Selected Works in Probability and Statistics

For futher volumes:
http://www.springer.com/series/8556

Richard A. Davis • Keh-Shin Lii • Dimitris N. Politis
Editors

Selected Works of Murray Rosenblatt

Springer

Editors
Richard A. Davis
Department of Statistics
Columbia University
Amsterdam Ave. 1255
New York, NY 10027
USA
davis.richarda@gmail.com

Keh-Shin Lii
Department of Statistics
University of California
Riverside, CA 92521, USA
ksl@stat.ucr.edu

Dimitris N. Politis
Department of Mathematics
University of California, San Diego
La Jolla, CA 92093-0112
USA
politis@euclid.ucsd.edu

ISBN 978-1-4419-8338-1 e-ISBN 978-1-4419-8339-8
DOI 10.1007/978-1-4419-8339-8
Springer New York Dordrecht Heidelberg London

Library of Congress Control Number: 2011925370

Printed on acid-free paper

Springer is part of Springer Science+Business Media (www.springer.com)

Preface to the Series

Springer's Selected Works in Probability and Statistics series offers scientists and scholars the opportunity of assembling and commenting upon major classical works in statistics, and honors the work of distinguished scholars in probability and statistics. Each volume contains the original papers, original commentary by experts on the subjects papers, and relevant biographies and bibliographies.

Springer is committed to maintaining the volumes in the series with free access on SpringerLink, as well as to the distribution of print volumes. The full text of the volumes is available on SpringerLink with the exception of a small number of articles for which links to their original publisher is included instead. These publishers have graciously agreed to make the articles freely available on their websites. The goal is maximum dissemination of this material.

The subjects of the volumes have been selected by an editorial board consisting of Anirban DasGupta, Peter Hall, Jim Pitman, Michael Sörensen, and Jon Wellner.

Murray Rosenblatt

Preface

Murray Rosenblatt was born in New York City on September 7, 1926. After completing an undergraduate degree in mathematics at the City College of New York, he entered graduate school in mathematics at Cornell in 1946. While at Cornell, he studied both physics and probability theory and benefited greatly from his contact and exposure to some of the great minds of the day including Feynman, Feller, and Kac. It was under the direction of Marc Kac that Murray wrote his PhD dissertation in 1949 on extensions of the famed Feynman-Kac integral.

Murray met his wife Ady in the late 1940s; they were married in Ithaca in 1949, and subsequently had two children, Karin and Daniel. After receiving his PhD, Murray joined the "Committee on Statistics" at the University of Chicago. He held academic positions at the University of Indiana and Brown University before settling in the University of California at San Diego (UCSD) in 1964. He remains at UCSD to date—as Distinguished Professor Emeritus since 1994—having survived the loss of his wife Ady in 2009 after a courageous battle with cancer. The Statistical Science interview [1] provides further insights into Murray's professional life and career; more details are also found in the short section on key events included in this volume.

During the second half of the 20th century, Murray was one of the most celebrated and leading figures in probability and statistics with particular emphasis on time series, Markov processes, and nonparametric function estimation. In addition to being a fellow of IMS and AAAS, he was a Guggenheim fellow twice (1965–66, 1971–72), and was elected to the National Academy of Sciences in 1984. Among his many contributions, Murray conducted seminal work on density estimation, central limit theorems under strong mixing, spectral domain methods, long memory processes and Markov processes.

During his long—and continuing—career, Murray has published over 130 papers, and 5 books. His earliest book, the time series monograph [2] written jointly with U. Grenander, was published in 1957 and remains a classic today. This book was the first to lay out a comprehensive and modern set of techniques for modeling time series backed up by rigorous arguments. It had wide appeal to researchers and practitioners alike. Many of the results in this book are as relevant today as when they first appeared more than 50 years ago. In tribute to Murray's seminal contributions, the Rosenblatt name has been attached to two notable terms, the Rosenblatt Transformation (based on his 1952 paper [3]) and the Rosenblatt Process—a term coined by Murad Taqqu. At the close of 2009, these terms have an astonishing 506K and 1870K search listings, respectively, in Google.

We are delighted that several leading experts agreed to provide commentary and reflections on various directions of Murray's research portfolio. Rick Bradley gives a nice overview of some of Murray's work related to the central limit theory under strong mixing conditions. Peter Bickel describes Murray's insight into their papers on global measures of density estimates. David Brillinger provides a behind-the-scenes look at his papers with Murray on higher-order spectra. Murad Taqqu describes how Murray's paper on independence and dependence [4] inspired him to look at a class of limit processes with long memory, the aforementioned Rosenblatt Processes. T.-C. Sun gives an interesting account of Murray's work on Markov processes on semi-groups and the connections to results by Lévy, and Kawada and Itô. Finally, Keh-Shin Lii, a long time collaborator of Murray's, describes some of their joint work and problems in deconvolution and non-Gaussian time series modeling.

Murray was an accomplished advisor with over 20 PhD students to his credit. He had an interesting style of advising that may have reflected his experience with his own mentor Marc Kac. In his interview [1], Murray says Kac "helped occasionally with suggestions, but sort of left you alone without saying you've got to do this or that so forth and so on. He let you to go your own way." Murray breaks down his own style of advising by "suggesting an area" and "if a student is bright enough to make his own way, why do you have to impose on him?" Of course, as two of the undersigned can attest, it was not always clear that we could make it on our own! In meetings during our graduate student days, Murray was fond of frequently uttering the "what if" refrain. He was not only trying to probe our level of understanding, but trying to push us to think more broadly beyond the conventional boundaries. Murray has extraordinary insight which, unfortunately, is something that is not readily passed on to one's students. It was often a revelation when after studying a topic, one would finally come to understand the insight Murray had expressed years earlier.

This volume is a celebration of Murray Rosenblatt's stellar research career that spans over six decades. While space limitations prevented us from publishing all of Murray's numerous research papers, we have attempted to gather in this volume some of his most interesting and influential papers. The task was difficult—and in the end rather subjective—but we received help and guidance from our six expert discussants and from Murray himself.

We would like to thank Anirban DasGupta, past editor of the IMS Collections and Lectures Notes—Monograph Series, who first proposed putting a volume together in honor of Murray. Even though IMS had to back out of its commitment to this series as a result of unanticipated financial concerns, Anirban, together with Jim Pitman, Peter Hall and Jon Wellner, have been stalwart supporters of these "Selected Works" projects. We are indebted to John Kimmel and Springer for stepping in and salvaging the "Selected Works Series" that honors some of the leading figures in our field.

New York — *Richard A. Davis*
Riverside — *Keh-Shin Lii*
La Jolla — *Dimitris N. Politis*

Bibliography

[1] D. R. Brillinger and R. A. Davis. A conversation with Murray Rosenblatt *Statistical Science*, 24:116–140.

[2] U. Grenander and M. Rosenblatt. *Statistical analysis of stationary time series*. John Wiley & Sons, New York, 1957.

[3] M. Rosenblatt. Remarks on a multivariate transformation. *Ann. Math. Statistics*, 23:470–472, 1952.

[4] M. Rosenblatt. Independence and dependence. In *Proc. 4th Berkeley Sympos. Math. Statist. and Prob., Vol. II*, pages 431–443. Univ. California Press, Berkeley, Calif., 1961.

Acknowledgements

This series of selected works is possible only because of the efforts and cooperation of many people, societies, and publishers. The series editors originated the series and directed its development. The volume editors spent a great deal of time compiling the previously published material and the contributors provided comments on the significance of the papers. The societies and publishers who own the copyright to the original material made the volumes possible and affordable by their generous cooperation:

American Institute of Physics
American Mathematical Society
American Statistical Association
Applied Probability Trust
Bernoulli Society
Cambridge University Press
Canadian Mathematical Society
Elsevier
Foundation of the Scandinavian Journal of Statistics
Indian Statistical Institute
Institute of Mathematical Statistics
International Chinese Statistical Association
International Statistical Institute
John Wiley and Sons
l'Institut Fourier
London Mathematical Society
New Zealand Statistical Association
Oxford University Press
Polish Academy of Sciences
Princeton University and the Institute for Advanced Studies
Springer
Statistical Society of Australia
University of California Press
University of Illinois, Department of Mathematics
University of Michigan, Department of Mathematics
University of North Carolina Press

Contents

Contributors

Peter Bickel Department of Statistics, University of California, Berkeley, California, 94720, E-mail: bickel@stat.berkeley.edu

Richard Bradley Department of Mathematics, Indiana University, Bloomington, Indiana, 47405, E-mail: bradleyr@indiana.edu

David R. Brillinger Department of Statistics, University of California, Berkeley, California, 94720, E-mail: brill@stat.berkeley.edu

Keh-Shin Lii Department of Statistics, University of California, Riverside, California, 92521, E-mail: ksl@stat.ucr.edu

T.-C. Sun Department of Mathematics, Wayne State University, Detroit, Michigan, 48202, E-mail: tsun@math.wayne.edu

Murad Taqqu Department of Mathematics, Boston University, Boston, Massachusetts, 02215, E-mail: murad@math.bu.edu

Murray Rosenblatt: Professional Timeline

1926	Born New York City, September 7, 1926
1946	B.S. City College of New York
1947	M.S. Cornell University
1949	PH.D. Cornell University thesis: "On distributions of certain Wiener functionals". Supervisor: Mark Kac
1949–1950	ONR postdoctoral grant (at Cornell University)
1950–1951	Instructor, Committee on Statistics, University of Chicago
1951–1955	Assistant professor, Committee on Statistics, Chicago
1953	University of Stockholm
1955	Department of Statistics, Columbia
1956–1959	Associate professor, Indiana University
1959–1964	Professor of Probability and Statistics, Brown University
1959	Brookhaven Lab visit
1960	—Madhav Heble, Indiana University, dissertation: "Linear estimation of regression coefficients, orthogonal matrix polynomials and application to multidimensional weakly stationary processes; interpolation and regression"
1960	—Perry Scheinok, Indiana University, dissertation: "The error on using the asymptotic variance and bias of spectrograph estimates for finite observation time"
1961	—Jack Hachigan, Indiana University, dissertation: "Some further results on functions of Markov processes"
1962	Symposium on Time Series, Brown University
1963	—T.-C. Sun, Brown University, dissertation: "A central limit theorem for non-linear functions of a normal stationary process"
1964	—Henry Krieger, Brown University, dissertation: "Toeplitz operators on locally compact abelian groups"
1964	—John van Ness, Brown University, dissertation: "Estimates of the bispectrum of stationary random processes"
1964–1994	Professor of Mathematics, University of California, San Diego
1965	—Mahendra Nadkarni, Brown University, dissertation: "Vector valued weakly stationary stochastic processes and factorization of matrix valued functions and strong mixing and uniformly ergodic gaussian processes"
1965	—Henry Richardson III, Brown University, dissertation: "Regression analysis when the least-squares estimate is not asymptotically efficient"
1965	Guggenheim fellowship, University College and Imperial College, London
1967–1968	Chair, Mathematics Department, UCSD
1970	—Seung Chay, USCD, dissertation: "On quasi-Markov random fields"
1970	Wald lecturer
1971	Guggenheim fellowship, University College and Imperial College, London

1972	—Michael Sturgeon, UCSD, dissertation: "The representation of invariant conservative Baire measures for certain Markov processes"
1975	—Keh-Shin Lii, UCSD, dissertation: "Density estimation and splines"
1978	—Richard Bradley, UCSD, dissertation: "Measures of dependence on stationary sequences of random variables"
1978	—Yue-Pok Mack, UCSD, dissertation: "k-nearest neighbor estimation"
1979	—Richard Davis, UCSD, dissertation: "Extremes of stationary processes"
1979	—Stuart Strait, UCSD, dissertation: "A quadratic measure of deviation of spectral estimates"
1979	Fellow at Churchill College, Cambridge
1984	—Larry Goldstein, UCSD, dissertation: "Extensions of stochastic approximation procedures"
1984	—Diane Marcus, UCSD, dissertation: "On the approximation of distributions of sums of independent summands by infinitely divisible distributions—n dimensional case"
1984	Member, National Academy of Sciences
1985	—Karen Messer, UCSD, dissertation: "Boundary effects of smoothing splines"
1991	—Bruce Wahlen, UCSD, dissertation: "A nonparametric measure of independence"
1993	—Michael Kramer, UCSD, dissertation: "The fluctuation of the gaussian likelihood for stationary random sequences"
1994	—Yukuang Chiu, UCSD, dissertation: "Topics on prediction and representation of stationary processes"
1998	—Anthony Gamst, UCSD, dissertation: "Stochastic Burgers flows"
1994–present	Professor Emeritus, University of California—San Diego
1994–1996,	Visiting Professor, Colorado State University (fall semesters)
2003	President, Emeriti Association, UCSD

Murray Rosenblatt: Bibliography

Books Authored

[B1] U. Grenander and M. Rosenblatt. *Statistical analysis of stationary time series*. John Wiley & Sons, New York, 1957.

[B2] M. Rosenblatt. *Random processes*. University Texts in the Mathematical Sciences. Oxford University Press, New York, 1962.

[B3] M. Rosenblatt and T. Czechowski. *Procesy stochastyczne*. Państwowe Wydaw. Naukowe, 1967. Polish translation of *Random Processes*.

[B4] M. Rosenblatt. *Markov Process: Structure and Asymptotic Behavior*. Die Grundlehren Der Mathematischen Wissenschaften, Band 184. Springer-Verlag, 1971.

[B5] M. Rosenblatt. *Random processes*. World Publishing Corporation, Beijing, 1974. Reprint of Springer-Verlag's 1974 ed. of *Random Processes*.

[B6] M. Rosenblatt. *Random processes*. Graduate Texts in Mathematics, No. 17. Springer-Verlag, New York, second edition, 1974.

[B7] U. Grenander and M. Rosenblatt. *Statistical analysis of stationary time series*. Chelsea Publishing Co., New York, second edition, 1984.

[B8] M. Rosenblatt. *Stationary sequences and random fields*. Birkhäuser Boston Inc., Boston, MA, 1985.

[B9] M. Rosenblatt. *Stochastic Curve Estimation*, volume 3 of *NSF-CBMS Regional Conference Series in Probability and Statistics*. Institute of Mathematical Statistics, 1991.

[B10] M. Rosenblatt. *Gaussian and non-Gaussian linear time series and random fields*. Springer Series in Statistics. Springer-Verlag, New York, 2000.

Books Edited

[E1] M. Rosenblatt, editor. *Proceedings of the Symposium on Time Series Analysis*, New York, 1963. John Wiley and Sons.

[E2] M. Rosenblatt and C. V. Atta, editors. *Statistical models and turbulence*. Lecture Notes in Physics, Vol. 12. Springer-Verlag, Berlin, 1972.

[E3] M. Rosenblatt, editor. *Studies in probability theory*, volume 18 of *MAA Studies in Mathematics*. Mathematical Association of America, Washington, D.C., 1978.

[E4] T. Gasser and M. Rosenblatt, editors. *Smoothing techniques for curve estimation*, volume 757 of *Lecture Notes in Mathematics*, Berlin, 1979. Springer.

[E5] M. Rosenblatt, editor. *Errett Bishop: reflections on him and his research*, volume 39 of *Contemporary Mathematics*, Providence, RI, 1985. American Mathematical Society.

[E6] D. Brillinger, P. Caines, J. Geweke, E. Parzen, M. Rosenblatt, and M. S. Taqqu, editors. *New directions in time series analysis. Part I*, volume 45 of *The IMA Volumes in Mathematics and its Applications*. Springer-Verlag, New York, 1992.

[E7] D. Brillinger, P. Caines, J. Geweke, E. Parzen, M. Rosenblatt, and M. S. Taqqu, editors. *New directions in time series analysis. Part II*, volume 46 of *The IMA Volumes in Mathematics and its Applications*. Springer-Verlag, New York, 1993.

[E8] P. M. Robinson and M. Rosenblatt, editors. Athens Conference on Applied Probability and Time Series An*alysis. Vol. II*, volume 115 of *Lecture Notes in Statistics*. Springer-Verlag, New York, 1996. Time series analysis, In memory of E. J. Hannan, Papers from the conference held in Athens, March 22–26, 1995.

Mathematical Articles

[M1] M. Rosenblatt. On a class of Markov processes. *Trans. Amer. Math. Soc.*, 71:120–135, 1951.

[M2] U. Grenander and M. Rosenblatt. On spectral analysis of stationary time series. *Proc. Nat. Acad. Sci. U. S. A.*, 38:519–521, 1952.

[M3] M. Rosenblatt. The behavior at zero of the characteristic function of a random variable. *Proc. Amer. Math. Soc.*, 3:498–504, 1952.

[M4] M. Rosenblatt. Limit theorems associated with variants of the von Mises statistic. *Ann. Math. Statistics*, 23:617–623, 1952.

[M5] M. Rosenblatt. On the oscillation of sums of random variables. *Trans. Amer. Math. Soc.*, 72:165–178, 1952.

[M6] M. Rosenblatt. Remarks on a multivariate transformation. *Ann. Math. Statistics*, 23:470–472, 1952.

Indicates article found in this volume.

[M7] K. A. Brownlee, J. L. Hodges, and M. Rosenblatt. The up-and-down method with small samples. Journal of the American Statistical Association, 48(262):262–277, 1953.

[M8] U. Grenander and M. Rosenblatt. Comments on statistical spectral analysis. Skand. Aktuarietidskr., 36:182–202, 1953.

[M9] U. Grenander and M. Rosenblatt. Statistical spectral analysis of time series arising from stationary stochastic processes. Ann. Math. Statistics, 24: 537–558, 1953.

[M10] J. L. Hodges, Jr. and M. Rosenblatt. Recurrence-time moments in random walks. Pacific J. Math., 3:127–136, 1953.

[M11] U. Grenander and M. Rosenblatt. An extension of a theorem of G. Szegö and its application to the study of stochastic processes. Trans. Amer. Math. Soc., 76:112–126, 1954.

[M12] U. Grenander and M. Rosenblatt. Regression analysis of time series with stationary residuals. Proc. Nat. Acad. Sci. U. S. A., 40:812–816, 1954.

[M13] M. Rosenblatt. An inventory problem. Econometrica, 22:244–247, 1954.

[M14] J. R. Blum and M. Rosenblatt. A class of stationary processes and a central limit theorem. Proc. Nat. Acad. Sci. U.S.A., 42:412–413, 1956.

[M15] U. Grenander and M. Rosenblatt. Some problems in estimating the spectrum of a time series. In Proceedings of the Third Berkeley Symposium on Mathematical Statistics and Probability, 1954–1955, vol. I, pages 77–93, Berkeley and Los Angeles, 1956. University of California Press.

[M16] M. Rosenblatt. A central limit theorem and a strong mixing condition. Proc. Nat. Acad. Sci. U. S. A., 42:43–47, 1956.

[M17] M. Rosenblatt. On the estimation of regression coefficients of a vector-valued time series with a stationary residual. Ann. Math. Statist., 27:99–121, 1956.

[M18] M. Rosenblatt. Remarks on some nonparametric estimates of a density function. Ann. Math. Statist., 27:832–837, 1956.

[M19] M. Rosenblatt. Some regression problems in time series analysis. In Proceedings of the Third Berkeley Symposium on Mathematical Statistics and Probability, 1954–1955, vol. I, pages 165–186, Berkeley and Los Angeles, 1956. University of California Press.

[M20] J. R. Blum and M. Rosenblatt. A class of stationary processes and a central limit theorem. Duke Math. J., 24:73–78, 1957.

[M21] M. Rosenblatt. The multidimensional prediction problem. In Proceedings of the National Academy of Sciences of the United States of America, volume 43, pages 989–992, November 15 1957.

[M22] M. Rosenblatt. A random model of the sea surface generated by a hurricane. J. Math. Mech., 6:235–246, 1957.

[M23] M. Rosenblatt. Some purely deterministic processes. J. Math. Mech., 6: 801–810, 1957.

Indicates article found in this volume.

[M24] J. R. Blum, H. Chernoff, M. Rosenblatt, and H. Teicher. Central limit theorems for interchangeable processes. Canad. J. Math., 10:222–229, 1958.

[M25] C. Burke and M. Rosenblatt. Consolidation of probability matrices. Bull. Inst. Internat. Statist., 36(3):7–8, 1958.

[M26] C. Burke and M. Rosenblatt. Consolidation of probability matrices. Bull. Inst. Internat. Statist., 36(3):7–8, 1958.

[M27] C. J. Burke and M. Rosenblatt. A Markovian function of a Markov chain. Ann. Math. Statist., 29:1112–1122, 1958.

[M28] M. Rosenblatt. A multi-dimensional prediction problem. Ark. Mat., 3: 407–424, 1958.

[M29] J. R. Blum and M. Rosenblatt. On the structure of infinitely divisible distributions. Pacific J. Math., 9:1–7, 1959.

[M30] M. Rosenblatt. Functions of a Markov process that are Markovian. J. Math. Mech., 8:585–596, 1959.

[M31] M. Rosenblatt. Stationary processes as shifts of functions of independent random variables. J. Math. Mech., 8:665–681, 1959.

[M32] M. Rosenblatt. Statistical analysis of stochastic processes with stationary residuals. In Probability and statistics: The Harald Cramér volume (edited by Ulf Grenander), pages 246–275. Almqvist & Wiksell, Stockholm, 1959.

[M33] M. Rosenblatt. Asymptotic distribution of eigenvalues of block Toeplitz matrices. Bull. Amer. Math. Soc., 66:320–321, 1960.

[M34] M. Rosenblatt. Limits of convolution sequences of measures on a compact topological semigroup. J. Math. Mech., 9:293–305, 1960.

[M35] M. Rosenblatt. The multidimensional prediction problem. In Statistical methods in radio wave propagation : proceedings of a symposium held at the University of California, Los Angeles 1958. Symposium Division Pergamon Pr., 1960.

[M36] M. Rosenblatt. Stationary Markov chains and independent random variables. J. Math. Mech., 9:945–949, 1960.

[M37] M. Rosenblatt. Some comments on narrow band-pass filters. Quart. Appl. Math., 18:387–393, 1960/1961.

[M38] J. R. Blum, J. Kiefer, and M. Rosenblatt. Distribution free tests of independence based on the sample distribution function. Ann. Math. Statist., 32: 485–498, 1961.

[M39] M. Rosenblatt. Independence and dependence. In Proc. 4th Berkeley Sympos. Math. Statist. and Prob., Vol. II, pages 431–443. Univ. California Press, Berkeley, Calif., 1961.

[M40] W. Freiberger, M. Rosenblatt, and J. W. Van Ness. Regression analysis of vector-valued random processes. J. Soc. Indust. Appl. Math., 10:89–102, 1962.

[M41] J. Hachigian and M. Rosenblatt. Functions of reversible Markov processes that are Markovian. J. Math. Mech., 11:951–960, 1962.

Indicates article found in this volume.

[M42] G. F. Newell and M. Rosenblatt. Zero crossing probabilities for Gaussian stationary processes. Ann. Math. Statist., 33:1306–1313, 1962.

[M43] M. Rosenblatt. Asymptotic behavior of eigenvalues of Toeplitz forms. J. Math. Mech., 11:941–949, 1962.

[M44] M. Rosenblatt and D. Slepian. *n*th order Markov chains with every N variables independent. J. Soc. Indust. Appl. Math., 10:537–549, 1962.

[M45] M. Heble and M. Rosenblatt. Idempotent measures on a compact topological semigroup. Proc. Amer. Math. Soc., 14:177–184, 1963.

[M46] M. Rosenblatt. Asymptotic behavior of eigenvalues for a class of integral equations with translation kernels. In Proc. Sympos. Time Series Analysis (Brown Univ., 1962), pages 316–326. Wiley, New York, 1963.

[M47] M. Rosenblatt. The representation of a class of two state stationary processes in terms of independent random variables. J. Math. Mech., 12:721–730, 1963.

[M48] M. Rosenblatt. Some results on the asymptotic behavior of eigenvalues for a class of integral equations with translation kernels. J. Math. Mech., 12:619–628, 1963.

[M49] M. Rosenblatt. Almost periodic transition operators acting on the continuous functions on a compact space. J. Math. Mech., 13:837–847, 1964.

[M50] M. Rosenblatt. Equicontinuous Markov operators. Teor. Verojatnost. i Primenen., 9:205–222, 1964.

[M51] M. Rosenblatt. Some nonlinear problems arising in the study of random processes. J. Res. Nat. Bur. Standards Sect. D, 68D:933–936, 1964.

[M52] M. Rosenblatt. Products of independent identically distributed stochastic matrices. J. Math. Anal. Appl., 11:1–10, 1965.

[M53] M. Rosenblatt and J. W. Van Ness. Estimation of the bispectrum. Ann. Math. Statist., 36:1120–1136, 1965.

[M54] M. Rosenblatt. Remarks on ergodicity of stationary irreducible transient Markov chains. Z. Wahrscheinlichkeitstheorie und Verw. Gebiete, 6: 293–301, 1966.

[M55] M. Rosenblatt. Functions of Markov processes. Z. Wahrscheinlichkeitstheorie und Verw. Gebiete, 5:232–243, 1966.

[M56] M. Rosenblatt. Remarks on higher order spectra. In Multivariate Analysis (Proc. Internat. Sympos., Dayton, Ohio, 1965), pages 383–389. Academic Press, New York, 1966.

[M57] D. R. Brillinger and M. Rosenblatt. Asymptotic theory of estimates of *k*th-order spectra. Proc. Nat. Acad. Sci. U.S.A., 57:206–210, 1967.

[M58] D. R. Brillinger and M. Rosenblatt. Asymptotic theory of estimates of *k*-th order spectra. In B. Harris, editor, Spectral Analysis Time Series (Proc. Advanced Sem., Madison, Wis., 1966), pages 153–188. John Wiley, New York, 1967.

Indicates article found in this volume.

[M59] D. R. Brillinger and M. Rosenblatt. Computation and interpretation of k-th order spectra. In B. Harris, editor, Spectral Analysis Time Series (Proc. Advanced Sem., Madison, Wis., 1966), pages 189–232. John Wiley, New York, 1967.

[M60] M. Rosenblatt. A strong mixing condition and a central limit theorem on compact groups. J. Math. Mech., 17:189–198, 1967.

[M61] M. Rosenblatt. Transition probability operators. In Proc. Fifth Berkeley Sympos. Math. Statist. and Probability (Berkeley, Calif., 1965/66), Vol. II: Contributions to Probability Theory, Part 2, pages 473–483. Univ. California Press, Berkeley, Calif., 1967.

[M62] M. Rosenblatt. Remarks on the Burgers equation. J. Mathematical Phys., 9:1129–1136, 1968.

[M63] M. Rosenblatt. Conditional probability density and regression estimators. In Multivariate Analysis, II (Proc. Second Internat. Sympos., Dayton, Ohio, 1968), pages 25–31. Academic Press, New York, 1969.

[M64] M. Rosenblatt. Stationary measures for random walks on semigroups. In Semigroups (Proc. Sympos., Wayne State Univ., Detroit, Mich., 1968), pages 209–220. Academic Press, New York, 1969.

[M65] M. Rosenblatt. Density estimates and Markov sequences. In Nonparametric Techniques in Statistical Inference (Proc. Sympos., Indiana Univ., Bloomington, Ind., 1969), pages 199–213. Cambridge Univ. Press, London, 1970.

[M66] M. Rosenblatt. A prediction problem and central limit theorems for stationary Markov sequences. In Proc. Twelfth Biennial Sem. Canad. Math. Congr. on Time Series and Stochastic Processes; Convexity and Combinatorics (Vancouver, B.C., 1969), pages 99–114. Canad. Math. Congr., Montreal, Que., 1970.

[M67] M. Rosenblatt. Curve estimates. Ann. Math. Statist., 42:1815–1842, 1971.

[M68] M. Rosenblatt. Central limit theorem for stationary processes. In Proceedings of the Sixth Berkeley Symposium on Mathematical Statistics and Probability (Univ. California, Berkeley, Calif., 1970/1971), Vol. II: Probability theory, pages 551–561, Berkeley, Calif., 1972. Univ. California Press.

[M69] M. Rosenblatt. Probability limit theorems and some questions in fluid mechanics. In Statistical models and turbulence (Proc. Sympos., Univ. California, La Jolla, Calif., 1971), pages 27–40. Lecture Notes in Phys., Vol. 12. Springer, Berlin, 1972.

[M70] M. Rosenblatt. Uniform ergodicity and strong mixing. Z. Wahrscheinlichkeitstheorie und Verw. Gebiete, 24:79–84, 1972.

[M71] M. Rosenblatt. Invariant and subinvariant measures of transition probability functions acting on continuous functions. Z. Wahrscheinlichkeitstheorie und Verw. Gebiete, 25:209–221, 1972/73.

Indicates article found in this volume.

[M72] P. Bickel and M. Rosenblatt. Two-dimensional random fields. In Multivariate analysis, III (Proc. Third Internat. Sympos., Wright State Univ., Dayton, Ohio, 1972), pages 3–15. Academic Press, New York, 1973.

[M73] P. J. Bickel and M. Rosenblatt. On some global measures of the deviations of density function estimates. Ann. Statist., 1:1071–1095, 1973.

[M74] M. Rosenblatt. Recurrent points and transition functions acting on continuous functions. Z. Wahrscheinlichkeitstheorie und Verw. Gebiete, 30:173–183, 1974.

[M75] P. J. Bickel and M. Rosenblatt. Corrections to: "On some global measures of the deviations of density function estimates" (Ann. Statist. **1** (1973), 1071–1095). Ann. Statist., 3(6):1370, 1975.

[M76] K. S. Lii and M. Rosenblatt. Asymptotic behavior of a spline estimate of a density function. Comput. Math. Appl., 1(2):223–235, 1975.

[M77] K. S. Lii and M. Rosenblatt. Asymptotic results on a spline estimate of a probability density. In Statistical inference and related topics (Proc. Summer Res. Inst. Statist. Inference for Stochastic Processes, Indiana Univ., Bloomington, Ind., 1974, Vol 2; dedicated to Z. W. Birnbaum), pages 77–85. Academic Press, New York, 1975.

[M78] M. Rosenblatt. The local behavior of the derivative of a cubic spline interpolator. J. Approximation Theory, 15(4):382–387, 1975.

[M79] M. Rosenblatt. Multiply schemes and shuffling. Math. Comput., 29:929–934, 1975.

[M80] M. Rosenblatt. A quadratic measure of deviation of two-dimensional density estimates and a test of independence. Ann. Statist., 3:1–14, 1975.

[M81] M. Rosenblatt. Note correcting a remark in a paper of Karl Bosch ("Notwendige und hinreichende Bedingungen dafür, dass eine Funktion einer homogenen Markoffschen Kette Markoffsch ist", (Z. Wahrscheinlichkeitstheorie und Verw. Gebiete **31** (1974/75), 199–202)). Z. Wahrscheinlichkeitstheorie und Verw. Gebiete, 33(3):219, 1975/76.

[M82] K. S. Lii, M. Rosenblatt, and C. V. Atta. Bispectral measurements in turbulence. J. Fluid Mech., 77, Pt. 1:45–62, 1976.

[M83] J. Rice and M. Rosenblatt. Estimation of the log survivor function and hazard function. Sankhyá Ser. A, 38(1):60–78, 1976.

[M84] M. Rosenblatt. Fractional integrals of stationary processes and the central limit theorem. J. Appl. Probability, 13(4):723–732, 1976.

[M85] M. Rosenblatt. On the maximal deviation of k-dimensional density estimates. Ann. Probability, 4(6):1009–1015, 1976.

[M86] M. Rosenblatt. Asymptotics and representation of cubic splines. J. Approximation Theory, 17(4):332–343, 1976.

[M87] K. N. Helland, K. S. Lii, and M. Rosenblatt. Bispectra of atmospheric and wind tunnel turbulence. In P. R. Krishnaiah, editor, Applications of Statistics. North Holland Publishing Co., 1977.

Indicates article found in this volume.

[M88] P. A. W. Lewis, L. H. Liu, D. W. Robinson, and M. Rosenblatt. Empirical sampling study of a goodness of fit statistic for density function estimation. In Multivariate analysis—IV: proceedings of the fourth International Symposium on Multivariate Analysis, pages 139–174, 1977.

[M89] M. Rosenblatt. Energy transfer for the Burgers' equation. Phys. Fluids, 21(10):1694–1697, 1978.

[M90] M. Rosenblatt. Dependence and asymptotic independence for random processes. In Studies in probability theory, volume 18 of MAA Stud. Math., pages 24–45. Math. Assoc. America, Washington, D.C., 1978.

[M91] K. N. Helland, K. S. Lii, and M. Rosenblatt. Bispectra and energy transfer in grid-generated turbulence. In Developments in statistics, Vol. 2, pages 123–155. Academic Press, New York, 1979.

[M92] K. N. Helland and M. Rosenblatt. Spectral variance estimation and the analysis of turbulence. *Phys. Fluids*, 22:819–823, 1979.

[M93] Y. P. Mack and M. Rosenblatt. Multivariate k-nearest neighbor density estimates. *J. Multivariate Anal.*, 9(1):1–15, 1979.

[M94] M. Rosenblatt. Global measures of deviation for kernel and nearest neighbor density estimates. In *Smoothing techniques for curve estimation (Proc. Workshop, Heidelberg, 1979)*, volume 757 of *Lecture Notes in Math.*, pages 181–190. Springer, Berlin, 1979.

[M95] M. Rosenblatt. Linearity and nonlinearity in time series: prediction. In *Proceedings of the 42nd session of the International Statistical Institute, Vol. 1 (Manila, 1979)*, volume 48, pages 423–434, 1979.

[M96] M. Rosenblatt. Some limit theorems for partial sums of quadratic forms in stationary Gaussian variables. *Z. Wahrsch. Verw. Gebiete*, 49(2):125–132, 1979.

[M97] M. Rosenblatt. Some remarks on a mixing condition. *Ann. Probab.*, 7(1):170–172, 1979.

[M98] M. Rosenblatt. Correction to: "Some remarks on a mixing condition" [Ann. Probab. **7** (1979), no. 1, 170–172; MR 80k:60071a]. *Ann. Probab.*, 7(6):1097, 1979.

[M99] M. Rosenblatt. Linear processes and bispectra. *J. Appl. Probab.*, 17(1): 265–270, 1980.

[M100] M. Rosenblatt. Some limit theorems for partial sums of stationary sequences. In *Multivariate analysis, V (Proc. Fifth Internat. Sympos., Univ. Pittsburgh, Pittsburgh, Pa., 1978)*, pages 239–248. North-Holland, Amsterdam, 1980.

[M101] M. Rosenblatt. Erratum: "Asymptotics and representation of cubic splines" [J. Approx. Theory **17** (1976), no. 4, 332–343; MR **54** #5682]. *J. Approx. Theory*, 28(2):184, 1980.

[M102] J. Rice and M. Rosenblatt. Integrated mean squared error of a smoothing spline. *J. Approx. Theory*, 33(4):353–369, 1981.

Indicates article found in this volume.

[M103] M. Rosenblatt. Limit theorems for Fourier transforms of functionals of Gaussian sequences. *Z. Wahrsch. Verw. Gebiete*, 55(2):123–132, 1981.

[M104] M. Rosenblatt. Polynomials in Gaussian variables and infinite divisibility? In *Contributions to probability*, pages 139–142. Academic Press, New York, 1981.

[M105] K. S. Lii, K. N. Helland, and M. Rosenblatt. Estimating three-dimensional energy transfer in isotropic turbulence. *J. Time Ser. Anal.*, 3(1):1–28, 1982.

[M106] K. S. Lii and M. Rosenblatt. Deconvolution and estimation of transfer function phase and coefficients for non-Gaussian linear processes. *Ann. Statist.*, 10(4):1195–1208, 1982.

[M107] J. Rice and M. Rosenblatt. Boundary effects on the behavior of smoothing splines. In *Statistics and probability: essays in honor of C. R. Rao*, pages 635–643. North-Holland, Amsterdam, 1982.

[M108] M. Rosenblatt. Corrections: "A quadratic measure of deviation of two-dimensional density estimates and a test of independence" [Ann Statist. **3** (1975), 1–14; MR **55** #1600]. *Ann. Statist.*, 10(2):646, 1982.

[M109] J. Rice and M. Rosenblatt. Smoothing splines: regression, derivatives and deconvolution. *Ann. Statist.*, 11(1):141–156, 1983.

[M110] M. Rosenblatt. Cumulants and cumulant spectra. In *Time series in the frequency domain*, volume 3 of *Handbook of Statist.*, pages 369–382. North-Holland, Amsterdam, 1983.

[M111] M. Rosenblatt. Linear random fields. In *Studies in econometrics, time series, and multivariate statistics*, pages 299–309. Academic Press, New York, 1983.

[M112] K. S. Lii and M. Rosenblatt. Remarks on non-Gaussian linear processes with additive Gaussian noise. In *Robust and nonlinear time series analysis (Heidelberg, 1983)*, volume 26 of *Lecture Notes in Statist.*, pages 185–197. Springer, New York, 1984.

[M113] K. S. Lii and M. Rosenblatt. Non-Gaussian linear processes, phase and deconvolution. In *Statistical signal processing (Annapolis, Md., 1982)*, volume 53 of *Statist. Textbooks Monogr.*, pages 51–58. Dekker, New York, 1984.

[M114] M. Rosenblatt. Asymptotic normality, strong mixing and spectral density estimates. *Ann. Probab.*, 12(4):1167–1180, 1984.

[M115] M. Rosenblatt. Short-range and long-range dependence. In H. A. David and H. T. David, editors, *Statistics, an appraisal: proceedings of a conference marking the 50th anniversary of the Statistical Laboratory, Iowa State University, Ames, Iowa, June 13-15, 1983*, pages 509–520. Iowa State University Press, 1984.

[M116] M. Rosenblatt. A two-dimensional smoothing spline and a regression problem. In *Limit theorems in probability and statistics, Vol. I, II (Veszprém, 1982)*, volume 36 of *Colloq. Math. Soc. János Bolyai*, pages 915–931. North-Holland, Amsterdam, 1984.

Indicates article found in this volume.

[M117] K. S. Lii and M. Rosenblatt. A fourth-order deconvolution technique for non-Gaussian linear processes. In *Multivariate analysis VI (Pittsburgh, Pa., 1983)*, pages 395–410. North-Holland, Amsterdam, 1985.

[M118] M. Rosenblatt and F. J. Samaniego. Julius R. Blum, 1922–1982. *Ann. Statist.*, 13(1):1–9, 1985.

[M119] K. S. Lii and M. Rosenblatt. Estimation of a transfer function in a non-Gaussian context. In *Function estimates (Arcata, Calif., 1985)*, volume 59 of *Contemp. Math.*, pages 49–51. Amer. Math. Soc., Providence, RI, 1986.

[M120] K.-S. Lii and M. Rosenblatt. Deconvolution of non-Gaussian linear processes with vanishing spectral values. *Proc. Nat. Acad. Sci. U.S.A.*, 83(2):199–200, 1986.

[M121] M. Rosenblatt. Convolution sequences of measures on the semigroup of stochastic matrices. In *Random matrices and their applications (Brunswick, Maine, 1984)*, volume 50 of *Contemp. Math.*, pages 215–220. Amer. Math. Soc., Providence, RI, 1986.

[M122] M. Rosenblatt. Parameter estimation for finite-parameter stationary random fields. *J. Appl. Probab.*, (Special Vol. 23A):311–318, 1986. Essays in time series and allied processes.

[M123] M. Rosenblatt. Prediction for some non-Gaussian autoregressive schemes. *Adv. in Appl. Math.*, 7(2):182–198, 1986.

[M124] M. Rosenblatt. Non-Gaussian sequences and deconvolution. In *Proceedings of the 1st World Congress of the Bernoulli Society, Vol. 2 (Tashkent, 1986)*, pages 349–353, Utrecht, 1987. VNU Sci. Press.

[M125] M. Rosenblatt. Scale renormalization and random solutions of the Burgers equation. *J. Appl. Probab.*, 24(2):328–338, 1987.

[M126] M. Rosenblatt. Remarks on limit theorems for nonlinear functionals of Gaussian sequences. *Probab. Theory Related Fields*, 75(1):1–10, 1987.

[M127] M. Rosenblatt. Some models exhibiting non-Gaussian intermittency. *IEEE Trans. Inform. Theory*, 33(2):258–262, 1987.

[M128] K.-S. Lii and M. Rosenblatt. Estimation and deconvolution when the transfer function has zeros. *J. Theoret. Probab.*, 1(1):93–113, 1988.

[M129] K.-S. Lii and M. Rosenblatt. Nonminimum phase non-Gaussian deconvolution. *J. Multivariate Anal.*, 27(2):359–374, 1988.

[M130] J. A. Rice and M. Rosenblatt. On frequency estimation. *Biometrika*, 75(3):477–484, 1988.

[M131] M. Rosenblatt. Comments on structure and estimation for non-gaussian linear processes. In E. Weginan, S. Schwartz, and J. Thomas, editors, *Topics in Non-Gaussian Signal Processing*, pages 88–97. Springer-Verlag, 1989.

[M132] M. Rosenblatt. Erratum: "Prediction for some non-Gaussian autoregressive schemes" [Adv. in Appl. Math. **7** (1986), no. 2, 182–198; MR0845375 (88a:62260)]. *Adv. in Appl. Math.*, 10(1):130, 1989.

Indicates article found in this volume.

[M133] M. Rosenblatt. Book Review: Correlation theory of stationary and random functions vol. I; Basic results, vol. II, Supplementary notes and references. *Bull. Amer. Math. Soc. (N.S.)*, 20(2):207–211, 1989.

[M134] M. Rosenblatt. A note on maximum entropy. In *Probability, statistics, and mathematics*, pages 255–260. Academic Press, Boston, MA, 1989.

[M135] F. J. Breidt, R. A. Davis, K.-S. Lii, and M. Rosenblatt. Nonminimum phase non-Gaussian autoregressive processes. *Proc. Nat. Acad. Sci. U.S.A.*, 87(1):179–181, 1990.

[M136] K. S. Lii and M. Rosenblatt. Asymptotic normality of cumulant spectral estimates. *J. Theoret. Probab.*, 3(2):367–385, 1990.

[M137] K. S. Lii and M. Rosenblatt. Cumulant spectral estimates: bias and covariance. In *Limit theorems in probability and statistics (Pécs, 1989)*, volume 57 of *Colloq. Math. Soc. János Bolyai*, pages 365–405. North-Holland, Amsterdam, 1990.

[M138] M. Rosenblatt. Discussion of Bartlett's paper "chance or chaos". *J. Roy. Stat. Soc. Ser. A.*, 153, 1990.

[M139] F. J. Breidt, R. A. Davis, K.-S. Lii, and M. Rosenblatt. Maximum likelihood estimation for noncausal autoregressive processes. *J. Multivariate Anal.*, 36(2):175–198, 1991.

[M140] R. A. Davis and M. Rosenblatt. Parameter estimation for some time series models without contiguity. *Statist. Probab. Lett.*, 11(6):515–521, 1991.

[M141] K. Helland, K. S. Lii, and M. Rosenblatt. Monte Carlo and turbulence. In *Nonparametric functional estimation and related topics (Spetses, 1990)*, volume 335 of *NATO Adv. Sci. Inst. Ser. C Math. Phys. Sci.*, pages 405–418. Kluwer Acad. Publ., Dordrecht, 1991.

[M142] M. Rosenblatt. Introductory remarks. In *Spatial stochastic processes*, volume 19 of *Progr. Probab.*, pages viii–x. Birkhäuser Boston, Boston, MA, 1991.

[M143] K.-S. Lii and M. Rosenblatt. An approximate maximum likelihood estimation for non-Gaussian non-minimum phase moving average processes. *J. Multivariate Anal.*, 43(2):272–299, 1992.

[M144] M. Rosenblatt. Gaussian and non-Gaussian linear sequences. In *New directions in time series analysis, Part I*, volume 45 of *IMA Vol. Math. Appl.*, pages 327–333. Springer, New York, 1992.

[M145] M. Rosenblatt and B. E. Wahlen. A nonparametric measure of independence under a hypothesis of independent components. *Statist. Probab. Lett.*, 15(3):245–252, 1992.

[M146] M. Kramer and M. Rosenblatt. The Gaussian log likelihood and stationary sequences. In *Developments in time series analysis*, pages 69–79. Chapman & Hall, London, 1993.

[M147] K.-S. Lii and M. Rosenblatt. Bispectra and phase of non-Gaussian linear processes. *J. Theoret. Probab.*, 6(3):579–593, 1993.

Indicates article found in this volume.

[M148] K.-S. Lii and M. Rosenblatt. Non-Gaussian autoregressive moving average processes. *Proc. Nat. Acad. Sci. U.S.A.*, 90(19):9168–9170, 1993.

[M149] M. Rosenblatt. The central limit theorem and Markov sequences. In *Doeblin and modern probability (Blaubeuren, 1991)*, volume 149 of *Contemp. Math.*, pages 171–177. Amer. Math. Soc., Providence, RI, 1993.

[M150] M. Rosenblatt. A note on prediction and an autoregressive sequence. In *Stochastic processes*, pages 291–295. Springer, New York, 1993.

[M151] M. Rosenblatt. Parameter estimation for some finite parameter stationary sequences. In *Probability, statistics and optimisation*, Wiley Ser. Probab. Math. Statist. Probab. Math. Statist., pages 213–218. Wiley, Chichester, 1994.

[M152] M. Rosenblatt. Prediction and non-Gaussian autoregressive stationary sequences. *Ann. Appl. Probab.*, 5(1):239–247, 1995.

[M153] K.-S. Lii and M. Rosenblatt. Maximum likelihood estimation for non-Gaussian nonminimum phase ARMA sequences. *Statist. Sinica*, 6(1):1–22, 1996.

[M154] K.-S. Lii and M. Rosenblatt. Non-Gaussian autoregressive sequences and random fields. In *Stochastic modelling in physical oceanography*, volume 39 of *Progr. Probab.*, pages 295–309. Birkhäuser Boston, Boston, MA, 1996.

[M155] M. Rosenblatt. The likelihood of an autoregressive scheme. In *Athens Conference on Applied Probability and Time Series Analysis, Vol. II (1995)*, volume 115 of *Lecture Notes in Statist.*, pages 352–362. Springer, New York, 1996.

[M156] M. Rosenblatt. Comments on estimation and prediction for autoregressive and moving average non-Gaussian sequences. In *Stochastic models in geosystems (Minneapolis, MN, 1994)*, volume 85 of *IMA Vol. Math. Appl.*, pages 353–358. Springer, New York, 1997.

[M157] M. Rosenblatt. Some simple remarks on an autoregressive scheme and an implied problem. *J. Theoret. Probab.*, 10(2):295–305, 1997. Dedicated to Murray Rosenblatt.

[M158] K.-S. Lii and M. Rosenblatt. Line spectral analysis for harmonizable processes. *Proc. Natl. Acad. Sci. USA*, 95(9):4800–4803 (electronic), 1998.

[M159] M. Rosenblatt. Non-Gaussian autoregressive and moving average schemes. In *Asymptotic methods in probability and statistics (Ottawa, ON, 1997)*, pages 731–737. North-Holland, Amsterdam, 1998.

[M160] K.-S. Lii and M. Rosenblatt. Spectral analysis for harmonizable processes. *Ann. Statist.*, 30(1):258–297, 2002.

[M161] K.-S. Lii and M. Rosenblatt. Correction: "Spectral analysis for harmonizable processes" [Ann. Statist. **30** (2002), no. 1, 258–297; mr1892664]. *Ann. Statist.*, 31(5):1693, 2003.

Indicates article found in this volume.

[M162] M. Rosenblatt. Non-Gaussian time series models. In *Time series analysis and applications to geophysical systems*, volume 139 of *IMA Vol. Math. Appl.*, pages 227–237. Springer, New York, 2004.

[M163] F. J. Breidt, R. A. Davis, N.-J. Hsu, and M. Rosenblatt. Pile-up probabilities for the Laplace likelihood estimator of a non-invertible first order moving average. In *Time series and related topics*, volume 52 of *IMS Lecture Notes Monogr. Ser.*, pages 1–19. Inst. Math. Statist., Beachwood, OH, 2006.

[M164] K.-S. Lii and M. Rosenblatt. Estimation for almost periodic processes. *Ann. Statist.*, 34(3):1115–1139, 2006.

[M165] M. Rosenblatt. An example and transition function equicontinuity. *Statist. Probab. Lett.*, 76(18):1961–1964, 2006.

[M166] K. S. Lii and M. Rosenblatt. Prolate spheroidal spectral estimates. *Statist. Probab. Lett.*, 78(11):1339–1348, 2008.

[M167] K.-S. Lii and M. Rosenblatt. Correction: "Estimation for almost periodic processes" [Ann. Statist. **34** (2006), no. 3, 1115–1139; mr2278353]. *Ann. Statist.*, 36(3):1508, 2008.

[M168] M. Rosenblatt. A comment on a conjecture of N. Wiener. *Statist. Probab. Lett.*, 79(3):347–348, 2009.

Interviews and Biographies

[I1] Publications of Murray Rosenblatt. *J. Theoret. Probab.*, 10(2):287–293, 1997.

[I2] T. C. Sun. Murray Rosenblatt: his contributions to probability and statistics. *J. Theoret. Probab.*, 10(2):279–286, 1997.

[I3] D. R. Brillinger and R. A. Davis. A conversation with Murray Rosenblatt. *Statistical Science*, 24(1):116–140, 2009.

Indicates article found in this volume.

Discussion of Bickel and Rosenblatt's work on Global Measures of Deviations for Density Estimates

By Peter Bickel

The paper [1] of Murray Rosenblatt's, which I coauthored, seems to have had a reasonably long shelf life. I believe Murray enjoyed working on this paper—I certainly did! The question posed by Murray was that of obtaining asymptotic approximations to the distribution of the maximum deviation of a kernel density estimate $\hat{f}_n$ of an unknown density f based on a sample $X_1, \ldots, X_n$. If a limit distribution existed for

$$\sigma_n \left[\max_{0 \le x \le 1} \left\{ \frac{|\hat{f}_n(x) - f(x)|}{\sqrt{\hat{f}_n(x)}} \right\} - \mu_n \right],$$

where μ_n, $\sigma_n > 0$ are normalizing constants not depending on f, then it is clear how to

(a) construct approximate uniform confidence bands for f of the form $\hat{f}_n \pm c_n \sqrt{\hat{f}_n}$.

(b) construct goodness of fit tests of point hypotheses $H : f = f_0$ using

$$\max_x \frac{|\hat{f}_n(x) - f_0(x)|}{\sqrt{f_0(x)}}$$

as a test statistic.

The heuristic that Murray advanced was the following. If $U_n(x)$ denotes the stochastic process defined by

$$U_n(x) \equiv \sqrt{nh_n} \frac{(\hat{f}_n(x) - f(x)}{\sqrt{\hat{f}(x)}},$$

then $U_n(x_1), \ldots, U_n(x_k)$ are asymptotically iid Gaussian and $U_n(x)$ behaves locally like a stationary Gaussian process. Hence one would expect that the statistic based on $\max_{0 \le x \le 1} U_n(x)$ would behave like the corresponding statistic for a stationary Gaussian process over an increasing stretch of the real line; namely the limit would have a Gumbel distribution. Murray and I were able to show this using strong embedding of the empirical process techniques.

We also studied another potential test statistic, the L_2 deviation, defined by

$$\int_0^1 \frac{(\hat{f}(x) - f_0(x))^2}{f_0(x)} dx.$$

We showed that under strong conditions, this statistic had a limiting Gaussian distribution consistent with its χ^2 like nature. Following this up in [5], Rosenblatt showed that this result held under very weak conditions in view of the representation of the statistic as a U statistic with a kernel dependent on the sample size. This was indeed a nice and useful observation.

Subsequently, we used our methods to consider the testing problem and examined the rate at which alternatives tend to the hypothesis with non-negligible power. Not surprisingly, the rate is less than $n^{-\frac{1}{2}}$ which is what one obtains for Kolmogorov-Smirnov and similar tests, although the quadratic deviation is consistently better. However, as results of Ingster and Suslina [4] show, minimax results for different smoothness classes of alternatives are obtained for tests which resemble quadratic density deviation measures. These results clearly depend on what class of alternatives is

R.A. Davis et al. (eds.), *Selected Works of Murray Rosenblatt*, Selected Works in Probability and Statistics, DOI 10.1007/978-1-4419-8339-8_1, © Springer Science+Business Media, LLC 2011

considered, $\{f : \|F - F_0\|_\infty \geq \delta\}$ where F is the cdf and f is the density is suitable for KS tests while $\{f : |f(x) - f(y)| \leq M|x - y|^\alpha$ all x, y, $\|f - f_0\|_2 \geq \delta\}$ is suitable for the test of [5].

Extensions of this work to the dependent and multivariate cases, and processes other than kernel density ones were developed quickly by Rosenblatt and others [6]. Murray and I only had one further paper in this area, [2], although we had certainly corresponded on [5] and others. The second paper was more probabilistic and derived results on the maxima of Gaussian random fields, which were later obtained more generally by Hogan and Siegmund [4] using a different method. We envisaged these results to be applied to the extension of the results of [1] to the multivariate case.

It is clear from all of the follow-up work that Rosenblatt's initial question was important and his intuition perfect.

References

[1] Bickel, P.J. and Rosenblatt M. (1973). On some global measures of the deviations of density function estimates, *Ann. Statist. 1*, 1071–1095.
[Correction: *Ann. Statist. 3* (1975), no. 6, 1370].

[2] Bickel P.J. and Rosenblatt M. (1973). Two-dimensional random fields, *Multivariate Analysis III*, 3–15, Academic Press NY.

[3] Hogan M.L. and Siegmund D. (1986). Large deviations for the maxima of some random fields, *Advances in Applied Probability 7*, 2–22.

[4] Ingster Y. I. and Suslina, I. A. (2000). Minimax nonparametric hypothesis testing for ellipsoids and Besov bodies. *ESAIM Probab. Statist. 4*, 53–135.

[5] Rosenblatt M. (1975). A quadratic measure of deviation of two-dimensional density estimates and a test of independence, *Ann. Statist. 3*, 1–14.

[6] Rosenblatt, M. (1976). On the maximal deviation of k-dimensional density estimates, *Ann. Probab. 4*, 1009–1015.

MURRAY ROSENBLATT'S CONTRIBUTIONS TO STRONG MIXING

By Richard C. Bradley

1. Introduction. Murray Rosenblatt's research has contributed much to the field of "strong mixing conditions," (i) by providing many results of his own in that field, and (ii) by inspiring a vast amount of research in that field by other people. This note will give just a small snapshot of (i) and an even smaller snapshot of (ii).

Within the broader field of probability theory, the field of "strong mixing conditions" seemed to "take off" with the publication in 1956 of the paper of Rosenblatt [17] titled "A central limit theorem and a strong mixing condition." Before that, weaker notions of "mixing" had been studied by Eberhard Hopf [9] and others in ergodic theory; and "strong" forms of "mixing" had been implicitly present in the work of Doeblin [4] involving what is now known as "Doeblin's condition" for Markov chains, and also in Doeblin's [5] work on continued fractions (see Iosifescu [13] for details on this latter work). But it was Rosenblatt's [17] paper — which introduced the "strong mixing" ("α-mixing") condition, showed how it "coordinated" well with a Bernstein-type blocking scheme and produced a central limit theorem under that condition — that really seemed to light a spark and lead to the development of "strong mixing conditions" as a field of its own. In particular, the 1956 paper inspired a considerable number of papers by other researchers in which variants of "strong mixing conditions" were proposed and used in various types of limit theorems. Those papers include, among many others, Ibragimov [10, 11], Kolmogorov and Rozanov [14], and Cogburn [2]. Subsequently, the field of "strong mixing conditions" exploded into an active and vibrant area of research.

Strong mixing conditions and the associated central limit theorems have enjoyed broad appeal beyond probability and into the mainstream statistics community. Applications of strong mixing include applications as diverse as block or stationary bootstrapping of time series (see e.g., Politis and Romano [16]); inference for linear time series; and parametric and nonparametric inference for nonlinear time series (see e.g., Fan and Yao [7]). Strong mixing is often the key tool for establishing asymptotic normality of various estimators of complicated time series models. In particular, inference for popular financial time series models such as GARCH and stochastic volatility models, often rely on the process being strongly mixing.

A short note such as this on Rosenblatt's work on strong mixing conditions cannot do justice to all of his own contributions to this field, and it cannot even begin to touch on the many key advances in this field that have been made by others. For details on the

R.A. Davis et al. (eds.), *Selected Works of Murray Rosenblatt*, Selected Works in Probability and Statistics, DOI 10.1007/978-1-4419-8339-8_2, © Springer Science+Business Media, LLC 2011

history of the development of the field, see e.g. [6] or the series [1]. This article will provide a sampling of just a few of the many contributions that Rosenblatt has made to this field, with a couple of examples of their direct inspiration of subsequent research by others.

To set the table for the discussion to follow, the rest of Section 1 here will be devoted to the definition of strong mixing and two other mixing conditions.

Suppose $(\Omega, \mathcal{F}, P)$ is a probability space. For any two σ-fields $\mathcal{A}$ and $\mathcal{B} \subset \mathcal{F}$, define the following three measures of dependence:

$$\alpha(\mathcal{A}, \mathcal{B}) := \sup_{A \in \mathcal{A}, B \in \mathcal{B}} |P(A \cap B) - P(A)P(B)|, \tag{1.1}$$

$$\phi(\mathcal{A}, \mathcal{B}) := \sup_{A \in \mathcal{A}, B \in \mathcal{B}, P(A)>0} |P(B|A) - P(B)|, \quad \text{and} \tag{1.2}$$

$$\rho(\mathcal{A}, \mathcal{B}) := \sup |\text{Corr}(f, g)| \tag{1.3}$$

where this last supremum is taken over all pairs of square-integrable random variables f and g such that f is $\mathcal{A}$-measurable and g is $\mathcal{B}$-measurable.

Now suppose $X := (X_k, k \in \mathbf{Z})$ is a strictly stationary sequence of random variables on our given probability space $(\Omega, \mathcal{F}, P)$. For each positive integer n, define the following three dependence coefficients:

$$\alpha(n) := \alpha(\sigma(X_k, k \le 0), \sigma(X_k, k \ge n)), \tag{1.4}$$

$$\phi(n) := \phi(\sigma(X_k, k \le 0), \sigma(X_k, k \ge n)), \quad \text{and} \tag{1.5}$$

$$\rho(n) := \rho(\sigma(X_k, k \le 0), \sigma(X_k, k \ge n)). \tag{1.6}$$

Here and below, the notation $\sigma(\ldots)$ means the σ-field of events generated by $(\ldots)$. The (strictly stationary) sequence X is said to be "strongly mixing" (or "α-mixing") if $\alpha(n) \to 0$ as $n \to \infty$, "ϕ-mixing" if $\phi(n) \to 0$ as $n \to \infty$, and "ρ-mixing" if $\rho(n) \to 0$ as $n \to \infty$. The "strong mixing" ("α-mixing") condition is the one that was introduced and used in a central limit theorem in the 1956 paper of Rosenblatt [17]. The "ϕ-mixing" condition was introduced by Ibragimov [10] and was studied by both Cogburn [2] and Ibragimov [11]. The "ρ-mixing" condition was introduced by Kolmogorov and Rozanov [14]. (The "maximal correlation coefficient" defined in (1.3) had been studied earlier in statistical contexts that did not involve "random sequences.") It is well known that ϕ-mixing implies ρ-mixing, that ρ-mixing implies strong mixing (α-mixing), and that the converses of both of those statements are false. For a formulation of other classic strong mixing conditions and their connections with the ones here, see e.g. [6] or [1, V1, Chapters 3 and 5].

2. Markov chains. An ongoing theme of research by many people is the question of what strong mixing properties are satisfied by random sequences with various other specific types of dependence structure. The books by Rosenblatt [20, 24, 26] gave (along with much other material) a treatment of different facets of this general question. In particular, strong mixing conditions for strictly stationary Markov chains (with, say, state space **R**) has been a topic of much research by Rosenblatt. A small selection of specific contributions of his on this topic will be briefly reviewed here.

Strong mixing conditions for Markov chains are treated extensively in Chapter 7 of Rosenblatt's 1971 book [20] on Markov processes. Along with much other information, that chapter gave what were, to the best of my knowledge, (1) the first known examples of strictly stationary Markov chains that are ϕ-mixing in the "usual" direction of time but *fail* to be ϕ-mixing in the "reversed" direction of time, and (2) the first known examples of strictly stationary ρ-mixing Markov chains such that for every positive integer n, the n-step (conditional) transition probability distributions are almost surely singular with respect to the marginal distribution.

Central limit theory under the ρ-mixing condition apparently started with central limit theorems of Rosenblatt [19] [20, Chapter 7] for square-integrable "instantaneous" functions of strictly stationary ρ-mixing Markov chains, and also with his paper [18] giving central limit theorems for some kernel-type estimators of marginal probability density for ρ-mixing Markov chains with an absolutely continuous marginal distribution. Shortly thereafter, the development of central limit theory for general (not necessarily Markovian) strictly stationary ρ-mixing sequences was promoted by Ibragimov [12]. (For more on that theory, see e.g. [1, V1, Chapter 11].)

In [18] and [20, Chapter 7], Rosenblatt used the Riesz Convexity (Interpolation) Theorem to prove the equivalence of a class of mixing conditions (including ρ-mixing) for Markov chains. That was apparently the first use of "interpolation theorems" of that type for comparison of different mixing conditions. This work subsequently inspired a more extensive use (not restricted to Markov chains) of various interpolation theorems to establish "covariance" and "cumulant" inequalities under various strong mixing conditions, and to compare different measures of dependence on which strong mixing conditions are based. For some of the main results and history of that later development, see [1, V1, Chapter 4], including the notes at the end of that chapter.

In [23], Rosenblatt showed that (in essence) an ultra-strong version of a "Markov regularity condition" that had been used in various papers, could not be satisfied by strictly stationary Markov chains other than the i. i. d. sequences. Around the same time, Rosenblatt conjectured to me that this condition could not be satisfied by *any* strictly stationary sequences, Markovian or not, except the i. i. d. sequences. Shortly thereafter, I was able to confirm the conjecture. For a detailed exposition, see [1, V2, Theorem 24.2].

3. A little known contribution. A remarkable paper of Rosenblatt [21] contains some ingenious fundamental insights in connection with the strong mixing condition for Markov chains. Unfortunately, this paper seems to have been practically unknown for at least thirty years after its publication. In this section, the content and the connections of this paper with the work of other researchers will be reviewed in some detail.

It will be useful to first formulate a classic result of Doeblin [4] (just the "aperiodic" case) involving what is now known as "Doeblin's condition" for Markov chains. It is well known that Doeblin's result can be reformulated in a different but equivalent way in terms of the ϕ-mixing condition (which was defined a couple of decades later), as follows:

Theorem 3.1 (Doeblin (reformulated)). *Suppose $X := (X_k, k \in \mathbf{Z})$ is a strictly stationary Markov chain which is ergodic, aperiodic, and satisfies $\phi(n) < 1$ for some $n \geq 1$. Then $\phi(n) \to 0$ (at least exponentially fast) as $n \to \infty$.*

We shall return to this theorem below, after picking up some other threads.

For strictly stationary random sequences, Cogburn [2] introduced, under the name "uniform ergodicity," a "Cesaro" variant of the strong mixing condition, and (along with other results) proved a central limit theorem involving this condition. The uniform ergodicity condition need not be formally defined here. Now (for strictly stationary sequences) strong mixing implies uniform ergodicity, and uniform ergodicity implies ergodicity. A natural question, pertinent to the development of limit theory for strictly stationary sequences, is how much weaker uniform ergodicity is than strong mixing. Rosenblatt [21] examined this question and proved the following result involving Markov chains:

Theorem 3.2 (Rosenblatt). *Suppose $X := (X_k, k \in \mathbf{Z})$ is a strictly stationary Markov chain which is "mixing" (in the ergodic-theoretic sense) and satisfies uniform ergodicity; then $\alpha(n) \to 0$ as $n \to \infty$ (strong mixing).*

In the same paper, Rosenblatt [21] also suggested that this result might still hold for general strictly stationary sequences (i.e., without the Markov assumption). That now seems to be an open question, with (so far) no decisive evidence either way.

Formally, Theorem 3.2 was the "goal" of Rosenblatt's [21] paper. However, the various arguments in this paper contained further peculiar and ingenious insights that have remained little known. In particular, Rosenblatt showed that if $X := (X_k, k \in \mathbf{Z})$ is a strictly stationary Markov chain that *fails* to satisfy strong mixing, then there exists a number $q \in (0, 1)$ such that the following condition holds: For every positive integer N and every $\theta > 0$, there exist events $G \in \sigma(X_k, k \leq 0)$ and $H \in \sigma(X_k, k \geq N)$ such that $q - \theta < P(H) < q + \theta$ and $P(G \Delta H) < \theta$. (Here Δ denotes symmetric difference. Obviously, adjusting the parameters a little, one can have $q - \theta < P(G) < q + \theta$ as well.) Rosenblatt also showed that this condition holds for *every* $q \in (0, 1)$ under the additional assumption of a trivial (past or) future "tail σ-field." Combining this fact with $q = 1/2$ and other arguments in Rosenblatt's [21] paper, implicitly yields the following theorem:

Theorem 3.3 (Rosenblatt). *Suppose $X := (X_k, k \in \mathbf{Z})$ is a strictly stationary Markov chain which is ergodic, aperiodic, and satisfies $\alpha(n) < 1/4$ for some $n \geq 1$. Then $\alpha(n) \to 0$ as $n \to \infty$.*

Now for any two σ-fields $\mathcal{A}$ and $\mathcal{B}$, one has by trivial arguments that $\alpha(\mathcal{A},\mathcal{B}) \le 1/4$ and $\phi(\mathcal{A},\mathcal{B}) \le 1$. Hence for a given strictly stationary sequence (Markovian or not), one has that $\alpha(n) \le 1/4$ and $\phi(n) \le 1$ for every positive integer n. Thus Theorem 3.3 is a nearly exact analog, for strong mixing, of Doeblin's result (transcribed to "ϕ-mixing") in Theorem 3.1. The only significant difference is that in the context of Theorem 3.1, the mixing rate has to be (at least as fast as) exponential, whereas in Theorem 3.3, the mixing rate need not be exponential (as was shown, for example, for the Markov chains studied by Davydov [3]). Yet while Theorem 3.1 receives frequent attention in the literature, Theorem 3.3 was apparently "absent from the radar screen" for three decades after the publication of [21].

For a detailed exposition of the material in Rosenblatt's [21] paper (including Theorems 3.2 and 3.3 and the other insights from that paper described here), see [1, V2, Sections 21.25-21.28 and 24.17-24.22].

4. Strong mixing conditions and estimation problems for random fields. At this point, we depart from Markov chains and turn to some contributions of Rosenblatt involving more general random sequences as well as random fields.

As Richard Olshen explained to me about 30 years ago, a major motivation behind Rosenblatt's [17] paper and the subsequent development of the field of strong mixing conditions, was to provide a way of proving limit theorems for use in statistical inference for data from time series that seemed to be "weakly dependent" but did not necessarily seem to "fit" more specific models of dependence such as Markov chains, martingales, Gaussian sequences, or ARMA models. Since then, in that spirit, a vast amount of research has been done by many researchers on statistical estimation of various kinds under strong mixing conditions. The 1985 book of Rosenblatt [24] was a prominent treatise on this topic. Further, much of that book gave a treatment of random fields (not just random sequences) satisfying strong mixing conditions, an ongoing topic of research of Rosenblatt and many others. We shall return to that book after providing some background information.

Suppose d is a positive integer, and $X := (X_k, k \in \mathbf{Z}^d)$ is a strictly stationary random field. For a given positive integer n, define

$$\alpha^*(n) := \sup \alpha(\sigma(X_k, k \in A), \sigma(X_k, k \in B)), \tag{4.1}$$

where the supremum is taken over all pairs of nonempty, disjoint sets $A, B \subset \mathbf{Z}^d$ such that

$$\text{dist}(A, B) := \min_{a \in A, b \in B} \|a - b\| \ge n. \tag{4.2}$$

Here, for a given $k \in \mathbf{Z}^d$, $\|k\|$ denotes the Euclidean norm of k. The (strictly stationary) random field X will be said to be "α^*-mixing" if $\alpha^*(n) \to 0$ as $n \to \infty$.

In the case $d = 1$ (the classic case of strictly stationary random sequences), α^*-mixing is a stronger condition than the strong mixing (α-mixing) condition (involving (1.4)), because in (4.1) the two index sets A and B are not restricted to "past" and "future" (as in (1.4)); they can instead be "interlaced," with each set having elements between ones in the other set.

Extending well known arguments and results of Kolmogorov and Rozanov [14] involving the usual strong mixing and ρ-mixing conditions in the case of stationary *Gaussian* sequences, Rosenblatt [22] showed that for a given stationary *Gaussian* random field $X := (X_k, k \in \mathbf{Z}^d)$, (i) the α^*-mixing condition is equivalent to the corresponding "ρ^*-mixing" condition (defined using in the right hand side of (4.1) the maximal correlation coefficient from (1.3)), and (ii) those equivalent conditions are satisfied if X has a continuous positive spectral density function. (An exposition of those results was also given in [24, pp. 73-77].)

Later, it became known that observation (i) holds for general (not necessarily Gaussian) strictly stationary random fields $X := (X_k, k \in \mathbf{Z}^d)$. (See [1, V3, Theorem 29.12(II)(A)] for the result, its proof, and its history.) We shall continue to use the term "α^*-mixing" (instead of "ρ^*-mixing") in the discussion below. Of course in contrast, for random sequences (the case $d = 1$), the strong mixing (α-mixing) condition (based on (1.4), with the two index sets being just "past" and "future") is strictly weaker than the ρ-mixing condition (based similarly on (1.6)).

(In a slightly different vein, the paper of Rosenblatt [22] also gave a central limit theorem for random fields under "martingale-like" assumptions similar to those used by Gordin [8] in a classic central limit theorem for random sequences.)

Now let us return to the 1985 book of Rosenblatt [24]. This book gives an extensive treatment of various estimation problems for strictly stationary α-mixing random sequences $X := (X_k, k \in \mathbf{Z})$ and strictly stationary α^*-mixing random fields $X := (X_k, k \in \mathbf{Z}^d)$. The problems considered there include the estimation of (under appropriate assumptions) autocovariances, spectral densities, higher-order spectral densities, marginal probability densities, and regression functions. The book contains a wealth of results and opened up a broad spectrum of new research problems.

In the formulation of most of the theorems in the 1985 book involving α-mixing or α^*-mixing, there is no assumption on the rate of convergence of $\alpha(n)$ or $\alpha^*(n)$ to 0; instead the "mixing rate" is allowed to be arbitrarily slow. To compensate, there typically are assumptions on higher-order moments and/or higher-order cumulants. For example, in a central limit theorem for some estimators of spectral density for strictly stationary random fields, Rosenblatt [24, p. 157, Theorem 7] assumes (in addition to α^*-mixing) the summability of cumulant functions of the X_k's up to order 8. Now the usual estimators of spectral density (such as ones based on periodograms) involve quadratic forms of the X_k's, and hence the natural basic moment assumption for a central limit theorem for those estimators is $EX_0^4 < \infty$, in order for the quadratic forms to have finite second moments. A couple of years after the publication of his book, Rosenblatt [25] posed the problem of whether in theorems such as the one alluded to here, one could get by with just the summability of cumulants of order 4 (instead of order 8), or even with just the assumption of finite fourth moments (of course in addition to the assumption of α^*-mixing).

In his 1994 Ph.D. thesis at Indiana University (see also [15, Theorem 2]), Curtis Miller gave an affirmative answer to that question for a class of estimators that involve "blocks" of the X_k's. Miller's result only assumes $EX_0^4 < \infty$; no assumption of moments of higher than fourth order and no further assumption on covariances or cumulants are required. (Miller formally used the "ρ^*-mixing" condition, but under strict stationarity that condition is equivalent to α^*-mixing as noted above.) The question is apparently still open for other estimators of spectral density, such as ones that involve averaging the periodograms over neighboring frequencies. Anyhow, this is just another particular example of the many ways in which the work of Rosenblatt on strong mixing conditions has directly inspired the research in the field by others, including young researchers.

Starting with his 1956 paper [17], Murray Rosenblatt has contributed a great deal to the field of strong mixing conditions, and has helped establish it as an active field of research. His work has directly inspired many contributions in that field by other researchers. Strong mixing conditions continue to be an active field of research that has important modeling and inference ramifications in a variety of applications.

Acknowledgement: My thanks go to Richard Davis, who made many helpful comments and suggestions and provided the third paragraph of this article.

References

[1] R.C. Bradley. *Introduction to Strong Mixing Conditions,* Volumes 1, 2, and 3. Kendrick Press, Heber City (Utah), 2007.

[2] R. Cogburn. Asymptotic properties of stationary sequences. *Univ. Calif. Publ. Statist.* 3 (1960) 99-146.

[3] Yu.A. Davydov. Mixing conditions for Markov chains. *Theor. Probab. Appl.* 18 (1973) 312-328.

[4] W. Doeblin. Sur les propriétés asymptotiques de mouvement régis par certains types de chaines simples. *Bull. Math. Soc. Roum. Sci.* 39 (1937), no. 1, 57-115, no. 2, 3-61.

[5] W. Doeblin. Remarques sur las théorie métriques des fractions continues. *Compositio Math.* 7 (1940) 353-371.

[6] P. Doukhan. *Mixing: Properties and Examples.* Springer, New York, 1995.

[7] J. Fan and Q. Yao. *Nonlinear Time Series: Nonparametric and Parametric Methods.* Springer, New York, 2003.

[8] M.I. Gordin. The central limit theorem for stationary processes. *Soviet Math. Doklady* 10 (1969) 1174-1176.

[9] E. Hopf. *Ergodentheorie.* Springer-Verlag, Berlin, 1937.

[10] I.A. Ibragimov. Some limit theorems for stochastic processes stationary in the strict sense. *Dokl. Akad. Nauk. SSSR* 125 (1959) 711-714.

[11] I.A. Ibragimov. Some limit theorems for stationary processes. *Theor. Probab. Appl.* 7 (1962) 349-382.

[12] I.A. Ibragimov. A note on the central limit theorem for dependent random variables. *Theor. Probab. Appl.* 20 (1975) 135-141.

[13] M. Iosifescu. Doeblin and the metric theory of continued fractions: A functional theoretic solution to Gauss' 1812 problem. In: *Doeblin and Modern Probability,* (H. Cohn, ed.), pp. 97-110. Contemporary Mathematics 149. American Mathematical Society, Providence (Rhode Island), 1993.

[14] A.N. Kolmogorov and Yu. A. Rozanov. On strong mixing conditions for stationary Gaussian processes. *Theor. Probab. Appl.* 5 (1960) 204-208.

[15] C. Miller. A CLT for the periodograms of a ρ^*-mixing random field. *Stochastic Process. Appl.* 60 (1995) 313-330.

[16] D.N. Politis and J.P. Romano. The stationary bootstrap. *J. Amer. Statist. Assoc.* 89 (1994) 1303-1313.

[17] M. Rosenblatt. A central limit theorem and a strong mixing condition. *Proc. Natl. Acad. Sci. USA* 42 (1956) 43-47.

[18] M. Rosenblatt. Density estimates and Markov sequences. In: *Nonparametric Techniques in Statistical Inference,* (M.L. Puri, ed.), pp. 199-210. Cambridge University Press, London, 1970.

[19] M. Rosenblatt. A prediction problem and central limit theorems for stationary Markov sequences. In: *Proceedings of the Twelfth Biennial Seminar of the Canadian Mathematical Congress on Time Series and Stochastic Processes; Convexity and Combinatorics (Vancouver, B.C., 1969),* (R. Pyke, ed.), pp. 99-114. Canadian Mathematical Congress, Montreal, 1970.

[20] M. Rosenblatt. *Markov Processes. Structure and Asymptotic Behavior.* Springer-Verlag, Berlin, 1971.

[21] M. Rosenblatt. Uniform ergodicity and strong mixing. *Z. Wahrsch. verw. Gebiete* 24 (1972) 79-84.

[22] M. Rosenblatt. Central limit theorems for stationary processes. *Proceedings of the Sixth Berkeley Symposium on Probability and Statistics,* Volume 2, pp. 551-561. University of California Press, Los Angeles, 1972.

[23] M. Rosenblatt. Some remarks on a mixing condition. *Ann. Probab.* 7 (1979) 170-172.

[24] M. Rosenblatt. *Stationary Sequences and Random Fields.* Birkhäuser, Boston, 1985.

[25] M. Rosenblatt. Private communication, 1987.

[26] M. Rosenblatt. *Gaussian and Non-Gaussian Linear Time Series and Random Fields.* Springer, New York, 2000.

Murray Rosenblatt and cumulant/higher-order/poly spectra

By David R. Brillinger

The bispectrum, or third order spectrum, of a stationary process has been around at least since the early 50s, for example it appeared in the paper Tukey (1953). It was studied in some detail in John Van Ness's thesis, "Estimates of the Bispectrum of Stationary Random Processes", supervised by Murray Rosenblatt. A 1965 Annals of Mathematical Statistics paper, "Estimation of the bispectra", Rosenblatt and Van Ness (1965) followed.

There is a Soviet history extending the concept. In particular Shiryaev (1989) writes: "In the late 1950s and early 1960s Kolmogorov suggested to his pupils V. P. Leonov and A. N. Shiryaev a series of problems related to the issue of nonlinear analysis of random processes (in particular, in radio technology) which brought about the techniques of calculating cumulants under nonlinear transformations, and the development of the theory of spectral analysis of high-order moments of stationary random processes."

I first met Murray in New Jersey in 1963. He was then consulting at Bell Telephone Laboratories Murray Hill under a Brown University program. In particular he was developing properties of a cepstrum estimate, the cepstrum being the Fourier transform of the log of a second-order spectrum. It had been developed as a tool for estimating how deep explosions, as opposed to earthquakes, were sited.

The study in the polyspectra paper, Brillinger (1965), started out as a young academic having fun. Independently of the Russian work I had found that cumulants were a pertinent tool. They were a multilinear operator with the property that if for a vector-valued variate some subset of its entries was independent of the complementary subset, then the joint cumulant of all was zero. This property takes one directly to a definition of mixing for general stationary processes. Perhaps the Russians already knew that type of result. At one point Tukey had mentioned the word polyspectra, and a connection was made. The 1965 paper surely led to my getting to collaborate with Murray later when we were both in London. Ady and Murray had come to London on a Guggenheim Fellowship in the Fall of 1965. On meeting Murray again I was surely a bit disrespectful going right up to Murray and saying something like, ``How about we do some work and write a paper together?" Murray replied with a remark like ``Sure.", and the papers Brillinger and Rosenblatt (1967a, b, c) resulted.

The first, Brillinger and Rosenblatt (1967a,) summarizes the work to be presented in (1967b) and (1967c). It is noteworthy for also seeking to develop a tensor-like notation for joint cumulant spectra after encouragement from John Tukey.

R.A. Davis et al. (eds.), *Selected Works of Murray Rosenblatt*, Selected Works in Probability and Statistics,
DOI 10.1007/978-1-4419-8339-8_3,

Brilllinger and Rosenblatt (1967b) sets down some asymptotic theory of estimates of a cumulant spectrum developed under an asymptotic independence or mixing assumption. The estimate is based on an empirical Fourier transform of a long stretch of data. A higher order periodogram is defined and smoothed to form the estimate. An important insight in the work is to base the estimates on the Fourier transforms evaluated at the Fourier frequencies. Large sample means and variances of the estimate were developed, with uniform error bounds. In developing the results substantial use was made of the theorem in Leonov and Shiryaev (1959). The mixing condition involved the absolute summability of the time domain joint cumulant functions. Asymptotic normality was shown by proving that the standardized cumulants of order greater that 2 tended to those of the normal, which is determined by its moments.

Brillinger and Rosenblatt (1967c) concerns properties and interpretations of cumulant spectra as well as details of computation and a scientific example. The thorny problem of aliasing is addressed with fundamental domains for the second, third and fourth order cases presented. Estimates of the second, third and fourth order spectra are computed for a long series of monthly relative sunspot numbers. A fast Fourier transform is employed in the computations.

Murray was an academic role model for me as my career progressed. He was someone to ask to write recommendation letters and to provide advice concerning the professional sides of academic life. For example, I remember being quite concerned when I found an error in a paper that I had published. Murray's reaction was, "Is that the first one?" There have been others since.

References.

D. R. Brillinger, "An introduction to polyspectra", Ann. Math. Statist. 36 (1965) 1351-1374.

D. R. Brillinger and M. Rosenblatt: "Asymptotic theory of estimates of kth order spectra," Proc. Nat. Acad. Sci., 57 (1967a), 206-210.

D. R. Brillinger and M. Rosenblatt: "Asymptotic theory of estimates of kth order spectra," Spectral Analysis of Time Series, John Wiley (1967b), 153-188.

D. R. Brillinger and M. Rosenblatt. "Computation and interpretation of kth order spectra," in Spectral Analysis of Time Series, John Wiley (1967c), 189-232.

M. Rosenblatt: "Remarks on higher order spectra," in Multivariate Analysis, Academic Press (1966), 383-389

M. Rosenblatt: "Cumulants and cumulant spectra," in Handbook of Statistics, vol. 3 (eds. D. Brillinger and P. Krishnaiah), (1983), 369-382.

M. Rosenblatt and J. W. Van Ness: "Estimation of the bispectra," Ann. Math. Stat., 36 (1965), 1120-1136.

A. N. Shiryaev, "Kolmogorov: life and creative activities." Ann Probab. 17 (1989) 866-944.

J. W. Tukey: "The spectral representation and transformation properties of the higher moments of stationary time series." Reprinted in The Collected Works of John W. Tukey, Vol. 1 (Edited by D. R. Brillinger), 165-184. Wadsworth, Belmont.

J. Van Ness "Estimates of the Bispectrum of Stationary Random Processes" 1964 Ph.D. Dissertation, Brown University.

Rosenblatt's Contributions to Deconvolution

By Keh-Shin Lii

A simple model of deconvolution can be described as observing $\{x(t)\}$ which is a convolution of a signal $\{s(t)\}$ with a filter $\{f(j)\}$, $x = s * f$. More specifically, we have

$$x(t) = \sum_{j=-\infty}^{\infty} f(j)s(t-j).$$

The problem of deconvolution is to recover $\{s(t)\}$ based on the output process $\{x(t)\}$. If the filter $\{f(j)\}$ is known then the problem is fairly straightforward. The blind deconvolution, in signal processing terminology, is to recover $\{s(t)\}$ based solely on $\{x(t)\}$ without knowing $\{f(j)\}$. Statisticians may be more interested in the estimation of $\{f(j)\}$ under certain conditions on $\{s(t)\}$ and $\{f(j)\}$. This problem and its many variations have very broad applications in signal processing, image restoration, geo-exploration, seismology, radio astronomy among others [11, 33, 38, 39].

Assume that the signal random variables $\{s(t)\}$ are independent and identically distributed with mean 0 and variance 1. Let the filter $\{f(j)\}$ be a sequence of real constants such that

$$\sum_{j=-\infty}^{\infty} f^2(j) < \infty$$

and $f(z) = \sum_j f(j)z^j$ be the z-transform corresponding to the process $\{x(t)\}$. Then

$$f(e^{-i\lambda}) = \sum_j f(j)e^{-ij\lambda} = |f(e^{-i\lambda})| \exp\{ih(\lambda)\}$$

is the frequency response function or the transfer function where $h(\lambda)$ is the phase function of the transfer function. If we know $f(e^{-i\lambda})$ for all $\lambda \in [0, 2\pi]$, then we can obtain $\{f(j)\}$ for all j. The modulus $f(j)$, $|f(j)|$, of the frequency response function can be obtained from

R.A. Davis et al. (eds.), *Selected Works of Murray Rosenblatt*, Selected Works in Probability and Statistics,
DOI 10.1007/978-1-4419-8339-8_4,

the spectral density of $\{x(t)\}$ which is

$$g(\lambda) = \frac{1}{2\pi}|f(e^{-i\lambda})|^2.$$

It is clear that if the $\{x(t)\}$ process or the random signal process $\{s(t)\}$ is Gaussian then the full probability structure of $\{x(t)\}$ is determined by $g(\lambda)$ or equivalently by $|f(e^{-i\lambda})|$ which is determined by the second order covariance property of the process. The phase information $h(\lambda)$ of $f(e^{-i\lambda})$ is not identifiable in the Gaussian case. Any hope of getting information on the phase function will require the process to be nonGaussian.

Murray Rosenblatt's interest and insight in this problem may have stemmed from his interest in the higher order spectra, especially higher-order cumulant spectra which is fundamental in dealing with nonGaussian processes [34, 4, 5, 19]. The kth-order cumulant spectral density of $\{x(t)\}$ is given by

$$b_k(\lambda_1, \lambda_2, \cdots, \lambda_{k-1}) = \frac{\gamma_k}{(2\pi)^{k-1}} f(e^{-i\lambda_1}) f(e^{-i\lambda_2}) \cdots f(e^{-i\lambda_{k-1}}) f(e^{i(\lambda_1+\lambda_2+\cdots+\lambda_{k-1})}),$$

where γ_k is the kth order cumulant of $\{s(t)\}$. Therefore the phase of the kth order spectrum $b_k(\lambda_1, \lambda_2, \cdots, \lambda_{k-1})$ is related to the phase $h(\lambda)$ of the the transfer function $f(e^{i\lambda})$.

In the paper [35], Murray laid out the basic idea on how the phase function $h(\lambda)$ can be identified up to a linear shift of $c\lambda$ using the phase of the bispectra of the process. It was noted that the same result holds by using any kth-order spectra with $k > 2$. A similar but different idea of using bispectra to obtain phase information of the transfer function is outlined in Brillinger [6]. Of course for different c's we will get different transfer functions and the deconvolution can not be realized.

The problem of indeterminancy of the linear shift of the phase was solved in [20], where it was shown that the phase function can be identified up to an integer shift $\tau\lambda$ with τ an integer. The basic idea is that the transfer function $f(e^{-i\lambda})$ has to be real when $\lambda = \pi$ and hence its phase at π has to be $\tau\pi$ with τ an integer. Since this integer phase shift corresponds to reindexing the noise sequence $\{s(t)\}$, the deconvolution problem is essentially solved in this case. If the linear process $\{x(t)\}$ is a finite parameter ARMA(p,q) process then there are other methods available to identify the phase function. In [20] a few other ideas were laid out to identify the correct ARMA model in terms of the location of the roots of the characteristic functions of the ARMA model. Again, the basic idea here is to match various higher order moments or cumulants of the process with the correct coeffients or roots.

The paper [20] seemed to have generated quite a bit interest in the signal proceesing community. The importance of the phase in signals has been noted in signal and image processing and geoexploration problems [31, 33, 38, 39]. Many papers were published using higher-order statistics and higher-order spectral analysis to analyze signals, images and systems. These activities resulted in the first 'Workshop on Higher-Order Spectral Analysis' held at Vail, Colorado in 1989 with a Proceedings of the Workshop and a 'Special Section on Higher Order Spectral' that appeared in the July 1990 issue of the IEEE Transactions on Acoustics, Speech and Signal Processing. Following the first workshop, four more 'workshops on higher-order statistics' were held in odd years with the location alternating betwwen Europe and North America until the last one in 1997. The last two Proceedings published in the 1995 and 1997 Workshops have more than 450 pages of articles each. The 1995 workshop resulted in IEEE Signal Processing September 1996 issue being the 'special issue on higher-order statistics'. A book [32] was published in 1993 on these topics.

If the output process $\{x(t)\}$ is observed with an independent Gaussian noise $\eta(t)$ it is shown in [21] that we can still estimate the filter $\{f(j)\}$ up to an unknown scale factor c. If the additive Gaussian noise $\eta(t)$ is white then we can estimate the filter consistently up to a time shift. A detailed discussion on the use of fourth-order cumulants to estimate the filter function for deconvolution is given in [22], where a discussion on the sample size relative to various orders of cumulants of $\{s(t)\}$ is given. In [23] it is demonstrated with a well-log data and water-gun signature that the deconvolution can be more effective using the Fourier transform of the tapered sample higher-order cumulants to obtain higher-order spectrum instead of using the smoothed higher-order periodogram. For this reason, theoretical properties of such higher-order spectral estimates are given in [25, 26]. The particular formula using the phase of a higher-order spectrum to obtain the phase of the transfer function in [20] utilized only a subset of the full higher-order spectral phase function. Another method which utilized the whole effective higher-order spectral phase function is given in [28] with better asymptotic convergence properties as a function of sample size. Generlization of the previous results to the deconvolution problem in autoregressive random fields is given in [30, 37].

In the deconvolution problem described above it is generally assumed that the spectral density of the process is strictly positive. In [24] our attention turned to the blind deconvolution problem when the transfer function has zeros, which do occur in geophysical investigations [31]. In [24] it is shown that if the zeros are finitely many and are of finite order then the transfer function can still be consistently estimated without the *minimum phase* assumption when the process is nonGaussian. A procedure is given so that the deconvolution can be effectively carried out. If the transfer function is zero in an interval for $\lambda \in (a, b)$ with $0 < a < b < \pi$, then the smallest mean square error in deconvolution that can be achieved is $(b - a)/2\pi$ even if we assume that the phase can be estimated. It was also shown that the phase is not identifiable if the transfer function is zero in an interval around π with positive length i.e., the process is band-limited. However if all higher order cumulants of $\{x(t)\}$ are available then the phase is identifiable up to a linear phase shit $\delta\lambda$ with δ a real number.

In the previous discussion the process is a general linear process and the approach to the identification of the phase information is the use of higher-order cumulant spectra which in general, is 'nonparametric' in the sense that the probability distribution of the signal process

$\{s(t)\}$ is not used explicitly other than that it is nonGaussian and that a certain kth order cumulant exists and is non-zero. Now if the process $\{x(t)\}$ can be represented by a finite parameter ARMA(p,q) process, then as noted before certain methods based on moments can be used to identify the phase information and in this case it is equivalent to finding the locations of the roots of the charateristic functions associated with the AR and MA parts of the process or eqivalently the coefficients of the ARMA model without the usual causal or invertible conditions under the Gaussian assumption. Maximum likelihood estimation of the parameters of an ARMA model under the Gaussian assumption has been discussed widely in the literature [9, 36]. Of course in this case the phase information is not available and the ARMA model is assumed to be causal and invertible. In [7] it is shown that given a nonGaussian probability density for the independent and identically distributed process $\{s(t)\}$, the stationary AR(p) process

$$x(t) = \phi_1 x(t-1) + \ldots + \phi_p x(t-p) + s(t)$$

is identifiable through a maximum likelihood procedure whether the process is causal or not. The idea is to reparametrize the model by decomposing the autoregressive polynomial into its causal and purely non-causal components and then analyzing the corresponding AR processes that resulted from this decomposition. The likelihood function can be approximated using these processes and the estimates of the parameters of the possibly non-causal AR process are the solutions to the likelihood equations. Similar results for possibly non-invertible MA(q) process are given in [27] and for general ARMA process in [29]. These results give possibly 'efficient' methods for blind deconvolution. If the nonGaussian probability distribution of the $\{s(t)\}$ process is unknown then it is demonstrated in the previous papers that a quasi-likelihood method can be used to estimate the parameters of the ARMA process by assuming that the unknown probability density is Laplace (two-sided exponentail) which leads to a least absolute deviation criterion. A modified least absolute deviation method is shown to be consistent when the input process in the AR(p) model has a stable law distribution with index $\alpha \in (1,2)$ [13].

Without the causal and invertible conditions for nonGaussian ARMA processes one is naturally led to an interesting subclass of processes called all-pass ARMA models. An ARMA model is all-pass if every root of the AR polynomial $\phi(z)$ is a reciprocal of a root of the MA polynomial $\theta(z)$ matched with multiplicity and vice versa. In such a case the corresponding filter is $f(e^{-i\lambda}) = \phi(e^{-i\lambda})/\theta(e^{-i\lambda})$ and $|f(e^{-i\lambda})|$ is a constant. This means that whatever the spectum of the input process $\{s(t)\}$, it will pass the ARMA filter unchanged except for a multiplicative constant, hence the name all-pass. So the spectral density of the process $\{x(t)\}$

$$g(\lambda) = \frac{1}{2\pi}\left|\frac{\phi(e^{-i\lambda})}{\theta(e^{-i\lambda})}\right|^2$$

is a constant. This means that the process $\{x(t)\}$ is uncorrelated or white but not independent if the input independent process is nonGaussian. Processes which are second order uncorrelated or white but with higher order dependence occur often in financial data. The estimation of parameters of such all-pass ARMA models using least absolute deviations is given in [8], using maximum likelihood in [1] and using rank based procedures in [2]. The rank based method can have the same asymtotic efficiency as maximum likelihood estimators and are robust to some distributional assumptions. There are deconvolution problems in signal processsing when the probability distribution of the signal process $\{s(t)\}$ is discrete with finitely many points of support such as in the finite alphabet transmission. These deconvolution problems were treated in [17, 12]. The finite tone image deconvolution or debluriing problems were treated in [18] where the distribution of the pixels is two-tone (or finite-tone) without the stationarity assumption. Methods using the maximization of the standardized cumulant of the deconvolved process to estimate the filter are given in [10].

Murray's interest in higher-order spectra began more than four decades ago [34], which ultimately led to the deconvolution problem [35]. Murray's fundamental contributions in both areas have had a long lasting impact on many aspects of statistical probems and applications as described in this brief article. His influence in the area of blind or noncausal deconvolution is still ongoing [3], and has expanded to many related problems in economics [15], in medicine [14], and in signal processing [16,40], among others.

References

[1] Andrews, B., Davis, R.A., and Breidt, F. J. (2006). "Maximum Likelihood Estimation for All-Pass Time Series Models". J. Multivariate Analysis. 97, 1638-1659

[2] Andrews, B., Davis, R.A., and Breidt, F. J. (2007). "Rank Estimation for All-Pass Time Series Models". Annals of Statistics, 35, 844-869.

[3] Andrews, B., Calder, M. and Davis, R.A. (2009). "Maximum likelihood estimation for α-stable autoregressive processes". Annals of Statistics, 37, 1946-1982.

[4] Brillinger, D. R., Rosenblatt, M. (1967a) "Asymptotic theory of k-th order spectra." In *Spectral Analysis of Time Series*, Ed. B. Harris, 153-188. New York, Wiley.

[5] Brillinger, D. R., Rosenblatt, M. (1967b) "Computation and interpretation of k-th order spectra." In *Spectral Analysis of Time Series*, Ed. B. Harris, 189-232. New York, Wiley.

[6] Brillinger, D. (1977). " The Identification of a Particular Nonlinear Time Series System," Biometrika, 64, 509-515.

[7] Breidt, F. J., Davis, R. A., Lii, K. S., and Rosenblatt, M. (1991). "Maximum Likelihood Estimation for Noncausal Autoregressive Processes", J. Multivariate Analysis, 36, 175-198.

[8] Breidt, F.J., Davis, R.A., and Trindade, A. (2001). "Least Absolute Deviation Estimation for All-Pass Time Series Models". Annals of Statistics 29, 919-946

[9] Brockwell. P. J., and Davis, R. A. (1987). Time Series: Theory and Methods. Springer-Verlag, New York.

[10] Cheng, Q. (1990). "Maximum standardized cumulant deconvolution of non-Gaussian linear processes", Annals of Statistics, 18, 1774-1783.

[11] Donoho, D. (1981). "On minimum entropy deconvolution", In Applied Time Series Analysis, II. Ed. by Findly, D. F. 565-608. Academic Press New York.

[12] Gamboa,F., and Gassiat, E. (1996). "Blind deconvolution of discrete linear systems", Annals of Statistics, 24, 1964-1981.

[13] Jian, H., and Pawitan, Y. (1999). "Consistent estimation for non-Gaussian non-causal autoregressive processes", J. of Time Series Analysis, 20, 417-423.

[14] Kannnan, G,, Milani, A. A., Panahi, I., and Briggs, R. (2008). "Equalizing secondary path effects using the periodicity of fMRI accoustic noise", 30th Annual Intrernational IEEE EMBS Conference, Cancouver, British Columbia, Canada, August 20-24, 2008, 25-28.

[15] Lanne, M., and Saikkonen, P. (2009). "Noncausal vector autoregression", Bank of Finland Research, Discussion papapers 18, 2009.

[16] Larue, A., Mars, J. I., and Jutten, C. (2006). "Frequency-domain blind deconvolution based on mutual information rate", IEEE Transactions on Signal Processing, 54, 1771-1781.

[17] Li, T. H. (1995). "Blind deconvolution of linear systems with multilevel stationary inputs", Annals of Statistics, 23 690-704.

[18] Li, T. H., and Lii, K. S. (2002). "A joint estimation approach for tw-tone image deblurring by blind deconvolution", IEEE Tansactions on Image Processing, 11, 847-858.

[19] Lii, K. S., Rosenblatt, M., and Van Atta, C. (1976). "Bispectra measurements in turbulance", Part 1, J. Fluid Mechanics, 77, 45-62.

[20] Lii, K. S., and Rosenblatt, M. (1982)."Deconvolution and estimation of the transfer function phase and coefficients for non-Gaussian linear processes," Ann. Statist., 10, 1195-1208.

[21] Lii, K. S., and Rosenblatt, M. (1984). "Remarks on nonGaussian linear processes with additive Gaussian noise". Lecture Notes in Statistics, V. 26, 185-197. In "Robust and Nonlinear Time Series Analysis" Ed. by Franke, J., Hardle, W., and Martin, D.

[22] Lii, K. S., and Rosenblatt, M. (1985). "A fourth order deconvolution technique for nonGaussian linear processes". In Krishnaiah, P. R. (eds), Multivariate Analysis VI, 395-410.

[23] Lii, K. S., and Rosenblatt, M. (1988). "Nonminimum phase non-Gaussian deconvolution", J. Multivariate Analysis, 27, 359-374.

[24] Lii, K. S., and Rosenblatt, M. (1988)."Estimation and Deconvolution When the Transfer Function Has Zeros," J. Theoretical Probability, 93-113.

[25] Lii, K. S., and Rosenblatt, M. (1990)."Cumulant spectral estimates: Bias and covariance," Proc. Third Hungarian Colloquium on Limit Theorems in Probability and Statistics (held at PECS), 365-405.

[26] Lii, K. S., and Rosenblatt, M. (1990)."Asymptotic normality of cumulant spectral estimates," J. Theoretical Probability, 3, 367-385.

[27] Lii, K. S., and Rosenblatt, M. (1992). "An Approximate Maximum Likelihood Estimation for Non-Gaussian Non-Minimum Phase Moving Average processes", J. Multivariate Analysis, 43, 272-299.

[28] Lii, K. S., and Rosenblatt, M. (1993). "Bispectra and phase of nonGaussian linear processes", J. Theoretcal Probability, 6, 579-593.

[29] Lii, K. S., and Rosenblatt, M. (1996). "Maximum Likelihood Estimation for NoGaussian NonMinimum Phase ARMA Sequences", Statistica Sinica, 6, 1-22.

[30] Lii, K. S., and Rosenblatt, M. (1996)."NonGaussian autoregressive sequences and random fields". In Adler, R. J., Muller, P., and Rozovskii, B. (Eds). Stochastic Modeling in Physical Oceanography, 295-309.

[31] Matsuoka, T., and Ulrych, T. J. (1984). "Phase Estimation Using Bispectrum", Proc. IEEE, 72, 1403-1411.

[32] Nikias, C. L., and Petropulu, A. P. (1993) Higher-order Spectra Analysis, a Nonlinear Signal Process Framework, Prentice Hall, Englewood Cliffs, NJ.

[33] Oppenheim, A. V., and Lim, J. S. (1981). " The importance of phase in Signals," Proc. IEEE, 69, 529-541.

[34] Rosenblatt, M., and Van Ness, J. W. (1965). "Estimation of bispectra", Annals of Mathematical Statistics, 36, 1120-1136.

[35] Rosenblatt, M. (1980). "Linear processes and Bispectra", J. Appl. Probab., 17, 265-270.

[36] Rosenblatt, M. (1985). Stationary Sequences and Random Fields. Birkhauser, Boston.

[37] Rosenblatt, M. (2000). Gaussian and Non-Gausian Linear Time Series and Random Fields. Springer, New York.

[38] Scargle, J. D. (1981). "Phase-sensitive deconvolution to model random processes with special reference to astronomical data". In Applied Time Series Analysis, II. Ed. by Findly, D. F. 549-564. Academic Press New York.

[39] Wiggins, R. A. (1978). "Minimum entropy deconvolution", Geoexploration 16, 21-35.

[40] Xia, B., and Zhang, L. (2007). " Blind deconvolution in nonminimum phase systems using cascade structure", EURASIP J. on Advances in Signal Processing, V. 2007, Article ID 48432, 10 pages.

Rosenblatt's Contributions to Random Walks on Compact Semigroups

By T. C. Sun

Murray Rosenblatt's interest in random walks on compact semigroups probably came from his work on representations of stationary processes as shifts of functions of independent random variables described in [11], where products of matrices were studied. His papers [12],[5],[13],[14] generalized the work of Lévy [2] on random walks on the circle and the work of Kawada and Itô [4] on random walks on compact groups. In [14], he also completely characterized the structure of the limit measures for the special case of compact semigrougs of $n \times n$ stochastic matrices.

To describe Rosenblatt's work in this area in more detail, we first introduce some definitions and basic facts about compact topological semigroups (see [15] and [3]).

1. Every compact topological semigroup S has a minimal two sided ideal K which is called the *kernel* of S.
2. The kernel K is a compact completely simple semigroup which has a Rees structure $X \times G \times Y$, where
   (a) X is a compact left-zero semigroup, Y is a compact right-zero semigroup and G is a compact group.
   (b) The topology on $X \times G \times Y$ is the product topology.
   (c) The binary operation on $X \times G \times Y$ is given by
   $$(x, g, y)(x', g', y') = (x, g\phi(x', y)g', y'),$$
   where $\phi(x', y) : X \times Y \to G$ is a continuous function which, in our case, can be chosen to be $\phi(x', y) = yx'$. The identity in each group $\{(x, g, y), g \in G\}$ is $(x, (yx)^{-1}, y)$.
3. The convolution of two regular probability measures ν and μ on S is defined by
   $$\nu * \mu(A) = \int \mu(s^{-1}A)\nu(ds) = \int \nu(s^{-1}A)\mu(ds).$$

R.A. Davis et al. (eds.), *Selected Works of Murray Rosenblatt*, Selected Works in Probability and Statistics, DOI 10.1007/978-1-4419-8339-8_5,

4. Let ν be a regular probability measure on S and denote its nth-fold convolution with itself by $\nu^{(n*)}$, i.e.,

$$\nu^{(2*)} = \nu * \nu, \quad \nu^{(n*)} = \nu^{((n-1)*)} * \nu.$$

5. Suppose $X_1, X_2, \cdots, X_n$ are independent and identically distributed random variables with values in S and common distribution ν on S. Then the distribution of $W_n = X_1 + X_2 + \cdots + X_n$ is $\nu^{(n*)}$, which has support contained in S and $\{W_n, n = 1, 2, \ldots\}$ is called a *random walk* on S.

In the following we shall always assume that ν is a regular probability measure on S and

$$\overline{\cup_{n=1}^{\infty} \text{supp}(\nu^{(n*)})} = S$$

or S is the compact semigroup generated by the support of ν, because the random walk can never exit the semigroup which is generated by the support of ν.

In [12], Rosenblatt first showed that the sequence $(1/n)\sum_{i=1}^{n} \nu^{(i*)}$ converges weakly to a measure μ on S satisfying,

$$\mu * \nu = \nu * \mu = \mu^{(2)} = \mu.$$

and where the support of μ is the kernel K of S. So, unlike the case of compact groups where the limit measure is the Haar measure on all of S, here in the case of compact semigroups the probability measure is absorbed into the kernel K of S in the limiting process.

In the next step, Heble and Rosenblatt in [5] determined the structure of this limit measure μ on $K = X \times G \times Y$. They showed that if a regular idempotent probability measure μ on S has a completely simple subsemigruoup $K = X \times G \times Y$ as its support, then μ is a product measure and

$$\mu = \alpha \times \chi \times \beta$$

where χ is the Haar measure of the group G and α, β are regular probability measures on X, Y respectively. So the components α and β of the product measure are the extra elements in the semigroup case. J. S. Pym also obtained this results in [10].

Later in [13], Rosenblatt found a necessary and sufficient condition for the sequence $\{\nu^{(n*)}\}$ to converge weakly. He showed that if ν is a regular probability measure on S whose support generates S, then the sequence $\{\nu^{(n*)}\}$ will not converge as $n \to \infty$ if and only if there is a proper closed subgroup G' (proper inclusion) of G such that $YX \subset G'$ and the support of ν is contained in

$$(X \times G' \times Y)^{-1}(X \times gG' \times Y),$$

where $g \notin G'$ and further, $\overline{\cup_{j=1}^{\infty} g^j G'} = G$. In the compact group case, Kawada and Itô's result says the sequence $\nu^{(n*)}$ will not converge if and only if there is a proper closed normal subgroup G' of G such that for some element $g \notin G$ the support of ν is contained in the coset gG'. This is because the support of $\nu^{(n*)}$ jumps from one coset of G' to another coset of G' as n changes. However the sequence $(1/n)\sum_{i=1}^{n} \nu^{(i*)}$ always converges.

In a more concrete setting, when the semigroup S is the set of all $m \times m$ stochastic matrices, one wonders what a completely simple kernel semigroup K would look like? In [14], Rosenblatt gave an answer about the structure of such a K as the following.

1. There is a partition of integers $\{1, 2, \ldots, m\}$ into disjoint classes of integers $T, C_1, C_2, \ldots, C_r$ of numbers $n_0, n_1, n_2, \ldots, n_r$ respectively with $\sum_{i=0}^{r} n_i = m$ and $T = \{1, 2, \ldots, n_0\}$, $C_1 = \{n_0 + 1, n_0 + 2, \ldots, n_0 + n_1\}, \ldots, C_r = \{n_0 + n_1 + \cdots + n_{r-1} + 1, \ldots, m\}$.

2. There are column vectors

$$\rho^{(\alpha)} = (\rho_j^{(\alpha)}; j \in T), \quad \rho_j^{(\alpha)} \geq 0, \quad \sum_{\alpha=1}^{r} \rho_j^{(\alpha)} = 1,$$

$$u^{(\alpha)} = (u_j^{(\alpha)}; j \in C_\alpha), \quad u_j^{(\alpha)} > 0, \quad \sum_{\alpha=1}^{r} u_j^{(\alpha)} = 1,$$

$$1^{(\alpha)} = (1; j \in C_\alpha)$$

3. H is the permutation group on $\{1, 2, \ldots, r\}$.

Then a completely simple kernel semigroup K in S contains all matrices of the form (for all $h \in H$),

$$\begin{pmatrix} 0 & \rho^{(h1)} u^{(1)\prime} & \cdots & \rho^{(hr)} u^{(r)\prime} \\ 0 & \delta_{1-h1} 1^{(1)} u^{(1)\prime} & \cdots & \delta_{1-hr} 1^{(1)} u^{(1)\prime} \\ \cdots & \cdots & \cdots & \cdots \\ 0 & \delta_{r-h1} 1^{(r)} u^{(1)\prime} & \cdots & \delta_{r-hr} 1^{(r)} u^{(r)\prime} \end{pmatrix},$$

where

1. the vectors $\{u^{(\alpha)}; \alpha = 1, 2, \cdots, r\}$ corresponds to Y,
2. the vectors $\{\rho^{(\alpha)}; \alpha = 1, 2, \cdots, r\}$ corresponds to X,
3. H corresponds to the group G,
4. the map $\phi(x, y)$ maps everything into the identity in G.

One may also wonder what the limit measures α and β would look like. In [15], in the Notes of Chapter 5, Rosenblatt gave a simple example of a measure ν on the semigroup S of 2×2 stochastic matrices whose support contains only two points. Each 2×2 stochastic matrix

$$\begin{pmatrix} a & 1-a \\ b & 1-b \end{pmatrix}$$

can be identified with a point (a, b). In this setting both the X and G components of the kernel $K = X \times G \times Y$ of S degenerate into one point and $Y = \{(a, a); a \in [0, 1]\}$. So $K = Y$ and $\mu = \beta$. With proper choice of the two points in the support of ν, the limit measure β on Y can be discrete, singularly continuous or absolutely continuous. In [17], T. Sun generalized this result and, in particular, showed geometrically how this sequence $\nu^{(n*)}$ of measures is attracted to the kernel $K = Y$. More examples in this direction can be found in [18] and [3].

To summarize, Rosenblatt's work completely generalizes Kawada and Ito's result from compact groups to compact semigroups. Of course the result on compact semigroups is much more complicated because of the appearance of the two components X and Y in addition to the group component G. However his work on the structure of the kernel K for the semigroup of $m \times m$ stochastic matrices at least reveals some mysteries about the 2 components X and Y.

Rosenblatt's seminal research in this area inspired a number of related papers. Here we mention just a few.

1. Mukherjea and Tserpes showed in [8] that any idempotent probability measure on a locally compact semigroup must have support which is a completely simple subsemigroup and the measure is a product measure. They generalized Pym, Heble and Rosenblatt's result from compact semigroups to locally compact semigroups.

2. After Rosenblatt's results, it was natural to study recurrent random walks on compact and locally compact semigroups [19], [9]. It is not surprising that any recurrent state must belong to some subgroup and, in fact, in a compact semigroup the set of recurrent states is exactly the kernel of the semigroup. Since a completely simple semigroup is really a union of isomorphic groups, should there be any type of harmonic analysis on a completely simple semigroup?

3. A study of products of random matrices on the semigroup of all $m \times m$ matrices with positive entries [6] and on the semigroup of all $m \times m$ matrices with real entries is contained in [1], [7]. In [16] Rosenblatt posed the question of determining the structure of a completely simple semigroup in these types of matrix semigroups. It seems that this question remains unanswered.

For more recent results on the random walks on semigroups we refer to the excellent reference book [3] by Hognas and Mukherjea.

References

[1] Bougerol, P., Tightness of products of random matrices and stability of linear stochastic systems, *Ann. Prob.* **15**, 40-74 (1987).

[2] Levy, P., L'addition des variables aleatoires definis sur une circonference, *Bull. Soc. Math. France*, **67**, 1-41 (1939).

[3] Hognas, G. and Mukherjea, A., *Probability Measures on Semigroups*, Plenum Press (1995).

[4] Kawada, Y. and Ito, K., On the probability distribution on a compact group, *Proc. Phys,-Math. Soc. of Japan*, **22** 977-998 (1940).

[5] Heble, M. and Rosenblatt, M., Idempotent measures on a topological semigroup, *Proc. Amer. Math. Soc.* **14**, 177-184 (1963).

[6] Kesten, H. and Spitzer, F., Convergence in distribution of products of random matrices, *Z. Wahr. verw. Gebiete* **67**, 363-383 (1984).

[7] Mukherjea, A., Tightness of products of i.i.d. random matrices II, *Ann. Prob.* **22**, 2223-2233 (1994).

[8] Mukherjea, A. and Tserpes, N., Idempotent measures on Locally compact semigroups, *Proc. Am. Math. Soc.* **29**, 143-150 (1971).

[9] Mukherjea, A., Sun, T. and Tserpes, N., Random walks on compact semigroups, *Proc. Am. Math. Soc.* **39**, 599-605 (1973).

[10] Pym, J., Idempotent measures on semigroups, *Pacific J. Math.* **12**, 685-698 (1962).

[11] Rosenblatt, M., Stationary processes as shifts of functions of independent random variables, *Jour. of Math. and Mech*, **8**, 665-682 (1959).

[12] Rosenblatt, M., Limits of convolution sequences of measures on a compact topological group, *Jour. of Math. and Mech.* **9**, 293-395 (1960).

[13] Rosenblatt, M., Equicontinuous Markov operators, *Theory of Prob. and Appl.*, **9** 205-222 (1964).

[14] Rosenblatt, M., Products of independent identically distributed stochastic matrices, *Jour. Math. Anal. Appl.* **11**, 1-10 (1965).

[15] Rosenblatt, M., *Markov Processes. Structure and Asymptotic Behavior*, Springer-Verlag (1971).

[16] Rosenblatt, M., Convolution of sequence of measures on the semigroup of stochastic matrices, 215-220, *Contemporary Math. vol. 50, Am. Math. Soc.* (1986).

[17] Sun, T., Limits of convolutions of probability measures on the set of 2×2 stochastic matrices. *Bull. Inst. of Math. Acad. Sinica*, **3**, 235-248 (1975).

[18] Sun, T., Random walks on semigroups, *Contemporary Math. vol. 50, Am. Math. Soc.* (1986).

[19] Sun, T., Mukherjea, A. and Tserpes, N., On recurrent random walks on semigroups, *Trans. Am. Math. Soc.* **185**, 213-227 (1973).

The Rosenblatt Process*

By Murad S. Taqqu

Abstract

This is a brief history of the Rosenblatt process, how it came about, the role it played, its properties and a detailed description of its various representations.

1 Introduction

I shall describe here the Rosenblatt process and the history behind it. The Rosenblatt process is important because it is one of the so-called "Hermite processes", which are limits of normalized sums of long-range dependent random variables. The simplest Hermite process is the Gaussian fractional Brownian motion. The Rosenblatt process is the simplest non-Gaussian Hermite process. It has continuous non-differentiable paths and is self-similar with stationary increments. It is Murray Rosenblatt who first conceived of it, and this is why I coined it the *Rosenblatt process* in [32].

I will start with Rosenblatt's strong mixing condition, quote the corresponding central limit theorem and describe Rosenblatt's two counterexamples. The second counterexample is of interest. It involves a limiting non-Gaussian distribution which is now called the Rosenblatt distribution. This marginal distribution has been expanded in [32] to the finite-dimensional distributions of a stochastic process called the Rosenblatt process. The distributions are not known in closed form and the corresponding characteristic functions are expressed in terms an expansion involving cumulants. There are, however, a number of integral representations which clarify the nature of the Rosenblatt process and which I will describe in the sequel. These include the time, spectral and finite time interval representations. There is also a wavelet representation due to Vladas Pipiras which extends to the Rosenblatt process the wavelet representation of fractional Brownian motion in [20]. I shall not describe this representation here but will direct the reader to [26], and in particular, to the first two sections which explain heuristically the ideas behind the expansion. This expansion is implemented in [1]. I will also go over some properties of the Rosenblatt process. The paper of Ciprian Tudor [35] provides additional information, in particular, about the stochastic calculus associated with the Rosenblatt process.

The paper is organized as follows. The strong mixing condition and Rosenblatt's corresponding central limit theorem are in introduced in Section 2. The second counterexample, mentioned above, is described in Section 3. The joint characteristic function of the Rosenblatt process and the associated cumulants are given in Section 4. The process is represented through Wiener-Itô stochastic integrals and these are introduced in Section 5. The corresponding stochastic integral representations for fractional Brownian motion and for the

*This work was partially supported by the NSF grant DMS-0706786 at Boston University.

R.A. Davis et al. (eds.), *Selected Works of Murray Rosenblatt*, Selected Works in Probability and Statistics, DOI 10.1007/978-1-4419-8339-8_6,

Rosenblatt process are discussed in Section 6 and Section 7 respectively. Section 8 describes some properties of the Rosenblatt processes and an Appendix provides details about the derivation of the characteristic function.

2 Strong mixing

The classical central limit theorem (CLT) states that the distribution of suitably normalized sums of independent and identically distributed random variables converges to the normal distribution. This result continues to hold if there is some dependence, for example, if the random sequence is m-dependent, that is, if random variables are independent when separated by m time lags. What if we increase the range of the dependence? More generally, will there be a natural boundary, that is, a dependence structure beyond which the CLT will not hold. This is unlikely to be the case as the notion of dependence is not linearly ordered.

Nevertheless, Murray Rosenblatt wanted to pursue A. Markov's idea "that one expects a central limit theorem to hold for $X_1, X_2, \ldots$ if the random variables behave more like independent random variables the further they are separated (assuming appropriate moments exist)". Rosenblatt focused on the following condition, known as the strong mixing condition, which he expresses in the following way in [29].

Definition 2.1 (Strong mixing condition) *Consider a strictly stationary sequence $\{X_k\}$. Let A and B be any events determined by conditions on the random variables X_k, $k \leq 0$ and X_k, $k \geq n$, respectively with $n > 0$. The process $\{X_k\}$ is said to satisfy the strong mixing condition if*

$$\sup_{A,B} |P(A \cap B) - P(A)P(B)| \leq d(n), \tag{1}$$

where the sup is over all such events A and B, and where d is a function of the positive integers n that decreases to zero as $n \to \infty$.

Since that time, research on mixing conditions has been extensive. See for example the surveys of Magda Peligrad [25] and Richard Bradley [7], [8], and the three volumes Bradley [9, 10, 11]. For the CLT to hold, often mixing conditions need to be accompanied by moments conditions. Typically, the slower $d(n) \to 0$ as $n \to \infty$, the more moments should exist.

Rosenblatt in [30] states the following central limit theorem for strictly stationary random variables. It is proved in [29] without the assumption of stationarity[1].

Theorem 2.2 (CLT for strong mixing sequences) *Suppose $\{X_j\}$ is a strictly stationary sequence with mean zero satisfying the strong mixing condition and let $S_n = \sum_{j=1}^n X_j$. Suppose also that the following two moment conditions:*

$$\mathbb{E}S_n^2 = \sigma_n^2 \to \infty \quad as \quad n \to \infty, \tag{2}$$

$$\mathbb{E}S_n^4 = O(\sigma_n^4) \quad as \quad n \to \infty \tag{3}$$

hold. Then as $n \to \infty$,

$$\frac{S_n}{\sigma_n} \xrightarrow{d} N(0,1) \tag{4}$$

In this theorem, there are no assumptions on the speed at which $d(n)$, defined in (1), tends to zero, but it is required that all moments up to order 4 exist. The following stronger result sheds light on Theorem 2.2 (see Denker [13], Theorem 3).

[1] In [29], the relation corresponding to (3) below is stated as $o(\sigma_n^4)$ but the proof holds with the weaker condition $O(\sigma_n^4)$ as stated in [6].

Theorem 2.3 *Suppose $\{X_j\}$ is a strictly stationary sequence with mean zero satisfying the strong mixing condition and condition (2). Then the CLT (4) holds if and only if*

$$\frac{S_n^2}{\sigma_n^2} \quad \textit{is uniformly integrable.} \tag{5}$$

Thus $\sigma_n^{-1} S_n$ tends to a N(0,1) distribution if, for example, $\mathbb{E}|S_n|^{2+\delta} = O(\sigma_n^{2+\delta})$ for some $\delta > 0$. Rosenblatt's Theorem 2.2 is stated with $\delta = 2$. For the history behind Theorem 2.3, see [9].

To get an idea of the limitations of Theorem 2.2, Rosenblatt provides two counterexamples in [30]. The first is $X_k = \epsilon_k - \epsilon_{k-1}$, where the ϵ_k are i.i.d., non-Gaussian random variables, with zero mean and finite fourth order moment. Then, $\sigma_n^2 = \mathbb{E}|\sum_{i=1}^n X_i|^2 = \mathbb{E}|\epsilon_n - \epsilon_0|^2 = 2\mathbb{E}\epsilon_0^2$ does not tend to infinity with n. While the strong mixing condition and condition (3) are satisfied, condition (2) fails, and in fact, the conclusion (4) fails as well.

3 Rosenblatt's counterexample

The second counterexample is more intriguing and is the basis of later developments which gave rise to the Rosenblatt process. Let $\{Y_k\}$ be a Gaussian sequence with mean zero, unit variance and covariance

$$r_k = \mathbb{E}Y_0 Y_k = (1+k^2)^{-D/2} \sim k^{-D} \quad \text{as} \quad k \to \infty$$

with $0 < D < 1/2$. Consider

$$X_k = Y_k^2 - 1.$$

and choose $\sigma > 0$. Then,

$$Z_n = \frac{\sigma}{n^{1-D}} \sum_{k=1}^n X_k \tag{6}$$

tends to a non-Gaussian distribution as $n \to \infty$. Since the X_k are dependent chi-squared random variables, the characteristic function of Z_n can be readily written and one can then let n tend to infinity. The detailed computation is instructive and can be found in the Appendix. Rosenblatt shows in [30] that the sequence Z_n converges in distribution to a random variable $Z(1)$ whose characteristic function is

$$\varphi(\theta) = \exp\left(\frac{1}{2}\sum_{k=2}^{\infty}(2i\theta\sigma)^k \frac{c_k}{k}\right), \tag{7}$$

where

$$c_k = \int_0^1 dx_1 \ldots \int_0^1 dx_k |x_1 - x_2|^{-D}|x_2 - x_3|^{-D} \ldots |x_{k-1} - x_k|^{-D}|x_k - x_1|^{-D}. \tag{8}$$

In this example, both conditions (2) and (3) are satisfied. Since the conclusion of the theorem is not satisfied, the sequence $\{X_k\}$ cannot be strong mixing. Since $X_k = Y_k^2 - 1$, the σ-fields generated by $\{X_k,\ k \le m\}$ and $\{X_k,\ k \ge n\}$ are respectively smaller than the σ-fields generated by $\{Y_k,\ k \le m\}$ and $\{Y_k,\ k \ge n\}$. Therefore $\{X_k\}$ not strong mixing implies that $\{Y_k\}$ is not strong mixing either. Rosenblatt, in this way, was able to characterize a non-trivial, non-strongly mixing Gaussian sequence.

The parameter σ is arbitrary. As we will see below, if one sets

$$\sigma = \left[\frac{1}{2}(1-2D)(1-D)\right]^{1/2}, \tag{9}$$

the limit of Z_n has unit variance. In fact, (7) is an expansion in cumulants. Indeed, if we set

$$\ln \varphi(\theta) = \sum_{k=2}^{\infty} (i\theta)^k \frac{\kappa_k}{k!} = \frac{1}{2} \sum_{k=2}^{\infty} (2i\theta\sigma)^k \frac{c_k}{k},$$

we can identify the cumulants as

$$\kappa_k = 2^{k-1}(k-1)!\sigma^k c_k.$$

With σ as in (9), the variance, which equals the second cumulant, becomes

$$\kappa_2 = 2c_2\sigma^2 = 2\sigma^2 \int_0^1 \int_0^1 |x_1 - x_2|^{-2D} dx_1 dx_2 = 2\sigma^2 \frac{2}{(1-2D)(2-2D)} = 1,$$

so that the distribution is standardized.

I have been intrigued by this counterexample. What is so special about the square transformation $X_k = Y_k^2 - 1$? Why is the usual central limit theorem failing? Why was D taken to satisfy $D < 1/2$? What about weak convergence to a stochastic process?

The answers turn out to be very interesting. The transformation $X_k = Y_k^2 - 1$ is not an arbitrary one: it involves in fact the Hermite polynomial of order 2, $H_2(y) = y^2 - 1$. The Hermite polynomials

$$H_0(y) = 1, \quad H_1(y) = y, \quad H_2(y) = y^2 - 1, \quad H_3(y) = y^3 - 3y, \ldots$$

span the L^2 space generated by the Gaussian sequence. This space contains any non-linear function $G(Y)$, where $Y \sim \mathrm{N}(0,1)$ with

$$\mathbb{E}G^2(Y) = \int_{\infty}^{\infty} G^2(y) \frac{e^{-\frac{y^2}{2}}}{\sqrt{2\pi}} dy < \infty.$$

One is then led to consider not only $X_k = Y_k^2 - 1 = H_2(Y_k)$, but also $X_k = H_m(Y_k)$ for $m \geq 2$, and more generally, $X_k = G(Y_k)$ with $\mathbb{E}G^2(Y_k) < \infty$. This is done in Taqqu, [32], [33], Dobrushin and Major [14], Major [16].

What about the condition on D? Observe that

$$\sum_{k=1}^{n} r_k = \sum_{k=1}^{n} (1+k^2)^{-D/2} \sim \frac{1}{1-D} n^{1-D} \to \infty \text{ as } n \to \infty.$$

The condition $\sum_{k=1}^{\infty} r_k = \infty$ is now known as "long-range dependence", "strong dependence" or "long memory". Since $0 < D < 1/2$, the $\{Y_k\}$ above are strongly dependent (they would be so even if $0 < D < 1$). What about the $\{X_k\}$? One nice property of the Hermite polynomials is their orthogonality. In fact,

$$\mathbb{E}H_{m_1}(Y_0)H_{m_2}(Y_k) = m_1!\ \mathbb{E}(Y_0Y_k)^{m_1}\delta_{m_1,m_2}.$$

In particular,

$$\mathbb{E}X_0X_k = 2(\mathbb{E}Y_0Y_k)^2 = 2r_k^2.$$

Since $0 < D < 1/2$,

$$\sum_{k=1}^{n} \mathbb{E}X_0X_k = 2\sum_{k=1}^{n}(1+k^2)^{-D} \sim \frac{2}{1-2D} n^{1-2D} \to \infty \text{ as } n \to \infty,$$

and

$$\sqrt{\mathbb{E}\left(\sum_{k=1}^{n} X_k\right)^2} = \sqrt{2\sum_{j=1}^{n-1}\sum_{k=1}^{j} \mathbb{E}X_0X_k} \sim \sqrt{\frac{2}{((1-2D)(1-D))}} n^{1-D} \text{ as } n \to \infty \tag{10}$$

which explains the normalization factor in (6) and the value of σ in (9).

Observe that $\{X_k\}$ has a weaker dependence than $\{Y_k\}$ since r_k^2 tends faster to 0 than r_k as $k \to \infty$. Nevertheless, the condition $0 < D < 1/2$ ensures that $\{X_k\}$ remains long-range dependent, that is, one has not only $\sum_{k=1}^{\infty} \mathbb{E}Y_0 Y_k = \infty$ but also $\sum_{k=1}^{\infty} \mathbb{E}X_0 X_k = \infty$. This would not be the case anymore if D were to be greater than $1/2$.

4 The joint characteristic function of the Rosenblatt process

What if one considered not random variables Z_n as defined in (6), but a stochastic process

$$Z_n(t) = \frac{\sigma}{n^{1-D}} \sum_{j=1}^{\lfloor nt \rfloor} X_j, \quad 0 \le t \le 1, \tag{11}$$

where $\lfloor \cdot \rfloor$ denotes the integer part. Observe that our previous Z_n is $Z_n(1)$. We shall then choose σ as in (9) so that $\lim_{n\to\infty} \mathbb{E}Z_n(1)^2 = 1$. The limiting characteristic function of the vector $(Z_n(t_1), \ldots, Z_n(t_p))$ is given in Taqqu [32]. Denoting it as the characteristic function of the limiting vector $(Z(t_1), \ldots, Z(t_n))$, it is[2]

$$\begin{aligned}\varphi(\theta_1, \ldots, \theta_p) &= \mathbb{E}e^{i(\theta_1 Z(t_1)+\cdots+\theta_p Z(t_p))} \\ &= \exp\left(\frac{1}{2}\sum_{k=2}^{\infty} \frac{(2i\sigma)^k}{k} \sum_{s_1,\ldots,s_k \in \{1,\ldots,p\}} \theta_{s_1}\ldots\theta_{s_k} S_D(t_{s_1}, \ldots, t_{s_k})\right),\end{aligned} \tag{12}$$

where

$$S_D(t_{s_1}, \ldots, t_{s_k}) = \int_0^{t_{s_1}} dx_1 \int_0^{t_{s_2}} dx_2 \ldots \int_0^{t_{s_k}} dx_k |x_1 - x_2|^{-D} |x_2 - x_3|^{-D} \ldots |x_{k-1} - x_k|^{-D} |x_k - x_1|^{-D}. \tag{13}$$

As noted in [32], since we have by the Cauchy-Schwarz inequality,

$$S_D \le \left\{\int_{-1}^{1} |y|^{-2D} dy\right\}^{1/2} \left\{\int_{-1}^{1} dy_1 \cdots \int_{-1}^{1} dy_{k-1} |y_1|^{-2D} \cdots |y_{k-1}|^{-2D}\right\}^{1/2} = \left\{\int_{-1}^{1} |y|^{-2D} dy\right\}^{k/2} =: C^k,$$

then for $|\theta_i| \le \epsilon$, $i = 1, \cdots, p$,

$$|\varphi(\theta_1, \ldots, \theta_p)| \le \exp\left(\frac{1}{2}\sum_{k=2}^{\infty} \frac{(2\sigma C p \epsilon)^k}{k}\right). \tag{14}$$

Thus the series in (12) converges for small values of ϵ, and this is enough to characterize the distribution (see [5], Theorem 30.1).

It is also convenient to consider the cumulants κ_k of the random variable $\sum_i^p \theta_i Z(t_i)$. These are $\kappa_1 = 0$ and

$$\kappa_k = 2^{k-1}(k-1)!\, \sigma^k \sum_{s_1,\ldots,s_k \in \{1,\ldots,p\}} \theta_{s_1}\ldots\theta_{s_k} S_D(t_{s_1}, \ldots, t_{s_k}), \quad k \ge 2, \tag{15}$$

and they characterize the finite-dimensional distribution of the vector $(Z(t_1), \ldots, Z(t_p))$ uniquely.

In fact, the sequence $Z_n(t)$ converges to $Z(t)$, not only in terms of the finite-dimensional distributions, but also as weak convergence of the probability measures in the space $D[0,1]$ endowed with the Skorohod topology (see [33]). In [32], the limiting process $Z(t)$ was called the Rosenblatt process and this name has been used since then.

[2] Replace the expression in [32] by the one given here.

5 Wiener-Itô stochastic integrals

The Rosenblatt process can be represented in terms of Wiener-Itô stochastic integrals. We shall now introduce them.

Let $\{B(t),\ t \in \mathbb{R}\}$ be a standard Brownian motion with $B(0) = 0$. A Wiener-Itô stochastic integral of order m is an integral of the form

$$I_m(f) = \int_{\mathbb{R}^m}' f(x_1, \ldots, x_m) dB(x_1) \ldots dB(x_m), \tag{16}$$

where f is a non-random kernel and $dB(x)$ can be interpreted as Gaussian noise with mean 0 and variance dx with $\mathbb{E}dB(x_1)dB(x_2) = 0$ if $x_1 \neq x_2$ and $\mathbb{E}(dB(x))^2 = dx$. The prime indicates that one does not integrate over the diagonals, that is, one always supposes $x_1 \neq x_2 \neq \cdots \neq x_m$. This ensures that $\mathbb{E}dB(x_1) \ldots dB(x_m) = 0$ and thus that $I_m(f)$ has mean 0. Moreover,

$$I_m(f) \stackrel{d}{=} I_m\{f_{sym}\} \tag{17}$$

where f_{sym} denotes the symmetrized version of f and where '$\stackrel{d}{=}$' means equality in distribution (either of the marginal or joint distributions, depending on the context). The integral (16) is defined for all f satisfying

$$\|f\|^2_{L^2(\mathbb{R}^m)} = \int_{\mathbb{R}^m} |f(x_1, \ldots, x_m)|^2 dx_1 \ldots dx_m < \infty,$$

that is, for all $f \in L^2(\mathbb{R}^m)$. To define the integral precisely, one first defines it for simple functions. Then one takes a limit in $L^2(\mathbb{R}^m)$, while ensuring that the isometry

$$\mathbb{E}(I_m(f))^2 = m!\|f\|^2_{L^2(\mathbb{R}^m)}$$

holds.

In the case $m = 2$, the distribution of $I_2(f)$ is determined by its cumulants. If f is symmetric, these are

$$\kappa_k(f) = 2^{k-1}(k-1)! \int_{\mathbb{R}^k} f(x_1, x_2) f(x_2, x_3) \ldots f(x_{k-1}, x_k) f(x_k, x_1) dx_1 \ldots dx_k \tag{18}$$

(see [15]). Circular integrals of this type appear in (8) and (13). See [24] for more information about multiple integrals including formulas for moments and cumulants.

It is also possible to view the integral (16) from the perspective of Parseval's formula, and write

$$I_m(f) \stackrel{d}{=} \int_{\mathbb{R}^m}'' f(\lambda_1, \ldots, \lambda_m) d\tilde{B}(\lambda_1) \ldots d\tilde{B}(\lambda_m), \tag{19}$$

where

$$\tilde{f}(\lambda_1, \ldots, \lambda_m) = \frac{1}{(2\pi)^{m/2}} \int_{\mathbb{R}^m} e^{i \sum_{j=1}^m x_j \lambda_j} f(x_1, \ldots, x_m) dx_1 \ldots dx_m$$

is the Fourier transform of f in $L^2(\mathbb{R}^m)$ and $d\tilde{B}(\lambda)$ is viewed as the complex-valued Fourier transform of $dB(x)$. Since $dB(x)$ is real, $d\tilde{B}(\lambda)$ will satisfy $d\tilde{B}(\lambda) = \overline{d\tilde{B}(-\lambda)}$ and thus, in the integral (19), the double prime on $\int''_{\mathbb{R}^m}$ means that one excludes integration of $\lambda_1 = \pm\lambda_2 = \cdots = \pm\lambda_m$. This is made precise in [33] where the equality in distribution between (16) and (19) is established as well as the relation

$$\mathbb{E}I_m(f)^2 = m! \int_{\mathbb{R}^m} |\tilde{f}(\lambda_1, \ldots, \lambda_m)|^2 d\lambda_1 \ldots d\lambda_m < \infty.$$

If one interprets the x's in (16) as "time" and the λ's in (19) as "frequency", then one has two representations for $I_m(f)$, one in time, and the other in frequency.

6 Stochastic representations of fractional Brownian motion

To better understand the representations of the Rosenblatt process which involve double integrals ($m = 2$) in (16) and (19), it is best to describe first the representation of *fractional Brownian motion*, since it involves only a single integral ($m = 1$).

Fractional Brownian motion $\{B_H(t),\ t \geq 0\}$ is characterized by the following properties:

- It is Gaussian with mean zero.
- It has stationary increments with $B_H(0) = 0$.
- It is self-similar with index $H \in (0,1)$, that is for any $a > 0$, $\{B_H(at),\ t \geq 0\}$ and $\{a^H B_H(t),\ t \geq 0\}$ have the same finite-dimensional distributions.

For more details, see for example [34]. These three characterizing conditions are very handy because in order to check that a given single Wiener-Itô stochastic integral ($m = 1$) represents fractional Brownian motion, it is sufficient to verify that the three conditions mentioned above are satisfied. Self-similarity and stationary increments imply moreover that

$$\mathbb{E}B_H^2(t) = \sigma^2 t^{2H} \tag{20}$$

$$\mathbb{E}B_H(t_1)B_H(t_2) = R_H(t_1,t_2) = \frac{\sigma^2}{2}\left\{t_1^{2H} + t_2^{2H} - |t_1 - t_2|^{2H}\right\}. \tag{21}$$

This process is "standardized" if $\sigma^2 = 1$.

If one sets

$$Y_k = B_H(k+1) - B_H(k),\ k \geq 0,$$

then $\{Y_k\}$ is Gaussian, stationary and has long-range dependence if $1/2 < H < 1$ because

$$r_k = \mathbb{E}Y_0Y_k \sim \sigma^2 H(2H-1)k^{2H-2},$$

and hence

$$\sum_{k=1}^{\infty} r_k = \infty.$$

Observe that $D = 2 - 2H \in (0,1)$. The increments $\{Y_k\}$ are called *fractional Gaussian noise* (see [31] and [34]).

6.1 Time representation of fractional Brownian motion

The following time representation of fractional Brownian motion is the most familiar one:

$$B_H(t) = C_1(H)\int_{\mathbb{R}}\left((t-x)_+^{H-1/2} - (-x)_+^{H-1/2}\right)dB(x), \quad t \geq 0, \tag{22}$$

where[3]

$$C_1(H) := \frac{(\Gamma(2H+1)\sin(\pi H))^{1/2}}{\Gamma(H+1/2)} = \left\{\frac{\sin(\pi H)}{\beta(H+1/2, H+1/2)}\right\}^{1/2} = \left\{\frac{2H}{(H-1/2)\beta(H-1/2, 2-2H)}\right\}^{1/2} \tag{23}$$

[3]The constant $C_1(H)$ here is the inverse of the constant $C_1(H)$ in [31] and [34].

ensures that $\mathbb{E}B_H^2(t) = t^{2H}$, that is, $\sigma^2 = 1$. For the value of the constant $C_1(H)$, see relation (9.3) in [34], Appendix A in [21] and also (25) below. Here

$$\beta(a,b) = \int_0^1 x^{a-1}(1-x)^{b-1}dx = \int_0^\infty x^{a-1}(1+x)^{-a-b}dx = \Gamma(a)\Gamma(b)/\Gamma(a+b), \ a,b > 0,$$

denotes the Beta function. Since there is a single integral in (22), the process $B_H(t)$ is Gaussian. Moreover, the form of the integrand is (22) also ensures that the process has stationary increments.

The self-similarity follows from the fact that the integrand scales and so does $dB(x)$ because $dB(ax) \stackrel{d}{=} a^{1/2}dB(x)$. Thus the representation (22) is that of fractional Brownian motion, but it is far from unique and different time representations are possible.

If $1/2 < H < 1$, one can express the kernel in (22) as an integral and obtain

$$B_H(t) = C_1(H)(H-1/2)\int_{\mathbb{R}}\left(\int_0^t (s-x)_+^{H-3/2}ds\right)dB(x). \tag{24}$$

This is a useful expression. Firstly, with (41) below, it can be used to verify that

$$\begin{aligned}\mathbb{E}B_H(1)^2 &= C_1(H)^2(H-1/2)^2\int_0^1\int_0^1 (s_1-x)_+^{H-3/2}(s_2-x)_+^{H-3/2}ds_1ds_2dx \\ &= C_1(H)^2(H-1/2)^2H^{-1}(2H-1)^{-1}\beta(H-1/2,2-2H) = 1, \end{aligned} \tag{25}$$

that is, the constant $C_1(H)$ ensures standardization.

Secondly (24) suggests a heuristic way of representing fractional Brownian motion. Although it is strictly speaking incorrect, it can be made precise by using "generalized functions". The idea is as follows. Since $B_H(t)$ has stationary increments, let us represent it as the integral of its "derivative". To do this, suppose $1/2 < H < 1$ and interchange the integral signs in (24), writing

$$B_H(t) \text{“} \stackrel{d}{=} \text{”} C_1(H)(H-1/2)\int_0^t\left(\int_{\mathbb{R}}(s-x)^{H-3/2}dB(x)\right)ds \tag{26}$$

$$= C_1(H)(H-1/2)\int_0^t B_H'(s)ds. \tag{27}$$

The change of order of integration in (26) is not justified and neither is the derivative $B_H'(t)$ well-defined because the integrand $(s-x)_+^{H-3/2}$ in (26) is not square integrable. Nevertheless (26) and (27) provide a physical interpretation of the time representation (22) and a better understanding of the corresponding representation for the Rosenblatt process.

6.2 Spectral representation of fractional Brownian motion

The spectral representation[4] for $B_H(t)$ is

$$B_H(t) \stackrel{d}{=} C_2(H)\int_{\mathbb{R}}\frac{e^{i\lambda t}-1}{i\lambda}\frac{1}{|\lambda|^{H-1/2}}d\tilde{B}(\lambda) \tag{28}$$

where[5]

$$C_2(H) := \left\{\frac{H(2H-1)}{2\Gamma(2-2H)\sin[(H-1/2)\pi]}\right\}^{1/2} = \left\{\frac{H\Gamma(2H)\sin H\pi}{\pi}\right\}^{1/2}$$

[4]In [31], the spectral representation is called the harmonizable representation.
[5]The constant $C_2(H)$ here is the inverse of the constant $C_2(H)$ in [31] and [34].

(see [33], Theorem 6.3 and [34], relation (9.8)). The term $(e^{i\lambda t}-1)/(i\lambda)$ is the Fourier transform of $\mathbf{1}_{[0,t]}$, yielding stationary increments and the term $c(\lambda)|\lambda|^{-(H-1/2)}|$, with $|c(\lambda)|=1$, can be viewed heuristically and up to a multiplicative constant as the Fourier transform of the kernel $s_+^{H-3/2}$ in (26) (see the proof of Lemma 6.2 in [33]). The equivalence between the time representation (22) and the spectral representation (28) follows from the discussion in Section 5.

6.3 Finite time interval representation of fractional Brownian motion

There is a third representation for fractional Brownian motion which involves only an integral from 0 to t. It is related to the so-called Molchan martingale and can be found in [23] and [27] where it is used for prediction. See Molchan [22] for a brief history and a discussion of the prediction problem. See also Mishura [21] who treats the case $0<H<1/2$ as well. We suppose here $1/2<H<1$. The representation is:

$$B_H(t) \stackrel{d}{=} \int_0^t K^H(t,x)dB(x), \tag{29}$$

where

$$\begin{aligned} K^H(t,x) &= C_3(H)x^{1/2-H}\int_x^t (u-x)^{H-3/2}u^{H-1/2}du, \quad 0<x<t, \\ &= C_3(H)x^{1/2-H}\int_0^t (u-x)_+^{H-3/2}u^{H-1/2}du, \quad 0<x<t, \end{aligned} \tag{30}$$

and

$$C_3(H) := \left(\frac{H(2H-1)}{\beta(2-2H,H-1/2)}\right)^{1/2} \tag{31}$$

(see [35]). Since

$$K^H(t,x) = K^H(t,x) - K^H(x,x) = \int_x^t \frac{\partial K^H}{\partial u}(u,x)du, \quad 0<x<t,$$

we can express (29) as

$$B_H(t) \stackrel{d}{=} \int_0^t \left[\int_x^t \frac{\partial K^H}{\partial u}(u,x)du\right] dB(x), \tag{32}$$

a form which will be handy when we will consider the corresponding extension to the Rosenblatt process. Observe that we are proceeding as in (24) by involving the derivative of the original kernel. Here

$$\frac{\partial K^H}{\partial u}(u,x) = C_3(H)\left(\frac{x}{u}\right)^{1/2-H}(u-x)_+^{H-3/2}, \quad 0<x<u, \tag{33}$$

so that one has

$$B_H(t) \stackrel{d}{=} C_3(H)\int_0^t\left[\int_x^t \left(\frac{x}{u}\right)^{1/2-H}(u-x)_+^{H-3/2}du\right] dB(x). \tag{34}$$

The self-similarity follows from

$$K^H(at,ax) = a^{H-1/2}K^H(t,x) \ \text{ for any } \ a>0. \tag{35}$$

The stationarity of the increments is not obvious, and since the process is Gaussian, it is best to compute the covariance directly. The relation

$$|u-v|^{2H-2} = \frac{(uv)^{H-1/2}}{\beta(2-2H,H-1/2)}\int_0^{u\wedge v}(u-x)_+^{H-3/2}(v-x)_+^{H-3/2}x^{1-2H}dx \tag{36}$$

for $u, v > 0$ is very handy (see (3.4) in [27]), because it shows that

$$\begin{aligned}
\mathbb{E}B_H(t_1)B_H(t_2) &= \int_0^{t_1\wedge t_2} K^H(t_1,x)K^H(t_2,x)dx \\
&= \frac{1}{H(2H-1)}\int_0^{t_1}\int_0^{t_2}|u-v|^{2H-2}dudv \\
&= R_H(t_1,t_2),
\end{aligned}$$

given in (21). Hence, the choice of $C_3(H)$ does ensure that $\mathbb{E}B_H^2(1) = 1$.

7 Stochastic representations of the Rosenblatt Process

The Rosenblatt process $\{Z(t),\ t \geq 0\}$ is H-self-similar with $1/2 < H < 1$, has zero mean and has stationary increments. It has therefore the same second order properties as fractional Brownian motion, namely,

$$\mathbb{E}Z^2(t) = \sigma^2 t^{2H}, \quad \mathbb{E}Z(t_1)Z(t_2) = R_H(t_1,t_2) = \frac{\sigma^2}{2}\left\{t_1^{2H} + t_2^{2H} - |t_1 - t_2|^{2H}\right\}.$$

It is, however, non-Gaussian and is expressed as a Wiener-Itô stochastic integral of order $m = 2$.

Remark. It is important to distinguish between the self-similarity index of fractional Brownian motion which we now call H_0 (instead of H), and the self-similarity index of the Rosenblatt process which we will denote by H.

7.1 Time representation of the Rosenblatt process

The time representation of the Rosenblatt process is

$$\begin{aligned}
Z(t) &= a_1(H_0)\int_{\mathbb{R}^2}'\left(\int_0^t (s-x_1)_+^{H_0-3/2}(s-x_2)_+^{H_0-3/2}ds\right)dB(x_1)dB(x_2) && (37) \\
&= A_1(H)\int_{\mathbb{R}^2}'\left(\int_0^t (s-x_1)_+^{-\frac{2-H}{2}}(s-x_2)_+^{-\frac{2-H}{2}}ds\right)dB(x_1)dB(x_2), && (38)
\end{aligned}$$

where

$$a_1(H_0) := \frac{\sqrt{(H_0-1/2)(4H_0-3)}}{\beta(H_0-1/2, 2-2H_0)} = \frac{\sqrt{(H/2)(2H-1)}}{\beta(H/2, 1-H)} =: A_1(H) \tag{39}$$

ensures that $\mathbb{E}Z^2(1) = 1$ (see relation (1.6) together with (6.1) in [33][6]). I have expressed the time representation in two ways: one, involving H_0 to indicate the analogy with the corresponding representation of fractional Brownian motion in (24), the other involving H, the self-similarity index of the Rosenblatt process. The relation between H_0 and H is

$$H_0 = \frac{H+1}{2} \tag{40}$$

obtained by comparing the exponent $H_0 - 3/2$ in (24) with the exponent $-(2-H)/2$ in (38). Observe also that

$$1/2 < H < 1 \Longleftrightarrow 3/4 < H_0 < 1.$$

[6] One sometimes replaces the integral $\int_{\mathbb{R}^2}'$ by the integral $\int_{\{x_1<x_2\}}$ which is smaller by a factor of 2!. This affects the normalization constants.

Moreover, the relation between the H in (38) and the D in (6) and (11) is

$$H = 1 - D.$$

The self-similarity with index H and the stationarity of the increments can be readily verified. To check that the process in (38) represents the Rosenblatt process, it is sufficient to show that the cumulants of $\sum_i^p \theta_i Z(t_i)$ are given by (15). This is easily done using (18) and the relation

$$\int_{\mathbb{R}} (u-x)_+^{\alpha-1}(v-x)_+^{\alpha-1}dx = \beta(\alpha, 1-2\alpha)|u-v|^{2\alpha-1} \tag{41}$$

valid for $0 < \alpha < 1/2$.

Interchanging the integrals in (37), we could write heuristically,

$$Z(t) = a_1(H_0)\int_0^t \left(\int_{\mathbb{R}^2}' (s-x_1)_+^{H_0-3/2}(s-x_2)_+^{H_0-3/2}dB(x_1)dB(x_2)\right) ds. \tag{42}$$

This is not well-defined because the function $f_s(x_1,x_2) = (s-x_1)_+^{-\frac{2-H}{2}}(s-x_2)^{-\frac{2-H}{2}}$ is not in $L^2(\mathbb{R}^2)$. If we ignore this fact and also suppress the prime in $\int_{\mathbb{R}^2}'$ which would have ensured mean 0 (if the process were well-defined), then we could view the Rosenblatt process as

$$Z(t) = a_1(H_0)\int_0^t \left(B'_{H_0}(s)\right)^2 ds \tag{43}$$

where B'_{H_0} denotes the "derivative" of fractional Brownian motion with self-similarity index given by (40).

7.2 Spectral representation of the Rosenblatt process

The corresponding spectral representation of the Rosenblatt process is

$$\begin{aligned} Z(t) &\stackrel{d}{=} a_2(H_0)\int_{\mathbb{R}^2}'' \frac{e^{i(\lambda_1+\lambda_2)t}-1}{i(\lambda_1+\lambda_2)}\frac{1}{|\lambda_1|^{H_0-1/2}}\frac{1}{|\lambda_2|^{H_0-1/2}}d\tilde{B}(\lambda_1)d\tilde{B}(\lambda_2) && (44) \\ &= A_2(H)\int_{\mathbb{R}^2}'' \frac{e^{i(\lambda_1+\lambda_2)t}-1}{i(\lambda_1+\lambda_2)}\frac{1}{|\lambda_1|^{H/2}}\frac{1}{|\lambda_2|^{H/2}}d\tilde{B}(\lambda_1)d\tilde{B}(\lambda_2), && (45) \end{aligned}$$

where

$$a_2(H_0) := \sqrt{\frac{(2H_0-1)(4H_0-3)}{2[2\Gamma(2-2H_0)\sin(\pi(H_0-1/2))]^2}} = \sqrt{\frac{H(2H-1)}{2[2\Gamma(1-H)\sin(H\pi/2)]^2}} =: A_2(H) \tag{46}$$

with (40) ensures that $\mathbb{E}Z^2(1) = 1$ (see [33], Theorem 6.3 and Section 5). Compare this spectral representation with the one of fractional Brownian motion in (28) and note that in view of (40), the powers of λ_i, $i = 1, 2$ are $H/2 = H_0 - 1/2$. Thus one has in the spectral domain, a structure consistent with the heuristic interpretation (43).

7.3 Finite time interval representation of the Rosenblatt process

Let us now turn to the third representation involving integration over the finite range $[0, t]$ only. We have seen that fractional Brownian motion has the representation (32). The corresponding representation of the

Rosenblatt process is

$$Z(t) \stackrel{d}{=} a_3(H_0)\int_{[0,t]^2}\left(\int_{x_1\vee x_2}^{t}\frac{\partial K^{H_0}}{\partial u}(u,x_1)\frac{\partial K^{H_0}}{\partial u}(u,x_2)du\right)dB(x_1)dB(x_2) \quad (47)$$

$$= A_3(H)\int_{[0,t]^2}\left(\int_{x_1\vee x_2}^{t}\frac{\partial K^{\frac{H+1}{2}}}{\partial u}(u,x_1)\frac{\partial K^{\frac{H+1}{2}}}{\partial u}(u,x_2)du\right)dB(x_1)dB(x_2), \quad (48)$$

where K^{H_0} is defined in (30), $H_0=(H+1)/2$ as in (40) and where

$$a_3(H_0) := \frac{1}{2H_0}\sqrt{\frac{2(4H_0-3)}{2H_0-1}} = \frac{1}{H+1}\sqrt{\frac{2(2H-1)}{H}} =: A_3(H) \quad (49)$$

with (40) ensures that $\mathbb{E}Z^2(1)=1$ (see [35]). The self-similarity follows from (35) which yields

$$\frac{\partial K^{H_0}}{\partial u}(au,ax) = a^{H_0-3/2}\frac{\partial K^{H_0}}{\partial u}(u,x),$$

and hence one gets $2(H_0-3/2)+1+1/2+1/2 = 2H_0-1 = H$ by applying (40). Again, the stationarity of the increments is not obvious. But it will follow by showing that $Z(t)$, as defined in (47), has the same distribution as the Rosenblatt process defined through the time representation (22). We will show in fact that they share the same cumulants. But let us compute first the covariance of $Z(t)$. Observe that

$$\begin{aligned}\int_0^{u\wedge v}\frac{\partial K^{H_0}}{\partial u}(u,x)\frac{\partial K^{H_0}}{\partial v}(v,x)dx &= C_3(H_0)^2\int_0^{u\wedge v}\left(\frac{x}{u}\right)^{1/2-H_0}\left(\frac{x}{v}\right)^{1/2-H_0}(u-x)_+^{3/2-H_0}(v-x)_+^{3/2-H_0}dx\\ &= C_3(H_0)^2\beta(2-2H_0,H_0-1/2)|u-v|^{2H_0-2}\\ &= H_0(2H_0-1)|u-v|^{2H_0-2}, \end{aligned} \quad (50)$$

by (31) and (36). Hence the covariance equals

$$\begin{aligned}&\mathbb{E}Z(t_1)Z(t_2)\\ &= 2(a_3(H_0))^2\int_0^{t_1}dx_1\int_0^{t_2}dx_2\left[\int_{x_1\vee x_2}^{t_1}\frac{\partial K^{H_0}}{\partial u}(u,x_1)\frac{\partial K^{H_0}}{\partial u}(u,x_2)du\right]\left[\int_{x_1\vee x_2}^{t_2}\frac{\partial K^{H_0}}{\partial v}(v,x_1)\frac{\partial K^{H_0}}{\partial v}(v,x_2)du\right]\\ &= 2(a_3(H_0))^2\int_0^{t_1}du\int_0^{t_2}dv\left[\int_0^{u\wedge v}\frac{\partial K^{H_0}}{\partial u}(u,x)\frac{\partial K^{H_0}}{\partial v}(v,x)dx\right]^2\\ &= 2[H_0(2H_0-1)]^2(a_3(H_0))^2\int_0^{t_1}du\int_0^{t_2}dv|u-v|^{H_0-4}\\ &= H(2H-1)\int_0^{t_1}du\int_0^{t_2}dv|u-v|^{2H-2}\\ &= R_H(t_1,t_2),\end{aligned}$$

by (47), (50) and (49). Thus, $Z(t)$, as defined in (47), has the same covariance as standard fractional Brownian motion, and in particular, we have $\mathbb{E}Z^2(t)=1$.

Moreover, the cumulants of $\sum_{i=1}^{p}\theta_i Z(t_i)$ are the same as the cumulants of the corresponding time representation (22), namely they are given by (15). To see that this is the case, apply again (18) and the relation (50), using computations similar to those performed in the computation of the covariance.

The Rosenblatt process is not differentiable, but as for fractional Brownian motion, one could view it heuristically as the integral of its derivative. One merely has to interchange the integrals in (47) and write

$$Z(t) \stackrel{d}{=} a_3(H_0) \int_0^t \left[\int'_{[0,u]^2} \frac{\partial K^{H_0}}{\partial u}(u, x_1) \frac{\partial K^{H_0}}{\partial u}(u, x_2) dB(x_1) dB(x_2) \right] du. \tag{51}$$

8 Some properties of the Rosenblatt process

The equalities in the previous sections hold in the sense of finite-dimensional distribution. In fact, the Rosenblatt process $\{Z(t),\ t \geq 0\}$ has an almost sure continuous path version since

$$\mathbb{E}|Z(t_1) - Z(t_2)|^2 = |t_1 - t_2|^{2H}$$

with $2H > 1$ and thus Kolmogorov's criterion holds (see [4], Theorem 12.4). But the Rosenblatt process is not differentiable in mean square since

$$\lim_{t \to 0} \mathbb{E} \left(\frac{Z(t)}{t} \right)^2 = \lim_{t \to 0} t^{2H-2} = \infty.$$

In fact, it is almost surely non differentiable. The argument is based on self-similarity and is identical to the one for fractional Brownian motion (see [19] and Proposition 1.7.1 in [3]).

We have seen that the Rosenblatt process is self-similar with index $1/2 < H < 1$ and has stationary increments. Therefore its increments

$$\Delta Z(k) = Z(k+1) - Z(k), \quad k \geq 0$$

have long-range dependence, that is,

$$\sum_{k=0}^{\infty} \mathbb{E}[(\Delta Z(0)) \Delta Z(k)] = \infty.$$

In fact, they have the exact same covariances as the increments of fractional Brownian motion, namely

$$\mathbb{E}[(\Delta Z(0)) \Delta Z(k)] = \frac{\sigma^2}{2} \left\{ |k+1|^2 - 2|k|^{2H} + |k-1|^2 \right\} \sim \sigma^2 H(2H-1) k^{2H-2},$$

as $k \to \infty$, and where $1/2 < H < 1$.

In Section 7, we presented several representations of the Rosenblatt process. While representations are very physical, they do not provide immediate information about the distributions. However, since the Rosenblatt process is represented as a double Wiener-Itô stochastic integral ($m = 2$), one can use a general result due to Major [17] on probability bounds for double integrals. Namely, there is a universal constant $C > 0$ such that

$$P[Z(1) > u] \leq C \exp\left\{ -\frac{1}{2} \left(\frac{u}{\sigma} \right) \right\} \quad u > 0, \tag{52}$$

where $\sigma^2 = \mathbb{E}Z(1)^2$. The result is proved in [18]. Observe, for example, that if Y is N(0,1), then $P[Y^2 > u] = P[|Y| > u^{1/2}]$ which is consistent with (52).

Albin [2] focuses on extreme values. One has clearly $\mathrm{P}[\sup_{t \in [0,1]} Z(t) > u] \geq \mathrm{P}[Z(1) > u]$, but Albin also shows that

$$\limsup_{u \to \infty} \frac{\mathrm{P}[\sup_{t \in [0,1]} Z(t) > u]}{u^{(1-H)/H} \mathrm{P}[Z(1) > u]} < \infty.$$

Observe that the factor $u^{(1-H)/H}$ tends to infinity with u. Observe also that the exponent $(1 - H)/H$ decreases from 1 to 0 as H increases from 1/2 to 1. Hence the greater H, the greater the similarity between $P[\sup_{t\in[0,1]} Z(t) > u]$ and $P[Z(1) > u]$ for large values of u.

The marginal distribution of $Z(t)$ has a probability density function. This follows from a general result on multiple integrals (see for example [12]).

We also remarked in Section 7, that the Rosenblatt process can be viewed heuristically as the integral of its derivative. However, as noted in Tudor [35], one can use this point of view to show that the Rosenblatt process can be approximated by a sequence of semi-martingales. Indeed, define $Z^{(\epsilon)}(t)$ by replacing each $K^{H_0}(u,x)$ in (51) by $K^{H_0}(u+\epsilon, x)$ where $\epsilon > 0$, that is,

$$Z^{(\epsilon)}(t) \stackrel{d}{=} a_3(H_0) \int_0^t \left[\int'_{[0,u]^2} \frac{\partial K^{H_0}}{\partial u}(u+\epsilon, x_1) \frac{\partial K^{H_0}}{\partial u}(u+\epsilon, x_2) dB(x_1)dB(x_2) \right] du. \tag{53}$$

We can express the integral $\int'_{[0,u]^2}$ as $2\int_{x_1<x_2}$ by using (17). We then observe that $Z^{(\epsilon)}(t)$ is the integral over $[0,t]$ of an adapted process and hence $Z^{(\epsilon)}(t)$ is a semi-martingale. Since $\mathbb{E}|Z^{(\epsilon)}(t) - Z(t)|^2 \to 0$ as $\epsilon \to 0$, one obtains that the Rosenblatt process can be approximated by a sequence of semi-martingales. Because $Z^{(\epsilon)}(t)$ is well-defined as the integral of its derivative, the semi-martingales are in fact processes with bounded variation.

9 Appendix

We provide here a proof of Rosenblatt's result, that the random sequence Z_n, $n \geq 1$ in (6) converges in distribution to a random variable Z(1) with characteristic function given by (7) and (8).

Let $d_n = n^{1-D}$ and let R_n denote the covariance matrix of the Gaussian vector $(Y_1, \ldots, Y_n)$. Each component has mean 0 and unit variance. Let $y' = (y_1, \ldots, y_n)$ denote the row vector and $|\cdot|$ a determinant. Then the characteristic function of Z_n is

$$\begin{aligned}
\mathbb{E}e^{i\theta Z_n} &= \int_{\mathbb{R}^n} \exp\left\{i\theta\sigma d_n^{-1}\sum_{j=1}^n (y_j^2 - 1)\right\} \frac{1}{\sqrt{2\pi|R_n|}} e^{-\frac{1}{2}y'R_n^{-1}y} d^n y \\
&= \exp\left\{-i\theta\sigma d_n^{-1} n\right\} \frac{1}{\sqrt{2\pi|R_n|}} \int_{\mathbb{R}^n} \exp\left\{-\frac{1}{2}y'[R_n^{-1} - 2i\theta\sigma d_n^{-1} I]y\right\} d^n y \\
&= \exp\left\{-i\theta\sigma d_n^{-1} n\right\} |R_n|^{-1/2} |R_n^{-1} - 2i\theta\sigma d_n^{-1} I|^{-1/2} \\
&= \exp\left\{-i\theta\sigma d_n^{-1} n\right\} |I - 2i\theta\sigma d_n^{-1} R_n|^{-1/2} \\
&= \exp\left\{\frac{1}{2}\sum_{j=1}^n [-2i\theta\sigma d_n^{-1} - \ln(1 - 2i\theta\sigma d_n^{-1}\lambda_{j,n})]\right\} \\
&= \exp\left\{\frac{1}{2}\Big[-2i\theta\sigma d_n^{-1} n - \sum_{j=1}^n \ln(1 - 2i\theta\sigma d_n^{-1}\lambda_{j,n})\Big]\right\},
\end{aligned}$$

where the $\lambda_{j,n}$, $j = 1, \ldots, n$ denote the eigenvalues of R_n. Let Tr denote the trace. Expanding the logarithm, we get

$$\ln(1 - 2i\theta\sigma d_n^{-1}\lambda_{j,n}) = -\sum_{k=1}^{\infty} \frac{(2i\theta\sigma d_n^{-1}\lambda_{j,n})^k}{k},$$

and hence

$$-\sum_{j=1}^{n}\ln(1-2i\theta\sigma d_n^{-1}\lambda_{j,n}) = 2i\theta\sigma d_n^{-1}\sum_{j=1}^{n}\lambda_{j,n} + \sum_{j=1}^{n}\sum_{k=2}^{\infty}\frac{(2i\theta\sigma d_n^{-1}\lambda_{j,n})^k}{k}.$$

Since R_n has 1 in its diagonals, $\sum_{j=1}^{n}\lambda_{j,n} = n$, and thus

$$\mathbb{E}e^{i\theta Z_n} = \exp\left\{\frac{1}{2}\sum_{k=2}^{\infty}\frac{(2i\theta\sigma)^k}{k}d_n^{-k}\sum_{j=1}^{n}\lambda_{j,n}^k\right\}, \tag{54}$$

where

$$\begin{aligned} d_n^{-k}\sum_{j=1}^{n}\lambda_{j,n}^k &= d_n^{-k}\ \mathrm{Tr}(R_n^k) \\ &= d_n^{-k}\sum_{i_1,i_2,\ldots,i_k=1}^{n} r(|i_1-i_2|)\ r(|i_2-i_3|)\ldots r(|i_{k-1}-i_k|)\ r(|i_k-i_1|) \to c_k \end{aligned}$$

as $n\to\infty$. The expansion (54) converges absolutely for $|\theta|<\epsilon$, ϵ small enough, as in (14). This proves (7) and (8). □

Observe that the $k=1$ term, $d_n^{-1}\sum_{j=1}^{n}\lambda_{j,n} = d_n^{-1}\mathrm{Tr}(R_n) = n^{D-1}n = n^D \to \infty$, but this term has been canceled and creates no trouble. The $k=2$ term provides the variance and hence the normalization d_n. Observe also that the expression (7) for the characteristic function is not valid for all values of $\theta\in\mathbb{R}$, but only for small values of θ. This, however, is enough to characterize the distribution.

This also suggests the following representation for the limiting random variable $Z(1)$ (see [28]). Let $\epsilon_j,\ j\ge 1$ be i.i.d. N(0,1) and let $\lambda_j,\ j\ge 1$ be such that

$$\sum_{j=1}^{\infty}\lambda_j^k = c_k, \tag{55}$$

for $k\ge 2$, where c_k is given by (8). Then one has

$$Z(1) \stackrel{d}{=} \sigma\sum_{j=1}^{\infty}\lambda_j(\epsilon_j^2-1), \tag{56}$$

because they both have the same characteristic function (7). Moreover,

$$\mathrm{Var}\Big\{\sigma\sum_{j=1}^{\infty}\lambda_j(\epsilon_j^2-1)\Big\} = 2\sigma^2\sum_{j=1}^{\infty}\lambda_j^2 = 2\sigma^2 c_2 < \infty,$$

and therefore the sum in (56) converges almost surely.

Acknowledgments. I would like to thank Magda Peligrad for comments on strong mixing and also Vladas Pipiras and Mark Veillette for interesting discussions.

References

[1] P. Abry and V. Pipiras. Wavelet-based synthesis of the Rosenblatt process. *Signal processing*, 86:2326–2339, 2006.

[2] J. M. P. Albin. On extremal theory for self-similar processes. *The Annals of Probability*, 26(2):743–793, 1998.

[3] F. Biagini, Y. Hu, B. Øksendal, and T. Zhang. *Stochastic calculus for fractional Brownian motion and applications.* Probability and its Applications. Springer-Verlag, London, 2008.

[4] P. Billingsley. *Convergence of Probability Measures.* Wiley, New York, 1968.

[5] P. Billingsley. *Probability and Measure.* Wiley, New York, 3rd edition, 1995.

[6] J.R. Blum and M. Rosenblatt. A class of stationary processes and a central limit theorem. *Proceedings of the National Academy of Sciences USA*, 42:412–413, 1956.

[7] R. C. Bradley. Basic properties of strong mixing conditions. In E. Eberlein and M. S. Taqqu, editors, *Dependence in Probability and Statistics*, Progress in Probability and Statistics, pages 165–192. Birkhäuser, 1986.

[8] R.C. Bradley. Basic properties of strong mixing conditions. A survey and some open questions. *Probability Surveys*, 2:107–144 (electronic), 2005. Update of, and a supplement to, the 1986 original.

[9] R.C. Bradley. *Introduction to strong mixing conditions. Vol. 1.* Kendrick Press, Heber City, UT, 2007.

[10] R.C. Bradley. *Introduction to strong mixing conditions. Vol. 2.* Kendrick Press, Heber City, UT, 2007.

[11] R.C. Bradley. *Introduction to strong mixing conditions. Vol. 3.* Kendrick Press, Heber City, UT, 2007.

[12] Yu. A. Davydov. Distributions of multiple Wiener-Itô integrals. *Theory of Probability and its Applications*, 35(1):51–62, 1990.

[13] M. Denker. Uniform integrability and the central limit theorem for strongly mixing processes. In Ernst Eberleein and Murad S. Taqqu, editors, *Dependence in probability and statistics*, volume 11 of *Progress in Probability and Statistics*, pages 269–289. Birkhäuser, Boston, MA, 1986.

[14] R. L. Dobrushin and P. Major. Non-central limit theorems for non-linear functions of Gaussian fields. *Zeitschrift für Wahrscheinlichkeitstheorie und verwandte Gebiete*, 50:27–52, 1979.

[15] R. Fox and M. S. Taqqu. Central limit theorems for quadratic forms in random variables having long-range dependence. *Probability Theory and Related Fields*, 74:213–240, 1987.

[16] P. Major. *Multiple Wiener-Itô Integrals*, volume 849 of *Springer Lecture Notes in Mathematics*. Springer-Verlag, New York, 1981.

[17] P Major. Tail behaviour of multiple random integrals and U-statistics. *Probab. Surv.*, 2:448–505 (electronic), 2005.

[18] P. Major. On a multivariate version of Bernstein's inequality. *Electron. J. Probab.*, 12, 2007.

[19] B. B. Mandelbrot and J. W. Van Ness. Fractional Brownian motions, fractional noises and applications. *SIAM Review*, 10:422–437, 1968.

[20] Y. Meyer, F. Sellan, and M. S. Taqqu. Wavelets, generalized white noise and fractional integration: the synthesis of fractional Brownian motion. *The Journal of Fourier Analysis and Applications*, 5(5):465–494, 1999.

[21] Y.S. Mishura. *Stochastic calculus for fractional Brownian motion and related processes*, volume 1929 of *Lecture Notes in Mathematics*. Springer-Verlag, Berlin, 2008.

[22] G. M. Molchan. Historical comments related to fractional Brownian motion. In P. Doukhan, G. Oppenheim, and M. S. Taqqu, editors, *Theory and Applications of Long-range Dependence*, pages 39–42. Birkhäuser, 2003.

[23] I. Norros, E. Valkeila, and J. Virtamo. An elementary approach to a Girsanov formula and other analytical results on fractional Brownian motions. *Bernoulli*, 15:571–587, 1999.

[24] G. Peccati and M. S. Taqqu. Moments, cumulants and diagram formulae for non-linear functionals of random measures. Preprint: arXiv:0811.1726v1 [math.PR], 2009.

[25] M. Peligrad. Recent advances in the Central Limit Theorem and its weak invariance principle for mixing sequences of random variables. In Ernst Eberlein and Murad S. Taqqu, editors, *Dependence in Probability and Statistics*, Progress in Probability and Statistics, pages 193–224. Birkhäuser, 1986.

[26] V. Pipiras. Wavelet-type expansion of the Rosenblatt process. *The Journal of Fourier Analysis and Applications*, 10(6):599–634, 2004.

[27] V. Pipiras and M. S. Taqqu. Are classes of deterministic integrands for fractional Brownian motion on an interval complete? *Bernoulli*, 7(6):873–897, 2001.

[28] A. Plikusas. Some properties of the multiple Itô integral. *Lithuanian Mathematical Journal*, 21(2):184–191, 1981.

[29] M. Rosenblatt. A central limit theorem and a strong mixing condition. *Proceedings of the National Academy of Sciences*, 42:43–47, 1956.

[30] M. Rosenblatt. Independence and dependence. In *Proceedings of the 4th Berkeley Symposium Mathematical Statistics and Probability*, pages 431–443. University of California Press, 1961.

[31] G. Samorodnitsky and M. S. Taqqu. *Stable Non-Gaussian Processes: Stochastic Models with Infinite Variance*. Chapman and Hall, New York, London, 1994.

[32] M. S. Taqqu. Weak convergence to fractional Brownian motion and to the Rosenblatt process. *Zeitschrift für Wahrscheinlichkeitstheorie und verwandte Gebiete*, 31:287–302, 1975.

[33] M. S. Taqqu. Convergence of integrated processes of arbitrary Hermite rank. *Zeitschrift für Wahrscheinlichkeitstheorie und verwandte Gebiete*, 50:53–83, 1979.

[34] M. S. Taqqu. Fractional Brownian motion and long-range dependence. In P. Doukhan, G. Oppenheim, and M. S. Taqqu, editors, *Theory and Applications of Long-range Dependence*, pages 5–38. Birkhäuser, 2003.

[35] C.A. Tudor. Analysis of the Rosenblatt process. *ESAIM Probability and Statistics*, 12:230–257, 2008.

ON SPECTRAL ANALYSIS OF STATIONARY TIME SERIES*

By Ulf Grenander and Murray Rosenblatt

Department of Mathematical Statistics, University of Stockholm, and Committee on Statistics, University of Chicago

Communicated by Saunders MacLane, April 30, 1952

1. The present statistical theory of analysis of stationary time series (e.g., extrapolation) has assumed complete knowledge of the covariance sequence or equivalently of the spectrum of the process. It is, therefore, of great importance to be able to estimate one of these. However, knowledge of the spectrum seems to yield greater immediate insight into the structure of the process. This seems to have first been noted in a fundamental paper by Bartlett.[1] An unpublished paper by Tukey[2] deals with some aspects of the problem of estimating the spectrum.

2. Consider a discrete parameter stochastic process x_t ($t = \ldots, -1, 0, 1, \ldots$) of the following form:

$$x_t = \sum_{\nu=-\infty}^{\infty} a_{t-\nu}\xi_\nu$$

where $\{a_n\}$ is a real sequence and

$$\sum_{-\infty}^{\infty} a_n^2 < \infty .$$

The ξ_ν's are assumed to be independent and identically distributed real-valued random variables with

$$E\xi_\nu = 0,\ E\xi_\nu^2 = \sigma^2,\ E\xi_\nu^4 = \mu_4.$$

The conditions cited above imply that the stochastic process x_t is strictly

R.A. Davis et al. (eds.), *Selected Works of Murray Rosenblatt*, Selected Works in Probability and Statistics, DOI 10.1007/978-1-4419-8339-8_7, © Springer Science+Business Media, LLC 2011

stationary. Such a process is known to have an absolutely continuous spectral distribution function $F(\lambda)$ with a density given by

$$F'(\lambda) = f(\lambda) = \frac{1}{2\pi}\left|\sum_{-\infty}^{\infty} a_\nu e^{i\nu\lambda}\right|^2 \sigma^2.$$

Proofs of the results cited below will require $f(\lambda)$ to satisfy a smoothness condition and the ξ_ν's to have moments up to the eighth order.

As the process is real-valued, the spectral density is symmetric about zero and hence one need only consider a spectral distribution function $F(\lambda)$ as the integral of the spectral density from zero to λ, $0 \leq \lambda \leq \pi$. A reasonable estimate of the spectral distribution function $F(\lambda)$ is

$$F_N^*(\lambda) = \int_0^\lambda I_N(l)dl$$

where

$$I_N(l) = \frac{1}{2\pi N}\left|\sum_1^N x_n e^{inl}\right|^2$$

is the so-called periodogram. Professor Doob suggested that we investigate the problem of giving confidence bands for the estimation of $F(\lambda)$ in terms of $F_N^*(\lambda)$. In the case of a normally distributed process, Theorem 3 yields a solution which is non-parametric with respect to the spectrum. In the general case, Theorem 1 and an appropriate estimate of $e = \frac{\mu_4}{\sigma^4} - 3$ can be used to obtain conservative bands non-parametric with respect to the spectrum. Theorem 4 can be used to test whether two independently drawn time series come from processes having the same spectrum.

3. The theorems are as follows:

THEOREM 1.

$$\lim_{N\to\infty} P\left\{\max_{0\leq\lambda\leq\pi}\left| N[F_N^*(\lambda) - F(\lambda)]\right| \leq \alpha\right\} = P\left\{\max_{0\leq\lambda\leq\pi}\left|Z(\lambda)\right| \leq \right\}$$

where $Z(\lambda)$ is a normal process whose distribution is completely determined by the mean value $EZ(\lambda) = 0$ and the covariance

$$EZ(\lambda)Z(\mu) = eF(\lambda)F(\mu) + 2\pi G(\text{min. }(\lambda, \mu)),$$

where

$$G(\lambda) = \int_0^\lambda f^2(l)dl.$$

COROLLARY. *If $e = 0$*

$$\lim_{N\to\infty} P\left\{\max_{0\leq\lambda\leq\pi}\left|\sqrt{N}[F_N^*(\lambda) - F(\lambda)]\right| \leq \alpha\right\} = P\left\{\max_{0\leq\lambda\leq 2\pi G(\pi)}\left|W(\lambda)\right| \leq \alpha\right\}$$

where $W(\lambda)$ is the Wiener process, i.e.,

$$EW(\lambda) = 0 \qquad EW(\lambda)W(\mu) = \min.\ (\lambda, \mu)$$

and $W(\lambda)$ is normally distributed.

Note that $e = 0$ when the ξ_ν are normally distributed.

THEOREM 2. *The statistic*

$$R_N = \frac{1}{2}\sum_{\nu=-kN^\alpha}^{kN^\alpha} \frac{c_\nu^2}{N^2}$$

converges to $2\pi G(\pi)$ in probability as $N \to \infty$ if $k > 0, 0 < \alpha < 1$. The c_ν's are the product sums

$$c_\nu = \sum_{1}^{N-|\nu|} x_n x_{n+|\nu|}.$$

THEOREM 3. *If $e = 0$*

$$\lim_{N\to\infty} P\left\{\frac{\max_{0\le\lambda\le\pi}\left|\sqrt{N}[F_N^*(\lambda) - F(\lambda)]\right|}{R_N^{1/2}} \le \alpha\right\} = P\left\{\max_{0\le\lambda\le 1}|W(\lambda)| \le \alpha\right\}.$$

It is well known that the probability on the right can be explicitly evaluated in a convenient form.

THEOREM 4. *If $e = 0$ and if $F_{N1}^*(\lambda)$, $F_{N2}^*(\lambda)$, R_{N1}, R_{N2} are the statistics defined above corresponding to two independently drawn samples from the same stochastic process, then*

$$\lim_{N\to\infty} P\left\{\frac{\max_{0\le\lambda\le\pi}\left|\sqrt{N}[F_{N1}^*(\lambda) - F_{N2}^*(\lambda)]\right|}{(R_{N1} + R_{N2})^{1/2}} \le \alpha\right\} = P\left\{\max_{0<\lambda\le\pi}|W(\lambda)| \le \alpha\right\}.$$

4. Although the plausibility of the basic Theorem 1 is apparent by heuristic arguments, the proof requires a long and formidable sequence of lemmas. The practical value of the theory cited above can only be judged after having completed an extensive computational program involving analyses of artificially generated time series. Some preliminary analyses have already been made and tables of the related distributions are being compiled. Considerable investigation must be made of the practical application of the theory. The proofs of the above theorems and considerable elaboration of the theory (e.g., solution of the corresponding problem for the spectral density) as well as aspects of the practical application of the theory will be contained in a forthcoming publication.

* This work was supported by a grant from the Office of Naval Research.

[1] Bartlett, M. S., "Periodogram Analysis and Continuous Spectra," *Biometrika*, **37**, (1950).

[2] Tukey, J. W., "Measuring Noise Color," unpublished manuscript.

REMARKS ON A MULTIVARIATE TRANSFORMATION[1]

By Murray Rosenblatt

University of Chicago

The object of this note is to point out and discuss a simple transformation[2] of an absolutely continuous k-variate distribution $F(x_1, \cdots, x_k)$ into the uniform distribution on the k-dimensional hypercube. A discussion of related transformations has been given by P. Lévy [1].

Let $X = (X_1, \cdots, X_k)$ be a random vector with distribution function $F(x_1, \cdots, x_k)$. Let $z = (z_1, \cdots, z_k) = Tx = T(x_1, \cdots, x_k)$, where T is the transformation considered. Then T is given by

$$z_1 = P\{X_1 \leqq x_1\} = F_1(x_1),$$

$$z_2 = P\{X_2 \leqq x_2 \mid X_1 = x_1\} = F_2(x_2 \mid x_1),$$

$$\vdots$$

$$z_k = P\{X_k \leqq x_k \mid X_{k-1} = x_{k-1}, \cdots, X_1 = x_1\} = F_k(x_k \mid x_{k-1}, \cdots, x_1).$$

One can readily show that the random vector $Z = TX$ is uniformly distributed on the k-dimensional hypercube, for

$$P\{Z_i \leqq z_i; i = 1, \cdots, k\}$$

$$= \int_{\{Z \mid Z_i \leqq z_i\}} \cdots \int d_{x_k} F_k(x_k \mid x_{k-1}, \cdots, x_1) \cdots d_{x_1} F_1(x_1)$$

$$= \int_0^{z_k} \cdots \int_0^{z_1} dz_1 \cdots dz_k = \prod_{i=1}^{k} z_i,$$

when $0 \leqq z_i \leqq 1$, $i = 1, \cdots, k$. Hence $Z_1, \cdots, Z_k$ are uniformly and independently distributed on [0, 1].

Let $X_{(i)} = (X_{1(i)}, \cdots, X_{k(i)})$, $i = 1, \cdots, n$, be a random sample of n vectors from a population with distribution function $F(x_1, \cdots, x_k)$ and let $G(x_1, \cdots, x_k)$ be the corresponding sample distribution function. It has been found that the probability distribution of the Kolmogorov-Smirnov statistic,

$$\max_{x_1, \cdots, x_k} \mid F(x_1, \cdots, x_k) - G(x_1, \cdots, x_k) \mid,$$

is not the same for all continuous F when $k > 1$ [2]. The same can be said for the multidimensional von Mises statistic

$$\int_{-\infty}^{\infty} \cdots \int_{-\infty}^{\infty} (F(x_1, \cdots, x_k) - G(x_1, \cdots, x_k))^2 \, d_{x_1 \cdots x_k} F(x_1, \cdots, x_k).$$

However, it would still be of interest to study the Kolmogorov-Smirnov and von Mises statistics for the case of sampling from a population uniformly distributed on the k-dimensional hypercube. One could test whether the $X_{(i)}$, $i = 1, \cdots, n$, are a sample from a population with distribution function $F(x_1, \cdots, x_k)$ by

[1] Work done under ONR contract.

[2] I have recently learned that J. H. Curtiss and I. R. Savage have also considered this transformation.

R.A. Davis et al. (eds.), *Selected Works of Murray Rosenblatt*, Selected Works in Probability and Statistics, DOI 10.1007/978-1-4419-8339-8_8, © Springer Science+Business Media, LLC 2011

using either the Kolmogorov-Smirnov or von Mises statistic to test whether the $Z_{(i)} = TX_{(i)}$ are a sample from a population uniformly distributed on the k-dimensional hypercube. Both statistics are invariant under permutation of the $X_{(i)}, i = 1, \cdots, n$.

The transformation is of some interest in setting up regions of equal probability mass in a chi-square test of goodness of fit. Assume again that one wishes to test whether the $X_{(i)}, i = 1, \cdots, n$, come from a population with absolutely continuous distribution function $F(x_1, \cdots, x_k)$. This, as remarked before, is equivalent to testing whether the $Z_{(i)} = TX_{(i)}, i = 1, \cdots, n$, come from a population with distribution function $z_1 \cdots z_k$, where $0 \leqq z_j \leqq 1, j = 1, \cdots, k$. Divide the k-dimensional hypercube into the N^k cells

$$C_{j_1 \cdots j_k} = \left\{ z \,\middle|\, \frac{j_i}{N} < z_i < \frac{j_i + 1}{N},\ i = 1, \cdots, k \right\},$$

where $j_1, \cdots, j_k = 0, 1, \cdots, N - 1$. Each of these cells has probability mass $1/N^k$ under the null hypothesis. Let $\nu_{j_1 \cdots j_k}$ be the number of transformed observations $Z_{(i)} = TX_{(i)}$ in $C_{j_1 \cdots j_k}$. The chi-square statistic

$$\sum \frac{\left(\nu_{j_1 \cdots j_k} - \frac{n}{N^k}\right)^2}{n/N^k}$$

can be used to test whether the sample $X_{(i)}$ comes from a population with distribution function $F(x_1, \cdots, x_k)$. A test for goodness of fit based on the chi-square statistic above is consistent as $n, N \to \infty$.

There are $k!$ transformations T of the type described above corresponding to the $k!$ ways in which one can number the coordinates $x_1, \cdots, x_k$. One might think this unsatisfactory because of a suspicion that the experimenter might carry out one of the procedures described above with each of the $k!$ transformations and then choose that particular transformation which yields the result he wishes to obtain. But this situation can arise in any case where there is a multitude of tests in the same context.

The transformation T can be written down explicitly in several cases. We shall write down the transformation when $F(x_1, \cdots, x_k)$ is a normal distribution with mean $M = (m_1, \cdots, m_k)$ and covariance matrix $\Lambda = \{\lambda_{ij}\}, i, j = 1, \cdots, k$. Let $\Lambda^{(r)} = \{\lambda_{ij}\}, i, j = 1, \cdots, r \leqq k$, and $\Lambda_{ij}^{(r)}$ be the cofactor of $\{\lambda_{ij}\}$ in $\Lambda^{(r)}$. The transformation T is then given by

$$F_1(x_1) = \Phi\left(\frac{x_1 - m_1}{\sqrt{\lambda_{11}}}\right),$$

$$F_2(x_2 \mid x_1) = \Phi\left(\frac{x_2 - m_2 + (\Lambda_{21}^{(2)}/\Lambda_{22}^{(2)})(x_1 - m_1)}{\sqrt{\Lambda^{(2)}/\Lambda_{22}^{(2)}}}\right),$$

$$\vdots$$

$$F_k(x_k \mid x_{k-1}, \cdots, x_1) = \Phi\left(\frac{x_k - m_k + \sum_{j=1}^{k-1} (\Lambda_{kj}/\Lambda_{kk})(x_j - m_j)}{\sqrt{\Lambda/\Lambda_{kk}}}\right).$$

Let $F(x_1, x_2)$ be a normal distribution with means m_1, m_2, variances σ_1^2, σ_2^2 and correlation coefficient ρ. The transformation can then be written as

$$F_1(x_1) = \Phi\left(\frac{x_1 - m_1}{\sigma_1}\right),$$

$$F_2(x_2 \mid x_1) = \Phi\left(\frac{x_2 - m_2 + \frac{\rho\sigma_1}{\sigma_2}(x_1 - m_1)}{\sigma_2\sqrt{1 - \rho^2}}\right).$$

REFERENCES

[1] P. Lévy, *Théorie de l'Addition des Variables Aléatoires*, Gauthier-Villars, Paris, 1937, pp. 71–73, 121–123.

[2] Paul B. Simpson, "Note on the estimation of a bivariate distribution function," *Annals of Math. Stat.*, Vol. 22 (1951), pp. 476–478.

ABSTRACTS OF PAPERS

(Abstracts of papers presented at the Eugene meeting of the Institute, June 19–21, 1952)

1. The Auditory Cortex—A Probability Model. Archie R. Tunturi, University of Oregon Medical School.

The role played by the brain in communication is well known, but in what manner the brain handles information is not understood. Some progress has been made in this direction by studying the anatomy and physiology of the auditory cortex in the anesthetized dog with controlled acoustic signals. Communication may be thought of as making a representation in a space of a representation in another space. In three of the four auditory areas (on one side of the brain), the entire frequency spectrum from 100 to 12800 cps is represented literally spacewise by groups of cells that respond only to a narrow range of frequencies. A special method increases the signal to noise ratio, by augmenting the electrical response of the cells, thereby permitting exact measurements of the characteristic frequency and intensity for each group of cells. This is similar to a narrow band filter, and does not reveal the effect of other frequencies on the information. The information capacity of the system can be inferred if it can be assumed that occurrence of the augmented response for the group of cells follows some probability function. These probabilities for all groups of cells can be assembled into a model representing the behavior of the system as a communication device. If there are 70 groups of cells between 100 and 12800 cps, the probability of any particular combination would be $1/2^{70}$, if the selections were equally probable. The effect of noise on this system will be considered. (Research sponsored in part by the Office of Naval Research.)

2. Testing Message Diffusion: The Utilization of Mathematical Models. Stuart C. Dodd, Richard J. Hill, and Susan Huffaker, University of Washington.

In connection with the study of interpersonal verbal communication, the Washington Public Opinion Laboratory designed an experimental procedure which yielded data on the temporal diffusion of thirty-three different messages in a population of 184 individuals. Data (including the recipient of each message, the initiator of communication, and the time of communication) were obtained on 5,522 separate instances of communication. The

STATISTICAL SPECTRAL ANALYSIS OF TIME SERIES ARISING FROM STATIONARY STOCHASTIC PROCESSES[1]

By Ulf Grenander and Murray Rosenblatt

University of Chicago

Summary. We consider time series which are realizations of a stochastic process. From the time series we construct various estimates of the spectral distribution function of the process (Section 3) and we study the sampling distributions of some functionals of these estimates (Sections 4–7). We then obtain confidence regions for the spectral distribution function and various tests of hypotheses in the normal case (Sections 8–10).

1. Introduction. Let us consider a real discrete parameter stochastic process' that is, a sequence of stochastic variables x_t, $t = \cdots, -1, 0, 1, \cdots$.[2] We introduce the quantities $Ex_t = m_t$, the mean value sequence, and the covariances $\rho_{s,t} = E(x_s - m_s)(x_t - m_t)$. The process $y_t = x_t - m_t$ is said to be stationary in the wide sense if $\rho_{s,t} = \rho_{s-t}$. Then it follows from a theorem of Herglotz [10] that $\rho_t = \int_{-\pi}^{\pi} e^{it\lambda}\, dF(\lambda)$ where $F(\lambda)$ is a bounded and nondecreasing function in $(-\pi, \pi)$. $F(\lambda)$ is called the *spectral distribution function* of the process as it can be said to describe the distribution of the spectral energy of the process (see Wold [22], p. 16). As x_t is real, $\rho_t = \rho_{-t}$ and the distribution of the spectral energy is symmetric about zero.

Since knowledge of the distribution of the spectral energy is equivalent to knowledge of the covariance sequence, it is a matter of convenience which to choose in analyzing the process. The reduced process y_t can be written as

$$y_t = \int_{-\pi}^{\pi} e^{it\lambda}\, dZ(\lambda) \tag{1.1}$$

[3] where $Z(\lambda)$ is an orthogonal process such that

$$\begin{gathered} EZ(\lambda) = 0 \\ E\,|\,Z(\lambda) - Z(\mu)\,|^2 = |\,F(\lambda) - F(\mu)\,|\,. \end{gathered} \tag{1.2}$$

(For a discussion of stochastic integrals used in this paper see J. L. Doob, *Stochastic Processes*, John Wiley and Sons, 1953.) Hence $dF(\lambda)$ can be interpreted as the variance of the stochastic amplitude $dZ(\lambda)$ corresponding to the harmonic $e^{it\lambda}$ in the Fourier expansion of y_t. $F(\lambda)$ seems to be more directly related to the structure of the process than the covariance sequence. Later on

Received 10/22/52.

[1] This work has been carried out with support of the Office of Naval Research.

[2] Any notation introduced in the paper will be adhered to in the remainder of the paper without any further explanation.

R.A. Davis et al. (eds.), *Selected Works of Murray Rosenblatt*, Selected Works in Probability and Statistics, 52
DOI 10.1007/978-1-4419-8339-8_9, © Springer Science+Business Media, LLC 2011

we shall find other reasons for preferring to deal with the spectrum instead of the covariances.

We can decompose $F(\lambda)$ into three parts, $F(\lambda) = F_a(\lambda) + F_d(\lambda) + F_s(\lambda)$, where $F_a(\lambda) = \int_{-\pi}^{\lambda} f(l)\, dl$, $F_d(\lambda) = \sum_{\lambda_\nu \leqq \lambda} \Delta F(\lambda_\nu)$, and $F_s(\lambda)$ is constant except on a set of Lebesgue measure zero. Here $f(\lambda) = F'(\lambda)$, the *spectral density*, is a nonnegative integrable function and the λ_ν's are the points of discontinuity of $F(\lambda)$ with associated jumps $\Delta F(\lambda_\nu) > 0$. We are going to deal with the absolutely continuous case, but will make a few remarks later on concerning possible discontinuities of the spectral distribution function.

A model of some practical interest is the following one. A linear filter $\mathfrak{F}$ can be characterized by its output a_t at time t generated by a unit input at time 0. It is not necessary for our purpose to demand that $a_t = 0$ for $t < 0$. Suppose we feed independent random impulses ξ_t, identically distributed with mean 0 and standard deviation 1, into the filter. Following Tukey we shall call a process of the type ξ_t "*pure white noise*" as contrasted with a more general process η_t with $E\eta_t = 0$, $E\eta_s\eta_t = \delta_{st}$ (called "*white noise*"), but where the η's are not necessarily independent. Note that in both cases one has a uniform spectrum, $F'(\lambda) = (2\pi)^{-1}$. The resulting output at time t is then

$$(1.3) \qquad y_t = \sum_{n=-\infty}^{\infty} a_n \xi_{t-n}.$$

If $\sum_{-\infty}^{\infty} a_\nu^2 < \infty$, this sum converges with probability 1 (see Levy [15], p. 139–142). If the input had been η_t we would only be able to state convergence in the mean square. In both cases the spectrum is absolutely continuous with spectral density (see Karhunen [12], p. 71)

$$f(\lambda) = \frac{1}{2\pi} \left| \sum_{n=-\infty}^{\infty} a_n e^{-in\lambda} \right|^2 .$$

More restrictive conditions will be imposed on the process later. A process of the type (1.3) is called a *linear process*.

On the other hand, if a real stationary process y_t has an absolutely continuous spectral distribution function $F(\lambda) = \int_{-\pi}^{\lambda} f(l)\, dl$, Doob (see *Proceedings of the Berkeley Symposium on Mathematical Statistics and Probability*, University of California Press, 1949, p. 327) has shown that

$$(1.4) \qquad y_t = \sum_{n=-\infty}^{\infty} a_n \eta_{t-n},$$

where we can choose $a_n = \int_{-\pi}^{\pi} e^{in\lambda} \sqrt{f(\lambda)}\, d\lambda$ and $E\eta_t = 0$, $E\eta_s\eta_t = \delta_{st}$. The η's are elements of the real Hilbert space spanned by the y's and an appropriately chosen set of real random variables (see [12], p. 42). They are then real and so are the a's. In the present paper we shall deal only with the linear processes. It would be interesting to extend the domain in which our results are valid to a more general class of processes.

A normally distributed process with an absolutely continuous spectrum is clearly linear as the η's appearing in (1.4) are then automatically independent and identically distributed.

2. Some methods that have been proposed for time series analysis. An important group of statistical problems arise in the following way. We observe a sample $x_1, x_2, \cdots, x_N$ (a time series) from the stochastic process x_t and want to make inferential statements concerning the covariances, or equivalently the spectrum of x_t. There has been a long history of work aimed at answering such problems. A considerable part of the literature has been devoted to testing the independence of the observations, usually in the case of normal time series (for a list of references see [14]). Other statisticians have studied what one could call *finite parameter schemes*. Two important schemes of this type are the autoregressive scheme (2.1) and the moving average (2.2)

$$\sum_{n=0}^{p} b_n y_{t-n} = \xi_t \tag{2.1}$$

$$y_t = \sum_{n=0}^{p} \beta_n \xi_{t-n} \tag{2.2}$$

where the b's and the β's are constants. Specifying a priori an order p for the process, they have given estimates for the coefficients and have devised tests for various hypotheses (for a list of references see [14]). Whittle [21] has gone further and has given tests for discriminating between finite parameter schemes of different orders without specifying the coefficients a priori.

Any linear process can be approximated as closely as is desired by a finite parameter scheme of sufficiently high order. In most practical situations, however, it does not seem realistic to take the order of the scheme as some number p, usually small, given a priori.

A great deal of the older literature has been devoted to the so-called periodogram analysis (see Kendall [13] for references). This was originally devised to deal with processes of the type

$$x_t = \sum_{\nu=1}^{m} (A_\nu \sin \lambda_\nu t + B_\nu \cos \lambda_\nu t) + \xi_t$$

where m, A_ν, B_ν, and λ_ν are unknown. Clearly x_t has a spectrum with a discrete component. To estimate the frequencies λ_ν the following statistic called the *periodogram*

$$\begin{aligned} I_N(\lambda) &= \frac{1}{2\pi N} \left| \sum_{\nu=1}^{N} x_\nu e^{-i\nu\lambda} \right|^2 \\ &= \frac{1}{2\pi N} \left[\left(\sum_{\nu=1}^{N} x_\nu \cos \nu\lambda \right)^2 + \left(\sum_{\nu=1}^{N} x_\nu \sin \nu\lambda \right)^2 \right] \end{aligned}$$

has been used. A rationale for this method is that $I_N(\lambda)$ can be shown to diverge to infinity as $N \to \infty$ if λ coincides with some frequency λ_ν. Hence if $I_N(\lambda)$ is large we suspect that λ is one of the frequencies λ_ν of the scheme. For a corresponding test of significance see Fisher [6].

Periodogram analysis is not immediately applicable to the case of an absolutely continuous spectrum. However, some work of the last few years has indicated that when it is properly modified, it is the most powerful method that has been found to work without very special assumptions concerning the covariance structure of the process. We would like to mention especially Bartlett [1], [2], and Tukey [20].

A brief discussion of some of the results of this paper can be found in [9].

3. Some preliminary considerations. Consider a process

$$y_t = \int_{-\pi}^{\pi} e^{-it\lambda}\, dZ(\lambda) = \sum_\nu Z_\nu e^{-it\lambda} + y'_t$$

where y'_t is the component of the process with a continuous spectrum and the other term is the one with a discrete spectrum. If we observe a complete realization y_t, $-\infty < t < \infty$, we can specify the sample value of any Z_ν with probability one. However, we cannot estimate $E\,|\,Z_\nu\,|^2 = \Delta F(\lambda_\nu)$ consistently unless $|\,Z_\nu\,|$ is constant with probability one. We note in passing that the model of random phases in which $Z_\nu = A_\nu e^{i\varphi_\nu}$, where the φ_ν's are independent and uniformly distributed in $(-\pi, \pi)$ and the A_ν's are constants, has this property and is not without interest (see Lévy [16], p. 114).

Although the periodogram is a legitimate tool for estimating the frequencies λ_ν of a discrete spectrum as has been remarked above, it cannot be used to estimate the spectral density of an absolutely continuous spectrum consistently [7]. Still the periodogram plays a fundamental role in our paper as our estimates are closely related to it.

Let y_t be a linear process and let the ξ's used in constructing the process have a fourth moment μ_4. We set

$$e = \mu_4 - 3.$$

Note that e is the fourth cumulant of the random variables ξ. The periodogram $I_N(\lambda)$ of the process y_t then is

$$\begin{aligned} I_N(\lambda) &= \frac{1}{2\pi N}\left|\sum_{t=1}^{N} y_t e^{-it\lambda}\right|^2 \\ &= \frac{C_0}{2\pi N} + \frac{1}{\pi N}\sum_{\nu=1}^{N-1} C_\nu \cos \nu\lambda \end{aligned}$$

where $C_\nu = \sum_{t=1}^{N-\nu} y_t y_{t+\nu}$. We are interested in statistics of the form

$$\Phi_N = \int_{-\pi}^{\pi} I_N(l)\phi(l)\, dl \tag{3.1}$$

where $\phi(l)$ is bounded, symmetric about zero, and has at most a finite number of discontinuities. Let

$$\Phi = \int_{-\pi}^{\pi} f(l)\phi(l)\, dl. \tag{3.2}$$

THEOREM 1. *Let Φ_N, Φ be defined as in (3.1), (3.2) and let Ψ_N, Ψ be defined analogously with weight function ψ. If the spectral density $f(l)$ is continuous*

$$\lim_{N\to\infty} E\Phi_N = \Phi, \qquad \lim_{N\to\infty} E\Psi_N = \Psi$$

and

$$\lim_{N\to\infty} N \operatorname{cov}(\Phi_N, \Psi_N) = e\Phi\Psi + 4\pi \int_{-\pi}^{\pi} f^2(l)\psi(l)\phi(l)\, dl. \tag{3.3}$$

PROOF. By definition of Φ_N

$$E\Phi_N = \int_{-\pi}^{\pi} EI_N(l)\phi(l)\, dl. \tag{3.4}$$

But using (1.1)

$$I_N(l) = \frac{1}{2\pi N}\left|\sum_1^N y_\nu e^{-i\nu l}\right|^2 = \frac{1}{2\pi N}\left|\int_{-\pi}^{\pi} e^{-i(\lambda-l)}\frac{1-e^{iN(\lambda-l)}}{1-e^{i(\lambda-l)}}\, dZ(\lambda)\right|^2$$

and hence from (1.2)

$$EI_N(l) = \frac{1}{2\pi N}\int_{-\pi}^{\pi}\left|e^{i(\lambda-l)}\frac{1-e^{iN(\lambda-l)}}{1-e^{i(\lambda-l)}}\right|^2 dE\mid Z(\lambda)\mid^2$$

$$= \frac{1}{2\pi N}\int_{-\pi}^{\pi}\frac{\sin^2\frac{N}{2}(\lambda-l)}{\sin^2\frac{\lambda-l}{2}} f(\lambda)\, d\lambda$$

which tends to $f(l)$ as $N\to\infty$. But the integrand in (3.4) is bounded by

$$\mid\phi(\lambda)\mid \max_\lambda f(\lambda)$$

so that we have $E\Phi_N \to \Phi$ as $N\to\infty$.

Let us now consider $N \operatorname{cov}(\Phi_N, \Psi_N)$. We have

$$N\operatorname{cov}(I_N(\lambda), I_N(\mu)) = \frac{1}{4\pi^2 N}\sum_{\alpha,\beta,\gamma,\delta=1}^{N}\operatorname{cov}(y_\alpha y_\beta, y_\gamma y_\delta)e^{i\lambda(\alpha-\beta)-i\mu(\gamma-\delta)}.$$

But

$$Ey_\alpha y_\beta y_\gamma y_\delta = \sum_{\nu_1,\nu_2,\nu_3,\nu_4=-\infty}^{\infty} a_{\alpha-\nu_1}a_{\beta-\nu_2}a_{\gamma-\nu_3}a_{\delta-\nu_4}E\xi_{\nu_1}\xi_{\nu_2}\xi_{\nu_3}\xi_{\nu_4}$$

$$= (\mu_4-3)\sum_{\nu=-\infty}^{\infty} a_{\alpha-\nu}a_{\beta-\nu}a_{\gamma-\nu}a_{\delta-\nu} + \sum_{\nu,\mu=-\infty}^{\infty} a_{\alpha-\nu}a_{\beta-\nu}a_{\gamma-\mu}a_{\delta-\mu}$$

$$+ \sum_{\nu,\mu=-\infty}^{\infty} a_{\alpha-\nu}a_{\beta-\mu}a_{\gamma-\mu}a_{\delta-\nu} + \sum_{\nu,\mu=-\infty}^{\infty} a_{\alpha-\nu}a_{\beta-\mu}a_{\gamma-\nu}a_{\delta-\mu}$$

so that

$$\operatorname{cov}(y_\alpha y_\beta, y_\gamma y_\delta) = (\mu_4 - 3) \sum_{\nu=-\infty}^{\infty} a_{\alpha-\nu} a_{\beta-\nu} a_{\gamma-\nu} a_{\delta-\nu} + \rho_{\alpha-\delta}\rho_{\beta-\gamma} + \rho_{\alpha-\gamma}\rho_{\beta-\delta}.$$

The fourth moments of Y_α can be shown to exist and the operations that have been carried out can be justified by a repeated application of Schwarz' inequality and Fatou's lemma. Hence

$$N \operatorname{cov}(\Phi_N, \Psi_N) = \frac{1}{4\pi^2 N}$$

$$\cdot \left\{ \sum_{\alpha,\beta,\gamma,\delta=1}^{N} \left[(\mu_4 - 3) \sum_{\nu=-\infty}^{\infty} a_{\alpha-\nu} a_{\beta-\nu} a_{\gamma-\nu} a_{\delta-\nu} + \rho_{\alpha-\delta}\rho_{\beta-\gamma} + \rho_{\alpha-\gamma}\rho_{\beta-\delta} \right] \phi_{\alpha-\beta} \bar{\psi}_{\gamma-\delta} \right\}$$

$$= S_1 + S_2 + S_3$$

where $\phi_\nu = \int_{-\pi}^{\pi} e^{i\nu\lambda}\phi(\lambda)\, d\lambda$ and ψ_ν is defined analogously.

We first deal with S_1. Note that

$$S_1 = \frac{\mu_4 - 3}{4\pi^2 N} \int_{-\pi}^{\pi} \int_{-\pi}^{\pi} \sum_{\nu=-\infty}^{\infty} \sum_{\alpha,\beta=1}^{N} a_{\alpha-\nu} a_{\beta-\nu} e^{i(\alpha-\beta)\lambda} \sum_{\gamma,\delta=1}^{N} a_{\gamma-\nu} a_{\delta-\nu} e^{-i(\gamma-\delta)\mu} \phi(\lambda)\psi(\mu)\, d\lambda\, d\mu$$

$$= \frac{\mu_4 - 3}{4\pi^2 N} \int_{-\pi}^{\pi} \int_{-\pi}^{\pi} \sum_{\nu=-\infty}^{\infty} \left| \sum_{\alpha=1}^{N} a_{\alpha-\nu} e^{i\alpha\lambda} \right|^2 \left| \sum_{\gamma=1}^{N} a_{\gamma-\nu} e^{i\gamma\mu} \right|^2 \phi(\lambda)\psi(\mu)\, d\lambda\, d\mu.$$

First we consider the special case

$$\phi(\lambda) = e^{-in\lambda}, \qquad \psi(\mu) = e^{-im\mu}.$$

The function

$$c_{n,m}(\lambda) = \sum_{k=-\infty}^{\infty} a_k a_{k-n} e^{ik\lambda} \sum_{l=-\infty}^{\infty} a_l a_{l-m} e^{-il\lambda}$$

is continuous. We have

$$\frac{1}{2\pi} \int_{-\pi}^{\pi} c_{n,m}(\lambda) e^{-i(\alpha-\gamma)\lambda}\, d\lambda = \sum_{\nu=-\infty}^{\infty} a_{\alpha-\nu} a_{\alpha-n-\nu} a_{\gamma-\nu} a_{\gamma-m-\nu}$$

so that

$$S_1(e^{-in\lambda} e^{-im\mu}) = \frac{e}{4\pi^2 N} (2\pi)^2 \sum_{\alpha=\max(1,1+m)}^{\min(N,N+m)} \sum_{\gamma=\max(1,1+n)}^{\min(N,N+n)} \frac{1}{2\pi} \int_{-\pi}^{\pi} c_{n,m}(\lambda) e^{-i(\alpha-\gamma)\lambda}\, d\lambda.$$

But then

$$S_1(e^{-in\lambda} e^{-im\mu}) = \frac{e}{2\pi N} \int_{-\pi}^{\pi} e^{\frac{i(n-m)\lambda}{2}} c_{n,m}(\lambda) \frac{\sin \frac{N-m}{2}\lambda \sin \frac{N-n}{2}\lambda}{\sin^2 \frac{\lambda}{2}}\, d\lambda$$

$$= \frac{e}{2\pi N} \int_{-\pi}^{\pi} e^{\frac{i(n-m)\lambda}{2}} c_{n,m}(\lambda)$$

$$\left[\frac{\sin^2 \frac{N-n}{2}\lambda}{\sin^2 \frac{\lambda}{2}} \cos \frac{n-m}{2}\lambda + \frac{1}{2} \frac{\sin (N-n)\lambda \sin \frac{n-m}{2}\lambda}{\sin^2 \frac{\lambda}{2}} \right] d\lambda.$$

Using [5] we see that the second term is dominated by constant $(\log N)/(N) \to 0$ as $N \to \infty$. The first term tends to $ec_{n,m}(0) = e\rho_n\rho_m$. Now let us consider a finite trigonometric polynomial

$$h(\lambda, \mu) = \sum_{n,m=-p}^{p} h_{n,m} e^{-in\lambda} e^{-im\mu}$$

in place of $\phi(\lambda)\psi(\lambda)$ in S_1. Then

$$\lim_{N\to\infty} S_1(h(\lambda, \mu)) = \frac{e}{4\pi^2} \sum_{n,m=-p}^{p} h_{n,m}\, \rho_n\, \rho_m = e \int_0^{\pi}\int_0^{\pi} h(\lambda, \mu) f(\lambda) f(\mu)\, d\lambda\, d\mu.$$

Now given any $\phi(\lambda)$ and $\psi(\mu)$, we can choose two finite trigonometric polynomials $h_1(\lambda, \mu)$, $h_2(\lambda, \mu)$ such that

$$h_1(\lambda, \mu) \leqq \phi(\lambda)\psi(\mu) \leqq h_2(\lambda, \mu), \qquad \int_{-\pi}^{\pi}\int_{-\pi}^{\pi} [h_2(\lambda, \mu) - h_1(\lambda, \mu)]\, d\lambda\, d\mu < \epsilon.$$

Then $S_1(\phi(\lambda)\psi(\mu))$ lies between $S_1(h_1(\lambda, \mu))$, $S_1(h_2(\lambda, \mu))$ and hence

$$\overline{\lim_{N\to\infty}}\ , \underline{\lim_{N\to\infty}}\ S_1(\phi(\lambda)\psi(\mu))$$

lies between $e \int_0^{\pi}\int_0^{\pi} h_1(\lambda, \mu) f(\lambda) f(\mu)\, d\lambda\, d\mu$ and $e \int_0^{\pi}\int_0^{\pi} h_2(\lambda, \mu) f(\lambda) f(\mu)\, d\lambda\, d\mu$. On letting $\epsilon \to 0$ we see that

$$\lim_{N\to\infty} S_1(\phi(\lambda)\psi(\mu)) = e \int_0^{\pi} \phi(\lambda) f(\lambda)\, d\lambda \int_0^{\pi} \psi(\mu) f(\mu)\, d\mu.$$

To evaluate the remainder $S_2 + S_3$ of cov (Φ_N, Ψ_N) we note that it would be the actual covariance if the process were normal. But in this case the limiting covariance has been evaluated in [7] when ϕ and ψ are even, although under the restriction that the spectral density has a continuous derivative. However, an argument similar to that just carried out above indicates that the result in [7] is valid for continuous $f(\lambda)$. This proves (3.3).

One should note that as the periodogram is an even function, it is sufficient to consider estimates of the form $\Phi_N = \int_0^{\pi} I_N(\lambda)\phi(\lambda)\, d\lambda$ corresponding to the weight function $\frac{1}{2}[\phi(\lambda) + \phi(-\lambda)]$. Expression (3.3) then becomes

$$\lim_{N\to\infty} N \operatorname{cov}(\Phi_N, \Psi_N) = e\Phi\Psi + 2\pi \int_0^{\pi} f^2(l)\phi(l)\psi(l)\, dl.$$

The following theorem gives an estimate of the speed with which $E\Phi_N$ converges to Φ.

THEOREM 2. *Suppose that $f(\lambda)$ has a bounded derivative. Then*

$$E\Phi_N = \Phi + 0(\log N/N).$$

PROOF. By the proof of Theorem 1

$$E\Phi_N - \Phi = \int_{-\pi}^{\pi} \frac{1}{2\pi N} \int_{-\pi}^{\pi} \frac{\sin^2 \frac{N}{2}(\lambda - l)}{\sin^2 \frac{\lambda - l}{2}} [f(l) - f(\lambda)]\phi(\lambda)\, d\lambda\, dl.$$

But we have uniformly $f(l) - f(\lambda) = O(l - \lambda)$ so that

$$E\Phi_N - \Phi = \frac{1}{2\pi N}\int_{-2\pi}^{2\pi} \frac{\sin^2 \frac{N}{2} x}{\sin^2 \frac{x}{2}} O(|x|)\sqrt{2}(2\pi - |x|)\, dx \leqq \frac{K \log N}{N}$$

according to [5].

We remark that if the function $f(\lambda)$ defined on the real line with period 2π has a bounded second derivative everywhere we get instead $E\Phi_N - \Phi = O(1/N)$.

We now prove three elementary lemmas that are required in the development of the main results of the paper. These results have been proved under conditions that are probably far from necessary. But we believe these conditions give an idea of the practical domain of applicability of the results.

In the remainder of the paper we shall assume that $E |\xi_\nu|^8 < \infty$.

LEMMA 1. *Consider the covariances* $c_\nu = \sum_{n=1}^{N-|\nu|} \xi_n \xi_{n+|\nu|}$. *Then* if $\nu, \mu \neq 0$

$$|E c_\nu c_{\nu+j} c_\mu c_{\mu+j}| \leqq \begin{cases} A_1 N & \text{if} \quad \mu \neq \nu \\ A_2 N^2 & \text{if} \quad \mu = \nu. \end{cases} \tag{3.5}$$

The reader should notice that C_ν refers to the y-process and c_ν to the ξ-process.

PROOF. It is clear that (3.5) is made up of terms of the type $E\prod_{i=1}^{8} \xi_{n_i}$. If one of the indices n_i is different from the rest this term vanishes. As each of the terms is bounded by $E |\xi_\nu|^8$, it is sufficient for our purpose to enumerate the nonvanishing terms. But we have restraints, say

$$n_2 = n_1 + a, \qquad n_4 = n_3 + b, \qquad n_6 = n_5 + c, \qquad n_8 = n_7 + d,$$

where we do not yet specify the integers a, b, c, d except that they should all be different from zero. We can then treat the eight variables in a completely symmetric way. Let us fix n_1. As $n_2 \neq n_1$, n_2 has to be equal to some other n_i, say n_3.

Now we separate two cases. As $n_5 \neq n_6$, it has to be equal to one of n_1, n_2, n_3, n_4 or to one of n_7, n_8. Let us consider the first alternative. Then n_7 has to be equal to one of $n_1, n_2, n_3, n_4, n_5, n_6$ and whichever we choose, we have

$$n_i = n_1 + \alpha_i, \qquad i = 2, 3, \cdots, 8. \tag{3.6}$$

As $1 \leqq n_i \leqq N$ this gives us at most N possibilities.

In the second alternative let us take $n_5 = n_7$. If any of n_1, n_2, n_3, n_4 is equal to any of n_5, n_6, n_7, n_8 we have again a system of restrictions of type (3.6) and hence again at most N possibilities so we can exclude this case. But $n_2 = n_3 \neq n_1$ and $n_6 \neq n_7 = n_5$ so that the only other way of getting a nonvanishing term is $n_2 = n_4$, $n_6 = n_8$, which requires that $a = b$, $c = d$. If that is the case we clearly get at most N^2 possibilities. The result now follows easily.

LEMMA 2. *The distribution of the variables* $(c_0 - N)/\sqrt{N}, c_1/\sqrt{N}, \cdots, c_k/\sqrt{N}$ *tends to the distribution of* $k + 1$ *normal and independent variables with mean zero and variances* $e + 2, 1, 1, \cdots, 1$.

PROOF. Introduce

$$z_j = t_0(\xi_j^2 - 1) + t_1 \xi_j \xi_{j+1} + \cdots + t_k \xi_j \xi_{j+k}$$

so that

$$U_N = \frac{1}{\sqrt{N}} \sum_{j=1}^{N} z_j = \frac{1}{\sqrt{N}} [t_0(c_0 - N) + t_1 c_1' + \cdots + t_k c_k']$$

where

$$c_\nu' = c_\nu + \sum_{j=N-\nu+1}^{N} \xi_j \xi_{j+\nu}, \qquad \nu > 0.$$

It is evident that $(c_\nu' - c_\nu)/(\sqrt{N}) \to 0$ in probability as $N \to \infty$. But z_j is a stationary $(k + 1)$-dependent sequence of random variables. Applying a theorem of Hoeffding and Robbins [11] we see that U_N is asymptotically normal with mean zero and variance $t_0^2(e + 2) + \sum_1^k t_\nu^2$. Hence

$$Ee^{iU_N} = E \exp\left\{it_0 \frac{c_0 - N}{\sqrt{N}} + i \sum_1^k t_\nu \frac{c_\nu'}{\sqrt{N}}\right\} \to e^{-\frac{1}{2}[t_0^2(e+2)+\Sigma_1^k t_\nu^2]}$$

which proves the lemma.

LEMMA 3. *If y_t is a normal process with a positive spectral density having an integrable second derivative, then a_ν can be chosen so that $a_\nu = O(1/\nu^2)$.*

PROOF. Choose a_ν so that $a_\nu = \int_{-\pi}^{\pi} e^{i\nu\lambda} \sqrt{f(\lambda)}\, d\lambda$ and integrate by parts twice. The result follows immediately.

4. Treatment of pure white noise.

THEOREM 3. *Consider the empirical spectral distribution function*

$$F_{N,\xi}^*(\lambda) = \int_0^\lambda I_{N,\xi}(l)\, dl$$

where

$$I_{N,\xi}(\lambda) = \frac{1}{2\pi N} \left| \sum_{\nu=1}^{N} \xi_\nu e^{-i\nu\lambda} \right|^2.$$

Then the limiting probability distribution of

$$\max_{0 \leq \lambda \leq \pi} \sqrt{N} \left| F_{N,\xi}^*(\lambda) - \frac{\lambda}{2\pi} \right|$$

as $N \to \infty$ is the same as that of

$$\max_{0 \leq \lambda \leq \pi} |\zeta(\lambda)|$$

where $\zeta(\lambda)$ is a normally distributed process with mean zero and covariance

$$E\zeta(\lambda)\zeta(\mu) = \frac{e\lambda\mu}{4\pi^2} + \frac{1}{2\pi} \min(\lambda, \mu).$$

PROOF. We have

$$\sqrt{N}\left[F^*_{N,\xi}(\lambda) - \frac{\lambda}{2\pi}\right] = \frac{\lambda}{2\pi}\left(\frac{c_0}{\sqrt{N}} - \sqrt{N}\right) + \sum_1^{N-1} \frac{c_\nu}{\sqrt{N}} \frac{\sin \nu\lambda}{\pi\nu}$$

$$= s_{N,k}(\lambda) + r_{N,k}(\lambda)$$

where $s_{N,k}(\lambda)$ consists of the first term and the $k - 1$ first summands of the sum. We shall show that with probability close to one, for sufficiently large k, $| r_{N,k}(\lambda) |$ is small uniformly in N, λ.

Let us consider

$$\left| \sum_{\nu=m}^{l} \frac{c_\nu}{\sqrt{N}} \frac{\sin \nu\lambda}{\nu} \right| \leqq \left| \sum_{\nu=m}^{l} \frac{c_\nu}{\sqrt{N}} \frac{e^{i\nu\lambda}}{\nu} \right|$$

where $0 < m < l < N$. But

$$\left| \sum_{\nu=m}^{l} \frac{c_\nu}{\sqrt{N}} \frac{e^{i\nu\lambda}}{\nu} \right|^2 \leqq 2 \sum_{j=0}^{l-m} \left| \sum_{\nu=m}^{l-j} \frac{c_\nu c_{\nu+j}}{N\nu(\nu + j)} \right|.$$

To get a bound for this sum we consider

$$E\left| \sum_{\nu=m}^{l-j} \frac{c_\nu c_{\nu+j}}{N\nu(\nu + j)} \right|^2 = \sum_{\nu,\mu=m}^{l-j} \frac{Ec_\nu c_{\nu+j} c_\mu c_{\mu+j}}{N^2\nu(\nu + j)\mu(\mu + j)}. \tag{4.1}$$

We know by Lemma 1 that

$$| Ec_\nu c_{\nu+j} c_\mu c_{\mu+j} | \leqq \begin{cases} A_2 N^2 & \text{if} \quad \nu = \mu \\ A_1 N & \text{if} \quad \nu \neq \mu. \end{cases}$$

Thus the expression in (4.1) is bounded (use the Schwarz inequality) by

$$A_2 \sum_{\nu=m}^{l-j} \frac{1}{\nu^2(\nu + j)^2} + \frac{A_1}{N} \sum_{\nu,\mu=m}^{l-j} \frac{1}{\nu(\nu + j)\mu(\mu + j)} \leqq \sum_{\nu=m}^{l-j} \frac{A_3}{\nu^2(\nu + j)^2}.$$

Now we choose $m = 2^p$ and $l = 2^{p+1}$. Then again using the Schwarz inequality we get

$$E \max_{0 \leqq \lambda \leqq \pi} \left| \sum_{\nu=2^p}^{2^{p+1}} \frac{c_\nu}{\sqrt{N}} \frac{\sin \nu\lambda}{\nu} \right| \leqq A_4 \sum_{j=0}^{2^p} \left[\sum_{\nu=2^p}^{2^{p+1}-j} \frac{1}{\nu^2(\nu + j)^2} \right]^{1/2}$$

$$\leqq A_5 \sum_{j=0}^{2^p} \frac{1}{2^{3p/2}} = A_5/2^{p/2}.$$

Let $k = 2^n$. Then with probability greater than $1 - A_5/2^{p/4}$

$$T_p = \max_{0 \leqq \lambda \leqq \pi} \left| \sum_{\nu=2^p}^{2^{p+1}} \frac{c_\nu}{\sqrt{N}} \frac{\sin \nu\lambda}{\nu} \right| < \frac{1}{2^{p/4}}.$$

Consequently

$$| r_{N,k}(\lambda) | \leqq \sum_{p=n}^{[\log N]+1} T_p \leqq A_6 \frac{1}{2^{n/4}}$$

with probability greater than $1 - A_6/2^{n/4}$. Choosing n sufficiently large $|r_{N,k}(\lambda)|$ is uniformly small.

We want to find the limiting value of

$$P\left(\max_{0 \leq \lambda \leq \pi} \sqrt{N}\left|F^*_{N,\xi}(\lambda) - \frac{\lambda}{2\pi}\right| \leq \alpha\right).$$

But from what we have just shown, we see that for any $\epsilon, \delta > 0$

$$P(\max_{0 \leq \lambda \leq \pi} |s_{N,k}(\lambda)| < \alpha - \epsilon) - \delta \leq P\left(\max_{0 \leq \lambda \leq \pi} \sqrt{N}\left|F^*_{N,\xi}(\lambda) - \frac{\lambda}{2\pi}\right| \leq \alpha\right)$$
$$\leq P(\max_{0 \leq \lambda \leq \pi} |s_{N,k}(\lambda)| < \alpha + \epsilon) + \delta$$

for k sufficiently large, uniformly in $N > N(\epsilon, \delta)$. But

$$s_{N,k}(\lambda) = \frac{c_0 - N}{2\pi\sqrt{N}}\lambda + \frac{1}{\pi}\sum_1^k \frac{c_\nu}{\sqrt{N}}\frac{\sin \nu\lambda}{\nu}$$

and we have shown in Lemma 2 that the joint distribution of $(c_0 - N)/\sqrt{N}$, $c_1/\sqrt{N}, \cdots, c_k/\sqrt{N}$ (k fixed) converges to the distribution of k independent normal variables with common mean zero and variances $e + 2, 1, 1, \cdots, 1$. Consider the related process

$$s_k(\lambda) = \frac{\gamma_0}{2\pi}\sqrt{e+2}\,\lambda + \sum_1^k \frac{\gamma_\nu}{\pi}\frac{\sin \nu\lambda}{\nu}$$

where the γ's are $N(0, 1)$ and independent. It is easily seen that

$$\lim_{N \to \infty} P(\max_{0 \leq \lambda \leq \pi} |s_{N,k}(\lambda)| \leq \alpha) = P(\max_{0 \leq \lambda \leq \pi} |s_k(\lambda)| \leq \alpha)$$

as the relevant point set in $(k + 1)$-space is closed [18]. But on letting k tend to infinity $s_k(\lambda)$ converges uniformly to

$$\zeta(\lambda) = \frac{\gamma_0}{2\pi}\sqrt{e+2}\,\lambda + \sum_1^\infty \frac{\gamma_\nu}{\pi}\frac{\sin \nu\lambda}{\nu}$$

(see Paley-Wiener [17], p. 148–151). But then

$$P(\max_{0 \leq \lambda \leq \pi} |\zeta(\lambda)| < \alpha - \epsilon) - \delta < P\left(\max_{0 \leq \lambda \leq \pi} \sqrt{N}\left|F^*_{N,\xi}(\lambda) - \frac{\lambda}{2\pi}\right| \leq \alpha\right)$$
$$< P(\max_{0 \leq \lambda \leq \pi} |\zeta(\lambda)| < \alpha + \epsilon) + \delta$$

and if $N \to \infty$ we can let $\delta, \epsilon \to 0$. Since the distribution function of $\max_{0 \leq \lambda \leq \pi} |\zeta(\lambda)|$ is continuous (see Section 7), this completes the proof.

5. Reduction to pure white noise. Let $G(\lambda) = \int_0^\lambda f^2(l)\,dl$.

THEOREM 4. *Let $f(\lambda)$ be a nonnegative absolutely continuous function. Then*

$$\lim_{N \to \infty} P\left\{\max_{0 \leq \lambda \leq \pi} \sqrt{N}\left|\int_0^\lambda f(l)\{2\pi I_{N,\xi}(l) - 1\}\,dl\right| \leq \alpha\right\} = P\{\max_{0 \leq \lambda \leq \pi} |\eta(\lambda)| < \alpha\}$$

where $\eta(\lambda)$ *is the normal process with,* $E\eta(\lambda) = 0$, $E\eta(\lambda)\eta(\mu) = eF(\lambda)F(\mu) + 2\pi G(\min(\lambda, \mu))$.

PROOF. Integrating by parts we have

$$\int_0^\lambda f(l)\{2\pi I_{N,\xi}(l) - 1\}\, dl = 2\pi f(\lambda)\left[F_{N,\xi}^*(\lambda) - \frac{\lambda}{2\pi}\right] - \int_0^\lambda 2\pi f'(l)\left[F_{N,\xi}^*(l) - \frac{l}{2\pi}\right] dl.$$

Given $\epsilon > 0$, by Theorem 3 we know that

$$\max_{0 \leq \lambda \leq \pi} \left| \sqrt{N}\left(F_{N,\xi}^*(\lambda) - \frac{\lambda}{2\pi}\right) - s_{N,k}(\lambda)\right| < \epsilon$$

with probability $1 - \epsilon$ uniformly in $N > N(\epsilon)$ for k sufficiently large. But then

$$\max_{0 \leq \lambda \leq \pi} \left| f(\lambda)\left[\sqrt{N}\left(F_{N,\xi}^*(\lambda) - \frac{\lambda}{2\pi}\right) - s_{N,k}(\lambda)\right] - \int_0^\lambda f'(l)\left[\sqrt{N}\left(F_{N,\xi}^*(l) - \frac{l}{2\pi}\right) - s_{N,k}(l)\right] dl \right| < B_1 \epsilon$$

with probability $1 - \epsilon$. However,

$$P\left\{\max_{0 \leq \lambda \leq \pi} \left| f(\lambda)s_{N,k}(\lambda) - \int_0^\lambda f'(l)s_{N,k}(l)\, dl\right| \leq \alpha\right\} \to P\left\{\max_{0 \leq \lambda \leq \pi} \left| f(\lambda)s_k(\lambda) - \int_0^\lambda f'(l)s_k(l)\, dl\right| \leq \alpha\right\}$$

as $N \to \infty$ as the relevant set is again closed in $(k + 1)$-space. We know that for any $\epsilon > 0$

$$\max_{0 \leq \lambda \leq \pi} | s_k(\lambda) - \zeta(\lambda) | < \epsilon$$

with probability $1 - \epsilon$ for sufficiently large k. Let

$$\eta(\lambda) = f(\lambda)\zeta(\lambda) - \int_0^\lambda f'(l)\zeta(l)\, dl.$$

In summary

$$P\left\{\max_{0 \leq \lambda \leq \pi} | \eta(\lambda) | < \alpha - \epsilon\right\} - \delta < P\left\{\max_{0 \leq \lambda \leq \pi} \sqrt{N}\left|\int_0^\lambda f(l)\{2\pi I_{N,\xi}(l) - 1\}\, dl\right| \leq \alpha\right\} < P\left\{\max_{0 \leq \lambda \leq \pi} | \eta(\lambda) | < \alpha + \epsilon\right\} + \delta$$

and on letting $N \to \infty$ so that $\epsilon, \delta \to 0$ we obtain the desired result. Note that the event $\max_{0 \leq \lambda \leq \pi} | \eta(\lambda)| \leq \alpha$ has a well defined probability since $\eta(\lambda)$ is a process with continuous sample functions with probability one.

THEOREM 5. *Let* $a_\nu = O(\nu^\beta)$ *where* $\beta < -3/2$. *Then given any* $\epsilon > 0$

$$\sqrt{N} \max_{0 \leq \lambda \leq \pi} \left| \int_0^\lambda [I_N(l) - 2\pi f(l) I_{N,\xi}(l)]\, dl \right| < \epsilon$$

with probability $1 - \epsilon$ *for* $N > N_\epsilon$.

PROOF. We have

$$2\pi\sqrt{N} \int_0^\lambda [I_N(l) - 2\pi f(l) I_{N,\xi}(l)]\, dl = \frac{1}{\sqrt{N}} \sum_{r,s=-\infty}^{\infty} a_r a_s d_{rs} \tag{5.1}$$

where

$$d_{r,s} = \sum_{n,m=1}^{N} \xi_{n-r}\, \xi_{m-s} \frac{e^{i(n-m)\lambda} - 1}{i(n-m)} - \sum_{\substack{n=1+r, N+r \\ m=1+s, N+s}} \xi_{n-r}\, \xi_{m-s} \frac{e^{i(n-m)\lambda} - 1}{i(n-m)}.$$

The coefficients of the terms when $n = m$ should be interpreted as λ. We note that there may be a lattice rectangle $R_{r,s}^{(N)}$ of points (n, m) common to both sums. Let us call the complement of $R_{r,s}^{(N)}$ with respect to the set consisting of all the lattice points in both summations $C_{r,s}^{(N)}$. Then we have

$$E \max_{0 \leq \lambda \leq \pi} | d_{r,s} | \leq \sum_{n,m \in C_{r,s}^{(N)}} g(n - m)$$

where

$$g(x) = \begin{cases} \dfrac{2}{|x|} & \text{if } x \neq 0 \\ \pi & \text{if } x = 0. \end{cases}$$

It is easily verified that

$$\sum_{n,m \in C_{r,s}^{(N)}} g(n - m) \leq \begin{cases} 2N \log N \text{ if } |r| > N \text{ or } |s| > N \\ 4(s + r) \log N \text{ otherwise.} \end{cases}$$

The expression in (5.1) is then bounded by

$$\frac{2}{\sqrt{N}} \sum_{s=-\infty}^{\infty} \sum_{|r|>N} | a_r a_s |\, 2N \log N + \frac{8}{\sqrt{N}} \sum_{s=-\infty}^{\infty} \sum_{|r| \leq N} | a_r a_s |\, r \log N$$

$$\leq \left(\sum_{s=-\infty}^{\infty} | a_s | \right) \left(4\sqrt{N} \log N \sum_{|r|>N} | a_r | + \frac{8 \log N}{\sqrt{N}} \sum_{|r| \leq N} r\, | a_r | \right).$$

Under the assumption made above, this expression tends to zero as $N \to \infty$ which proves the theorem.

6. Treatment of the general case. Let $F_N^*(\lambda) = \int_0^\lambda I_N(l)\, dl$.

THEOREM 6. *Let*

1. $f(\lambda)$ *be absolutely continuous*
2. $a_\nu = 0(\nu^\beta)$, $\beta < -3/2$.

Then

$$\lim_{N\to\infty} P\{\max_{0\leqq\lambda\leqq\pi} \sqrt{N}\mid F_N^*(\lambda) - F(\lambda)\mid \leqq \alpha\} = P\{\max_{0\leqq\lambda\leqq\pi} \mid \eta(\lambda)\mid \leqq \alpha\}$$

where $\eta(\lambda)$ *is the process defined in Section* 5.

The proof follows immediately from Theorems 3, 4 and 5. An important special case is contained in the following corollary.

COROLLARY 1. *When* y_t *is a normal process with a positive spectral density with an integrable second derivative, Theorem* 6 *reduces to the following statement:*

$$\lim_{N\to\infty} P\{\max_{0\leqq\lambda\leqq\pi} \sqrt{N}\mid F_N^*(\lambda) - F(\lambda)\mid \leqq \alpha\}$$

(6.1)
$$= \sum_{k=-\infty}^{\infty} (-1)^k[\Phi((2k+1)\alpha/x) - \Phi((2k-1)\alpha/x)] = \Delta(\alpha/x)$$

where $\Phi(u)$ *is the normal distribution function and* $x = \sqrt{2\pi G(\pi)}$.

PROOF. Lemma 3 implies that the assumptions of Theorem 6 are satisfied.

As $e = 0$ the process $\eta(\lambda)$ reduces to the Wiener process with the following changed scale of time: $t = 2\pi G(\lambda)$, $0 \leqq \lambda \leqq \pi$. The reader may note that (6.1) is the probability that a particle in Brownian motion on the line is not absorbed by the barriers α, $-\alpha$ in the time interval 0, $2\pi G(\pi)$, given that the particle starts from 0 at time 0 (see Sommerfeld [19], pp. 74–79).

7. The limiting distribution in the case of pure white noise. When the process y_t is pure white noise (so that $f(\lambda) \equiv$ constant, say 1) the limiting probability we are interested in is given by

(7.1)
$$P\{\max_{0\leqq\lambda\leqq\pi} \mid \zeta(\lambda)\mid \leqq \alpha\}$$

where $\zeta(\lambda)$ is the normal process with

$$E\zeta(\lambda) = 0$$

$$E\zeta(\lambda)\zeta(\mu) = e\lambda\mu + 2\pi \min(\lambda, \mu).$$

Now (7.1) has been explicitly evaluated when $e = 0$ (see 6.1). However, the limiting distribution can still be evaluated when $e \neq 0$.

THEOREM 7.

$$P\{\max_{0\leqq\lambda\leqq\pi} \mid \zeta(\lambda)\mid \leqq \alpha\} = \sum_{k=-\infty}^{\infty} (-1)^k e^{(-k^2\alpha^2/\pi^2)(1-(\gamma/2))} \cdot\left[\Phi\left(\frac{\sqrt{\gamma}\alpha}{\pi}\left(k+\frac{1}{\gamma}\right)\right) - \Phi\left(\frac{\sqrt{\gamma}\alpha}{\pi}\left(k-\frac{1}{\gamma}\right)\right)\right]$$

where $\gamma = e + 2$.

PROOF. Let $\tau = \lambda/\pi$. Let X be a normal random variable with mean zero and variance one. Let $X(\tau)$ be a normal process with

$$EX(\tau) = 0$$

$$EX(\tau)X(\tau') = \min(\tau, \tau') - \tau\tau',$$

that is, $X(\tau)$ is the Wiener process conditioned so that $X(1) = 0$. X and $X(\tau)$ are independent. Consider the process $u(\tau) = \sqrt{\gamma}\pi\tau X + \pi\sqrt{2}X(\tau)$. The process $u(\tau)$ has the same probability distribution as $\zeta(\lambda)$ so that

$$P\{\max_{0 \leqq \lambda \leqq \pi} |\zeta(\lambda)| \leqq \alpha\} = P\{\max_{0 \leqq \tau \leqq 1} |u(\tau)| \leqq \alpha\}.$$

But

$$P\{\max_{0 \leqq \tau \leqq 1} |u(\tau)| \leqq \alpha\}$$

$$(7.2) \qquad = \int_{-\infty}^{\infty} P\left\{\frac{-\alpha - \sqrt{\gamma}\pi\tau x}{\pi\sqrt{2}} \leqq X(\tau) \leqq \frac{\alpha - \sqrt{\gamma}\pi\tau x}{\pi\sqrt{2}}; 0 \leqq \tau \leqq 1\right\} \varphi(x)\, dx$$

where $\varphi(x)$ is the normal frequency function. Let

$$Y(t) = (t+1)X\left(\frac{t}{t+1}\right), \qquad 0 \leqq t < \infty.$$

Then $Y(t)$ is the Wiener process (see [4]), $0 \leqq t < \infty$. We have then to consider

$$(7.3) \qquad P\{-a - (a+b)t \leqq y(t) \leqq a + (a-b)t; 0 \leqq t < \infty\}$$

where $a = \alpha/\pi\sqrt{2}$, $b = \sqrt{\gamma}x/\sqrt{2}$. The integrand in (7.2) is clearly zero unless $|\sqrt{\gamma}x| \leqq \alpha/\pi$. But supposing this inequality to hold, Doob [4] has evaluated (7.3) as

$$\sum_{m=1}^{\infty} \{e^{-2[(2m-1)^2a^2-(2m-1)ba]} + e^{-2[(2m-1)^2a^2+(2m-1)ba]} - e^{-2[(2m)^2a^2-2mba]} - e^{-2[(2m)^2a^2+2mba]}\}$$

$$= 2\sum_{n=1}^{\infty} (-1)^{n+1} e^{-2n^2a^2} \cosh 2nba.$$

Thus (7.1) is equal to

$$\Phi\left(\frac{\alpha}{\pi\sqrt{\gamma}}\right) - \Phi\left(\frac{-\alpha}{\pi\sqrt{\gamma}}\right) - 2\sum_{n=1}^{\infty} (-1)^{n+1} e^{-(n^2\alpha^2/\pi^2)} \int_{-\alpha/\pi\sqrt{\gamma}}^{\alpha/\pi\sqrt{\gamma}} \varphi(x) \cosh n \frac{\sqrt{\gamma}\alpha}{\pi} x\, dx$$

$$= \Phi\left(\frac{\alpha}{\pi\sqrt{\gamma}}\right) - \Phi\left(\frac{-\alpha}{\pi\sqrt{\gamma}}\right) - \sum_{n=1}^{\infty} (-1)^{n-1} e^{-(n^2[(\alpha^2/\pi^2)-(\gamma\alpha^2/2\pi^2)])}$$

$$\left\{\Phi\left(\frac{n\sqrt{\gamma}\alpha}{\pi} + \frac{\alpha}{\pi\sqrt{\gamma}}\right) - \Phi\left(\frac{n\sqrt{\gamma}\alpha}{\pi} - \frac{\alpha}{\pi\sqrt{\gamma}}\right) + \Phi\left(\frac{-n\sqrt{\gamma}\alpha}{\pi} + \frac{\alpha}{\pi\sqrt{\gamma}}\right)\right.$$

$$\left. - \Phi\left(-\frac{n\sqrt{\gamma}\alpha}{\pi} - \frac{\alpha}{\pi\sqrt{\gamma}}\right)\right\} = \sum_{n=-\infty}^{\infty} (-1)^n e^{-(n^2\alpha^2/\pi^2)(1-(\gamma/2))}$$

$$\cdot \left[\Phi\left(\frac{\sqrt{\gamma}\alpha}{\pi}\left(n + \frac{1}{\gamma}\right)\right) - \Phi\left(\frac{\sqrt{\gamma}\alpha}{\pi}\left(n - \frac{1}{\gamma}\right)\right)\right].$$

This expression parallels (6.1) and will be used in a later publication in which we will study some applications of the theory.

8. Statistical applications. The corollary of Theorem 6 lends itself to important statistical applications. First, however, we will have to estimate $G(\pi)$ which is in general unknown. We note that $G(\pi) = (4\pi)^{-1}\sum_{-\infty}^{\infty} \rho_\nu^2$. Of course the C_ν/N are consistent estimates of ρ_ν. It is natural to consider an estimate of the form (8.1). One reason for this particular choice is that in practice one will try to avoid the cumbersome calculation of all the lagged product sums C_ν.

LEMMA 4. *If* $a_\nu = 0(\nu^\beta)$, $\beta < -1$, *the statistic*

$$G^*(\pi) = \frac{1}{4\pi N^2}\left[C_0^2 + 2\sum_{\nu=1}^{[KN^\alpha]} C_\nu^2\right], \quad 0 < \alpha < 1, \quad K > 0, \tag{8.1}$$

is a consistent estimate of $G(\pi)$.

PROOF.

$$G^*(\pi) = \frac{1}{4\pi N^2}\left\{C_0^2 + 2\sum_{\nu=1}^{m} C_\nu^2\right\} + \frac{1}{2\pi N^2}\sum_{\nu=m+1}^{[KN^\alpha]} C_\nu^2. \tag{8.2}$$

But we see using (3.5) that the last sum is less than or equal to

$$\sum_{\nu=m+1}^{[KN^\alpha]}\sum_{j,k=1}^{N}\sum_{\beta_1\beta_2=-\infty}^{\infty}\{| a_{j-\beta_1}a_{j+\nu-\beta_1}a_{k-\beta_2}a_{k+\nu-\beta_2}| + | a_{j-\beta_1}a_{j+\nu-\beta_2}a_{k-\beta_1}a_{k+\nu-\beta_2}|$$
$$+ | a_{j-\beta_1}a_{j+\nu-\beta_2}a_{k-\beta_2}a_{k+\nu-\beta_1}|\}(4+\mu_4) = \left(\sum\nolimits_1 + \sum\nolimits_2 + \sum\nolimits_3\right)(4+\mu_4).$$

Let the r_ν be understood to be the covariances of the process $\sum_{\nu=-\infty}^{\infty} | a_{n-\nu} | \xi_\nu$. The spectral density $(2\pi)^{-1} | \sum | a_\nu | e^{i\nu\lambda} |^2$ of this process is continuous and hence quadratically integrable, so that $\sum_{-\infty}^{\infty} r_\nu^2 < \infty$. But then

$$\left|\sum\nolimits_1\right| = \sum_{\nu=m+1}^{[KN^\alpha]} N^2 r_\nu^2 \leq N^2 \sum_{\nu=m+1}^{\infty} r_\nu^2$$

so that $|\sum_1|/N^2 < \epsilon$ for m sufficiently large uniformly in N. Also we get

$$\frac{1}{N^2}\left|\sum\nolimits_2\right| = \frac{1}{N^2}\sum_{\nu=m+1}^{[KN^\alpha]}\sum_{j,k=1}^{N} r_{j-k}^2 \leq \frac{1}{N}\sum_{m+1}^{[KN^\alpha]}\sum_{n=-\infty}^{\infty} r_n^2 \leq \frac{KN^\alpha+1}{N}\sum_{-\infty}^{\infty} r_n^2 \to 0$$

as $N \to \infty$. The third sum is handled in almost the same way but using Schwarz' inequality. We now choose m so large that the last sum in (8.2) is less than δ with probability larger than $1 - \delta$. The term in brackets in (8.2) consists of a fixed number of terms each of which converges in probability to the corresponding term in the expression for $G(\pi)$. This proves the lemma.

The corollary of Theorem 6 shows that in the normal domain the asymptotic distribution of $\max_\lambda \sqrt{N} \mid F_N^*(\lambda) - F(\lambda) \mid$ depends only on one parameter $G(\pi)$. This together with Lemma 4 enables us to construct confidence bands for $F(\lambda)$.

THEOREM 8. *Suppose that* y_t *is a normal process and that its spectral density is positive and has an integrable second derivative. Then*

$$F_N^*(\lambda) - \alpha\sqrt{\frac{2\pi G^*(\pi)}{N}} \leq F(\lambda) \leq F_N^*(\lambda) + \alpha\sqrt{\frac{2\pi G^*(\pi)}{N}} \tag{8.3}$$

holds with a probability tending to $\Delta(\alpha)$ *as* N *tends to infinity.*

The proof follows immediately from Lemma 4 and Corollary 1 as $\Delta(\alpha)$ is continuous.

Clearly Theorem 8 also gives a test of significance for the simple hypothesis of a completely specified spectrum. A more interesting case, however, is the following.

THEOREM 9. *From a stochastic process satisfying Theorem 8 we observe two independent time series of length* N_1 *and* N_2. *Compute the corresponding estimates* $F^*_{N_1}(\lambda)$, $F^*_{N_2}(\lambda)$, $G^*_1(\pi)$, $G^*_2(\pi)$. *Then, assuming that* $N_1/N_2 \to c > 0$ *and putting* $N = 2N_1N_2/(N_1 + N_2)$,

$$(8.4) \qquad \frac{\max_{0 \leq \lambda \leq \pi} \sqrt{N} \mid F^*_{N_1}(\lambda) - F^*_{N_2}(\lambda) \mid}{\sqrt{G^*_1(\pi) + G^*_2(\pi)}} > \alpha,$$

holds with a probability tending to $1 - \Delta(\alpha)$ *as* N *tends to infinity under the hypothesis* $F_1(\lambda) = F_2(\lambda)$.

Theorems 8 and 9 are somewhat restricted in their applicability as they stand, since in many contexts $m_t = Ex_t$ is not identically zero. We shall consider the case

$$m_t = d_1 \varphi_t^{(1)} + d_2 \varphi_t^{(2)} + \cdots + d_p \varphi_t^{(p)},$$

where $\varphi_t^{(1)}, \varphi_t^{(2)}, \cdots, \varphi_t^{(p)}$ are given sequences and the regression coefficients d_ν are unknown. To avoid unnecessary complications we will confine ourselves to the case $p = 2$, which illustrates the general situation. We have to introduce the following condition which prevents the two regression variables $\varphi_t^{(1)}$ and $\varphi_t^{(2)}$ from becoming linearly dependent in the limit

$$(8.5) \qquad R = \overline{\lim_{N \to \infty}} \frac{\left(\sum_1^N \varphi_t^{(1)} \varphi_t^{(2)} \right)^2}{\sum_1^N \varphi_t^{(1)2} \sum_1^N \varphi_t^{(2)2}} < 1.$$

We will use the least square estimates d_1^*, d_2^* of d_1, d_2 as they have been shown to be asymptotically efficient [8].

THEOREM 10. *Under the conditions of Theorems* 8 *and* 9 *and* (8.5), *formulas* (8.3) *and* (8.4) *remain valid if* $F^*_N(\lambda)$ *is computed using* $x_t - d_1^* \varphi_t^{(1)} - d_2^* \varphi_t^{(2)}$ *in place of* y_t.

PROOF. It is sufficient to prove that

$$\max_{0 \leq \lambda \leq \pi} \sqrt{N} \left| \frac{1}{2\pi N} \int_0^\lambda \left| \sum_{\nu=1}^N (x_\nu - d_1^* \varphi_\nu^{(1)} - d_2^* \varphi_\nu^{(2)}) e^{-i\nu l} \right|^2 dl - \frac{1}{2\pi N} \int_0^\lambda \left| \sum_1^N y_\nu e^{-i\nu l} \right|^2 dl \right|$$

tends to zero in probability as $N \to \infty$. The expression inside the absolute value sign is

$$\frac{1}{2\pi N} \sum_{\nu,\mu=1}^N [-2(d_1^* - d_1)\varphi_\nu^{(1)} y_\mu - 2(d_2^* - d_2)\varphi_\nu^{(2)} y_\mu + (d_1^* - d_1)^2 \varphi_\nu^{(1)} \varphi_\mu^{(1)} + (d_2^* - d_2)^2 \varphi_\nu^{(2)} \varphi_\mu^{(2)} + 2(d_1^* - d_1)(d_2^* - d_2)\varphi_\nu^{(1)} \varphi_\mu^{(2)}] \cdot \frac{\sin(\nu - \mu)\lambda}{\nu - \mu}$$

$$= \Sigma_1 + \Sigma_2 + \Sigma_3 + \Sigma_4 + \Sigma_5,$$

where the same convention as before is used in interpreting the terms with $\nu = \mu$. The least square estimates d_1^*, d_2^* are unbiased linear estimates with coefficients depending only upon the $\varphi_\nu^{(1)}$'s and $\varphi_\nu^{(2)}$'s. A simple argument using the fact that the spectral density is bounded shows that

$$(8.6)\quad \operatorname{var} d_1^* \leqq 2\pi \max_{0\leqq\lambda\leqq\pi} f(\lambda) \frac{\sum \varphi_\nu^{(2)^2}}{\sum \varphi_\nu^{(1)^2} \sum \varphi_\nu^{(2)^2} - (\sum \varphi_\nu^{(1)}\varphi_\nu^{(2)})^2}$$

$$\leqq \frac{2\pi \max_{0\leqq\lambda\leqq\pi} f(\lambda)}{1-R} \frac{1}{\sum_1^N \varphi_\nu^{(1)^2}}.$$

Now

$$\pi\sqrt{N} \max_{0\leqq\lambda\leqq\pi} \left| \sum\nolimits_1 \right| \leqq \frac{1}{\sqrt{N}} \left\{ \left| d_1^* - d_1 \right| \sum_{p=1}^{N-1} \frac{1}{p} \left| \sum_{\nu=1}^{N-p} y_\nu \varphi_{\nu+p}^{(1)} \right| \right.$$

$$\left. + \pi \left| d_1^* - d_1 \right| \cdot \left| \sum_{\nu=1}^{N} y_\nu \varphi_\nu^{(1)} \right|^2 + \left| d_1^* - d_1 \right| \sum_{p=-N+1}^{1} \frac{1}{|p|} \left| \sum_{\nu=1-p}^{N} y_\nu \varphi_{\nu+p}^{(1)} \right| \right\}.$$

But we have

$$(8.7)\quad E\left[\sum_{\nu=1}^{N-p} y_\nu \varphi_{\nu+p}^{(1)}\right]^2 \leqq 2\pi \max_{0\leqq\lambda\leqq\pi} f(\lambda) \sum_{\nu=1}^{N} \varphi_\nu^{(1)^2}.$$

We know from (8.6) that with as large a probability as desired $|d_1^* - d_1|$ is less than $k/\sqrt{\sum_{\nu=1}^N \varphi_\nu^{(2)^2}}$. Then it follows from Schwarz' inequality and (8.7) that with large probability

$$\pi\sqrt{N} \max_{0\leqq\lambda\leqq\pi} \left| \sum\nolimits_1 \right| < K' \log N/\sqrt{N} \to 0.$$

$\sum_2$ can be handled in the same way. Now

$$2\pi\sqrt{N} \max_{0\leqq\lambda\leqq\pi} \left| \sum\nolimits_3 \right| = \frac{(d_1^* - d_1)^2}{\sqrt{N}} \int_0^\lambda \left| \sum_1^N \varphi_\nu^{(1)} e^{i\nu l} \right|^2 dl \leqq \frac{(d_1^* - d_1)^2}{\sqrt{N}} \pi \sum_1^N \varphi_\nu^{(1)^2}$$

and the expectation of this tends to zero as $N^{-\frac{1}{2}}$. $\sum_4$ and $\sum_5$ can be treated in a similar manner which proves the theorem.

An important special case is $\varphi_\nu \equiv 1$, which corresponds to a constant unknown mean value of the process. Another situation of some interest arises when the spectrum of the process has a discrete component with frequencies $\lambda_1, \lambda_2, \cdots, \lambda_p$. There we take the φ_ν's as trigonometric functions with these frequencies.

9. Alternative estimates of the spectral distribution function. It is clear that Theorems 6, and 8–10 are still valid if the estimate $F_N^*(\lambda)$ is replaced by a truncated estimate

$$F_{N,tr}^*(\lambda) = \frac{C_0\lambda}{2\pi N} + \frac{1}{\pi N}\sum_{\nu=1}^{h_N} C_\nu \frac{\sin \nu\lambda}{\nu} = \frac{1}{2\pi}\int_0^\lambda \int_{-\pi}^{\pi} I_N(l) \frac{\sin (h_N + \frac{1}{2})(l-\mu)}{\sin (l - \mu/2)} dl\, d\mu$$

where $h_N \to \infty$ as $N \to \infty$, as they can be proved in exactly the same manner since the weighting factors are 1 from $\nu = 0$ up to $\nu = h_N$ and 0 from that point on. Note that the estimate $F^*_{N,tr}(\lambda)$, in general, is not nondecreasing with probability 1 which may at times be an unpleasant feature. We can choose $h_N = [kN^\alpha]$ as in Lemma 4. This reduces the computational work considerably as one then only needs to compute the C_ν's for $\nu \leqq [kN^\alpha]$.

We shall now consider a class of estimates of the spectral distribution function with nondecreasing weight functions. Let $W_N(\lambda) = \int_{-\pi}^{\lambda} w_N(l)\,dl$ where $w_N(l) \geqq 0$ in $(-\pi, \pi)$, $\int_{-\pi}^{\pi} w_N(l)\,dl = 1$ and

$$\lim_{N\to\infty} W_N(\lambda) = \begin{cases} 0 & \text{if } \lambda < 0 \\ 1 & \text{if } \lambda > 0. \end{cases}$$

Let

$$F^*_N(\lambda, W_N) = \int_0^{\pi} I_N(l) W_N(\lambda - l)\,dl$$

$$F(\lambda, W_N) = \int_0^{\pi} f(l) W_N(\lambda - l)\,dl.$$

We prove the following theorem.

THEOREM 11. *Under the conditions of Theorem* 6

$$\lim_{N\to\infty} P\{\max_{0\leqq\lambda\leqq\pi} \sqrt{N} \mid F^*_N(\lambda, W_N) - F(\lambda, W_N) \mid \leqq \alpha\} = P\{\max_{0\leqq\lambda\leqq\pi} \mid \eta(\lambda) \mid \leqq \alpha\}.$$

PROOF. On integrating by parts we have

$$\begin{aligned}\sqrt{N}[F^*_N(\lambda, W_N) - F(\lambda, W_N)] &= \sqrt{N} \int_0^{\pi} [I_N(l) - f(l)]\,dl W_N(\lambda - \pi) \\ &\quad + \int_0^{\pi} \sqrt{N} \int_0^{l} [I_N(\mu) - f(\mu)]\,d\mu\, w_N(\lambda - l)\,dl.\end{aligned} \tag{9.1}$$

By Theorem 5 we can replace (9.1) by

$$\begin{aligned}&\sqrt{N} \int_0^{\pi} f(l)[2\pi I_{N,\xi}(l) - 1]\,dl W_N(\lambda - \pi) \\ &\quad + \int_0^{\pi} \sqrt{N} \int_0^{l} [2\pi I_{N,\xi}(\mu) - 1] f(\mu)\,d\mu\, w_N(\lambda - l)\,dl\end{aligned} \tag{9.2}$$

committing an error of at most $\epsilon > 0$ uniformly in λ with probability $1 - \epsilon$. However, on integrating by parts twice we obtain

$$\begin{aligned}&\sqrt{N} 2\pi[F^*_{N,\xi}(\pi) - \tfrac{1}{2}] f(\pi) W_N(\lambda - \pi) - \int_0^{\pi} \sqrt{N} 2\pi \left[F^*_{N,\xi}(l) - \frac{l}{2\pi}\right] \\ &\qquad \cdot \frac{d}{dl}(f(l) W_N(\lambda - l))\,dl.\end{aligned} \tag{9.3}$$

But

$$\sqrt{N} \max_{0 \le \lambda \le \pi} \left| F^*_{N,\xi}(\lambda) - \frac{\lambda}{2\pi} - \frac{1}{\sqrt{N}} s_{N,m}(\lambda) \right| < \epsilon$$

with probability $1 - \epsilon$ where m is a large but fixed number. We can then replace (9.3) by

$$\int_0^\pi f(l) W_N(\lambda - l) s'_{N,m}(l)\, dl = \frac{c_0 - N}{2\pi\sqrt{N}} \int_0^\pi f(l) W_N(\lambda - l)\, dl$$

$$+ \frac{1}{\pi} \sum_1^m \int_0^\pi \frac{c_\nu}{\sqrt{N}} \cos \nu l f(l) W_N(\lambda - l)\, dl$$

with an error of at most ϵ uniformly in λ. But then reasoning as in Theorem 4 we get

$$\lim_{N \to \infty} P\left\{ \max_{0 \le \lambda \le \pi} \left| \int_0^\pi f(l) W_N(\lambda - l) s'_{N,m}(l)\, dl \right| \le \alpha \right\}$$

$$= P\left\{ \max_{0 \le \lambda \le \pi} \left| \int_0^\lambda f(l) s'_m(l)\, dl \right| \le \alpha \right\}$$

making use of the fact that

$$\max_{0 \le \lambda \le \pi} \left| \int_0^\pi \cos \nu l f(l) W_N(\lambda - l)\, dl - \int_0^\lambda \cos \nu l f(l)\, dl \right|$$

$$\le \max_{0 \le \lambda \le \pi} f(\lambda) \left\{ \int_{-\pi}^0 W_N(l)\, dl + \int_0^\pi | W_N(l) - 1 |\, dl \right\} \to 0$$

as $N \to \infty$. But

$$\max_{0 \le \lambda \le \pi} \left| \int_0^\lambda f(l) s'_m(l)\, dl - \eta(\lambda) \right| < \epsilon$$

with large probability if m is sufficiently large and the theorem follows immediately.

We are usually more interested in estimating $F(\lambda)$ than $F(\lambda, W_N)$. The following corollary enables us to do this.

COROLLARY 2. *Theorems* 6, 8–10 *remain valid when* $F^*_N(\lambda)$ *is replaced by* $F^*_N(\lambda, W_N)$ *if*

$$\int_{-\pi}^0 W_N(\lambda)\, d\lambda + \int_0^\pi [1 - W_N(\lambda)]\, d\lambda = o(N^{-\frac{1}{2}}). \tag{9.4}$$

PROOF. The proof follows immediately as

$$\max_{0 \le \lambda \le \pi} \sqrt{N}\, | F(\lambda) - F(\lambda, W_N) |$$

$$\le \sqrt{N} \max_{0 \le \lambda \le \pi} f(\lambda) \left\{ \int_{-\pi}^0 W_N(\lambda)\, d\lambda + \int_0^\pi [1 - W_N(\lambda)]\, d\lambda \right\} = o(1)$$

It is clear from the comment above that the time series analyst has a large class of estimates of the spectral distribution function at his disposal. At present we are not able to claim that certain estimates are preferable to others. We hope to investigate such problems in a later paper.

It is worthwhile noting that the Fejér kernel

$$\frac{1}{2\pi h_N} \frac{\sin^2 \frac{h_N}{2} \lambda}{\sin^2 \frac{\lambda}{2}} = w_N(\lambda)$$

satisfies (9.4) if $\log h_N/h_N = o(N^{-\frac{1}{2}})$, so that we can choose the truncation point h_N as $h_N = [kN^\alpha]$, $\frac{1}{2} < \alpha < 1$. Note that the estimate corresponding to this kernel is

$$F_N^*(\lambda, W_N) = \frac{C_0}{2\pi N} + \frac{1}{\pi N} \sum_{\nu=1}^{h_N} C_\nu \frac{\sin \nu\lambda}{\nu} \left(1 - \frac{\nu}{h_N}\right).$$

This estimate of the spectral distribution function is closely related to an estimate of the spectral density given by Bartlett [1]. It is nondecreasing and does not require the computation of all the C_ν's.

10. Alternative statistics and the corresponding limit theorems. In some instances one might prefer to consider statistics other than $\max_{0 \leqq \lambda \leqq \pi} | F_N^*(\lambda) - F(\lambda) |$ in analyzing time series. We shall therefore consider some alternative statistics in this section.

Consider the linear space consisting of continuous functions $c = c(\lambda)$ on $0 \leqq \lambda \leqq \pi$ with the norm $\| c \| = \sup_{0 \leqq \lambda \leqq \pi} | c(\lambda) |$. Consider a functional $\varphi(c)$ uniformly continuous in this topology.

THEOREM 12. *Under the conditions of Theorem* 6 *we have*

$$\lim_{N \to \infty} P\{\varphi(\sqrt{N}[F_N^*(\lambda) - F(\lambda)]) \leqq \alpha, 0 \leqq \lambda \leqq \pi\} = P\{\varphi(\eta(\lambda)) \leqq \alpha, 0 \leqq \lambda \leqq \pi\}.$$

PROOF. Writing

$$\sqrt{N}\, [F_N^*(\lambda) - F(\lambda)] = s_{N,k}(\lambda) + r_{N,k}(\lambda)$$

as before we note that $\| r_{N,k}(\lambda) \| < \epsilon$ if k is chosen sufficiently large. Hence we commit only a small error by considering instead the probability of the event

$$\{\varphi(s_{N,k}(\lambda)) \leqq \alpha,\ 0 \leqq \lambda \leqq \pi\}$$

which is a closed set in $(k + 1)$-space. This probability converges to

$$P\{\varphi(s_k(\lambda)) \leqq \alpha,\ 0 \leqq \lambda \leqq \pi\}$$

as $N \to \infty$. But we can choose k so large that $\| s_k(\lambda) - \eta(\lambda) \| < \delta$ with probability $1 - \delta$. This together with the uniform continuity of $\varphi(c)$ proves the theorem.

As an example we can choose

$$\varphi(c) = \left\{ \int_0^\lambda | c(\lambda) |^p \, d\mu(\lambda) \right\}^{1/p}, \qquad p \geqq 1,$$

where $\mu(\lambda)$ is bounded and nondecreasing. This will give a statistic of the von Mises type.

11. Acknowledgement. We are indebted to J. L. Doob who suggested the problem that led to the investigation resulting in this paper.

REFERENCES

[1] M. S. Bartlett, "Smoothing periodograms from time series with continuous spectra," *Nature*, Vol. 161 (1948), pp. 686–687.

[2] M. S. Bartlett, "Periodogram analysis and continuous spectra," *Biometrika*, Vol. 37 (1950), pp. 1–16.

[3] H. Cramér, "On harmonic analysis in certain functional spaces," *Ark. Mat. Astr. Fys.*, Vol. 28B (1942), p. 12.

[4] J. L. Doob, "Heuristic approach to the Kolmogorov-Smirnov theorems," *Ann. Math. Stat.*, Vol. 20 (1949), pp. 393–403.

[5] L. Fejér, "Lebesguesche konstanten und divergente fourier-reihen," *J. Reine Angew. Math.*, Vol. 138 (1910), pp. 22–53.

[6] R. A. Fisher, "Tests of significance in harmonic analysis," *Proc. Roy. Soc. London, Ser. A*, Vol. 125 (1929), pp. 54–59.

[7] U. Grenander, "On empirical spectral analysis of stochastic processes," *Ark. Mat.*, Vol. 1 (1951), pp. 503–531.

[8] U. Grenander, "On the estimation of regression coefficients in the case of an autocorrelated disturbance," unpublished manuscript.

[9] U. Grenander and M. Rosenblatt, "On spectral analysis of stationary time series," *Proc. Nat. Acad. Sci.*, U. S. A., Vol. 38 (1952), pp. 519–521.

[10] G. Herglotz, "Über Potenzreihen mit positivem reelem Teil im Einheitskreis," *Sitzgsber. Sächs. Akad. Wiss.*, Vol. 63 (1911), p. 501.

[11] W. Hoeffding and H. Robbins, "The central limit theorem for dependent random variables," *Duke Math. J.*, Vol. 15 (1948), pp. 773–780.

[12] K. Karhunen, "Uber lineare Methoden in der Wahrscheinlichkeitsrechnung," *Ann. Acad. Sc. Fenn.* Ser. A I, Math.-Phys. No. 37, 79 pp. 1947.

[13] M. G. Kendall, *The Advanced Theory of Statistics*, Vol. 2, C. Griffith and Co., London, 1946.

[14] T. Koopmans, *Statistical Inference in Dynamic Economic Models*, John Wiley and Sons, 1950.

[15] P. Lévy, *Théorie de l'Addition des Variables Aléatoires*, Gauthier-Villars, Paris, 1937.

[16] P. Lévy, *Les Processus Stochastiques et le Mouvement Brownien*, Gauthier-Villars, Paris, 1948.

[17] R. E. A. C. Paley and N. Wiener, *Fourier Transforms in the Complex Domain*, Amer. Math. Soc., New York, 1934.

[18] H. Scheffé, "A useful convergence theorem for probability distributions," *Ann. Math. Stat.*, Vol. 18 (1947), pp. 434–438.

[19] A. Sommerfeld, *Partial Differential Equations in Physics*, Academic Press, New York, 1949.

[20] J. Tukey, "Measuring noise color," unpublished manuscript.

[21] P. Whittle, *Hypothesis Testing in Time Series Analysis*, Uppsala, 1951.

[22] H. Wold, *A Study in the Analysis of Stationary Time Series*, Almqvist and Wiksells boktr., Uppsala, 1938.

RECURRENCE-TIME MOMENTS IN RANDOM WALKS

J. L. HODGES, JR. AND M. ROSENBLATT

1. Introduction. Consider an irreducible time-homogeneous Markov chain with discrete time. The recurrence-time moments of the states of such stochastic processes are studied. We point out that if the recurrence time of one state has its first k moments finite, then the recurrence times of all the other states have their first k moments finite. We then specialize and investigate the recurrence-time moments of random walks. The main result of the paper consists of exhibiting random walks whose first $k-1$ recurrence-time moments exist and whose higher moments are infinite, for $k = 1, 2, \cdots$. A comparison theorem is derived that permits the moment properties of recurrence times of a large class of random walks to be determined.

2. Preliminary considerations. We begin with the following:

DEFINITION. By the *index* of the random variable X, denoted by $I(X)$, we shall mean the largest integer k such that $E(X^k) < \infty$. If all moments of X are finite, we write $I(X) = \infty$. Clearly $I(X) \geq 0$. We shall consider only nonnegative random variables.

LEMMA 1.[1] (a) *If $X_1, \cdots, X_n$ are independent random variables, then*

$$I(X_1 + \cdots + X_n) = \min\{I(X_1), \cdots, I(X_n)\}.$$

(b) *If X has the geometric distribution*

$$P(X = n) = p(1-p)^{n-1} \qquad (p > 0),$$

then $I(X) = \infty$.

(c) *If $N, X_1, X_2, \cdots$ are independent random variables, N is positive integer valued, $X_1, X_2, \cdots$ are identically distributed, and then*

---

[1] The proof of these simple facts is omitted.

Received April 28, 1952. This work has been carried out under ONR contract.
Pacific J. Math. 3 (1953), 127-136

R.A. Davis et al. (eds.), *Selected Works of Murray Rosenblatt*, Selected Works in Probability and Statistics, 74
DOI 10.1007/978-1-4419-8339-8_10, © Springer Science+Business Media, LLC 2011

$$I(X_1 + \cdots + X_N) = \min\{I(N), I(X_1)\}.$$

We adopt the terminology and notation for Markov chains used by Feller [1, Chap. 15]. Let $E_1, E_2, \cdots$ be a denumerable set of states. We assume the chain is irreducible; i.e., for every pair E_i, E_j there exists $n(i, j)$ such that transition from E_i to E_j in $n(i, j)$ steps has positive probability. Let R_i be the random number of steps for first return to E_i, and let R_{ij} be the random number of steps for first passage from E_i to E_j, starting at E_i. It is well known [1] that if, for any j, $P(R_j < \infty) = 1$, then this holds true for all j, and we shall assume this to be the case; i.e., our process is assumed to be recurrent.

LEMMA 2.[2] (a) *$I(R_i)$ has a constant value, say $I(C)$, for all $i = 1, 2, \cdots$.*
(b) *For every pair $i \neq j$,*

$$I(R_i) = \min\{I(R_{ij}), I(R_{ji})\}.$$

This lemma is related to the well-known result that in a recurrent irreducible time-homogeneous Markov chain, if the expected time for first return to any state is finite, then the same holds for all states. In the language of [1, Chap. 15], if any state is null, then every state is null. The lemma extends this result from first moment to arbitrary moments.

Examples of Markov chains of index k are to be found in a class of Markov chains considered by Feller [1]. Let $X(n)$ be the Markov chain, and let

$$p_j = P\{X(n) = E_{j+1} \mid X(n-1) = E_j\},$$

$$q_j = 1 - p_j = P\{X(n) = E_1 \mid X(n-1) = E_j\},\ j = 1, 2, \cdots.$$

Then

$$P(R_1 = n) = (1 - p_n)\prod_{i=1}^{n-1} p_i = \prod_{i=1}^{n-1} p_i - \prod_{i=1}^{n} p_i.$$

We obviously obtain a Markov chain with $E(R_1^k) < \infty$, $E(R_1^{k+i}) = \infty$, where $i \geq 1$, if the p_i are such that

$$\prod_1^{n-1} p_i - \prod_1^{n} p_i = \frac{C}{n^{k+1+\epsilon}} \qquad (0 < \epsilon < 1),$$

[2] After obtaining this result we learned that it had been obtained earlier by K. L. Chung and by R. N. Snow, in a more general form (unpublished). We omit the proof.

where

$$C \sum_{1}^{\infty} \frac{1}{n^{k+1+\epsilon}} = 1.$$

3. Comparison theorem for random walks. We shall now specialize from Markov chains to random walks. There is no loss of generality in taking the states of the walk to be consecutive integers, and we shall let p_i denote the probability of transition from i to $i+1$. To ensure irreducibility, we shall require $0 < p_i < 1$ for all interior states. To apply Lemma 2, we must have the walk recurrent, so we shall assume that the boundaries, if any, are reflecting, and that (if the walk is unbounded in one or both directions) the probability of escape to infinity is 0. Denote the walk by W and its index by $I(W)$.

THEOREM 1. (a) *$I(R_{ij})$ has a constant value, say I^R, for all $i > j$, and a constant value, say I^L, for all $i < j$.*

(b) *Further,* $I(W) = \min\{I^L, I^R\}$.

Proof. We shall first show that $I(R_{i+1,i}) = I(R_{i,i-1})$. Consider a walk starting at i and indefinitely prolonged. It is certain eventually to reach $i-1$; let M denote the number of times the walk is at i, including the start, before reaching $i-1$; M has the distribution

$$P(M = m) = (1 - p_i)\, p_i^{m-1};$$

and given $M = m$, the walk consists of $m-1$ steps from i to $i+1$, $m-1$ first passages from $i+1$ to i (denote these by $R^{(1)}_{i+1,i}, \cdots, R^{(m-1)}_{i+1,i}$), and a terminal step from i to $i-1$. Thus

$$R_{i,i-1} = M + R^{(1)}_{i+1,i} + \cdots + R^{(M)}_{i+1,i}.$$

Apply Lemma 1 (b), (c), (a).

To complete the proof of (a), represent a walk from i to j, $i > j$, as the sum

$$R_{i,j} = R_{i,i-1} + R_{i-1,i-2}, + \cdots + R_{j+1,j},$$

and apply Lemma 1 (a). As for (b), this follows from Lemma 2 (b).

It is clear that in a walk over finitely many states, the index is ∞. (See [1, problem 8, p. 345].) Similarly, passages away from a reflecting boundary have moments of all orders. The interest attaches to passages away from an unbounded side, and by virtue of Theorem 1 (b) we may consider the two sides separately. We may therefore, without loss of generality, consider only passages to the left in semi-infinite walks unbounded on the right, and we may take the states to be nonnegative integers. We now approach the following general question: given the sequence of transition probabilities $p_1, p_2, \cdots$ of such a walk, to determine the index of first passages from right to left.

This question is somewhat analogous to the question of determining the minimum integer k such that the series $\sum X_i^k$ is convergent. As in the convergence of series, all that matters is the performance of $\{p_i\}$ in the tail—any finite number of terms may be changed or suppressed without altering the result. We also have a "comparison" theorem.

THEOREM 2. *Consider two random walks, W and V, with transition probabilities $\{p_i\}$ and $\{q_i\}$, respectively. If $p_i \leq q_i$ for every $i = 1, 2, \cdots$, then $I(W) \geq I(V)$.*

Proof. We may assume that the walks begin at the same state. Refer both walks to the same infinite process $X_1, X_2, \cdots$, where $X_1, X_2, \cdots$ are mutually independent random variables, each uniformly distributed over $(0, 1)$. If the walk W is at state j for trial i, it proceeds to the right if and only if $X_i < p_j$, and similarly for walk V, which proceeds to the right if and only if $X_i < q_j$.

Consider $X_1, X_2, \cdots$ fixed at observed values $x_1, x_2, \cdots$. Observe that W and V always differ by an even number and that this difference changes by either 0 or 2 on each step. We shall show that walk W can never be to the right walk V. If it were, there would have to be a first time on which this occurred, and on the preceding time, say trial i, the walks would coincide, say at state j. But $p_j \leq q_j$, so that $x_i < p_j$ and $x_i \geq q_j$ are contradictory.

As an application, we remark: if lim sup $p_i < 1/2$, then the index is ∞; and if lim inf $p_i > 1/2$, when the walk is nonrecurrent. This fact follows from comparison with the classical walk with constant transition probability p.

The fact just cited indicates that the interest will lie in those sequences $\{p_i\}$ for which lim sup $p_i \geq 1/2$ and lim inf $p_i \leq 1/2$. We shall in the next section investigate a class of walks in which lim $p_i = 1/2$, and these will serve, together with the comparison theorem, to handle a large class of problems.

4. A class of random walks. We now consider a class of random walks which are related to the ordinary unbiased coin tossing. The random walk has as its possible states the integers r, $r + 1$, $r + 2$, $\cdots$. Let

$$p_k = P\{X(n) = k+1 \mid X(n-1) = k\},$$

$$1 - p_k = q_k = P\{X(n) = k-1 \mid X(n-1) = k\},$$

$$q_{k,n} = P\{\text{going from } k \text{ to } r \text{ for the first time in } n \text{ steps}\} \qquad (k > r).$$

Then

$$q_{k,n+1} = p_k q_{k+1,n} + q_k q_{k-1,n} \qquad (k > r).$$

The state r will be an absorbing barrier. This implies that

$$q_{r,0} = 1, \text{ while } q_{r,n} = 0 \text{ if } n > 0.$$

We shall set $q_{r-1,n} = 0$ for all n as a convenient convention. Note that

$$q_{r+1,1} = q_{r+1}, \text{ while } q_{k,1} = 0 \text{ if } k > r+1;$$

$$q_{k,0} = 0 \text{ if } k > r.$$

Define

$$G_k(s) = \sum_{n=0}^{\infty} q_{k,n} s^n \qquad (k = r-1, r, r+1, \cdots),$$

and observe that

$$G_{r-1}(s) = 0, \quad G_r(s) = 1.$$

Let

$$p_k = \frac{1}{2}\left(1 - \frac{r}{k}\right), \quad q_k = \frac{1}{2}\left(1 + \frac{r}{k}\right) \qquad (k = r, r+1, \cdots).$$

Note that $r = 0$ gives us the ordinary case of unbiased coin tossing.

The generating functions satisfy the following equations:

$$G_k(s) = p_k s G_{k+1}(s) + q_k s G_{k-1}(s) \qquad (k > r),$$

$$G_r(s) = p_r s G_{r+1}(s) + q_r s G_{r-1}(s) + 1 = q_r s G_{r-1}(s) + 1.$$

Let

$$G(t, s) = \sum_{k=r}^{\infty} G_k(s)t^k =$$

$$\frac{s}{2}\left[\frac{G(t, s)-t^r}{t} + tG(t, s)\right] + t^r + \frac{sr}{2}\left[\int_0^t G(t, s)\,dt - \int_0^t \frac{G(t, s)-t^r}{t^2}\,dt\right].$$

We now differentiate with respect to t $(0 < t < 1)$, use $[1-(s/2)(1/t+t)]^{r+1}$ as an integrating factor, and integrate the differential equation between the limits t and $\sigma = [1-(1-s^2)^{1/2}]/s$, where $0 < t < \sigma$. We then obtain

$$(1) \qquad -G(t, s)\left[1-\frac{s}{2}\left(\frac{1}{t}+t\right)\right]^{r+1}$$

$$= \int_t^{\sigma}\left[\frac{s\tau^{r-2}}{2} + r\tau^{r-1}\right]\left[1-\frac{s}{2}\left(\frac{1}{\tau}+\tau\right)\right]^r d\tau.$$

Let ${}_c\mu_k$ denote the cth factorial moment of the first passage time from k to r. It is easily seen that the ${}_c\mu_k$ (c fixed), if they are finite, cannot increase with k more than exponentially. Therefore, if ${}_c\mu_k < \infty$ for some positive c and some $k > r$, then

$$\sum_{k=r}^{\infty} {}_c\mu_k\, t^{k-c}$$

is convergent for sufficiently small t. Therefore

$$\lim_{s \to 1-} \frac{d^c}{ds^c} G(t, s) = \sum_{k=r}^{\infty} {}_c\mu_k\, t^{k-c}$$

exists for sufficiently small t if and only if ${}_c\mu_k < \infty$ for some $k > r$ (and hence for all $k > r$).

It is now easily shown that the first r derivatives of the integral on the right of (1) with respect to s are bounded in the closed interval $(t, 1)$, and hence have finite limits as $s \longrightarrow 1-$. The $(r+1)$st derivative, however, contains a term with the factor $d\sigma/ds$, which diverges as $s \longrightarrow 1-$.

The same techniques can be used to prove the following result:

THEOREM 3. *If one has a one-sided random walk with*

$$p_k = \frac{1}{2}\left(1 - \frac{\alpha}{k}\right), \quad q_k = \frac{1}{2}\left(1 + \frac{\alpha}{k}\right)$$

$\alpha > 0$ $(k = \alpha, \alpha + 1, \cdots)$ *then* ${}_r\mu_k < \infty$ *if* $r - 1/2 < \alpha \leq r + 1/2$ *and* $k > \alpha$, *while* ${}_c\mu_k = \infty$ *for* $c > r$ *and* $k > \alpha$.

We shall now consider walks with $p_n > 1/2$. It is clear from Theorem 2 that for such walks the index is 0. The interest now attaches to the question of whether first passages are certain to occur.

LEMMA 3.[3] *Consider a random walk over the states* $0, 1, 2, \cdots$. *Assume* $p_k \geq \epsilon$ *for some* $\epsilon > 0$. *The walk is recurrent if and only if the series*

$$\sum_{k=1}^{\infty} \frac{q_1 q_2 \cdots q_k}{p_1 p_2 \cdots p_k}$$

is divergent.

Proof. Consider first the finite walk over $0, 1, 2, \cdots, n$, and let $P_k^{(n)}$ denote the probability that the walk, starting at k, reaches n before reaching 0. Clearly

$$P_0^{(n)} = 0, \; P_n^{(n)} = 1,$$

and

$$P_k^{(n)} = p_k P_{k+1}^{(n)} + q_k P_{k-1}^{(n)} \qquad (k = 1, 2, \cdots, n-1).$$

We may solve for $P_2^{(n)}, P_3^{(n)}, \cdots, P_{n-1}^{(n)}$ in terms of $P_1^{(n)}$, getting

$$q_k\left(P_k^{(n)} - P_{k-1}^{(n)}\right) = p_k\left(P_{k+1}^{(n)} - P_k^{(n)}\right) \qquad (k = 1, 2, \cdots, n-1),$$

and hence

$$P_k^{(n)} = (1 + b_1 + b_2 + \cdots + b_{k-1})\, P_1^{(n)},$$

where $b_j = (q_1 q_2 \cdots q_j)/(p_1 p_2 \cdots p_j)$. Now suppose $\sum_{j=1}^{\infty} b_j = \infty$. Since

[3] We were informed by the referee that this lemma and the following theorem have also been obtained by T. E. Harris.

$$P_1^{(n)} = P_{n-1}^{(n)} \Bigg/ \left(1 + \sum_{k=1}^{n-1} b_k\right) \leq 1 \Bigg/ \left(1 + \sum_{k=1}^{n-1} b_k\right),$$

we must have $P_1^{(n)} \longrightarrow 0$ as $n \longrightarrow \infty$. We conclude that the walk is recurrent.

Conversely, suppose $\sum_{k=1}^{\infty} b_k < \infty$. Then $\sum_{k=1}^{\infty} b_k = b_0$, where b_0 is some positive number. Using the fact that

$$P_{n-1}^{(n)} = p_{n-1} + q_{n-1} P_{n-2}^{(n)},$$

we have

$$\left(1 + \sum_{k=1}^{n-1} b_k\right) P_1^{(n)} = p_{n-1} + q_{n-1} \left(1 + \sum_{k=1}^{n-2} b_k\right) P_1^{(n)},$$

and therefore

$$P_1^{(n)} = 1 \Bigg/ \left[1 + \sum_{k=1}^{n-2} b_k + b_{n-1}/ p_{n-1}\right].$$

Thus

$$\lim_{n \to \infty} P_1^{(n)} = 1/b_0 > 0,$$

whence the walk is nonrecurrent.

As an illustration, consider a random walk over the nonnegative integers with reflecting barrier 0 and

$$p_k = \frac{1}{2}\left(1 + \frac{1}{\beta k}\right), \qquad (k = 1, 2, \cdots ; \ \beta > 1).$$

THEOREM 4. *The walk is recurrent if and only if* $\beta \geq 2$.

Proof. Let

$P_k = P$ {walk never reaches 0, given that it starts at k} $(k = 1, 2, \cdots)$.

We have

$$P_1 = \frac{1}{2}\left(1+\frac{1}{\beta}\right)P_2, \quad P_k = \frac{1}{2}\left(1-\frac{1}{\beta k}\right)P_{k-1} + \frac{1}{2}\left(1+\frac{1}{\beta k}\right)P_{k+1} \qquad (k = 2, 3, \cdots).$$

Rewriting, we have

$$P_2 - P_1 = \frac{\beta-1}{\beta+1}P_1, \quad P_{k+1} - P_k = \frac{\beta k-1}{\beta k+1}(P_k - P_{k-1}) \qquad (k = 2, 3, \cdots).$$

Thus

$$P_k = P_1\left\{1 + \frac{\beta-1}{\beta+1} + \frac{\beta-1}{\beta+1}\cdot\frac{2\beta-1}{2\beta+1} + \cdots + \left(\frac{\beta-1}{\beta+1}\cdots\frac{(k-1)\beta-1}{(k-1)\beta+1}\right)\right\}.$$

Recalling the preceding lemma, we see that the recurrence of the walk is equivalent to the divergence of

$$1 + \frac{\beta-1}{\beta+1} + \frac{\beta-1}{\beta+1}\cdot\frac{2\beta-1}{2\beta+1} + \frac{\beta-1}{\beta+1}\cdot\frac{2\beta-1}{2\beta+1}\cdot\frac{3\beta-1}{3\beta+1} + \cdots = f(\beta).$$

We easily see that

$$f(2) = 1 + \frac{1}{3} + \frac{1}{5} + \cdots = \infty,$$

so the walk is recurrent when $\beta = 2$. That it is recurrent for $\beta > 2$ then follows from the lemma. The series $f(\beta)$ $(1 < \beta < 2)$, may be shown to be convergent as follows. For $1 < \beta < 2$,

$$f(\beta) = 1 + \frac{\beta-1}{\beta+1} + \frac{\beta-1}{2\beta+1}\left(1-\frac{2-\beta}{\beta+1}\right) + \frac{\beta-1}{3\beta+1}\left(1-\frac{2-\beta}{\beta+1}\right)\left(1-\frac{2-\beta}{2\beta+1}\right) + \cdots$$

$$< 1 + \frac{1}{2} + \frac{1}{3}\left(1-\frac{2-\beta}{3}\right) + \frac{1}{4}\left(1-\frac{2-\beta}{3}\right)\left(1-\frac{2-\beta}{5}\right) + \cdots.$$

But

$$\left(1-\frac{2-\beta}{3}\right)\left(1-\frac{2-\beta}{5}\right)\cdots\left(1-\frac{2-\beta}{2k+1}\right) < e^{-(2-\beta)(1/3+1/5+\cdots+1/2k+1)}$$

$$< e^{[-(2-\beta)/4]\log(2k+1)} = (2k+1)^{-(2-\beta)/4}.$$

Thus

$$f(\beta) < 1 + \frac{1}{2} + \frac{1}{3} \frac{1}{3^{(2-\beta)/4}} + \frac{1}{4} \frac{1}{5^{(2-\beta)/4}} + \cdots < \infty.$$

The preceding class of examples of random walks together with the comparison theorem (Theorem 2) now permit us to determine the moment properties of a large class of random walks. For example, the one-sided walks with

$$p_n = \frac{1}{2}\left(1 + \frac{\alpha}{n^\beta}\right) \qquad (n = 1, 2, \cdots)$$

have the same moment index as the walk with $p_n = 1/2$ if $\alpha > 0$, $\beta > 1$, while if $\beta < 1$, $\alpha < 0$ the index is infinity (all moments exist). For $\alpha > 0$, the walk will be certain to return if $\beta < 1$, but not if $\beta > 1$.

REFERENCE

1. W. Feller, *An introduction to probability theory and its applications*, Wiley, New York, 1950.

UNIVERSITY OF CHICAGO

A CLASS OF STATIONARY PROCESSES AND A CENTRAL LIMIT THEOREM

BY J. R. BLUM AND MURRAY ROSENBLATT

1. **Summary.** A certain class of stationary processes is discussed. It is shown that each process in the class has an absolutely continuous spectrum. Under some moment conditions, it is shown that such processes satisfy the central limit theorem.

2. **Introduction.** Let $\eta = (\cdots, \eta_{-1}, \eta_0, \eta_1, \cdots)$ be a doubly infinite sequence of independent identically distributed random variables. Let T be the translation operator defined on η by

$$T\eta = \eta' = (\cdots, \eta_0, \eta_1, \eta_2, \cdots). \tag{2.1}$$

Let B be the Borel field generated by η and let $g = g(\eta)$ be a function defined on B such that $E\{|g|^2\} < \infty$.

Given any such function g, we define a stationary process $\{X_n\}$ where

$$X_n = g(T^n\eta), \qquad n = 0, \quad \pm 1, \cdots. \tag{2.2}$$

We assume without loss of generality that $E\{X_n\} \equiv 0$. Let $r_s = E\{X_n X_{n+s}\}$. Then $r_s = \int_{-\pi}^{\pi} e^{is\lambda} dF(\lambda)$, where $F(\lambda)$ is the spectral distribution function of the process. In §3 the spectral distribution function of any process of the form (2.2) is shown to be absolutely continuous. Finally it is shown in §4 that under some additional assumptions on the moment structure of the process the central limit theorem is applicable.

3. **Absolute continuity of the spectral distribution function.** In this section we prove

THEOREM I. *If $F(\lambda)$ is the spectral distribution function of a process of the form (2.2), $F(\lambda)$ is absolutely continuous.*

Proof. In proving the theorem we assume that each η_i is uniformly distributed on the unit interval. It is easily seen that this may be done without loss of generality. For if ξ is uniformly distributed on the unit interval, we can always construct a monotone function $\varphi(\xi)$ such that the distribution of $\varphi(\xi)$ is the same as that of any η_i, and consequently as far as the probability structure of the process $\{X_n\}$ is concerned, we may assume that each η_i is uniformly distributed [0, 1].

Received May 29, 1956; in revised form, November 19, 1956. In this work, the authors were sponsored by the Office of Ordnance Research, U. S. Army, under Contract No. DA-33-008-ORD-965, and the Office of Naval Research under Contract No. Nonr-908(10).

R.A. Davis et al. (eds.), *Selected Works of Murray Rosenblatt*, Selected Works in Probability and Statistics, DOI 10.1007/978-1-4419-8339-8_11, © Springer Science+Business Media, LLC 2011

For each integer n define a process $\{X_{n,m}\}$, $m = 1, 2, \cdots$, by

$$X_{n,m} = E\{X_n \mid \eta_{n-m+1}, \cdots, \eta_{n+m-1}\}. \tag{3.1}$$

It is easily verified that for n fixed, the process $\{X_{n,m}\}$ is a martingale. Furthermore $E\{|X_{n,m}|^2\} \leq E\{|X_n|^2\} = E\{|g(\eta)|^2\} < \infty$. It then follows from well-known theorems on martingales that $X_{n,m}$ converges in mean square and with probability one as m approaches infinity. To see that $\lim_{m\to\infty} X_{n,m} = X_n$, note that $X_{n,m}$ is the projection of X_n on the space generated by all measurable functions of $\eta_{n-m+1}, \cdots, \eta_{n+m-1}$. Since X_n lies in the space generated by the entire η sequence, it follows that $\lim_{m\to\infty} E\{|X_n - X_{n,m}|^2\} = 0$. (See e.g. Doob [1; 331, Theorem 4.3].

Let $\theta_j(u)$; $j = 0, 1, \cdots$ be an orthonormal sequence of polynomials where $\theta_j(u)$ is of degree j on $[0, 1]$ i.e., $\int_0^1 \theta_j(u)\, \theta_k(u) du = \delta_{jk}$ where δ_{jk} is the Kronecker delta. The functions $\{\theta_j(u)\}$ are complete in $L^2[0, 1]$. Let $(u_1, \cdots, u_k)$ be a point in the k-dimensional unit cube U_k. The sequence of products $\{\theta_{j_1}(u_1) \cdots \theta_{j_k}(u_k)\}$ with $j_1, \cdots, j_k$ nonnegative integers is a complete orthonormal system for $L^2[U_k]$. To get an orthonormal system for the infinite-dimensional unit cube U we take the function which is identically one and the system of products $\{\theta_{j_1}(u_{m_1}) \cdots \theta_{j_k}(u_{m_k})\}$ where $m_1 < m_2 < \cdots < m_k$ and $k = 1, 2, \cdots$. This system is clearly orthonormal. Now let k be an arbitrary positive integer. From the discussion above it follows that we can expand $X_{0,k}$ in the form

$$X_{0,k} = \sum_{j=1}^{2k-1} \sum_{-k+1 \leq m_1 < \cdots < m_j \leq k-1} \sum_{n_1, \cdots, n_j = 1}^{\infty} C(\bar{n}_j, \bar{m}_j)\varphi_{\bar{n}_j}(\eta_{\bar{m}_j}) \tag{3.2}$$

where $\bar{m}_j = (m_1, \cdots, m_j)$, $\bar{n}_j = (n_1, \cdots, n_j)$, and

$\varphi_{\bar{n}_j}(\eta_{\bar{m}_j}) = \theta_{n_1}(\eta_{m_1}) \cdots \theta_{n_j}(\eta_{m_j})$. Since $E\{|X_{0,k}|^2\} \leq E\{|X_0|^2\} < \infty$

it follows that

$$\sum_{j=1}^{\infty} \sum_{m_1 < m_2 < \cdots < m_j} \sum_{n_1, \cdots, n_j = 1}^{\infty} C(\bar{n}_j, \bar{m}_j)^2 < \infty.$$

Now $X_0 = \lim_{k\to\infty} X_{0,k}$ with probability one so that

$$X_0 = \sum_{j=1}^{\infty} \sum_{m_1 < \cdots < m_j} \sum_{n_1, \cdots, n_j = 1}^{\infty} C(\bar{n}_j, \bar{m}_j)\varphi_{\bar{n}_j}(\eta_{\bar{m}_j}) \tag{3.3}$$

with probability one Similarly we have

$$X_n = \sum_{j=1}^{\infty} \sum_{m_1 < \cdots < m_j} \sum_{n_1, \cdots, n_j = 1}^{\infty} C(\bar{n}_j, \bar{m}_j - n)\varphi_{\bar{n}_j}(\eta_{\bar{m}_j}) \tag{3.4}$$

where $\bar{m}_j - n = (m_1 - n, \cdots, m_j - n)$.

Now let k be a positive integer, $\bar{n} = (n_1, \cdots, n_k)$ be a k-tuplet of positive integers, and $\bar{m} = (m_1, \cdots, m_k)$ be a k-tuplet of integers with $m_1 < \cdots < m_k$.

Given any integer t, we define $X_t(\bar{m}, \bar{n})$ by

$$X_t(\bar{m}, \bar{n}) = \sum_{j=-\infty}^{\infty} C(\bar{n}, \bar{m} + j - t)\varphi_{\bar{n}}(\eta_{\bar{m}+j}) \tag{3.5}$$

where $\bar{m} + l = (m_1 + l, \cdots, m_k + l)$ for any integer l. $X_t(\bar{m}, \bar{n})$ is a moving average, and such processes are known to have an absolutely continuous spectrum [1; 499]. Now let $(\bar{m}', \bar{n}')$ be different from $(\bar{m}, \bar{n})$ in the sense that $\bar{n} \neq \bar{n}'$ or that $\bar{m} - \bar{m}'$ is not a vector with equal coordinates. Then it is easily seen that the processes $X_t(\bar{m}, \bar{n})$ and $X_t(\bar{m}', \bar{n}')$ are mutually orthogonal. There is a denumerable collection of such distinct processes and we enumerate them. Let $\{X_t^{(i)}\}$ be the i-th process and let $f^{(i)}(\lambda)$ be its spectral density. Consider the process $\sum_{i=1}^n X_t^{(i)}$ and let $f_n(\lambda)$ be its spectral density. From the mutual orthogonality of the processes $\{X_t^{(i)}\}$ it follows that $f_n(\lambda) = \sum_{i=1}^n f^{(i)}(\lambda)$. Since the $f^{(i)}(\lambda)$ are nonnegative, $f_n(\lambda)$ is non-decreasing in n. Let $f(\lambda) = \lim_{n\to\infty} f_n(\lambda)$. Then we have

$$\begin{aligned}\int_{-\pi}^{\pi} f(\lambda)\, d\lambda &= \int_{-\pi}^{\pi} \lim_{n\to\infty} f_n(\lambda)\, d\lambda \leq \lim_{n\to\infty} \int_{-\pi}^{\pi} f_n(\lambda)\, d\lambda \\ &= \lim_{n\to\infty} E\left\{\left| \sum_{i=1}^{n} X_t^{(i)} \right|^2\right\} = E\{| g |^2\} < \infty .\end{aligned} \tag{3.6}$$

Hence $f(\lambda)$ is almost everywhere finite and integrable. An easy computation shows that $f(\lambda)$ is the spectral density of the process $\{X_t\}$ and the theorem is proved.

It follows from the theorem just proved that if h is any square integrable function defined on the probability space generated by $X = (\cdots, X_{-1}, X_0, X_1, \cdots)$ and if $\{U_n\}$ is the process defined by $U_n = h(T^n X)$ then the process $\{U_n\}$ also has an absolutely continuous spectrum. It would be of some interest to determine if the converse of this proposition is also true. That is, if $\{X_n\}$ is strictly stationary and if $\{X_n\}$ and any process $\{U_n\}$ obtained from $\{X_n\}$ in the manner indicated above has an absolutely continuous spectrum, does it follow that X_n is of the form $g(T^n\eta)$ where g is square integrable and $\eta = (\cdots, \eta_{-1}, \eta_0, \eta_1, \cdots)$ is a doubly infinite sequence of independent identically distributed random variables?

4. **The central limit theorem.** In this section it is shown that under some additional moment conditions, the processes we have discussed satisfy the central limit theorem. As before let $X_{n,k} = E\{X_n \mid \eta_{n-k+1}, \cdots, \eta_{n+k-1}\}$ when $k = 1, 2, \cdots$, and let $X_{n,0} = 0$. Furthermore, let $V_{n,k} = X_{n,k+1} - X_{n,k}$, and for any integer s let $a_{s,k,k'} = E\{V_{n,k}V_{n+s,k'}\}$. Then we have

THEOREM 2. *If*

$$E\{| X_n |^{\alpha}\} < \infty \quad \textit{for some} \quad \alpha > 2, \tag{4.1}$$

$$\sum_{s=-\infty}^{\infty} \sum_{k,k'=1}^{\infty} | a_{s,k,k'} | < \infty, \tag{4.2}$$

and

$$\sum_{s=-\infty}^{\infty} \sum_{k,k'=1}^{\infty} a_{s,k,k'} \neq 0 \tag{4.3}$$

then $N^{-\frac{1}{2}} \sum_{i=1}^{N} X_i$ *is asymptotically normal with mean zero and variance* $2\pi f(0)$, *where* $f(\lambda)$ *is the spectral density of the process* $\{X_n\}$.

Proof. For fixed k it can easily be verified that the process $\{X_{n,k}\}$ is a stationary $(2k-1)$-dependent process. Marsaglia [2] has given conditions under which such a process satisfies the central limit theorem. Let $s_{N,k}^2 = E\{| \sum_{i=1}^{N} X_{i,k} |^2\}$. Then we can state Marsaglia's conditions in the way we shall use them:

$$E\{| X_{n,k} |^2\} < B < \infty, \qquad k = 1, 2, \cdots, \tag{4.4}$$

$$\frac{N}{s^2_{N,k}} < B, \qquad N = 1, 2, \cdots, \tag{4.5}$$

$$\lim_{N\to\infty} \frac{N^{1/\alpha} E\{| X_{n,k} |^\alpha\}^{1/\alpha}}{s_{N,k}} = 0, \quad \text{for some} \quad \alpha > 2. \tag{4.6}$$

We first show that these conditions are satisfied for k sufficiently large. Condition (4.4) follows immediately from the fact that $E\{| X_{n,k} |^2\} \leq E\{| g |^2\} < \infty$. To prove (4.5) we write

$$\begin{aligned}
\frac{s_{N,k}^2}{N} &= \frac{1}{N} E\left\{\left(\sum_{i=1}^{N} \sum_{j=0}^{k-1} V_{i,j}\right)^2\right\} \\
&= \frac{1}{N} \sum_{j,j'=0}^{k-1} \sum_{i,i'=1}^{N} a_{i'-i,j,j'} \\
&= \frac{1}{N} \sum_{j,j'=0}^{k-1} \sum_{u=-N+1}^{N-1} a_{u,j,j'}(N - | u |),
\end{aligned}$$

where $u = i' - i$. Hence

$$\lim_{N\to\infty} \frac{s_{N,k}^2}{N} = \sum_{j,j'=0}^{k-1} \sum_{u=-\infty}^{\infty} a_{u,j,j'} .$$

Given hypothesis (4.3) we can clearly choose k so large that (4.5) is satisfied. Jensen's inequality is used to obtain $E\{| X_{nk} |^\alpha\} \leq E\{| X_n |^\alpha\}$ for all k. Condition (4.6) follows immediately.

Now let $r_j = E\{X_n X_{n+j}\}$, $r_{j,k} = E\{X_{n,k} X_{n+j,k}\}$, and let $f_k(\lambda)$ be the spectral density of the process $\{X_{n,k}\}$. Then

$$f(\lambda) = \frac{1}{2\pi} \sum_{j=-\infty}^{\infty} r_j e^{-ij\lambda}$$

and

$$f_k(\lambda) = \frac{1}{2\pi} \sum_{j=-\infty}^{\infty} r_{j,k} e^{-ij\lambda}.$$

Hence

$$| f_k(\lambda) - f(\lambda) | \leq \frac{1}{2\pi} \sum_{j=-\infty}^{\infty} | r_{j,k} - r_j |.$$

For each j we have $\lim_{k\to\infty} r_{j,k} = r_j$. Furthermore it is easily established that

$$\sum_{j\geq j_0} | r_j | \leq \sum_{j\geq j_0} \sum_{i,i'=0}^{\infty} | a_{j,i,i'} |$$

and

$$\sum_{j\geq j_0} | r_{j,k} | \leq \sum_{j\geq j_0} \sum_{i,i'=0}^{\infty} | a_{j,i,i'} | \text{ uniformly in } k.$$

Applying hypothesis (4.2) we see that

$$\lim_{k\to\infty} \frac{1}{2\pi} \sum_{j=-\infty}^{\infty} | r_{j,k} - r_j | = 0.$$

Thus it is clearly seen that $\lim_{k\to\infty} f_k(\lambda) = f(\lambda)$ uniformly in λ. Since

$$f(0) = \frac{1}{2\pi} \sum_{j=-\infty}^{\infty} r_j = \frac{1}{2\pi} \sum_{j=-\infty}^{\infty} \sum_{k,k'=0}^{\infty} a_{j,k,k'} > 0,$$

we may choose k so large that $f_k(0) > 0$ and so that Marsaglia's theorem applies.

Then

$$\lim_{N\to\infty} P\left\{(2\pi N f_k(0))^{-\frac{1}{2}} \sum_{i=1}^{N} X_{i,k} \leq \alpha\right\} = (2\pi)^{-\frac{1}{2}} \int_{-\infty}^{\alpha} e^{-u^2/2}\, du.$$

Now

$$\begin{aligned} E\left\{\left| (Nf_k(0))^{-\frac{1}{2}} \sum_{i=1}^{N} (X_i - X_{i,k}) \right|^2\right\} &= \frac{1}{Nf_k(0)} E\left\{\left| \sum_{i=1}^{N} \sum_{j=k+1}^{\infty} V_{n,j} \right|^2\right\} \\ &\leq \frac{1}{Nf_k(0)} \sum_{i,i'=1}^{N} \sum_{j,j'=k+1}^{\infty} | a_{i'-i,j,j'} | \\ &= \frac{1}{f_k(0)} \sum_{j,j'=k+1}^{\infty} \sum_{u=-N}^{N} | a_{u,j,j'} | \left(1 - \frac{|u|}{N}\right) \end{aligned}$$

where $u = i' - i$. On letting N approach infinity in the last sum we obtain

$$\frac{1}{f_k(0)} \sum_{j,j'=k+1}^{\infty} \sum_{u=-\infty}^{\infty} | a_{u,j,j'} |.$$

Clearly we may make this sum arbitrarily small by choosing k large enough.

Since $\lim_{k\to\infty} f_k(0) = f(0)$ we obtain

$$\lim_{N\to\infty} P\left\{(2\pi N f(0))^{-\frac{1}{2}} \sum_{i=1}^{N} X_i \leq \alpha\right\} = (2\pi)^{-\frac{1}{2}} \int_{-\infty}^{\alpha} e^{-u^2/2}\, du,$$

which is the desired result.

References

1. J. L. Doob, *Stochastic Processes*, New York, 1953.
2. G. Marsaglia, *Iterated limits and the central limit theorem for dependent variables*, Proceedings of the American Mathematical Society, vol. 5(1954), pp. 987–991.

Indiana University

3^{-p}, either $F(nh) = 0$ or $\Phi_h(nh) = 0$. Thus the sum on the right-hand side of equation (3) is identically zero. Combining equation (3) with equation (2), we have the desired result.

* The research for this paper was supported by the United States Air Force under Contract No. AF18(600-685) monitored by the Office of Scientific Research.

[1] A discussion of the problem along with the necessary references will be found in Harry Pollard, "The Harmonic Analysis of Bounded Functions," *Duke Math. J.*, **20**, 499–512, 1953.

[2] See *ibid.*

[3] See S. Bochner, *Fouriersche Integrale* (Leipzig, 1932), p. 33.

[4] The Levitan polynomial. See N.I. Achieser, *Approximationstheorie* (Berlin, 1953), p. 146.

A CENTRAL LIMIT THEOREM AND A STRONG MIXING CONDITION

BY M. ROSENBLATT

COLUMBIA UNIVERSITY

Communicated by P. A. Smith, November 10, 1955

Introduction.—This paper presents a central limit theorem for a sequence of dependent random variables

$$X_1, X_2, \ldots .$$

The assumptions required are the usual assumptions on second and $2 + \delta$ order moments and a strong mixing condition. The theorem is of interest for two reasons. All general central limit theorems for dependent random variables formalize in some sense a heuristic notion of A. Markoff to the effect that one expects a central limit theorem to hold for $X_1, X_2, \ldots$, if the random variables behave more like independent random variables the farther they are separated (assuming that appropriate moments exist). An interesting discussion of this intuitive notion is given in S. Bernstein's paper on the central limit theorem.[1] The strong mixing condition used in this paper seems to be a more intuitively appealing formalization of this notion than most others. The condition is also of interest because it is a strong version of the mixing condition encountered in ergodic theory (see Hopf,[2] p. 35).

The Mixing Condition.—The mixing condition is assumed to hold for sets of the form

$$a_{k_r} < X_{k_r} \leq b_{k_r}, \qquad r = 1, \ldots, s, \qquad (1)$$

where $k_1 < \ldots < k_s$. Given any two such sets A and B, introduce the following notion of distance $d(A, B)$ between them. Consider the smallest interval containing the indices of the random variables in terms of which A is defined. Call this interval $I(A)$. In the case of set (1), this interval is $[k_1, k_s]$. The distance $d(A, B)$ is set equal to the distance between the intervals $I(A)$ and $I(B)$. Note that $d(A, B)$ does not satisfy the usual properties of a distance; for example, one may have $d(A, B) > d(A, C) + d(B, C)$ for some set C. The strong mixing condition is satisfied by the sequence of random variables $X_1, X_2, \ldots$, if

DOI 10.1007/978-1-4419-8339-8_12,

$$|P(A \cap B) - P(A)\,P(B)| < f(d(A, B)) \tag{2}$$

for all sets A, B of form (1), where $f(n)$ is a function defined for $n = 0, 1, 2, \ldots$. and decreasing to zero as $n \to \infty$.

The Central Limit Theorem.—Assume that the mean values $EX_k \equiv 0$, as one could in any case subtract the mean values EX_k from the random variables X_k, respectively. The assumptions are

1. $$E\left|\sum_{j=a}^{b} X_j\right|^2 \sim h(b-a)$$

as $b - a \to \infty$, where $h(m) \uparrow \infty$ as $m \to \infty$;

$$E\left|\sum_{j=a}^{b} X_j\right|^{2+\delta} = o(h(b-a)^{1+\delta/2}$$

as $b - a \to \infty$ for some $\delta > 0$.

2. The strong mixing condition (2) is satisfied by the sequence of random variables $X_1, X_2, \ldots$.

Introduce the following notation:

$$S_n = \sum_{j=1}^{n} X_j,$$

$$U_r = \sum_{j=(r-1)(p_n+q_n)+1}^{rp_n+(r-1)q_n} X_j,$$

$$V_r = \sum_{j=rp_n+(r-1)q_n+1}^{r(p_n+q_n)} X_j, \quad r = 1, \ldots, k,$$

where $k(p_n + q_n) = n$. The numbers k, p_n, q_n will be chosen in such a way that $p_n, q_n, k \to \infty$ and $q_n/p_n \to 0$ as $n \to \infty$.

We expect to show that

$$\frac{S_n}{\sqrt{kh(p_n)}}$$

is asymptotically normally distributed. Now

$$E^{1/2}\left|\sum_{r=1}^{k} \frac{V_r}{\sqrt{kh(p_n)}}\right|^2 \leq \sum_{r=1}^{k} \frac{E^{1/2}|V_r|^2}{\sqrt{kh(p_n)}}$$

$$\sim \frac{(kh(q_n)}{h(p_n))^{1/2}}.$$

Choose the sequences q_n, p_n, k such that

$$\frac{kh(q_n)}{h(p_n)} \to 0.$$

Then

$$\sum_{r=1}^{k} \frac{V_r}{\sqrt{kh(p_n)}} \to 0$$

in probability as $n \to \infty$. It can therefore be neglected in further computations. We shall later see that this can be done.

Set

$$G_{r,n}(x) = P\left\{\frac{U_r}{\sqrt{kh(p_n)}} \le x\right\},$$

and let the set

$$\left\{m_r\delta < \frac{U_r}{\sqrt{kh(p_n)}} \le (m_r + 1)\delta\right\} = A(r, n, m_r, \delta),$$

where m_r is an integer. Note that

$$\sum_{(m_1 + \ldots + m_k + k) \le x} P\left(\bigcap_{r=1}^{k} A(r, n, m_r, \delta)\right) \le P\left(\sum_{r=1}^{k} \frac{U_r}{\sqrt{kh(p_n)}} \le x\right) \le \sum_{(m_1 + \ldots + m_k)\delta \le x} P\left(\bigcap_{r=1}^{k} A(r, n, m_r, \delta)\right) \quad (3)$$

LEMMA 1

$$P\left(\max_{r=1,\ldots,k} \left|\frac{U_r}{\sqrt{kh(p_n)}}\right| > t_k\right) < \epsilon$$

if $t_k = (k/\epsilon)^{1/2}$.

Now

$$E\left|\max_{r=1,\ldots,k}\left|\frac{U_r}{\sqrt{kh(p_n)}}\right|\right|^2 \le E\left|\sum_{r=1}^{k}\left|\frac{U_r}{\sqrt{kh(p_n)}}\right|\right|^2 \le k.$$

The result is obtained by an application of the Tchbycheff inequality.

LEMMA 2

$$\left|\sum_{(m_1 + \ldots + m_k)\delta \le x} P\left(\bigcap_{r=1}^{k} A(r, n, m_r, \delta)\right) - \sum_{(m_1 + \ldots + m_k)\delta \le x} \prod_{r=1} P(A(r, n, m_r, \delta))\right| \le k\left(\frac{2t_k}{\delta}\right)^k f(q_n) + 2\epsilon.$$

The probability contributed by all the sets $\bigcap_{r=1}^{k} A(r, n, m_r, \delta)$ for which max $|U_r/\sqrt{kh(p_n)}| > t_k$ is at most ϵ. Consider the sets $\bigcap_{r=1}^{k} A(r, n, m_r, \delta)$ for which max $|U_r/\sqrt{kh(p_n)}| \le t_k$. By repeated application of condition (2), one can see that

$$\left|P\left(\bigcap_{r=1}^{k} A(r, n, m_r, \delta)\right) - \prod_{r=1}^{k} P(A(r, n, m_r, \delta))\right| \le kf(q_n).$$

Since there are $(2t_k/\delta)^k$ sets of this form, we obtain the desired inequality.

Let

$$G_{1,n} * \ldots * G_{k,n}(x)$$

be the convolution of $G_{1,n}(x), \ldots, G_{k,n}(x)$. Note that

$$G_{1,n} * \ldots * G_{k,n}(x) \leq \sum_{(m_1 + \ldots + m_k)\delta \leq x} \prod_{r=1}^{k} P(A(r, n, m_r, \delta)) \leq G_{1,n} * \ldots * G_{k,n}(x + k\delta) \tag{4}$$

and

$$G_{1,n} * \ldots * G_{k,n}(x - k\delta) < \sum_{(m_1 + \ldots + m_k + k)\delta \leq x} \prod_{r=1}^{k} P(A(r, n, m_r, \delta)) \leq G_{1,n} * \ldots * G_{k,n}(x). \tag{5}$$

Now $G_{1,n} * \ldots * G_{k,n}(x)$ tends to the normal distribution as $n \to \infty$ if $k \to \infty$. We want to let $k, p_n, q_n \to \infty$, $kp_n \sim n$, and $k\delta \to 0$ in such a way that

$$k \left(\frac{2t_k}{\delta}\right)^k f(q_n) \to 0 \tag{6}$$

$$\frac{kh(q_n)}{h(p_n)} \to 0. \tag{7}$$

Set $\delta = 1/k^2$. Then $k\,\delta \to 0$. Since

$$t_k = \left(\frac{k}{\epsilon}\right)^{1/2},$$

it follows that

$$k \left(\frac{2t_k}{\delta}\right)^k = k^{5/2k + 1} \frac{2^k}{\epsilon^{1/2k}} \leq k^{5k} C^k,$$

where $C = 2/\epsilon^{1/2}$. The function f can always be chosen in such a way that $f(n) > 1/n$ for all n. If k is chosen so that

$$k \leq [-\log f(q_n)]^{1/2}, \tag{8}$$

clearly condition (6) will be satisfied.

Now

$$h(n) \sim E\left|\sum_{j=1}^{n} X_j\right|^2 \leq k^2 h\left(\frac{n}{k}\right)$$

by the Minkowski inequality. But then

$$\frac{k^2}{h(n)} \geq \frac{1}{h(p_n)}.$$

Thus

$$\frac{kh(q_n)}{h(p_n)} \leq \frac{k^3 h(q_n)}{h(n)}.$$

Condition (4) will then be satisfied if

$$h(q_n) = o\left(\frac{h(n)}{k^3}\right). \tag{9}$$

Sequences k, p_n, q_n, $\to \infty$, $kp_n \sim n$ can be chosen so that conditions (8) and (9) are satisfied and hence conditions (6) and (7) are satisfied. On using inequalities (3), (4), and (5) and Lemma 2, we see that

$$\sum_{r=1}^{k} \frac{U_r}{\sqrt{kh(p_n)}}$$

is asymptotically normally distributed. Since

$$\sum_{r=1}^{k} \frac{V_r}{\sqrt{kh(p_n)}} \to 0$$

in probability as $n \to \infty$, it follows that

$$\frac{S_n}{\sqrt{kh(p_n)}}$$

is asymptotically normally distributed.

Remarks.—The result obtained includes the result of Höffding and Robbins.[3] It would be of very great interest to see how much stronger the notion of a strong mixing condition is than that of an ordinary mixing condition in the case of a strictly stationary process.

[1] S. Bernstein, "Sur l'extension du théorème limite du calcul des probabilités aux sommes de quantités dépendentes," *Math. Ann.*, **97**, 1–59, 1927.

[2] E. Hopf, *Ergodentheorie* ("Ergebnisse" series). J. Springer, Berlin, 1937.

[3] W. Höffding and H. Robbins, "The Central Limit Theorem for Dependent Random Variables," *Duke Math. J.*, **15**, 773–780, 1948.

REMARKS ON SOME NONPARAMETRIC ESTIMATES OF A DENSITY FUNCTION[1]

By Murray Rosenblatt[2]

University of Chicago

1. Summary. This note discusses some aspects of the estimation of the density function of a univariate probability distribution. All estimates of the density function satisfying relatively mild conditions are shown to be biased. The asymptotic mean square error of a particular class of estimates is evaluated.

2. Estimates of the density function. Let $X_1, \cdots, X_n$ be independent and identically distributed random variables with continuous density function $f(y)$. Let $S(y; X_1, \cdots, X_n)$ be an estimate of $f(y)$. The function $S(y; x_1, \cdots, x_n)$ is assumed to be jointly Borel measurable in $(y, x_1, \cdots, x_n)$. It is also assumed that

$$S(y; x_1, \cdots, x_n) \geqq 0,$$

since $f(y) \geqq 0$.

It can easily be shown that

$$S(y; X_1, \cdots, X_n)$$

is not an unbiased estimate of $f(y)$. Suppose to the contrary that

$$ES(y; X_1, \cdots, X_n) \equiv f(y) \tag{1}$$

for all continuous f and all y. Condition (1) implies that for each y,

$$ES(y; X_1, \cdots, X_n) < \infty.$$

Assume that $S(y; x_1, \cdots, x_n)$ is a symmetric function of $x_1, \cdots, x_n$, since the symmetrized n-tuple $(X_1, \cdots, X_n)$ is a sufficient statistic for the problem. But then

$$\int_a^b S(y; X_1, \cdots, X_n)\, dy \tag{2}$$

is a symmetric estimate of

$$F(b) - F(a) = \int_a^b f(y)\, dy.$$

Moreover, (2) is an unbiased estimate of $F(b) - F(a)$, since

$$E \int_a^b S(y; X_1, \cdots, X_n)\, dy = \int_a^b ES(y; X_1, \cdots, X_n)\, dy$$
$$= \int_a^b f(y)\, dy = F(b) - F(a)$$

Received April 27, 1955.

[1] Research carried out at the Statistical Research Center, University of Chicago, under the sponsorship of the Statistics Branch, Office of Naval Research.

The comments of R. R. Bahadur have been very helpful.

[2] Now at Indiana University.

DOI 10.1007/978-1-4419-8339-8_13, © Springer Science+Business Media, LLC 2011

by Fubini's theorem. However, the only unbiased estimate of $F(b) - F(a)$ symmetric in the observations $X_1, \cdots, X_n$ is $F_n(b) - F_n(a)$, where $F_n(y)$ is the sample distribution function. This follows immediately from the fact that the symmetrized n-tuple $(X_1, \cdots, X_n)$ is a complete statistic [2]. Thus,

$$F_n(b) - F_n(a) = \int_a^b S(y; X_1, \cdots, X_n)\, dy$$

for all a and b and almost all $X_1, \cdots, X_n$. But then $F_n(y)$ is absolutely continuous in y for almost all $X_1, \cdots, X_n$, which is impossible.

One need not require $S(y; x_1, \cdots, x_n)$ to be nonnegative. An assumption like

$$\int_a^b E|S(y; X_1, \cdots, X_n)|\, dy < \infty$$

for some two values, a, b, with $a < b$, would lead to the same conclusion, that is, that there are no unbiased estimates $S(y; X_1, \cdots, X_n)$ of $f(y)$ satisfying this condition.

3. The difference quotient of the sample distribution function. An obvious estimate of $f(y)$ is the difference quotient

$$S(y; X_1, \cdots, X_n) = f_n(y) = \frac{F_n(y + h) - F_n(y - h)}{2h}$$

of the sample distribution function $F_n(y)$, where $h = h_n$ is a function of the sample size n and approaches zero as $n \to \infty$. The asymptotic behavior of this estimate as $n \to \infty$ is examined in terms of its mean square error. Fix and Hodges have used an estimate of this form in their discussion of a nonparametric discrimination problem [1].

Now,

$$EF_n(y) = F(y),$$

$$E[F_n(y)F_n(y')] = \frac{1}{n} F(\min(y, y')) + \frac{n-1}{n} F(y)F(y'),$$

so that

$$\operatorname{cov}(F_n(y), F_n(y')) = \frac{1}{n} [F(\min(y, y')) - F(y)F(y')].$$

But then,

$$\operatorname{cov}(f_n(y), f_n(y')) = \frac{1}{4h^2 n} [F(\min(y + h, y' + h)) - F(y + h)F(y' + h)$$
$$- F(\min(y + h, y' - h)) + F(y + h)F(y' - h) - F(\min(y - h, y' + h))$$
$$+ F(y - h)F(y' + h) + F(\min(y - h, y' - h)) - F(y - h)F(y' - h)].$$

On setting $y = y'$,

$$\sigma^2(f_n(y)) = \frac{1}{4h^2 n} [F(y + h) - F(y - h) + (F(y + h) - F(y - h))^2].$$

Now consider the behavior of $f_n(y)$ where y is fixed as $n \to \infty$ and $h \to 0$. The mean square error

$$E|f_n(y) - f(y)|^2 = \sigma^2(f_n(y)) + (Ef_n(y) - f(y))^2$$

$$= \frac{1}{4h^2n}[F(y+h) - F(y-h) + (F(y+h) - F(y-h))^2]$$

$$+ \left[\frac{1}{2h}(F(y+h) - F(y-h)) - f(y)\right]^2$$

is a reasonable measure of how good an estimate $f_n(y)$ is of $f(y)$ locally at y. The density function f is assumed to be sufficiently regular for the following evaluation of the mean square error to be carried through. It will be enough to assume that the first three derivatives of f exist at y. Then

$$F(y+h) - F(y-h) = \int_{y-h}^{y+h} f(u)\,du$$

$$= \int_{y-h}^{y+h} [f(y) + (u-y)f'(y)$$

$$+ \tfrac{1}{2}(u-y)^2 f''(y) + 0\,|u-y|^3]\,du$$

$$= 2hf(y) + \tfrac{1}{3}f''(y)h^3 + 0(|h|^4).$$

Assume that $f''(y) \neq 0$. Then

$$(Ef_n(y) - f(y))^2 \sim \left(\frac{h^2}{6} f''(y)\right)^2 = \frac{h^4}{36}|f''(y)|^2$$

as $h \to 0$. The variance of the estimate

$$\sigma^2(f_n(y)) \sim \frac{f(y)}{2hn}$$

as $h \to 0$. The asymptotic mean square error

$$E|f_n(y) - f(y)|^2 \sim \frac{f(y)}{2hn} + \frac{h^4}{36}|f''(y)|^2 + o\left(\frac{1}{hn} + h^4\right) \tag{3}$$

as $h \to 0$ and $n \to \infty$. The question of an optimal choice of $h = h_n$ as a function of n now arises. If h is set equal to $kn^{-\alpha}$, $\alpha > 0$, it is easily seen from (3) that the optimal choice of α is $\alpha = \frac{1}{5}$. The optimal value of k is then the one minimizing

$$\frac{f(y)}{2k} + \frac{k^4}{36}|f''(y)|^2.$$

This value of k is

$$k = \left[\frac{9}{2}\frac{f(y)}{|f''(y)|^2}\right]^{1/5}.$$

With this choice of k and α, we find that

$$E|f_n(y) - f(y)|^2 \sim \tfrac{5}{4}9^{-1/5}2^{-4/5}f(y)^{4/5}|f''(y)|^{2/5}n^{-4/5}.$$

The choice of k would be based on guesses as to the magnitude of $f(y)$, $f''(y)$.

One is led to a choice of h as a function of n that is independent of y by considering a global measure of how good f_n is as an estimate of f. The integrated mean square error

$$\int_{-\infty}^{\infty} E\,|f_n(y) - f(y)|^2\,dy$$

is a simple measure of this type. Let $f(y)$, $f''(y)$ be bounded continuous functions that are square integrable. One is then led to the following asymptotic expression

$$\int_{-\infty}^{\infty} E\,|f_n(y) - f(y)|^2\,dy \sim \frac{1}{2hn} + \frac{h^4}{36}\int_{-\infty}^{\infty}|f''(y)|^2\,dy + o\left(\frac{1}{hn} + h^4\right)$$

as $h \to 0$ and $n \to \infty$. The optimal choice of h as a function of n is

$$h = kn^{-1/5},$$

where k is now

$$k = \left[\frac{9}{2\int_{-\infty}^{\infty}|f''(y)|^2\,dy}\right]^{1/5},$$

and

$$\int_{-\infty}^{\infty} E\,|f_n(y) - f(y)|^2\,dy \sim 2^{-4/5}9^{-1/5}\frac{5}{4}\left[\int_{-\infty}^{\infty}|f''(y)|^2\,dy\right]^{1/5} n^{-4/5}$$

as $n \to \infty$.

4. A class of estimates of the density function. The discussion of the previous section suggests that the following class of estimates will be of interest. Let $w_n(u)$ be a nonnegative function such that

$$\int_{-\infty}^{\infty} w_n(u)\,du = 1.$$

The sequence of functions $\{w_n(u)\}$ is chosen so that the total mass concentrates in the neighborhood of zero as $n \to \infty$; that is, given any $\epsilon > 0$,

$$\int_{|u|<\epsilon} w_n(u)\,du \to 1$$

as $n \to \infty$. Corresponding to each sequence of weight functions $\{w_n(u)\}$ of this type, there is an estimate

$$f_n(y) = \int_{-\infty}^{\infty} w_n(y - u)\,dF_n(u) = \frac{1}{n}\sum_{j=1}^{\infty} w_n(y - X_j).$$

Now

$$Ef_n(y) = \int_{-\infty}^{\infty} w_n(y - x)\, dF(x) = \int_{-\infty}^{\infty} w_n(y - x) f(x)\, dx,$$

and

$$\operatorname{cov}\,(f_n(y), f_n(y')) = \frac{1}{n}\left[\int_{-\infty}^{\infty} w_n(y - x) w_n(y' - x) f(x)\, dx - \left(\int_{-\infty}^{\infty} w_n(y - x) f(x)\, dx\right)^2\right].$$

On setting $y = y'$, we have

$$\sigma^2(f_n(y)) = \frac{1}{n}\left[\int_{-\infty}^{\infty} w_n^2(y - x) f(x)\, dx - \left(\int_{-\infty}^{\infty} w_n(y - x) f(x)\, dx\right)^2\right].$$

Note that all estimates of this form are themselves density functions; that is'

$$f_n(y) \geqq 0,$$

and

$$\int_{-\infty}^{\infty} f_n(y)\, dy = 1.$$

An estimate $f_n(y)$ with any desired regularity properties can be obtained by choosing a weight function $w_n(u)$ with these same regularity properties. Thus, $f_n(y)$ will be analytic if $w_n(u)$ is.

As an example, consider

$$w_n(u) = \frac{1}{h} w\left(\frac{u}{h}\right),$$

where $h = h_n \rightarrow 0$ as $n \rightarrow \infty$, and

$$\int_{-\infty}^{\infty} w(u)\, du = 1.$$

The estimate discussed in the previous section is obtained on setting

$$w(u) = \begin{cases} \frac{1}{2} & \text{when } |u| < 1, \\ 0 & \text{otherwise.} \end{cases}$$

The function $w(u)$ is assumed to be square integrable. Then

$$\sigma^2(f_n(y)) = \frac{1}{n}\left[\frac{1}{h}\int_{-\infty}^{\infty} w^2(u) f(y + hu)\, du - \left(\int_{-\infty}^{\infty} w(u) f(y + hu)\, du\right)^2\right],$$

and

$$Ef_n(y) - f(y) = \int w(u) f(y + hu)\, du - f(y).$$

It is clear that

$$\int w(u)f(y + hu)\, du \rightarrow f(y),$$

and that

$$\int w^2(u)f(y + hu)\, du \rightarrow f(y) \int w^2(u)\, du$$

as $h \rightarrow 0$ for every continuous density function f. Hence,

$$\sigma^2(f_n(y)) \sim \frac{1}{nh} f(y) \int w^2(u)\, du.$$

The integral

$$\int w(u)\, |u|^3\, du$$

is assumed to be finite, and $f(y)$ is assumed to have continuous derivatives of the first three orders in the following computation of the bias. The bias of the estimate is then

$$Ef_n(y) - f(y) = \int w(u)[f(y + hu) - f(y)]\, du$$

$$= hf'(y) \int w(u)u\, du + \tfrac{1}{2}h^2f''(y) \int w(u)u^2\, du + 0(|h|^3).$$

It is now clear that it would be advantageous to have

$$\int w(u)u\, du = 0.$$

This condition will be satisfied if $w(u)$ is symmetric about zero. Using the same sort of argument as was used in the last section, it is easily seen that the mean square error of these estimates can be made no smaller than $0(n^{-4/5})$ for all admissible f. It would be very interesting to find out whether there are other estimates $f_n(y)$ with an asymptotic behavior of the order $1/n$ as $n \rightarrow \infty$.

REFERENCES

[1] E. Fix and J. L. Hodges, Jr., *Discriminatory Analysis, Nonparametric Discrimination: Consistency Properties*, USAF School of Aviation Medicine, Project No. 21–49–004, Report No. 4.

[2] E. Lehmann, "Notes on the theory of estimation," University of California, Berkeley (1950).

R.A. Davis et al. (eds.), *Selected Works of Murray Rosenblatt*, Selected Works in Probability and Statistics, DOI 10.1007/978-1-4419-8339-8_14, © Springer Science+Business Media, LLC 2011

SOME REGRESSION PROBLEMS IN TIME SERIES ANALYSIS

MURRAY ROSENBLATT[1]
COLUMBIA UNIVERSITY AND NEW YORK UNIVERSITY

1. Introduction

Estimates of the regression coefficients which are unbiased and linear in the observations are discussed in this paper. The residual is assumed to be a stationary process. Two specific estimates are discussed, the least-squares estimate and the Markov estimate. I call the estimate which is computed under the assumption that the residual is an orthogonal process the least-squares estimate. The Markov estimate is the linear unbiased estimate with minimal covariance matrix. The basic assumptions made in the paper are discussed in section 2 and are held to throughout the paper. In section 3 some remarks about the approximation of a continuous positive definite matrix-valued function by finite trigonometric forms are made. These remarks are used in section 4 to obtain the main results about the asymptotic behavior of the covariance matrices of the least-squares and Markov estimates. The next section discusses the many interesting cases in which the least-squares estimate is asymptotically as good as the Markov estimate. The first really systematic discussion of some of these problems was given by U. Grenander [1]. Further work was carried out by U. Grenander and M. Rosenblatt in [2], [3], and [4]. The author considers some of these problems in the case of a vector-valued time series in [5]. Some of the results of this paper are a generalization of some of those obtained in [5].

A few cases in which the least-squares estimate is not asymptotically efficient in the class of linear unbiased estimates are discussed in sections 5 and 7. Some small sample computations for a linear regression with a residual which is a first order autoregressive scheme are carried out in section 6 to test the asymptotic theory.

2. Assumptions and notation

I assume that the observed process y_t is a vector-valued process (a k-vector)

$$y_t = x_t + m_t\,, \qquad t = \cdots, -1, 0, 1, \cdots, \tag{2.1}$$

where $m_t = Ey_t$ is the mean value sequence and x_t, $Ex_t \equiv 0$, is the sequence of residuals. The residual x_t is assumed to be weakly stationary, that is, the covariances

$$r_{t\,\tau} = r_{t-\tau} = Ex_t x_\tau' = E(y_t - m_t)(y_\tau - m_\tau)'\ ^{2} \tag{2.2}$$

depend only on the difference $t - \tau$. For mathematical convenience, in sections 3 and 4, I assume that the components of the vector observations are complex valued. The real-

Based in part on research supported by the Office of Naval Research at the Statistical Research Center, University of Chicago.

[1] Now at Indiana University.

[2] x_t is column vector. Given a matrix A, A' denotes the conjugated transpose of A.

valued case is the one of statistical interest and will be examined in detail later on. The mean value m_t is assumed to be of a regression form

$$m_t = \beta_1\varphi_t^{(1)} + \beta_2\varphi_t^{(2)} + \cdots + \beta_s\varphi_t^{(s)}, \tag{2.3}$$

where the regression vector sequences

$$\varphi_1^{(j)}, \varphi_2^{(j)}, \cdots, \varphi_n^{(j)}, \qquad j = 1, \cdots, s, \tag{2.4}$$

are assumed known and the regression coefficients $\beta_1, \cdots, \beta_s$ are unknown. I shall discuss the problem of estimating the regression coefficients by unbiased estimates linear in the observations

$$y_1, \cdots, y_n. \tag{2.5}$$

Two specific linear unbiased estimates will be discussed in some detail, the "least-squares" estimate and the Markov estimate. The least-squares estimate is obtained by treating the residuals as if they were orthogonal, that is,

$$Ex_t x_\tau' = \delta_{t,\tau} I \tag{2.6}$$

where $\delta_{t,\tau}$ is the Kronecker delta and I is the identity matrix. The Markov estimate is the optimal linear unbiased estimate in the sense of minimal covariance matrix of the estimate.

The covariance sequence r_t of a weakly stationary process has the representation

$$r_t = \int_{-\pi}^{\pi} e^{it\lambda} dF(\lambda) \tag{2.7}$$

where $F(\lambda)$ is a nondecreasing matrix-valued $(k \times k)$ function, that is, $\Delta F(\lambda) \geqq 0$.[3] I shall assume that $F(\lambda)$ is absolutely continuous, that is,

$$F(\lambda) = \int_{-\pi}^{\lambda} f(\mu)\, d\mu \tag{2.8}$$

so that

$$r_t = \int_{-\pi}^{\pi} e^{it\lambda} f(\lambda)\, d\lambda. \tag{2.9}$$

The function $F(\lambda)$ is called the spectral distribution function of the process while $f(\lambda)$ is called the spectral density of the process. The spectral density is a nonnegative function of λ since

$$f(\lambda) = \frac{dF(\lambda)}{d\lambda} \geqq 0. \tag{2.10}$$

I shall assume that $f(\lambda)$ is a continuous function of λ [each element of $f(\lambda)$ is a continuous function of λ] and that $f(\lambda)$ is nonsingular for all λ.

For convenience I introduce the following notation. Let

$$y = \begin{pmatrix} y_1 \\ \vdots \\ y_n \end{pmatrix}, \qquad x = \begin{pmatrix} x_1 \\ \vdots \\ x_n \end{pmatrix}. \tag{2.11}$$

[3] Given a square matrix A, $A \geqq 0$ means that the corresponding quadratic form is positive semidefinite. $A > 0$ means that the corresponding quadratic form is positive definite.

Let

$$\varphi_t = (\varphi_t^{(1)}, \cdots, \varphi_t^{(s)}) \tag{2.12}$$

and

$$\varphi = \begin{pmatrix} \varphi_1 \\ \cdot \\ \cdot \\ \cdot \\ \varphi_n \end{pmatrix}. \tag{2.13}$$

Equation (2.1) can then be rewritten in the form

$$y = \varphi\beta + x\,. \tag{2.14}$$

The matrix R is the covariance matrix of y (or x). Here

$$\beta = \begin{pmatrix} \beta_1 \\ \cdot \\ \cdot \\ \cdot \\ \beta_s \end{pmatrix}. \tag{2.15}$$

The matrix R is nonsingular since $f(\lambda)$ is continuous and nonsingular for all λ. The matrix $\varphi'\varphi$ is also assumed to be nonsingular.

The "least-squares" estimate β_L is the vector β that minimizes the quadratic form

$$(y - \varphi\beta)'(y - \varphi\beta) \tag{2.16}$$

and is given by

$$\beta_L = [\varphi'\varphi]^{-1}\varphi' y\,. \tag{2.17}$$

It is clearly an unbiased estimate and has covariance matrix

$$E(\beta_L - \beta)(\beta_L - \beta)' = [\varphi'\varphi]^{-1}\varphi' R\varphi[\varphi'\varphi]^{-1}\,. \tag{2.18}$$

The linear unbiased estimate with minimal covariance matrix or Markov estimate is given by

$$\beta_M = [\varphi' R^{-1}\varphi]^{-1}\varphi' R^{-1} y\,. \tag{2.19}$$

Its covariance matrix is given by

$$E(\beta_M - \beta)(\beta_M - \beta)' = [\varphi' R^{-1}\varphi]^{-1}\,. \tag{2.20}$$

These remarks on the least-squares and Markov estimates are well known.

The techniques used in the paper can be considered a sort of generalized harmonic analysis. In order to carry out the analysis, various assumptions on the asymptotic behavior of the regression vectors are introduced. These assumptions are broad enough to include most of the usual types of regression, such as polynomial and trigonometric regression. They do not include the case of exponential regression.

Let

$$\Phi_n^{(j)} = \sum_{t=1}^{n} \varphi_t^{(j)\prime}\varphi_t^{(j)}, \qquad j = 1, \cdots, s\,. \tag{2.21}$$

It is assumed that

$$\Phi_n^{(j)} \to \infty \tag{2.22}$$

as $n \to \infty$, $j = 1, \cdots, s$. It is also assumed that

$$\lim_{n\to\infty} \frac{\Phi_{n+h}^{(j)}}{\Phi_n^{(j)}} = 1 \tag{2.23}$$

for every fixed h. Consider the k-vectors

$$\varphi_t^{(j)} = \begin{pmatrix} {}_1\varphi_t^{(j)} \\ \cdot \\ \cdot \\ \cdot \\ {}_k\varphi_t^{(j)} \end{pmatrix} \qquad \begin{matrix} j = 1, \cdots, s \\ t = 1, \cdots, n. \end{matrix} \tag{2.24}$$

Let the limits

$$\lim_{n\to\infty} \sum_{t=1}^{n} \frac{{}_p\varphi_{t+h}^{(j)}\, {}_q\varphi_t^{(l)}}{\sqrt{\Phi_n^{(j)}\Phi_n^{(l)}}} = {}_{jl}m_h^{(p,\, q)} \tag{2.25}$$

exist, $j, l = 1, \cdots, s$ and $p, q = 1, \cdots, k$ [if $t < 0$ set ${}_q\varphi_t^{(l)} = 0$]. Set

$${}_{jl}M_h = \{{}_{jl}m_h^{(p,\, q)}; \qquad p, q = 1, \cdots, k\} \tag{2.26}$$

and

$$M_h = \{{}_{jl}M_h; \qquad j, l = 1, \cdots, s\}. \tag{2.27}$$

The matrices M_h, $h = \cdots, -1, 0, 1, \cdots$ form a positive definite sequence, that is, given any finite collection of ks-vectors $\{a_\nu\}$

$$\sum_{\nu,\, \mu} a_\nu' M_{\nu-\mu} a_\mu \geqq 0. \tag{2.28}$$

It then follows that the matrices M_h have the representation

$$M_h = \int_{-\pi}^{\pi} e^{ih\lambda} dM(\lambda) \tag{2.29}$$

where $M(\lambda)$ is a nondecreasing matrix-valued $(ks \times ks)$ function of λ. In accordance with the notation introduced in (2.26) and (2.27) I write

$$M(\lambda) = \{{}_{jl}M(\lambda); \qquad j, l = 1, \cdots, s\} \tag{2.30}$$

and

$$\begin{aligned} {}_{jl}M(\lambda) &= \{{}_{jl}M_{pq}(\lambda); \qquad p, q = 1, \cdots, k\}, \\ M_{pq}(\lambda) &= \{{}_{jl}M_{pq}(\lambda); \qquad j, l = 1, \cdots, s\}. \end{aligned} \tag{2.31}$$

Note that

$$M_0 = \int_{-\pi}^{\pi} dM(\lambda) = M(\pi) - M(-\pi). \tag{2.32}$$

It will be convenient to introduce some additional notation. Given a $k \times k$ matrix f and a form $M = \{{}_{ij}m_{uv};\ i, j = 1, \cdots, s;\ u, v = 1, \cdots, k\}$, let

$$(f \cdot M) = \left\{\sum_{i,\, j} f_{ij}\, {}_{uv}m_{ij}; \qquad u, v = 1, \cdots, s\right\}. \tag{2.33}$$

By the integral

$$(2.34) \qquad \int_{-\pi}^{\pi} [f(\lambda) \cdot dM(\lambda)]$$

I shall mean

$$(2.35) \qquad \left\{ \int_{-\pi}^{\pi} \sum_{i,j} f_{ij}(\lambda)\, d_{pq}M_{ij}(\lambda)\,; \qquad p, q = 1, \cdots, s \right\}.$$

The matrix

$$(2.36) \qquad T = (I \cdot M_0)$$

is assumed to be nonsingular. This means that the vector sequences

$$(2.37) \qquad \varphi_1^{(j)}, \cdots, \varphi_n^{(j)}, \qquad j = 1, \cdots, s,$$

are asymptotically linearly independent in a sense which is relevant in this context. Conditions (2.22) and (2.23) are introduced to ensure that the estimates of the regression coefficients converge to the true regression coefficients in the mean square.

It will also be convenient to introduce the matrix

$$(2.38) \qquad D_n = \begin{bmatrix} \Phi_n^{(1)1/2} & & 0 \\ & \ddots & \\ 0 & & \Phi_n^{(s)1/2} \end{bmatrix}.$$

3. Remarks on approximation

The arguments used to obtain results on the asymptotic behavior of the covariance matrices of the least-squares and Markov estimates are approximation arguments. They make use of finite trigonometric polynomials that uniformly approximate the matrix-valued spectral density $f(\lambda)$.

LEMMA 1. *Let $f(\lambda)$ be a continuous positive definite matrix-valued $(k \times k)$ function of λ. Given any $\epsilon > 0$, there is a positive definite matrix-valued trigonometric polynomial*

$$(3.1) \qquad g(\lambda) = \sum_{u=-p}^{p} g_u e^{iu\lambda}$$

with coefficients g_u Hermitian $k \times k$ matrices such that

$$(3.2) \qquad \epsilon z'z > z'[f(\lambda) - g(\lambda)]z > -\epsilon z'z$$

for every k-vector z.

There are finite trigonometric polynomials $g_{ij}(\lambda)$ such that

$$(3.3) \qquad g_{ij}(\lambda) = \overline{g_{ji}(\lambda)}$$

and

$$(3.4) \qquad |f_{ij}(\lambda) - g_{ij}(\lambda)| < \delta, \qquad \delta < 0,$$

where $i, j = 1, \cdots, k$. But then on setting $g(\lambda) = \{g_{ij}(\lambda); i, j = 1, \cdots, k\}$,

$$(3.5) \qquad z'[f(\lambda) - g(\lambda)]z \leqq \sum \delta |z_i| |z_j| \leqq k\delta z'z.$$

If $\delta > 0$ is chosen sufficiently small

$$(3.6) \qquad z'[f(\lambda) - g(\lambda)]z < \epsilon z'z.$$

One can show similarly that

$$-\epsilon z'z < z'[f(\lambda) - g(\lambda)]z . \tag{3.7}$$

Note that p is the maximal order of the polynomials $p_{ij}(\lambda)$.

LEMMA 2. *Let $f(\lambda)$ be a continuous positive definite matrix-valued $(k \times k)$ function of λ. Given any $\epsilon > 0$ sufficiently small, there are positive definite matrix-valued trigonometric polynomials*

$$\begin{aligned} g(\lambda) &= \sum_{u=-p}^{p} g_u e^{iu\lambda} \\ h(\lambda) &= \sum_{u=-p}^{p} h_u e^{iu\lambda} \end{aligned} \tag{3.8}$$

with coefficients g_u, h_u Hermitian $k \times k$ matrices such that

$$0 < f(\lambda) - \epsilon I < g(\lambda) < f(\lambda) < h(\lambda) < f(\lambda) + \epsilon I \tag{3.9}$$

where I is the identity matrix $(k \times k)$.

Since $f(\lambda)$ is a positive continuous function of λ, for all sufficiently small $\epsilon > 0$

$$0 < f(\lambda) - \epsilon I \tag{3.10}$$

and clearly

$$f(\lambda) - \epsilon I < f(\lambda) . \tag{3.11}$$

By lemma 1, there is a trigonometric polynomial $g(\lambda)$ such that

$$-\tfrac{1}{2}\epsilon I < f(\lambda) - \tfrac{1}{2}\epsilon I - g(\lambda) < \tfrac{1}{2}\epsilon I . \tag{3.12}$$

But then

$$0 < f(\lambda) - \epsilon I < g(\lambda) < f(\lambda) . \tag{3.13}$$

One can similarly show that there is a trigonometric polynomial $h(\lambda)$ such that

$$f(\lambda) < h(\lambda) < f(\lambda) + \epsilon I . \tag{3.14}$$

LEMMA 3. *Let*

$$f(\lambda) = a_0 + a_1 \cos \lambda + \cdots + a_p \cos p\lambda + b_1 \sin \lambda + \cdots + b_p \sin p\lambda \tag{3.15}$$

be a positive definite matrix-valued $(k \times k)$ function of λ with the coefficients a_ν, b_ν $(k \times k)$ matrices. Then $f(\lambda)$ can be written in the form

$$f(\lambda) = \frac{1}{2\pi} \left(\sum_{j=0}^{p} c_j e^{-ij\lambda} \right) \left(\sum_{j=0}^{p} c_j e^{-ij\lambda} \right) \tag{3.16}$$

where c_0 is nonsingular and

$$\int_{-\pi}^{\pi} e^{-il\lambda} \left(\sum_{j=0}^{p} c_j e^{-ij\lambda} \right)^{-1} d\lambda = 0 \tag{3.17}$$

where $l > 0$.

Since $f(\lambda)$ is a positive definite matrix-valued $(k \times k)$ function of λ, there is a k-dimensional weakly stationary process x_t, $Ex_t \equiv 0$, having $f(\lambda)$ as its spectral density. The covariance matrices

$$r_t = Ex_{t+\tau}x_\tau' = \int_{-\pi}^{\pi} e^{it\lambda} f(\lambda)\, d\lambda \tag{3.18}$$

are the null matrix if $|t| > p$. Let Px_t be the projection of x_t on the closed linear manifold spanned by $x_{t-1}, x_{t-2}, \cdots$. Consider $\eta_t = x_t - Px_t$. The η_t are orthogonal to each other, that is, $E\eta_t\eta_\tau' = 0$ when $t \neq \tau$. Since x_t and η_t are stationary I can write

$$x_t = \int_{-\pi}^{\pi} e^{it\lambda} dZ_x(\lambda), \qquad \eta_t = \int_{-\pi}^{\pi} e^{it\lambda} dZ_\eta(\lambda) \tag{3.19}$$

where $z_x(\lambda)$, $z_\eta(\lambda)$ are processes with orthogonal increments

$$\begin{aligned} EdZ_x(\lambda)\, dZ_x(\mu)' &= \delta_{\lambda\mu} f(\lambda)\, d\lambda \\ EdZ_\eta(\lambda)\, dZ_\eta(\mu)' &= \frac{1}{2\pi}\, \delta_{\lambda\mu} N d\lambda . \end{aligned} \tag{3.20}$$

Here $N = E\eta_t\eta_t'$. Because $r_t = 0$ for $t < -p$ it is clear that

$$x_t = \int_{-\pi}^{\pi} e^{it\lambda} dZ_x(\lambda) = \sum_{j=0}^{p} h_j \eta_{t-j} = \int_{-\pi}^{\pi} e^{it\lambda} \left(\sum_{j=0}^{p} h_j e^{-ij\lambda}\right) dZ_\eta(\lambda), \tag{3.21}$$

and $h_0 = I$. On approximating the characteristic function of the set $[-\pi, \lambda]$ in the mean square by linear combinations of the exponentials exp $it\lambda$, equation

$$\int_{-\pi}^{\lambda} dZ_x(\lambda) = \int_{-\pi}^{\lambda} \sum_{j=0}^{p} h_j e^{-ij\lambda} dZ_\eta(\lambda) \tag{3.22}$$

is obtained. On taking the covariance matrix of both sides of equation (3.22)

$$F_x(\lambda) = \frac{1}{2\pi} \int_{-\pi}^{\lambda} \left(\sum_{j=0}^{p} h_j e^{-ij\lambda}\right) N \left(\sum_{j=0}^{p} h_j e^{-ij\lambda}\right)' d\lambda \tag{3.23}$$

or

$$f(\lambda) = \frac{1}{2\pi} \left(\sum_{j=0}^{p} h_j e^{-ij\lambda}\right) N \left(\sum_{j=0}^{p} h_j e^{-ij\lambda}\right)' \tag{3.24}$$

is obtained. Since $f(\lambda)$ is nonsingular, the matrix N must be nonsingular. Norm the η_t's so as to get

$$\xi_t = N^{-1/2}\eta_t . \tag{3.25}$$

The ξ_t's are an orthonormal process, that is,

$$E\xi_t\, \xi_\tau' = \delta_{t,\tau} I . \tag{3.26}$$

I can now write

$$x_t = \sum_{j=0}^{p} c_j\, \xi_{t-j} \tag{3.27}$$

where $c_0 = N^{1/2}$. By using the argument that led to equation (3.22) one can see that

$$f(\lambda) = \frac{1}{2\pi} \left(\sum_{j=0}^{p} c_j e^{-ij\lambda}\right) \left(\sum_{j=0}^{p} c_j e^{-ij\lambda}\right)' . \tag{3.28}$$

Since ξ_t is weakly stationary

$$\xi_t = \int_{-\pi}^{\pi} e^{it\lambda} dZ_\xi(\lambda) = \int_{-\pi}^{\pi} e^{it\lambda} \left(\sum_{j=0}^{p} c_j e^{-ij\lambda} \right)^{-1} dZ_x(\lambda) \tag{3.29}$$

$$E dZ_\xi(\lambda)\, dZ_\xi(\mu)' = \delta_{\lambda\mu} \frac{1}{2\pi} I d\lambda. \tag{3.30}$$

By our construction ξ_t is in the linear manifold spanned by $x_t, x_{t-1}, \cdots$ so that

$$\int_{-\pi}^{\pi} e^{-il\lambda} \left(\sum_{j=0}^{p} c_j e^{-ij\lambda} \right)^{-1} d\lambda = 0 \tag{3.31}$$

when $l > 0$.

4. The asymptotic covariance matrices

In obtaining the asymptotic form of the covariance matrices of the least-squares and Markov estimates, it will be convenient to deal with

$$D_n E(\beta_L - \beta)(\beta_L - \beta)' D_n = D_n [\varphi'\varphi]^{-1} D_n D_n^{-1} \varphi' R \varphi D_n^{-1} D_n [\varphi'\varphi]^{-1} D_n \tag{4.1}$$

and

$$D_n E(\beta_M - \beta)(\beta_M - \beta)' D_n = D_n [\varphi' R^{-1} \varphi]^{-1} D_n . \tag{4.2}$$

THEOREM 1. *Under the conditions on the spectrum of the process x_t given in section 2 and the conditions on the regression vectors specified there*

$$\lim_{n\to\infty} D_n E(\beta_L - \beta)(\beta_L - \beta)' D_n = 2\pi T^{-1} \int_{-\pi}^{\pi} [f(-\lambda) \cdot dM(\lambda)] T^{-1}. \tag{4.3}$$

The conditions on the regression vectors specified in section 2 imply that

$$D_n^{-1} [\varphi'\varphi] D_n^{-1} \to T \tag{4.4}$$

as $n \to \infty$ and the limiting $s \times s$ matrix T is nonsingular. I therefore need only consider the asymptotic behavior of

$$D_n^{-1} \varphi' R \varphi D_n^{-1} . \tag{4.5}$$

Given any sufficiently small $\epsilon > 0$, by lemma 2 there are finite matrix-valued $(k \times k)$ trigonometric polynomials

$$\begin{aligned} g(\lambda) &= \frac{1}{2\pi} \sum_{u=-p}^{p} g_u e^{iu\lambda} \\ h(\lambda) &= \frac{1}{2\pi} \sum_{u=-1}^{p} h_u e^{iu\lambda} \end{aligned} \tag{4.6}$$

such that

$$0 < f(\lambda) - \epsilon I < g(\lambda) < f(\lambda) < h(\lambda) < f(\lambda) + \epsilon I . \tag{4.7}$$

Let G, H be the covariance matrices of y if x_t has the spectral densities $g(\lambda)$, $h(\lambda)$, respectively. Then

$$\varphi' G \varphi < \varphi' R \varphi < \varphi' H \varphi . \tag{4.8}$$

I shall obtain the limit of (4.5) as $n \to \infty$ when the covariance matrix of x is G. The matrix

$$\varphi' G \varphi = \sum_{t,\tau=1}^{n} \varphi_t' \, g_{t-\tau} \, \varphi_\tau . \tag{4.9}$$

A typical element of (4.9) is therefore of the form

$$\frac{\sum_{t,\tau=1}^{n} \varphi_t^{(p)\prime} g_{t-\tau} \varphi_\tau^{(q)}}{[\Phi_n^{(p)} \Phi_n^{(q)}]^{1/2}} . \tag{4.10}$$

Since there are only a finite number of nonzero g_u's, the limit of expression (4.10) is

$$2\pi \sum_{i,j=1}^{k} \int_{-\pi}^{\pi} g_{ij}(-\lambda) \, d_{pq} M_{ij}(\lambda) . \tag{4.11}$$

Thus

$$\lim_{n \to \infty} D_n^{-1} \varphi' G \varphi D_n^{-1} = 2\pi \int_{-\pi}^{\pi} [g(-\lambda) \cdot dM(\lambda)] . \tag{4.12}$$

Similarly

$$\lim_{n \to \infty} D_n^{-1} \varphi' H \varphi D_n^{-1} = 2\pi \int_{-\pi}^{\pi} [h(-\lambda) \cdot dM(\lambda)] . \tag{4.13}$$

On letting $\epsilon \to 0$, it is clear that

$$\lim_{n \to \infty} D_n^{-1} \varphi' R \varphi D_n^{-1} = 2\pi \int_{-\pi}^{\pi} [f(-\lambda) \cdot dM(\lambda)] . \tag{4.14}$$

Note that the expression

$$\int_{-\pi}^{\pi} [f(-\lambda) \cdot dM(\lambda)] \tag{4.15}$$

is nonsingular for $f(\lambda) > \epsilon I$ if $\epsilon > 0$ is sufficiently small. But then

$$\int_{-\pi}^{\pi} [f(-\lambda) \cdot dM(\lambda)] > \epsilon \int_{-\pi}^{\pi} [I \cdot dM(\lambda)] = \epsilon (I \cdot M_0) \tag{4.16}$$

which is nonsingular.

THEOREM 2. *Under the conditions on the spectrum of x_t and the regression vectors assumed in section* 2

$$\lim_{n \to \infty} D_n E(\beta_M - \beta)(\beta_M - \beta)' D_n = 2\pi \left(\int_{-\pi}^{\pi} [f^{-1}(-\lambda) \cdot dM(\lambda)] \right)^{-1} . \tag{4.17}$$

By lemmas 2 and 3, for every sufficiently small $\epsilon > 0$ there are finite trigonometric polynomials

$$g(\lambda) = \sum_{u=0}^{p} g_u e^{-iu\lambda}$$
$$h(\lambda) = \sum_{u=0}^{p} h_u e^{-iu\lambda} \tag{4.18}$$

with g_0, h_0 nonsingular and satisfying relation (3.17) and such that

$$0 < f(\lambda) - \epsilon I < \frac{1}{2\pi} g^{-1}(\lambda) g^{-1}(\lambda)' < f(\lambda) < \frac{1}{2\pi} h^{-1}(\lambda) h^{-1}(\lambda)' < f(\lambda) + \epsilon I .$$

Let G, H be the covariance matrices of the vector x if the process x_t has the spectral densities $(1/2\pi)g^{-1}(\lambda)g^{-1}(\lambda)'$, $(1/2\pi)h^{-1}(\lambda)h^{-1}(\lambda)'$, respectively.

In obtaining our result it will be sufficient to consider the asymptotic behavior of $D_n^{-1}\varphi'R^{-1}\varphi D_n^{-1}$. Now

$$0 < \varphi' H^{-1}\varphi < \varphi' R^{-1}\varphi < \varphi' G^{-1}\varphi \,. \tag{4.19}$$

Assume for the moment that x_t has the spectral density $(1/2\pi)g^{-1}(\lambda)g^{-1}(\lambda)'$ and consider the asymptotic behavior of $D_n^{-1}\varphi'G^{-1}\varphi D_n^{-1}$ as $n \to \infty$. Now carry out a Gramm-Schmidt orthogonalization procedure with respect to $x_1, \cdots, x_n$ starting with x_1. The first p equations are

$$\begin{aligned} d_{11}x_1 &= \xi_1 \\ d_{21}x_1 + d_{22}x_2 &= \xi_2 \\ &\cdots \\ d_{p1}x_1 + \cdots + d_{pp}x_p &= \xi_p \,. \end{aligned} \tag{4.20}$$

The residual ξ's are orthonormal. At the next step I have

$$g_p x_1 + \cdots + g_0 x_{p+1} = \xi_{p+1} = \eta_{p+1} \tag{4.21}$$

since x_t satisfies the difference equation

$$\sum_{u=0}^{p} g_u \, x_{t-u} = \eta_t \tag{4.22}$$

with the η_t's orthonormal. From then on we have the equation

$$g_p x_k + \cdots + g_0 \, x_{k+p} = \xi_{k+p} = \eta_{k+p} \,, \qquad k = 1, \cdots, n-p \tag{4.23}$$

Let the matrix

$$\Delta = \begin{pmatrix} d_{11} & & & & & \\ d_{21} d_{22} & & & & 0 & \\ \cdot \ \cdot \ \cdot & & & & & \\ d_{p1} d_{p2} & \cdots & d_{pp} & & & \\ g_p \ g_{p-1} & \cdots & g_1 & g_0 & & \\ & g_p & \cdots & g_2 & g_1 & g_0 \\ & & \cdot & & & \cdot \\ & 0 & & \cdot & & \cdot \\ & & & \cdot & & \cdot \end{pmatrix} . \tag{4.24}$$

Then

$$\Delta x = \xi = \begin{pmatrix} \xi_1 \\ \cdot \\ \cdot \\ \cdot \\ \xi_n \end{pmatrix} \tag{4.25}$$

where ξ is a vector of orthonormal random vectors. Δ is a nonsingular transformation taking $x_1, \cdots, x_n$ into $\xi_1, \cdots, \xi_n$. On taking the covariance matrix of both sides of equation (4.25) I have

$$\Delta G \Delta' = I \tag{4.26}$$

so that

$$G^{-1} = \Delta'\Delta \,. \tag{4.27}$$

Let $\{g_{u,v}^{-1}; u, v = 1, \cdots, n\} = G^{-1}$. Then

$$g_{\nu\mu}^{-1} = \sum_{u=-\infty}^{\infty} g'_{\nu+u}\, g_{\mu+u} \tag{4.28}$$

if ν or μ is greater than p. Here g_ν is understood to be zero if $\nu < 0$ or $\nu > p$. But

$$\sum_{u=-\infty}^{\infty} g'_{\nu+u}\, g_{\mu+u} = \frac{1}{2\pi} \int_{-\pi}^{\pi} e^{i(\nu-\mu)\lambda} g(\lambda)' g(\lambda)\, d\lambda \,. \tag{4.29}$$

Now the (p, q)th element of $D_n^{-1}\varphi' G^{-1} \varphi D_n^{-1}$ is

$$\frac{\sum_{t,\tau=1}^{n} \varphi_t^{(p)\prime} g_{t,\tau}^{-1} \varphi_\tau^{(q)}}{(\Phi_n^{(p)} \Phi_n^{(q)})^{1/2}} = \sum_{m=0}^{p} \sum_{v=1}^{n-m} \frac{\phi_{m+v}^{(p)\prime} \gamma_m \varphi_v^{(q)}}{\sqrt{\Phi_n^{(p)} \Phi_n^{(q)}}} + \sum_{m=-1}^{-p} \sum_{u=m+1}^{n} \frac{\varphi_u^{(p)\prime} \gamma_m \varphi_{u-m}^{(q)}}{\sqrt{\Phi_n^{(p)} \Phi_n^{(q)}}} + \delta_n \,. \tag{4.30}$$

Here $\gamma_m = (1/2\pi) \int_{-\pi}^{\pi} e^{im\lambda} g(\lambda)' g(\lambda) d\lambda$.

Now δ_n is the sum of at most $4p^2$ terms of the form

$$\frac{\varphi_u^{(p)\prime} g_{u,v}^{-1} \varphi_v^{(q)}}{\sqrt{\Phi_n^{(p)} \Phi_n^{(q)}}} \,. \tag{4.31}$$

Since every element of $g_{u,v}^{-1}$ is smaller in absolute value than δ^{-1} for some small but fixed $\delta > 0$, the elements of

$$\frac{\varphi_u^{(p)\prime} g_{u,v}^{-1} \varphi_v^{(q)}}{\sqrt{\Phi_n^{(p)} \Phi_n^{(q)}}} \tag{4.32}$$

converge to zero as $n \to \infty$. But then

$$\lim_{n\to\infty} \frac{\sum_{t,\tau=1}^{n} \varphi_t^{(p)\prime} g_{t,\tau}^{-1} \varphi_\tau^{(q)}}{(\Phi_n^{(p)} \Phi_n^{(q)})^{1/2}} = \frac{1}{2\pi} \int_{-\pi}^{\pi} \sum_{i,j=1}^{k} \{ g(-\lambda)' g(-\lambda) \}_{ij} d_{pq} M_{ij}(\lambda) \,. \tag{4.33}$$

I have now shown that

$$\lim_{n\to\infty} D_n^{-1} \varphi' G^{-1} \varphi\, D_n^{-1} = \int_{-\pi}^{\pi} [g(-\lambda)' g(-\lambda) \cdot dM(\lambda)] \,. \tag{4.34}$$

In the same way one can show that

$$\lim_{n\to\infty} D_n^{-1} \varphi' H^{-1} \varphi\, D_n^{-1} = \int_{-\pi}^{\pi} [h(-\lambda)' h(-\lambda) \cdot dM(\lambda)] \,. \tag{4.35}$$

On letting $\epsilon \downarrow 0$ the desired result

$$\lim_{n\to\infty} D_n^{-1} \varphi' R^{-1} \varphi\, D_n^{-1} = \frac{1}{2\pi} \int_{-\pi}^{\pi} [f^{-1}(-\lambda) \cdot dM(\lambda)] \tag{4.36}$$

is obtained. The matrix (4.36) can be seen to be nonsingular by using the argument used at the end of the proof of theorem 1.

5. Asymptotic efficiency

It is interesting to investigate those types of regression for which the least-squares estimate is asymptotically efficient in the class of linear unbiased estimates for all admissible spectra $f(\lambda)$. In most cases the covariance matrix R is unknown and a reasonably large sample size is required to get adequate precision in estimating it. For this reason it would be convenient if one could use the least-squares estimate instead of the Markov estimate since the least-squares estimate does not require knowledge of R. Even if R is known, it may be difficult to compute R^{-1} which is required for the Markov estimate. In view of the results already obtained, the least-squares estimate will be asymptotically efficient if

$$T^{-1}\int_{-\pi}^{\pi}[f(-\lambda)\cdot dM(\lambda)]\,T^{-1}\int_{-\pi}^{\pi}[f^{-1}(-\lambda)\cdot dM(\lambda)]=I \tag{5.1}$$

for all admissible $f(\lambda)$. The case of interest is that in which the process x_t and the regression vectors have real components. Because of this $f(\lambda)$ and $M(\lambda)$ must satisfy the additional restraints

$$\begin{aligned} f(\lambda) &= \overline{f(-\lambda)} \\ dM(\lambda) &= \overline{dM(-\lambda)}. \end{aligned} \tag{5.2}$$

When $k = 1$, asymptotic efficiency of the least-squares estimate has been discussed in U. Grenander [1] and U. Grenander and M. Rosenblatt [2], [3].

In the one-dimensional case it is convenient to set

$$T(\lambda)=M(\lambda+)-M(-\lambda-) \tag{5.3}$$

and rewrite (5.1) in the form

$$T^{-1}\int_{0-}^{\pi}f(\lambda)\,dT(\lambda)\,T^{-1}\int_{0-}^{\pi}f^{-1}(\lambda)\,dT(\lambda)=I. \tag{5.4}$$

Equation (5.4) is satisfied for all positive continuous $f(\lambda)$ if and only if $T(\lambda)$ increases only at a finite number of points $0 \leqq \lambda_1 < \cdots < \lambda_q \leqq \pi$, $q < s$, and the jumps

$$T_i=\Delta T(\lambda_i) \tag{5.5}$$

satisfy the relations

$$T_i T^{-1} T_j=\delta_{ij} T_i. \tag{5.6}$$

These conditions are satisfied if one has a polynomial regression

$$m_t=\beta_0+\cdots+\beta_{s-1}t^{s-1}, \tag{5.7}$$

a trigonometric regression

$$m_t=\beta_1\cos t\lambda_1+\cdots+\beta_{s_1}\cos t\lambda_{s_1}+\beta_{s_1+1}\sin t\lambda_1+\cdots+\beta_{s_1+s_2}\sin t\lambda_{s_2} \tag{5.8}$$

(with the points λ_i distinct), or more generally a mixed polynomial and trigonometric regression

$$m_t=\sum_{u=1}^{q}\sum_{v=1}^{{}_1s_u}{}_1\beta_{u,v}t^{v-1}\cos t\lambda_u+\sum_{u=1}^{q}\sum_{v=1}^{{}_2s_u}{}_2\beta_{u,v}t^{v-1}\sin t\lambda_u \tag{5.9}$$

(with points λ_i distinct) [3]. Obviously the sine terms in this last regression form disappear if $\lambda_i = 0$. Notice that these regression sequences include most of those used in standard statistical work. It is easy to construct a regression sequence where the least-squares estimate is not asymptotically efficient. Consider

$$m_t = \beta\,(a_0 + a_1 \cos t\lambda_1 + \cdots + a_k \cos t\lambda_k)\,, \quad 0 < \lambda_1 < \cdots < \lambda_k\,, \tag{5.10}$$

where the constants a_i are known. Such a regression has a form similar to the pulse trains encountered in communication theory. In the case of such a regression $T(\lambda)$ increases only at the points $0, \lambda_1, \cdots, \lambda_k$. The jump of $T(\lambda)$ at 0 is

$$\frac{a_0^2}{a_0^2 + \frac{1}{2}\sum_{j=1}^{k} a_j^2} \tag{5.11}$$

and the jump at $\lambda_j, j = 1, \cdots, k$, is

$$\frac{\frac{1}{2}a_j^2}{a_0^2 + \frac{1}{2}\sum_{j=1}^{k} a_j^2}\,. \tag{5.12}$$

The asymptotic efficiency of the least-squares estimate β_L in the class of linear unbiased estimates is

$$\frac{\left(a_0^2 + \frac{1}{2}\sum_{j=1}^{k} a_j^2\right)^2}{\left[a_0^2 f(0) + \frac{1}{2}\sum_{j=1}^{k} a_j^2 f(\lambda_j)\right]\left[\frac{a_0^2}{f(0)} + \frac{1}{2}\sum_{j=1}^{k}\frac{a_j^2}{f(\lambda_j)}\right]} \leqq 1\,. \tag{5.13}$$

In section 7 I shall discuss the question of how much additional information about the spectrum $f(\lambda)$ is required to construct an estimate with the same asymptotic mean square error as the Markov estimate.

In the case of multidimensional time series, new phenomena arise. Consider first the case of a polynomial regression. If each component of the time series has a polynomial regression, that is,

$$_i m_t = \sum_{k=1}^{s} {}_i\beta_k\, t^{k-1}\,, \qquad i = 1, \cdots, k\,, \tag{5.14}$$

the least-squares estimate β_L is still asymptotically efficient. However, if the different components have polynomial regressions of different orders, the least-squares estimate is no longer asymptotically efficient [5]. A simple example is that in which the mean value of the first coordinate of a two-dimensional time series is unknown while the mean value of the second coordinate is known to be zero. Then

$$m_t = \beta\varphi_t\,, \qquad \varphi_t = \begin{pmatrix}1\\0\end{pmatrix}. \tag{5.15}$$

The function $M(\lambda)$ increases only at zero and the jump at zero

$$\Delta M(0) = \begin{pmatrix}\Delta M_{11}(0) & \Delta M_{12}(0)\\ \Delta M_{21}(0) & \Delta M_{22}(0)\end{pmatrix} = \begin{pmatrix}1 & 0\\ 0 & 0\end{pmatrix}. \tag{5.16}$$

Thus

$$(5.17)\qquad \begin{aligned}\int_{-\pi}^{\pi}[f(-\lambda)\cdot dM(\lambda)] &= f_{11}(0)\\ \int_{-\pi}^{\pi}[f^{-1}(-\lambda)\cdot dM(\lambda)] &= \frac{f_{22}(0)}{f_{11}(0)f_{22}(0)-f_{12}^2(0)}.\end{aligned}$$

The asymptotic efficiency of the least-squares estimate is

$$(5.18)\qquad 1-\frac{f_{12}^2(0)}{f_{11}(0)f_{22}(0)}.$$

If there is a mixed polynomial and trigonometric regression (5.9) not only must the same regression form occur in each component, but one must also have ${}_1s_u = {}_2s_u$ if $\lambda_u \neq 0$. Thus, the least-squares estimate will be asymptotically efficient in the case of a regression

$$(5.19)\qquad \begin{aligned} m_t &= \beta_1\varphi_t^{(1)}+\beta_2\varphi_t^{(2)}+\beta_3\varphi_t^{(3)}+\beta_4\varphi_t^{(4)}\\ &= \beta_1\begin{pmatrix}\cos t\lambda\\ 0\end{pmatrix}+\beta_2\begin{pmatrix}\sin t\lambda\\ 0\end{pmatrix}+\beta_3\begin{pmatrix}0\\ \cos t\lambda\end{pmatrix}+\beta_4\begin{pmatrix}0\\ \sin t\lambda\end{pmatrix}, \qquad \lambda\neq 0,\end{aligned}$$

but not in the case of a regression

$$(5.20)\qquad m_t=\beta_1\varphi_t^{(1)}+\beta_2\varphi_t^{(2)}=\beta_1\begin{pmatrix}\cos t\lambda\\ 0\end{pmatrix}+\beta_2\begin{pmatrix}0\\ \cos t\lambda\end{pmatrix}, \qquad \lambda\neq 0.$$

It is worthwhile examining this last regression in a little more detail. The function $M(\lambda)$ increases at two points, λ and $-\lambda$. The jumps

$$(5.21)\qquad \begin{aligned}\Delta M_{11}(\lambda) &= \Delta M_{11}(-\lambda) = \begin{pmatrix}\frac{1}{2} & 0\\ 0 & 0\end{pmatrix}\\ \Delta M_{12}(\lambda) &= \Delta M_{12}(-\lambda) = \begin{pmatrix}0 & \frac{1}{2}\\ 0 & 0\end{pmatrix}\end{aligned}$$

$$(5.22)\qquad \begin{aligned}\Delta M_{21}(\lambda) &= \Delta M_{21}(-\lambda) = \begin{pmatrix}0 & 0\\ \frac{1}{2} & 0\end{pmatrix}\\ \Delta M_{22}(\lambda) &= \Delta M_{22}(-\lambda) = \begin{pmatrix}0 & 0\\ 0 & \frac{1}{2}\end{pmatrix}.\end{aligned}$$

Thus

$$(5.23)\qquad \int_{-\pi}^{\pi}[f(-\lambda)\cdot dM(\lambda)] = \begin{pmatrix} f_{11}(\lambda) & \operatorname{Re} f_{12}(\lambda)\\ \operatorname{Re} f_{12}(\lambda) & f_{22}(\lambda)\end{pmatrix}$$

and

$$(5.24)\qquad \int_{-\pi}^{\pi}[f^{-1}(-\lambda)\cdot dM(\lambda)] = \frac{1}{f_{11}(\lambda)f_{22}(\lambda)-|f_{12}(\lambda)|^2}\begin{pmatrix} f_{22}(\lambda) & -\operatorname{Re} f_{12}(\lambda)\\ -\operatorname{Re} f_{12}(\lambda) & f_{11}(\lambda)\end{pmatrix}.$$

It is worthwhile noting that one does have asymptotic efficiency of the least-squares estimate if $\operatorname{Im} f_{12}(\lambda) = 0$.

In the multidimensional case $k \geqq 2$ set

$$(5.25)\qquad N(\lambda) = \{T^{-1/2}\,{}_{jl}M(\lambda)T^{-1/2};\qquad j,l=1,\cdots,s\}.$$

Then equation (5.1) can be rewritten

$$\int_{-\pi}^{\pi} [f(-\lambda) \cdot dN(\lambda)] \int_{-\pi}^{\pi} [f^{-1}(-\lambda) \cdot dN(\lambda)] = I. \tag{5.26}$$

I have not been able to get simple necessary and sufficient conditions on $N(\lambda)$ for equation (5.26) to be satisfied for all admissible $f(\lambda)$. However, one can get simple conditions of this type when $N(\lambda)$ is known to increase only at zero. This corresponds to the interesting case of a polynomial regression. Equation (5.26) can then be written as

$$(f \cdot N)(f^{-1} \cdot N) = I, \tag{5.27}$$

where

$$f = f(0) > 0 \tag{5.28}$$

and

$$N = \Delta N \geqq 0. \tag{5.29}$$

Set

$$N_{ij} = \{{}_{pq}N_{ij}; \qquad p, q = 1, \cdots, s\} \tag{5.30}$$

and

$$W_{ij} = N_{ij} + N_{ji}, \qquad i \neq j. \tag{5.31}$$

Since $f = f(0)$ and $N = \Delta N(0)$, the elements of f and N are all real.

THEOREM 3. *Let f and N be positive definite and positive semidefinite symmetric matrices, respectively. The equation*

$$(f \cdot N)(f^{-1} \cdot N) = I \tag{5.32}$$

is valid for all positive definite symmetric f if and only if

$$\begin{aligned}
&N_{ii}N_{jj} = \delta_{ij}N_{ii}; && i, j = 1, \cdots, k, \\
&\sum N_{ii} = I \\
&W_{ij}W_{ik} = N_{jj}W_{jk} = W_{jk}N_{kk}, && j \neq k, \\
&\quad W_{ij}^2 = N_{ii} + N_{jj} \\
&W_{ij}W_{kl} = 0, && i \neq k, l; \quad j \neq k, l; \\
&W_{ij}N_{kk} = N_{kk}W_{ij} = 0, && i, j \neq k.
\end{aligned} \tag{5.33}$$

Consider first the case in which f is a diagonal matrix

$$f = \begin{pmatrix} \lambda_1 & & 0 \\ & \ddots & \\ 0 & & \lambda_k \end{pmatrix}, \qquad \lambda_i > 0. \tag{5.34}$$

Equation (5.32) then becomes

$$\sum \lambda_i N_{ii} \sum \lambda_j^{-1} N_{jj} = I. \tag{5.35}$$

This equation is valid for all positive λ_i if and only if

$$N_{ii} N_{jj} = \delta_{ij} N_{ii} \tag{5.36}$$

and $\sum N_{ii} = I$. Since N is positive semidefinite and $N_{ii}N_{jj} = \delta_{ij}N_{ii}$, it follows that

$$W_{ij} N_{kk} = N_{kk} W_{ij} = 0 \tag{5.37}$$

if $i, j \neq k$ and

$$W_{ij} W_{kl} = 0 \tag{5.38}$$

if $i \neq k, l$ and $j \neq k, l$. Now consider the case in which

$$f = \begin{pmatrix} f_1 & f_2 & 0 \\ f_2 & f_3 & \\ 0 & & I \end{pmatrix} \tag{5.39}$$

and is positive definite. On differentiating equation (5.32) with respect to f_2 twice, equation

$$W_{12}^2 = N_{11} + N_{22} \tag{5.40}$$

is obtained. On differentiating equation (5.32) first with respect to f_1 and then with respect to f_2, equation

$$N_{11} W_{12} = W_{12} N_{22} \tag{5.41}$$

is obtained. Now let

$$f = \begin{pmatrix} f_1 & f_2 & f_3 & \\ f_2 & f_4 & f_5 & 0 \\ f_3 & f_5 & f_6 & \\ & 0 & & I \end{pmatrix} \tag{5.42}$$

and be positive definite. Differentiate equation (5.32) with respect to f_1, f_2, and f_5. I now get

$$M_{11} T_{13} = T_{12} T_{23}. \tag{5.43}$$

All the other equations (5.33) are obtained by taking some subscripts other than 1, 2, 3 or interchanging subscripts.

By using the conditions (5.33), one can readily verify that equation (5.32) is satisfied with any positive definite symmetric f.

6. Some computations

It is worthwhile looking at a process with a stationary residual of a special and simple form to see how good the asymptotic theory considered is for finite samples. Consider a process

$$y_t = x_t + \beta_1 + \beta_{2t} \tag{6.1}$$

where the residual x_t is a first order autoregressive scheme with covariances

$$r_n = \frac{\text{[illegible]}\,\rho^{|n|}}{1 - \rho^2}, \qquad -1 < \rho < 1. \tag{6.2}$$

Here ρ is the correlation coefficient of the scheme. The regression coefficients β_1, β_2 of the scheme are unknown and to be estimated. The covariance matrices of the least-squares estimate β_L and the Markov estimate β_M are considered. I have already noted that

$$E(\beta_L-\beta)(\beta_L-\beta)'=(\varphi'\varphi)^{-1}\varphi'R\varphi(\varphi'\varphi)^{-1} \tag{6.3}$$

and

$$E(\beta_M-\beta)(\beta_M-\beta)'=(\varphi'R^{-1}\varphi)^{-1}. \tag{6.4}$$

Here when the sample size is n

$$\varphi'=\begin{pmatrix}1 & 1\cdots 1\\ 1 & 2\cdots n\end{pmatrix} \tag{6.5}$$

so that

$$(\varphi'\varphi)^{-1}=\begin{pmatrix}\dfrac{2(2n+1)}{n(n-1)} & \dfrac{-6}{n(n-1)}\\[2ex] \dfrac{-6}{n(n-1)} & \dfrac{12}{n(n^2-1)}\end{pmatrix}. \tag{6.6}$$

Straightforward but tedious manipulations lead to

$$\varphi'R^{-1}\varphi=\begin{bmatrix}(n-2)(1-\rho)^2+2(1-\rho) & \dfrac{(n-1)n}{2}(1-\rho)^2+(n+\rho)(1-\rho)\\[2ex] \dfrac{(n-1)n}{2}(1-\rho)^2 & \dfrac{(n-1)n(2n-1)}{6}(1-\rho)^2\\ \qquad+(n+\rho)(1-\rho) & \qquad+n^2(1-\rho)-\rho^2+n\rho\end{bmatrix}. \tag{6.7}$$

One can similarly show that

$$\varphi'R\varphi=\begin{bmatrix}\dfrac{n}{(1-\rho)^2}-\dfrac{2\rho(1-\rho^n)}{(1-\rho^2)(1-\rho)^2} & \dfrac{n(n+1)}{2(1-\rho)^2}-\dfrac{(n+1)\rho(1-\rho^n)}{(1-\rho^2)(1-\rho)^2}\\[2ex] \dfrac{n(n+1)}{2(1-\rho)^2}-\dfrac{(n+1)\rho(1-\rho^n)}{(1-\rho^2)(1-\rho)^2} & \dfrac{n(n+1)(2n+1)}{6(1-\rho)^2}-\dfrac{n(n+1)\rho}{(1+\rho)(1-\rho)^3}\\ & -\dfrac{2\rho^{n+2}(n+1)}{(1+\rho)(1-\rho)^4}+\dfrac{2\rho^2(1-\rho^{n+1})}{(1+\rho)(1-\rho)^5}\end{bmatrix}. \tag{6.8}$$

The covariance matrix as given by the asymptotic theory is

$$\frac{1}{(1-\rho)^2}\begin{pmatrix}\dfrac{4}{n} & -\dfrac{6}{n^2}\\[2ex] \dfrac{-6}{n^2} & \dfrac{12}{n^3}\end{pmatrix} \tag{6.9}$$

to the first order.

The (i,j)th elements of the covariance matrices of both the least-squares and Markov estimates, $i, j=1, 2$, are given in Table I for the sample sizes $n=10, 15, 20, 50$ and correlation coefficients $\rho=-.8, -.6, \cdots, .8$. The approximation suggested by asymptotic theory is also given.

TABLE I

COVARIANCE MATRICES OF THE LEAST-SQUARES AND MARKOV ESTIMATES (AND AN ASYMPTOTIC APPROXIMATION OF THE COVARIANCE MATRICES) OF A LINEAR REGRESSION, RESIDUAL FIRST-ORDER AUTOREGRESSIVE

ρ		Matrix Elements		
		(1, 1)	(1, 2)=(2, 1)	(2, 2)
			$n=10$	
+.2	(a)	0.65196	−0.091312	.016602
	(b)	0.64406	−0.090046	.016372
	(c)	0.62500	−0.093750	.018750
−.2	(a)	0.35898	−0.052116	.0094753
	(b)	0.34109	−0.049425	.0089863
	(c)	0.27778	−0.041666	.0083333
+.4	(a)	0.99189	−0.13464	.024880
	(b)	0.94826	−0.12785	.023245
	(c)	1.11111	−0.16667	.033333
−.4	(a)	0.29685	−0.043812	.0079656
	(b)	0.27749	−0.040614	.0073844
	(c)	0.20408	−0.030612	.0061224
+.6	(a)	1.69108	−0.21501	.039090
	(b)	1.54892	−0.19421	.035311
	(c)	2.50000	−0.37500	.075000
−.6	(a)	0.27142	−0.040923	.0074402
	(b)	0.22345	−0.032950	.0059909
	(c)	0.15625	−0.023438	.0046876
+.8	(a)	3.48150	−0.35878	.065229
	(b)	3.17040	−0.32391	.058893
	(c)	10.00000	−1.50000	.30000
−.8	(a)	0.32541	−0.051326	.0093318
	(b)	0.18379	−0.027257	.0049559
	(c)	0.12346	−0.018518	.0037037
			$n=15$	
+.2	(a)	0.42883	− .040942	.0051178
	(b)	0.42456	− .040469	.0050587
	(c)	0.41667	− .041667	.0055556
−.2	(a)	0.21958	− .021498	.0026873
	(b)	0.21716	− .021227	.0026534
	(c)	0.18519	− .018519	.0024692
+.4	(a)	0.68983	− .064546	.0080684
	(b)	0.66268	− .061576	.0076970
	(c)	0.74075	− .074075	.0098767
−.4	(a)	0.17508	− .017363	.0021704
	(b)	0.16642	− .016383	.0020478
	(c)	0.13606	− .013606	.0018141
+.6	(a)	1.29297	− .11604	.014505
	(b)	1.18142	− .10428	.013034
	(c)	1.66669	− .16669	.022223

(a) Least-squares, (b) Markov, and (c) Asymptotic.

TABLE I—*Continued*

ρ		Matrix Elements (1, 1)	(1, 2)=(2, 1)	(2, 2)
		n=15—*Continued*		
−.6	(a)	0.15221	− .015363	.0019204
	(b)	0.13159	− .013023	.0016278
	(c)	0.10417	− .010417	.0013889
+.8	(a)	3.16763	− .24715	.030894
	(b)	2.76763	− .21008	.026261
	(c)	6.66675	− .66675	.088890
−.8	(a)	0.15817	− .016409	.0020513
	(b)	0.10665	− .010598	.0013247
	(c)	0.082305	− .0082305	.0010974
		n=20		
+.2	(a)	0.31940	− .023130	.0022033
	(b)	0.31679	− .022912	.0021821
	(c)	0.31250	− .023438	.0023438
−.2	(a)	0.15777	− .011650	.0011097
	(b)	0.15640	− .011532	.0010983
	(c)	0.13889	− .010417	.0010417
+.4	(a)	0.52774	− .037659	.0035871
	(b)	0.50996	− .036167	.0034445
	(c)	0.55556	− .041667	.0041667
−.4	(a)	0.12341	− .0092078	.00087707
	(b)	0.11854	− .0087887	.00083702
	(c)	0.10204	− .0076530	.00076530
+.6	(a)	1.03889	− .071960	.0068545
	(b)	0.95601	− .065169	.0062066
	(c)	1.25000	− .093750	.0093750
−.6	(a)	0.10438	− .0079058	.00075303
	(b)	0.092934	− .0069183	.00065888
	(c)	0.078126	− .0058595	.00058595
+.8	(a)	2.84008	− .17755	.016913
	(b)	2.45270	− .14856	.014148
	(c)	5.00000	− .37500	.037500
−.8	(a)	0.10458	− .0081673	.00077794
	(b)	0.074810	− .0055867	.00053206
	(c)	0.061728	− .0046296	.00046296
		n=50		
+.2	(a)	0.12612	− .0037310	.000146311
	(b)	0.12565	− .0037140	.000145648
	(c)	0.12500	− .0037500	.000150000
−.2	(a)	0.058456	− .0017433	.000068360
	(b)	0.058229	− .0017352	.000068045
	(c)	0.055555	− .0016667	.000066666
+.4	(a)	0.21807	− .0064149	.000251554
	(b)	0.21442	− .0062866	.000246535
	(c)	0.2222	− .0066667	.000266667

TABLE I—*Continued*

ρ		Matrix Elements (1, 1)	(1, 2)=(2, 1)	(2, 2)
		$n=50$—*Continued*		
−.4	(a)	0.044081	−.0013209	.000051799
	(b)	0.043302	−.0012933	.000050719
	(c)	0.040816	−.0012245	.000048979
+.6	(a)	0.46697	−.013595	.00053312
	(b)	0.44575	−.012856	.00050415
	(c)	0.50000	−.015000	.00060000
−.6	(a)	0.035245	−.0010643	.000041737
	(b)	0.033457	−.0010010	.000039256
	(c)	0.031250	−.00093751	.000037500
+.8	(a)	1.61368	−.045419	.0017810
	(b)	1.43485	−.039365	.0015437
	(c)	2.00000	−.060000	.0024000
−.8	(a)	0.031137	−.00095752	.000037549
	(b)	0.026625	−.00079766	.000031281
	(c)	0.024691	−.00074074	.000029629

7. Some special examples

In section 5 a few special but interesting types of regression sequences were considered where the least-squares estimate of the regression coefficient was not asymptotically efficient in the class of linear unbiased estimates. I now consider two of these regression sequences to find out what information about the spectrum $f(\lambda)$ is required to construct an estimate of the regression coefficient with the same asymptotic mean square error as the Markov estimate.

The first example is that of a one-dimensional process

$$y_t = x_t + \beta\varphi_t \tag{7.1}$$

where x_t is stationary and

$$\varphi_t = a_0 + a_1 \cos t\lambda_1 + \cdots + a_k \cos t\lambda_k\,, \qquad 0 < \lambda_1 < \cdots < \lambda_k\,. \tag{7.2}$$

Note that

$$\Phi_n = \sum_{t=1}^{n} \varphi_t^2 \sim n\left(a_0^2 + \frac{1}{2}\sum_{j=1}^{k} a_j^2\right). \tag{7.3}$$

An estimate of β which has the same asymptotic mean square error as the Markov estimate is

$$\beta^* = \frac{1}{n}\left(\frac{a_0^2}{f(0)} + \frac{1}{2}\sum_{j=1}^{k}\frac{a_j^2}{f(\lambda_j)}\right)\sum_{t=1}^{n} y_t\left(\frac{a_0}{f(0)} + \sum_{j=1}^{k}\frac{a_j}{f(\lambda_j)}\cos t\lambda_j\right) \tag{7.4}$$

for

$$E\beta^* = \beta + O\left(\frac{1}{n}\right) \tag{7.5}$$

and

$$\sigma^2(\beta^*) \sim \frac{1}{n}\left(\frac{a_0^2}{f(0)} + \frac{1}{2}\sum_{j=1}^{k}\frac{a_j^2}{f(\lambda_j)}\right)^{-1}. \tag{7.6}$$

Notice that the only information about the spectrum $f(\lambda)$ required for the construction of β^* is knowledge of the ratios

$$\frac{f(0)}{f(\lambda_1)}, \ldots, \frac{f(0)}{f(\lambda_k)}. \tag{7.7}$$

The second example is a two-dimensional process

$$\nu_t = x_t + \beta\varphi \tag{7.8}$$

with x_t stationary and

$$\varphi_t = \begin{pmatrix}1\\0\end{pmatrix}. \tag{7.9}$$

An estimate of β which has the same asymptotic mean square error as the Markov estimate is

$$\beta^* = \frac{1}{n}\sum_{t=1}^{n}\left({}_1y_t - \frac{f_{12}(0)}{f_{22}(0)}\,{}_2y_t\right) \tag{7.10}$$

where ${}_1y_t$, ${}_2y_t$ are the components of y_t. Note that

$$E\beta^* = \beta \tag{7.11}$$

and

$$\sigma^2(\beta^*) \sim \frac{1}{n}\left[f_{11}(0) - \frac{f_{12}^2(0)}{f_{22}(0)}\right]. \tag{7.12}$$

8. Final remarks

There are many interesting open problems. It is clear that one ought to be able to obtain analogues of the results obtained thus far in the case of a continuous time parameter. It is likely that such a program would require heavier tools.

The results obtained thus far have an immediate implication for various types of nonstationary processes, specifically processes which are integrals or sums of stationary processes. Consider as an example

$$Z_t = \sum_{\tau=1}^{t} x_\tau \tag{8.1}$$

where x_t is a stationary process. Results on estimation of regression coefficients with Z_t as a residual can be obtained from corresponding results with x_t as a residual.

A much more detailed investigation of specific types of regression sequences would be worthwhile pushing through. It is worthwhile noting that all the main results obtained can be derived for processes with a vector time parameter in the same way.

I wish to thank M. de Groot and the computing staff of the Statistical Research Center at the University of Chicago for carrying out the computations in section 6.

REFERENCES

[1] U. GRENANDER, "On the estimation of regression coefficients in the case of an autocorrelated disturbance," *Annals of Math. Stat.*, Vol. 25 (1954), pp. 252–272.

[2] U. GRENANDER and M. ROSENBLATT, "An extension of a theorem of G. Szegö and its application to the study of stochastic processes," *Trans. Amer. Math. Soc.*, Vol. 76 (1954), pp. 112–126.

[3] ———, "Regression analysis of time series with stationary residuals," *Proc. Nat. Acad. Sci.*, Vol. 40 (1954), pp. 812–816.

[4] ———, A monograph on time series analysis, to be published.

[5] M. ROSENBLATT, "On the estimation of regression coefficients of a vector-valued time series with a stationary residual," to be published in *Annals of Math. Stat.*

Some Purely Deterministic Processes

MURRAY ROSENBLATT

1. Introduction. The linear prediction problem as it arises in the case of stationary processes has attracted much attention. The problem has been studied most intensively when the prediction error is known to be positive. It is of some interest to consider a few simple situations in which the prediction error is zero, especially as a situation of this sort arises in NEUMANN's theoretical model of storm-generated ocean waves [3]. I shall, unfortunately, not be able to discuss the problem that arises in the context of NEUMANN's model in any detail.

Let x_t , $Ex_t \equiv 0$, be a weakly stationary process, that is,

$$r_{t,\tau} = Ex_t x_\tau = r_{t-\tau}$$

depends only on the time difference $t - \tau$. Our process is assumed to be real-valued. The time parameter t may either range over the real numbers or it may range over the integers. The first case is the continuous parameter case and the second is that of a discrete parameter. Examples of both continuous parameter and discrete parameter processes will be discussed. The discrete parameter processes discussed will be of much greater interest than the continuous parameter processes.

2. Preliminary Remarks. If x_t, $Ex_t \equiv 0$, is a weakly stationary process, it has a random Fourier representation of the form

$$(1) \qquad x_t = \int_{-\infty}^{\infty} e^{it\lambda}\, dZ(\lambda), \qquad E\, dZ(\lambda)\, \overline{dZ(\mu)} = \delta_{\lambda\mu}\, dF(\lambda)$$

in the continuous parameter case $-\infty < t < \infty$, and a random Fourier representation

$$(2) \qquad x_t = \int_{-\pi}^{\pi} e^{it\lambda}\, dZ(\lambda), \qquad E\, dZ(\lambda)\, \overline{dZ(\mu)} = \delta_{\lambda\mu}\, dF(\lambda)$$

in the discrete parameter case $t = \cdots, -1, 0, 1, \cdots$. Here $\delta_{\lambda\mu}$ is the Kronecker delta

$$\delta_{\lambda\mu} = \begin{cases} 1 & \text{if } \lambda = \mu \\ 0 & \text{if } \lambda \neq \mu. \end{cases}$$

Journal of Mathematics and Mechanics, Vol. 6, No. 6 (1957).

R.A. Davis et al. (eds.), *Selected Works of Murray Rosenblatt*, Selected Works in Probability and Statistics, 124
DOI 10.1007/978-1-4419-8339-8_15, © Springer Science+Business Media, LLC 2011

Note that $dZ(\lambda)$ is the random amplitude of the harmonic $e^{i\lambda}$. The covariances

$$r_t = \int_{-\infty}^{\infty} e^{it\lambda}\, dF(\lambda), \qquad -\infty < t < \infty, \tag{3}$$

in the continuous parameter case. In the discrete parameter case

$$r_t = \int_{-\pi}^{\pi} e^{it\lambda}\, dF(\lambda), \qquad t = \cdots, -1, 0, 1, \cdots. \tag{4}$$

$F(\lambda)$ is a bounded nondecreasing function and is called the spectral distribution function of the process x_t. Note that $dF(\lambda)$ is the variance of the random amplitude $dZ(\lambda)$,

$$dF(\lambda) = E\,|dZ(\lambda)|^2.$$

The case of greatest interest is that in which $F(\lambda)$ is absolutely continuous, that is,

$$F(\lambda) = \int_{-\infty}^{\lambda} f(\mu)\, d\mu$$

in the continuous parameter case, and

$$F(\lambda) = \int_{-\pi}^{\lambda} f(\mu)\, d\mu$$

in the discrete parameter case. The function $f(\lambda) \geqq 0$ is called the spectral density of the process x_t.

In the linear prediction problem, the parameter t is thought of as time. The process x_t has been observed over the time $t \leqq \tau$ and one wishes to predict s time units into the future ($t = \tau$ is thought of as the present) by a predictor that is linear in the observations. Thus, one wishes to predict $x_{\tau+s}$, $s > 0$ by a predictor $px_{\tau+s}$ which is linear in the observations x_t, $t \leqq \tau$. The desired predictor is the one that minimizes the mean square error of prediction

$$E\,|x_{\tau+s} - px_{\tau+s}|^2.$$

Note that this prediction problem is somewhat unrealistic as it assumes knowledge of the entire past of the process. Nonetheless, in certain situations the solution to this problem will lead to reasonable procedures in the more realistic problem in which only a finite amount of the past of the process is known. A process x_t is called "purely deterministic" if the mean square error of prediction is zero in this idealized problem, that is, if one can predict perfectly by linear techniques when the entire past of the process is known.

Some specific discrete parameter purely deterministic processes will be discussed. A finite amount of the past of the process x_t, $\tau - m \leqq t \leqq \tau$, is assumed known. The prediction error in these cases is positive since only a finite amount of the past is known. The rate at which the prediction error decreases to zero as $m \to \infty$ is obtained. The discussion of these discrete parameter processes is based on the helpful comments of G. Szegö and some computations he made.

3. Continuous Parameter Processes. Let x_t, $-\infty < t < \infty$, be a continuous parameter process. Assume that its spectral distribution function $F(\lambda)$ satisfies the condition

$$\int_{-\infty}^{\infty} e^{t|\lambda|}\, dF(\lambda) < \infty \tag{5}$$

for some $t > 0$. It is then clear that the process x_t is purely deterministic (see J. L. Doob [1] p. 584). Let the kth derivative of x_t, if it exists, be denoted by $x_t^{(k)}$. The derivative $x_t^{(1)}$ is the limit in the mean square of $(x_{t+h} - x_t)/h$ as $h \to 0$, that is, the random variable $\hat{x}$ such that

$$E\left|\hat{x} - \frac{x_{t+h} - x_t}{h}\right|^2 \to 0$$

as $h \to 0$. Condition (5) implies that

$$x = x_t^{(1)}$$

exists. The derivative $x_t^{(2)}$ is defined similarly in terms of $x_t^{(1)}$. By proceeding recursively in this manner one can define $x_t^{(k)}$ in terms of $x_t^{(k-1)}$. Condition (5) implies that all the derivatives $x_t^{(k)}$ exist and are given by

$$x_t^{(k)} = \int_{-\infty}^{\infty} (i\lambda)^k e^{it\lambda}\, dZ(\lambda).$$

Let

$$\int_{-\infty}^{\infty} e^{T|\lambda|}\, dF(\lambda) < \infty, \qquad T > 0. \tag{6}$$

This implies that

$$E\left|x_t - \sum_{k=0}^{\infty} \frac{x_\tau^{(k)}(t-\tau)^k}{k!}\right|^2 = 0$$

if $|t - \tau| < \frac{1}{2}T$, or that

$$x_t = \sum_{k=0}^{\infty} \frac{x_\tau^{(k)}(t-\tau)^k}{k!} \tag{7}$$

if $|t - \tau| < \frac{1}{2}T$. In particular, if $x_0^{(k)}$, $k = 0, 1, 2, \cdots$, is known,

$$x_t = \sum_{k=0}^{\infty} \frac{x_0^{(k)} t^k}{k!}$$

if $|t| < \frac{1}{2}T$. Given that we now know x_t, $|t| < \frac{1}{2}T$, by applying (7) to t values in the range $|t| < \frac{1}{2}T$ we see that x_t is determined in the range $|t| < T$. By applying formula (7) repeatedly x_t is determined for all t. Thus, *knowledge of* $x_0^{(k)}$, $k = 0, 1, 2, \cdots$, *when condition* (5) *is satisfied, determines the past and future of* x_t *completely*. This means that knowledge of the process x_t over any small interval $|t| < \epsilon$, $\epsilon > 0$, determines the full history of the process. This is unrealistic as in practice there is always some noise perturbing the process of interest so that condition (5) is not satisfied. Nonetheless, the study of this idealized situation implies that in practical work one might be able to use a truncated power series expansion with approximations of the derivatives inserted in the expansion. The point at which the power series is truncated and the type of approximation of the derivatives to be used (one might use difference quotients) would be dictated by some *a priori* knowledge available about the form of the spectrum of the process and the perturbing noise.

4. Prediction Error for Discrete Parameter Processes when the Spectral Density is Zero on an Interval. Let x_t , $t = \cdots , -1, 0, 1, \cdots$, be a discrete parameter weakly stationary process with mean value $Ex_t \equiv 0$. The process is assumed to have an absolutely continuous spectral distribution function with spectral density $f(\lambda)$. It will be convenient for us to let λ vary from zero to 2π instead of $-\pi$ to π. Further let $f(\lambda)$ be continuous and positive on the closed interval $[\frac{1}{2}\pi - \alpha, \frac{1}{2}\pi + \alpha]$ and zero elsewhere. Thus $f(\lambda)$ is zero on a set of length $2\pi - 2\alpha$. This may disturb the reader a little as such a spectral density corresponds in general to a complex-valued process since it is not symmetric about π, and real-valued processes are of interest. However, this can easily be remedied by adding $\frac{1}{2}\pi$ to λ. This convention is held to because of the convenience in the later discussion. The prediction error for such a spectral density is the same as in the corresponding problem with λ shifted by $\frac{1}{2}\pi$ units.

We would like to consider $f(\lambda)$ as a function on the unit circle $| z | = 1$ in the complex plane. This can be accomplished by setting

$$f(\lambda) = f\left(\frac{1}{i} \log (e^{i\lambda})\right) = W(e^{i\lambda}).$$

$W(e^{i\lambda})$ is now a positive continuous function on the arc $z = e^{i\lambda}$, $\frac{1}{2}\pi - \alpha \leqq \lambda \leqq \frac{1}{2}\pi + \alpha$, of the unit circle. The error in predicting x_0 , given that x_{-1} , x_{-2} , $\cdots$, x_{-n} are known, is

$$\begin{aligned} \sigma_n^2 &= \min_{c_1, \cdots, c_n} E \left| x_0 - \sum_{j=1}^{n} c_j x_{-j} \right|^2 \\ &= \min_{c_1, \cdots, c_n} \int_{-\pi}^{\pi} \left| e^{in\lambda} - \sum_{j=1}^{n} c_j e^{i(n-j)\lambda} \right|^2 f(\lambda)\, d\lambda \\ &= \min_{c_1, \cdots, c_n} \int_{\frac{1}{2}\pi - \alpha}^{\frac{1}{2}\pi + \alpha} \left| e^{in\lambda} - \sum_{j=1}^{n} c_j e^{i(n-j)\lambda} \right|^2 W(e^{i\lambda})\, d\lambda. \end{aligned}$$

This means that the prediction error σ_n^2 is equal to

$$\min_{p_n} \int_{\frac{1}{2}\pi - \alpha}^{\frac{1}{2}\pi + \alpha} |p_n(e^{i\lambda})|^2 \; W(e^{i\lambda})\, d\lambda$$

where $p_n(z)$ is a polynomial of degree n in z with the coefficient of z^n one. In our problem the process x_t is purely deterministic since

$$\int_{-\pi}^{\pi} \log f(\lambda)\, d\lambda = -\infty .$$

But we want to study the rate at which the prediction error σ_n^2 approaches zero as $n \to \infty$. In evaluating the rate at which $\sigma_n^2 \to 0$ as $n \to \infty$ we will lean heavily on some results of G. SZEGÖ that are given in Chapter 16 of his monograph *Orthogonal Polynomials*. The crucial results will be given in our discussion.

SZEGÖ considers the following problem. $W(z)$ is a positive continuous function defined on a Jordan curve C in the complex plane. C is the boundary of a simply connected region T in the complex z-plane containing $z = \infty$ as an interior point. Let L be the length of the curve C. Now let $p_n(z)$ be a polynomial of degree n in z with the coefficient of z^n one. One wishes to obtain

$$\min \frac{1}{L} \int_C |p_n(z)|^2\, W(z)\, |dz|$$

over the class of such polynomials. In our problem the curve C is the arc of the unit circle $e^{i\lambda}$, $\frac{1}{2}\pi - \alpha \leqq \lambda \leqq \frac{1}{2}\pi + \alpha$, and $W(e^{i\lambda}) = f((1/i) \log (e^{i\lambda}))$ on this curve. T consists of all the points of the plane off the arc.

Define the inner product of two functions $h(z)$, $g(z)$ where z is on C by

$$(h, g) = \frac{1}{L} \int_C h(z)\overline{g(z)} W(z)\, |dz| .$$

If the system $1, z, z^2, \cdots$ is orthogonalized by the Gramm-Schmidt orthogonalization procedure, one obtains an orthonormal set of polynomials $\phi_n(z)$,

$$\frac{1}{L} \int_C \phi_n(z)\overline{\phi_m(z)} W(z)\, |dz| = \delta_{nm}\,, \qquad n, m = 0, 1, 2, \cdots ,$$

where $\phi_n(z)$ is a polynomial of degree n with the coefficient of z^n real and positive. Let k_n be the coefficient of z^n in $\phi_n(z)$. Then

$$\frac{1}{\sigma_n^2} = \frac{L}{k_n^2} = \frac{2\alpha}{k_n^2}$$

in our case since $L = 2\alpha$.

Let

$$z = \eta(\mu) = a\mu + a_0 + a_1\mu^{-1} + a_2\mu^{-2} + \cdots$$

be the analytic function, regular and one-valued for $|\mu| > 1$, which maps $|\mu| > 1$ conformally onto the region T, preserving the point at infinity and the direction there. The function $\eta(\mu)$ is uniquely determined and a is called the capacity of the curve C. Let $\mu = N(z)$ be the inverse function of $z = \eta(\mu)$. Now let $W(z)$ be the positive continuous function on C. Then $W\{\eta(e^{-i\theta})\}$ is positive and continuous on the unit circle $\mu = e^{i\theta}$, $0 \leqq \theta \leqq 2\pi$. Let

$$D(W\eta; \mu) = \exp\left\{\frac{1}{4\pi} \int_0^{2\pi} \log\, [W\{\eta(e^{-it})\}]\, \frac{1 + \mu e^{-it}}{1 - \mu e^{-it}}\, dt\right\}.$$

On substituting into the analytic function $D(W\eta;\mu)$ the function $\mu = \{N(z)\}^{-1}$ we obtain the analytic function $\Delta(z)$. $\Delta(z)$ is regular in T including $z = \infty$. Moreover $\Delta(z) \neq 0$, $\Delta(\infty)$ is real and positive, and

$$\lim_{z\to z_0} |\Delta(z)|^2 = W(z_0)$$

where z_0 is a point on C and z approaches z_0 from within the region T. SZEGÖ reduces the minimum problem on the curve C to the problem on the full unit circle by the mapping given above (see pages 355–356, 270 of SZEGÖ [4]).

We now cite two interesting results of SZEGÖ (see pages 365–367 [4]). If $\Delta(z)$ is regular in the closed exterior of C (the closure of T)

$$\phi_n(z) \cong \left(\frac{L}{2\pi}\right)^{\frac{1}{2}} \frac{\{N'(z)\}^{\frac{1}{2}}\{N(z)\}^n}{\Delta(z)} \tag{8}$$

when z is in the closure of T. Moreover

$$\frac{1}{k_n^2} \cong \frac{2\pi}{L} \{\Delta(\infty)\}^2 a^{2n+1}.$$

Now let us see what we can derive from these general results in our problem. In our problem

$$z = \eta(\mu) = \mu \sin\frac{\alpha}{2} + i\cos^2\frac{\alpha}{2} + \frac{\cos^2\frac{\alpha}{2}\sin\frac{\alpha}{2}}{\mu + i\sin\frac{\alpha}{2}}$$

maps the region $|\mu| > 1$ onto the complex z-plane cut by the circular arc $z = e^{i\lambda}$, $\frac{1}{2}\pi - \alpha \leqq \lambda \leqq \frac{1}{2}\pi + \alpha$, $0 < \alpha < \pi$. The function

$$\mu = N(z) = \frac{-(i-z) - \sqrt{(i-z)^2 + 4i\sin^2\frac{\alpha}{2}}}{2\sin\frac{\alpha}{2}}.$$

Here

$$\begin{aligned}\Delta(\infty) = D(W\eta;\{N(\infty)\}^{-1}) &= D(W\eta;0)\\ &= \exp\left\{\frac{1}{4\pi}\int_0^{2\pi} \log\left[W\{\eta(e^{-it})\}\right] dt\right\}\\ &= \exp\left\{\frac{1}{4\pi}\int_0^{2\pi} \pi \log W\left\{\frac{e^{-2it}}{i}\frac{\sin\frac{\alpha}{2} + ie^{it}}{\sin\frac{\alpha}{2} - ie^{-it}}\right\} dt\right\}\end{aligned}$$

where

$$W(e^{i\lambda}) = f(\lambda), \qquad \tfrac{1}{2}\pi - \alpha \leqq \lambda \leqq \tfrac{1}{2}\pi + \alpha.$$

Note that the capacity of the arc in our problem is $a = \sin \frac{1}{2}\alpha$. Thus

$$\sigma_n^2 \cong 2\pi\{\Delta(\infty)\}^2 \left(\sin \frac{\alpha}{2}\right)^{2n+1}.$$

By using (8) we can also obtain information about the orthogonal polynomials $\phi_n(z)$ as $n \to \infty$ (or equivalently the predictors). However, we shall not pursue this topic any further. We have already evaluated the asymptotic behavior of the prediction error and we see that it approaches zero as fast as the $(2n + 1)^{\text{th}}$ power of a positive number less than one.

Note that the result obtained can be used to obtain upper and lower bounds for the rate at which the prediction error approaches zero when the spectral density is a continuous function that is zero on a set of positive measure and positive on a set of positive measure. It is clear that in such a case the prediction error approaches zero as fast as the $(2n + 1)^{\text{th}}$ power of a positive number less than one.

5. Prediction Error for Some Discrete Parameter Processes Having Spectra with a High Order Contact with Zero. The spectral density that NEUMANN (3) suggests as the theoretical spectrum of a storm-generated ocean surface is positive except for one point. However, the order of contact of the spectral density with zero at this one point is so high that the corresponding stationary process is purely deterministic. Thus far, no one has been able to discuss the prediction problem with such a spectral density in any detail.

G. SZEGÖ has suggested looking at a discrete parameter weakly stationary process x_t, $Ex_t \equiv 0, t = \cdots, -1, 0, 1, \cdots$, that has a spectral density positive everywhere except for one point. At this one point it has a very high contact with zero that is not exactly of the same character as that shown by NEUMANN's spectral density. Nonetheless, it is close enough to give some small insight into the asymptotic behavior of the prediction error.

The computations of this section were carried out by G. SZEGÖ. Throughout this discussion we shall refer to SZEGÖ's book *Orthogonal Polynomials* and a paper of his on POLLACZEK's polynomials.

Let the spectral density of the process x_t be

$$f(\lambda; a) = W(\cos \lambda; a)\,|\sin \lambda| = \frac{e^{(2\lambda-\pi)\varphi(\lambda)}}{\cos \lambda(\pi\varphi(\lambda))}$$

$$f(-\lambda; a) = f(\lambda; a), \qquad 0 \leqq \lambda \leqq \pi,$$

where

$$\varphi(\lambda) = \frac{\text{actn } \lambda}{2}$$

Here a is a fixed parameter. Note that

$$f(\lambda; a) \sim 2 \exp\left\{-\frac{\pi}{|\lambda|}\right\} |\sin \lambda|$$

as $\lambda \to 0$ so that $f(\lambda)$ has a very high order contact with zero at $\lambda = 0$. Let $\phi_n(z) = k_n z^n + \cdots + l_n$ be the orthogonal polynomial of degree n with respect to $f(\lambda)$, $z = e^{i\lambda}$, that is

$$\int_{-\pi}^{\pi} \phi_n(e^{i\lambda})\overline{\phi_m(e^{i\lambda})} f(\lambda)\, d\lambda = \delta_{nm} .$$

We know that the prediction error one step ahead, given observations at n successive time points, is

$$\sigma_n^2 = \frac{1}{k_n^2}.$$

The evaluation of the asymptotic behavior of k_n^{-1} as $n \to \infty$ is required.

The asymptotic behavior of k_n is evaluated by studying a closely related family of orthogonal polynomials on the interval $(-1, 1)$. Let

$$(9) \qquad g(x, z) = g(\cos\lambda, z) = \sum_{n=0}^{\infty} P_n(x; a) z^n = (1 - ze^{i\lambda})^{-\frac{1}{2}+i\varphi(\lambda)}(1 - ze^{-i\lambda})^{-\frac{1}{2}-i\varphi(\lambda)}, \qquad x = \cos\lambda,$$

be the generating function of the polynomials $P_n(x)$ of degree n. Let

$$p_n(x) = p_n x^n + \cdots = \left(n + \frac{a+1}{2}\right)^{\frac{1}{2}} P_n(x).$$

The polynomials $p_n(x)$ are orthonormal with respect to the weight function

$$W(x; a) = W(\cos\lambda; a) = \frac{e^{(2\lambda-\pi)\varphi(\lambda)}}{\cosh(\pi\varphi(\lambda))}$$

on the interval $-1 \leqq x \leqq 1$. Replace z by z/x in formula (9) and let $x \to \infty$ (see [5] p. 731–732). One then obtains

$$\sum_{n=0}^{\infty} \frac{p_n}{\left(n + \frac{a+1}{2}\right)^{\frac{1}{2}}} z^n = (1 - 2z)^{-\frac{1}{2}} \exp\left\{a \sum_{m=1}^{\infty} \frac{z^m}{m} 2^{m-1}\right\} = (1 - 2z)^{-(a+1)/2}.$$

But then

$$\frac{p_n}{\left(n + \frac{a+1}{2}\right)^{\frac{1}{2}}} = (-1)^n \begin{bmatrix} -\frac{a+1}{2} \\ n \end{bmatrix} 2^n = \frac{2^n}{\Gamma\left(\frac{a+1}{2}\right)} \frac{\Gamma\left(n + \frac{a+1}{2}\right)}{\Gamma(n+1)} \sim \frac{2^n}{\Gamma\left(\frac{a+1}{2}\right)} n^{(a-1)/2}$$

so that

$$p_n \sim \frac{2^n}{\left(\frac{a+1}{2}\right)} n^{\frac{1}{2}a}.$$

Also $p_n(1) \sim n^{\frac{1}{4}} \exp\{2\sqrt{an}\}$ (see [5] p. 733). Now let

$$q_n = q_n x^n + \cdots$$

be the orthonormal family of polynomials with weight function $(1 - x^2)W(x; a)$ on the interval $-1 \leqq x \leqq 1$. Then it is known that

$$q_n(x) = cQ_n(x)$$

where

$$Q_n(x) = \frac{1}{1-x^2}\begin{vmatrix} p_n(x) & p_{n+2}(x) \\ p_n(1) & p_{n+2}(1) \end{vmatrix} = p_{n+2}p_n(1)x^n + \cdots = \frac{p_{n+2}}{p_n}\, p_n(x) + \cdots$$

(see [4] pp. 28–30). Now

$$\int_{-1}^{1} (1-x^2)W(x)(Q_n(x))^2\, dx = \int_{-1}^{1} \begin{vmatrix} p_n(x) & p_{n+2}(x) \\ p_n(1) & p_{n+2}(1) \end{vmatrix} W(x)\,\frac{p_{n+2}}{p_n}\, p_n(1)p_n(x)\, dx$$

$$= \frac{p_{n+2}}{p_n}\, p_n(1)p_{n+2}(1)$$

so that

$$c = \sqrt{\frac{p_n}{p_{n+2}}}\,[p_n(1)p_{n+2}(1)]^{-\frac{1}{2}}$$

and

$$q_n = \sqrt{p_n p_{n+2}}\left[\frac{p_n(1)}{p_{n+2}(1)}\right]^{\frac{1}{2}} \sim 2p_n\ .$$

The orthonormal polynomials $p_n(x)$, $q_n(x)$ can be expressed in terms of the orthonormal polynomials $\phi_n(z)$, $x = \frac{1}{2}(z + z^{-1})$, $z = e^{i\lambda}$,

$$p_n(x) = (2\pi)^{-\frac{1}{2}}\left\{1 + \frac{\phi_{2n}(0)}{k_{2n}}\right\}^{-\frac{1}{2}}\{z^{-n}\phi_{2n}(z) + z^n\phi_{2n}(z^{-1})\}$$

$$q_n(x) = \left(\frac{2}{\pi}\right)^{\frac{1}{2}}\left\{1 - \frac{\phi_{2n+2}(0)}{k_{2n+2}}\right\}^{-\frac{1}{2}} \frac{z^{-n-1}\phi_{2n+2}(z) - z^{n+1}\phi_{2n+2}(z^{-1})}{z - z^{-1}}$$

(see [4] p. 287). Thus

$$p_n = (2\pi)^{-\frac{1}{2}}\left(1 + \frac{l_{2n}}{k_{2n}}\right)^{-\frac{1}{2}} 2^n(k_{2n} + l_{2n}) = (2\pi)^{-\frac{1}{2}}2^n k_{2n}^{\frac{1}{2}}(k_{2n} - l_{2n})^{\frac{1}{2}}$$

and

$$q_n = \left(\frac{2}{\pi}\right)^{\frac{1}{2}}\left(1 - \frac{l_{2n+2}}{k_{2n+2}}\right)^{-\frac{1}{2}} 2^{n+1}(k_{2n+2} - l_{2n+2})$$

$$= \left(\frac{2}{\pi}\right)^{\frac{1}{2}} 2^{n+1}k_{2n+2}^{\frac{1}{2}}(k_{2n+2} - l_{2n+2})^{\frac{1}{2}}.$$

By (11.3.6) on p. 283 [4] we know that $|l_n| < k_n$. But

$$k_{2n}^{\frac{1}{2}}(k_{2n} + l_{2n})^{\frac{1}{2}} \sim \frac{(2\pi)^{\frac{1}{2}}}{\Gamma\left(\frac{a+1}{2}\right)} n^{\frac{1}{2}a} = \gamma_n$$

$$k_{2n}^{\frac{1}{2}}(k_{2n} - l_{2n})^{\frac{1}{2}} \sim \gamma_n .$$

Thus $k_{2n} \sim \gamma_n$. Now $k_{2n-1} \sim \gamma_n$. Therefore

$$k_n \sim \frac{\pi^{\frac{1}{2}} 2^{1-a/2}}{\Gamma\left(\frac{a+1}{2}\right)} n^{\frac{1}{2}a}$$

and

$$\sigma_n^2 \sim n^{-a}.$$

6. Conclusion. It is worthwhile contrasting the asymptotic behavior of the prediction error σ_n^2 for the processes dealt with in sections 4 and 5. In section 4 the spectral density is positive and continuous except for an interval of length $2\pi - 2\alpha$, $\pi > \alpha > 0$, where the spectral density is zero. The prediction error σ_n^2 approaches zero geometrically

$$\sigma_n^2 \sim \left(\sin \frac{\alpha}{2}\right)^{2n+1}. \tag{10}$$

In section 5 the spectral density $f(\lambda; a)$ is positive away from $\lambda = 0$ and has a very high order of contact with zero at $\lambda = 0$

$$f(\lambda; a) \sim 2 \exp\left\{-\frac{\pi}{|\lambda|}\right\} |\sin \lambda|.$$

The prediction error σ_n^2 approaches zero as $n \to \infty$ but at a slower rate than (10)

$$\sigma_n^2 \sim n^{-a}.$$

This paper was prepared for the Office of Naval Research under contract Nonr 285(17) at New York University. It may be reproduced in whole or in part for any purpose of the United States Government.

References

[1] J. L. Doob, *Stochastic Processes*, John Wiley, New York, 1953.
[2] U. Grenander, On Toeplitz forms and stationary processes, *Arkiv for Matematik* 1 (1951).
[3] W. J. Pierson, Jr., *Wind Generated Gravity Waves*, Advances in Geophysics, Vol. 2, Academic Press, New York, 1955.
[4] G. Szegö, *Orthogonal Polynomials*, Amer. Math. Soc. Colloq. Pub. Vol. 23, 1939.
[5] G. Szegö, On certain special sets of orthogonal polynomials, *Proc. Amer. Math. Soc.* 1, pp. 731–737 (1950).

Functions of a Markov Process that are Markovian

M. ROSENBLATT*

Summary. In this paper we are primarily concerned with discrete time parameter Markov processes $\{X(n)\}$, $n = 0, 1, 2, \cdots$, with stationary transition mechanism. The processes $\{Y(n)\} = \{f(X(n))\}$ generated by a given many-one function f and the processes $\{X(n)\}$ with a fixed stationary transition mechanism are constructed. The processes $\{Y(n)\}$ are in a one-to-one correspondence with the possible initial distributions of $\{X(n)\}$. The object of the paper is to determine conditions under which $\{Y(n)\}$ is Markovian, whatever the initial distribution of $\{X(n)\}$. Necessary and sufficient conditions for the new processes $\{Y(n)\}$ to be Markovian are obtained under the assumption that the family of measures corresponding to the fixed transition mechanism (of $\{X(n)\}$) is dominated [4]. The conditions are expressed, of course, in terms of the function $f(\cdot)$ and the transition mechanism. Generally the processes $\{Y(n)\}$ do not have stationary transition mechanism. The conditions simplify in the case of a continuous time parameter Markov chain. Some of the discussions may at times have the flavor of those used in considering the concept of sufficiency [4]. Some aspects of the problem discussed in the paper are touched on in [2].

Basic Concepts. *Let* Λ *be a space of points* x and $\mathcal{B}$ *a Borel field of* Λ *sets.* The sets consisting of single points $\{x\}$, $x \in \Lambda$, are assumed to be in $\mathcal{B}$. A function $P(\cdot) = P_0(\cdot)$ of $B \in \mathcal{B}$ and a family of functions $P_n(\cdot, \cdot)$ $(n = 0, 1, \cdots)$ of $x \in \Lambda$, $B \in \mathcal{B}$, satisfying the following recursive conditions are assumed given:

(i) $P(B) = P_0(B)$ *is a probability measure in* $\mathcal{B}$.

(ii) $P_0(x, B)$ *for almost any fixed* x (P_0 *measure*) *is a probability measure in* $\mathcal{B}$ *and for every* B ($\in \mathcal{B}$) *is a measurable function with respect to* $\mathcal{B}$ *in* x.

(iii) *Let* $P_n(\cdot)$, $P_n(\cdot, \cdot)$ *be given where* $P_n(B)$ *is probability measure in* $\mathcal{B}$ *and where* $P_n(x, B)$ *is a probability measure in* $\mathcal{B}$ *for almost every fixed* x (P_n *measure*) *and for every* $B \in \mathcal{B}$ *is a measurable function with respect to* $\mathcal{B}$ *in* x. *Then*

$$(1) \qquad P_{n+1}(B) = \int_\Lambda P_n(dx)P_n(x, B);$$

*This research was supported by the Office of Naval Research under Contract Onr 908 (10). Reproduction in whole or in part is permitted for any purpose of the United States Government.

Journal of Mathematics and Mechanics, Vol. 8, No. 4 (1959).

R.A. Davis et al. (eds.), *Selected Works of Murray Rosenblatt*, Selected Works in Probability and Statistics, DOI 10.1007/978-1-4419-8339-8_16, © Springer Science+Business Media, LLC 2011

and $P_{n+1}(x, B)$ *is a probability measure in* $\mathcal{B}$ *for almost every* x (P_{n+1} *measure*) *and for every* $B \in \mathcal{B}$ *is a measurable function with respect to* $\mathcal{B}$ *in* x.

Define a probability measure in the space Ω of sequences $w = (x_0, x_1, \cdots)$, $x_i \in \Lambda$, as follows. Let $X_n(w)$, $n = 0, 1, 2, \cdots$, be the nth coordinate function, that is, $X_n(w) = x_n$ if w is the point $(x_0, x_1, \cdots)$. Set

$$P(X_0(w) \in B_0) = P_0(B_0), \qquad B_0 \in \mathcal{B}, \tag{2}$$

and, more generally,

$$\begin{aligned} &P(X_0(w) \in B_0, \cdots, X_n(w) \in B_n) \\ &\quad = \int_{B_0} P_0(dx_0) \int_{B_1} P_0(x_0, dx_1) \cdots \int_{B_{n-1}} P_{n-1}(x_{n-1}, B_n), \qquad B_i \in \mathcal{B}. \end{aligned} \tag{3}$$

The sequence of functions $\{X_n(w)\}$ *is called the Markov process with initial distribution* $P_0(\cdot)$ *and stochastic transition functions* (or, more loosely, transition probabilities) $P_n(\cdot, \cdot)$, $n = 0, 1, \cdots$.

Notice that a Markov process, as introduced above, may have a non-stationary stochastic transition function (not independent of the time n) and our interest is in functions $Y(n) = f(X(n))$ of Markov processes $X(n)$ with stationary transition functions. Since we may be led to processes with non-stationary transition mechanism, the more general definition above has been given.

As already remarked, the Markov process is said to have a *stationary transition function* if $P_n(x, B) = P(x, B)$ (independent of n). Consider the family of Markov processes with stationary transition function generated by the fixed transition function $P(x, B)$ and any initial distribution $P(B)$. Since we speak about any initial distribution $P(\cdot)$, a more stringent condition has to be imposed on $P(x, B)$. $P(x, B)$ must be *assumed to be a probability measure in* $\mathcal{B}$ *for every fixed* $x \in \Lambda$. The family of Markov processes introduced will be referred to as the *family of Markov processes generated by* $P(x, B)$.

Let Λ' *be a space of points* y *with a Borel field* $\mathcal{B}'$ *of* Λ' *sets*. Let $f(\cdot)$ *be a many-one function from* Λ *onto* Λ' *measurable* (*with respect to* $\mathcal{B}$). The field Λ' is also assumed to contain the sets consisting of single points $\{y\}$, $y \in \Lambda'$. Given a function $f(\cdot)$ and a stationary transition function $P(x, B)$, we are interested in finding conditions under which all the processes $Y(n) = f(X(n))$, derived from $f(\cdot)$ and the family of processes generated by $P(x, B)$, are Markovian. Notice that the sub-Borel field of $\mathcal{B}$ induced by f,

$$f^{-1}f\mathcal{B} = \{f^{-1}fB \mid B \in \mathcal{B}\} = f^{-1}\mathcal{B}',$$

is of importance. The results may be phrased in terms of $f^{-1}f\mathcal{B}$ or $\mathcal{B}'$, depending on which is more convenient.

We shall want to obtain stochastic transition functions (in the sense given above) for the derived processes $Y(n)$. This will mean that the conditional probabilities computed have to be consistent in such a way that conditional probability measures (the stochastic transition functions) can be manufactured from them. This is possible if there is a one-to-one mapping of the range space Λ'(of f) into the real line taking the sets $\mathcal{B}'$ into Borel sets (of the real line) with the inverse images of Borel sets of the family $\mathcal{B}'$ (see [1, 3, 5]). This assumption is made throughout this paper. This includes the case where Λ' is a Borel set of countably dimensional Euclidean space with $\mathcal{B}'$ the Borel field of Borel subsets of Λ'.

A Dominated Stochastic Transition Function. Consider the family of Markov processes generated by the stochastic transition function $P(x, B)$, $x \varepsilon \Lambda$, $B \varepsilon \mathcal{B}$. The transition function is a family of probability measures indexed by the points $x \varepsilon \Lambda$. The family of measures $P(x, f^{-1}B')$, $x \varepsilon \Lambda$, $B' \varepsilon \mathcal{B}'$, on $f^{-1}\mathcal{B}' = f^{-1}f\mathcal{B}$ is induced by the transition function and the function f. Necessary and sufficient conditions for the processes $Y(n) = f(X(n))$, derived from $f(\cdot)$ and the Markov processes $X(n)$ generated by $P(x, B)$, to be Markovian will be obtained under the assumption that the measures $P(x, f^{-1}B')$ are dominated by a sigma-finite measure (see [4] for a discussion of this terminology).

If the family of measures $P(x, f^{-1}B')$ is dominated, it is well known that one can find an equivalent countable subset of measures $P(x_i, f^{-1}B')$, $x_i \varepsilon \Lambda$, $i = 1, 2, \cdots$ [4]. The dominating (in fact, equivalent) measure can be taken, and shall be taken, as the probability measure

$$\mu(f^{-1}B') = \sum_{i=1}^{\infty} \alpha_i P(x_i, f^{-1}B'), \qquad \alpha_i > 0, \qquad i = 1, 2, \cdots .$$

Call a set $S \varepsilon f^{-1}\mathcal{B}'$ a single entry set if $\mu(S) > 0$ and there is a point $y \varepsilon \Lambda'$ such that $P(x, S) = 0$ if $x \not\varepsilon f^{-1}\{y\}$. Note that since $\mu(S) > 0$ (μ equivalent to the family of measures $P(x, f^{-1}B')$) there are points $x \varepsilon f^{-1}\{y\}$ such that $P(x, S) > 0$. *S is a maximal single entry set if it is a single entry set and is not contained in a single entry set of larger μ measure.* We shall identify a single entry set with all sets differing from it by at most a set of μ measure zero. *The point $y \varepsilon \mathcal{B}'$ from which one can enter a single entry set in one step with positive probability is called a single entry point.*

Lemma 1. *If the measures $P(x, f^{-1}B')$, $x \varepsilon \Lambda$, $B' \varepsilon \mathcal{B}'$, are dominated, there are at most a countable number of disjoint maximal single entry sets and corresponding single entry points.*

This follows immediately from the fact that μ is a finite measure. Lemma 1 ensures the existence of a unique (up to sets of μ measure zero) countable maximal collection of maximal single entry sets. *Let the single entry sets of this maximal collection be $S_1, S_2, \cdots$. Let $M = \Lambda - \bigcup S_i$.*

Lemma 2. *Under the hypothesis of Lemma 1, if $\mu(M) = 0$ then the processes $Y(n) = f(X(n))$ are Markov processes. The space Λ' is then the countable set of single entry points.*

Let y_i be the single entry point leading into the maximal single entry set

S_i . Since $\mu(M) = 0$ any point $x \varepsilon \Lambda$ must fall in some set $f^{-1}\{y_i\}$, $j = 1, 2, \cdots$. The space Λ' is therefore the countable set of points y_i . Every set $f^{-1}\{y_i\}$ for which $\mu(f^{-1}\{y_i\}) > 0$ is a single entry set. In verifying that the processes $Y(n)$ are Markovian it is enough to look at the probabilities

$$(4)\qquad P(Y(j) = y_{m(j)}\ ; j = 0, \cdots, n) = \int_{f^{-1}\{y_{m(0)}\}} P(dx_0) \cdot \int_{f^{-1}\{y_{m(1)}\}} P(x_0, dx_1) \cdots \int_{f^{-1}\{y_{m(n-1)}\}} P(x_{n-1}, f^{-1}\{y_{m(n)}\})$$

where $P(\cdot)$ is the initial probability distribution. Let

$$(5)\qquad \begin{aligned} P^{(1)}(x, B) &= P(x, B), \\ P^{(n+1)}(x, B) &= \int_\Lambda P^{(n)}(x, dx')P(x', B), \qquad n = 1, 2, \cdots . \end{aligned}$$

It is apparent that under the above assumptions

$$(6)\qquad \begin{aligned} P(Y(j) = y_{m(j)}\ ; j = 0, \cdots, n) &= P(Y(n) = y_{m(n)}) \\ &= \int_\Lambda P(dx_0)P^{(n)}(x_0, f^{-1}\{y_{m(n)}\}) \end{aligned}$$

if (4) is positive. Notice that we need only consider points y_i for which $P(f^{-1}\{y_i\}) > 0$. Let $Q(\cdot)$ be the initial distribution (on $\mathfrak{B}'$) and $Q_n(\cdot\ ,\cdot)$ the stochastic transition functions (of $y \varepsilon \Lambda'$ and $B' \varepsilon \mathfrak{B}'$) of $Y(n)$ that are to be constructed. Clearly $Q(\{y_i\}) = P(f^{-1}\{y_i\})$. If $P_n(f^{-1}\{y_i\}) > 0$ let

$$(7)\qquad \begin{aligned} Q_n(y_i, \{y_k\}) &= P(Y(n+1) = y_k \mid Y(n) = y_i) \\ &= \begin{cases} P_{n+1}(f^{-1}\{y_k\})/P_n(f^{-1}\{y_i\}) \\ \qquad\qquad \text{if } y_i \text{ is the single entry point for the set of } f^{-1}\{y_k\} \\ 0 \qquad \text{otherwise.} \end{cases} \end{aligned}$$

It is then clear that $Y(n)$ is a Markov process with initial distribution $Q(\cdot)$ and transition functions $Q_n(\cdot, \cdot)$ since (3) is satisfied with $Q(\cdot)$, $Q_n(\cdot, \cdot)$, Λ', $\mathfrak{B}'$ in place of $P(\cdot)$, $P_n(\cdot, \cdot)$, Λ, $\mathfrak{B}$. Since this is true for any initial distribution $P(\cdot)$ the lemma has been proved.

The Markovian property of $Y(n)$ was trivially verified when $\mu(M) = 0$. The interesting case is clearly that in which $\mu(M) > 0$. There are then single entry sets if and only if $\mu(M) < 1$ (assuming μ to be a probability measure). If S is a single entry set, we may be able to enter S from a point $x \varepsilon M$ outside the single entry sets in a finite number of steps with positive probability. If this is possible, there is an integer n and there are points $y_{m(1)}, \cdots, y_{m(n)} \varepsilon \mathfrak{B}'$ with the following properties. The sets $f^{-1}\{y_{m(j)}\}$, $j = 2, \cdots, n$, are single entry sets while $f^{-1}\{y_{m(1)}\}$ is not. Further, the point $y_{m(j)}$ is the single entry point for $f^{-1}\{y_{m(j+1)}\}$, $j = 1, \cdots, n - 1$, and the point $y_{m(n)}$ is the single entry point for S. Given any single entry set S into which entry is possible from points $x \varepsilon M$ with positive probability (that is, there is an integer $n' \geqq 1$ such that $P^{(n')}(x, S)$ for some $x \varepsilon M$), the integer n and the points $y_{m(j)}$ as specified above are uniquely determined. We shall then say that *a thread of length n consisting of the points* $y_{m(1)}, \cdots, y_{m(n)} \varepsilon \mathcal{B}'$ *enters* S.

Theorem 1. *Let the measures* $P(x, f^{-1}B')$, $x \varepsilon \Lambda$, $B' \varepsilon \mathcal{B}'$, *be dominated. A necessary and sufficient condition for the processes* $Y(n) = f(X(n)$, *derived from* $f(\cdot)$ *and the family of Markov processes* $X(n)$ *generated by* $P(x, B)$, *to be Markovian is given as follows:*

(a) *There is a function* $R(x, B)$ *of* $x \varepsilon M$, $B \varepsilon f^{-1}\mathcal{B}'$ *that is measurable in* x *with respect to* $f^{-1}\mathcal{B}'$ *for each* $B \varepsilon f^{-1}\mathcal{B}'$ *and satisfies*

$$\int_A P(x_0, dx_1)P(x_1, B) = \int_A P(x_0, dx_1)R(x_1, B) \tag{8}$$

for every $x_0 \varepsilon \Lambda$ *and* $A, B \varepsilon f^{-1}\mathcal{B}'$ *where* $A \subset M$.

(b) *If* S *is a maximal single entry set with a thread of length* n *entering it, there is a function* ${}_nR(x, B)$ *of* $x \varepsilon S$, $B \varepsilon f^{-1}\mathcal{B}'$ *that is measurable in* x *with respect to* $f^{-1}\mathcal{B}'$ *for each* $B \varepsilon f^{-1}\mathcal{B}'$ *and satisfies*

$$\int_A P^{(n+1)}(x_0, dx_{n+1})P(x_{n+1}, B) = \int_A P^{(n+1)}(x_0, dx_{n+1})\, {}_nR(x_{n+1}, B) \tag{9}$$

for all $x_0 \varepsilon \Lambda$, $A, B \varepsilon f^{-1}\mathcal{B}'$, $A \subset S$.

First consider the necessity of the conditions. If $\mu(M) = 0$ the conditions are vacuously satisfied. Assume therefore that $\mu(M) > 0$ and let the necessity of condition (a) be our concern. Take any two points $x, x' \varepsilon \Lambda$ and the corresponding measures $P(x, B)$, $P(x', B)$ on the sets $B \varepsilon f^{-1}\mathcal{B}'$, $B \subset M$. Let $M(x, x') \subset M$ be the set on which these two measures are mutually absolutely continuous in M. Set

$$P(x, A, B) = \int_A P(x, dx')P(x', B), \qquad A, B \varepsilon f^{-1}\mathcal{B}', \qquad A \subset M, \tag{10}$$

and let $P(x, df^{-1}y, B)/P(x, df^{-1}y)$ denote the Radon-Nikodym derivative of (10) with respect to $P(x, A)$ on the Borel field $f^{-1}\mathcal{B}'$ restricted to subsets of M. We want to show that

$$\frac{P(x, df^{-1}y, B)}{P(x, df^{-1}y)} = \frac{P(x', df^{-1}y, B)}{P(x', df^{-1}y)} \tag{11}$$

almost everywhere with respect to the dominating μ measure on $M(x, x')$ for each $B \varepsilon f^{-1}\mathcal{B}'$. Let us first consider points $x, x' \varepsilon \Lambda$ with $x' \not\varepsilon f^{-1} f\{x\}$. Let $P(\cdot)$ be the initial distribution with mass α, $0 < \alpha < 1$, at x and mass $\beta = 1 - \alpha$ at x'. Then

$$\begin{aligned} P(Y(2) \varepsilon fB \mid Y(1) = y) &= \frac{\alpha P(x, df^{-1}y, B) + \beta P(x', df^{-1}y, B)}{\alpha P(x, df^{-1}y) + \beta P(x', df^{-1}y)} \\ &= \frac{\alpha g(x, f^{-1}y, B) + \beta g(x', f^{-1}y, B)}{\alpha g(x, f^{-1}y) + \beta g(x', f^{-1}y)} \end{aligned} \tag{12}$$

where $g(x, f^{-1}y, B), g(x', f^{-1}y)$ are the Randon-Nikodym derivatives of $P(x, A, B)$, $P(x, A)$ respectively with respect to $\mu(A)$, $A, B \in f^{-1}\mathcal{B}$, $A \subset M$. Since the $Y(n)$ process is Markovian, whatever the initial distribution of $X(n)$, we must, in particular, have

$$P(Y(2) \in fB,\ Y(1) \in fA \mid Y(0) = fx) = \int_A P(x, dx')P(x', B) \tag{13}$$
$$= \int_{fA} P(x, df^{-1}y)P(Y(2) \in fB \mid Y(1) = y)$$

for $A, B \in f^{-1}\mathcal{B}'$, $A \subset M(x, x')$. But this cannot happen for every α unless (11) holds almost everywhere with respect to μ measure on $M(x, x')$. Thus (11) is valid for pairs of points $x, x' \in \Lambda$ with $x' \not\in f^{-1} f\{x\}$. Now consider a pair $x, x' \in \Lambda$ with $x' \in f^{-1} f\{x\}$ and the set $M(x, x')$ as defined above. Assume that $\mu(M(x, x')) > 0$ since otherwise (11) is obviously valid. Consider all points $x'' \in \Lambda$, $x'' \not\in f^{-1} f\{x\}$ and the corresponding sets $M(x'', x)$. There is an integer $n > 0$ such that for some such x'', $\mu(M(x'', x)M(x', x)) > (\frac{1}{2})^n$. Choose the smallest such integer n and a corresponding point $x'' \not\in f^{-1} f\{x\}$ and call them n_1, x_1. Assume that one has points $x_i \not\in f^{-1} f\{x\}$ and corresponding integers n_i, $i = 1, \cdots, m$, $n_1 \leqq n_2 \leqq \cdots \leqq n_m$, with

$$\mu\Big(M(x, x')M(x, x_i) - \bigcup_{j=1}^{i-1} M(x, x_j)\Big) > \frac{1}{2^{n_i}}, \qquad i = 1, \cdots, m. \tag{14}$$

If

$$\mu\Big(M(x, x') - \bigcup_{j=1}^{m} M(x, x_j)\Big) > 0 \tag{15}$$

there is an integer $n \geqq n_m$ and a corresponding point $x'' \not\in f^{-1} f\{x\}$ with

$$\mu\Big(M(x, x')M(x, x'') - \bigcup_{j=1}^{m} M(x, x_j)\Big) > \frac{1}{2^n}. \tag{16}$$

Choose the smallest such n and a corresponding point x'' and call them n_{m+1}, x_{m+1}. If this construction stops in a finite number of steps,

$$\mu\Big(M(x, x') - \bigcup_{j=1}^{m} M(x, x_j)\Big) = 0 \tag{17}$$

where m is the number of steps. Otherwise we have an infinite sequence of x_j's with corresponding $n_j \to \infty$ as $j \to \infty$. Moreover,

$$\mu\Big(M(x, x') - \bigcup_{j=1}^{\infty} M(x_j, x)\Big) = 0 \tag{18}$$

for otherwise we would have a contradiction. Now $P(x, df^{-1}y, B)/P(x, df^{-1}y)$, $P(x', df^{-1}y, B)/P(x', df^{-1}y)$ are both equal to $P(x_j, df^{-1}y, B)/P(x_j, df^{-1}y)$, $B \in f^{-1}\mathcal{B}'$, on $M(x, x')M(x_j, x)$ and hence equal to each other there with respect to μ measure. Because of (17) or (18) they are therefore equal almost everywhere on $M(x, x')$ with respect to μ measure. Assume that we have taken the equivalent measure

$$\mu(B) = \sum_{j=1}^{\infty} \alpha_j P(x_j', B), \qquad \alpha_j > 0, \qquad \sum \alpha_j = 1, \qquad B \in f^{-1}\mathcal{B}',$$

where $\{P(x_j', B)\}$ is an equivalent countable subset of the collection of measures $\{P(x, B), x \in \Lambda\}$, $B \in f^{-1}\mathcal{B}'$. Now consider the Radon-Nikodym derivative of

$$\int_A \sum \alpha_j P(x_j', dx) P(x, B)$$

with respect to $\mu(A)$, $A, B \in f^{-1}\mathcal{B}'$, and call it $R(x, B)$. Now it is clear that $R(x, B)$ is measurable in x with respect to $f^{-1}\mathcal{B}'$ for each $B \in f^{-1}\mathcal{B}'$. Further,

$$\frac{P(x_1, df^{-1}y, B)}{P(x_1, df^{-1}y)} = R(x_2, B), \qquad x_2 \in f^{-1}\{y\}, \tag{19}$$

almost everywhere with respect to $P(x, A)$ measure, $A \in f^{-1}\mathcal{B}'$, $A \subset M$. But this is condition (a).

An argument paralleling the one given above for the necessity of condition (a) suffices for the necessity of condition (b) if S is a maximal single entry set with a thread of length n entering it. The function ${}_nR(x, B)$ can be taken as the Radon-Nikodym derivative of

$$\int_A \sum_j \alpha_j P^{(n+1)}(x_j', dx) P(x, B) \tag{20}$$

with respect to $\sum_j \alpha_j P^{(n+1)}(x_j', A)$, $A, B \in f^{-1}\mathcal{B}'$, $A \subset S$.

We now consider the sufficiency of conditions (a), (b). If $\mu(M) = 0$ conditions (a), (b) are vacuously satisfied and Lemma 2 tells us that the new process $Y(n)$ is Markovian. Let us therefore consider $\mu(M) > 0$. Let $P(\cdot)$ be the initial distribution of $X(n)$. The initial distribution of $Y(n)$ is then $Q(B') = P(f^{-1}B')$, $B' \in \mathcal{B}'$. Now consider how to construct the stochastic transition functions $Q_n(\cdot, \cdot)$ of the $Y(n)$ process. If $y \in fM$ let $Q_n(y, B') = R(x, f^{-1}B')$, $n \geqq 1$, where $x \in f^{-1}\{y\}$ and B' is any set in $\mathcal{B}'$. Set $Q_0(y, B')$ equal to the Radon-Nikodym derivative of $P(Y(0) \in A', Y(1) \in B')$ with respect to $P(Y(0) \in A')$, $A', B' \in \mathcal{B}'$, $y \in \Lambda'$. If $y \notin fM$, $f^{-1}\{y\}$ is contained in some single entry set S. If there is no thread of length $m < n$ entering S let

$$Q_n(y, B') = \int_{f^{-1}\{y\}} P_n(dx) P(x, f^{-1}B') / P_n(f^{-1}\{y\}), \qquad B' \in \mathcal{B}', \tag{21}$$

if $P_n(f^{-1}\{y\}) > 0$. If there is a thread of length $m < n$ entering S, let

$$Q_n(y, B') = {}_mR(x, f^{-1}B'), \qquad B' \in \mathcal{B}', \tag{22}$$

where $x \in f^{-1}\{y\}$. The $Q_n(\cdot, \cdot)$'s have been set equal to conditional probabilities. Since we have assumed that there is a one-to-one mapping of the range space

Λ' into the real line taking the sets of $\mathcal{B}'$ into Borel sets of the real line and conversely, the $Q_n(\cdot, \cdot)$ can be chosen so that they are conditional probability measures. The process $Y(n)$ must be shown to be Markovian with the specification of the stochastic transition functions $Q_n(\cdot, \cdot)$ as given above, namely

$$P(Y(0) \in B_0', \cdots, Y(n) \in B_n') = \int_{B_0'} Q(dy_0) \cdot \int_{B_1'} Q_0(y_0, dy_1) \cdots \int_{B_{n-1}'} Q_{n-1}(y_{n-1}, B_n'), \qquad B_j' \in \mathcal{B}'. \tag{23}$$

It is enough to show that relation (23) holds for sets of the form

$$(Y(j) \in fB_j ; j = 0, \cdots, n), \qquad B_j \in \mathcal{B}, \tag{24}$$

where for each j separately either $B_j \subset M$ or else B_j is a single entry set, since every set $(Y(j) \in fB_j ; j = 0, \cdots, n)$, $B_j \in \mathcal{B}$, is the sum of a countable number of sets of the form (24). However, it is apparent from the construction of the stochastic transition functions $Q_n(\cdot, \cdot)$ that (23) holds for sets of the form (24). The proof of the sufficiency is complete.

Probably the most interesting context covered in Theorem 1 is that in which there are no single entry sets. Now the original process $X(n)$ has stationary transition mechanism. Suppose we insist that the new process $Y(n)$ (assuming it to be Markovian) also have stationary transition mechanism (which might conceivably depend on the initial distribution $P(\cdot)$ of the $X(n)$ process). Condition (8) simplifies since stationarity of the transition mechanism implies that the Radon-Nikodym derivative of

$$\int_{f^{-1}A'} P(dx)P(x, f^{-1}B') \tag{25}$$

with respect to $P(f^{-1}A')$, A', $B' \in \mathcal{B}'$, whatever the initial distribution $P(\cdot)$, must be equal to $Q(y, B')$ almost everywhere with respect to the μ measure. Corollary 1 follows.

Corollary 1. *Assume that the hypotheses of Theorem 1 are satisfied and that there are no single entry sets. A necessary and sufficient condition for the processes $Y(n)$, derived from $f(\cdot)$ and the processes $X(n)$, to be Markovian and have stationary transition mechanism ($Q_n(\cdot, \cdot) = Q(\cdot, \cdot)$) is given by*

$$P(x, f^{-1}B') \quad \textit{measurable in } x \tag{26}$$

with respect to $\mathcal{B}'$ for every $B' \in \mathcal{B}'$.

However, condition (26) of Corollary 1 is not generally valid if one allows single entry sets. The following example illustrates this. Consider a Markov chain $X(n)$ (Λ a countable set which is, for convenience, labeled by the integers) with transition probability matrix $P = (p_{ij})$. The function f collapses the states (elements of Λ) into three distinct points y_1, y_2, y_3 ($\Lambda' = \{y_1, y_2, y_3\}$)

in such a way that at least two $X(n)$ states are mapped into each y point. Let

$$p_{ij} = \begin{cases} c\,\delta_{ij} & \text{if } i, j \in f^{-1}\{y_1\}, \quad 0 < c < 1, \\ 0 & \text{if } i \notin f^{-1}\{y_1\}, \quad j \in f^{-1}\{y_1\} \end{cases} \tag{27}$$

while

$$P(i, f^{-1}\{y_\beta\}) = \sum_{j \in f^{-1}\{y_\beta\}} p_{ij} = Q(y_\alpha, \{y_\beta\}) \tag{28}$$

for $i \in f^{-1}\{y_\alpha\}$, $\alpha, \beta = 2, 3$. The remaining p_{ij} ($i \in f^{-1}\{y_1\}$, $j \notin f^{-1}\{y_1\}$) are arbitrary aside from the non-negativity condition and

$$P(i, \Lambda - f^{-1}\{y_1\}) = 1 - c, \qquad i \in f^{-1}\{y_1\}. \tag{29}$$

The processes $Y(n) = f(X(n))$ are Markovian whatever the initial distribution $P(\{i\}) = p_i$ of $X(n)$. Further, the transition probability matrix of $Y(n)$ is stationary, whatever the initial distribution p_i of $X(n)$. Of course, it does depend on the initial distribution of $X(n)$ since

$$\begin{aligned} &P(X(n+1) \in f^{-1}\{y_2\} \mid X(n) \in f^{-1}\{y_1\}) \\ &\qquad = \sum_{i \in f^{-1}\{y_1\}} p_i P(i, f^{-1}\{y_2\}) \Big/ \sum_{i \in f^{-1}\{y_1\}} p_i\,. \end{aligned} \tag{30}$$

If we insist that the transition mechanism of $Y(n)$ not only be stationary, but also independent of the initial distribution $P(\cdot)$ of $X(n)$, we are again led to (26) as a necessary and sufficient condition. Condition (26) arises in some of the results of [2, 6, 7, 8, 9].

Notice that condition (8) implies that the transition mechanism of $Y(n)$ is almost stationary. It is, in fact, then stationary for $n \geqq 1$. There are simple conditions that imply (8), for example, condition (26). The assumption that the measures $P(x, B)$ are of the form

$$P(x, B) = \int_B f(x, z)\nu(dz), \tag{31}$$

where ν is a probability measure on $\mathfrak{B}$ and $f(x, z)$ is measurable in (x, z) on the product space $\Lambda \times \Lambda$ with respect to the Borel field $\mathfrak{B} \times f^{-1}\mathfrak{B}'$, is enough to ensure (8). It is reasonable to ask whether (26) or (31) or trivial variants of them are the only conditions that imply (8). A simple discussion indicates that this is not the case generally.

Suppose that the function $P(x, B)$ is given as a function in $x \in \Lambda$ and $B \in f^{-1}\mathfrak{B}'$. It is measurable for each fixed $B \in f^{-1}\mathfrak{B}'$ in x with respect to $\mathfrak{B}$ and a probability measure for each fixed $x \in \Lambda$ on $f^{-1}\mathfrak{B}'$. We wish to find the family of stochastic transition functions $P(\cdot, \cdot)$ of $x \in \Lambda$ and $B \in \mathfrak{B}$ which are consistent with $P(x, B)$, $B \in f^{-1}\mathfrak{B}'$, that is, they agree with the latter on the restricted field $f^{-1}\mathfrak{B}'$ and, further, condition (8) is satisfied. Assume that the given family of measures $P(x, B)$, $x \in \Lambda$, $B \in f^{-1}\mathfrak{B}'$, is equivalent to the probability measure $\bar{\mu}$ on $f^{-1}\mathfrak{B}'$. Call μ a refinement of $\bar{\mu}$ if it is a probability measure on $\mathfrak{B}$ and agrees with $\bar{\mu}$

on $f^{-1}\mathcal{B}'$. The following corollary characterizes the family of stochastic transition functions with corresponding sets of measures dominated by a probability measure μ and consistent with $P(x, B)$, $B \in f^{-1}\mathcal{B}'$.

Corollary 2. *Let $P(x, B)$, $x \in \Lambda$, $B \in f^{-1}\mathcal{B}'$, be given and be dominated by the probability measure $\bar{\mu}$ on $f^{-1}\mathcal{B}'$. Then $P(\cdot, \cdot)$ is a dominated* (by a probability measure) *stochastic transition function consistent* (so that (8) is satisfied) *with $P(x, B)$, $B \in f^{-1}\mathcal{B}'$, if and only if it is a transition function that can be expressed in the form*

$$P(x, B) = \int_B (1 + a(x, z))g(x, z)\mu(dz), \qquad x \in \Lambda, \qquad B \in \mathcal{B}, \tag{32}$$

where μ is a refinement of $\bar{\mu}$, where $g(x, z) \geqq 0$ is a function of $\Lambda \times \Lambda$ measurable with respect to $\mathcal{B} \times f^{-1}\mathcal{B}'$ and $a(x, z)$ is a function on $\Lambda \times \Lambda$ measurable with respect to $\mathcal{B} \times \mathcal{B}$ and satisfying

$$a(x, z) \geqq -1, \qquad x, z \in \Lambda, \tag{33}$$

$$\int_{f^{-1}A'} a(x, z)g(x, z)\mu(dz) = 0, \tag{34}$$

$$\int_{f^{-1}A'} a(x, z)g(x, z)\mu(dz)P(z, f^{-1}B') = 0 \tag{35}$$

for all $x \in \Lambda$ and A', $B' \in \mathcal{B}'$.

Suppose $P(\cdot, \cdot)$ is a dominated stochastic transition function consistent with $P(x, B)$, $B \in f^{-1}\mathcal{B}'$. Let μ be a probability measure on $\mathcal{B}$ equivalent to the family of measures generated by $P(\cdot, \cdot)$. Since $P(\cdot, \cdot)$ agrees with $P(x, B)$, $B \in f^{-1}\mathcal{B}'$, on $B \in f^{-1}\mathcal{B}'$, μ can be taken to be a refinement of $\bar{\mu}$. Let $g(x, z)$ be the Radon-Nikodym derivative of $P(x, B)$, $B \in f^{-1}\mathcal{B}'$, with respect to $\bar{\mu}$. The function $g(x, z)$ is defined on $\Lambda \times \Lambda$ and measurable with respect to $\mathcal{B} \times f^{-1}\mathcal{B}'$. Let $h(x, z)$ be the Radon-Nikodym derivative of $P(x, B)$, $B \in \mathcal{B}$, with respect to

$$\int_B g(x, z)\mu(dz), \qquad B \in \mathcal{B}. \tag{36}$$

Set $a(x, z) = h(x, z) - 1$. Then $P(\cdot, \cdot)$ has the representation (32) and $g(x, z) \geqq 0$. We only have to verify conditions (33–35). But $h(x, z) = 1 + a(x, z) \geqq 0$. Condition (34) follows from

$$\int_B h(x, z)g(x, z)\mu(dz) = \int_B g(x, z)\mu(dz), \qquad B \in f^{-1}\mathcal{B}', \tag{37}$$

and (35) follows from (8). On the other hand, if $P(\cdot, \cdot)$ has a representation of the form (32) it is clearly a stochastic transition function consistent with $P(x, B)$, $x \in \Lambda$, $B \in f^{-1}\mathcal{B}'$. Actually (34) is just the special case of (35) obtained by taking $B' = \Lambda'$. It was written out because we made use of it to get (35).

Corollary 2 provides us with a method of constructing transition functions consistent with a specification of $P(x, B)$, $x \varepsilon \Lambda$, $B \varepsilon f^{-1}\mathcal{B}'$. The specific transition function $P(\cdot, \cdot)$ obtained will depend on the choice of the refinement μ of $\bar{\mu}$ and the function $a(x, z)$. There is always one possible choice of $a(x, z)$, namely $a(x, z) \equiv 0$. Generally, there will be many other nontrivial choices of $a(x, z)$ but this will, of course, depend on how strong the restriction (35) is (see the example in section 3 of [2]). The condition $a(x, z) \equiv 0$ amounts to the repreesentation (31).

The statement of Theorem 1 in the special case of a countable state space is given in [10]. It is clear that the notion of a maximal single entry set does not make sense if the condition that the measures $P(x, f^{-1}B')$, $x \varepsilon \Lambda$, $B' \varepsilon \mathcal{B}'$, be dominated is relaxed. The question of determining when a function $Y(n) = f(X(n))$ of a stationary Markov process $X(n)$ (not any initial distribution) is Markovian was brought up in [2]. It is clear that (8) provides a sufficient condition broader than (26).

Continuous Parameter Markov Chains. Now consider what happens in the case of a continuous parameter Markov chain $X(t)$ with stationary transition mechanism, $0 \leqq t < \infty$. Let the state space Λ be labeled by the positive integers. The chain is assumed to have a decent transition probability function

$$\begin{aligned} P(t) &= (p_{ij}(t); i, j = 1, 2, \cdots) \\ &= (P(X(t + \tau) = j \mid X(\tau) = i)), \qquad 0 \leqq t, \qquad \tau < \infty, \end{aligned}$$

continuous in t, $0 \leqq t < \infty$, with

$$\lim_{t \downarrow 0} P(t) = I. \tag{38}$$

Let $S_\alpha = f^{-1}\{y_\alpha\}$, $\alpha = 1, 2, \cdots$, be the disjoint sets of states generated by the function f. Necessary and sufficient conditions for $Y(t) = f(X(t))$ to be Markovian, whatever the initial distribution $p_i = P(X(0) = i)$, $i = 1, 2, 3, \cdots$, of $X(t)$, are obtained. We shall say that S_α is (is not) a single entry set at time $t > 0$ according as to whether it is (is not) a single entry set with respect to the matrix $P(t)$ and its powers (in the sense of the previous section). If S_α is not a single entry set at time $t > 0$, it is not a single entry set for all $\tau \geqq t$. Suppose S_α is a single entry set at time $t > 0$. Then S_α can only be entered from itself in time $\tau \leqq t$ because of (38). It must then be a single entry set for all $\tau \geqq t$. For if it were not, there would be a maximal $T < \infty$ such that for $t < T$ S_α is a single entry set and for $t > T$ S_α is not a single entry set. Then for every $\epsilon > 0$ there is an $i \not\varepsilon S_\alpha$ such that

$$p_{i, S_\alpha}(T + \epsilon) = \sum_{j \varepsilon S_\alpha} p_{ij}(T + \epsilon) > 0. \tag{39}$$

But

$$p_{i, S_\alpha}(T + \epsilon) = \sum_j p_{ij}(T - \epsilon) p_{j, S_\alpha}(2\epsilon) = \sum_{j \varepsilon S_\alpha} p_{ij}(T - \epsilon) p_{j, S_\alpha}(2\epsilon) = 0 \tag{40}$$

if $\epsilon < \frac{1}{2}T$, a contradiction. Thus there are only two possibilities. Either S_α is a single entry set for all t or else it is not a single entry set for all t, $\alpha = 1, 2, \cdots$.

If S_α is a single entry set for all t

$$p_{i,S_\alpha}(t) \equiv 0 \tag{41}$$

for all $i \not\in S_\alpha$. Suppose S_α is not a single entry set for all t. Then

$$\sum_{j \in S_\beta} p_{ij}(t) p_{j,S_\beta}(\tau) / p_{i,S_\alpha}(t) = Q_{S_\alpha,S_\beta}(t, \tau), \tag{42}$$

$t, \tau > 0$, for all $i \in S_\alpha$ and all β. As $t \downarrow 0$ the left side of (42) approaches $p_{i,S_\beta}(\tau)$. But the right side of (42) depends on i only through S_α. Thus

$$p_{i,S_\beta}(\tau) = Q_{S_\alpha,S_\beta}(\tau) \tag{43}$$

for all $i \in S_\alpha$ and we have the following result.

Theorem 2. *Let $X(t)$, $0 \leqq t < \infty$, be a continuous parameter Markov chain with stationary transition probability function $P(t) = (p_{ij}(t);\ i, j = 1, 2, \cdots)$. Let $P(t)$ be continuous in t, $0 \leqq t < \infty$, with*

$$\lim_{t \downarrow 0} P(t) = I. \tag{44}$$

The process $Y(t) = f(X(t))$ will be Markovian, whatever the initial distribution of $X(t)$, if and only if each set S_α of states generated by f separately satisfies one or the other of the following two conditions for all t:

$$\text{(i)} \quad p_{i,S_\alpha}(t) \equiv 0 \quad \textit{for all} \quad i \not\in S_\alpha, \tag{45}$$

$$\text{(ii)} \quad p_{i,S_\beta}(t) = Q_{S_\alpha,S_\beta}(t) \quad \textit{for all} \quad i \in S_\alpha \quad \textit{and all} \quad \beta. \tag{46}$$

It would be interesting to find out under what conditions the results of Theorem 2 hold in the case of a more general state space for the processes $X(t)$.

References

[1] R. R. Bahadur, Sufficiency and statistical decision functions, *Ann. Math. Stat.*, **25** (1954), pp. 423–462.

[2] C. J. Burke & M. Rosenblatt, A Markovian function of a Markov chain, *Ann. Math. Stat.*, **29** (1958), pp. 1112–1122.

[3] J. L. Doob, *Stochastic Processes*, New York (1952).

[4] P. R. Halmos & L. J. Savage, Application of the Radon-Nikodym theorem to the theory of sufficient statistics, *Ann. Math. Stat.*, **20** (1949), pp. 225–241.

[5] M. Loeve, *Probability Theory*, New York (1955).

[6] B. Rankin, *The concept of enchainment—a relation between stochastic processes*, unpublished (1955).

[7] D. Rosenblatt, On aggregation and consolidation in linear systems, to be published in the *Naval Research Logistics Quarterly*.

[8] D. Rosenblatt, *On aggregation and consolidation in linear systems*, Technical Report C, prepared under Contract Nonr-1180 (00) (NR-047-012) 1956.

[9] D. Rosenblatt, Abstracts of "On aggregation and consolidation in finite substochastic systems I, II, III, IV," *Ann. Math. Stat.*, **28** (1957), pp. 1060–1061.

[10] M. Rosenblatt, Abstract of "Functions of Markov Chains," *Ann. Math. Stat.*, **29** (1958), p. 1291.

Indiana University
Bloomington, Indiana

Stationary Processes as Shifts of Functions of Independent Random Variables

M. ROSENBLATT*

1. Summary. Let x_n, $n = 0, \pm 1, \pm 2, \cdots$, be a strictly stationary process. Two closely related problems are posed with respect to the structure of strictly stationary processes. In the first problem we ask whether one can construct a random variable $\xi_n = g(x_n, x_{n-1}, \cdots)$, a function of $x_n, x_{n-1}, \cdots$, that is independent of the past, that is, independent of $x_{n-1}, x_{n-2}, \cdots$. Such a sequence of random variables $\{\xi_n\}$ is a sequence of independent and identically distributed random variables. Further, given such a construction, is x_n a function of ξ_n, $\xi_{n-1}, \cdots$. Necessary and sufficient conditions for such a representation are obtained in the case where x_n is a finite state Markov chain with the positive transition probabilities in any row of the transition probability matrix $P = (p_{ij})$ of x_n distinct (Section 3). Such a representation is comparatively rare for a finite state Markov chain. In the second problem, the assumption that the independent and identically distributed ξ_n's be functions of x_n, $x_{n-1}, \cdots$ is removed. We ask whether for some such family $\{\xi_n\}$ there is a process $\{y_n\}$, $y_n = g(\xi_n, \xi_{n-1}, \cdots)$, with the same probability structure as $\{x_n\}$. This is shown to be the case for every ergodic finite state Markov chain with nonperiodic states (Section 4). Sufficient conditions for such representations in the case of a general strictly stationary process are obtained in Section 5.

2. Introduction. There are problems on the representation of strictly stationary processes that are motivated by certain results in the linear prediction problem and remarks of P. LÉVY.

In the linear prediction problem one considers a weakly stationary process x_n, $n = 0, \pm 1, \cdots$, that is, a process x_n with

$$\mathrm{E}x_n^2 < \infty, \qquad \mathrm{E}x_n x_m = r_{n-m}, \qquad \mathrm{E}x_n \equiv 0. \tag{1}$$

Under certain conditions on the spectrum of the process (the spectral distribution function $F(\lambda)$ absolutely continuous with log $F'(\lambda)$ integrable), the process x_n can be represented as a one-sided weighted moving average of orthonormal random variables ξ_n defined on the present and past of the process. More explicitly,

*This research was supported by the Office of Naval Research. Reproduction in whole or in part is permitted for any purpose of the United States Government.

Journal of Mathematics and Mechanics, Vol. 8, No. 5 (1959).

R.A. Davis et al. (eds.), *Selected Works of Murray Rosenblatt*, Selected Works in Probability and Statistics, 146
DOI 10.1007/978-1-4419-8339-8_17, © Springer Science+Business Media, LLC 2011

ξ_n is an element of the closed linear manifold (closure in mean square) generated by x_m, $m \leqq n$, with

$$E\xi_n = 0, \qquad E\xi_n\xi_m = \delta_{n,m} \tag{2}$$

and x_n has a representation

$$x_n = \sum_{k=0}^{\infty} a_k\xi_{n-k}, \qquad \sum a_k^2 < \infty \tag{3}$$

(see [1]). This suggests a corresponding representation problem with orthogonality replaced by independence and linear relationship by functional dependence.

Before going on to present the problem, it should be noted that any statement made in the paper is to be understood modulo sets of measure zero, where the measure is the probability measure implicit in the problem at hand. Thus if $g(\cdot)$ is said to be measurable with respect to a Borel field of sets $\mathcal{B}$ we understand that it is equal almost everywhere to a function strictly measurable with respect to $\mathcal{B}$. Further, two Borel fields $\mathcal{B}$, $\mathcal{B}'$ will be said to be the same if given any set $B \in \mathcal{B}$ there is a set $B' \in \mathcal{B}'$ with the measure of the symmetric difference of B and B' zero, and if $\mathcal{B}'$ has the same property relative to $\mathcal{B}$. Clearly the two Borel fields are not the same strictly, but they are the same modulo sets of measure zero and that is of interest to us.

Let x_n, $n = 0, \pm 1, \cdots$, be a strictly stationary process with T the corresponding one-step shift operator. $\mathcal{B}_n$ is the Borel field of sets generated by x_n, x_{n-1}, $\cdots$.

Problem 1. (a) *When can one find a random variable ξ_0 measurable with respect to $\mathcal{B}_0$, independent of x_{-1}, x_{-2}, $\cdots$, and such that $\mathcal{B}_0 = \mathcal{B}_{-1} \times \mathcal{A}_0$ (product field) where $\mathcal{A}_0$ is the Borel field generated by ξ_0?*

(b) If (a) is satisfied, the random variables $\xi_n = T^n\xi_0$, $n = 0, \pm 1, \cdots$, are a sequence of independent, identically distributed random variables. *Let $\mathcal{A}_n$ be the Borel field generated by ξ_n. Is x_n measurable with respect to $\cdots \times \mathcal{A}_{n-1} \times \mathcal{A}_n$?*

A second problem is suggested by some remarks by P. LÉVY [5]. Let x_1, x_2, $\cdots$ be a sequence of random variables. Under rather broad conditions P. LÉVY has shown that one can find a sequence of independent, uniformly distributed random variables ξ_1, ξ_2, $\cdots$ and Borel functions $y_1 = g_1(\xi_1)$, $y_2 = g_2(\xi_2, \xi_1)$, $\cdots$ of ξ_1, ξ_2, $\cdots$ such that y_1, y_2, $\cdots$ have the same probability structure as x_1, x_2, $\cdots$.

Problem 2. Let x_n, $n = 0, \pm 1, \cdots$ be a strictly stationary process. *Can one find a sequence of independent, uniformly distributed* (*on* [0, 1]) *random variables ξ_n, $n = 0, \pm 1, \cdots$, and a Borel function $g(\xi) = g(\xi_0, \xi_{-1}, \cdots)$, $\xi = (\cdots, \xi_{-1}, \xi_0, \xi_1, \cdots)$, such that*

$$y_n = g(T^n\xi), \qquad n = 0, \pm 1, \pm 2, \cdots, \tag{4}$$

has the same probability structure as x_n, $n = 0, \pm 1, \cdots$? Here T is the shift operator, that is,

$$T\xi = T(\cdots, \xi_{-1}, \xi_0, \xi_1, \cdots) = (\cdots, \xi_0, \xi_1, \xi_2, \cdots). \tag{5}$$

Notice that g is a one-sided function of the sequence of random variables ξ_n.

If problem 1 is soluble, problem 2 is soluble. The converse is, however, not true. Notice that the solutions to these problems have nothing to do with the existence or nonexistence of moments of the process considered. They are problems about the Borel fields $\mathcal{B}_n$ generated by the process and the probability measure of the process on these fields. *Let* $\bigcap \mathcal{B}_n$ *be the Borel field of sets common to all the Borel fields* $\mathcal{B}_n$, $n = 0, \pm 1, \cdots$.

Lemma 1. *If problem 2 is soluble, the only functions measurable with respect to* $\bigcap \mathcal{B}_n$ *are the constant functions.*

Consider any random variable z, $E|z| < \infty$, measurable with respect to the Borel field generated by a stationary process $\{y_n\}$ for which problem 2 is soluble. We can assume that the random variables y_n are bounded. Just consider any monotone one-one function $h(\cdot)$ mapping $(-\infty, \infty)$ into $(0, 1)$. The new process $\{\bar{y}_n\}$, $\bar{y}_n = h(y_n)$, is bounded and has the same Borel fields $\mathcal{B}_n$ as $\{y_n\}$.

Since the random variables y_n are bounded, given any fixed $\epsilon > 0$ we can find a polynomial $p(y_n, \cdots, y_{n-k})$ in an appropriately large but finite number of random variables $y_n, \cdots, y_{n-k}$ such that

$$E|z - p| < \epsilon. \tag{6}$$

Since the y_n's are one-sided functions of the ξ_n's (independent and uniformly distributed $[0, 1]$), there is a Borel function $q(\xi_n, \cdots, \xi_{n-j})$ in $\xi_n, \cdots, \xi_{n-j}$ for j sufficiently large such that

$$E|q - p| < \epsilon. \tag{7}$$

But then

$$E|z - q| < 2\epsilon. \tag{8}$$

Now

$$E(q|\cap\mathcal{B}_n) = E(q|\mathcal{B}_m) = E(q|\xi_m, \xi_{m-1}, \cdots) = Eq \tag{9}$$

for $m < n - j$. But $E(z \mid \mathcal{B}_m) \to E(z \mid \bigcap \mathcal{B}_n)$ as $m \to -\infty$ in absolute mean by a martingale convergence theorem [1], [6]. Further,

$$E|E(z|\mathcal{B}_m) - E(q|\mathcal{B}_m)| \leqq E|z - q| < 2\epsilon. \tag{10}$$

This implies that

$$E|E(z|\cap\mathcal{B}_n) - E(z)| < 4\epsilon. \tag{11}$$

Since this is true for every $\epsilon > 0$, $E(z \mid \bigcap \mathcal{B}_n) = E(z)$ with probability one. Now consider any function w measurable with respect to $\bigcap \mathcal{B}_n$. We can always assume that $E|w| < \infty$, since otherwise one could find a one-one monotone function $h(\cdot)$ such that $z = h(w)$ has finite first moment. If z is a constant almost everywhere so is w. But clearly z is a constant almost everywhere since $E(z| \cap \mathcal{B}_n) = Ez$. The proof is complete.

It is an open question as to whether the converse of Lemma 1 is true. We shall look at this question in a few special but interesting cases. The necessary condition for the representation of problem 2 obtained in Lemma 1 is, of course, a necessary condition for the representation of problem 1. *Let* $\{x_n\}$ *be a strictly stationary process with the only functions measurable with respect to* $\cap \mathcal{B}_n$ *the constant functions.* It is natural to *call such processes purely nondeterministic processes.*

From this point on *assume that the conditional probability distribution function of the process* $\{x_n\}$,

$$(12) \qquad F(a_n|a_{n-1}, a_{n-2}, \cdots) = P[x_n \leqq a_n|x_{n-1} = a_{n-1}, x_{n-2} = a_{n-2}, \cdots],$$

is a Borel measurable function of the real variables $a_n, a_{n-1}, \cdots$. Given this weak assumption, we shall look for conditions on the conditional probability distribution that imply that part (a) of problem 1 is satisfied. The following convenient notion of equivalence of two distribution functions F, G is introduced. *F and G are equivalent if there is a one-one function mapping the jumps of F onto the jumps of G and preserving the size of the jumps.* Notice that any two continuous distribution functions F, G are equivalent according to this definition.

Lemma 2. *There is a random variable ξ_n satisfying part* (a) *of problem* 1 *if and only if the distribution functions $F(a_n \mid a_{n-1}, a_{n-2}, \cdots)$ (as functions of a_n) are equivalent for almost all $a_{n-1}, a_{n-2}, \cdots$ (with respect to the probability measure of the process $\{x_n\}$).*

Since ξ_n is independent of $x_{n-1}, x_{n-2}, \cdots$ the conditional probability distribution of ξ_n given $x_{n-1}, x_{n-2}, \cdots$ is independent of $x_{n-1}, x_{n-2}, \cdots$. Moreover, since $\mathcal{B}_n = \mathcal{B}_{n-1} \times \mathcal{A}_n$ the conditional probability distribution of ξ_n given $x_{n-1}, x_{n-2}, \cdots$ must with probability one be equivalent to the conditional probability distribution of x_n given $x_{n-1}, x_{n-2}, \cdots$. The probability distribution functions $F(a_n \mid a_{n-1}, \cdots)$ must therefore be equivalent for almost all a_{n-1}, $a_{n-2}, \cdots$.

Conversely, assuming the equivalence of the distribution functions $F(a_n \mid a_{n-1}, a_{n-2}, \cdots)$ for almost all $a_{n-1}, a_{n-2}, \cdots$ we must show how to construct a random variable ξ_n with the desired properties. If the distribution functions $F(a_n \mid a_{n-1}, \cdots)$ are continuous for almost all $a_{n-1}, a_{n-2}, \cdots$ there is no difficulty. Simply set

$$(13) \qquad \xi_n = F(x_n|x_{n-1}, x_{n-2}, \cdots),$$

and an argument of P. LÉVY [5] indicates that ξ_n is uniformly distributed [0, 1] and independent of $x_{n-1}, x_{n-2}, \cdots$. Further,

$$(14) \qquad x_n = F^{-1}(\xi_n|x_{n-1}, x_{n-2}, \cdots)$$

with probability one. Let us therefore consider what happens if the distribution functions $F(a_n \mid a_{n-1}, a_{n-2}, \cdots)$ have with probability one a discrete part. Let the jumps in the discrete part, ordered in terms of magnitude, be $p_1 \geqq p_2 \geqq \cdots$. For convenience assume there is strict inequality among the p_i. An obvious elaboration of the argument given in the case of strict inequality holds in the more general context with equality allowed. The magnitudes of the jumps $p_1, p_2, \cdots$ with probability one do not depend on $a_{n-1}, a_{n-2}, \cdots$ because of the equivalence of the conditional probability distributions. Let A_j be the set of points

$$A_j = \{(a_n, a_{n-1}, \cdots) | \Delta F(a_n | a_{n-1}, \cdots) = p_j\} \tag{15}$$

where ΔF denotes the magnitude of the jump (if there is any) of the conditional distribution function at the argument indicated. We associate a set of integers $S((a_n, a_{n-1}, \cdots))$ with every point $(a_n, a_{n-1}, \cdots) \not\in \bigcup A_j$. The integer $j \in S((a_n, a_{n-1}, \cdots))$ if there is a set A_j and a point $(a_n', a_{n-1}, a_{n-2}, \cdots) \in A_j$ with $a_n' < a_n$. Of course, the set $S((a_n, a_{n-1}, \cdots))$ may be the empty set. Set

$$\xi_n = \begin{cases} j & \text{if} \quad (x_n, x_{n-1}, \cdots) \in A_j \\ F(x_n | x_{n-1}, \cdots) - \sum\limits_{j \in S((x_n, x_{n-1}, \cdots))} p_j & \text{if} \quad (x_n, x_{n-1}, \cdots) \not\in \bigcup A_j . \end{cases} \tag{16}$$

Then ξ_n is independent of $x_{n-1}, x_{n-2}, \cdots$. The random variable ξ_n is uniformly distributed on $(0, 1 - \sum p_j)$ with density function one and assumes the integral value j with probability p_j. Further, the construction indicates that x_n is given with probability one by the knowledge of $\xi_n, x_{n-1}, x_{n-2}, \cdots$.

There is a case in which part (a) of problem 1 is satisfied trivially and yet there is no hope of having part (b) satisfied. Suppose $F(a_n \mid a_{n-1}, \cdots)$ is for almost all $a_{n-1}, a_{n-2}, \cdots$ a pure jump function with a jump of size one. Then x_n is determined with probability one given $x_{n-1}, x_{n-2}, \cdots$ so that $\mathcal{B}_n = \mathcal{B}_{n-1} = \cdots$ for all n. It is natural to *call such a process a purely deterministic process*. Such a process will satisfy the necessary condition of Lemma 1 only if the random variables x_n are a constant c with probability one. Since we are interested in problems 1 and 2, the purely deterministic processes are to be avoided.

3. Markov Chains and Problem 1.

We shall not be able to obtain very broad results in the general context spoken of in Section 2. The discussion in the general context of strictly stationary processes is relegated to a later section (Section 5) because more detailed and enlightening results can be obtained for strictly stationary Markov chains with a finite number of states.

Let $\{x_n\}$ be a strictly stationary Markov chain with transition probability matrix

$$\mathrm{P} = (p_{ij}), \qquad p_{ij} = \mathrm{P}[x_{n+1} = j | x_n = i], \qquad i, j = 1, 2, \cdots, \tag{17}$$

and stationary vector of probabilities

$$p_i = \mathrm{P}[x_n = i] > 0. \tag{18}$$

It is already clear from Lemma 2 that we shall not be able to construct a random variable ξ_n measurable with respect to $\mathcal{B}_n$ satisfying the conditions of part (a) of problem 1 unless the probability distributions $\{p_{ij}\,;\, j = 1, 2, \cdots\}$ indexed by i are equivalent. An alternative way of stating this is as follows. Given any i, i' there is a permutation $M(i, i')$ of the integers j such that

$$p_{ij} = p_{i', M(i,i')j} \tag{19}$$

for all j. Such Markov chains are admittedly rather special. Nonetheless they are interesting and give some insight into problem 1. We shall *call such Markov chains uniform chains* because the row probability distributions are the same except for permutation. Let the positive probability masses common to the distributions $\{p_{ij}\,, j = 1, 2, \cdots\}$ be $q_1 \geqq q_2 \geqq \cdots > 0$ in order of magnitude, $\sum q_i = 1$. When the q_i's are all distinct one can characterize the random variable ξ_n simply.

Lemma 3. *If x_n is a stationary uniform Markov chain with q_i's distinct, ξ_n is uniquely determined (up to a one-one transformation modulo sets of measure zero) and is of the form*

$$\xi_n = \xi(x_n\,, x_{n-1}) = k \quad \textit{for} \quad (x_n\,, x_{n-1}) \quad \textit{such that} \quad p_{x_{n-1}\,, x_n} = q_k\,. \tag{20}$$

Without too much abuse of notation we can write $\xi_n = \xi(x_n\,, x_{n-1}\,, \cdots)$. It is already clear that ξ must take m distinct values ($m \leqq \infty$ the number of distinct q's) with probability one, and we label the values 1, 2, $\cdots$ for convenience. Let the inverse function of ξ, as a function of x_n for fixed $x_{n-1}\,, x_{n-2}\,, \cdots$, be $\eta(\xi, x_{n-1}\,, x_{n-2}\,, \cdots)$. Since ξ_n is independent of $x_{n-1}\,, x_{n-2}\,, \cdots$

$$\begin{aligned} &\mathrm{P}[\xi_n = j \mid x_{n-1} = i_1\,, x_{n-2} = i_2\,, \cdots] \\ &\qquad = \mathrm{P}[\xi_n = j] = p_{i_1\,, \eta(j, i_1\, i_2\,, \cdots)} > 0. \end{aligned} \tag{21}$$

But the q_i's are distinct, and therefore for fixed $i_1\,, j$ the function $\eta(j, i_1\,, i_2\,, \cdots)$ must be independent of $i_2\,, i_3\,, \cdots$. Thus $\eta(j, i_1\,, i_2\,, \cdots) = \eta(j, i_1)$ and hence $\xi(x_n\,, x_{n-1}\,, \cdots) = \xi(x_n\,, x_{n-1})$. It is clear that ξ must be constant on the set

$$\{(i, j) \mid p_{ij} = q_k\}, \tag{22}$$

and for convenience we set $\xi = k$ on this set. Any other ξ satisfying the conditions of part (a) of problem 1 must be a one-one function of this ξ.

The random variable ξ_n was seen to be essentially uniquely determined when the q_i's are distinct. It is easy to see that it is not uniquely determined if the q_i's are not distinct. For example, if the transition probability matrix is

$$P = \begin{pmatrix} a & a & b \\ a & b & a \\ b & a & a \end{pmatrix}, \qquad a, b > 0, \qquad a \neq b, \tag{23}$$

we can take

$$\xi(x_n, x_{n-1}) = \begin{cases} 0 & \text{if } (x_n, x_{n-1}) = (0,0), (0,1), (1,2) \\ 1 & \text{if } (x_n, x_{n-1}) = (1,0), (2,1), (2,2) \\ 2 & \text{otherwise} \end{cases} \tag{24}$$

or, alternatively,

$$\xi(x_n, x_{n-1}) = \begin{cases} 0 & \text{if } (x_n, x_{n-1}) = (0,0), (2,1), (2,2) \\ 1 & \text{if } (x_n, x_{n-1}) = (1,0), (0,1), (1,2) \\ 2 & \text{otherwise.} \end{cases} \tag{25}$$

In fact, ξ need no longer be a function of only x_n, x_{n-1}, for one can take

$$\xi(x_n, x_{n-1}, x_{n-2}) = \begin{cases} 0 & \text{if } (x_n, x_{n-1}, x_{n-2}) = (0,0,0), (1,0,1), \\ & \quad (0,0,2), (0,1,0), (2,1,1), (0,1,2), \\ & \quad (1,2,0), (2,2,1), (1,2,2) \\ 1 & \text{if } (x_n, x_{n-1}, x_{n-2}) = (1,0,0), (0,0,1), \\ & \quad (1,0,2), (2,1,0), (0,1,1), (2,1,2), \\ & \quad (2,2,0), (1,2,1), (2,2,2) \\ 2 & \text{otherwise.} \end{cases} \tag{26}$$

Let us restrict ourselves to uniform Markov chains with distinct q_i's and try to find out under what circumstances the answer to problem 1 is in the affirmative. We have already seen that a positive answer to problem 2, and hence to problem 1 (making use of Lemma 1), can be given only if all the functions measurable with respect to $\bigcap \mathcal{B}_n$ are the constant functions. This means that *the set of states of the Markov chain must be an irreducible closed set of ergodic* (that is, nonperiodic) *states* [2]. An interesting necessary and sufficient condition for a positive answer to problem 1 can be given in algebraic terms.

Let M_k be the matrix

$$M_k = \{e_{ij}(k)\} \tag{27}$$

where

$$e_{ij}(k) = \begin{cases} 1 & \text{if } p_{ij} = q_k \\ 0 & \text{otherwise.} \end{cases} \tag{28}$$

Given the Markov chain $\{x_n\}$ there are m corresponding matrices M_k. Each matrix M_k has precisely one element equal to one in each row and all other elements zero. No two M_k's have a one in the same entry. Every product of a finite number of the M_k's also has these properties. The family of matrices $\{M_k\}$ corresponds to a family of mappings of the set of states of the process into the same set of states induced by the process $\{x_n\}$. Consider the semigroup (under multiplication) generated by the matrices M_k. This semigroup may or may not contain a matrix with a column of ones. If it does, we *call the semigroup point collapsing.*

Theorem 1. *Let $\{x_n\}$ be a finite state stationary uniform Markov chain with q_i's distinct. Let $\{M_k\}$ be the family of mapping matrices generated by $\{x_n\}$. The answer to problem 1 is positive if and only if the semigroup generated by $\{M_k\}$ is point collapsing.*

First assume the semigroup generated by the family of matrices $\{M_k\}$ is point collapsing. There is then a finite product $M_{k_1} \cdots M_{k_s}$ of matrices of this family (with possible duplication) which has all the elements in the ith column one. Now the random variable ξ_n satisfying the conditions of part (a) of problem 1 is uniquely determined by Lemma 3 and is given by

$$\xi_n = \xi(x_n, x_{n-1}) = k \quad \text{if} \quad e_{x_{n-1},x_n}(k) = 1. \tag{29}$$

Consider an event A_j of the form

$$A_j = \{\xi_j = k_1, \cdots, \xi_{j+s} = k_s\}. \tag{30}$$

If A_j occurs, it is clear that $x_{j+s} = i$ with probability one. The events A_{n-s}, $A_{n-2s}, \cdots$ are independent, and hence by the lemma of Borel-Cantelli at least one of the events will occur with probability one. Thus at least one of the events A_j, $j \leqq n - s + 1$, must occur with probability one. Let T_k be the mapping induced by M_k, that is, $T_k i = j$ where j is the unique j such that $e_{ij}(k) = 1$. We now construct a function $g(\xi_n, \xi_{n-1}, \cdots)$ such that $x_n = g(\xi_n, \xi_{n-1}, \cdots)$ with probability one. Consider the sequences $(\cdots, \xi_{n-1}, \xi_n)$. Let S_j be the set of sequences $(\cdots, \xi_{n-1}, \xi_n)$ for which some A_i, $i \leqq n - s + 1$, occurs and A_j is the A_i with largest index, $i \leqq n - s + 1$, that occurs. The sets S_j, $j \leqq n - s + 1$, are disjoint and $P(\overline{\cup S_j}) = 0$ ($\bar{S}$ denotes the complement of S) as already remarked above. Let

$$g(\xi_n, \xi_{n-1}, \cdots) = \begin{cases} T_{\xi_n} \cdots T_{\xi_{j+s+1}} i & \text{if} \quad (\cdots, \xi_{n-1}, \xi_n) \in S_j \\ 0 & \text{otherwise.} \end{cases} \tag{31}$$

It is then clear that $x_n = g(\xi_n, \xi_{n-1}, \cdots)$ with probability one since $(\cdots, \xi_{n-1}, \xi_n) \in S_j$ for some $j \leqq n - s + 1$ with probability one. Thus the answer to problem 1 is positive.

Now, on the contrary, assume that the semigroup generated by the family of matrices $\{M_k\}$ is not point collapsing. Every matrix of the semigroup has a certain number of distinct columns containing a one as a column entry. The minimal number of distinct columns among all matrices of the semigroup is certainly greater than one (let us say $r > 1$) since the semigroup is not point collapsing. But this means that knowledge of all the ξ_i's, $i \leqq n$, can tell no more than that x_n takes one of r distinct values. The answer to problem 1 must therefore be negative.

One might hope for a positive answer to problem 1 if there is a random variable ξ_n satisfying part (a) of the problem and the necessary condition of Lemma 1 is satisfied. However, by making use of Theorem 1 a large number of processes satisfying these conditions for which the answer to problem 1 is negative can be constructed.

We give the simplest example exhibiting this interesting sort of pathology. Let $\{x_n\}$ be the strictly stationary Markov chain with the two states 1, 2, the stationary instantaneous distribution

$$\mathrm{P}[x_n = 1] = \mathrm{P}[x_n = 2] = \tfrac{1}{2} \tag{32}$$

and transition probability matrix

$$P = \begin{bmatrix} p & q \\ q & p \end{bmatrix}, \qquad 0 < p,\ q = 1 - p < 1, \qquad p \neq q.$$

Now

$$M_1 = \begin{bmatrix} 1 & 0 \\ 0 & 1 \end{bmatrix}, \qquad M_2 = \begin{bmatrix} 0 & 1 \\ 1 & 0 \end{bmatrix} \tag{33}$$

so that

$$\xi_n = \xi(x_n, x_{n-1}) = \begin{cases} 1 & \text{if } (x_{n-1}, x_n) = (1, 1), (2, 2) \\ 2 & \text{otherwise} \end{cases}$$

is measurable with respect to $\mathcal{B}_n$ and yet independent of $x_{n-1}, x_{n-2}, \cdots$. By Lemma 3 it is essentially the only random variable with these properties. Further, $\{x_n\}$ is purely nondeterministic. Nonetheless x_n is not a function of $\xi_n, \xi_{n-1}, \cdots$ since $\mathrm{E}(x_n \mid \xi_n, \xi_{n-1}, \cdots) = \mathrm{E}x_n$.

One would like to give a neat characterization of the point collapsing semigroups generated by the matrices $\{M_k\}$ induced by a uniform Markov chain. This does not appear to be possible. However, one can easily sketch the details of a computational procedure that would determine whether or not such a semigroup is point collapsing. Let $N < \infty$ be the number of states of the finite state uniform Markov chain $\{x_n\}$. Let $\{T_k\}$ be the family of mappings of the integers $1, \cdots, N$ into the integers $1, \cdots, N$ determined by the matrices $\{M_k\}$. If the semigroup is point collapsing some T_k must map the N states onto $N_1 < N$ states. Given N_1 states, consider the ensemble of all distinct N_1-tuples (including the original one) that can be obtained from this N_1-tuple by applying a finite number of the T_k mappings. Some one of these N_1-tuples must be collapsed further onto $N_2 < N_1$ states by some T_k mapping if the semigroup is point collapsing. Continue in this manner until the original set of states is collapsed onto one state. If, at some stage in the construction, all the distinct N_j-tuples ($N_j > 1$) obtained cannot be contracted, the semigroup is not point collapsing. If N is not small, this is admittedly a very tedious procedure.

4. Markov Chains and Problem 2. The answer to problem 1 is generally negative, even in the case of a finite state purely nondeterministic Markov chain. However, the answer to problem 2 for such Markov chains will be shown to be positive. The following theorem will be proved.

Theorem 2. *Let $\{x_n\}$ be a stationary purely nondeterministic Markov chain. Let there be an $\epsilon > 0$ and a finite number of states M such that*

$$\sum_{j \varepsilon M} p_{ij} = p_{i,M} > \epsilon \tag{34}$$

for all i. A representation of the type mentioned in problem 2 holds for such a process.

Notice that this result holds for all ergodic finite state stationary Markov chains with nonperiodic states. Actually Theorem 2 is proved by enriching the probability space of the $\{x_n\}$ process in an appropriate way.

Introduce the Markov process y_n , $1 < y_n < \infty$, with stationary probability density

$$p(y|y') = p_{ij} \quad \text{if} \quad y' \varepsilon (i, i+1], \qquad y \varepsilon (j, j+1], \qquad i, j = 1, 2, \cdots , \tag{35}$$

and stationary instantaneous probability distribution

$$\mathrm{P}[y_n \varepsilon A] = \sum p_i m(A \cap (i, i+1]), \tag{36}$$

where $p_i = \mathrm{P}[x_n = i]$. We shall construct a process on the probability space of the y_n process with the same probability structure as that of the x_n process. In spite of a possible abuse of notation we shall call this new process the x_n process. The process x_n can be taken as $x_n = g(y_n)$ where $g(y) = j$ if $y \varepsilon (j, j+1]$ (see [7]). The probability space of the y_n process is rich enough so that one can construct random variables $\alpha_n = G(y_n \mid y_{n-1})$ independent of y_{n-1} , y_{n-2} , $\cdots$ with α_n uniformly distributed on (0, 1). $G(y \mid y')$ is a Borel function of y, y' and is so constructed that, for fixed y_{n-1} , $G(y_n \mid y_{n-1})$ is with probability one a one-one function of y_n . We can therefore write $y_n = G^{-1}(\alpha_n \mid y_{n-1})$. It will be convenient to introduce the associated family of transformations T_α , $0 < \alpha < 1$, where

$$T_\alpha y = G^{-1}(\alpha|y). \tag{37}$$

Notice that it is equivalent to give either $G(y \mid y')$ or the corresponding family of transformations $\{T_\alpha\}$ due to the essentially one-one character of G. Such a family of random variables α_n can be introduced, for example, by taking $G(y|y') = F(y \mid y')$ where $F(y \mid y')$ is the conditional distribution function of y_n given y_{n-1} . Note that in this case T_α for each α, $0 < \alpha < 1$, maps each interval $(j, j+1]$ into a single point. There are other families of transformations T_α , associated with random variables α_n of the above form, that have this property. We shall construct a convenient family T_α for our purposes in the proof of Theorem 2.

By (34) for some fixed $\epsilon > 0$ there is a finite set of states which we can take as 1, $\cdots$, M (by relabeling the states if necessary) such that $\sum_{j=1}^{M} p_{ij} > \epsilon$

for all i. Since the Markov chain $\{x_n\}$ is irreducible and the states are nonperiodic, there must be two paths from state one to state one of positive probability with lengths that are relatively prime. Add all the states passed through in these paths to the states $1, \cdots, M$. Further, add all the other states passed through in a path of minimal length and positive probability from state i to state j, $i, j = 1, \cdots, M$. Call the collection of states included thus far S. We can then connect any two states $i, j \in S$ by a path of positive probability containing only states of S. If the states of S are $N(S)$ in number it may be convenient to relabel the states of the process so that they appear as the states $1, 2, \cdots, N = N(S)$. Notice that the matrix of transition probabilities of the process $\{x_n\}$, as restricted to the states of S, is irreducible and aperiodic, that is, some sufficiently high power of the restricted matrix has all its elements positive.

Let S' be any subset of S and $N(S')$ the number of states in S'. *Call S' a contracting set if there is a set of states $S'' \subset S$ with $N(S'') < N(S')$ such that for each $i \in S'$ there is a $j \in S''$ with $p_{ij} > 0$.* The set $S'' \subset S$ can be reached from $S' \subset S$ if for each $i \in S'$ there is a $j \in S''$ with $p_{ij} > 0$. *S'' is accessible from S' if for some k there is a sequence of subsets (of S) $S_1 = S'$, $S_2, \cdots, S_k = S''$ such that S_j can be reached from S_{j-1}, $j = 2, \cdots, k$.* If $N(S') > 1$, $S' \subset S$, there is a set $S'' \subset S$ with $N(S'') < N(S')$ that is accessible from S'. For suppose there were not. Every m-tuple of states of S is accessible from every other m-tuple of states of S because of the irreducible and nonperiodic character of the transition matrix when restricted to the states of S. If there were no set of states S'' with the desired property no m-tuple with $m = N(S') > 1$ could be contracted. Take any two distinct states $i, j \in S$. Since they are in some m-tuple and no m-tuple is contracting, there is no state $k \in S$ such that $p_{ik}, p_{jk} > 0$. This means that for any $i \in S$ there is one and only one $j = j(i)$ such that $p_{ij} > 0$. But this implies that the matrix of transition probabilities, as restricted to the states of S, is periodic, a contradiction.

Start with the whole set of states $S_0 = S = \{1, 2, \cdots, N\}$. This set is contracting. Let $S_1 \subset S$ be a set of states with $N(S_1) < N(S)$ that is accessible from S_0 in a minimal number of steps. Let $S_2 \subset S$ be a set with $N(S_2) < N(S_1)$ that is accessible from S_1 in a minimal number of steps. Continue the construction until a set, say S_k, with one element is obtained. Let $M_1, \cdots, M_a$ be sets of states with $M_1 = S_0$, $M_2 = S_1, \cdots, M_{2+r_1} = S_2, \cdots, M_{2+r_1+r_2} = S_3, \cdots, M_a = S_k$ where $a = 2 + r_1 + \cdots + r_k$ and M_{j+1} can be reached from M_j. In other words, the M_j's are the sets of states passed through in going from S_0 to S_k. The number of steps in going from S_j to S_{j+1} can be at most $N^2 = N(S)^2$ in number (see [8]). Therefore the total number of steps from M_1 to M_a can be at most N^3. Let the minimal positive transition probability p_{ij}, $i, j \in S$, be δ (> 0). Given any state $i \in M_j$, let $l = f_j(i)$ be the state of lowest index in M_{j+1} for which $p_{il} > 0$. Further, given any state $i \notin S$. there is a state $l = f(i) \in S$ such that $p_{il} > \epsilon/N$. Let $\delta' = \min(\epsilon/N, \delta)$.

Now consider the Markov process $\{y_n\}$, constructed at the beginning of our proof, and the Markov chain $x_n = g(y_n)$. We now set up random variables

$\alpha_n = G(y_n \mid y_{n-1})$, uniformly distributed and independent of $y_{n-1}, y_{n-2}, \cdots$ with the associated family of one-one transformations T_α satisfying conditions now to be specified. Consider disjoint subintervals (γ_i, δ_i), $i = 0, 1, \cdots, a$, of (0, 1) with $0 < \delta_i - \gamma_i < \delta'/(N^3 + 1)$. The transformations T_α are to be set up (and can be set up) so that they satisfy the following conditions. Let

$$\text{(38)} \qquad \begin{aligned} T_\alpha(i, i+1] &\subset (f(i), f(i)+1] \quad \text{if} \quad i \not\in S, \\ T_\alpha(i, i+1] &\subset (1, N+1] \qquad \text{if} \quad i \in S \end{aligned}$$

when $\gamma_0 < \alpha < \delta_0$. Thus if $\gamma_0 < \alpha_n < \delta_0$ we know that $1 < y_n \leqq N + 1$ (or equivalently $x_n \in S$). Further, let

$$\text{(39)} \qquad T_\alpha(i, i+1] \subset (f_j(i), f_j(i)+1]$$

for all $i \in M_j$ if $\gamma_j < \alpha < \delta_j$. Moreover, let the family of transformations T_α, $0 < \alpha < 1$, map each interval $(i, i + 1]$, $i = 1, 2, \cdots$, into a single point. The point will generally depend on α. Now if the event A_j occurs,

$$\text{(40)} \qquad A_j = \{\gamma_0 < \alpha_j < \delta_0, \cdots, \gamma_a < \alpha_{j+a} < \delta_a, \gamma_0 < \alpha_{j+a+1} < \delta_0\},$$

we know with probability one the value of y_{j+a+1}. For the first $a + 1$ conditions imply that $i < y_{j+a} \leqq i + 1$ where i is the single state of M_a. Thus $y_{j+a+1} = T_{\alpha_{j+a+1}}(i, i + 1]$. The random variables $\{\alpha_j\}$ are independent uniformly distributed random variables. Just as in the proof of Theorem 1, the Borel-Cantelli lemma implies that at least one of the events A_j, $j \leqq n - a - 1$, occurs with probability one. Let E_j be the set of sequences $(\cdots, \alpha_{n-1}, \alpha_n)$ for which some A_i, $i \leqq n - a - 1$, occurs and A_j is the A_i with largest index, $i \leqq n - a - 1$, that occurs. The events E_j are disjoint, and with probability one every sequence $(\cdots, \alpha_{n-1}, \alpha_n)$ lies in some E_j, $j \leqq n - a - 1$. Let

$$\text{(41)} \qquad h(\alpha_n, \alpha_{n-1}, \cdots) = \begin{cases} T_{\alpha_n} \cdots T_{\alpha_{j+a+1}}(i, i+1] & \text{if} \quad (\cdots, \alpha_{n-1}, \alpha_n) \in E_j \\ 0 & \text{otherwise.} \end{cases}$$

Then $y_n = h(\alpha_n, \alpha_{n-1}, \cdots)$ with probability one and hence

$$x_n = g(h(\alpha_n, \alpha_{n-1}, \cdots))$$

with probability one. The proof of Theorem 2 is complete.

5. Stochastic Processes. We now consider some conditions sufficient (but not necessary) for a positive answer to problem 1, and hence problem 2 in the case of a general strictly stationary stochastic process $\{x_n\}$. There are several conditions that will be introduced to avoid unduly tedious measure theoretic complications. We have already seen that there are difficulties in obtaining random variables ξ_n satisfying the conditions of part (a) of problem 1 if the conditional distribution function $F(a_n \mid a_{n-1}, \cdots)$ has a discrete component with positive probability. To avoid this assume that *$F(a_n \mid a_{n-1}, \cdots)$ is continuous in a_n with probability one.* Further, *let the random variables x_n of the process be* *bounded between zero and one.* This last assumption is on restriction since it can always be brought about by considering a process $z_n = h(x_n)$ where h is a one-one monotone function mapping $(-\infty, \infty)$ into $(0, 1)$. Further, *let* $F(a_n \mid a_{n-1}, \cdots)$ *be a strictly increasing function of* a_n, $0 < a_n < 1$, *with probability one.* By Lemma 2 there is a random variable ξ_n satisfying part (a) of problem 1. Such a random variable is, in fact, given by

$$\xi_n = F(x_n|x_{n-1}, x_{n-2}, \cdots). \tag{42}$$

This is not the only possible construction of such a random variable. The conditions to be specified will be enough to ensure that the process x_n is of the form

$$x_n = g(\xi_n, \xi_{n-1}, \cdots); \tag{43}$$

in other words problem 1 has a positive answer with this particular construction (see formula (42)) of such a random variable. In a report KALLIANPUR & WIENER [3] discuss some work in this direction. The conjectured result is that a representation of the form (43) holds under the above conditions and under the assumption that $\{x_n\}$ is purely nondeterministic. The argument given implies that the conjectured representation can be established in terms of the specific random variables (42). We shall later give an example for which a representation of the form (43) is valid, but not in terms of the random variables (42). This indicates that generally one has to consider constructing random variables ξ_n satisfying part (a) of problem 1 which are not of the form (42).

Let μ be the measure on half infinite sequences $(a_n, a_{n-1}, \cdots)$, $0 < a_i < 1$, induced by the probability measure of the process $\{x_n\}$. Let T_α, $0 < \alpha < 1$, be the family of transformations of the set of sequences $(a_n, a_{n-1}, \cdots)$ into the real numbers a_{n+1}, $0 < a_{n+1} < 1$, induced by the conditional distribution function, that is,

$$T_\alpha(a_n, a_{n-1}, \cdots) = F^{-1}(\alpha|a_n, a_{n-1}, \cdots). \tag{44}$$

Notice that

$$x_n = T_{\xi_n}(x_{n-1}, x_{n-2}, \cdots) \tag{45}$$

with probability one where ξ_n is the uniformly distributed random variable

$$\xi_n = F(x_n|x_{n-1}, \cdots) \tag{46}$$

independent of $x_{n-1}, x_{n-2}, \cdots$. If $x_n = g(\xi_n, \xi_{n-1}, \cdots)$ then clearly $x_n = \mathrm{E}(x_n \mid \xi_n, \xi_{n-1}, \cdots)$ and hence

$$x_n - \mathrm{E}(x_n|\xi_n, \xi_{n-1}, \cdots, \xi_{n-k}) \to 0 \tag{47}$$

in probability as $k \to \infty$ or equivalently

$$\mathrm{E}\,|x_n - \mathrm{E}(x_n|\xi_n, \cdots, \xi_{n-k})| \to 0 \tag{48}$$

because of the boundedness of the random variables x_n. Because of the stationarity of the $\{x_n\}$ process, condition (48) could equally well have been written

$$\text{E}\,|x_n - \text{E}(x_n|\xi_n, \cdots, \xi_1)| \to 0 \tag{49}$$

as $n \to \infty$. To avoid cumbersome formulas we introduce the following notation. Let

$$\begin{aligned} \bar{x}_m &= (x_m, x_{m-1}, \cdots), \\ T_{\xi_m}\bar{x}_{m-1} &= T_{\xi_m}(x_{m-1}, x_{m-2}, \cdots), \\ \{T_{\xi_m}\bar{x}_{m-1}\} &= (T_{\xi_m}\bar{x}_{m-1}, x_{m-1}, x_{m-2}, \cdots). \end{aligned} \tag{50}$$

Theorem 3. *The condition*

$$\text{E}\,|x_n - \text{E}(x_n|\xi_n, \cdots, \xi_1)| \to 0 \quad as \quad n \to \infty \tag{51}$$

is satisfied if and only if there is a set S of $\bar{y}_0$ points of μ measure one such that

$$T_{\xi_n}\{T_{\xi_{n-1}}\{\cdots\{T_{\xi_1}\bar{y}_0\}\cdots\}\} - T_{\xi_n}\{T_{\xi_{n-1}}\{\cdots\{T_{\xi_1}\bar{y}_0'\}\cdots\}\} \to 0 \tag{52}$$

in $\xi_1, \cdots, \xi_n, \cdots$ measure for all $\bar{y}_0, \bar{y}_0' \in S$ as $n \to \infty$.

Suppose there is a set S of μ measure one satisfying condition (52). Let $\bar{z}_0 \in S$. Then for all $\bar{x}_0 \in S$ the measure

$$m\{((\xi_1, \cdots, \xi_n)||T_{\xi_n}\{\cdots\{T_{\xi_1}\bar{x}_0\}\cdots\} - T_{\xi_n}\{\cdots\{T_{\xi_1}\bar{z}_0\}\cdots\}| > \epsilon)\} \to 0$$

as $n \to \infty$ for any fixed $\epsilon > 0$. But then

$$\begin{aligned} \text{P}\{|x_n - T_{\xi_n}\{\cdots\{T_{\xi_1}\bar{z}_0\}\cdots\}| > \epsilon\} = \int m\{((\xi_1, \cdots, \xi_n)||T_{\xi_n}\{\cdots\{T_{\xi_1}\bar{x}_0\}\cdots\} \\ - T_{\xi_n}\{\cdots\{T_{\xi_1}\bar{z}_0\}\cdots\}| > \epsilon)\}\, d\mu(\bar{x}_0) \to 0 \end{aligned}$$

as $n \to \infty$ and hence

$$x_n - T_{\xi_n}\{\cdots\{T_{\xi_1}\bar{z}_0\}\cdots\} \to 0$$

in probability as $n \to \infty$. On the other hand,

$$\begin{aligned} &\text{E}\,|\text{E}(x_n|\xi_n, \cdots, \xi_1) - T_{\xi_n}\{\cdots\{T_{\xi_1}\bar{z}_0\}\cdots\}| \\ &\quad = \text{E}\,|\text{E}(T_{\xi_n}\{\cdots\{T_{\xi_1}\bar{x}_0\}\cdots\} - T_{\xi_n}\{\cdots\{T_{\xi_1}\bar{z}_0\}\cdots\}|\xi_n, \cdots, \xi_1)| \\ &\quad \leqq \text{E}(\text{E}(|T_{\xi_n}\{\cdots\{T_{\xi_1}\bar{x}_0\}\cdots\} - T_{\xi_n}\{\cdots\{T_{\xi_1}\bar{z}_0\}\cdots\}||\xi_n, \cdots, \xi_1)) \\ &\quad = \text{E}\,|x_n - T_{\xi_n}\{\cdots\{T_{\xi_1}\bar{z}_0\}\cdots\}| \to 0. \end{aligned}$$

But then $x_n - \text{E}(x_n|\xi_n, \cdots, \xi_1) \to 0$ as $n \to \infty$ in probability.

Now suppose that $x_n - \text{E}(x_n | \xi_n, \cdots, \xi_1) \to 0$ in probability as $n \to \infty$ and condition (52) is not satisfied. Then given a sufficiently small $\epsilon > 0$ there are sets S_1, S_2 with $\mu(S_1), \mu(S_2) > \delta(\epsilon) > 0$ such that for $\bar{x}_0 \in S_1$, $\bar{x}_0' \in S_2$

$$m\{((\xi_1, \cdots, \xi_n)||T_{\xi_n}\{\cdots\{T_{\xi_1}\bar{x}_0\}\cdots\} - T_{\xi_n}\{\cdots\{T_{\xi_1}\bar{x}_0'\}\cdots\}| > \epsilon)\} > \delta(\epsilon)$$

for an infinite number of values of n. However,

$$x_n - \mathrm{E}(x_n | \xi_n, \cdots, \xi_1) \to 0$$

implies that for almost all $\bar{x}_0$ (μ measure)

$$T_{\xi_n}\{\cdots\{T_{\xi_1}\bar{x}_0\}\cdots\} - \mathrm{E}(T_{\xi_n}\{\cdots\{T_{\xi_1}\bar{x}_0'\}\cdots\}|\xi_n, \cdots, \xi_1) \to 0$$

in $\xi_1, \cdots, \xi_n, \cdots$ measure, a contradiction.

Corollary 1. *Let the family of transformations T_α, $0 < \alpha < 1$, be such that, given any two $\bar{x}_0$, $\bar{x}_0'$ with $0 < x_{-j}, x_{-j}' < 1$,*

$$\mathrm{E}\ |T_{\xi_1}\bar{x}_0 - T_{\xi_1}\bar{x}_0'| \leqq \sum_{j=0}^{\infty} \beta_j\ |x_{-j} - x_{-j}'| \tag{53}$$

for some sequence of positive constants β_j with $\sum \beta_j < 1$. Then $\{x_n\}$ has a representation of form (43). *The expectation in formula* (53) *is understood to be an average with respect to the uniformly distributed random variable ξ.*

It is clear that

$$\mathrm{E}\ |T_{\xi_n}\{\cdots\{T_{\xi_1}\bar{x}_0\}\cdots\} - T_{\xi_n}\{\cdots\{T_{\xi_1}\bar{x}_0'\}\cdots\}| \leqq (\sum \beta_j)^n$$

so that

$$T_{\xi_n}\{\cdots\{T_{\xi_1}\bar{x}_0\}\cdots\} - T_{\xi_n}\{\cdots\{T_{\xi_1}\bar{x}_0'\}\cdots\} \to 0$$

as $n \to \infty$ in probability for all $\bar{x}_0$, $\bar{x}_0'$ and hence the hypothesis of Theorem 3 is satisfied.

Corollary 2. *Suppose there is a unique ergodic distribution corresponding to the transition probability measure generated by the conditional probability distribution $F(a_n \mid a_{n-1}, \cdots)$ and that the Doeblin condition* (D) ([1], [6]) *is satisfied by this transition probability measure. Further, given any two points $\bar{x}_0$, $\bar{x}_0' \in S$* (see Theorem 3) *with $0 < x_{-j} < x_{-j}' < 1$ for all j, let*

$$T_\alpha \bar{x}_0 < T_\alpha \bar{x}_0' \tag{54}$$

for all α, $0 < \alpha < 1$. Then $\{x_n\}$ has a representation of the form (43).

Notice that

$$x_n - x_n' = T_{\xi_n}\{\cdots\{T_{\xi_1}\bar{x}_0\}\cdots\} - T_{\xi_n}\{\cdots\{T_{\xi_1}\bar{x}_0'\}\cdots\} \leqq 0$$

if $\bar{x}_0 < \bar{x}_0'$, that is, if $x_{-j} < x_{-j}'$ for all j ($\leqq 0$). However, the hypothesis of the corollary implies that x_n, x_n' have a common limiting distribution, say $F(a)$. But

$$\mathrm{P}\{x_n' > a, x_n \leqq a\} + \mathrm{P}\{x_n' \leqq a\} = \mathrm{P}\{x_n' \leqq a\}$$

so that

$$\mathrm{P}\{x_n' > a, x_n \leqq a\} \to 0$$

at every point a ($F(a)$ is continuous since $F(a_n \mid a_{n-1}, \cdots)$ is continuous in a_n).

Given any $\epsilon > 0$ we can find $k(\epsilon)$ points $b_1 < \cdots < b_k$ such that

$$\bigcup_{j=1}^{k} \{x_n \leqq b_j , x_n' > b_j\} \supset \{x_n' - x_n > \epsilon\}.$$

But then

$$P\{x_n' - x_n > \epsilon\} \to 0$$

for every $\epsilon > 0$ as $n \to \infty$, and hence $x_n' - x_n \to 0$ in probability as $n \to \infty$. This argument has been carried out for any two points $\bar{x}_0$, $\bar{x}_0'$ with $\bar{x}_0 < \bar{x}_0'$. However, given any two points $\bar{x}_0$, $\bar{x}_0'$ (without such an order relation) we can find a third point $\bar{x}_0''$ such that $\bar{x}_0'' < \bar{x}_0, \bar{x}_0'$ and hence $x_n' - x_n \to 0$ for x_n', x_n derived from any two points $\bar{x}_0'$, $\bar{x}_0$. Now simply apply Theorem 3.

We now construct the simple example mentioned at the beginning of this section to show that even though a strictly stationary process $\{x_n\}$ may not be a one-sided function $x_n = g(\xi_n, \xi_{n-1}, \cdots)$ of the random variables (42), it may have such a representation in terms of random variables ξ_n' satisfying part (a) of problem 1 and defined in a different and more suitable manner. Set up a sequence of points b_i, $i = 0, \pm 1, \cdots$, with $0 < \cdots < b_i < b_{i+1} < \cdots < 1$. Let $S \subset (0, 1)$ be

$$S = \bigcup_{i=-\infty}^{\infty} \{y | b_{2i} < y \leqq b_{2i+1}\} \tag{55}$$

and $\bar{S} = (0, 1) - S$. Let $\{x_n\}$ be the strictly stationary Markov process with transition probability density

$$p(y|y') = \begin{cases} p_i/(b_{i+1} - b_i) & \text{if} \quad b_i < y \leqq b_{i+1}, \quad y' \in S, \\ p_i/(b_i - b_{i-1}) & \text{if} \quad b_{i-1} < y \leqq b_i, \quad y' \in \bar{S}, \end{cases} \tag{56}$$

where the p_i's are distinct positive numbers with $\sum p_i = 1$. Consider the family of transformations T_α generated by the conditional distribution function of the process $\{x_n\}$. If $\sum_{j=-\infty}^{i-1} p_j < \alpha \leqq \sum_{j=-\infty}^{i} p_j$, then $b_i < T_\alpha x \leqq b_{i+1}$ when $x \in S$ and $b_{i-1} < T_\alpha x \leqq b_i$ when $x \in \bar{S}$. Thus if we start out with two points x, x' with $x \in S$, $x' \in \bar{S}$ their transforms $T_{\xi_n} \cdots T_{\xi_1} x$, $T_{\xi_n} \cdots T_{\xi_1} x'$ will never be both in S or both in $\bar{S}$. Moreover, at any stage n we can separate the transforms by a quantity bounded below by a positive number by specifying an appropriate condition of the form $\beta < \xi_n < \beta'$. Thus the condition of Theorem 3 is not satisfied and hence $\{x_n\}$ does not have a representation of form (43) with the ξ_n's given by (42). Nonetheless the process $\{x_n\}$ is of the same type as the process $\{y_n\}$ constructed in the proof of Theorem 2. By using the construction given in Theorem 2 we can generate random variables ξ_n' satisfying the conditions of part (a) of problem 1 and such that $x_n = g(\xi_n', \xi_{n-1}', \cdots)$.

In conclusion, it is worthwhile noting that problem 1 can be rephrased in the terminology of information theory [4]. The original process is passed through a deterministic channel that manufactures independent random variables ξ_n

independent of the past, that is, independent of $x_{n-1}, x_{n-2}, \cdots$. We ask under what circumstances the original process $\{x_n\}$ can be expressed in terms of the output of the channel $\{\xi_n\}$ with the time sense preserved, that is, when $x_n = g(\xi_n, \xi_{n-1}, \cdots)$ for some Borel function g.

REFERENCES

[1] J. L. DOOB, *Stochastic Processes*, New York (1953).

[2] W. FELLER, *An Introduction to Probability Theory and its Applications*, New York (1950).

[3] G. KALLIANPUR & N. WEINER, *Non-linear prediction*, Technical Report No. 1 (1956), Office of Naval Research, Cu-2-56-Nonr-266, (39)-CIRMIP, Project NR-047-015.

[4] A. I. KHINCHIN, *Mathematical Foundations of Information Theory*, New York (1957).

[5] P. LÉVY, *Theorie de l'addition des variables aleatoires*, Paris (1937).

[6] M. LOEVE, *Probability Theory*, New York (1955).

[7] M. ROSENBLATT, Functions of a Markov process that are Markovian, *Journal of Mathematics and Mechanics*, **8** (1959), pp. 585–596.

[8] H. WIELANDT, Unzerlegbare, nicht negative Matrizen, *Math. Zeit.*, **52** (1950), pp. 642–648.

Indiana University*
Bloomington, Indiana

*The author is now located in the Division of Applied Mathematics, Brown University.

ASYMPTOTIC DISTRIBUTION OF EIGENVALUES OF BLOCK TOEPLITZ MATRICES

M. ROSENBLATT[1]

Communicated by Edwin Hewitt, June 1, 1960

Let $g(\lambda)$, $-\pi \leqq \lambda \leqq \pi$, be a $p \times p$ $(p=1, 2, \cdots)$ matrix-valued Hermitian function. Further $g(\lambda)$ is bounded on $[-\pi, \pi]$, that is, its elements are bounded on $[-\pi, \pi]$. The Fourier coefficients

$$a_k = \frac{1}{2\pi}\int_{-\pi}^{\pi} e^{ik\lambda} g(\lambda)\,d\lambda, \qquad k = 0, \pm 1, \cdots, \tag{1}$$

are then bounded in k. We call the $np \times np$ matrix

$$A_n = (a_{j-k};\, j, k = 1, \cdots, n)$$

(an $n \times n$ matrix of the $p \times p$ blocks a_{j-k}) the nth section block Toeplitz matrix generated by $g(\lambda)$. Notice that the block Toeplitz matrix A_n is generally not Toeplitz. Our interest is in obtaining the asymptotic distribution of eigenvalues of A_n as $n \to \infty$. The proof is suggested by an argument given in the one-dimensional case $(p=1)$ (see [**3**]) and is based on results in the multidimensional prediction problem [**5**].

If the real number α is sufficiently small in absolute value $f(\lambda) = [I_p + \alpha g(\lambda)]$ is positive definite for all λ and bounded (I_p is the identity matrix of order p). Let $R_n = I_{np} + \alpha A_n$ be the nth section block Toeplitz matrix generated by $f(\lambda)$. Further denote the (i, j)th block element ($p \times p$ matrix), $i, j = 1, \cdots, n$, of the inverse R_n^{-1} of R_n by ${}_n r_{i,j}^{(-1)}$. The basic result on the determinant of the prediction error covariance matrix in the multidimensional prediction problem [**5**] tells us that

$$\lim_{n\to\infty} \det \left({}_n r_{11}^{(-1)}\right)^{-1} = (2\pi)^p \exp\left\{\frac{1}{2\pi}\int_{-\pi}^{\pi} \log \det\left(\frac{f(\lambda)}{2\pi}\right) d\lambda\right\}$$

since $f(\lambda)/2\pi$ can be regarded as the spectral density function of a p-vector weakly stationary stochastic process. However,

$$\det \left({}_n r_{11}^{(-1)}\right)^{-1} = \det (R_n)/\det (R_{n-1}) = \sigma_n^2$$

(see [**1**, p. 21]). Let $\lambda_{\nu,n}$, $\nu = 1, \cdots, np$, be the eigenvalues of A_n.

[1] This research was supported in part by the U. S. Army Signal Corps.

R.A. Davis et al. (eds.), *Selected Works of Murray Rosenblatt*, Selected Works in Probability and Statistics, 163
DOI 10.1007/978-1-4419-8339-8_18, © Springer Science+Business Media, LLC 2011

Now

$$\det(R_n) = \prod_{\nu=1}^{np}(1 + \alpha\lambda_{\nu,n})$$

and

$$\lim \log \sigma_n^2 = \lim \frac{1}{n}\sum_{k=1}^{n} \log \sigma_k^2 = \lim \frac{1}{n}\sum_{\nu=1}^{np} \log(1 + \alpha\lambda_{\nu,n}).$$

Thus

$$\lim_{n\to\infty} \frac{1}{n}\sum_{\nu=1}^{np} \log(1 + \alpha\lambda_{\nu,n}) = \frac{1}{2\pi}\int_{-\pi}^{\pi} \log \det(I_p + \alpha g(\lambda))d\lambda.$$

On taking $s_{n,k} = \sum_{\nu=1}^{np} \lambda_{\nu,n}^k$, it is readily seen that

$$\lim \sum_{k=1}^{\infty} \frac{(-1)^{k-1}}{k} \frac{s_{n,k}}{pn} \alpha^k = \sum_{k=1}^{\infty} \frac{(-1)^{k-1}}{k} \frac{1}{2\pi p}\int_{-\pi}^{\pi} \operatorname{tr}(g^k(\lambda))\alpha^k d\lambda$$

for $|\alpha|$ sufficiently small ($\operatorname{tr}(A)$ denotes the trace of A). This cannot hold unless

$$\lim_{n\to\infty} \frac{s_{n,k}}{pn} = \frac{1}{2\pi p}\int_{-\pi}^{\pi} \operatorname{tr}(g^k(\lambda))d\lambda. \tag{2}$$

Let $\mu_1(\lambda), \cdots, \mu_p(\lambda)$ *be the eigenvalues of* g *at* λ, let us say for convenience in order of magnitude. Then relation (2) implies *that the asymptotic distribution of eigenvalues* $\lambda_{\nu,n} = 1, \cdots, pn$, *is given by*

$$\lim_{n\to\infty} \frac{\text{number of eigenvalues} \leqq a}{pn} = \frac{1}{p}\sum_{j=1}^{p} P[\mu_j(X) \leqq a] \tag{3}$$

where X *is a random variable uniformly distributed on* $[-\pi, \pi]$. Aside from the interest in this result for its own sake, it is clear that it suggests a number of good approximations for the joint distribution of spectral estimates in the case of multidimensional normal stationary processes [**2**; **4**].

References

1. F. R. Gantmacher, *The theory of matrices*, vol. 1, New York, Chelsea Publishing Co., 1959.

2. N. R. Goodman, *On the joint estimation of the spectra, cospectrum and quadrature spectrum of a two-dimensional Gaussian process*, Scientific Paper No. 10, Engineering Statistics Laboratory, New York University, 1957, p. 168.

3. U. Grenander and G. Szegö, *Toeplitz forms and their applications*, University of California Press, 1958.

4. M. Rosenblatt, *Statistical analysis of stochastic processes with stationary residuals*, The H. Cramér Volume, Stockholm, Almquist and Wiksell, 1959.

5. N. Wiener and P. Masani, *The prediction theory of multivariate stochastic processes*, Acta Math. vol. 98 (1957) pp. 111–150.

Brown University

Limits of Convolution Sequences of Measures on a Compact Topological Semigroup

M. ROSENBLATT

Introduction. Limit properties of the convolution sequence of a regular measure on a compact topological semigroup are examined in this paper. Similar questions, as they arise in the case of a compact group, were examined by KAWADA & ITO [4]. Recently BELLMAN [1] and GRENANDER [3] considered special limit theorems for products of independent identically distributed random operators. Such problems are closely related to those in this paper. It should be noted that similar questions arise when considering the structure of stationary stochastic processes [7]. Various results on compact semigroups are used in characterizing the class of limit measures [5], [6], [8].

Basic Concepts. *Let S be a compact Hausdorff space. Take $\mathcal{B}$ as the Borel field generated by the open sets of the topology.* Further, *let S be a topological semigroup.* This means that there is a product for pairs of elements of S and that the product operation is continuous with respect to the given topology (see [5], [6]). Let $\mathcal{B} \times \mathcal{B}$ be the Borel field in the product space $S \times S$ generated by the Borel field $\mathcal{B}$ in the component spaces. The set

$$A_B = \{(x, y) \mid xy \in B\} \in \mathcal{B} \times \mathcal{B} \tag{1}$$

if $B \in \mathcal{B}$ since it is true for open sets B. We can define the product measure $\nu \times \mu$ on $\mathcal{B} \times \mathcal{B}$ in the usual manner [2] and the convolution $\nu * \mu$ of two measures as

$$\nu * \mu(B) = \int c_B(xy)\, d(\nu \times \mu)((x, y)) \tag{2}$$

for $B \in \mathcal{B}$, where c_B is the characteristic function of B.

In Lemmas 1 and 2 some simple properties of a measure on a topological space are obtained. No use is made of the semigroup property.

Lemma 1. *Let ν be a measure on a Borel field $\mathcal{G}$ of a topological space. Let $\mathcal{F}$ be the collection of sets $E \in \mathcal{B}$ such that for any $\epsilon > 0$ there are open and closed sets*

This research was supported by the Office of Naval Research under contract Nonr 908(10). Reproduction in whole or in part is permitted for any purpose of the United States Government.

Journal of Mathematics and Mechanics, Vol. 9, No. 2 (1960).

R.A. Davis et al. (eds.), *Selected Works of Murray Rosenblatt*, Selected Works in Probability and Statistics, 165
DOI 10.1007/978-1-4419-8339-8_19, © Springer Science+Business Media, LLC 2011

O, C such that

$$C \subset E \subset O \tag{3}$$

and

$$\nu(O - E) < \epsilon, \qquad \nu(E - C) < \epsilon. \tag{4}$$

Then $\mathfrak{F}$ is a monotone field.

Consider any two sets E_1, $E_2 \in \mathfrak{F}$. Given any $\epsilon > 0$ there are open sets O_i and closed sets C_i,

$$C_i \subset E_i \subset O_i, \qquad i = 1, 2, \tag{5}$$

with

$$\nu(O_i - E_i) < \tfrac{1}{2}\epsilon, \qquad \nu(E_i - C_i) < \tfrac{1}{2}\epsilon. \tag{6}$$

But

$$\begin{aligned} &\nu(O_1 \cup O_2 - E_1 \cup E_2) \leqq \nu((O_1 - E_1) \cup (O_2 - E_2)) < \epsilon, \\ &\nu(E_1 \cup E_2 - C_1 \cup C_2) \leqq \nu((E_1 - C_1) \cup (E_2 - C_2)) < \epsilon. \end{aligned} \tag{7}$$

Thus $E_1 \cup E_2 \in \mathfrak{F}$.

We now show that $E_1 - E_2 \in \mathfrak{F}$. There is an open set $O_{1,2}$ and a closed set C_2 such that

$$E_1 \cup E_2 \subset O_{1,2}, \qquad C_2 \subset E_2, \tag{8}$$

with

$$\nu(O_{1,2} - E_1 \cup E_2) < \tfrac{1}{2}\epsilon, \qquad \nu(E_2 - C_2) < \tfrac{1}{2}\epsilon. \tag{9}$$

Now $O_{1,2} - C_2$ is open and $E_1 - E_2 \subset O_{1,2} - C_2$. Further,

$$O_{1,2} - C_2 - (E_1 - E_2) = (O_{1,2} - E_1 \cup E_2) \cup (E_2 - C_2) \tag{10}$$

and hence

$$\nu(O_{1,2} - C_2 - (E_1 - E_2)) < \epsilon. \tag{11}$$

Note that the empty set $\emptyset$ and the whole space Ω obviously belong to $\mathfrak{F}$. One can reduce the problem of finding an approximating closed set within $E_1 - E_2$ to the argument given above by complementation. Thus $E_1 - E_2 \in \mathfrak{F}$.

Now consider a monotone sequence of sets $E_n \in \mathfrak{F}$, $E_n \uparrow E$. We shall show that $E \in \mathfrak{F}$. Let n be so large that $\nu(E - E_n) < \frac{1}{2}\epsilon$. There is then a closed set $C \subset E_n$ with $\nu(E_n - C) < \frac{1}{2}\epsilon$. But then $\nu(E - C) < \epsilon$. Given each E_n there is an open set O_n, $E_n \subset O_n$, with $\nu(O_n - E_n) < \epsilon/2^{n+1}$. Now $O = \bigcup O_n$ is open, $E \subset \bigcup O_n$, and

$$\nu(O - E) \leqq \nu(\bigcup (O_n - E_n)) < \epsilon. \tag{12}$$

Thus $\mathfrak{F}$ is a monotone field.

If ν is a measure on the Borel field $\mathcal{G}$ and the monotone field $\mathfrak{F}$ of Lemma 1 *is identical with $\mathcal{G}$, we say that ν is regular on $\mathcal{G}$.*

Lemma 2. *If ν, μ are regular measures on $\mathcal{B}$, the product measure $\nu \times \mu$ is regular on $\mathcal{B} \times \mathcal{B}$. If the product measure $\nu \times \mu$ is regular, the factor measures ν, μ are regular.*

Clearly the product sets with projections on the factor spaces elements of $\mathcal{B}$ have the property of Lemma 1. Therefore the σ-field $\mathcal{B} \times \mathcal{B}$ generated by these sets has this property.

Conversely let $\nu \times \mu$ be regular and $B \in \mathcal{B}$. Then there is a closed set $C \in \mathcal{B} \times \mathcal{B}$, $C \subset B \times \Omega$ with $\nu \times \mu\ (B \times \Omega - C) < \epsilon$. Let $C' \in \mathcal{B} \times \mathcal{B}$ be the set of all points whose x projection is the same as the x projection of some point of C. The set C is closed since $f((x, y)) = x$ is continuous. Let the x projection of the set C' be $C'' \in \mathcal{B}$. Then $C'' \subset B$ and

$$\nu(B - C'') = \nu \times \mu(B \times \Omega - C') \leqq \nu \times \mu(B \times \Omega - C). \tag{13}$$

Lemma 3. *If ν, μ are regular measures on $\mathcal{B}$, the convolution $\nu * \mu$ is regular (on $\mathcal{B}$).*

Let $B \in \mathcal{B}$. Set

$$A_B = \{(x, y) \mid xy \in B\}. \tag{14}$$

Now $A_B \in \mathcal{B} \times \mathcal{B}$, and hence by the regularity of $\nu \times \mu$ there is a closed set $C' \subset A_B$ with

$$\nu \times \mu(A_B - C') < \epsilon.$$

Let

$$C = \{z \mid z = xy \quad \text{for some pair} \quad (x, y) \in C'\}. \tag{15}$$

It is clear that $C' \subset A_C \subset A_B$. Thus

$$\nu * \mu(B - C) = \nu \times \mu(A_B - A_C) < \epsilon. \tag{16}$$

Further, C is a closed set since it is the image of a continuous function on the closed set C'. We can find an open set containing B with the desired properties by complementation.

Let us now consider the family of regular probability measures $\mathfrak{M}$ on the compact topological semigroup S. We shall say that *the sequence of regular probability measures μ_n on S converges to a regular probability measure μ, $\mu_n \to \mu$, if*

$$\int f\, d\mu_n \to \int f\, d\mu \tag{17}$$

for every continuous function f on S.

Since the continuous functions on a compact space are bounded, it is clear that it is enough to have convergence for continuous functions f, $0 \leqq f \leqq 1$.

Let I be the closed unit interval [0, 1]. Consider the product space $\prod_f I_f$ where the coordinate spaces I_f are indexed by continuous functions f, $0 \leqq f < 1$, with the usual Tychonov topology [2]. $\prod I_f$ is a compact space since I_f is compact. Each measure $\mu \in \mathfrak{M}$ can be considered as a point in $\prod I_f$ with component in the fth coordinate space $I_f = \int f\, d\mu$. Let us consider a limit point $c \in \prod I_f$ of the family of measures $\mathfrak{M}$. Clearly $c_f \geqq 0$ for all f and $c_1 = 1$ (1 is the function $f \equiv 1$). Further, given any two functions f_1, f_2, $0 \leqq f_1, f_2, f_1 + f_2 \leqq 1$, and any $\epsilon > 0$ there is a measure $\mu \in \mathfrak{M}$ in the neighborhood

$$\{c' \mid |c'_{f_1} - c_{f_1}| < \epsilon, |c'_{f_2} - c_{f_2}| < \epsilon, |c'_{f_1+f_2} - c_{f_1+f_2}| < \epsilon\}. \tag{18}$$

But this implies that $c_{f_1+f_2} = c_{f_1} + c_{f_2}$. Let $c_f = L(f)$. $L(f)$ is a functional on the continuous functions f, $0 \leqq f \leqq 1$. We can similarly show that $c_{\alpha f} = \alpha c_f$ for $\alpha \geqq 0$ if $0 \leqq \alpha f, f \leqq 1$. Further, $|c_f| \leqq ||f||$ if $0 \leqq f \leqq 1$ (absolute supremum norm). We now extend L to all continuous f. First suppose $f \geqq 0$. Then there is a positive constant α such that $0 \leqq \alpha f \leqq 1$. Let $L(f) = L(\alpha f)/\alpha$. Thus $L(f)$ is defined on all continuous nonnegative f. If f is not nonnegative, introduce $f^+ = \max(0, f)$ and $f^- = f - f^+$. Then f^+, $-f^-$ are nonnegative continuous functions. Let $L(f) = L(f^+) - L(-f^-)$. L is linear on the continuous functions since $L(\alpha f) = \alpha L(f)$, $L(f_1 + f_2) = L(f_1) + L(f_2)$. L is positive, that is, $L(f) \geqq 0$ if $f \geqq 0$. Further, $L(f) \leqq ||f||$ for $f \geqq 0$. But then

$$\begin{aligned} |L(f)| = |L(f^+) - L(-f^-)| &\leqq \max(L(f^+), L(-f^-)) \\ &\leqq \max(||f^+||, ||-f^-||) = ||f|| \end{aligned} \tag{19}$$

for all continuous f, and thus $||L|| \leqq 1$. By the Riesz representation theorem [2] there is a regular probability measure μ on S (since $L(1) = 1$) such that $L(f) = \int f\, d\mu$. Therefore $c \in \mathfrak{M}$ and $\mathfrak{M}$, as a point set in $\prod I_f$, is closed. Thus $\mathfrak{M}$ is compact.

Lemma 4. *The set of regular probability measures $\mathfrak{M}$ on S is a compact set in $\prod I_f$.*

Now let us investigate the continuity of the convolution operation on the set of regular probability measures. First consider a continuous function $f(x, y)$ on $S \times S$. It is easy to see that the set of finite sums of products of continuous functions of the form

$$\sum_{i=1}^{N} g_i(x)h_i(y) \tag{20}$$

with $g_i(x)$, $h_i(y)$ continuous on S is an algebra of functions that separate points. For if $(x, y) \neq (x', y')$ either $x \neq x'$ or $y \neq y'$. Suppose $x \neq x'$. Take a continuous function $g(x)$ that separates x, x' and let $h(y) \equiv 1$. Then $g(x)h(y) = g(x)$ separates (x, y), (x', y'). Since $S \times S$ is compact, by the Stone-Weierstrass theorem any continuous function $f(x, y)$ on $S \times S$ can be uniformly approximated by a function of the form (20) with g_i, h_i continuous on S. We shall now prove Lemma 5.

Lemma 5. *The convolution operation is continuous on the probability measures* $\mathfrak{M}$.

It will be enough to show that the inverse image of a set in the subbasis for regular probability measures is open. Let us consider two probability measures $\nu, \mu \in \mathfrak{M}$. A subbasis neighborhood of $\nu * \mu$ is of the form

$$\left\{c \mid \left|\int f \, d\, \nu * \mu - c\right| < \epsilon\right\} \tag{21}$$

where f is a continuous function, $0 \leqq f \leqq 1$. Now

$$\int f(x) \, d\, \nu * \mu(x) = \int f(xy) \, d\, \nu \times \mu((x, y)). \tag{22}$$

But f is a continuous function of (x, y), and hence there is an approximating function

$$u(x, y) = \sum_{i=1}^{N} g_i(x) h_i(y) \tag{23}$$

with h_i , g_i continuous on S such that

$$|u(x, y) - f(x, y)| < \tfrac{1}{4}\epsilon. \tag{24}$$

Then

$$\left|\int [u(x, y) - f(xy)] \, d\, \nu \times \mu((x, y))\right| < \tfrac{1}{4}\epsilon. \tag{25}$$

Let $M < \infty$ be such that $|g_i|, |h_i| < M$ for all i. Consider the following neighborhoods of ν, μ respectively:

$$\begin{aligned} N(\nu) &= \left\{\eta \mid \left|\int g_i(x) \, d\eta(x) - \int g_i(x) \, d\nu(x)\right| < \delta; i = 1, \cdots, N\right\}, \\ N(\mu) &= \left\{\eta \mid \left|\int h_i(x) \, d\eta(x) - \int h_i(x) \, d\mu(x)\right| < \delta; i = 1, \cdots, N\right\}. \end{aligned} \tag{26}$$

If $\nu' \in N(\nu)$, $\mu' \in N(\mu)$ then

$$\begin{aligned} &\left|\int u(x, y) \, d\, \nu \times \mu((x, y)) - \int u(x, y) \, d\, \nu' \times \mu'((x, y))\right| \\ &\leqq \sum_{i=1}^{N} \left|\int g_i(x) h_i(y) [d\, \nu \times \mu((x, y)) - d\, \nu' \times \mu'((x, y))]\right| \leqq 2N\, \delta M, \end{aligned} \tag{27}$$

and thus if $\delta < \epsilon/4NM$,

$$\left|\int f \, d\, \nu * \mu - \int f \, d\, \nu' * \mu'\right| < \epsilon. \tag{28}$$

The following theorem is an immediate consequence of the results obtained thus far.

Theorem 1. *The family of regular probability measures $\mathfrak{M}$ on S with the convolution operation as a multiplicative operation is a compact Hausdorff topological semigroup.*

Limit Properties of Convolution Sequences. Let ν be a regular probability measure on S. We shall now consider limit properties of the sequence of measures $\nu^{(1)} = \nu, \nu^{(2)} = \nu * \nu, \cdots, \nu^{(n+1)} = \nu^{(n)} * \nu, \cdots$ generated from ν by convolution. A classical probability problem of this sort arises in considering the limit properties of sums of independent, identically distributed random variables. The values taken on by the random variables (since they are numerical-valued) commute with each other. In our case, we are interested in limit properties of products of independent, identically distributed elements of a semigroup (possibly a semigroup of operators) and a non-abelian semigroup is of considerable interest to us.

The notion of the spectrum of a regular measure ν on S is of particular interest. *The spectrum $\Sigma(\nu)$ of a measure $\nu \in \mathfrak{M}$ consists of the points $s \in S$ such that any open set O containing s has positive ν measure.* The spectrum is clearly a closed set. Given any two subsets A, B of the semigroup S let $AB = \{ss' \mid s \in A, s' \in B\}$. Now the closure of the set of elements $\bigcup_n (\Sigma(\nu))^n$ is a closed semigroup and hence is compact. This is the relevant semigroup insofar as the sequence of convolutions $\nu^{(n)}$ is concerned. Let us call this subsemigroup *the semigroup generated by the spectrum of ν*. Since this semigroup is of importance in considering the limit properties of the sequence $\nu^{(n)}$, we shall take $S = \overline{\bigcup_n (\Sigma(\nu))^n}$ ($\bar{A}$ is the closure of A).

An element e of S is called an idempotent if $e^2 = e$. Every compact Hausdorff semigroup has at least one idempotent [5]. By a left (right) ideal of S we mean a nonvacuous subset L (R) of S such that $SL \subset L$ ($RS \subset R$). If M is both a left and right ideal of S, M is called a (two-sided) ideal of S. The minimal two-sided ideal K of S exists and is given by $K = \bigcap_e SeS$ [5]. *The minimal ideal K is called the kernel of the semigroup S.* K is closed and hence compact [5].

We shall say that *μ is a limit measure of the sequence $\nu^{(n)}$, $n = 1, 2, \cdots$, if there is a subsequence n_k such that*

$$\mu = \lim_{k \to \infty} \nu^{(n_k)}.$$

Theorem 2. *Let ν be a regular probability measure on the compact Hausdorff semigroup S. Every limit measure μ of the sequence $\nu^{(n)}$ (ν regular) has all its mass concentrated on the kernel of the semigroup generated by the spectrum of ν.*

It is clear that we can take $S = \overline{\bigcup (\Sigma(\nu))^n}$. Let k be an element of the kernel K of the semigroup generated by the spectrum of ν. Let $N(k)$ be any neighborhood of k. Then for m sufficiently large there are elements $s_1, \cdots, s_m \in \Sigma(\nu)$ such that $\prod_{i=1}^m s_i \in N(k)$. By the continuity of the multiplicative operation there are neighborhoods $N(s_i)$ of s_i, $i = 1, \cdots, m$, such that $\prod_{i=1}^m N(s_i) \subset N(k)$. Let O be any open set containing K. Then there must be a neighborhood of k, $N(k)$, such that $SN(k)$, $N(k)S \subset O$. For given any point $s \in S$ there are neighborhoods $N_{1s}(k)$, $N_{2s}(k)$ of k and a neighborhood $N(s)$ of s such that $N_{1s}(k)N(s)$, $N(s)N_{2s}(k) \subset O$ since $SK = KS = K$. Let $N_s(k) = N_{1s}(k) \cap N_{2s}(k)$. Then $N(s)N_s(k)$, $N_s(k)N(s) \subset O$. The open sets $N(s)$ cover S, and hence there are a finite number of s_i^1's, $i = 1, \cdots, n$, such that the $N(s_i^1)$ cover S (because of the compactness of S). $N(k) = \bigcap_1^n N_{s_i^1}(k)$ is a neighborhood of k such that $SN(k)$, $N(k)S \subset O$. Given this neighborhood $N(k)$ of k choose a finite number of elements $s_1, \cdots, s_m$ with corresponding neighborhoods $N(s_i)$, $i = 1, \cdots, m$, such that $\prod_{i=1}^m N(s_i) \subset N(k)$. Now consider the infinite product space of sequences $(t_1, t_2, \cdots)$ of elements t_i of S with product measure generated by ν. Now $\nu(N(s_i)) > 0$, $i = 1, \cdots, m$. By the lemma of Borel-Cantelli there is at least one block of m successive elements from $N(s_1), \cdots, N(s_m)$ respectively in any such sequence with measure one. Take n sufficiently large so that the measure of the set with a block of m such successive elements in truncated sequences of length n is greater than $1 - \epsilon$. It then follows that $\nu^{(n+1)}(O) > 1 - \epsilon$. But this implies that for any limit measure μ of the sequence $\nu^{(n)}$, $\mu(O) = 1$. However, this is true for any open set O containing K. Therefore $\mu(K) = 1$.

Lemma 6. *Let $f((x, y))$ be a continuous function on $S \times S$ where S is a compact Hausdorff space. Then the family of functions $f_y(x) = f((x, y))$ defined on S and indexed by $y \in S$ is equicontinuous for all $y \in S$.*

To say that the family $\{f_y(x)\}$ is equicontinuous is equivalent to stating that for any given $\epsilon > 0$ and any $x \in S$ there is a neighborhood $N(x)$ of x such that for all $x' \in N(x)$

$$|f_y(x') - f_y(x)| < \epsilon \tag{29}$$

independently of y. Suppose the family is not equicontinuous. Then for some $\epsilon > 0$ and some $x \in S$ there is no neighborhood $N(x)$ of x such that $|f_y(x') - f_y(x)| < \epsilon$ for all $x' \in N(x)$ and all y. Index the neighborhoods N_α of x by α. Given any neighborhood N_α of x there is a point $(x_\alpha, y_\alpha) \in S \times S$, $x_\alpha \in N_\alpha$, such that $|f_{y_\alpha}(x_\alpha) - f_{y_\alpha}(x)| \geqq \epsilon$. The indices α are partially ordered if we set $\alpha \leqq \alpha'$ when $N_{\alpha'} \subset N_\alpha$. Further, the indices are a directed set since any finite subset of them has an upper bound. Thus $\{x_\alpha\}$ and $\{(x_\alpha, y_\alpha)\}$ are generalized sequences of elements in S and $S \times S$ respectively [2]. Further, $\lim_\alpha x_\alpha = x$. The generalized sequence (x_α, y_α) has a cluster (limit) point (x, y). But then $|f_y(x) - f_y(x)| = |0| = 0 \geqq \epsilon > 0$, a contradiction.

Theorem 3. *Let ν be a regular probability measure on the compact Hausdorff semigroup S. There is then a unique limit measure of the averaged convolution sequence*

$$\mu = \lim_n \frac{1}{n} \sum_{j=1}^n \nu^{(j)}. \tag{30}$$

Further,

$$\mu * \nu = \nu * \mu = \mu^{(2)} = \mu. \tag{31}$$

Consider the regular probability measure ν as an operator T on the family of continuous functions f on S, that is,

$$(Tf)(x) = \int f(xy)\, d\nu(y).$$

The family of functions $f_y(x) = f(xy)$ is an equicontinuous family of functions by Lemma 6, and hence T maps the continuous functions into continuous functions. Consider iterates of the operator T, T^n, acting on f. Then

$$(T^n f)(x) = \int (T^{n-1}f)(xy)\, d\nu(y) = \int f(xy)\, d\nu^{(n)}(y). \tag{32}$$

Let

$$\bar{T}^n = \frac{1}{n}\sum_{j=1}^{n} T^j. \tag{33}$$

The sequence of functions $(T^n f)(x)$ is an equicontinuous family of functions since $f_y(x) = f(xy)$ is equicontinuous. This implies that $(\bar{T}^n f)(x)$ is equicontinuous and hence $\{\bar{T}^n f\}$ is strongly compact (sequentially). Further, $\|T\| = 1$ and $T^n f/n \to 0$. Therefore $\bar{T}^n \to \bar{T}$ strongly where $\bar{T}$ is linear,

$$\|\bar{T}\| = 1, \qquad \bar{T}T = T\bar{T} = \bar{T}^2 = \bar{T}, \tag{34}$$

and $\bar{T}$ is positivity preserving. Hence there is a regular probability measure μ such that

$$\int \bar{T}f(x)\, d\nu(x) = \int f(x)\, d\mu(x). \tag{35}$$

But then

$$\frac{1}{n}\int f(x)\, d\Big(\sum_{j=1}^{n} \nu^{(j)}(x)\Big) \to \int f(x)\, d\mu(x) \tag{36}$$

for all continuous f, and hence $(1/n)\sum_{j=1}^{n} \nu^{(j)} \to \mu$.

Limit Measures on the Kernel of the Semigroup. Consider limit measures of the kind discussed in Theorem 3. These measures are the idempotents in the semigroup of probability measures on S. For simplicity, we restrict ourselves to those measures ν in this semigroup of measures whose spectrum generates S. By Theorem 2 the limit measure μ of $(1/n)\sum_1^n \nu^{(j)}$ is concentrated on the kernel K of S. In Theorem 4 we shall show that the spectrum of a limit measure μ is K. The detailed structure of such measures will be examined in this section.

However, we will require certain basic results on the structure of the kernel K of a compact Hausdorff semigroup. When studying K, it is natural to look at it with the relative topology. K is a completely simple semigroup, that is, $KkK = K$ for all $k \in K$ and K has a minimal left and a minimal right ideal [8]. Every compact completely simple semigroup can be represented as the product space $T \times X \times Y$ of a compact topological group T and compact Hausdorff spaces X and Y where the multiplication is given by

$$(t, x, y)(t', x', y') = (t\varphi(x, y')t', x', y) \tag{37}$$

with φ a continuous function on the product space $X \times Y$ into T [8]. We shall therefore identify K with such a space $T \times X \times Y$ and the accompanying φ function.

Theorem 4. *Let S be a compact Hausdorff semigroup and ν a regular measure on S whose spectrum generates S. The spectrum of the limit measure $\mu = \lim (1/n) \sum_1^n \nu^{(i)}$ is K, the kernel of S.*

Let k be any element of K. Let $N(k)$ be a neighborhood of k. It is enough to show that $\mu(N(k)) > 0$. Any point $k \in K$ has the form $k = (t, x, y)$. Let $x = x(k)$, $y = y(k)$ be the x, y coordinates of k. Further, let $N_{x,y}(k)$ be the collection of all points k (in K, of course) with x, y coordinates the same as those of points in $N(k) \cap K$. Take k' to be an element in the spectrum of μ. There is then a neighborhood $N_1(k)$ such that

$$N_1(k)k'N_1(k) \subset N_{x,y}(k). \tag{38}$$

This follows from the continuity of the multiplicative operation. But then $\mu(N_{x,y}(k)) > 0$ for any neighborhood $N(k)$ of k. For

$$\mu(N_{x,y}(k)) = \int c_{N_{x,y}(k)}(s_1 s_2 s_3)\, d\nu^{(n)}(s_1)\, d\mu(s_2)\, d\nu^{(n)}(s_3) \tag{39}$$

and for some n

$$\nu^{(n)}(N_{x,y}(k)) > 0. \tag{40}$$

Thus there is a point $k'' \in N_{x,y}(k)$ in the μ spectrum.

Given the neighborhood $N(k)$ there is a point $k' \in N(k)$ with a corresponding point $k'' \in N_{x,y}(k)$, $x(k'') = x(k')$, $y(k'') = y(k')$, that is, in the μ spectrum. There are then points k_1'', k_2'' such that $k_1''k''k_2'' = k'$. This implies that there are neighborhoods $N(k_1'')$, $N(k'')$, $N(k_2'')$ of k_1'', k'', k_2'' such that

$$N(k_1'')N(k'')N(k_2'') \subset N(k') \subset N(k) \tag{41}$$

by the continuity of the multiplicative operation and the structure of K. Then all of K is the spectrum of μ since

$$\nu^{(n)} * \mu * \nu^{(m)} = \mu. \tag{42}$$

The proof is complete.

Let us now consider a product measure $\bar{\mu}$ on K of the form

$$\bar{\mu} = \chi \times \alpha \times \beta \tag{43}$$

where χ is the normed Haar measure ($\chi(T) = 1$) of the compact topological group T and α and β are regular probability measures on X, Y respectively.

We shall show that such a measure is an idempotent on K. First consider a set $G \varepsilon \mathcal{B}$ which is a product set, that is, $G = A \times B \times C$ where A, B, C are sets of the Borel fields on T, X, Y respectively. Let $s^{-1}G = \{s' \mid ss' \varepsilon G\}$. But

$$\bar{\mu}^{(2)}(G) = \int \bar{\mu}(s^{-1}G)\bar{\mu}\,(ds) = \int_{y(s)\varepsilon C} \bar{\mu}(s^{-1}G)\bar{\mu}\,(ds). \tag{44}$$

Now

$$\bar{\mu}(s^{-1}G) = \int_B \int_Y \chi([t\varphi(x(s), y'')]^{-1}A)\,d\alpha(x')\,d\beta(y'') = \chi(A)\alpha(B) \tag{45}$$

by the invariance of Haar measure. Thus

$$\bar{\mu}^{(2)}(G) = \int_{y(s)\varepsilon C} \chi(A)\alpha(B)\bar{\mu}\,(ds) = \chi(A)\alpha(B)\beta(C) = \bar{\mu}(G). \tag{46}$$

Since $\bar{\mu}^{(2)}(G) = \bar{\mu}(G)$ for all product sets, the relation holds for all G in the field of finite sums of product sets and hence for all G in the Borel field $\mathcal{B}$ generated by these sets.

Lemma 7. *Let K be a completely simple compact semigroup. Any measure $\bar{\mu}$ of the form* (43) *is an idempotent measure on K.*

We shall now show that for a fairly wide class of completely simple compact semigroups, the $\bar{\mu}$ measures are the only idempotent measures.

Theorem 5. *Let K be a completely simple compact semigroup with the function $\varphi(x, y)$ of the form $\varphi(x, y) = \varphi_1(x)\varphi_2(y)$. Then every idempotent measure μ with spectrum the whole semigroup is a $\bar{\mu}$ measure.*

Assume that the function $\varphi(x, y) = \varphi_1(x)\varphi_2(y)$. Introduce the following function ψ of $k = (t, x, y)$:

$$\psi(k) = \varphi_2(y)\,t\varphi_1(x). \tag{47}$$

Now $\psi(k)$ is a continuous function of k since φ is. Suppose μ is an idempotent measure on K with its spectrum K. Let ν be the derived measure

$$\nu(A) = \mu(\psi(k)\,\varepsilon\,A) \tag{48}$$

on the compact group T. Here A is an element of the Borel field on T. Since μ is an idempotent measure on K, it follows that ν is an idempotent measure on T [4]. The spectrum of ν is the whole group T since the spectrum of μ is the whole semigroup K. However, this implies that ν is the normed Haar measure of T. Let

$$\begin{aligned} \nu_1(A, C) &= \mu(k \mid t\varphi_1(x)\,\varepsilon\,A,\ y\,\varepsilon\,C), \\ \nu_2(A, B) &= \mu(k \mid \varphi_2(y)t\,\varepsilon\,A,\ x\,\varepsilon\,B). \end{aligned} \tag{49}$$

Now if μ is an idempotent measure $\mu^{(3)} = \mu$, and hence

(50) $$\nu(A \times B \times C) = \int \nu_2(dt, B)\nu(t'^{-1}At)\nu_1(dt', C) = \nu_2(\cdot, B)\nu_1(\cdot, C)\nu(A)$$

since ν is the Haar measure of T. Thus μ is a $\bar{\mu}$ measure.

Corollary. *If S is a compact Abelian Hausdorff semigroup, the kernel of S is an Abelian group. Any idempotent measure is concentrated on the kernel. If the spectrum of the measure is the kernel, it is the normed Haar measure of the kernel.*

This is almost an immediate consequence of Theorem 4.

We shall now show that the result of Theorem 5 is true for a general finite completely simple semigroup. This makes one suspect that Theorem 5 very likely holds without the restriction on $\varphi(x, y) = \varphi_1(x)\varphi_2(y)$.

Theorem 6. *Let K be a finite completely simple semigroup. Every idempotent measure μ with spectrum the whole semigroup is a $\bar{\mu}$ measure.*

Since μ is idempotent

(51) $$\mu(k') = \sum_k \mu(k'k^{-1})\mu(k)$$

where $k'k^{-1} = \{a \mid ak = k'\}$. Note that $k'k^{-1}$ is the null set except when $x(k) = x(k')$. Thus

(52) $$\mu(k''k^{-1}) = \sum_{k'} \mu(k''k^{-1}k'^{-1})\mu(k').$$

But $k''k^{-1} = \{a \mid ak = k''\}$, $k''k^{-1}k'^{-1} = \{b \mid bk'k = k''\} = k''k'''^{-1}$ where $k''' = k'k$, so that

(53) $$\mu(k''k^{-1}) = \sum_{k'} \mu(k''k'^{-1})\mu(k'k^{-1}).$$

We can then regard $\mu(kk'^{-1}) = p(k \mid k')$ as the conditional probability of going from k' to k for a Markov chain with state space the states k. The measure μ assigns positive measure to each point k. Notice that there is an irreducible class of states for the Markov chain with transition probabilities $\mu(kk'^{-1})$ corresponding to each distinct x value. By the classical results on Markov chains

(54) $$p(k \mid k') = \begin{cases} p(k) & \text{if } \ x(k') = x(k) \\ 0 & \text{otherwise.} \end{cases}$$

This tells us that

(55) $$\mu(k''k'^{-1}) = \mu(\bigcup_x (t''t'^{-1}\varphi(x, y')^{-1}, x, y''))$$

is independent of t', y' if $x(k'') = x(k')$. But expression (55) is then independent of t'', t', y'. Then

(56) $$\mu(k') = \sum_{x(k')=x(k)} \mu(k'k^{-1})\mu(k) = \sum_{x(k)=x'} a(x, y')\mu(s) = b(x', y').$$

Thus

$$\mu((\cdot, x', y')) = \sum_{x'', t, y} \mu((\cdot, x'', y'))\mu((t, x', y)) = a(x')b(y') \tag{57}$$

and

$$\mu((t, x, y)) = \frac{1}{n} a(x)b(y) \tag{58}$$

where n is the number of elements in the finite group T.

Notice that if K contains exactly one minimal left (right) ideal, the idempotent measure μ is an r^* (l^*) invariant measure in the sense of ROSEN [6].

A Special Class of Semigroups. In certain contexts it is of especial interest to determine whether the semigroup S is such that all elements of the kernel K are left (right) annihilators, that is, $kS = k$ ($Sk = k$) for every $k \in K$. This is of particular interest in certain structure problems for stationary processes (see [7]). We shall show that in this case not only do the averaged convolutions converge, but also the convolution sequence itself.

Theorem 7. *Let S be a compact Hausdorff semigroup with $kS = k$ ($Sk = k$) for every $k \in K$, the kernel of S. If the semigroup generated by the spectrum of the regular measure ν on S is S,*

$$\lim_{n\to\infty} \nu^{(n)} = \mu \tag{59}$$

exists with μ having K as its spectrum.

Suppose $kS = k$ for all $k \in K$. Let

$$\mu = \lim_{n\to\infty} \frac{1}{n} \sum_{j=1}^{n} \nu^{(j)}. \tag{60}$$

We shall show that $\mu = \lim_n \nu^{(n)}$. Let $N(k)$ be a neighborhood of $k \in K$. For sufficiently large n $\nu^{(n)}(N(k)) > 0$. Let $K_N = \{k' \in K \mid k' \in N(k)\}$. Take $C \subset K_N$ a closed set of points such that $\mu(C) > \mu(K_N) - \epsilon$, $\epsilon > 0$. Given a point $k' \in C$, there is a neighborhood $N(k')$ of k' such that $N(k')S \subset N(k)$. For, given each point $s \in S$, there are neighborhoods $N(s)$ of s and $N_s(k')$ of k' such that $N_s(k')$ $N(s) \subset N(k)$. Since S is compact there are a finite number of points s_i such that $N(s_i)$ cover S. Let $N(k') = \bigcap N_{s_i}(k')$. It is then clear that $N(k')S \subset N(k)$. Now $\bigcup_{k' \in C} N(k')S \subset N(k)$. Let $N = \bigcup_{k' \in C} N(k')$. Then for some sufficiently large n, $\nu^{(n)}(N) > \mu(K_N) - \epsilon$. But then for all $m > n$

$$\nu^{(m)}(N(k)) = \int \nu^{(n)}(N(k)s^{-1})\nu^{(m-n)}\,(ds) \geqq \nu^{(n)}(N) > \mu(K_N) - \epsilon. \tag{61}$$

Thus

$$\varliminf_{m\to\infty} \nu^{(m)}(N(k)) \geqq \mu(N(k)). \tag{62}$$

This is not possible unless

$$\lim_{m\to\infty} \nu^{(m)}(N(k)) = \mu(N(k)). \tag{63}$$

Thus $\lim \nu^{(m)} = \mu$. The proof is essentially the same when $Sk = k$ for every $k \varepsilon K$.

A very simple but interesting application of Theorem 7 can be given. Let ν be a regular measure on the $n \times n$ probability matrices, that is, matrices with nonnegative elements and row sums one. Suppose the spectrum of ν contains a matrix with identical rows. The assumption of Theorem 6 is then satisfied by the semigroup of matrices S generated by the spectrum of ν. Thus $\lim \nu^{(m)}$ exists and is concentrated on the kernel K of S which consists of matrices with identical rows. Let $a_1, \cdots, a_n, \sum a_i = 1$, be the random elements of the common row for the limit measure concentrated on the matrices of K. Then the characteristic function

$$\begin{aligned}
\varphi(z_1, \cdots, z_{n-1}) &= \mathrm{E} \exp\left\{i \sum_{j=1}^{n-1} z_j a_j\right\} \\
&= \mathrm{E} \exp\left\{i \sum_1^{n-1} z_j \left(\sum_{k=1}^{n-1} a_k p_{kj} + \left[1 - \sum_1^{n-1} a_k\right] p_{nj}\right)\right\} \\
&= \mathrm{E}\left[\mathrm{E}\left(\exp\left\{i \sum_1^{n-1} z_j p_{nj}\right\} \exp\left\{i \sum_1^{n-1} z_j \sum_1^{n-1} (p_{kj} - p_{nj}) \mid p_{kj}\right\}\right)\right] \\
&= \int \varphi\left(\sum_1^{n-1} z_j(p_{1j} - p_{nj}), \cdots, \sum_1^{n-1} z_j(p_{n-1,j} - p_{nj})\right) \\
&\qquad \cdot \exp\left\{i \sum_1^{n-1} z_j p_{nj}\right\} d\nu(p).
\end{aligned} \tag{64}$$

BIBLIOGRAPHY

[1] R. BELLMAN, Limit theorems for noncommutative operations I, *Duke Math. Journ.* **21**, (1954).

[2] N. DUNFORD & J. T. SCHWARTZ, *Linear Operators, Part I*, 1958 (New York).

[3] U. GRENANDER, Some nonlinear problems in probability theory, *H. Cramér Volume* (1959).

[4] Y. KAWADA & K. ITO, On the probability distribution on a compact group I, *Proc. Phys.-Math. Soc. of Japan*, **22** (1940), pp. 977–998.

[5] K. NUMAKURA, On bicompact semigroups, *Math. Journ. of Okayama Univ.* (1952), pp. 99–108.

[6] W. G. ROSEN, On invariant means over compact semigroups, *Proc. Amer. Math. Soc.*, **7** (1956), pp. 1076–1082.

[7] M. ROSENBLATT, Stationary processes as translates of functions of independent random variables, *Journ. Math. and Mech.*, **8** (1959), pp. 665–682.

[8] A. D. WALLACE, The Rees-Suschkewitsch structure theorem for compact simple semigroups, *Proc. Nat. Acad. Sci. U.S.A.*, **42** (1956), pp. 430–432.

Indiana University
Bloomington, Indiana

Addendum to

"Limits of Convolution Sequences of Measures on a Compact Topological Semigroup"*

M. ROSENBLATT

It should be noted that in Theorem 5 the functions $\varphi_1(x)$, $\varphi_2(y)$ are assumed to be functions on X, Y respectively into the group T.

---

* *Journal of Mathematics and Mechanics*, 9 (1960), 293–305.

R.A. Davis et al. (eds.), *Selected Works of Murray Rosenblatt*, Selected Works in Probability and Statistics, DOI 10.1007/978-1-4419-8339-8_20, © Springer Science+Business Media, LLC 2011

INDEPENDENCE AND DEPENDENCE

M. ROSENBLATT
BROWN UNIVERSITY

1. Introduction

A stochastic process is commonly used as a model in studying the behavior of a random system through time. It will be convenient for us to take the stochastic process $\{x_t\}$ as *discrete in time* $t = \cdots, -1, 0, 1, \cdots$. Processes of independent random variables are the simplest and most completely understood. It is, however, clear that these are extremely limited in scope as models and one must have recourse to dependent processes (the random variables x_t not independent) in order to have any power in description. For simplicity, *let us* further *restrict ourselves to* processes that are stable through time, *stationary processes*. For such processes the probabilities of events shifted through time remain the same, that is, the probability

$$P\{x_{t_1+h} \leqq a_1, \cdots, x_{t_n+h} \leqq a_n\} \tag{1}$$

is independent of h. Such models occur fairly often in the physical sciences. If mean properties of the process are to be capable of being estimated reasonably well from part of a realization of the process, some form of asymptotic independence for blocks of random variables of the process that are widely separated must be satisfied. This is, in effect, the gist of many of the results in ergodic theory. Two types of interesting problems are posed. *The first* of these is *concerned with reasonable notions of asymptotic independence* and *what types of processes satisfy them. The second is that of characterizing those processes* $\{x_t\}$ *that can be constructed out of independent processes by a function and its shifts*, that is,

$$x_t = f(\cdots, \xi_{t-1}, \xi_t, \xi_{t+1}, \cdots) \tag{2}$$

where $\{\xi_t\}$ is a process of independent random variables. Neither of these questions have elicited satisfactory answers. However, there are some small results that do give insights into the problems. The object of this paper is a presentation and discussion of a few of these limited results.

2. Mixing

Ergodicity itself might be thought of as a form of asymptotic independence. However, the most obvious formulation of asymptotic independence is the

This research was supported by the Office of Naval Research.

stronger *mixing condition*. Suppose T is the shift (in time) operator associated with the stationary process. The process is said to be mixing if

$$P\{A \cap T^n B\} \to P\{A\}P\{B\} \tag{3}$$

as $n \to \infty$ for any two events A, B. The normal stationary processes are the simplest stationary processes and Maruyama [6] has obtained a simple and intuitive necessary and sufficient condition for such a process to be mixing. To verify the mixing property, it is enough to verify property (3) for any two events A, B determined by conditions on a finite number of the random variables x_t.

MARUYAMA'S THEOREM. *A necessary and sufficient condition for a normal stationary process to be mixing is that the covariance* $r_t = \text{Cov}\,(x_{\tau+t}, x_\tau)$ *tend to zero as* $|t| \to \infty$.

It is clear that no generality is lost by assuming that the mean of the process is zero. The necessity of the condition that $r_t \to 0$ as $|t| \to \infty$ is then almost completely apparent. We approximate the random variables $x_{\tau+t}, x_\tau$ by step functions in the following way. Let

$$A(k, \epsilon) = \{\omega | k\epsilon \leqq x_\tau(\omega) < (k+1)\epsilon\} \tag{4}$$

and

$$c_{A(k,\epsilon)}(\omega) = \begin{cases} 1 & \text{if} \quad \omega \in A(k, \epsilon) \\ 0 & \text{otherwise.} \end{cases} \tag{5}$$

Set

$$x_\tau^{(N,\epsilon)}(\omega) = \sum_{k=-N}^{K=N} k\epsilon c_{A(k,\epsilon)}(\omega). \tag{6}$$

Then for sufficiently large N

$$\begin{aligned} &E|x_\tau^{(N,\epsilon)}(\omega) - x_\tau(\omega)| = E|x_\tau^{(N,\epsilon)}(T^t\omega) - x_{t+\tau}(\omega)| < \delta, \\ &|r_t - Ex_\tau^{(N,\epsilon)}(T^t\omega)x_\tau^{(N,\epsilon)}(\omega)| \leqq E|x_{t+\tau}(\omega)x_\tau(\omega) - x_\tau^{(N,\epsilon)}(T^t\omega)x_\tau^{(N,\epsilon)}(\omega)| < \delta. \end{aligned} \tag{7}$$

Since the mean of the process is zero, for N sufficiently large

$$|Ex_\tau^{(N,\epsilon)}(\omega)| = |Ex_\tau^{(N,\epsilon)}(T^t\omega)| < \delta. \tag{8}$$

However by the mixing condition

$$\begin{aligned} Ex_\tau^{(N,\epsilon)}(T^t\omega)x_\tau^{(N,\epsilon)}(\omega) &= \sum_{k,j=-N}^{N} k\epsilon j\epsilon E c_{A(k,\epsilon)}(T^t\omega)c_{A(j,\epsilon)}(\omega) \\ &= \sum_{k,j=-N}^{N} kj\epsilon^2 P[A(j,\epsilon) \cap T^t A(k,\epsilon)] \\ &\to \sum_{k,j=-N}^{N} kj\epsilon^2 P[A(j,\epsilon)]P[A(k,\epsilon)] \\ &= [Ex_\tau^{(N,\epsilon)}(\omega)]^2. \end{aligned} \tag{9}$$

But this limit relation coupled with the inequalities (7) indicate that

$$\overline{\lim_{|t|\to\infty}}\, |r_t| < 3\delta \tag{10}$$

and since this is true for any δ, we have $\lim_{|t|\to\infty} r_t = 0$.

The sufficiency of this condition follows just as easily. Let A, B be determined by conditions on blocks of k successive random variables (these could in fact be taken to be the same blocks if k is sufficiently large). Then

$$(11)\qquad P\{A \cap T^n B\} = (2\pi)^{-k}|M|^{-1/2} \int_A \int_B \exp\left\{-\frac{1}{2}(u, v)M^{-1}(u, v)'\right\} du\, dv,$$

where u, v are two row k-vectors and (u, v) the $2k$-vector formed by joining them. The matrix M is the covariance matrix of two blocks of k successive random variables of the process separated by $n - k$ random variables. Write M in the form

$$(12)\qquad M = \begin{pmatrix} R_1 & R_2 \\ R_2' & R_1 \end{pmatrix}$$

where R_1 is the $k \times k$ covariance matrix of a block of k successive random variables and R_2 the $k \times k$ matrix of covariances of the one block of k with the other block removed from it by $n - k$ random variables (R_2' is the transpose of R_2). The fact that $r_t \to 0$ as $|t| \to \infty$ implies that M^{-1} exists. It is in fact given by

$$(13)\qquad M^{-1} = \begin{pmatrix} A & B \\ B' & C \end{pmatrix}$$

with $A = (R_1 - R_2R_1^{-1}R_2')^{-1}$, $C = (R_1 - R_2'R_1^{-1}R_2)^{-1}$, $B = -R_1^{-1}R_2C = -AR_2R_1^{-1}$. The matrix R_1 is fixed since k is fixed. But R_2 is a $k \times k$ matrix whose entries are the covariances r_t with $|t| > n - k$. They approach zero uniformly as $n \to \infty$ so that R_2 approaches the null matrix. Thus

$$(14)\qquad M^{-1} \to \begin{pmatrix} R_1^{-1} & 0 \\ 0 & R_1^{-1} \end{pmatrix}$$

$$(15)\qquad P\{A \cap T^n B\} \to (2\pi)^{-k}|R_1|^{-1} \int_A \int_B \exp\left\{-\frac{1}{2}[uR_1^{-1}u' + vR_1^{-1}v']\right\} du\, dv$$
$$= P\{A\}P\{B\}.$$

Stronger mixing conditions are sometimes invoked in deriving results like central limit theorems for dependent processes. The following condition is an example of such a strong mixing condition: *Let A, B be any events determined by conditions on the random variables x_k, $k \leq m$, and x_k, $k \geq n$, respectively with $n > m$. The process $\{x_k\}$ is said to satisfy the strong mixing condition M if*

$$(16)\qquad |P\{A \cap B\} - P\{A\}P\{B\}| \leq d(n - m)$$

for all such events A, B and some function d where d is a function on the positive integers n that decreases to zero as $n \to \infty$. Notice that such a mixing condition is a uniform mixing condition in what we might call the distance $n - m$ between the events A, B.

Let $\{x_k\}$ be a stationary process with mean zero, $Ex_k \equiv 0$, that satisfies the strong mixing condition M. Further let the process satisfy the two moment conditions

$$E\left|\sum_{j=1}^{n} x_j\right|^2 = \sum_{k=-n}^{n} (n - |k|) r_k = h(n) \to \infty \qquad \text{as} \quad n \to \infty, \tag{17}$$

$$\begin{aligned} E\left|\sum_{j=1}^{n} x_j\right|^4 &= \sum_{j_i=1}^{n} E x_{j_1} x_{j_2} x_{j_3} x_{j_4} \\ &= \sum_{j_i=1}^{n} [r_{j_1-j_2} r_{j_3-j_4} + r_{j_1-j_3} r_{j_2-j_4} + r_{j_1-j_4} r_{j_2-j_3} + P_{j_2-j_1, j_3-j_1, j_4-j_1}] \\ &= 3[h(n)]^2 + \sum_{|a_i| \leq n} P_{a_1, a_2, a_3} N(a_1, a_2, a_3) \\ &= O[h(n)]^2. \end{aligned} \tag{18}$$

The sums $\sum_{j=1}^{n} x_j$ can then be shown to be asymptotically and nontrivially normally distributed when suitably normed as $n \to \infty$. Of course, the statement of nontrivialness means that the limiting normal distribution does not have variance zero. The proof of this limit theorem can be found in [7]. The following two examples in part at least indicate the limits of the central limit theorem cited above.

In the first example the process $\{x_k\}$ is taken to be

$$x_k = \epsilon_k - \epsilon_{k-1} \tag{19}$$

where the ϵ_k are independent, identically distributed nonnormal random variables with finite 4th moment and vanishing 1st moment. Since $\sum_{k=1}^{n} x_k = \epsilon_n - \epsilon_0$, it is readily seen that $h(n)$ does not diverge as $n \to \infty$. Thus, we get a nonnormal limiting distribution, namely that of $\epsilon_1 - \epsilon_0$, without norming the sum at all. The process certainly satisfies the strong mixing condition and moment condition (18) but not moment condition (17). The difficulty arises because the process $\{x_k\}$ has too little spectral mass in the vicinity of $\lambda = 0$. Its spectral density is $f(\lambda) = (1/2\pi)|1 - \exp(-i\lambda)|^2$.

The second example has greater interest. Consider a normal process $\{y_k\}$ with mean zero and covariance sequence $r_k = (1 + k^2)^{-\gamma}$ with $\gamma > 0$. The spectral density of the process $\{y_k\}$ is

$$a(\lambda) = \sum_{k=-\infty}^{\infty} b(\lambda + 2k\pi) \tag{20}$$

where

$$b(\lambda) = b(|\lambda|) = \{\Gamma(\gamma)\}^{-2} e^{-|\lambda|} \int_0^{\infty} \mu^{\gamma-1} (\mu + |\lambda|)^{\gamma-1} e^{-2\mu} \, d\mu. \tag{21}$$

The spectral density $a(\lambda)$ is continuous and bounded away from zero if $|\lambda| > \epsilon > 0$. If $\gamma < 1/2$, it has a singularity of the form $|\lambda|^{2\gamma-1}$ in the vicinity of $\lambda = 0$. However, it should be noted that the process is in any case, purely nondeterministic and therefore mixing in the ordinary sense. Let the process $\{x_k\}$ be of the form $x_k = y_k^2 - 1$. The covariance sequence of the process $\{x_k\}$ is $r_k = 2(1 + k^2)^{-2\gamma}$. We shall be interested in processes for which $\gamma < 1/4$. The spectral density $f(\lambda)$ of $\{x_k\}$ has a singularity of the form $|\lambda|^{4\gamma-1}$ in the neighborhood of $\lambda = 0$ as can be seen from (21). We shall see that $n^{-1+2\gamma} \sum_{k=1}^{n} x_k$ has a limiting nonnormal distribution as $n \to \infty$. For the characteristic function of $n^{-1+2\gamma} \sum_{k=1}^{n} x_k$ is

$$|I - 2itn^{-1+2\gamma}R|^{-1/2} \exp\{-in^{2\gamma}tr_0\} = \exp\left\{\sum_{k=2}^{\infty} (2itn^{-1+2\gamma})^k \operatorname{sp}(R^k)/k\right\} \tag{22}$$

and

$$\begin{aligned}(n^{-1+2\gamma})^k &\operatorname{sp}(R^k) \\ &= (n^{-1+2\gamma})^k \sum_{i_j=1}^{n} r_{i_1-i_2} r_{i_2-i_3} \cdots r_{i_k-i_1} \\ &\to \underbrace{\int_0^1 \cdots \int}_{k} |x_1 - x_2|^{-2\gamma} |x_2 - x_3|^{-2\gamma} \cdots |x_k - x_1|^{-2\gamma} \, dx_1 \cdots dx_k \\ &= c_k > 0,\end{aligned} \tag{23}$$

as $n \to \infty$ (here $\operatorname{sp}(R)$ denotes the trace of the matrix R). The characteristic function of the limiting distribution is

$$\exp\left\{\sum_{k=2}^{\infty} (2it)^k \frac{c_k}{k}\right\} \tag{24}$$

the characteristic function of a nonnormal distribution. The curious aspect of this example is that moment conditions 1 and 2 are satisfied. We must therefore conclude that the strong mixing condition is not satisfied by $\{x_k\}$. But this means that the normal process $\{y_k\}$ does not satisfy a strong mixing condition, since $\{x_k\}$ is derived from it by an instantaneous nonlinear filter, even though $\{y_k\}$ is purely nondeterministic and hence mixing. *This is rather surprising.* It certainly indicates that *there would be a great deal of interest in a result that would characterize the Gaussian stationary processes that satisfy the strong mixing condition discussed in this section.*

3. Stationary processes generated by independent random variables

Most of the stationary processes that arise in practice are generated from independent identically distribution random variables by a linear or nonlinear operation and its shifts. Let $\eta = (\cdots, \eta_{-1}, \eta_0, \eta_1, \cdots)$ be a doubly infinite sequence of independent random variables with a common distribution function. Take T as the shift operator defined by

$$T\eta = \eta' = (\cdots, \eta_0, \eta_1, \eta_2, \cdots), \tag{25}$$

that is, the nth component of $\eta', (T\eta')_n = \eta_{n+1}$. The processes we are thinking of are generated in the following way. Let g be a Borel function of an infinite number of variables such that the random variable $g(\eta)$ has finite second moment $E|g(\eta)|^2 < \infty$. A stationary process $\{x_n\}$ is generated by taking

$$x_n = g(T^n\eta), \qquad n = 0, \pm 1 \cdots, \tag{26}$$

We can take $E\{x_n\} \equiv 0$ without any loss of generality. In this section we shall prove a result of Doob and Leibler [3] indicating that *the spectral distribution function of any process of this type is absolutely continuous*. The form of the proof given is that in [2].

In proving this result it is enough to assume that the η_j are uniformly distributed on the unit interval. For if η is not uniformly distributed, one can find a monotone function φ such that $\varphi(\xi)$ has the same distribution as η if ξ is uniformly distributed on the unit interval.

For each integer $m = 1, 2, \cdots$ we define a process $\{x_{n,m}\}$ by

$$x_{n,m} = E\{x_n | \eta_{n-m+1}, \cdots, \eta_{n+m-1}\}. \tag{27}$$

For fixed n, $\{x_{n,m}\}$ is a martingale. It follows from well-known theorems on martingales that $x_{n,m}$ converges to x_n in mean square and with probability one as $m \to \infty$.

Let $\theta_j(u_j)$, $j = 0, 1, \cdots$, be a sequence of orthonormal polynomials on the unit interval $[0, 1]$ with $\theta_j(u)$ of degree j precisely. It is well-known that the functions $\{\theta_j(u)\}$ are complete in $L^2[0, 1]$. The family of products $\{\theta_{j_1}(u_1) \cdots \theta_{j_k}(u_k)\}$ is a complete orthonormal system for $L^2[U_k]$ where U_k is the k-dimensional unit cube. To get a complete orthonormal system for the infinite-dimensional cube U we take the function identically one and the system of products $\{\theta_{j_1}(u_{m_1}) \cdots \theta_{j_k}(u_{m_k})\}$ where $m_1 < m_2 < \cdots < m_k$, $j_1, \cdots, j_k \geqq 1$ and $k = 1, 2, \cdots$. Let k be an arbitrary positive integer. From the discussion given above it follows that $x_{0,k}$ can be expanded in the form

$$x_{0,k} = \sum_{j=1}^{2k-1} \sum_{-k+1 \leqq m_1 < \cdots < m_j \leqq k-1} \sum_{n_1, \cdots, n_j = 1}^{\infty} C(\bar{n}_j, \bar{m}_j) \varphi_{\bar{n}_j}(\eta_{\bar{m}_j}) \tag{28}$$

where $\bar{m}_j = (m_1, \cdots, m_j)$, $\bar{n}_j = (n_1, \cdots, n_j)$ and $\varphi_{\bar{n}_j}(\eta_{\bar{m}_j}) = \theta_{n_1}(\eta_{m_1}) \cdots \theta_{n_j}(\eta_{m_j})$. Since $E\{|x_{0,k}|^2\} \leqq E\{|x_0|^2\} < \infty$ it follows that

$$\sum_{j=1}^{\infty} \sum_{m_1 < m_2 < \cdots < m_j} \sum_{n_1, \cdots, n_j = 1} C(\bar{n}_j, \bar{m}_j)^2 < \infty. \tag{29}$$

Since $x_0 = \lim_{k \to \infty} x_{0,k}$ with probability one it follows that

$$x_0 = \sum_{j=1}^{\infty} \sum_{m_1 < \cdots < m_j} \sum_{n_1, \cdots, n_j = 1}^{\infty} C(\bar{n}_j, \bar{m}_j) \varphi_{\bar{n}_j}(\eta_{\bar{m}_j}) \tag{30}$$

with probability one. The identity

$$x_n = \sum_{j=1}^{\infty} \sum_{m_1 < \cdots < m_j} \sum_{n_1, \cdots, n_j = 1}^{\infty} C(\bar{n}_j, \bar{m}_j - n) \varphi_{\bar{n}_j}(\eta_{\bar{m}_j}) \tag{31}$$

with $\bar{m}_j - n = (m_1 - n, \cdots, m_j - n)$ follows in a similar manner.

Let k be a positive integer, $\bar{n} = (n_1, \cdots, n_k)$ a k-tuplet of positive integers and $\bar{m} = (m_1, \cdots, m_k)$ a k-tuplet of integers with $m_1 < \cdots < m_k$. For any integer t let $x_t(\bar{m}, \bar{n})$ be given by

$$x_t(\bar{m}, \bar{n}) = \sum_{j=-\infty}^{\infty} C(\bar{n}, \bar{m} + j - t) \varphi_{\bar{n}}(\eta_{\bar{m}+j}) \tag{32}$$

where $\overline{m} + l = (m_1 + l, \cdots, m_k + l)$. $\{x_t(\overline{m}, \overline{n})\}$ is a moving average and such processes have an absolutely continuous spectrum. We shall distinguish $(\overline{m}', \overline{n}')$ from $(\overline{m}, \overline{n})$ if $\overline{n} \neq \overline{n}'$ or $\overline{m} - \overline{m}'$ is not a vector with equal coordinates. Then the processes $x_t(\overline{m}, \overline{n})$ and $x_t(\overline{m}', \overline{n}')$ are orthogonal if $(\overline{m}, \overline{n})$ and $(\overline{m}', \overline{n}')$ are distinct in the sense indicated above. There is an enumerable collection of such processes. Let $\{x_t^{(i)}\}$ be the ith process with $f^{(i)}(\lambda)$ its spectral density. Because of the orthogonality of the processes $\{x_t^{(i)}\}$ the spectral density $f_n(\lambda)$ of $\sum_{i=1}^n x_t^{(i)}$ is given by $\sum_{i=1}^n f^{(i)}(\lambda)$. The sequence $f_n(\lambda)$ is nondecreasing in n since the $f^{(i)}(\lambda)$ are nonnegative. Let $f(\lambda) = \lim_{n\to\infty} f_n(\lambda)$. Then

$$\begin{aligned}(33) \qquad \int_{-\pi}^{\pi} f(\lambda)\, d\lambda &= \int_{-\pi}^{\pi} \lim_{n\to\infty} f_n(\lambda)\, d\lambda \leqq \lim_{n\to\infty} \int_{-\pi}^{\pi} f_n(\lambda)\, d\lambda \\ &= \lim_{n\to\infty} E\left\{\left|\sum_{i=1}^{n} x_t^{(i)}\right|^2\right\} = E\{|g|^2\} < \infty.\end{aligned}$$

The function $f(\lambda)$ is finite almost everywhere and integrable. An easy computation indicates that $f(\lambda)$ is the spectral density of the process $\{x_t\}$ and the desired result is established. Analogous results for continuous parameter processes have been considered by Anzai [1], Kakutani [5] and Itô [4]. The importance of such processes has been stressed in the work of Wiener [10].

It follows from the result obtained above that if h is any square integrable function defined on the probability space of $x = (\cdots, x_{-1}, x_0, x_1, \cdots)$ and if $\{V_n\}$ is the process given by $V_n = h(T^n x)$, then $\{V_n\}$ has an absolutely continuous spectrum. One would like to know whether one can characterize such processes in terms of independent random variables with a representation like that of (26). At present this question is unresolved. In fact, not much work has been done on the question of determining those processes that can be given a representation of the form (26). Some insight into this question is given by the discussion in the remainder of this paper.

4. Structural questions for stationary processes

We now pose two interesting structural questions for stationary processes. Let $\{x_t\}$, with $t = 0, \pm 1, \cdots$, be a stationary process and T the corresponding one-step shift operator. Further, let $\mathcal{B}_n$ be the Borel field generated by $x_n, x_{n-1}, \cdots$. The first problem can be conveniently split into two parts.

PROBLEM 1. (a) *When can one find a random variable ξ_0 measurable with respect to $\mathcal{B}_0$, independent of $x_{-1}, x_{-2}, \cdots$ and such that $\mathcal{B}_0 = \mathcal{B}_{-1} \times \mathcal{A}_0$* (product field) *where $\mathcal{A}_0$ is the Borel field generated by ξ_0?* As we shall see, the answer to this part is almost immediate and rather simple. The second part is the nontrivial question.

(b) Notice that if (a) is satisfied, the random variables $\xi_n = T^n\xi_0$, $n = 0, \pm 1, \cdots$ form a sequence of independent, identically distributed random variables. Let $\mathcal{A}_n$ be the Borel field generated by ξ_n. *Is x_n measurable with respect to $\cdots \times \mathcal{A}_{n-1} \times \mathcal{A}_n$?*

This problem can be easily rephrased in the language of information theory. The random variable ξ_n can be regarded as the output of a noiseless channel ξ that $\{x_n\}$ is sent through, that is, $\xi_n = \xi(x_n, x_{n-1}, \cdots)$. In part (a) we ask that the output at time n, ξ_n, be independent of the past of the input $(x_{n-1}, x_{n-2}, \cdots)$ and yet that ξ_n coupled with knowledge of the past of the input be enough to determine the input x_n at present. Thus we can write $x_n = f(\xi_n, x_{n-1}, x_{n-2}, \cdots)$ and by iteration x_n as a function of $\xi_n, \xi_{n-1}, \cdots, \xi_{n-k}, x_{n-k-1}, x_{n-k-2}, \cdots$ for any finite k. In part (b) one asks whether one can go to the limit so that x_n can be considered a function of $\xi_n, \xi_{n-1}, \cdots$ alone.

Essentially the second question can be regarded as a reformulation of the first question where the condition that ξ be a noiseless channel is relaxed.

PROBLEM 2. Can one find a sequence of independent uniformly distributed (on $[0, 1]$) random variables ξ_n, with $n = 0, \pm 1, \cdots$ and a Borel function $g(\xi) = g(\xi_0, \xi_{-1}, \cdots)$, $\xi = (\cdots, \xi_{-1}, \xi_0, \xi_1, \cdots)$ such that

$$y_n = g(T^n\xi) \qquad n = 0, \pm 1, \pm 2, \cdots \tag{34}$$

has the same probability structure as x_n, with $n = 0, \pm 1, \cdots$? Here T is the shift operator on ξ. Of course, a positive answer to problem 1 implies that problem 2 can be answered in the affirmative.

Let $\bigcap \mathcal{B}_n$ be the Borel field of sets common to all the Borel fields $\mathcal{B}_n$, with $n = 0, \pm 1, \cdots$. Call a stationary process $\{x_i\}$ purely nondeterministic if the only functions measurable with respect to $\bigcap \mathcal{B}_n$ are the constant functions. This means that there is no information in the infinite past of the process. A little reflection indicates that if problem 1 or problem 2 are soluble for $\{x_n\}$, it must be a purely nondeterministic process (see [8]).

As remarked before, the answer to part (a) of problem 1 is not difficult. Let

$$F(a_n|a_{n-1}, a_{n-2}, \cdots) = P\{x_n \leqq a_n|x_{n-1} = a_{n-1}, x_{n-2} = a_{n-2}, \cdots\} \tag{35}$$

be the conditional probability distribution function of $\{x_n\}$. Assume that (35) is a Borel measure function of the real variables $a_n, a_{n-1}, \cdots$. Given two distribution functions F and G, we shall say they are equivalent if there is a one-one function mapping the jumps of F and G onto each other and preserving the size of the jumps. In [8] it is shown that there is a random variable ξ_n satisfying part (a) of problem 1 if and only if the distribution functions $F(a_n|a_{n-1}, \cdots)$ (as functions of a_n) are equivalent for almost all $a_{n-1}, a_{n-2}, \cdots$ (with respect to the probability measure of the process $\{x_n\}$).

Let us assume that part (a) of problem 1 is satisfied and inquire under what conditions part (b) is fulfilled. The remarks of section 3 indicate that there is no loss in generality in assuming the processes $\{x_n\}$ and $\{\xi_n\}$ bounded. Naturally take $\xi(x_n, x_{n-1}, \cdots)$ as a Borel measurable function of its arguments. Since $\mathcal{B}_0 = \mathcal{B}_{-1} \times \mathcal{A}_0$ it is clear that there is a family of transformations S_α on half infinite sequences into the real numbers induced by ξ such that

$$x_n = S_{\xi_n}(x_{n-1}, x_{n-2}, \cdots) \tag{36}$$

with probability one where ξ_n is independent of $x_{n-1}, x_{n-2}, \cdots$ and is given by $\xi_n = \xi(x_n, x_{n-1}, \cdots)$. To say that x_n is measurable with respect to $\cdots \times \mathcal{A}_{n-1} \times \mathcal{A}_n$ is equivalent to stating that $x_n = g(\xi_n, \xi_{n-1}, \cdots)$ where g is a Borel measurable function. But this will hold if and only if

$$x_n - E(x_n | \xi_n, \xi_{n-1}, \cdots, \xi_{n-k}) \to 0 \tag{37}$$

in probability as $k \to \infty$ making use of the boundedness of the $\{x_n\}$ process. Let μ be the measure on half infinite sequences induced by the probability measure of the process $\{x_n\}$. For convenience we introduce the following notation. Let

$$\begin{aligned} \bar{x}_m &= (x_m, x_{m-1}, \cdots) \\ S_{\xi_m}\bar{x}_{m-1} &= S_{\xi_m}(x_{m-1}, x_{m-2}, \cdots) \\ \{S_{\xi_m}\bar{x}_{m-1}\} &= (S_{\xi_m}\bar{x}_{m-1}, x_{m-1}, x_{m-2}, \cdots). \end{aligned} \tag{38}$$

We now give an argument proving the following theorem.

THEOREM. *The condition that*

$$x_n - E(x_n | \xi_n, \cdots, \xi_1) \to 0 \tag{39}$$

in probability as $n \to \infty$ is satisfied if and only if there is a set of points M of $\bar{y}_0$ points of μ measure one such that

$$S_{\xi_n}\{S_{\xi_{n-1}}\{\cdots \{S_{\xi_1}\bar{y}_0\} \cdots\}\} - S_{\xi_n}\{S_{\xi_{n-1}}\{\cdots \{S_{\xi_1}\bar{y}_0'\} \cdots\}\} \to 0 \tag{40}$$

in $\xi_1, \cdots, \xi_n, \cdots$ measure for all $\bar{y}_0, \bar{y}_0'$ in M as $n \to \infty$.

Assume there is a set M of μ measure one satisfying (40). Take $\bar{z}_0 \in M$. For all $\bar{x}_0 \in M$ and any fixed $\epsilon > 0$ the measure

$$m\{(\xi_1, \cdots, \xi_n) \big| |S_{\xi_n}\{\cdots \{S_{\xi_1}\bar{x}_0\} \cdots\} - S_{\xi_n}\{\cdots \{S_{\xi_1}\bar{z}_0\} \cdots\}| > \epsilon\} \to 0 \tag{41}$$

as $n \to \infty$. This implies that

$$\begin{aligned} &P\{|x_n - S_{\xi_n}\{\cdots \{S_{\xi_1}\bar{z}_0\} \cdots\}| > \epsilon\} \\ &\quad = \int m\{(\xi_1, \cdots, \xi_n) \big| |S_{\xi_n}\{\cdots \{S_{\xi_1}\bar{x}_0\} \cdots\} \\ &\qquad\qquad - S_{\xi_n}\{\cdots \{S_{\xi_1}\bar{z}_0\} \cdots\}| > \epsilon\} \, d\mu\,(\bar{x}_0) \to 0 \end{aligned} \tag{42}$$

as $n \to \infty$ so that

$$x_n - S_{\xi_n}\{\cdots \{S_{\xi_1}\bar{z}_0\} \cdots\} \to 0 \tag{43}$$

in probability as $n \to \infty$. On the other hand

$$\begin{aligned} &E\big|E(x_n | \xi_n, \cdots, \xi_1) - S_{\xi_n}\{\cdots \{S_{\xi_1}\bar{z}_0\} \cdots\}\big| \\ &\quad = E\big|E(S_{\xi_n}\{\cdots \{S_{\xi_1}\bar{x}_0\} \cdots\} - S_{\xi_n}\{\cdots \{S_{\xi_1}\bar{z}_0\} \cdots\} | \xi_n, \cdots, \xi_1)\big| \\ &\quad \leq E[E(|S_{\xi_n}\{\cdots \{S_{\xi_1}\bar{x}_0\} \cdots\} - S_{\xi_n}\{\cdots \{S_{\xi_1}\bar{z}_0\} \cdots\}| \,|\xi_n, \cdots, \xi_1)] \\ &\quad = E|x_n - S_{\xi_n}\{\cdots \{S_{\xi_1}\bar{z}_0\} \cdots\}| \to 0. \end{aligned} \tag{44}$$

Therefore $x_n - E(x_n | \xi_n, \cdots, \xi_1) \to 0$ as $n \to \infty$ in probability.

Now assume that $x_n - E(x_n | \xi_n, \cdots, \xi_1) \to 0$ in probability as $n \to \infty$ but that (40) is not satisfied. Given a sufficiently small $\epsilon > 0$ there are then sets M_1, M_2 with $\mu(M_1)$, $\mu(M_2) > \delta(\epsilon) > 0$ such that for $\bar{x}_0 \in M_1$, $\bar{x}_0' \in M_2$

$$m\{(\xi_1, \cdots, \xi_n) | \, |S_{\xi_n}\{\cdots \{S_{\xi_1}\bar{x}_0\} \cdots\} - S_{\xi_n}\{\cdots \{S_{\xi_1}\bar{x}_0'\} \cdots\}| > \epsilon\} > \delta(\epsilon) \tag{45}$$

for an infinite number of values of n. However

$$x_n - E(x_n|\xi_n, \cdots, \xi_1) \to 0 \tag{46}$$

implies that for almost all $\bar{x}_0$ (μ measure)

$$S_{\xi_n}\{\cdots \{S_{\xi_1}\bar{x}_0\} \cdots\} - E(S_{\xi_n}\{\cdots \{S_{\xi_1}\bar{x}_0'\} \cdots\}|\xi_n, \cdots, \xi_1) \to 0 \tag{47}$$

in $\xi_1, \cdots, \xi_n$ measure, a contradiction.

What results exist on these two problems are rather scattered. The most coherent set of results are those on Markov chains. Let us therefore consider problem 1 for Markov chains. Let the transition probability matrix of the chain be $P = (p_{ij})$. It has already been remarked that part (a) of problem 1 has an affirmative answer if and only if the conditional distribution functions are equivalent for almost every past. This means that the row probability distributions in the transition matrix must be the same except for permutation. Let the positive probability masses common to the distributions $\{p_{ij}, j = 1, 2, \cdots\}$ be $q_1 \geqq q_2 \geqq \cdots > 0$ in order of magnitude, $\sum q_i = 1$. We call such Markov chains uniform chains. When the q_i are all distinct one can characterize the random variable ξ_n simply. *If $\{x_n\}$ is a stationary uniform Markov chain with the q_i distinct, ξ_n is uniquely determined* (up to a one-one transformation modulo sets of measure zero) *and is of the form*

$$\xi_n = \xi(x_n, x_{n-1}) = k \quad \textit{for} \quad (x_n, x_{n-1}) \quad \textit{such that} \quad p_{x_{n-1},x_n} = q_k. \tag{48}$$

This is implied by the following argument. Write $\xi_n = \xi(x_n, x_{n-1}, \cdots)$. Now ξ must take m distinct values (where $m \leqq \infty$ is the number of distinct q) with probability one. Label the values 1, 2, $\cdots$ for convenience. Let the inverse function of ξ as a function of x_n for fixed $x_{n-1}, x_{n-2}, \cdots$ be $\eta(\xi, x_{n-1}, x_{n-2}, \cdots)$. Since ξ_n is independent of $x_{n-1}, x_{n-2}, \cdots$

$$P\{\xi_n = j|x_{n-1} = i_1, x_{n-2} = i_2, \cdots\} = P\{\xi_n = j\} = p_{i_1,\eta(j,i_1,i_2,\cdots)}. \tag{49}$$

Since the q_i are distinct, the function $\eta(j, i_1, i_2, \cdots)$ for fixed i_1, j must be independent of $i_2, i_3, \cdots$. Therefore $\eta(j, i_1, i_2, \cdots) = \eta(j, i_1)$ so that $\xi(x_n, x_{n-1}, \cdots) = \xi(x_n, x_{n-1})$. The function ξ must be constant on the set

$$\{(i, j)|p_{ij} = q_k\} \tag{50}$$

and for convenience we set $\xi = k$ on this set.

The remarks just made indicate that the answer to part (a) of problem 1 is rarely in the affirmative in the case of Markov chains even though the chain is purely nondeterministic. Nonetheless it is of some interest to find out *when part* (b) *of problem* 1 *is satisfied for stationary uniform Markov chains with the q_i distinct.* We have already noted that processes for which problem 1 (or 2) has a positive answer must be purely nondeterministic. *In the case of Markov chains this means that the set of all states must be an irreducible closed set of persistent states.* An interesting necessary and sufficient condition for a positive answer to problem 1 can be given in algebraic terms. Let M_k be the matrix

(51) $$M_k = \{e_{ij}(k)\}$$

where

(52) $$e_{ij}(k) = \begin{cases} 1 & \text{if } p_{ij} = q_k \\ 0 & \text{otherwise.} \end{cases}$$

There are m such distinct matrices M_k. Each such matrix M_k has precisely one element equal to one in each row and all other elements zero. Notice that every product of a finite number of the M_k also has these properties. No two distinct M_k have a one in the same location. The matrices $\{M_k\}$ correspond to a family of mappings of the set of states of the process into itself induced by the transition probability matrix P. First consider the semigroup generated (by finite operations) from the $\{M_k\}$. Notice that the matrices of this semigroup are of the same type, that is, have precisely one element equal to one in each row and all the rest zero. Complete this semigroup by adding to it all limit matrices. Call this semigroup the *completed semigroup generated by* $\{M_k\}$. We shall say that this semigroup is *point collapsing* if it has as an element a matrix with all the elements in some column one. A simple application of the theorem of this section indicates that *a purely nondeterministic stationary uniform Markov chain with distinct* q_i *has a representation of the type indicated in problem* 1 *if and only if the completed semigroup generated by the* $\{M_k\}$ *is point collapsing*. Since the proof is almost immediate we do not go through the details.

It is easy to give examples of purely nondeterministic Markov chains for which 1(a) is satisfied and 1(b) is not. The simplest such example is the Markov chain with the two states 1, 2 the stationary instantaneous distribution

(53) $$P\{x_n = 1\} = P\{x_n = 2\} = \frac{1}{2}$$

and the transition probability matrix

(54) $$P = \begin{pmatrix} p & q \\ q & p \end{pmatrix}, \qquad 0 < p,\, q = 1 - p < 1,\, p \neq q.$$

Here

(55) $$M_1 = \begin{pmatrix} 1 & 0 \\ 0 & 1 \end{pmatrix}, \quad M_2 = \begin{pmatrix} 0 & 1 \\ 1 & 0 \end{pmatrix},$$

so that

(56) $$\xi_n = \xi(x_n, x_{n-1}) = \begin{cases} 1 & \text{if } (x_{n-1}, x_n) = (1, 1), (2, 2) \\ 2 & \text{otherwise.} \end{cases}$$

Notice that x_n is not a function of $\xi_n, \xi_{n-1}, \cdots$ since $E(x_n | \xi_n, \xi_{n-1}, \cdots) = Ex_n$.

It is generally a tedious computation even in the case of a finite state uniform Markov chain to find out whether the completed semigroup generated by the $\{M_k\}$ is point collapsing. One might hope for a simple criterion that would provide either a necessary or a sufficient condition for a positive answer to problem 1 in terms perhaps of the invariant vector (stationary instantaneous distribution) of the Markov chain. In the case of the simple example given above, we note that the stationary distribution is uniform. Perhaps this would provide such a criterion. We give two simple examples to show that a satisfactory criterion must be somewhat more elaborate than a simple statement about uniformity of the invariant distribution. Consider a Markov chain with transition probability matrix

$$P = \begin{pmatrix} q_1 & q_2 & q_3 & q_4 \\ q_1 & q_2 & q_4 & q_3 \\ q_2 & q_1 & q_3 & q_4 \\ q_2 & q_1 & q_4 & q_3 \end{pmatrix} \tag{57}$$

with the q_i positive and distinct and $q_1 + q_2 = q_3 + q_4 = 1/2$. The invariant distribution is clearly uniform. Nonetheless, a simple computation indicates that the semigroup generated by the $\{M_k\}$ is point collapsing so that the answer to problem 1 is positive. A second example is provided by the Markov chain with transition matrix

$$P = \begin{pmatrix} q_1 & q_3 & q_2 & q_4 \\ q_1 & q_3 & q_2 & q_4 \\ q_2 & q_4 & q_3 & q_1 \\ q_2 & q_4 & q_3 & q_1 \end{pmatrix} \tag{58}$$

with the q_i positive and distinct. If the q_i are appropriately chosen, the invariant distribution will have distinct elements. Nonetheless the answer to problem 1 is negative since the semigroup generated by the $\{M_k\}$ is not point collapsing.

For reasons of clarity it is worthwhile seeing what one can say about the problem in terms of the concepts associated with the coding problem of information theory, even though it is trivial. Let us consider the capacity of the channel $\xi = \xi(x_n, x_{n-1})$ assuming that we are dealing with finite state processes (let us say m states). Since the channel is a noiseless channel with finite memory and essentially one-one given knowledge of x_{n-1}, it follows that the average entropy of an input, the input and output jointly, and the output alone are all the same. But this implies that the channel capacity is the maximum of average entropies for all possible inputs. Thus, the channel capacity is $m \log m$. Nonetheless, as we have noted there are inputs $\{x_n\}$ with average entropy smaller than $m \log m$ that cannot be reconstructed from the output. On reflection, the reason is clear. Here we do not allow coding. The channel ξ is determined so as to generate independent random variables from the input where the input is to be directly transmitted without coding. Given $\xi_n, \cdots, \xi_{n-k}$ the knowledge lacking to reproduce x_n is that of x_{n-k-1}. For some processes we recover this information as $k \to \infty$ and for others we don't. Problem 1 is that of determining the processes for which this information is recoverable.

The second problem has a more natural answer in the case of Markov chains. In fact, in [9] it is shown that the purely nondeterministic stationary Markov chains are precisely the Markov chains for which a representation of the form (19) is possible. In the case of Markov chains this is equivalent to a positive answer to problem 2. Clearly this is the best one can hope for. It is an interesting and open question as to whether the same is true of all purely nondeterministic stationary processes with respect to problem 2. Essentially the proof proceeds by first passing the Markov chain $\{x_n\}$ through a channel with noise and without memory so as to enlarge the probability space. The output of this channel is a Markov process $\{y_n\}$ from which one can immediately reproduce $\{x_n\}$. A deterministic channel ξ is then determined in terms of $\{y_n\}$. This channel manufactures independent random variables and it is in terms of this output that $\{x_n\}$ is constructed.

In obtaining the results cited above, the simplicity of the state space of the process (the fact that it is countable) plays an important role. In the case of a Markov process with a conditional probability distribution $F(a_n|a_{n-1})$ continuous in a_n for almost every past, the state space is continuous and one can immediately construct a noiseless channel satisfying part (a) of problem 1 by setting

$$\xi_n = \xi(x_n, x_{n-1}) = F(x_n|x_{n-1}). \tag{59}$$

Here ξ_n is independent of the past $x_{n-1}, x_{n-2}, \cdots$ and yet ξ_n and x_{n-1} determine x_n. Nonetheless, even if $\{x_n\}$ is purely nondeterministic, it is generally not true that problem 1 has an affirmative answer for this channel. In fact there may be some other channel $\bar{\xi}$ satisfying 1(a) for which problem 1 has an affirmative answer even though the answer is negative for ξ (see the example on page 680 of [8]). Therefore, one must generally take care in choosing the proper ξ. A discussion of some scattered results in the case of continuous state space stationary processes can be found in [8].

I would like to thank D. Slepian and S. P. Lloyd for their helpful comments.

REFERENCES

[1] H. Anzai, "A remark on spectral measures of the flow of Brownian motion," *Osaka Math. J.*, Vol. 1 (1949), pp. 95–97.

[2] J. R. Blum and M. Rosenblatt, "A class of stationary processes and a central limit theorem," *Duke Math. J.*, Vol. 24 (1957), pp. 73–78.

[3] J. L. Doob and R. A. Leibler, "On the spectral analysis of a certain transformation," *Amer. J. Math.*, Vol. 65 (1943), pp. 263–272.

[4] K. Itô, "Spectral type of the shift transformation of differential processes with stationary increments," *Trans. Amer. Math. Soc.*, Vol. 81 (1956), pp. 253–263.

[5] S. Kakutani, "Determination of the spectrum of the flow of Brownian motion," *Proc. Nat. Acad. Sci. U.S.A.*, Vol. 36 (1950), pp. 319–323.

[6] G. Maruyama, "The harmonic analysis of stationary stochastic processes," *Mem. Fac. Sci. Kyushu Univ., Ser. A*, Vol. 4 (1949), pp. 45–106.

[7] M. Rosenblatt, "A central limit theorem and a mixing condition," *Proc. Nat. Acad. Sci. U.S.A.*, Vol. 42 (1956), pp. 43–47.

[8] ———, "Stationary processes as shifts of functions of independent random variables," *J. Math. Mech.*, Vol. 8 (1959), pp. 665–682.

[9] ———, "Stationary Markov chains and independent random variables," *J. Math. Mech.*, Vol. 9 (1960), pp. 945–950.

[10] N. Wiener, "The homogeneous chaos," *Amer. J. Math.*, Vol. 60 (1939), pp. 897–936.

R.A. Davis et al. (eds.), *Selected Works of Murray Rosenblatt*, Selected Works in Probability and Statistics, DOI 10.1007/978-1-4419-8339-8_21, © Springer Science+Business Media, LLC 2011

CHAPTER 21

Asymptotic Behavior of Eigenvalues for a Class of Integral Equations with Translation Kernels*

M. Rosenblatt, *Division of Applied Mathematics*
Brown University

1. INTRODUCTION

The results obtained in this chapter may be of some interest from the point of view of analysis. However, they have an immediate interpretation in terms of certain representation theorems for stationary random processes on a finite time interval, and this provided part of the motivation for the investigation. Our interest is in finite interval translation kernel integral equation eigenvalue problems, that is, in the integral equation

$$\int_{-T}^{T} r(t-\tau)\, \phi(\tau)\, d\tau = \lambda\phi(t). \tag{1}$$

Here $\phi(t)$ is an eigenfunction and λ the associated eigenvalue. The principal result of the investigation runs as follows. Let $r(t)$ be a positive definite even function on $[-2T, 2T]$ of the form

$$r(t) = (-1)^k c|t|^{2k-1} + b(t), \qquad c > 0, \tag{2}$$

with $b(t)$ a function $2k - 1$ times continuously differentiable whose $(2k - 1)$st derivative is absolutely continuous ($k = 1, 2, \cdots$). We can then show that the jth eigenvalue λ_j of the integral equation (1) is asymptotically the same as

$$2(2k-1)!c\left(\frac{2T}{j\pi}\right)^{2k} \tag{3}$$

as $j \to \infty$.

The interest from the point of view of random processes is easy to explain. The positive definite function $r(t)$ can be considered as the covariance function

* This research was supported by the Office of Naval Research. Reproduction in whole or in part is permitted for any purpose of the United States Government.

316

of a real-valued random process $X(t)$ weakly stationary on $[-T, T]$; that is,

$$\begin{aligned} EX(t) &\equiv 0 \\ EX(t)\,X(\tau) &= r(t-\tau),\ |t|,\ |\tau| \leqslant T. \end{aligned} \tag{4}$$

The Karhunen-Loève theorem [7] then tells us that $X(t)$ has the following representation in mean square:

$$X(t) = \sum_j \lambda_j\, \phi_j(t) X_j, \tag{5}$$

where the $\phi_j(t)$'s are the eigenfunctions of the integral equation (1), the λ_j's are the corresponding eigenvalues, and the X_j's are orthonormal random variables given by

$$\begin{aligned} X_j &= \frac{1}{\lambda_j} \int_{-T}^{T} X(\tau)\, \phi_j(\tau)\, d\tau \\ EX_jX_k &= \delta_{jk}. \end{aligned} \tag{6}$$

The eigenfunctions $\phi_j(t)$ are, of course, assumed to be normalized over the interval $[-T, T]$. Estimates on the rate of decay of the eigenvalues are helpful in obtaining information on the speed of convergence in the representation (5). It is also useful in determining the sample function regularity of the process.

The result obtained is reminiscent also of the relationship between the tail behavior of a distribution function and the local behavior at zero of the corresponding characteristic function (see [3]). Assume that $r(t)$ is the restriction to $[-2T, 2T]$ of a positive definite symmetric function on $(-\infty, \infty)$. We then know that $r(t)$ has the Fourier-Stieltjes representation

$$r(t) = \int_{-\infty}^{\infty} e^{it\lambda}\, dF(\lambda), \tag{7}$$

where $F(\lambda)$ is a nondescreasing function of bounded variation on $(-\infty, \infty)$. Of course, $dF(\lambda) = dF(-\lambda)$. Then $1 - r(t) \cong c|t|^\alpha$ at zero if and only if $1 - F(\lambda) \cong c/\lambda^{1+\alpha}$ as $\lambda \to \infty$. Note that the asymptotic behavior of the eigenvalues λ_j [see (3)] when $r(t)$ satisfies (2) (here $\alpha = 2k - 1$) is essentially the same as the tail behavior of F as $\lambda \to \infty$. Our assumptions are such that the main irregularity of $r(t)$ is at $t = 0$ and $r(t)$ is by comparison smooth away from zero. Recently related results have been obtained for the special case in which F is absolutely continuous with $f(\lambda) = F'(\lambda)$, a rational function of λ[1]. A few remarks made later on in the chapter indicate that we can no longer expect the correspondence between asymptotic behavior of λ_j and tail behavior of F if $r(t)$ has irregularities as bad as that at $t = 0$ away from zero. Our results are in the case $\alpha = 2k - 1$ with k a positive integer, Corresponding results can be obtained for α nonintegral by means of more detailed estimation procedures and they have recently been written up by the author [8].

The derivations use the variational characterization of eigenvalues and a number of simple comparison lemmas for eigenvalues.

2. COMPARISON LEMMAS FOR EIGENVALUES

The estimates we obtain for the eigenvalues of the integral operators we deal with depend on a few well-known results that follow almost immediately from the classic variational definition of the eigenvalues. These results are stated and discussed for the sake of completeness.

Consider a completely continuous symmetric operator A on a Hilbert space H mapping H into itself. Let μ_k^+, μ_k^-, $k = 1, 2, \cdots$, be the nonnegative and nonpositive eigenvalues of A on H in nonincreasing and nondecreasing order of magnitude, respectively. We shall understand by μ_k^+

$$\mu_k^+ = \inf_{\phi_1, \cdots, \phi_{k-1}} \sup_{\substack{\phi \perp \phi_1, \cdots, \phi_{k-1} \\ \|\phi\| = 1,}} (A\phi, \phi) \tag{8}$$

where $(\cdot, \cdot)$ and $\|\cdot\|$ are the inner product and norm of an element in H. Here $\phi \perp \phi'$ means $(\phi, \phi') = 0$. Further, if there is no element $\phi \perp \phi_1, \cdots, \phi_{k-1}$ with $\|\phi\| = 1$, the supremum in formula (8) should be identified with zero. This is not quite the usual definition of eigenvalue. It always leads formally to an infinite number of eignevalues, even in the case of a finite dimensional Hilbert space. However, if H is n-dimensional $(n < \infty)$, then $\mu_k^+ = 0$ for all $k > n$. The eigenvalues μ_k^- are given by a formula analogous to (8), namely

$$\mu_k^- = \sup_{\phi_1, \cdots, \phi_{k-1}} \inf_{\substack{\phi \perp \phi_1, \cdots, \phi_{k-1} \\ \|\phi\| = 1.}} (A\phi, \phi) \tag{9}$$

Note that μ_k^+, μ_k^-, $\cdots$, runs through all the eigenvalues as defined in the conventional way [5] and that all the remaining eigenvalues as defined through (8) and (9) are zero.

Lemma 1. *Let A_1, A_2 be two symmetric completely continuous transformations and set $A = A_1 + A_2$. Denote the nth nonnegative eigenvalue of A_1, A_2, and A* [as given by (8)] *by $\mu_{1,n}^+$, $\mu_{2,n}^+$, and μ_n^+, respectively, and the nth nonpositive eigenvalue* [as given by (9)] by *$\mu_{1,n}^-$, $\mu_{2,n}^-$, and μ_n^-, respectively.* Then

$$\mu_{p+q-1}^+ \leqslant \mu_{1,p}^+ + \mu_{2,q}^+ \tag{10}$$

and

$$\mu_p^+ \geqslant \mu_{1,p+q-1}^+ + \mu_{2,q}^-. \tag{11}$$

The proof of this simple but extremely useful lemma can be found in [5]. It enables us to get upper and lower bounds for μ_p^+. Corresponding upper and lower bounds for μ_p^- can be written down.

We are also interested in a result that will let us compare corresponding eigenvalues for the two integral equations

$$\int_{-T_i}^{T_i} r(t, \tau)\phi(\tau)\, d\tau = \lambda\phi(t), \qquad i = 1, 2, \tag{12}$$

where $r(t, \tau)$ is Hermitian and square integrable in t, τ with $0 < T_1 < T_2$. A result of this type is stated in a more general context that may be of some independent interest.

Lemma 2. *Let A be a symmetric completely continuous operator on a Hilbert space H mapping H into itself. Let H_1 be a sub-Hilbert space of H and P the projection operator onto H_1. μ_k^+, μ_k^- $(\lambda_k^+, \lambda_k^-)$, $k = 1, 2, \cdots$ are the nonnegative and nonpositive eigenvalues of A $(PA = PAP)$ on H (H_1) in nonincreasing and nondecreasing order of magnitude, respectively. Then*

$$\begin{aligned} \mu_k^+ &\geqslant \lambda_k^+ \\ \mu_k^- &\leqslant \lambda_k^-, \end{aligned} \tag{13}$$

$k = 1, 2, \cdots$.

A discussion of the application of this lemma can be found in [6]. Its proof is so simple that we give it here. It is enough to show that $\mu_k^+ \geqslant \lambda_k^+$. Given any $\epsilon > 0$, there are $\phi_1, \cdots, \phi_{k-1} \in H$ such that

$$\sup_{\substack{\phi \perp \phi_1, \cdots, \phi_{k-1} \\ \|\phi\| = 1}} (A\phi, \phi) < \mu_k^+ + \epsilon \tag{14}$$

where the supremum is understood to be zero if there is no ϕ, $\|\phi\| = 1$, orthogonal to $\phi_1, \cdots, \phi_{k-1}$. If there is a $\phi \in H_1$, $\|\phi\| = 1$, orthogonal to $P\phi_1, \cdots, P\phi_{k-1}$, it is certainly orthogonal to $\phi_1, \cdots, \phi_{k-1}$. Hence

$$\sup_{\substack{\phi \perp P\phi_1, \cdots, P\phi_{k-1} \\ \phi \in H_1, \|\phi\| = 1}} (PA\phi, \phi) \leqslant \sup_{\substack{\phi \perp \phi_1, \cdots, \phi_{k-1} \\ \|\phi\| = 1}} (A\phi, \phi). \tag{15}$$

This implies that $\lambda_k^+ \leqslant \mu_k^+$ and the proof is complete.

In the integral equations (12) H is the Hilbert space of square integrable functions ϕ on $[-T_2, T_2]$, H_1 is the subspace of such square integrable functions ϕ equal to zero when $T_1 < |t| < T_2$, and

$$(A\phi)(t) = \int_{-T_2}^{T_2} r(t, \tau)\phi(\tau)\, d\tau. \tag{16}$$

3. SOME BOUNDS ON THE ASYMPTOTIC BEHAVIOR OF EIGENVALUES

Crude bounds on the asymptotic behavior of eigenvalues of finite interval translation kernel integral equations are obtained in this section. The basic idea is that of extending the kernel so that it is a section of a sufficiently smooth periodic function and then using one of the comparison lemmas of the last section. The following elementary lemma is derived first:

Lemma 3. *Let $r(t)$ be a function of period $2T$ symmetric about zero that is k times continuously differentiable ($k = 0, 1, 2, \cdots$) and whose kth derivative $r^{(k)}(t)$ is absolutely continuous. The eigenvalues λ_j* (in order of decreasing absolute magnitude) *of the integral equation*

$$\int_{-T}^{T} r(t - \tau)\phi(\tau)\, d\tau = \lambda\phi(t) \tag{16'}$$

are then $o(j^{-k-1})$ as $j \to \infty$.

The eigenfunctions of the integral equation (16′) are

$$\exp\left(\frac{\pi j t i}{T}\right) \qquad j = 0, \pm 1, \cdots, \tag{17}$$

since $r(t)$ is periodic with period $2T$. The eigenvalues are simply given by the Fourier coefficients

$$c_j = \int_{-T}^{T} r(t) \exp\left(\frac{\pi j t i}{T}\right) dt. \tag{18}$$

Clearly all the derivatives of $r(t)$ that exist have period $2T$. The odd and even derivatives of $r(t)$ are odd and even functions, respectively, about zero. Hence $r^{(m)}(T) = r^{(m)}(-T)$ for $0 \leqslant m \leqslant k$. On integrating by parts, we find that

$$\begin{aligned} c_j &= \left(\frac{T}{\pi j i}\right)^{k+1} (-1)^{k+1} \int_{-T}^{T} r^{(k+1)}(t) \exp\left(\frac{\pi j t i}{T}\right) dt \\ &= o(j^{-k-1}). \end{aligned} \tag{19}$$

The condition of periodicity is discarded in the following lemma:

Lemma 4. *Let $r(t)$ be an even function k times continuously differentiable ($k = 0, 1, 2, \cdots$.) whose kth derivative is absolutely continuous. The eigenvalues λ_j of the integral equation (16) are $o(j^{-k-1})$ as $j \to \infty$. If $r(t)$ is infinitely differentiable, the eigenvalues λ_j are $o(j^{-m})$ as $j \to \infty$ for every $m > 0$.*

Consider the function $r(t)$ on $[-2T, 2T]$. We can find an even function $r^*(t)$ k times continuously differentiable with kth derivative absolutely continuous that is periodic with period $4(T + \epsilon)$ for some $\epsilon > 0$ such that $r^*(t) = r(t)$ on $|t| \leqslant 2T$. The eigenvalues μ_j of the integral equation

$$\int_{-2(T+\epsilon)}^{2(T+\epsilon)} r^*(t - \tau)\, \phi(\tau)\, d\tau = \mu\phi(t) \tag{20}$$

are $o(j^{-k-1})$ by Lemma 3 as $j \to \infty$. However, by comparison Lemma 2 it then follows that the eigenvalues λ_j of the original integral equation (16) are $o(j^{-k-1})$ as $j \to \infty$. The statement in Lemma 4 about infinitely differentiable $r(t)$ follows immediately.

4. BERNOULLI POLYNOMIALS AND PERIODIC FUNCTIONS

Properties of the Bernoulli polynomials and periodic functions are useful in obtaining detailed estimates on the asymptotic behavior of eigenvalues of a certain class of finite translation kernel integral equations. The Bernoulli polynomials on the interval [0, 1] (see [2]) are recursively generated by setting

$$B_0(x) \equiv 1$$

and

$$B'_{k+1}(x) = B_k(x)$$

where

$$\int_0^1 B_k(x)\, dx = 0, \qquad k = 1, 2, \cdots.$$

The first few polynomials can be easily written down as follows

$$B_0(x) \equiv 1, \qquad B_1(x) = x - \frac{1}{2}, \qquad B_2(x) = \frac{x^2}{2} - \frac{x}{2} + \frac{1}{12}$$

$$B_3(x) = \frac{x^3}{3!} - \frac{x^2}{4} + \frac{x}{12}$$

$$B_4(x) = \frac{x^4}{4!} - \frac{x^3}{2 \cdot 3!} + \frac{x^2}{24} - \frac{1}{(24 \cdot 30)}$$

$$\cdots .$$

We are interested in the Bernoulli polynomials of even order. If the Fourier series of period unity of the polynomial $B_{2k}(x)$ is written out, we obtain the following expansion:

$$P_{2k}(x) = (-1)^{k-1} \sum_{n=1}^{\infty} \frac{\cos 2n\pi x}{2^{2k-1}(n\pi)^{2k}}. \tag{21}$$

Note that except for the factor $(-1)^{k-1}$ we have a positive definite function. Consider

$$r_k(t) = (-1)^{k-1} P_{2k}\left(\frac{t}{2T}\right) = (-1)^{k-1} B_{2k}\left(\frac{|t|}{2T}\right)$$

$$= 2T \sum_{n=1}^{\infty} \frac{\cos (n\pi t/T)}{T(2n\pi)^{2k}}, \qquad |t| < 2T \tag{22}$$

as the kernel of the integral equation

$$\int_{-T}^{T} r_k(t - \tau)\, \phi(\tau)\, d\tau = \lambda\phi(t), \qquad |t| < T.$$

As we have already remarked, the kernel $r_k(t)$ is positive definite. The eigenfunctions of the integral equation are

$$\begin{aligned} \phi_{2j-1}(t) &= T^{-\frac{1}{2}} \cos \frac{j\pi t}{T} \\ \phi_{2j}(t) &= T^{-\frac{1}{2}} \sin \frac{j\pi t}{T}, \end{aligned} \tag{23}$$

$j = 1, 2, \cdots$, with corresponding eigenvalues

$$\lambda_{2j-1} = \lambda_{2j} = \frac{2T}{(2j\pi)^{2k}}. \tag{24}$$

Since the only odd power of the Bernoulli polynomials $B_{2k}(x)$ is x^{2k-1}, the kernel $r_k(t)$, except for the term

$$\frac{(-1)^k|t|^{2k-1}}{2\cdot(2k-1)!(2T)^{2k-1}},$$

is a polynomial of degree $2k$ in t in $|t| < 2T$ with only the even powers of t appearing.

Theorem 1. *Let $\rho(t)$ be a positive definite even function on $[-2T, 2T]$ of the form*

$$\rho(t) = (-1)^k c|t|^{2k-1} + b(t) \tag{25}$$

with $c > 0$, $b(t)$ a function $2k - 1$ times continuously differentiable whose $(2k - 1)$st derivative is absolutely continuous ($k = 1, 2, \cdots$). Then the jth eigenvalue λ_j of the integral equation

$$\int_{-T}^{T} \rho(t-\tau)\, \phi(\tau)\, d\tau = \lambda\phi(t) \tag{26}$$

is asymptotically the same as

$$2(2k-1)!c\left(\frac{2T}{j\pi}\right)^{2k} \tag{27}$$

as $j \to \infty$.

Now

$$\rho(t) = 2c(2k-1)!(2T)^{2k-1}r_k(t) + d(t), \tag{28}$$

where $r_k(t)$ is given by (22) and $d(t)$ is an even function $2k - 1$ times continuously differentiable whose $(2k - 1)$st derivative is absolutely continuous. By Lemma 4 the translation integral equation of the form (26) with $d(t - \tau)$ as kernel has eigenvalues $\mu_j = o(j^{-2k})$ as $j \to \infty$. Further, the eigenvalues η_j of the translation integral equation with kernel $2c(2k - 1)!(2T)^{2k-1}r_k(t - \tau)$ are given precisely by $2c(2k - 1)!(2T)^{2k-1}$ times (24) in view of our earlier discussion. Comparison Lemma 1 tells us that

$$\lambda_{p+q-1} \leqslant \eta_p + \mu_q^+$$

and

$$\lambda_p \geqslant \eta_{p+q-1} + \mu_q^-.$$

Since $\mu_j = o(j^{-2k})$ as $j \to \infty$, we can find a positive integer-valued function $q = q(p)$ such that $q(p) = o(p)$ and $\mu_{q(p)} = o(p^{-2k})$ as $p \to \infty$. It then immediately follows that the eigenvalues λ_j are asymptotically the same as (27) as $j \to \infty$.

With this theorem, we can easily obtain the asymptotic behavior of eigenvalues for the integral equation (26) when $\rho(t)$ is of the form

$$\rho(t) = \frac{1}{\pi}\int_{-\infty}^{\infty} e^{it\lambda}(1 + \lambda^{2k})^{-1}\, d\lambda. \tag{29}$$

The integral (29) can be readily evaluated by a contour integration and is given by

$$\rho(t) = -\frac{i}{k}\sum_{j=0}^{k-1} z_j e^{itz_j}, \qquad t > 0,$$

with $\rho(t) = \rho(-t)$, where

$$z_j = \exp\{[\pi i(j + \tfrac{1}{2})]/k\}, \qquad j = 0, 1, \cdots, k-1.$$

Now

$$\rho(t) = -\frac{i}{k}\sum_{n=0}^{\infty}\frac{(it)^n}{n!}\sum_{j=0}^{k-1} z_j^{n+1}$$

where

$$\sum_{j=0}^{k-1} z_j^{n+1} = \begin{cases} 0 \text{ if } n+1 \text{ is even and not a multiple of } 2k \\ -i\left[\sin\dfrac{\pi(n+1)}{2k}\right]^{-1} \quad \text{if } n+1 \text{ is odd} \\ (-1)^s k \quad \text{if } n+1 = 2sk. \end{cases}$$

Thus $\rho(t)$, as given by (29), satisfies the assumptions of the theorem with $c = [(2k-1)!]^{-1}$; hence the eigenvalues λ_j are asymptotically

$$\lambda_j \cong 2\left(\frac{2T}{j\pi}\right)^{2k}$$

as $j \to \infty$. This suggests the following result for positive definite functions $\rho(t)$ of the form

$$\rho(t) = \int_{-\infty}^{\infty} e^{itx} f(x)\, dx \tag{30}$$

with $f(x) \geqslant 0$.

Theorem 2. *Consider the integral equation* (26) *with* $\rho(t)$ *of the form* (30) *where* $f(x)$ *is even with*

$$f(x) \cong \frac{c}{x^{2k}}, \qquad c > 0, \tag{31}$$

as $x \to \infty$. *The eigenvalues* λ_j *of the integral equation* (26) *are asymptotically of the form*

$$\lambda_j \cong 2\pi c\left(\frac{2T}{j\pi}\right)^{2k}. \tag{32}$$

Given any $\epsilon > 0$ there is a $K(\epsilon) > 0$ large enough so that for $|x| > K(\epsilon)$

$$\frac{c-\epsilon}{1+x^{2k}} < f(x) < \frac{c+\epsilon}{1+x^{2k}}.$$

Let

$$f_k(x) = \begin{cases} f(x) & \text{if } |x| > K(\epsilon) \\ 0 & \text{otherwise} \end{cases} \tag{33}$$

and

$$ {}_1f(x) = \begin{cases} \dfrac{c+\epsilon}{1+x^{2k}} & \text{if } |x| > K(\epsilon) \\ 0 & \text{otherwise} \end{cases} \tag{34} $$

$$ {}_2f(x) = \begin{cases} \dfrac{c-\epsilon}{1+x^{2k}} & \text{if } |x| > K(\epsilon) \\ 0 & \text{otherwise.} \end{cases} \tag{35} $$

Call the Fourier transforms of f_k, ${}_1f$, ${}_2f$ derived by formula (30) $\rho_k(t)$, ${}_1\rho(t)$, ${}_2\rho(t)$, respectively. Since f_k is bandlimited, $\rho_k(t)$ is analytic in t, and the eigenvalues λ_j of the integral equation (26) with ρ_k as kernel approach zero faster than any inverse power of j as $j \to \infty$ by Lemma 4. Lemma 1 implies that the eigenvalues ${}_1\lambda_j$ and ${}_2\lambda_j$ of integral equation (26) with kernels ${}_1\rho + \rho_k$, ${}_2\rho + \rho_k$, respectively, behave like

$$ \begin{aligned} {}_1\lambda_j &\cong 2\pi(c+\epsilon)\left(\frac{2T}{j\pi}\right)^{2k} \\ {}_2\lambda_j &\cong 2\pi(c-\epsilon)\left(\frac{2T}{j\pi}\right)^{2k} \end{aligned} \tag{36} $$

as $j \to \infty$. But the eigenvalues λ_j of the integral equation (26) with kernel ρ satisfy the inequality

$$ {}_2\lambda_j \leqslant \lambda_j \leqslant {}_1\lambda_j. \tag{37} $$

The inequality (37) with the asymptotic estimates (36) immediately imply the theorem.

It is curious that the kernel $(-1)^{k-1} P_{2k}(x)$ used to prove Theorem 1 is at the limit for which one can expect a result like that cited in the theorem to hold. If we consider this kernel over the interval $|x| \leqslant 2$ instead of $|x| \leqslant 1$ [it is positive definite on the infinite line $(-\infty, \infty)$], we find that the eigenvalues of the corresponding translation kernel integral equation are

$$ \lambda_{2j-1} = \lambda_{2j} = \frac{1}{2^{2k-1}(j\pi)^{2k}}, \tag{38} $$

whereas Theorem 1 would lead us to expect that

$$ \lambda_j \cong \left(\frac{2}{j\pi}\right)^{2k} \tag{39} $$

as $j \to \infty$. The actual eigenvalues are easily read off, since essentially they are the Fourier coefficients of the function. Here the asymptotic behavior of the eigenvalues is the same type as that in Theorem 1 but the coefficient is not that given in the theorem. The reason why the result of Theorem 1 does not hold is given in the introduction. The function $r(t)$ of Theorem 1 has a jump in its $(2k-1)$st derivative at $t = 0$ but nowhere else in $[-2T, 2T]$. However $(-1)^{k-1} P_{2k}(t)$ not only has a jump at $t = 0$ but also at $t = \pm 1$ in

$|t| \leqslant 2$. In this example we have seen that the asymptotic behavior is the same type as that given in Theorem 1 but the coefficient is off. However, even more erratic examples indicating the limits of Theorem 1 can be constructed by considering positive definite functions $r(t)$ with singularities of the same type as that at the origin distributed more widely in the interval $[-2T, 2T]$. Let

$$r(t) = \sum_j c_j e^{2\pi i j t}$$

with

$$c_j = c_{-j}$$

$$c_j = \begin{cases} \dfrac{1}{j^{1+\beta}} & \text{if } j = [n^\alpha] \text{ for some integer } n > 0,\ \alpha > 1, \\ 0 & \text{otherwise,} \end{cases}$$

where $j \geqslant 0$. Here $[x]$ is the largest integer less than or equal to x. The nonzero eigenvalues are essentially the Fourier coefficients if $T = \frac{1}{2}$, hence

$$\lambda_j \sim \frac{1}{j^{1+\beta}}.$$

If $F(x) = \sum_{j \leqslant x} c_j$ then $1 - F(x) \sim 1/x^{\beta/\alpha}$, hence

$$r(t) \sim c|t|^{\beta/\alpha} + b(t),$$

where $b(t)$ is smoother than $|t|^{\beta/\alpha}$ at $t = 0$. If we set $\beta/\alpha = 2k - 1$, we shall see that the eigenvalues λ_j decrease at a rate faster than that suggested by Theorem 1.

5. CRUDE UPPER BOUNDS FOR EIGENVALUES OF ANALYTIC KERNELS

Let $r(t)$ be a positive definite function on $(-\infty, \infty)$ that is the restriction to the real axis of a function analytic in the entire complex t-plane. Such a function, of course, has a power series representation

$$r(t) = \sum_{k=0}^{\infty} c_k t^k \tag{40}$$

valid for all t. We shall obtain crude upper bounds for the eigenvalues of (1) for an analytic kernel in terms of this power-series representation. Suppose we wish to obtain an upper bound for λ_j. Let

$$r_1(t) = \sum_{k=0}^{j-2} c_k t^k$$

$$r_2(t) = r(t) - r_1(t).$$

Call the integral operator generated by $r(t)$ as a translation kernel on $[-T, T]$ A and the corresponding integral operators generated by $r_1(t)$, $r_2(t)$, the operators A_1, A_2, respectively. The eigenvalues of A in order of magnitude are λ_j. The nonnegative eigenvalues of A_1, A_2 are $\lambda_{1,j}^+$ and $\lambda_{2,j}^+$, in order of magnitude. Of course $A = A_1 + A_2$. By Lemma 1

$$\lambda_j \leqslant \lambda_{1,j}^+ + \lambda_{2,1}^+.$$

However, A_1 has at most $j - 1$ nonzero eigenvalues so that $\lambda_{1,j}^+ = 0$. Further

$$\lambda_{2,1}^+ \leqslant \max_{|t| \leqslant T} \int_{-T}^{T} \sum_{k=j-1}^{\infty} |c_k| \, |t - \tau|^k \, d\tau. \tag{41}$$

Thus (41) provides an upper bound for λ_j. It may be of some value to use this estimation procedure for two analytic kernels. First let us consider $r(t) = \exp(-t^2)$. Here

$$\lambda_{2j} \leqslant \sum_{k=j}^{\infty} \frac{(2T)^{2k+1}}{k!(2k+1)}.$$

The kernel $r(t) = \sin t/t$ has been of especial interest [4]. In this case

$$\lambda_{2j} \leqslant \sum_{k=j}^{\infty} \frac{(2T)^{2k+1}}{(2k+1)!(2k+1)}.$$

In both cases it should be noted that the eigenvalues go to zero faster than geometrically, that is, given any c with $0 < c < 1$, for sufficiently large k:

$$\lambda_k < c^k.$$

ACKNOWLEDGMENT

Theorem 2 of this paper can be regarded as extension of an unpublished result of A. Beurling (see the footnote in [9]). I should like to thank H. O. Pollak and D. Slepian for many helpful discussions.

REFERENCES

1. Capon, J. Asymptotic eigenfunctions and eigenvalues of a homogeneous integral equation. *IRE Trans. Inform. Theory*, **8**, 2–4 (1962).
2. Franklin, P. *A Treatise on Advanced Calculus*. New York: Wiley, 1940.
3. Pitman, E. J. G. Some theorems on characteristic functions of probability distributions. *Proc. Fourth Berkeley Symp. Math. Statist. Prob.*, **2**, 393–402 (1961).
4. Pollak, H. O., and Slepian, D. Prolate spheroidal wave functions, Fourier analysis and uncertainty I. *Bell System Tech. J.*, **40**, 43–64 (1961).
5. Riesz, F., and Sz.-Nagy, B. *Leçons d'analyse functionelle*. Budapest, 1952.
6. Rosenbloom, P. *Linear Partial Differential Equations*, Surveys in Applied Mathematics V. New York: Wiley, 1958.
7. Rosenblatt, M. *Random Processes*. New York: Oxford University Press, 1962.
8. Rosenblatt, M. Some results on the asymptotic behavior of eigenvalues for a class of integral equations with translation kernels. To be published in the *J. of Math. and Mech.* (1963).
9. Slepian, D. Fluctuations of random noise power. *Bell System Tech. J.*, **37**, 163–184 (1957).

*Asymptotic Behavior of Eigenvalues of Toeplitz Forms**

M. ROSENBLATT

1. Introduction. Let $F(\lambda) = \{F_{jk}(\lambda); j, k = 1, \cdots, s\}$ be an $s \times s$ matrix-valued function of bounded variation on $[-\pi, \pi]$. By this we mean that every complex-valued element $F_{jk}(\lambda)$ is of bounded variation. Further, let every difference $F(\lambda_1) - F(\lambda_2)$ be Hermitian. It will be convenient and in no way less general to take $F(-\pi) = 0$, the null matrix. Let

$$r_k(dF) = \frac{1}{2\pi} \int_{-\pi}^{\pi} e^{ik\lambda} \, dF(\lambda), \qquad k = 0, \pm 1, \cdots, \tag{1}$$

be the sequence of Fourier-Stieltjes coefficients of F. The sequence $r_k(dF)$ is a sequence of $s \times s$ matrices. Let the Hermitian matrix

$$R_n(dF) = \{r_{j-k}(dF); j, k = 1, \cdots, n\} \tag{2}$$

be the $ns \times ns$ block Toeplitz matrix generated by dF. Since F is of bounded variation, its derivative

$$f(\lambda) = \frac{dF(\lambda)}{d\lambda} \tag{3}$$

exists almost everywhere and is Hermitian. Let $\mu_1(\lambda) \leqq \mu_2(\lambda) \leqq \cdots \leqq \mu_s(\lambda)$ be the eigenvalues of $f(\lambda)$ in order of magnitude. The functions $\mu_j(\lambda)$, $j = 1, \cdots, s$, are measurable (see Gyires [3]). Let $\mu_{j,n}$, $j = 1, \cdots, ns$, be the eigenvalues of $R_n(dF)$ and $N_n(x)$ the number of eigenvalues of $R_n(dF)$ less than or equal to x. We show that

$$\lim_{n\to\infty} \frac{N_n(x)}{ns} = \frac{1}{2\pi s} \sum_{j=1}^{s} m\{\lambda \mid \mu_j(\lambda) \leqq x\} \tag{4}$$

at every point of continuity of the right hand side, where m is Lebesgue measure on $[-\pi, \pi]$. The case of Toeplitz matrices arises when $s = 1$ and was treated by Szegö and Grenander & Szegö (see [2]). The result obtained in this paper is a generalization of that given by Gyires [3] and Rosenblatt [6] for block Toeplitz matrices.

* The results presented in this paper were obtained in the course of research carried out under grant NSF–G19046 of the National Science Foundation.

R.A. Davis et al. (eds.), *Selected Works of Murray Rosenblatt*, Selected Works in Probability and Statistics, 205
DOI 10.1007/978-1-4419-8339-8_22, © Springer Science+Business Media, LLC 2011

A crude upper bound is obtained for the maximal eigenvalue of Toeplitz matrices generated by an absolutely continuous F. This upper bound is used to obtain a result on the asymptotic distribution of covariance estimates of normal stationary processes.

2. Nondecreasing $F(\lambda)$. We first prove the result on the asymptotic distribution of eigenvalues of block Toeplitz forms in the case of nondecreasing $F(\lambda)$. The function $F(\lambda)$ is said to be nondecreasing if $F(\lambda_1) - F(\lambda_2)$ is a positive definite matrix whenever $\lambda_1 \geqq \lambda_2$.

Lemma 1. *Let $F(\lambda)$ be an $s \times s$ nondecreasing matrix-valued function of bounded variation on $[-\pi, \pi]$. The limiting distribution of the eigenvalues of the block Toeplitz matrix $R_n(dF)$ as $n \to \infty$ is given by*

$$\frac{1}{2\pi s} \sum_{j=1}^{s} m\{\lambda \mid \mu_j(\lambda) \leqq x\}. \tag{5}$$

Let $G_n(x)$ be the distribution function of the eigenvalues of $R_n(dF)$, that is,

$$G_n(x) = \frac{N_n(x)}{sn}. \tag{6}$$

We need only consider the distribution functions $G_n(x)$ on $[0, \infty)$. Now

$$\int_0^\infty x \, dG_n(x) = \frac{1}{ns} \sum_{j=1}^{ns} \mu_{j,n} = \frac{1}{ns} \operatorname{sp} (R_n(dF)) = \frac{1}{s} \operatorname{sp} (r_0(dF)) \tag{7}$$

where sp (M) denotes the trace of the matrix M. Notice that

$$y[1 - G_n(y)] \leqq \int_y^\infty x \, dG_n(x) \leqq \frac{1}{s} \operatorname{sp} (r_0) = c, \tag{8}$$

so that

$$1 - G_n(y) \leqq \frac{c}{y}. \tag{9}$$

Consider any weak limit $G(y)$ of a subsequence of the sequence $G_n(y)$. Clearly any such limit must also satisfy

$$1 - G(y) \leqq \frac{c}{y} \tag{10}$$

and hence is a distribution function. Let z be a nonnegative number. The basic result on the determinant of the prediction error covariance matrix in the multidimensional prediction problem tells us that

$$\begin{aligned} \frac{1}{ns} \sum_{j=1}^{ns} \log (1 + z\mu_{j,n}) &= \int_0^\infty \log (1 + zx) \, dG_n(x) \\ &\to \frac{1}{2\pi s} \int_{-\pi}^{\pi} \log \det (I + zf(\lambda)) \, d\lambda = \frac{1}{2\pi s} \sum_{j=1}^{s} \int_{-\pi}^{\pi} \log (1 + z\mu_j(\lambda)) \, d\lambda \end{aligned} \tag{11}$$

as $n \to \infty$ (see [5] and [6]). On integrating by parts we see that

$$\int_0^\infty \log(1+zx)\,dG_n(x) = \log(1+zx)(G_n(x)-1)\Big|_0^\infty \tag{12}$$
$$+ \int_0^\infty \frac{z}{1+zx}(1-G_n(x))\,dx = \int_0^\infty \frac{z}{1+zx}(1-G_n(x))\,dx.$$

Relations (11) and (12) imply that for any weak limit $G(y)$ of a subsequence of $G_n(y)$

$$\int_0^\infty \frac{z}{1+zx}(1-G(x))\,dx = \frac{1}{2\pi s}\sum_{j=1}^{s}\int_{-\pi}^{\pi} \log(1+z\mu_j(\lambda))\,d\lambda. \tag{13}$$

The distribution function (5) obviously satisfies (13). To complete the proof for nondecreasing F we only have to show that G is uniquely determined by the transform (13). Consider the uniqueness problem for the transform

$$\int_0^\infty \frac{z}{1+zx}H(x)\,dx, \qquad z \geqq 0, \tag{14}$$

where H is of bounded variation and satisfies $|H(x)| \leqq k/(1+x)$ for some constant k. This is equivalent to the uniqueness problem for the Stieltjes transform

$$\int_0^\infty \frac{1}{z+x}H(x)\,dx, \qquad z > 0. \tag{15}$$

A proof of uniqueness is easily given. Suppose

$$\int_0^\infty \frac{1}{z+x}H(x)\,dx \equiv 0 \tag{16}$$

for all $z > 0$. Equation (16) can be rewritten as

$$\int_0^\infty e^{-zu}\int_0^\infty e^{-xu}H(x)\,dx\,du \equiv 0. \tag{17}$$

But this implies that

$$\int_0^\infty e^{-xu}H(x)\,dx \equiv 0. \tag{18}$$

It follows that $H(x) \equiv 0$ almost everywhere. The proof of Lemma 1 is complete.

It is readily seen that the conclusion of Lemma 1 holds for every F of bounded variation such that $F(\lambda) + a(\lambda+\pi)I$ is nondecreasing for some real value a. It similarly holds for every F of bounded variation such that $F(\lambda) + a(\lambda+\pi)I$ is nonincreasing for some real value a.

3. General F. In this section it is shown that the conclusion of Lemma 1 holds for general Hermitian F of bounded variation. The extension will depend on the following remarks. Given two $n \times n$ Hermitian matrices A, B with

$A - B \geqq 0$ (difference positive definite), the eigenvalues λ_j and μ_j of A and B respectively (assume the eigenvalues are indexed in order of magnitude) satisfy the corresponding inequalities, that is,

$$\lambda_j \geqq \mu_j, \qquad j = 1, \cdots, n. \tag{19}$$

This follows readily from the minimax property of eigenvalues (see HILBERT & COURANT [1]).

Theorem 1. *Let $F(\lambda)$ $(F(-\pi) = 0)$ be an $s \times s$ Hermitian matrix-valued function of bounded variation on $[-\pi, \pi]$. The limiting distribution of the eigenvalues of the block Toeplitz matrix $R_n(dF)$ as $n \to \infty$ is given by*

$$\frac{1}{2\pi s} \sum_{j=1}^{s} m\{\lambda \mid \mu_j(\lambda) \leqq x\}.$$

We first prove the desired result for such a function F whose singular part is nondecreasing. Let $f(\lambda) = dF(\lambda)/d\lambda$. The singular part of F is given by $F(\lambda) - \int_{-\pi}^{\lambda} f(\mu)\, d\mu$. Let

$$S_a = \{\lambda \mid f(\lambda) + aI \geqq 0\}, \tag{20}$$

and $\bar{S}_a$ be the complement of S_a. The sets S_a are nondecreasing as $a \to \infty$ and such that the Lebesgue measure $m(S_a) \to 2\pi$ as $a \to \infty$. This follows from the fact that $f(\lambda)$ is integrable, that is, every element of f is integrable. Given any value λ, let

$$f^+(\lambda) = \sum_{u,v=1}^{s} s\, |f_{u,v}(\lambda)| \cdot I. \tag{21}$$

Note that

$$f(\lambda) \leqq f^+(\lambda) \tag{22}$$

($A \geqq B$ means that $A - B \geqq 0$). Inequality (22) is to be understood as holding almost everywhere where the elements of the matrices are finite and hence well defined. Let

$$F^a(\lambda) = F(\lambda) - \int_{-\pi}^{\lambda} f(\mu)\, d\mu + \int_{-\pi}^{\lambda} f^a(\mu)\, d\mu \tag{23}$$

where

$$f^a(\lambda) = \begin{cases} f(\lambda) & \text{on } S_a \\ f^+(\lambda) & \text{on } \bar{S}_a. \end{cases} \tag{24}$$

Because of (22) it follows that

$$R_n(dF) \leqq R_n(dF^a) \tag{25}$$

for all a. Further, $R_n(dF^a)$ is nonincreasing as $a \to \infty$. Let $G^{(n)}(x)$, $G_a^{(n)}(x)$ be the distribution functions of the eigenvalues of $R_n(dF)$ and $R_n(dF^a)$ respectively.

It then follows that

$$G_a^{(n)}(x) \leqq G^{(n)}(x) \tag{26}$$

for all n, a, and x. The integrability of $f(\lambda)$, $f^+(\lambda)$, inequality (25), and the fact that $m(\bar{S}_a) \to 0$ as $a \to \infty$ imply that for every $\epsilon > 0$ there is an $a(\epsilon) > 0$ such that for $a > a(\epsilon)$

$$\begin{aligned} 0 &\leqq \frac{1}{sn} \operatorname{sp}\left(R_n(dF^a) - R_n(dF)\right) \\ &= \int_{-\infty}^{\infty} x \, d[G_a^{(n)}(x) - G^{(n)}(x)] = \int_{-\infty}^{\infty} [G^{(n)}(x) - G_a^{(n)}(x)] \, dx < \epsilon \end{aligned} \tag{27}$$

independently of n. By the remarks at the end of section 2, it follows that $G_a^{(n)}(x)$ has the limiting distribution

$$G_a(x) = \frac{1}{2\pi s} \sum_{j=1}^{s} m\{\lambda \mid \mu_j^a(\lambda) \leqq x\} \tag{28}$$

as $n \to \infty$ where the $\mu_j^a(\lambda)$ are the eigenvalues of $f^a(\lambda)$. Let $H(x)$ be the limit of any subsequence of the functions $G^{(n)}(x)$. It follows that

$$G_a(x) \leqq H(x) \tag{29}$$

for all x and a. Inequality (29) implies that

$$0 \leqq \int_{-\infty}^{\infty} [H(x) - G_a(x)] \, dx \leqq \epsilon \tag{30}$$

for all $a > a(\epsilon)$. The family of functions $G_a(x)$ is nondecreasing at each x as $a \to \infty$ and has as limit as $a \to \infty$

$$G(x) = \frac{1}{2\pi s} \sum_{j=1}^{s} m\{\lambda \mid \mu_j(\lambda) \leqq x\}. \tag{31}$$

Thus

$$H(x) \geqq G(x) \tag{32}$$

and

$$0 \leqq \int_{-\infty}^{\infty} [H(x) - G(x)] \, dx \leqq \epsilon \tag{33}$$

for every ϵ. But this implies that $H(x) = G(x)$. Since this is true for the limit $H(x)$ of any subsequence of $G^{(n)}(x)$, there is a limiting distribution which is given by $G(x)$. The proof is complete for F with a nondecreasing singular part.

We now complete the proof of Theorem 1 by showing that (5) is the asymptotic distribution of eigenvalues of $R_n(dF)$ for a general Hermitian F of bounded variation. The singular part of F is

$$H(\lambda) = F(\lambda) - \int_{-\pi}^{\lambda} f(\mu) \, d\mu \tag{34}$$

where $f(\lambda) = dF(\lambda)/d\lambda$. Let

$$H^+(\lambda) = \sum_{u,v=1}^{s} s \int_{-\pi}^{\lambda} |dH_{u,v}(\mu)| \cdot I. \tag{35}$$

Set

$$F_1(\lambda) = \int_{-\pi}^{\lambda} f(\mu)\, d\mu + H^+(\lambda), \qquad F_2(\lambda) = \int_{-\pi}^{\lambda} f(\mu)\, d\mu - H^+(\lambda). \tag{36}$$

F_1 has a nondecreasing singular part H^+ and F_2 a nonincreasing singular part $-H^+$. Further,

$$R_n(dF_1) \geqq R_n(dF) \geqq R_n(dF_2) \tag{37}$$

for all n. Since $R_n(dF_1)$, $R_n(dF_2)$ have the same asymptotic eigenvalue distribution (5) by the result on the asymptotic distribution of eigenvalues for a function with a nondecreasing singular part given in the last paragraph, it follows from (37) that $R_n(dF)$ has the asymptotic eigenvalue distribution (5).

4. Asymptotic distribution of covariance estimates. It will be useful to have an upper bound on the maximal eigenvalue of the ordinary Toeplitz matrix $R_n(f)$ generated by a nonnegative function $f \in L^2$ in our derivation of the asymptotic distribution of the covariance estimates of a normal stationary process. There are refined estimates of $\lambda_{n,n}$ when f is a bounded function satisfying certain regularity conditions (see [2]). However, we wish to allow f to be unbounded. Let

$$G(x) = \frac{1}{2\pi} m\{\lambda \mid f(\lambda) \leqq x\}.$$

Lemma 2. *The maximal eigenvalue $\lambda_{n,n}$ of the Toeplitz matrix $R_n(f)$ generated by $f \in L$ is bounded above by*

$$n \int_{b(n)}^{\infty} x\, dG(x) \tag{38}$$

where $b(n)$ is the supremum of the values α such that $1 - G(\alpha) > 1/n$. If $f \in L^2$, this upper bound is of smaller order than $n^{\frac{1}{2}}$ as $n \to \infty$.

The eigenvalue $\lambda_{n,n}$ is the maximal value attained by the quadratic form cR_nc' with c an n-vector subject to the restraint $cc' = 1$. But

$$cR_nc' = \frac{1}{2\pi} \int_{-\pi}^{\pi} |c_n(e^{i\lambda})|^2 f(\lambda)\, d\lambda, \qquad c_n(e^{i\lambda}) = \sum_{j=1}^{n} c_j e^{ij\lambda} \tag{39}$$

and the absolute maximum of $|c_n(e^{i\lambda})|^2$ subject to the restraint $cc' = 1$ is n. Let

$$S_x = \{\lambda \mid f(\lambda) \geqq x\}. \tag{40}$$

We certainly overestimate $\lambda_{n,n}$ by replacing $|c_n(e^{i\lambda})|^2$ by a function equal to n on the set $S_{b(n)}$. Now

$$\frac{1}{2\pi} n \int_{S_{b(n)}} f(\lambda) \, d\lambda = n \int_{b(n)}^{\infty} x \, dG(x).$$

If $f \varepsilon L^2$ then $\int x^2 \, dG(x) < \infty$ and by the Schwarz inequality

$$n \int_{b(n)}^{\infty} x \, dG(x) \leqq n \left[\int_{b(n)}^{\infty} dG(x) \int_{b(n)}^{\infty} x^2 \, dG(x) \right]^{1/2} = o(n^{1/2}).$$

An upper bound for the maximal eigenvalue of a block Toeplitz matrix can be obtained by similar techniques.

We now consider the asymptotic distribution of the estimates

$$r_k^* = \frac{1}{N} \sum_{j=1}^{N-|k|} X_j X_{j+k}, \qquad k = 0, 1, \cdots, s, \tag{41}$$

of the covariances r_k, $k = 0, 1, \cdots, s$, of a normal stationary process $\{X_k\}$.

Theorem 2. *Let $\{X_k ; k = 0, \pm 1, \cdots\}$ be a normal stationary process with mean $\mathrm{E}\, X_k \equiv 0$ and covariances*

$$r_k = \operatorname{cov}(X_j, X_{j+k}) = \frac{1}{2\pi} \int_{-\pi}^{\pi} e^{ik\lambda} f(\lambda) \, d\lambda \qquad k = 0, \pm 1, \cdots, \tag{42}$$

with $f \varepsilon L^2$. Then

$$\sqrt{N}\, (r_k^* - \mathrm{E}\, r_k^*), \qquad k = 0, 1, \cdots, s, \tag{43}$$

are asymptotically jointly normally distributed with mean zero and covariances

$$c_{j,k} = \frac{1}{\pi} \int_{-\pi}^{\pi} \cos j\lambda \cos k\lambda \, f^2(\lambda) \, d\lambda, \qquad j, k = 0, 1, \cdots, s. \tag{44}$$

The joint characteristic function of the random variables (43) is

$$\begin{aligned} \mathrm{E}\left(\exp\left\{ \sum_{k=0}^{s} it_k (r_k^* - \mathrm{E}\, r_k^*) \sqrt{N} \right\} \right) &= \exp\left\{ -\sum_{k=0}^{s} it_k \sqrt{N}\, \mathrm{E}\, r_k^* \right\} (2\pi)^{-1/2N} |R_N|^{-1/2} \\ &\quad \cdot \int_{-\infty}^{\infty} \cdots \int \exp\{ ixT_N x'/\sqrt{N} - \tfrac{1}{2} x R_N^{-1} x' \} \, dx \\ &= |I - 2iR_N T_N/\sqrt{N}|^{-1/2} \exp\left\{ -\sum_{k=0}^{s} it_k \sqrt{N}\, \mathrm{E}\, r_k^* \right\} \end{aligned} \tag{45}$$

where x is a row N-vector, $R_N = R_N(f)$, and $T_N = \{t_{j,k} ; j, k = 1, \cdots, N\}$ is the Toeplitz matrix with $t_{i,i} \equiv t_0$, $t_{i,i\pm k} = \frac{1}{2} t_{|k|}$ if $0 < k \leqq s$, and all other elements zero. Now expression (45) can be written as

$$\exp\left\{ -\tfrac{1}{2} \sum_{j=1}^{N} [\log(1 - 2i\mu_{j,N}/\sqrt{N}) + 2\, i\mu_{j,N}/\sqrt{N}] \right\}$$

where the $\mu_{j,N}$ are the eigenvalues of $R_N^{1/2} T_N R_N^{1/2}$. Here $R_N^{\frac{1}{2}}$ is the positive definite square root of R_N. However,

$$\mu_{j,N} = O(\max_j \alpha_{j,N} \sum_j |t_j|),$$

where the $\alpha_{j,N}$ are the eigenvalues of R_N. By Lemma 2 the eigenvalues of R_N are $o(N^{\frac{1}{2}})$ uniformly in j so that $\mu_{j,N}/\sqrt{N} = o(1)$ uniformly in j. Therefore (45) can be given as

$$\exp\left\{- \operatorname{sp}(R_N T_N)^2/N + O\left(\sum_{j=1}^{N} |\mu_{j,N}|^3/N^{3/2}\right)\right\}. \tag{46}$$

But

$$\sum |\mu_{j,N}|^3 \leqq \left(\sum \mu_{j,N}^2 \sum \mu_{j,N}^4\right)^{1/2} = [\operatorname{sp}(R_N T_N)^2 \operatorname{sp}(R_N T_N)^4]^{1/2}. \tag{47}$$

Furthermore

$$0 \leqq \operatorname{sp}(R_N T_N)^{2k} \leqq \operatorname{sp}(M_N)^{2k}, \qquad k = 1, 2, \cdots, \tag{48}$$

where M_N is the $N \times N$ Toeplitz matrix with elements

$$m_{i-j} = \sum_{|u| \leq s} |r_{i-j-u}| \cdot |t_u|.$$

Notice that M_N is generated by a real-valued function $g \in L^2$ since $m_k = m_{-k}$ and $\sum |r_k|^2 < \infty$. The fact that $\sum m_k^2 < \infty$ implies that $\operatorname{sp}(M_N^2) = O(N)$. Since $g \in L^2$, Lemma 2 indicates that the eigenvalues $\eta_{j,N}$ of M_N are all uniformly $o(N^{\frac{1}{2}})$. But then

$$\operatorname{sp}(M_N)^4 = \sum \eta_{j,N}^4 = o(N) \operatorname{sp}(M_N)^2 = o(N^2). \tag{49}$$

This implies that

$$\sum_{j=1}^{N} |\mu_{j,N}|^3/N^{3/2} = o(1) \tag{50}$$

as $N \to \infty$. A simple computation indicates that

$$\begin{aligned} \lim_{N\to\infty} \frac{1}{N} \operatorname{sp}(R_N T_N)^2 &= \sum_{j,k=-s}^{s} \tfrac{1}{4}(1 + \delta_{j,0})(1 + \delta_{k,0}) t_{|j|} t_{|k|} \frac{1}{2\pi} \\ &\quad \cdot \int_{-\pi}^{\pi} e^{i(j-k)\lambda} f^2(\lambda)\, d\lambda = \sum_{j,k=0}^{s} \tfrac{1}{2} t_j t_k c_{j,k} \end{aligned} \tag{51}$$

where $c_{j,k}$ is given by (44). The proof of asymptotic normality is complete by the correspondence between characteristic functions and distribution functions, for the limit of (45) as $N \to \infty$

$$\exp\left\{-\tfrac{1}{2} \sum_{j,k=0}^{s} c_{j,k} t_j t_k\right\}$$

is the characteristic function of $s + 1$ jointly normal random variables with mean zero and covariances $c_{j,k}$.

Some results related to Theorem 2 are discussed in [4]. Relatively simple examples can be constructed to show that one need no longer have asymptotic normality if $f \notin L^2$ (see [7]). In fact, given any p with $1 \leqq p < 2$ one can construct an $f \in L^p$ but not in L^2 such that one does not have asymptotic normality of the covariance estimates.

References

[1] R. Courant & D. Hilbert, *Methods of Mathematical Physics*, Vol. 1, New York, 1953.

[2] U. Grenander & G. Szegö, *Toeplitz Forms and Their Applications*, Berkeley, 1958.

[3] B. Gyires, Uber die Spuren der verallgemeinten Toeplitzschen Matrizes, *Publ. Math.-Debrecen*, **8** (1961), pp. 93–116.

[4] E. J. Hannan, *Time Series Analysis*, New York, 1960.

[5] P. Masani & N. Wiener, The prediction theory of multivariate stochastic processes, *Acta. Math.*, **98** (1957), pp. 111–150.

[6] M. Rosenblatt, Asymptotic distribution of eigenvalues of block Toeplitz matrices, *Bull. Amer. Math. Soc.*, **66** (1960), pp. 320–321.

[7] M. Rosenblatt, Independence and dependence, *Proceedings of the 4th Berkeley Symposium on Probability and Statistics*, **2** (1961), pp. 431–443.

Brown University
Providence, R. I.

ESTIMATION OF THE BISPECTRUM[1]

BY M. ROSENBLATT AND J. W. VAN NESS

University of California, San Diego, and Stanford University

1. Summary. Recently interest has arisen in statistical applications of the bispectrum of stationary random processes. (The bispectrum can be thought of as the Fourier transform of the third-order moment function of the process.) The principal area of statistical harmonic analysis to receive attention previous to this time has been second-order (i.e. spectral) theory on which there is a vast literature. However, the spectrum is most useful in problems of a "linear nature" (see discussion beginning on p. viii of Blackman and Tukey [2]) and provides insufficient information in nonlinear problems.

A desire to study phenomena of a nonlinear character has attracted attention to the higher order theory. Such was the case, for example, in a recent study by Hasselmann, Munk and MacDonald [8] where the bispectrum is used in connection with oceanographic problems, among which, as the authors state, a number of interesting phenomena such as surf beats, wave breaking, and the energy transfer between wave components can be explained only by the nonlinearity of the wave motion.

The bispectrum therefore provides a first glimpse at the nonlinear effects. It is the purpose of the present paper to discuss estimating the weighted and unweighted bispectral density given a set of observations of the process. The relevant properties (consistency and asymptotic unbiasedness) of the estimates are derived for certain general classes of processes.

2. Introduction. The indexed set of random variables, $\{X_t\}$, is called a continuous parameter random process if the index set, T, is the real line and it is called discrete parameter if T is the positive and negative integers. For our purposes we take X_t to have mean zero and finite sixth-order moments and to be real-valued sixth-order weakly stationary all of which implies that for all t,

$$\begin{aligned} Ex_t &= 0, \\ EX_tX_{t+\nu} &= m_2(t, t+\nu) = r(\nu), \\ EX_tX_{t+\nu_1}X_{t+\nu_2} &= m_3(t, t+\nu_1, t+\nu_2) = r_3(\nu_1, \nu_2), \qquad (2.1)\\ &\vdots \\ EX_tX_{t+\nu_1}\cdots X_{t+\nu_5} &= m_6(t, t+\nu_1, \cdots, t+\nu_5) = r_6(\nu_1, \cdots, \nu_5). \end{aligned}$$

Consider $r(\nu)$ and $r_3(\nu_1, \nu_2)$ to be in L_1 (or l_1) and in the continuous parameter case to be continuous. It is well known from second-order harmonic theory that

Received 24 June 1964; revised 17 February 1965.

[1] This research was supported in part by the Office of Naval Research under Contract Nonr 562(29) at Brown University.

R.A. Davis et al. (eds.), *Selected Works of Murray Rosenblatt*, Selected Works in Probability and Statistics, 214
DOI 10.1007/978-1-4419-8339-8_23, © Springer Science+Business Media, LLC 2011

$r(\nu)$ has a Fourier-Stieltjes representation in terms of the spectral distribution function, $F(\lambda)$. Take $F(\lambda)$ to be absolutely continuous with a continuous density, $f(\lambda)$; then this representation becomes

$$r(\nu) = \int_{-\infty}^{\infty} e^{i\nu\lambda} f(\lambda)\, d\lambda. \tag{2.2}$$

(Throughout this paper, all formulae shall be written for the continuous parameter case; the discrete case can be obtained by easy modifications.) Further

$$f(\lambda) = (2\pi)^{-1} \int_{-\infty}^{\infty} e^{-i\nu\lambda} r(\nu)\, d\nu. \tag{2.3}$$

Analogously, in third-order harmonic theory we can define the bispectral density function, $g(\lambda_1, \lambda_2)$, as

$$g(\lambda_1, \lambda_2) = (2\pi)^{-2} \int_{-\infty}^{\infty} \int_{-\infty}^{\infty} \exp\,(-i\nu_1\lambda_1 - i\nu_2\lambda_2) r_3(\nu_1, \nu_2)\, d\nu_1\, d\nu_2 \tag{2.4}$$

and, assuming $g(\lambda_1, \lambda_2)\ \varepsilon\ L_1(R_2)$,

$$r_3(\nu_1, \nu_2) = \int_{-\infty}^{\infty} \int_{-\infty}^{\infty} \exp\,(i\nu_1\lambda_1 + i\nu_2\lambda_2) g(\lambda_1, \lambda_2)\, d\lambda_1\, d\lambda_2\,. \tag{2.5}$$

(For a discussion of the existence of the higher order spectral representations under some conditions, see Blanc-Lapierre and Fortet [3] and Sinai [20].) This then is the harmonic representation of the third-order moment function under the above assumptions.

Since the process is real, the following symmetries occur in the third-order functions

$$r_3(\nu_1, \nu_2) = r_3(\nu_2, \nu_1) = r_3(-\nu_1, \nu_2 - \nu_1), \tag{2.6}$$

$$g(\lambda_1, \lambda_2) = g(\lambda_2, \lambda_1) = g(\lambda_1, -\lambda_1 - \lambda_2) = \overline{g(-\lambda_1, -\lambda_2)}. \tag{2.7}$$

The symmetries (2.6) imply that $r_3(\nu_1, \nu_2)$ is completely specified over the entire plane by its values in any *one* of the six sectors, ⓐ through ⓕ, shown in Fig. 1. These sectors include their boundaries so that, for example, sector ⓐ is

$$0 \leqq \nu_1 < \infty, \quad 0 \leqq \nu_2 \leqq \nu_1\,. \tag{2.8}$$

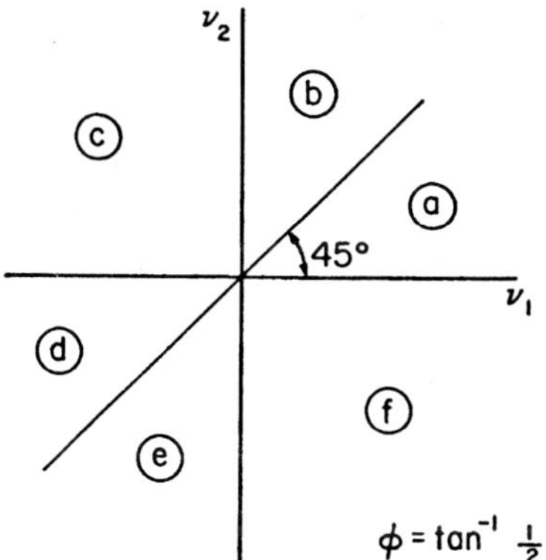

FIG. 1

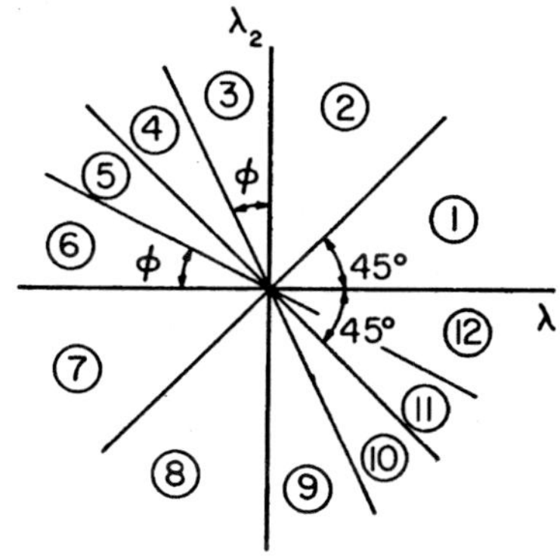

FIG. 2

Similarly $g(\lambda_1, \lambda_2)$ is, by (2.7), completely specified by its values in any one of the twelve sectors (including boundaries) shown in Fig. 2. A later paper will discuss the slightly more complicated situation which arises in the discrete parameter case.

Recall from second-order theory that the process permits the representation

$$X_t = \int_{-\infty}^{\infty} e^{it\lambda}\, dZ(\lambda), \tag{2.9}$$

where $Z(\lambda)$ is a random process of orthogonal increments with (in the case of a real process)

$$\begin{aligned} E\, dZ(\lambda) &= 0, \\ E\, dZ(\lambda_1)\, dZ(\lambda_2) &= \delta(\lambda_1 + \lambda_2)\, dF(\lambda_1), \\ \delta(x) &= 1 \qquad \text{if } x = 0 \\ &= 0 \qquad \text{if } x \neq 0. \end{aligned} \tag{2.10}$$

Equation (2.10) yields an interpretation of the spectrum. When there exists a Fourier-Stieltjes representation of the third-order moment function in terms of a complex-valued bispectral distribution function of bounded variation, a similar result holds:

$$E\, dZ(\lambda_1)\, dZ(\lambda_2)\, dZ(\lambda_3) = \delta(\lambda_1 + \lambda_2 + \lambda_3)\, dG(\lambda_1, \lambda_2). \tag{2.11}$$

More particularly a real process (implies $dZ(\lambda) = \overline{dZ(-\lambda)}$) has the real representation

$$X_t = \int_0^{\infty} \cos t\lambda\, dZ_1(\lambda) + \int_0^{\infty} \sin t\lambda\, dZ_2(\lambda) \tag{2.12}$$

where $dZ_1(\lambda) = 2 \operatorname{Re} dZ(\lambda)$ and $dZ_2(\lambda) = -2 \operatorname{Im} dZ(\lambda)$. It is interesting that these functions allow us to obtain the following insight into the role of the real and imaginary parts of the bispectrum. In these relations note that when $E\, dZ_{j_1}(\lambda_1)\, dZ_{j_2}(\lambda_2)\, dZ_{j_3}(\lambda_3)$, $j_i = 1, 2$, is nonzero, it is equal to some multiple of the real part of dG if an *odd* number of the j_i's equal 1 and equal to some multiple of the imaginary part of dG if an *even* number of the j_i's equal 1.

$$\begin{aligned} &E\, dZ_1(\lambda_1)\, dZ_1(\lambda_2)\, dZ_1(\lambda_3) \\ &\quad = 2 \operatorname{Re} [dG(\lambda_1, \lambda_2)\delta(\lambda_1 + \lambda_2 - \lambda_3) + dG(\lambda_1, \lambda_3)\delta(\lambda_1 - \lambda_2 + \lambda_3) \\ &\qquad + dG(\lambda_2, \lambda_3)\delta(-\lambda_1 + \lambda_2 + \lambda_3) + dG(0, 0)\delta(\lambda_1)\delta(\lambda_2)\delta(\lambda_3)] \\ &E\, dZ_2(\lambda_1)\, dZ_2(\lambda_2)\, dZ_2(\lambda_3) \\ &\quad = -2 \operatorname{Im} [dG(\lambda_1, \lambda_2)\delta(\lambda_1 + \lambda_2 - \lambda_3) + dG(\lambda_1, \lambda_3)\delta(\lambda_1 - \lambda_2 + \lambda_3) \\ &\qquad + dG(\lambda_2, \lambda_3)\delta(-\lambda_1 + \lambda_2 + \lambda_3)] \\ &E\, dZ_1(\lambda_1)\, dZ_1(\lambda_2)\, dZ_2(\lambda_3) \\ &\quad = 2 \operatorname{Im} [dG(\lambda_1, \lambda_2)\delta(\lambda_1 + \lambda_2 - \lambda_3) - dG(\lambda_1, \lambda_3)\delta(\lambda_1 - \lambda_2 + \lambda_3) \end{aligned} \tag{2.13}$$

$$- dG(\lambda_2, \lambda_3)\delta(-\lambda_1 + \lambda_2 + \lambda_3)]$$

$$E\, dZ_1(\lambda_1)\, dZ_2(\lambda_2)\, dZ_2(\lambda_3)$$
$$= 2 \text{ Re } [dG(\lambda_1, \lambda_2)\delta(\lambda_1 + \lambda_2 - \lambda_3) + dG(\lambda_1, \lambda_3)\delta(\lambda_1 - \lambda_2 + \lambda_3)$$
$$- dG(\lambda_2, \lambda_3)\delta(-\lambda_1 + \lambda_2 + \lambda_3) - dG(0, 0)\delta(\lambda_1)\delta(\lambda_2)\delta(\lambda_3)].$$

It is also worth noting that the symmetries indicate that the integrals in Equations (2.4) and (2.5) can be taken over one (or more) of the sectors instead of over the entire plane.

3. Estimation. Given observations, x_t, for $0 \leqq t \leqq N$, natural estimates for the bispectral density, $g(\lambda_1, \lambda_2)$, are those based on a function, $g_N(\lambda_1, \lambda_2)$, which is analogous to the periodogram of second-order theory (see, for example, Rosenblatt [17] for a discussion of periodogram theory). The usual estimate for $r_3(\nu_1, \nu_2)$ is

$$\rho_N(\nu_1, \nu_2) = N^{-1}\int_{D_N(\nu_1,\nu_2)} x_t x_{t+\nu_1} x_{t+\nu_2}\, dt \tag{3.1}$$

where the interval D_N restricts x_t, $x_{t+\nu_1}$, and $x_{t+\nu_2}$ to the domain in which they are defined (The set Φ is the empty set.):

$$D_N(\nu_1, \nu_2) = \Phi \qquad \text{if } |\nu_1| \text{ or } |\nu_2| \text{ or } |\nu_1 - \nu_2| > N$$
$$= [-\min[0, \nu_1, \nu_2], N - \max[0, \nu_1, \nu_2]] \qquad \text{otherwise.}$$

It is intuitively plausible that if $r_3(\nu_1, \nu_2)$ is replaced by $\rho_N(\nu_1, \nu_2)$ in (2.4), the resulting function is an estimate of $g(\lambda_1, \lambda_2)$, thus

$$g_N(\lambda_1, \lambda_2) = (2\pi)^{-2}\int_{-N}^{N}\int_{-N}^{N} \exp(-i\nu_1\lambda_1 - i\nu_2\lambda_2)\rho_N(\nu_1, \nu_2)\, d\nu_1\, d\nu_2 . \tag{3.2}$$

There are two customary requirements made of such estimates:

(i) the estimate should be asymptotically unbiased,

(ii) the variance of the estimate should go to zero as $N \to \infty$.

Theorem 1 states that under certain conditions on the process, $\{X_t\}$, $\rho_N(\nu_1, \nu_2)$ has both properties. However, $g_N(\lambda_1, \lambda_2)$, like the periodogram, has only Property (i) and not Property (ii). Prompted by the results of second-order theory, one corrects this difficulty by considering weighted estimates of the form

$$g_N^*(W) = \int_{-\infty}^{\infty}\int_{-\infty}^{\infty} W(\mu_1, \mu_2)g_N(\mu_1, \mu_2)\, d\mu_1\, d\mu_2 \tag{3.3}$$

where $W(\mu_1, \mu_2)$ is a "bispectral averaging function" to be defined later. By Theorems 2 and 3 below, it is seen that under certain conditions on W and the process, $g_N^*(W)$ is an asymptotically unbiased estimate of

$$g(W) = \int_{-\infty}^{\infty}\int_{-\infty}^{\infty} W(\mu_1, \mu_2)g(\mu_1, \mu_2)\, d\mu_1\, d\mu_2 \tag{3.4}$$

and that $\sigma^2(g_N^*(W)) \to 0$ as $N \to \infty$. Therefore $g_N^*(W)$ suffices as an estimate of (3.4) but it obviously is not an asymptotically unbiased estimate of $g(\lambda_1, \lambda_2)$.

To get an estimate of the bispectral density itself with both Properties (i) and (ii), a *sequence* of weight functions, $\{W_N(\mu_1, \mu_2)\}$, can be used so that

$$(3.5)\quad g_N^*(\lambda_1, \lambda_2) = \int_{-\infty}^{\infty}\int_{-\infty}^{\infty} W_N(\mu_1 - \lambda_1, \mu_2 - \lambda_2) g_N(\mu_1, \mu_2)\, d\mu_1\, d\mu_2 .$$

Appropriate choice of the weight functions will give $g_N^*(\lambda_1, \lambda_2)$ Property (ii). As N increases, the weight functions should narrow the region averaged over (that is tend to behave more and more like a two-variate δ-function with the "spike" at the origin); this tends to give the estimate Property (i). A proper rate of concentration by the weight functions must be chosen so that (under certain conditions on the process and on $\{W_N\}$) both properties can be obtained simultaneously (see Theorems 4 and 5).

4. Hypotheses characteristic of this paper. Fourth and sixth-order moments arise during the investigation of the covariance properties of the estimates. It is necessary to make some kind of assumption on these moments. A very advantageous assumption involves the use of cumulant (semi-invariant) functions (see Stratonovich ([21], Chapter 1) and Leonov and Shiryaev [11] for a discussion of cumulant functions). These functions are defined in a manner completely analogous to the way ordinary cumulants are defined for a single random variable.

Let a random vector $\eta = (\eta_1, \cdots, \eta_k)$ be given. Its characteristic function is

$$(4.1)\qquad \varphi_\eta(\alpha_1, \cdots, \alpha_k) = E \exp (i\alpha_1\eta_1 + \cdots + i\alpha_k\eta_k).$$

Assume $E|\eta_j|^n < \infty$; then the mixed moments

$$(4.2)\qquad m^{(\nu_1,\ldots,\nu_k)} = E\eta_1^{\nu_1} \cdots \eta_k^{\nu_k}$$

exist for all $\nu_1, \cdots, \nu_k$ such that $\nu_j \geqq 0$ and $\nu_1 + \cdots + \nu_k \leqq n$. Consequently φ_η has the Taylor expansion

$$(4.3)\quad \varphi_\eta(\alpha_1, \cdots, \alpha_k) = \sum_{\nu_1+\ldots+\nu_k\leqq n} (i^{\nu_1+\ldots+\nu_k}/\nu_1 ! \cdots \nu_k !) m^{(\nu_1,\ldots,\nu_k)} \alpha_1^{\nu_1} \cdots \alpha_k^{\nu_k} + o((|\alpha_1| + \cdots + |\alpha_k|)^n)$$

where the sum is over all non-negative $\nu_1, \cdots, \nu_k$ whose sum does not exceed n. Furthermore, log φ_η has a Taylor expansion exactly as in (4.3) except with $m^{(\nu_1,\ldots,\nu_k)}$ replaced by the coefficient $s^{(\nu_1,\ldots,\nu_k)}$. The quantities $s^{(\nu_1,\ldots,\nu_k)}$ are called the cumulants of the vector η. Also $m^{(\nu_1,\ldots,\nu_k)}$ can be expressed as a polynomial in the $s^{(\gamma_1,\ldots,\gamma_k)}$, $0 \leqq \gamma_j \leqq \nu_j$, $j = 1, \cdots, k$ (and vice versa) (see Leonov and Shiryaev ([11], p. 320)). For a process with zero mean, the two expressions needed in the following are

$$(4.4)\qquad m_4(\nu_1, \cdots, \nu_4) = s_4(\nu_1, \cdots, \nu_4) + \{s_2(\nu_1, \nu_2)s_2(\nu_3, \nu_4)\}_3$$

and

$$(4.5)\quad m_6(\nu_1, \cdots, \nu_6) = s_6(\nu_1, \cdots, \nu_6) + \{s_3(\nu_1, \nu_2, \nu_3)s_3(\nu_4, \nu_5, \nu_6)\}_{10} + \{s_2(\nu_1, \nu_2)s_4(\nu_3, \cdots, \nu_6)\}_{15} + \{s_2(\nu_1, \nu_2)s_2(\nu_3, \nu_4)s_2(\nu_5, \nu_6)\}_{15}$$

where (for $\eta = (X_{\nu_1}, \cdots, X_{\nu_6})$) for example

$$s_2(\nu_1, \nu_2) = s^{(1,1,0,0,0,0)},$$
$$s_2(\nu_1, \nu_3) = s^{(1,0,1,0,0,0)},$$
$$s_4(\nu_1, \nu_2, \nu_5, \nu_6) = s^{(1,1,0,0,1,1)}, \text{ etc.},$$

and where the notation $\{\cdot\}_j$ denotes the sum of all j *different* terms obtained by interchanging the arguments of the terms in brackets (the order of the arguments of the s_j being immaterial) (see Tables I, II and III below.) Thus (4.4) is

$$m_4(\nu_1, \cdots, \nu_4) = s_4(\nu_1, \cdots, \nu_4) + s_2(\nu_1, \nu_2)s_2(\nu_3, \nu_4) + s_2(\nu_1, \nu_3)s_2(\nu_2, \nu_4) + s_2(\nu_1, \nu_4)s_2(\nu_2, \nu_3).$$

Note that in the case of zero mean,

$$(4.6)\qquad \begin{aligned} s_1(\nu) &= 0,\\ s_2(\nu_1, \nu_2) &= m_2(\nu_1, \nu_2),\\ s_3(\nu_1, \nu_2, \nu_3) &= m_3(\nu_1, \nu_2, \nu_3). \end{aligned}$$

Due to stationarity we write

$$(4.7)\qquad \begin{aligned} s_2(t, t + \nu) &= \xi_2(\nu),\\ &\vdots\\ s_6(t, t + \nu_1, \cdots, t + \nu_5) &= \xi_6(\nu_1, \cdots, \nu_5). \end{aligned}$$

Then *the basic assumption on the process used in the following is that* $\xi_4(\nu_1, \nu_2, \nu_3) \,\varepsilon\, L_1(R_3)$ *and* $\xi_6(\nu_1, \cdots, \nu_5) \,\varepsilon\, L_1(R_5)$. This is a large class of processes which includes, for example, normal processes (trivially) and linear processes. More significantly, it includes all k-step dependent processes (see Hoeffding and Robbins [9]) (linear or nonlinear) which have sixth-order moments. The assumption that $\xi_4 \,\varepsilon\, L_1(R_3)$ was used by Magness [12] and Parzen [13] in works on spectral theory.

5. Third-order moment estimation. It was stated in Section 3 that, under certain restrictions, $\rho_N(\nu_1, \nu_2)$ is an asymptotically unbiased estimate of $r_3(\nu_1, \nu_2)$ and that $\sigma^2(\rho_N(\nu_1, \nu_2)) \to 0$ as $N \to \infty$. The specific result is

THEOREM 1. *If* (a) $\{X_t\}$, $EX_t = 0$, *is a real, 6th-order weakly stationary process,*
(b) $r(\nu)$, $r_3(\nu_1, \nu_2)$, $\xi_4(\nu_1, \nu_2, \nu_3)$, $\xi_6(\nu_1, \cdots, \nu_5)$, $g(\lambda_1, \lambda_2) \,\varepsilon\, L_1$, *and*
(c) $\gamma(N)$ *is a real function that is* $o(N)$, *then*

$$(5.1)\quad (1)\qquad \lim_{N\to\infty} \gamma(N)|E\rho_N(\nu_1, \nu_2) - r_3(\nu_1, \nu_2)| = 0,$$

$$(2)\quad \lim_{N\to\infty} N \operatorname{cov}[\rho_N(\nu_1, \nu_2), \rho_N(\nu_3, \nu_4)]$$
$$= \int_{-\infty}^{\infty} dy[r_6(\nu_1, \nu_2, y, y + \nu_3, y + \nu_4) - r_3(\nu_1, \nu_2)r_3(\nu_3, \nu_4)]$$
$$(5.2)\qquad = \int_{-\infty}^{\infty} dy[\xi_6(\nu_1, \nu_2, y, y + \nu_3, y + \nu_4)$$
$$+ \{m_3(0, \nu_1, \nu_2)m_3(y, y + \nu_3, y + \nu_4)\}_{10-1}$$

$$+ \{m_2(0, \nu_1)s_4(\nu_2, y, y + \nu_3, y + \nu_4)\}_{15}$$

$$+ \{m_2(0, \nu_1)m_2(\nu_2, y)m_2(y + \nu_3, y + \nu_4)\}_{15}]$$

where

$$\{m_3(0, \nu_1, \nu_2)m_3(y, y + \nu_3, y + \nu_4)\}_{10-1}$$

$$= \{m_3(0, \nu_1, \nu_2)m_3(y, y + \nu_3, y + \nu_4)\}_{10} - r_3(\nu_1, \nu_2)r_3(\nu_3, \nu_4),$$

(3) *Equation* (5.2) *can be written in the "frequency domain" provided* ξ_4 *and* ξ_6 *have Fourier transforms which are suitably regular* (*see* [18]).

PROOF. Assertion (1) is immediate and (2) is almost as immediate but will be proved in order to introduce some notation.

$$NE[\rho_N(\nu_1, \nu_2) - E\rho_N(\nu_1, \nu_2))(\rho_N(\nu_3, \nu_4) - E\rho_N(\nu_3, \nu_4))]$$

$$(5.3) \quad = N^{-1} \int_{D_N(\nu_1,\nu_2)} dt \int_{D_N(\nu_3,\nu_4)} d\tau [m_6(t, t + \nu_1, t + \nu_2, \tau, \tau + \nu_3, \tau + \nu_4) - r_3(\nu_1, \nu_2)r_3(\nu_3, \nu_4)].$$

Let $y = \tau - t$, $t = t$ to get

$$(5.4) \quad \int_{-N}^{N} dy[C_N(\nu_1, \nu_2, \nu_3, \nu_4, y)/N][r_6(\nu_1, \nu_2, y, y + \nu_3, y + \nu_4) - r_3(\nu_1, \nu_2)r_3(\nu_3, \nu_4)]$$

where C_N is defined as follows (see Fig. 3):

1. Construct the set $D_N(\nu_1, \nu_2) \times D_N(\nu_3, \nu_4) = \hat{D}_N$,
2. take the intersection of this set with the line $\tau - t = y$ and call that segment $\hat{C}_N$,
3. then $C_N(\nu_1, \nu_2, \nu_3, \nu_4, y)$ is equal to the length of the projection of $\hat{C}_N$ onto either axis.

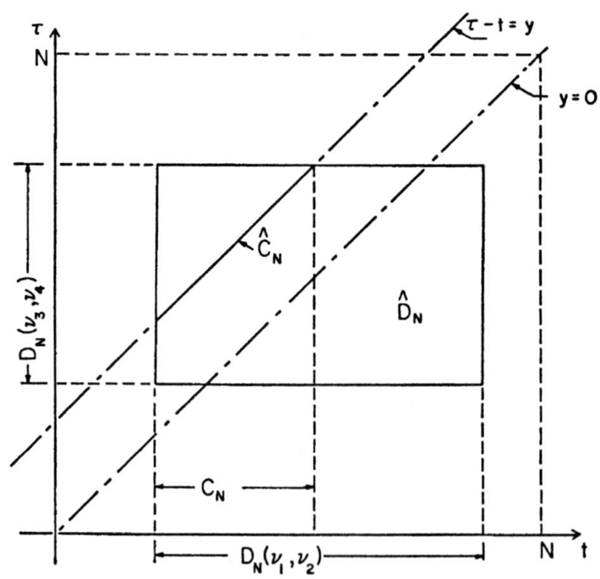

FIG. 3

Note that C_N can be written out analytically in terms of its arguments but the expression is cumbersome and not as enlightening as Fig. 3. Also, $0 \leqq C_N/N \leqq 1$ and $C_N/N \to 1$ as $N \to \infty$ uniformly on any fixed finite set of C_N's arguments. Using (4.5), (5.4) becomes

$$\begin{aligned}\int_{-N}^{N} dy(C_N/N)[\xi_6(\nu_1, \nu_2, y + \nu_3, y + \nu_4)& \\ + \{m_3(0, \nu_1, \nu_2)m_3(y, y + \nu_3, y + \nu_4)\}_{10-1}& \\ + \{m_2(0, \nu_1)s_4(\nu_2, y, y + \nu_3, y + \nu_4)\}_{15}& \\ + \{m_2(0, \nu_1)m_2(\nu_2, y)m_2(y + \nu_3, y + \nu_4)\}_{15}].& \qquad (5.5)\end{aligned}$$

Tables I, II, and III display the individual terms (later called minor terms) which make up (5.5). Looking at each of these 40 terms separately, it is obvious

TABLE I

The ten terms which sum to give the expression $\{s_3(0, \nu_1, \nu_2)s_3(y, y + \nu_3, y + \nu_4)\}_{10}$.

minor term number	$s_3(\cdot, \cdot, \cdot)s_3(\cdot, \cdot, \cdot)$	
1	$r_3(\nu_1, \nu_2)$	$r_3(\nu_3, \nu_4)$
2	$r_3(\nu_1, y)$	$r_3(y + \nu_3 - \nu_2, y + \nu_4 - \nu_2)$
3	$r_3(\nu_2, y)$	$r_3(y + \nu_3 - \nu_1, y + \nu_4 - \nu_1)$
4	$r_3(\nu_1, y + \nu_3)$	$r_3(y - \nu_2, y + \nu_4 - \nu_2)$
5	$r_3(\nu_1, y + \nu_4)$	$r_3(y - \nu_2, y + \nu_3 - \nu_2)$
6	$r_3(\nu_2, y + \nu_3)$	$r_3(y - \nu_1, y + \nu_4 - \nu_1)$
7	$r_3(\nu_2, y + \nu_4)$	$r_3(y - \nu_1, y + \nu_3 - \nu_1)$
8	$r_3(y, y + \nu_3)$	$r_3(\nu_2 - \nu_1, y + \nu_4 - \nu_1)$
9	$r_3(y, y + \nu_4)$	$r_3(\nu_2 - \nu_1, y + \nu_3 - \nu_1)$
10	$r_3(y + \nu_3, y + \nu_4)$	$r_3(\nu_2 - \nu_1, y - \nu_1)$

TABLE II

The fifteen terms which sum to give the expression $\{s_2(0, \nu_1)s_4(\nu_2, y, y + \nu_3, y + \nu_4)\}_{15}$.

minor term number	$m_2(\cdot, \cdot)s_4(\cdot, \cdot, \cdot)$	
1	$r(\nu_1)$	$\xi_4(\nu_2 - y, \nu_3, \nu_4)$
2	$r(\nu_2)$	$\xi_4(\nu_1 - y, \nu_3, \nu_4)$
3	$r(\nu_3)$	$\xi_4(\nu_1, \nu_2, \nu_4 + y)$
4	$r(\nu_4)$	$\xi_4(\nu_1, \nu_2, \nu_3 + y)$
5	$r(y)$	$\xi_4(\nu_2 - \nu_1, y + \nu_3 - \nu_1, y + \nu_4 - \nu_1)$
6	$r(\nu_2 - \nu_1)$	$\xi_4(-y, \nu_3, \nu_4)$
7	$r(\nu_4 - \nu_3)$	$\xi_4(\nu_1, \nu_2, y)$
8	$r(y - \nu_1)$	$\xi_4(\nu_2, y + \nu_3, y + \nu_4)$
9	$r(y - \nu_2)$	$\xi_4(\nu_1, y + \nu_3, y + \nu_4)$
10	$r(y + \nu_3)$	$\xi_4(\nu_1 - y, \nu_2 - y, \nu_4)$
11	$r(y + \nu_4)$	$\xi_4(\nu_1 - y, \nu_2 - y, \nu_3)$
12	$r(y + \nu_3 - \nu_1)$	$\xi_4(\nu_2, y, y + \nu_4)$
13	$r(y + \nu_3 - \nu_2)$	$\xi_4(\nu_1, y, y + \nu_4)$
14	$r(y + \nu_4 - \nu_1)$	$\xi_4(\nu_2, y, y + \nu_3)$
15	$r(y + \nu_4 - \nu_2)$	$\xi_4(\nu_1, y, y + \nu_3)$

TABLE III

The fifteen terms which sum to give the expression $\{s_2(0, \nu_1)s_2(\nu_2, y)s_2(y + \nu_3, y + \nu_4)\}_{15}$.

minor term number	$s_2(\cdot, \cdot)s_2(\cdot, \cdot)s_2(\cdot, \cdot)$		
1	$r(\nu_1)$	$r(\nu_2 - y)$	$r(\nu_3 - \nu_4)$
2	$r(\nu_2)$	$r(\nu_1 - y)$	$r(\nu_3 - \nu_4)$
3	$r(\nu_1)$	$r(y + \nu_3 - \nu_2)$	$r(\nu_4)$
4	$r(\nu_2)$	$r(y + \nu_3 - \nu_1)$	$r(\nu_4)$
5	$r(\nu_1)$	$r(y + \nu_4 - \nu_2)$	$r(\nu_3)$
6	$r(\nu_2)$	$r(y + \nu_4 - \nu_1)$	$r(\nu_3)$
7	$r(y)$	$r(\nu_2 - \nu_1)$	$r(\nu_4 - \nu_3)$
8	$r(y)$	$r(y + \nu_3 - \nu_1)$	$r(y + \nu_4 - \nu_2)$
9	$r(y)$	$r(y + \nu_3 - \nu_2)$	$r(y + \nu_4 - \nu_1)$
10	$r(y + \nu_3)$	$r(\nu_2 - \nu_1)$	$r(\nu_4)$
11	$r(y + \nu_4)$	$r(\nu_2 - \nu_1)$	$r(\nu_3)$
12	$r(y + \nu_3)$	$r(y - \nu_1)$	$r(y + \nu_4 - \nu_2)$
13	$r(y + \nu_3)$	$r(y - \nu_2)$	$r(y + \nu_4 - \nu_1)$
14	$r(y + \nu_4)$	$r(y - \nu_1)$	$r(y + \nu_3 - \nu_2)$
15	$r(y + \nu_4)$	$r(y - \nu_2)$	$r(y + \nu_3 - \nu_1)$

that the expression in square brackets is absolutely integrable with respect to y on $(-\infty, \infty)$. Thus writing $\int_{-N}^{N} dy$ as $\int_{-\infty}^{\infty} dy\chi_N(y)$ where χ_N is the characteristic function of the interval $[-N, N]$, (5.5) can be looked upon as the infinite integral of a sequence of functions, $\chi_N(y)(C_N/N)[\cdots]$, which (by the properties of C_N) tend to zero pointwise and are dominated by the fixed $L_1(-\infty, \infty)$ function given by the absolute value of the expression in square brackets. Consequently, Lebesgue's dominated convergence theorem gives the result.

Q.E.D.

6. Estimation of the weighted bispectral density. Next consider estimates of the form (3.3),

$$g_N^*(W) = \int_{-\infty}^{\infty} \int_{-\infty}^{\infty} W(\mu_1, \mu_2)g_N(\mu_1, \mu_2)\, d\mu_1\, d\mu_2 . \tag{3.3}$$

The following conditions will be placed on W.

DEFINITION 1. A bispectral averaging function, $W(\mu_1, \mu_2)$, of order $\alpha > 0$ is

(i) real-valued,

(ii) $\varepsilon L_1 \cap L_2$, and

(iii) its Fourier transform (which is bounded and L_2),

$$w(\nu_1, \nu_2) = (2\pi)^{-2} \int_{-\infty}^{\infty} \int_{-\infty}^{\infty} \exp(-i\nu_1\mu_1 - i\nu_2\mu_2)W(\mu_1, \mu_2)\, d\mu_1\, d\mu_2 \tag{6.1}$$

is such that $|w(\nu_1, \nu_2)| = O(|\nu_1| + |\nu_2|)^{-\alpha}$.

Condition (iii) could be weakened. Note that (iii) implies that there are constants K_1 and $b > 0$ such that for all $|\nu_1|$ and/or $|\nu_2| \geqq b$,

$$|w(\nu_1, \nu_2)| \leqq K_1(|\nu_1| + |\nu_2|)^{-\alpha}. \tag{6.2}$$

THEOREM 2. *If* (a) $\{X_t\}$ *is* 3rd-*order weakly stationary, real,*

(b) $r_3(\nu_1, \nu_2)$, $g(\lambda_1, \lambda_2) \varepsilon L_1(R_2)$, *and*

(c) $W(\mu_1, \mu_2)$ *is a bispectral averaging function of order* $s + \epsilon$, $\epsilon > 0$, $0 < s < 1$, *then* $\lim_{N\to\infty} N^s |Eg_N^*(W) - g(W)| = 0$.

PROOF. By (3.2), (3.3), and (6.1)

$$N^s |Eg_N^*(W) - g(W)|$$

$$= N^s \left|E \int_{-N}^{N} \int_{-N}^{N} w(\nu_1, \nu_2)[\rho_N(\nu_1, \nu_2) - r_3(\nu_1, \nu_2)]\, d\nu_1\, d\nu_2 - \int\int_{|\nu_1| \text{or} |\nu_2| > N} w(\nu_1, \nu_2) r_3(\nu_1, \nu_2)\, d\nu_1\, d\nu_2\right|. \tag{6.3}$$

The second integral on the left hand side of (6.3) tends to zero in absolute value by the boundedness of $w(\nu_1, \nu_2)$ and the fact that $r_3(\nu_1, \nu_2) \varepsilon L_1$. The first integral is

$$\leqq N^s \left|\int_{-N}^{N} \int_{-N}^{N} w(\nu_1, \nu_2)[1 - [c_N(\nu_1, \nu_2)/N] - 1] r_3(\nu_1, \nu_2)\, d\nu_1\, d\nu_2\right| \tag{6.4}$$

where

$$c_N(\nu_1, \nu_2) = \min [\hat{c}_N(\nu_1, \nu_2), N]$$
$$\hat{c}_N(\nu_1, \nu_2) = \max [|\nu_1|, |\nu_2|, |\nu_1 - \nu_2|]. \tag{6.5}$$

Since $c_N \leqq |\nu_1| + |\nu_2|$ and by (6.2), the integrand of (6.4), $[c_N(\nu_1, \nu_2)/N^{1-s}]w(\nu_1, \nu_2) r_3(\nu_1, \nu_2)$, is bounded in absolute value on $(-\infty, \infty)$ by $K_2 |r_3(\nu_1, \nu_2)|$ for some K_2. This integrand converges pointwise to zero so that again by Lebesgue's dominated convergence theorem the result is proved.

Q.E.D.

Property (i) of Section 3, the asymptotic unbiasedness, is therefore shown. Now turn to Property (ii) of Section 3.

THEOREM 3. *If* (a) $\{X_t\}$ *is* 6th-*order weakly stationary, real,*

(b) $r(\nu)$, $r_3(\nu_1, \nu_2)$, $\xi_4(\nu_1, \nu_2, \nu_3)$, $\xi_6(\nu_1, \cdots, \nu_5)$, $g(\nu_1, \nu_2) \varepsilon L_1$, *and*

(c) $W_j(\mu_1, \mu_2)$, $j = 1, 2$, *are bispectral averaging functions of order* $1 + \epsilon_j$, $\epsilon_j > 0$, *then*

(1) $\lim_{N\to\infty} N \operatorname{cov} [g_N^*(W_1), g_N^*(W_2)]$

$$= \int_{-\infty}^{\infty} \cdots \int_{-\infty}^{\infty} \int_{-\infty}^{\infty} d\nu_1 \cdots d\nu_4\, dy w_1(\nu_1, \nu_2)\overline{w_2(\nu_3, \nu_4)}$$
$$\cdot [\xi_6(\nu_1, \nu_2, y, y + \nu_3, y + \nu_4) + \{m_3(0, \nu_1, \nu_2) m_3(y, y + \nu_3, y + \nu_4)\}_{10-1}$$
$$+ \{m_2(0, \nu_1) s_4(\nu_2, y, y + \nu_3, y + \nu_4)\}_{15}$$
$$+ \{m_2(0, \nu_1) m_2(\nu_2, y) m_2(y + \nu_3, y + \nu_4)\}_{15}], \tag{6.6}$$

(2) *Equation* (6.6) *is written in the "time domain"; the result can be written in the "frequency domain"* (*see* [18]).

PROOF. Recalling the proof of Theorem 1,

$$NE[(g_N^*(W_1) - Eg_N^*(W_1))(g_N^*(W_2) - Eg_N^*(W_2))]$$
$$= \int_{-N}^{N} \cdots \int_{-N}^{N} \int_{-N}^{N} w_1(\nu_1, \nu_2)\overline{w_2(\nu_3, \nu_4)} [C_N(\nu_1, \cdots, \nu_4, y)/N]$$

$$(6.7)\qquad \cdot[\xi_6(\nu_1, \nu_2, y, y+\nu_3, y+\nu_4) + \{m_3(0, \nu_1, \nu_2)m_3(y, y+\nu_3, y+\nu_4)\}_{10-1}$$
$$+ \{m_2(0, \nu_1)s_4(\nu_2, y, y+\nu_3, y+\nu_4)\}_{15}$$
$$+ \{m_2(0, \nu_1)m_2(\nu_2, y)m_2(y+\nu_3, y+\nu_4)\}_{15}]\, d\nu_1 \cdots d\nu_4\, dy.$$

It remains to be shown that the integrand with the (C_N/N) term replaced by 1 is absolutely integrable on $(-\infty, \infty)$. To do this, consider the right hand side of (6.7) as made up of four "major terms" or 40 "minor terms", where the first major term and the first minor term are identical and the remaining minor terms are enumerated in accorance with Tables I, II, and III.

Each minor term is treated separately. Minor term 1 is obviously summable since w_1 and w_2 are bounded and $\xi_6 \varepsilon L_1(R_5)$. The remaining minor terms are grouped according to the major terms within which they are contained.

LEMMA 1. *The hypotheses of Theorem* 3 *imply that*

(1)

$$(6.8)\qquad \int_{-\infty}^{\infty} \cdots \int_{-\infty}^{\infty} \int_{-\infty}^{\infty} |\{m_3(0, \nu_1, \nu_2)m_3(y, y+\nu_3, y+\nu_4)\}_{10-1}\, w_1(\nu_1, \nu_2)\overline{w_2(\nu_3, \nu_4)}|\, d\nu_1 \cdots d\nu_4\, dy < \infty,$$

(2)

$$(6.9)\qquad \int_{-\infty}^{\infty} \cdots \int_{-\infty}^{\infty} \int_{-\infty}^{\infty} |\{m_2(0, \nu_1)s_4(\nu_2, y, y+\nu_3, y+\nu_4)\}_{15}\, w_1(\nu_1, \nu_2)\overline{w_2(\nu_3, \nu_4)}|\, d\nu_1 \cdots d\nu_4\, dy < \infty,$$

(3)

$$(6.10)\qquad \int_{-\infty}^{\infty} \cdots \int_{-\infty}^{\infty} \int_{-\infty}^{\infty} |\{m_2(0, \nu_1)m_2(\nu_2, y)m_2(y+\nu_3, y+\nu_4)\}_{15}\, w_1(\nu_1, \nu_2)\overline{w_2(\nu_3, \nu_4)}|\, d\nu_1 \cdots d\nu_4\, dy < \infty.$$

PROOF. The proof follows using the integrability properties of r_2, r_3 and ξ_4 and (6.1) and (6.2) (see [18] and [19]). Q.E.D.

The previous statements and Lemma 1 show that there is an L_1 upper bound for all N on the integrand. Therefore the properties of (C_N/N) and Lebesgue's dominated convergence theorem give (6.6). Q.E.D.

7. Estimation of the bispectral density. The most interesting properties are exhibited by the third form of estimate introduced in Section 3,

$$(3.5)\qquad g_N^*(\lambda_1, \lambda_2) = \int_{-\infty}^{\infty} \int_{-\infty}^{\infty} W_N(\mu_1 - \lambda_1, \mu_2 - \lambda_2) g_N(\mu_1, \mu_2)\, d\mu_1\, d\mu_2 .$$

Properties (i) and (ii) of Section 3 will be discussed not for the general form (3.5) but for a subclass of estimates described by first defining a bispectral estimating kernel.

DEFINITION 2. A symmetric bispectral estimating kernel, $w(\nu_1, \nu_2)$, of order $\alpha > 0$ is

(i) real-valued, εL_2,

(ii) $w(\nu_1, \nu_2) \leqq M_1 < \infty$,

(iii) $w(\nu_1, \nu_2) = w(\nu_2, \nu_1) = w(-\nu_1, \nu_2 - \nu_1)$ (same symmetries as those of $r_3(\nu_1, \nu_2)$),

(iv) for every $\epsilon > 0$ there exists an M_2 such that uniformly in ν_2,

$$\int_{|\nu_1|>M_2} |w(\nu_1, \nu_2)| \, d\nu_1 \leqq \epsilon,$$

(v) and

$$w^{(\alpha)} = \sup_P \lim_{\nu_1, \nu_2 \to 0} [|1 - w(\nu_1, \nu_2)|/(|\nu_1| + |\nu_2|)^\alpha],$$

where $P \equiv$ all possible paths, is well-defined and finite.
Letting $\{B_N\}$ be a sequence of positive constants tending to zero as $N \to \infty$, the estimates then discussed are of the form

$$(7.1) \quad g_N^*(\lambda_1, \lambda_2) = (2\pi)^{-2} \int_{-N}^{N} \int_{-N}^{N} \exp(-i\lambda_1\nu_1 - i\lambda_2\nu_2) \cdot w(B_N\nu_1, B_N\nu_2) \rho_N(\nu_1, \nu_2) \, d\nu_1 \, d\nu_2 .$$

By Condition (i) above, $w(\nu_1, \nu_2)$ is a Fourier transform in the L_2 sense,

$$(7.2) \quad w(\nu_1, \nu_2) = \int_{-\infty}^{\infty} \int_{-\infty}^{\infty} \exp(-i\nu_1\lambda_1 - i\nu_2\lambda_2) W(\lambda_1, \lambda_2) \, d\lambda_1 \, d\lambda_2 ,$$

Using this, (7.1) can be rewritten as

$$(7.3) \; g_N^*(\lambda_1, \lambda_2) = \int_{-\infty}^{\infty} \int_{-\infty}^{\infty} B_N^{-2} W[(\mu_1 - \lambda_1)/B_N), (\mu_2 - \lambda_2)/B_N] g_N(\mu_1, \mu_2) \, d\mu_1 \, d\mu_2$$

which is of the form (3.5). Parzen [13] uses an analogous estimate for the spectral density. The rate at which $B_N \to 0$ governs the rate of concentration of the weight functions.

Condition (iii) of Definition 2 implies that $g_N^*(\lambda_1, \lambda_2)$ has the same symmetries as $g(\lambda_1, \lambda_2)$. It is not necessary in proving the following theorems, but is included merely to permit the statement and proof of Theorem 5 to be written more compactly. A discussion of the nonsymmetric case including a statement of the theorems is in [19].

For convenience of notation, define the generalized qth weighted (with respect to some bounded weight function h) bispectral derivative, $g^{(q)}(\mu_1, \mu_2)$, by

$$(7.4) \quad g^{(q)}(\mu_1, \mu_2, h) = (2\pi)^{-2} \int_{-\infty}^{\infty} \int_{-\infty}^{\infty} \exp(-i\mu_1\nu_1 - i\mu_2\nu_2)(|\nu_1| + |\nu_2|)^q r_3(\nu_1, \nu_2) h(\nu_1, \nu_2) \, d\nu_1 \, d\nu_2 .$$

Then the bias of (7.3) satisfies the following.

THEOREM 4. *If* (a) $\{X_t\}$ *is* 3rd-*order weakly stationary, real,*

(b) $r_3(\nu_1, \nu_2)$, $g(\lambda_1, \lambda_2) \, \varepsilon \, L_1(R_2)$,

(c) $\int_{-\infty}^{\infty} \int_{-\infty}^{\infty} (|\nu_1| + |\nu_2|)^q r_3(\nu_1, \nu_2) \, d\nu_1 \, d\nu_2 < \infty$, $q > 0$,

(d) $w(\nu_1, \nu_2)$ *is a bispectral estimating kernel of order* $\alpha > q$, *and*

(e) B_N *chosen such that as* $N \to \infty$

(i) $B_N \to 0$,

(ii) $B_N N \to \infty$ *if* $q \leqq 1$
$B_N^q N \to \infty$ *if* $q > 1$

then

$$(7.5)\quad \lim_{N\to\infty} B_N^{-q}\,|Eg_N^*(\mu_1,\mu_2) - g(\mu_1,\mu_2)| = 0 \qquad \text{if } \alpha > q$$
$$= |g^{(q)}(\mu_1,\mu_2, w^{(q)})| \qquad \text{if } \alpha = q$$

where in the case $\alpha = q$, $w(\nu_1, \nu_2)$ *is assumed to have the property that for almost all* (ν_1, ν_2),

$$(7.6)\quad \lim_{N\to\infty} (1 - w(B_N\nu_1, B_N\nu_2))/(|B_N\nu_1| + |B_N\nu_2|)^q = w^{(q)}(\nu_1,\nu_2).$$

PROOF. Break up the bias expression into three terms,

$$\begin{aligned} &B_N^{-q}(Eg_N^*(\mu_1,\mu_2) - g(\mu_1,\mu_2)) \\ &\quad = [B_N^{-q}/(2\pi)^2][\textstyle\int_{-N}^{N}\int_{-N}^{N} \exp(-i\mu_1\nu_1 - i\mu_2\nu_2)(w(B_N\nu_1, B_N\nu_2) - 1)r_3 \\ (7.7)\quad &\qquad \cdot(\nu_1,\nu_2)\,d\nu_1\,d\nu_2 \\ &\quad - \textstyle\int_{-N}^{N}\int_{-N}^{N} \exp(-i\mu_1\nu_1 - i\mu_2\nu_2)\{\min[\max(|\nu_1|, |\nu_2|, |\nu_1 - \nu_2|), N]/N\} \\ &\qquad \cdot w(B_N\nu_1, B_N\nu_2) r_3(\nu_1,\nu_2)\,d\nu_1\,d\nu_2 \\ &\quad - \textstyle\int\int_{|\nu_1|\text{or}|\nu_2|>N} \exp(-i\mu_1\nu_1 - i\mu_2\nu_2) r_3(\nu_1,\nu_2)\,d\nu_1\,d\nu_2]. \end{aligned}$$

The third term is

$$\leqq (NB_N)^{-q} \int\int_{|\nu_1|\text{or}|\nu_2|>N} (|\nu_1| + |\nu_2|)^q\,|r_3(\nu_1,\nu_2)|\,d\nu_1\,d\nu_2 \to 0$$

by (c) and (e). The second term of (7.7) is

$$(7.8)\quad \leqq (M_1/B_N^q N) \int_{-N}^{N}\int_{-N}^{N} (|\nu_1| + |\nu_2|)|r_3(\nu_1,\nu_2)|\,d\nu_1\,d\nu_2\,.$$

First suppose $q \leqq 1$; then (7.8) is

$$\leqq [M_1/(NB_N)^q] \int_{-N}^{N}\int_{-N}^{N} [(|\nu_1| + |\nu_2|)^{1-q}/N^{1-q}](|\nu_1| + |\nu_2|)^q\,|r_3(\nu_1,\nu_2)|\,d\nu_1\,d\nu_2 \to 0$$

by (c) and (e). Next suppose $q > 1$; then (7.8) is

$$\leqq (M_1/NB_N^q) \int_{-N}^{N}\int_{-N}^{N} (|\nu_1| + |\nu_2|) r_3(\nu_1,\nu_2)\,d\nu_1\,d\nu_2 \to 0$$

again by (c) and (e). Therefore

$$\begin{aligned} &\lim_{N\to\infty} B_N^{-q}\,|Eg_N^*(\mu_1,\mu_2) - g(\mu_1,\mu_2)| \\ (7.9)\quad &= (2\pi)^{-2} \lim_{N\to\infty} \left|\textstyle\int_{-N}^{N}\int_{-N}^{N} \exp(-i\mu_1\nu_1 - i\mu_2\nu_2)(|\nu_1| + |\nu_2|)^q r_3(\nu_1,\nu_2)\right. \\ &\qquad \left.\cdot[(w(B_N\nu_1, B_N\nu_2) - 1)/(|B_N\nu_1| + |B_N\nu_2|)^\alpha](|B_N\nu_1| + |B_N\nu_2|)^{\alpha-q}\,d\nu_1\,d\nu_2\right|. \end{aligned}$$

Since $w(\nu_1, \nu_2)$ is a bispectral estimating kernel of order α, there exists a constant K_3 such that the integrand of (7.9) is bounded by $K_3(|\nu_1| + |\nu_2|)^q\,|r_3(\nu_1,\nu_2)|$ which is integrable. Q.E.D.

Secondly, the asymptotic behavior of the variance can be described under the assumptions listed in the next theorem. In the proof of this theorem the following lemmas are needed. Lemma 2 is motivated by the Riemann-Lebesgue lemma. The proofs are simple and are omitted (see [18]).

LEMMA 2. *The hypotheses of Theorem 5 imply that*

(1) $\int_{-M}^{M}\int_{-M}^{M} d\nu_1\, d\nu_2 \exp[-i\nu_1\mu_1 - i\nu_2(\mu_2/B_N)]w(B_N\nu_1, \nu_2)r(\nu_1) \to 0$ *as* $N \to \infty$ *unless* $\mu_2 = 0$,

(2) $\int_{-M}^{M} \cdots \int_{-M}^{M}\int_{-M}^{M} d\nu_1 \cdots d\nu_4\, dy \exp\{-i\nu_1(\mu_1/B_N) - i\nu_2\mu_2 + i\nu_3\mu_3 + i\nu_4 \cdot(\mu_4/B_N)\}w(\nu_1, B_N\nu_2 + \nu_4)w(B_N\nu_3 + \nu_1, \nu_4)r(y)r(y + \nu_3)r(y - \nu_2) \to 0$ *unless* $\mu_1 = \mu_4 = 0$.

LEMMA 3. *If* (a) $\{\gamma(M, N)\}$ *is a sequence of constants,*

(b) $\gamma(M) = \lim_{N\to\infty} \gamma(M, N)$ *exists for all* M,

(c) $\gamma = \lim_{M\to\infty} \gamma(M)$ *exists, and*

(d) *for any* $\epsilon > 0$ *there exists an* $M_0(\epsilon)$ *such that for all* $N > M > M_0$, $|\gamma(M, N) - \gamma(N, N)| < \epsilon$,

then $\lim_{N\to\infty} \gamma(N, N) = \gamma$.

THEOREM 5. *If* (a) $\{X_t\}$ *is* 6th-*order weakly stationary, real,*

(b) $r(\nu)$, $r_3(\nu_1, \nu_2)$, $\xi_4(\nu_1, \nu_2, \nu_3)$, $\xi_6(\nu_1, \cdots, \nu_5)$, $g(\lambda_1, \lambda_2) \in L_1$,

(c) $w(\nu_1, \nu_2)$ *is a symmetric bispectral estimating kernel,*

(d) B_N *is a sequence of positive constants such that*

(i) $B_N \to 0$ *as* $N \to \infty$

(ii) $B_N^2 N \to \infty$ *as* $N \to \infty$,

(e) $w(\nu_1, \nu_2)$ *is continuous* a.e. *and for* $a < \infty$, $w(B_N a, \nu_2) \to w(0, \nu_2)$ *for almost all* ν_2, *and*

(f) *for brevity in writing results,* (μ_1, μ_2) *and* (μ_3, μ_4) *are taken in their first sections of definition as shown in Fig.* 2, i.e. $0 \leqq \mu_1, \mu_3 < \infty$, $0 \leqq \mu_2 \leqq \mu_1$, $0 \leqq \mu_4 \leqq \mu_3$, *then*

$$\begin{aligned}
&\lim_{N\to\infty} NB_N^2 \operatorname{cov}[g_N^*(\mu_1, \mu_2), g_N^*(\mu_3, \mu_4)] \\
(7.10)\qquad &= (2\pi)^{-1}[f(\mu_1)f(\mu_2)f(\mu_1 + \mu_2)f(\mu_3)f(\mu_4)f(\mu_3 + \mu_4)]^{1/2} \\
&\quad \cdot\{w_1\delta(\mu_2)\delta(\mu_4)[1 + 2\delta(\mu_1)][1 + 2\delta(\mu_3)] \\
&\quad + w_2\delta(\mu_1 - \mu_3)\delta(\mu_2 - \mu_4)[1 + \delta(\mu_1 - \mu_4) + 4\delta(\mu_1)\delta(\mu_2)]\},
\end{aligned}$$

where $w_1 = [\int_{-\infty}^{\infty} w(0, \nu)\, d\nu]^2$, $w_2 = \int_{-\infty}^{\infty}\int_{-\infty}^{\infty} w^2(\nu_1, \nu_2)\, d\nu_1\, d\nu_2$.

PROOF. The proof will be given in a series of lemmas and individual discussions of major and minor terms. Recalling Theorems 1 and 3, write

$$\begin{aligned}
&NB_N^2 \operatorname{cov}[g_N^*(\mu_1, \mu_2), g_N^*(\mu_3, \mu_4)] \\
&= (B_N^2/(2\pi)^4)\int_{-N}^{N} \cdots \int_{-N}^{N} d\nu_1 \cdots d\nu_4 \exp(-i\mu_1\nu_1 - i\mu_2\nu_2 + i\mu_3\nu_3 + i\mu_4\nu_4) \\
&\quad \cdot w(B_N\nu_1, B_N\nu_2)w(B_N\nu_3, N_N\nu_4)\int_{-N}^{N} dy(C_N(\nu_1, \cdots, \nu_4, y))/N \\
(7.11)\qquad &\quad \cdot[\xi_6(\nu_1, \nu_2, y, y + \nu_3, y + \nu_4) + \{m_3(0, \nu_1, \nu_2)m_3(y, y + \nu_3, y + \nu_4)\}_{10-1} \\
&\quad + \{m_2(0, \nu_1)s_4(\nu_2, y, y + \nu_3, y + \nu_4)\}_{15} \\
&\quad + \{m_2(0, \nu_1)m_2(\nu_2, y)m_2(y + \nu_3, y + \nu_4)\}_{15}].
\end{aligned}$$

The boundedness of w and (C_N/N) and the $L_1(R_5)$ property of ξ_6 imply the first major term is $O(B_N^2)$. Further, by also using the integrability properties of r_2, r_3, ξ_4 and w, the second and third major terms are seen to be $O(B_N)$. By these statements

$$\lim_{N\to\infty} NB_N^2 \operatorname{cov}[g_N^*(\mu_1, \mu_2), g_N^*(\mu_3, \mu_4)]$$
$$= \lim_{N\to\infty} (B_N^2/(2\pi)^4)\int_{-N}^{N} \cdots \int_{-N}^{N}\int_{-N}^{N} d\nu_1 \cdots d\nu_4\, dy$$
$$\cdot\exp(-i\mu_1\nu_1 - i\mu_2\nu_2 + i\mu_3\nu_3 + i\mu_4\nu_4)w(B_N\nu_1, B_N\nu_2)$$
$$\cdot w(B_N\nu_3, B_N\nu_4)(C_N(\nu_1, \cdots, \nu_4, y)/N)\{m_2(0, \nu_1)m_2(\nu_2, y)m_2(y+\nu_3, y+\nu_4)\}_{15}.$$

Look at the 15 minor terms, listed by Table III, separately. The first term yields

$$\lim_{N\to\infty} (B_N^2/(2\pi)^4)\int_{-N}^{N} \cdots \int_{-N}^{N}\int_{-N}^{N} d\nu_1 \cdots d\nu_4\, dy$$
$$\cdot\exp(-i\mu_1\nu_1 - i\mu_2\nu_2 + i\mu_3\nu_3 + i\mu_4\nu_4)w(B_N\nu_1, B_N\nu_2)w(B_N\nu_3, B_N\nu_4) \tag{7.12}$$
$$\cdot(C_N(\nu_1, \cdots, \nu_4, y)/N)r(\nu_1)r(\nu_2 - y)r(\nu_3 - \nu_4).$$

Let $\hat{y} = y - \nu_2$, $\hat{\nu}_4 = \nu_4 - \nu_3$, $\hat{\nu}_2 = B_N\nu_2$, $\hat{\nu}_3 = B_N\nu_3$; then (7.2) becomes

$$\lim_{N\to\infty} (2\pi)^{-4}\int_{-N}^{N} d\nu_1\int_{-NB_N}^{NB_N} d\hat{\nu}_2\int_{-NB_N}^{NB_N} d\hat{\nu}_3 \int_{-N-(\hat{\nu}_3/B_N)}^{N-(\hat{\nu}_3/B_N)} d\hat{\nu}_4\int_{-N-(\hat{\nu}_2/B_N)}^{N-(\hat{\nu}_2/B_N)} d\hat{y}$$
$$\cdot\exp\{-i\nu_1\mu_1 - i\hat{\nu}_2(\mu_2/B_N) + i\hat{\nu}_3[(\mu_3+\mu_4)/B_N] + i\hat{\nu}_4\mu_4\}w(B_N\nu_1, \hat{\nu}_2) \tag{7.13}$$
$$\cdot w(-\hat{\nu}_3, B_N\hat{\nu}_4)[C_N(\nu_1, (\hat{\nu}_2/B_N), (\hat{\nu}_3/B_N), \hat{\nu}_4 + (\hat{\nu}_3/B_N), \hat{y} + (\hat{\nu}_2/B_N))/N]$$
$$\cdot r(\nu_1)r(\hat{y})r(\hat{\nu}_4).$$

An inspection of Fig. III reveals that (see [18] or [19])

$$[C_N(\nu_1, (\nu_2/B_N), (\nu_3/B_N), \nu_4 + (\nu_3/B_N), y + (\nu_2/B_N))/N] \to 1 \text{ as } N \to \infty. \tag{7.14}$$

Denote by $\square_M$, the j-dimensional hypercube centered at the origin with sides of length $2M$ parallel to the j axes. The dimension, j, will be obvious from the context. Also let $\square_M'$ denote the complement of $\square_M$ in R_j. We have the following easy lemma (see [18] or [19]).

LEMMA 4. *The hypotheses of Theorem 5 imply that for any $\epsilon > 0$, there is an $M_2(\epsilon)$ independent of N such that for all $M > M_2$ and for all N,*

$$\int\int\int\int\int_{\square_M'} |w(B_N\nu_1, \nu_2)w(-\nu_3, B_N\nu_4)r(\nu_1)r(y)r(\nu_4)|\, d\nu_1 \cdots d\nu_4\, dy < \epsilon.$$

Let

$$\gamma_1(M, N) = (2\pi)^{-4}\int\int\int\int\int d\nu_1 \cdots d\nu_4\, dy \exp\{-i\nu_1\mu_1 - i\nu_2(\mu_2/B_N)$$
$$+ i\nu_3[(\mu_3+\mu_4)/B_N] + i\nu_4\mu_4\}w(B_N\nu_1, \nu_2)w(-\nu_3, B_N\nu_4)r(\nu_1)r(y)r(\nu_4),$$
$$\gamma_1(M) = \lim_{N\to\infty}\gamma_1(M, N).$$

Lemma 4 and (7.14) indicate that the limit (7.13) is the same as $\lim_{N\to\infty}\gamma_1(N, N)$. Lemma 2 says that $\gamma_1(M) = 0$ unless $\mu_2 = \mu_3 + \mu_4 = 0$. Using the boundedness and continuity properties of w, Lebesgue's convergence theorem, and Lemma 3;

$$\lim_{N\to\infty} \gamma_1(N, N) = (2\pi)^{-1} w_1 f(0) f(\mu_1) f(\mu_3) \delta(\mu_2) \delta(\mu_3 + \mu_4).$$

Minor terms 2 through 7 and 10 and 11 follow by similar arguments. Minor term 8 offers a slight variation. Put $\hat{\nu}_3 = \nu_3 - \nu_1$, $\hat{\nu}_2 = \nu_2 - \nu_4$, $\hat{\nu}_1 = B_N \nu_1$, $\hat{\nu}_4 = B_N \nu_4$ to get (omitting the "hats")

$$\begin{aligned}
&(2\pi)^{-4} \lim_{N\to\infty} \int_{-NB_N}^{NB_N} \int_{-NB_N}^{NB_N} d\nu_1\, d\nu_4 \int_{-N-(\nu_4/B_N)}^{N-(\nu_4/B_N)} d\nu_2 \int_{-N-(\nu_1/B_N)}^{N-(\nu_1/B_N)} d\nu_3 \\
&\cdot \exp\{i\nu_1[(\mu_1 - \mu_3)/B_N] - i\nu_2\mu_2 + i\nu_3\mu_3 + i\nu_4[(\mu_4 - \mu_2)/B_N]\} \\
&\cdot w(\nu_1, B_N\nu_2 + \nu_4) w(B_N\nu_3 + \nu_1, \nu_4) \qquad (7.15) \\
&\cdot \int_{-N}^{N} dy [C_N((\nu_1/B_N), \nu_2 + (\nu_4/B_N), \nu_3 + (\nu_1/B_N), (\nu_4/B_N), y)/N] \\
&\cdot r(y) r(y + \nu_3) r(y - \nu_2).
\end{aligned}$$

Again it is seen that $(C_N/N) \to 1$ as $N \to \infty$. Corresponding to Lemma 4 is the following:

LEMMA 5. *The hypotheses of Theorem* 5 *imply that for any* $\epsilon > 0$, *there is an* $M_3(\epsilon)$ *independent of* N *such that for all* $M > M_3(\epsilon)$ *and for all* $N > M$,

$$\int\int\int\int\int_{\square_M} d\nu_1 \cdots d\nu_4\, dy |w(\nu_1, B_N\nu_2 + \nu_4) w(B_N\nu_3 + \nu_1, \nu_4) \cdot r(y) r(y + \nu_3) r(y - \nu_2)| < \epsilon.$$

Proof: The lemma follows using Schwarz's inequality (see [18] or [19]). Q.E.D. Define

$$\begin{aligned}
\gamma_8(M, N) = {} & (2\pi)^{-4} \int_{-M}^{M} \cdots \int_{-M}^{M} \int_{-M}^{M} d\nu_1 \cdots d\nu_4\, dy \exp\{i\nu_1[(\mu_1 - \mu_3)/B_N] - i\nu_2\mu_2 \\
& + i\nu_3\mu_3 + i\nu_4[(\mu_4 - \mu_2)/B_N]\} w(\nu_1, B_N\nu_2 + \nu_4) w(B_N\nu_3 + \nu_1, \nu_4) \\
& \cdot r(y) r(y + \nu_3) r(y - \nu_2), \\
\gamma_8(M) = {} & \lim_{N\to\infty} \gamma_8(M, N).
\end{aligned}$$

Lemma 2, Lebesgue's theorem, and Lemma 3 gives minor term 8 as

$$\lim_{N\to\infty} \gamma_8(M, N) = (2\pi)^{-1} w_2 f(\mu_2) f(\mu_3) f(\mu_2 + \mu_3) \delta(\mu_1 - \mu_3) \delta(\mu_2 - \mu_4).$$

Certain generalizations of Theorem 5 could readily be obtained. For example, two weight functions, w_1 and w_2, could be used; different (but interrelated) sequences, $\{B_N^{(1)}\}$ and $\{B_N^{(2)}\}$, could also be used; and as pointed out earlier certain restrictions on w and the process could be relaxed slightly.

8. Further results. A later paper (part of the results to appear in this paper are contained in Rosenblatt and Van Ness [18].) will discuss the estimates in the discrete parameter case, in particular their asymptotic distribution which (for certain processes) turns out to be complex normal with independent real and imaginary parts. In the second and third-order results the assumptions that the moment functions $r(\nu)$ and $r_3(\nu_1, \nu_2)$ are in L_1 were crucial to the estimation of their transforms. When we go to higher order theory the assumption that $r_n(\nu_1, \cdots, \nu_{n-1}) \in L_1$, $n \geqq 4$ is too restrictive. This difficulty can be overcome by estimating the Fourier transforms of the cumulant functions, $\xi_n(\nu_1, \cdots, \nu_{n-1})$, instead of the transforms of the moment functions. (This is actually what we've done in the second and third-order cases since $r(\nu) = \xi(\nu)$ and $r_3(\nu_1, \nu_2) = \xi_3(\nu_1, \nu_2)$.) As stated earlier, the assumption $\xi_n(\nu_1, \cdots, \nu_{n-1}) \varepsilon L_1(R_{n-1})$ is not so restrictive.

REFERENCES

[1] Bartlett, M. S. (1955). *An Introduction to Stochastic Processes*. Cambridge Press.

[2] Blackman, R. B. and Tukey, J. W. (1959). *The Measurement of Power Spectra from the Point of View of Communications Engineering*. Dover, New York.

[3] Blanc-Lapierre, A. and Fortet, R. (1953). *Théorie des Fonctions Aléatoires*. Masson et Cie, Paris.

[4] Doob, J. L. (1953). *Stochastic Processes*. Wiley, New York.

[5] Gnedenko, B. V. and Kolmogorov, A. D. (1954). *Limit Distributions for Sums of Independent Random Variables*. Addison-Wesley, Cambridge.

[6] Grenander, U. and Rosenblatt, M. (1957). *Statistical Analysis of Stationary Time Series*. Wiley, New York.

[7] Hannan, E. J. (1960). *Time Series Analysis*. Methuen, London.

[8] Hasselman, K., Munk, W. and MacDonald, G. (1963). Bispectrum of ocean waves. *Proc. Symp. Time Series Analysis* (M. Rosenblatt, ed.). 125–139. Wiley, New York.

[9] Hoeffding, W. and Robbins, H. (1948). The central limit theorem for dependent random variables. *Duke Math. J.* **15** 773–780.

[10] Jenkins, G. M. (1961). General considerations in estimation of spectra. *Technometrics* **3** 319–329.

[11] Leonov, V. P. and Shiryaev, A. N. (1959). On a method of calculation of semi-invariants. (English translation). *Theor. Prob. Appl.* **4** 319–329.

[12] Magness, T. A. (1954). Spectral response of a quadratic device to non-Gaussian noise. *J. Appl. Phys.* **25** 1357–1365.

[13] Parzen, E. (1957). On consistent estimates of the spectrum of a stationary time series. *Ann. Math. Statist.* **28** 329–348.

[14] Parzen, E. (1961). Mathematical considerations in estimation of spectra. *Technometrics* **3** 167–190.

[15] Parzen, E. (1961). An approach to time series analysis. *Ann. Math. Statist.* **32** 951–989.

[16] Rosenblatt, M. (1959). Statistical analysis of stochastic processes with stationary residuals. *Probability and Statistics—The Harald Cramér Volume* (Ulf Grenander, ed.). 125–139. Wiley, New York.

[17] Rosenblatt, M. (1962). *Random Processes*. Oxford, New York.

[18] Rosenblatt, M. and Van Ness, J. W. (1964). Estimates of the bispectrum of stationary random processes. Report Nonr 562(29) 11, Division of Applied Mathematics, Brown Univ.

[19] Rosenblatt, M. and Van Ness, J. W. (1964). Estimation of the bispectrum. Report Nonr 562(29)13, Division of Applied Mathematics, Brown Univ.

[20] Sinai, Ya. G. (1963). On the properties of spectra of ergodic dynamical systems. *Dokl. Akad. Nauk, CCCP* **150** 1235–1237.

[21] Stratonovich, R. L. (1963). *Topics in the Theory of Random Noise*, **1** (English translation). Gordon and Breach, New York.

[22] Tukey, J. W. (1959). An introduction to the measurement of spectra. *Probability and Statistics—The Harald Cramér Volume* (Ulf Grenander, ed.). 300–330. Wiley, New York.

[23] Tukey, J. W. (1959). The estimation of power spectra and related quantities. *On Numerical Approximation* (R. E. Langer, ed.). 389–411. Univ. of Wisconsin Press.

JOURNAL OF MATHEMATICAL PHYSICS VOLUME 9, NUMBER 7 JULY 1968

Remarks on the Burgers Equation

M. ROSENBLATT
University of California, San Diego, California

(Received 21 November 1967)

Periodic and aperiodic solutions of the Burgers equation $u_t + uu_x = \mu u_{xx}$, $\mu > 0$, are studied in this paper. A harmonic analysis of the solutions is carried out and the form of the spectrum is estimated for large time. Corresponding estimates of energy decay are also made. In Burgers' work on this equation, the case in which $\mu \downarrow 0$ with t fixed, and one then lets $t \to \infty$, is studied. In our investigation, a fixed value of $\mu > 0$ is taken and then one lets $t \to \infty$. A similar analysis is also carried out for an irrotational solution of a similar 3-dimensional system of equations. For large time and moderate wavenumbers there is, to the first order, a drift of spectral mass from low wavenumbers to higher wavenumbers. Comments are also made on the asymptotic distribution of a class of random solutions.

INTRODUCTION

Periodic and aperiodic solutions of the Burgers equation are studied in this paper. A harmonic analysis of the solutions is carried out and the form of the spectrum is estimated asymptotically for large time. Corresponding estimates of energy decay are also made. A similar analysis is also carried out for an irrotational solution of a similar 3-dimensional system of equations. For large time and moderate wavenumbers there is, to the first order, a drift of spectral mass from low wavenumbers to higher wavenumbers. Comments are also made on the asymptotic distribution of a class of random solutions of the equations.

1. THE BURGERS EQUATION

Burgers introduced the equation

$$u_t + uu_x = \mu u_{xx}, \quad \mu > 0, \tag{1.1}$$

as a model equation for $t > 0$, $-\infty < x < \infty$, in order to get some insight into turbulence.[1] Hopf[2] and Cole[3] showed that the study of this equation could be reduced to that of the heat equation. Specifically, if $u_0(x) = u(0, x)$ is continuous and

$$\int_0^x u_0(\xi)\, dx = o(x^2) \tag{1.2}$$

as $|x| \to \infty$, then

$$u(x, t) = \frac{\displaystyle\int_{-\infty}^{\infty} \frac{x - y}{t} \exp\left[-\frac{1}{2\mu} F(x, y, t)\right] dy}{\displaystyle\int_{-\infty}^{\infty} \exp\left[-\frac{1}{2\mu} F(x, y, t)\right] dy,} \tag{1.3}$$

with

$$F(x, y, t) = \frac{(x - y)^2}{2t} + \int_0^y u_0(y)\, dy,$$

is the unique solution of the initial-value problem. Letting

$$\varphi_0(y) = \exp\left[-\frac{1}{2\mu}\int_0^y u_0(\eta)\, d\eta\right],$$

$$\varphi(x, t) = (4\pi\mu t)^{-\frac{1}{2}} \int_{-\infty}^{\infty} \varphi_0(y) \exp\left[-\frac{(x - y)^2}{4\mu t}\right] dy, \tag{1.4}$$

it is clear that $\varphi(x, t)$ is a nonnegative solution of the heat equation

$$\varphi_t = \mu\varphi_{xx} \tag{1.5}$$

and that

$$u(x, t) = -2\mu \frac{\partial}{\partial x} \log \varphi(x, t). \tag{1.6}$$

Because of the continuity of u_0,

$$\lim_{\substack{t\to 0\\ y\to x}} u(t, y) = u_0(x). \tag{1.7}$$

It is sometimes convenient to consider solutions of Burgers equation on the circle (or equivalently periodic solutions of the equation). One then has a corresponding result on existence and uniqueness of the solution without condition (1.2). A standard argument indicates that the solutions of (1.1), of period 2π with continuous initial data $u_0(x)$ and

$$\int_0^{2\pi} u_0(x)\, dx = 0, \tag{1.8}$$

are given in terms of solutions $\varphi(x, t)$ of the heat equation with period 2π,

$$\varphi(x, t) = \sum y_j \exp(-j^2\mu t) \exp(ijx), \tag{1.9}$$

with initial condition

$$\varphi(x, 0) = \varphi_0(x) = \sum y_j \exp(ijx) \tag{1.10}$$

given by (1.4) in terms of $u_0(x)$. Again, φ is positive so that $y_j = \bar{y}_{-j}$ and $y_0 > 0$. Then

$$u(x, t) = -2\mu \frac{\partial}{\partial x} \log \varphi(x, t).$$

[1] G. K. Batchelor, *The Theory of Homogeneous Turbulence* (Cambridge University Press, 1953).
[2] E. Hopf, Commun. Pure Appl. Math. 3, 201 (1950).
[3] J. D. Cole, Quart. Appl. Math. 9, 226 (1951).

R.A. Davis et al. (eds.), *Selected Works of Murray Rosenblatt*, Selected Works in Probability and Statistics, DOI 10.1007/978-1-4419-8339-8_24, © Springer Science+Business Media, LLC 2011

It will sometimes be convenient to look at random solutions of the Burgers equation that are stationary in x. We look in particular at some stationary solutions obtained from solutions of the heat equation that are stationary in x. If φ is a solution of the heat equation on the circle $[0, 2\pi)$ (or, equivalently, a periodic solution with period 2π) it takes the form (1.9). Stationarity up to second-order moments implies that the coefficients y_j in representation (1.10) are orthogonal

$$Ey_j\bar{y}_k = \delta_{j-k}a_j \geq 0, \quad \sum a_j < \infty.$$

Stationarity up to second-order moments for solutions of the heat equation on the full real line implies that one can write

$$\varphi(x, t) = \int_{-\infty}^{\infty} e^{ix\lambda}e^{-\mu\lambda^2 t}\, dY(\lambda), \qquad (1.11)$$

where $Y(\lambda)$ is a process with orthogonal increments

$$E\, dY(\lambda)\, d\bar{Y}(\mu) = \delta_{\lambda-\mu}\, dF(\lambda), \qquad (1.12)$$

with F a bounded nondecreasing function. The sequence $\{a_j\}$ and function $F(\lambda)$ correspond to the second-order spectral decomposition of the process in the periodic and general case.

2. PERIODIC SOLUTIONS

Consider a periodic solution of the Burgers equation with period 2π and spatial mean zero; that is, condition (1.8) is satisfied. For continuous initial data, it has already been noted that the solution can be given in terms of a positive solution $\varphi(x, t)$ of the heat equation (1.5) with $y_0 > 0$. The continuity of initial data implies that

$$\sum_j |y_j|^2 < \infty. \qquad (2.1)$$

In view of (1.6), the solution $u(x, t)$ of the Burgers equation can be written,

$$u(x, t) = -2\mu \frac{\sum_j y_j(ij) \exp(-j^2\mu t) \exp(ijx)}{\sum_j y_j \exp(-j^2\mu t) \exp(ijx)} = \sum_k Z_k(t) \exp(ikx). \qquad (2.2)$$

If we formally look at the Burgers equation in the Fourier domain,

$$Z_k'(t) + \sum_j Z_{k-j}(t)Z_j(t)ij = -\mu k^2 Z_k(t) \qquad (2.3)$$

is obtained for the Fourier coefficients $Z_k(t)$ of $u(x, t)$. This equation is certainly satisfied for sufficiently large t. We shall obtain estimates of $Z_k(t)$ with error term valid for large t and moderate k from the representation (2.2). These estimates will be used to show that, in the range of large t and moderate k, there is a unidirectional transfer of spectral mass from wavenumbers $1, 2, \cdots, k$ to $k + 1$ to the first order. This is consistent with some of the heuristic arguments employed in the discussion of nonlinear phenomena like turbulence (see Refs. 1–3).

Make the strong assumption

$$\sum_{j\neq 0} |y_j| < y_0, \qquad (2.4)$$

or the stronger assumption (for a later computation)

$$\sum_{j\neq 0} |y_j|\, j^2 < y_0, \qquad (2.4')$$

so that the expansion

$$u(x, t) = 2\mu \frac{\psi_x}{y_0 - \psi} = 2\mu \sum_{j=0}^{\infty} y_0^{-j-1} \frac{\partial}{\partial x} \psi^{j+1} \Big/ (j + 1) \qquad (2.5)$$

is valid, where

$$\varphi(x, t) = y_0 - \psi. \qquad (2.6)$$

Even if (2.4) or (2.4′) are not satisfied, they will be satisfied at some later time t by $y_j \exp(-j^2\mu t)$ so that the qualitative features of our conclusions are valid for any periodic solution satisfying (1.8).

Now

$$Z_k(t) = 2\mu ik \sum_{j=0}^{\infty} \frac{1}{j + 1} \sum_{\substack{s_1+\cdots+s_{j+1}=k \\ s_i\neq 0}} y_{s_1} \cdots y_{s_{j+1}} \times \exp[-(s_1^2 + \cdots + s_{j+1}^2)\mu t] y_0^{-j-1} \qquad (2.7)$$

and

$$Z_k'(t) = 2\mu^2 ik \sum_{j=0}^{\infty} \frac{1}{(j + 1)} \times \sum_{\substack{s_1+\cdots+s_{j+1}=k \\ s_i\neq 0}} y_{s_1} \cdots y_{s_{j+1}}(-s_1^2 - \cdots - s_{j+1}^2) \times \exp[-(s_1^2 + \cdots + s_{j+1}^2)\mu t] y_0^{-j-1}. \qquad (2.8)$$

Our aim is now to give the principal terms for $Z_k(t)$, $Z_k'(t)$ together with bounds for the error terms. First consider $Z_k(t)$ with $k \geq 1$ [$Z_{-k}(t) = \bar{Z}_k(t)$ since the solutions are real]. The principal term for large t is

$$2\mu i(y_1/y_0)^k \exp(-k\mu t) \qquad (2.9)$$

corresponding to the term in the inner summation of (2.7) with $j + 1 = k$ and $s_1 = s_2 = \cdots = s_k = 1$. All the other terms of the inner summation corresponding to $j+1=k$ are in total bounded by

$$2\mu k\Big[\sum_{s\neq 0}|y_s/y_0|\exp(-s^2\mu t)\Big]^{k-1} \times\Big[\sum_{s\neq 0,1}|y_s/y_0|\exp(-s^2\mu t)\Big]. \quad (2.10)$$

The total of the terms of the inner summation corresponding to fixed $j+1$ with $j+1>k$ is bounded by

$$2\mu k\Big[\sum_{s\neq 0}|y_s/y_0|\exp(-s^2\mu t)\Big]^{j+1} \quad (2.11)$$

Under assumption (2.4), the total of the terms of the inner summation corresponding to fixed $j+1$ with $j+1<k$ is bounded by

$$2\mu k\exp\{-M_{j+1}\mu t\}, \quad (2.12)$$

where

$$M_{j+1}=\min_{\substack{s_1+\cdots+s_{j+1}=k\\ s_i\neq 0}}\{s_1^2+\cdots+s_{j+1}^2\}. \quad (2.13)$$

However, $M_{j+1}\geq k+2\{k-(j+1)\}$ if $j+1<k$. Thus, for large t,

$$|Z_k(t)-2\mu i(y_1/y_0)^k e^{-k\mu t}|<\mu kC\exp[-(k+1)\mu t], \quad (2.14)$$

where C is an absolute constant. Under assumption (2.4′) a similar argument implies that $Z_k'(t)$ is to the first order

$$-2\mu^2 ik(y_1/y_0)^k e^{-k\mu t} \quad (2.15)$$

and that

$$|Z_k'(t)+2\mu^2 ik(y_1/y_0)e^{-k\mu t}|<\mu^2 kC\exp[-(k+1)\mu t], \quad (2.16)$$

with C an absolute constant. Let us now look at Eq. (2.3) with the estimates given by (2.14) and (2.16). Notice that

$$\Big|\sum_{\substack{j<0\\ \text{or } j>k}} Z_{k-j}(t)Z_j(t)ij\Big|\leq Ck^2e^{-\mu t}e^{-(k+1)\mu t}, \quad (2.17)$$

where C is an absolute constant. This implies that

$$Z'(t)+\sum_{j=1}^{k-1}Z_{k-j}(t)Z_j(t)ij+\mu k^2Z_k(t) = O\{(k^3e^{-\mu t})\exp[-(k+1)\mu t]\}, \quad (2.18)$$

as $t\to\infty$ where the constant, corresponding to the O order of magnitude estimate, is independent of k. At the very least, if $k^3\exp(-\mu t)=o(1)$ as $t\to\infty$, the error estimate on the right of Eq. (2.18) is negligible compared to the individual terms on the left of the equation since each of those terms is precisely of order of magnitude $\exp(-k\mu t)$. It follows that we have a drift of spectral mass from lower wavenumbers $1\leq j<k$ to k through the convolution term

$$\sum_{j=1}^{k-1}Z_{k-j}(t)Z_j(t)ij, \quad (2.19)$$

as given in a number of heuristic discussions of turbulence. This is enough to counterbalance the dissipative term $-\mu k^2Z_k(t)$ to such an extent that $Z_k(t)$ is of order of magnitude $\exp(-k\mu t)$ [see (2.14)] rather than $\exp(-k^2\mu t)$, as it would be if the convolution transfer term did not enter. Notice that all three terms of Eq. (2.3) are of order $\exp(-k\mu t)$.

Cole[4] has recently obtained estimates related to (2.14).

3. ENERGY ESTIMATES

We first consider a real deterministic solution of the Burgers equation on the real line. *Let u be a solution with u_t, u_x bounded and $u\in L$,*

$$\lim_{|x|\to\infty}u(x)=0. \quad (3.1)$$

The assumptions on u imply that $u\in L^2$. On multiplying the Burgers equation by u and integrating with respect to x, the following is obtained:

$$\begin{aligned}\int_{-N}^{N}[uu_t+u^2u_x]\,dx &= \frac{1}{2}\frac{d}{dt}\int_{-N}^{N}u^2\,dx+\tfrac{1}{3}[u^3(N)-u^3(-N)]\\ &=\mu\int_{-N}^{N}uu_{xx}\,dx\\ &=\mu uu_x\Big|_{-N}^{N}-\mu\int_{-N}^{N}u_x^2\,dx. \end{aligned}\quad (3.2)$$

The relation

$$\frac{1}{2}\frac{d}{dt}\int u^2\,dx=-\mu\int u_x^2\,dx \quad (3.3)$$

is obtained in the limit as $N\to\infty$ and describes the rate of decay of energy.

A similar argument can be used to give a corresponding result in the case of a random solution of the Burgers equation stationary in x. It will be more than enough to *assume that moments of u and its partial derivatives u_t, u_x,* and *u_{xx} exist up to fourth order*. Multiply the Burgers equation by u and take expectations to obtain

$$E\{uu_t+u_xu^2\}=\mu E(uu_{xx}). \quad (3.4)$$

The simple equalities

$$Eu_xu^2=\frac{1}{3}\frac{\partial}{\partial x}Eu^3=0$$

[4] J. D. Cole (unpublished).

and

$$\frac{\partial}{\partial x} Euu_x = Eu_x^2 + Euu_{xx} = 0$$

imply that

$$\frac{1}{2}\frac{d}{dt} Eu^2 = -\mu Eu_x^2. \tag{3.5}$$

We shall get estimates of the energy and the rate of decay of energy when t is large for a class of solutions of the Burgers equation in both the deterministic solution of the Burgers equation of the form $u = -2\mu\varphi_x(c+\varphi)^{-1}$, where $\max_x |\varphi| < c$ and

$$\varphi(x, t) = \int \exp(ix\lambda - \mu t\lambda^2) h(\lambda)\, d\lambda, \tag{3.6}$$

where $h \in L$ and is bounded. Now

$$\int |u(x,t)|^2\, dx = 4\mu^2 \int \left|\frac{\varphi_x}{c+\varphi}\right|^2 dx$$

and, since $|\varphi| \to 0$ as $t \to \infty$,

$$\int |u(x,t)|^2\, dx \cong 4\mu^2 c^{-2} \int |\varphi_x|^2\, dx, \tag{3.7}$$

as $t \to \infty$. The energy content of the solution u of the Burgers equation given above is the same as that of the solution $2\mu c^{-1}\varphi_x$ of the heat equation as $t \to \infty$. The rate of decay of total energy is governed by $\int |u_x|^2\, dx$ and

$$u_x(x,t) = -2\mu[(c+\varphi)\varphi_{xx} - \varphi_x^2](c+\varphi)^{-2}. \tag{3.8}$$

Thus

$$\int |u_x(x,t)|^2\, dx \cong 4\mu^2 c^{-4} \int |(c+\varphi)\varphi_{xx} - \varphi_x^2|^2\, dx,$$

as $t \to \infty$. Since

$$\int |\varphi_{xx}|^2\, |c+\varphi|^2\, dx \cong c^2 \int |\varphi_{xx}|^2\, dx$$

as $t \to \infty$ and

$$\int (c+\varphi)\varphi_{xx}\varphi_x^2\, dx = \int \varphi\varphi_{xx}\varphi_x^2\, dx = -\frac{1}{3}\int \varphi_x^4\, dx,$$

it follows that

$$\int |u_x|^2\, dx \cong 4\mu^2 c^{-2} \int |\varphi_{xx}|^2\, dx + \tfrac{20}{3}\mu^2 c^{-4} \int |\varphi_x|^4\, dx, \tag{3.9}$$

as $t \to \infty$. Since

$$\sup_x |\varphi_x(x,t)| \le \int \exp(-\mu t\lambda^2)\, |\lambda|\, |h(\lambda)|\, d\lambda = O[(\mu t)^{-1}]$$

as $t \to \infty$, then

$$\int |\varphi_x(t,x)|^4\, dx \le C(\mu t)^{-2} \int \lambda^2 \exp(-2\mu t\lambda^2)\, |h(\lambda)|^2\, d\lambda$$

as $t \to \infty$ with C a constant. If $h(\lambda) \cong k\lambda^\alpha$, $\alpha \ge 0$, and $k \ne 0$, as $\lambda \downarrow 0$, we shall have the second term on the right of relation (3.9) of smaller order of magnitude than the first as $t \to \infty$, so that

$$\int |u_x|^2\, dx \cong 4\mu^2 c^{-2} \int |\varphi_{xx}|^2\, dx \tag{3.10}$$

as $t \to \infty$. The asymptotic rate of decay of energy will then be the same as for the solution $2\mu c^{-1}\varphi_x$ of the heat equation. Notice that our discussion of total energy and its rate of decay for large t depends on the low-frequency range of λ in the neighborhood of zero.

Similar conclusions can be obtained for a random solution $u = -2\mu\varphi_x(c+\varphi)^{-1}$, where

$$\varphi(x,t) = \int \exp(ix\lambda - \mu t\lambda^2)\, dZ(\lambda), \tag{3.11}$$

with $\varphi(x, 0)$ a bounded process stationary in x, of absolute value less than c and mean zero. Here $Z(\cdot)$ is the random spectral function corresponding to $\varphi(x, 0)$. We shall assume that $\varphi(x, 0)$ has second-order spectral density $f(\lambda)$ and fourth-order cumulant spectral density $g(\lambda_1, \lambda_2, \lambda_3)$. These will be well defined and continuous if the second-order moments and fourth-order cumulants are integrable as functions of the one and three linearly independent differences of spatial arguments respectively (see Ref. 5 for a discussion of higher-order spectra). As before, $|\varphi| \to 0$ as $t \to \infty$, so that

$$Eu^2 \cong 4\mu^2 c^{-2} E\varphi_x^2 = 4\mu^2 c^{-2} \int \lambda^2 \exp(-2\mu t\lambda^2) f(\lambda)\, d\lambda \tag{3.12}$$

as $t \to \infty$. The rate of dissipation of energy is given by (3.5), with u_x given by (3.8). Then

$$Eu_x^2 \cong 4\mu^2 c^{-4} E\, |(c+\varphi)\varphi_{xx} - \varphi_x^2|^2,$$

with

$$E\, |(c+\varphi)\varphi_{xx} - \varphi_x^2|^2 \cong c^2 E\varphi_{xx}^2 + \tfrac{5}{3} E\varphi_x^4.$$

Now, $E\varphi_x^4 = c_4(\varphi_x) + 3\sigma^4(\varphi_x)$, where c_4 is the fourth cumulant of φ_x and $\sigma^2(\varphi_x)$ is the variance of φ_x. Thus

$$\begin{aligned} E\varphi_x^4 = {} & 3\left[\int \lambda^2 \exp(-2\mu t\lambda^2) f(\lambda)\, d\lambda\right]^2 \\ & + \int \lambda_1 \lambda_2 \lambda_3 (-\lambda_1 - \lambda_2 - \lambda_3) \\ & \times \exp\{-\mu t[\lambda_1^2 + \lambda_2^2 + \lambda_3^2 + (\lambda_1 + \lambda_2 + \lambda_3)^2]\} \\ & \times g(\lambda_1, \lambda_2, \lambda_3)\, d\lambda_1\, d\lambda_2\, d\lambda_3. \end{aligned} \tag{3.13}$$

[5] Ya. G. Sinai, Theory Probability Appl. (USSR) 8, 429 (1963).

Assuming that f and g are bounded and that f is not zero in the neighborhood of zero,

$$E\varphi_{xx}^2 \sim (\mu t)^{-\frac{5}{2}},$$

while

$$E\varphi_x^4 = O[(\mu t)^{-3}],$$

as $t \to \infty$. In that case the principal contribution to the decay of energy would be expected from the term $E\varphi_{xx}^2$ when t is large.

4. A SOLUTION ON THE REAL LINE

In this section we shall again look at a solution of the Burgers equation of the form $u = 2\mu\Psi'_x(c - \Psi')^{-1}$ with $\max_x |\Psi'| < c$ and Ψ' given by (3.6). The function h is assumed to be bounded, real, and continuous with $h(0) = c > 0$. Our object is to get an estimate of the Fourier transform $f(\lambda, t)$ (with respect to x) of $u(x, t)$ at a fixed $\lambda \neq 0$ as $t \to \infty$. This is a limited parallel to the computation carried out in Sec. 2 for a periodic solution. Notice that as t gets large this will fall outside of both the energy-containing and energy-dissipative range for which $\lambda \sim (\mu t)^{-\frac{1}{2}}$ as $t \to \infty$ by arguments like those given in the section on energy estimates.

Now

$$u(x, t) = 2\mu \sum_{k=0}^{\infty} \frac{y_0^{-k-1} \dfrac{\partial}{\partial x} \Psi'^{k+1}}{(k + 1)}. \tag{4.1}$$

Consider the Fourier transform of Ψ'^k $(k > 1)$

$$(2\pi)^{-(k-1)} \int \exp\left\{-\mu t\left[\left(\lambda - \sum_1^{k-1} \lambda_i\right)^2 t + \sum_1^{k-1} \lambda_i^2\right]\right\}$$
$$\times h\left[\lambda - \sum_1^{k-1} \lambda_i\right] \prod_1^{k-1} h(\lambda_i)\, d\lambda_1 \cdots d\lambda_{k-1}. \tag{4.2}$$

Set

$$\lambda_i = \frac{\lambda}{k} + (2\mu t)^{-\frac{1}{2}} z_i.$$

Integral (4.2) can then be rewritten as

$$(2\pi)^{-(k-1)} \exp(-\mu t\lambda^2/k)(2\mu t)^{-(k-1)/2}$$
$$\times \int \exp\left\{-\frac{1}{2}\left[\left(\sum_1^{k-1} z_i\right)^2 + \sum_1^{k-1} z_i^2\right]\right\} h\left[\frac{\lambda}{k} - (2\mu t)^{-\frac{1}{2}} \sum_1^{k-1} z_i\right]$$
$$\times \prod_1^{k-1} h\left(\frac{\lambda}{k} + (2\mu t)^{-\frac{1}{2}} z_i\right) dz_1 \cdots dz_{k-1}. \tag{4.3}$$

This suggests the estimate

$$\exp(-\mu t\lambda^2/k)\left\{h\left(\frac{\lambda}{k}\right)\right\}^k (4\pi\mu t)^{-(k-1)/2} k^{-\frac{1}{2}} \tag{4.4}$$

for integral (4.3) to the first order and therefore

$$2\mu i\lambda \sum_{k=1}^{\infty} \exp(-\mu t\lambda^2/k)\left[y_0^{-1} h\left(\frac{\lambda}{k}\right)\right]^k (4\pi\mu t)^{-(k-1)/2} k^{-\frac{3}{2}} \tag{4.5}$$

as an estimate of $f(\lambda, t)$. For $\lambda \sim (\mu t)^{-1}$, it is clear that only the term corresponding to $k = 1$ is of interest as $t \to \infty$. Consider a fixed $\lambda > 0$ and assume that

$$h(\lambda) = h_0 + \lambda h_1 + o(\lambda), \quad \lambda > 0, \tag{4.6}$$

in the neighborhood of $\lambda = 0$. The following approximation is obtained by taking into account those terms in the summation (4.5) for values of k in the neighborhood of $k_{\max}$, which maximizes the typical term

$$f(\lambda, t) \sim 2\mu i\lambda \exp\left(-\lambda\left\{\mu t\left[\tfrac{1}{2}\log\left(\frac{4\pi\mu t y_0^2}{h_0^2}\right)\right]\right\}^{\frac{1}{2}}\right.$$
$$\left. + O[\log(\mu t)]\right). \tag{4.7}$$

5. AN OVERSIMPLIFIED VERSION OF THE NAVIER–STOKES EQUATIONS

A three-dimensional example like the Burgers equation can be obtained by looking at the Navier–Stokes system of equations without the continuity equation and the pressure terms, and by looking for solutions that are gradients. We wish to consider Cauchy's problem for the equations

$$\frac{\partial \mathbf{v}}{\partial t} + (\mathbf{v} \cdot \nabla)\mathbf{v} = \mu\Delta\mathbf{v}, \quad \mu > 0, \tag{5.1}$$

where $\mathbf{v} = (v_1, v_2, v_3)$ is a 3-vector depending on 3 spatial variables and time, and the initial vector $\mathbf{v}_0 = \mathbf{v}(x, 0)$ satisfies curl $\mathbf{v}_0 = 0$. This system is briefly mentioned at the end of Cole's paper[3] but is studied at some length in the paper of Kuznetsov and Rozhdestvenskii.[6] The fact that curl $\mathbf{v}_0 = 0$ implies that a solution with this initial specification is irrotational for all time. By introducing a scalar function φ satisfying

$$\mathbf{v} = -2\mu\nabla \log \varphi, \tag{5.2}$$

the initial-value problem for Eq. (5.1) is reduced to that of

$$\frac{\partial \varphi}{\partial t} = \mu\Delta\varphi. \tag{5.3}$$

If we set $\varphi(x, t) = y_0 - \Psi'$, with $\max|\varphi| < y_0$ as before,

$$-2\mu \log \varphi - 2\mu \log y_0 = 2\mu \sum_{j=0}^{\infty} y_0^{-j-1}\Psi'^{j+1}/(j + 1). \tag{5.4}$$

[6] N. N. Kuznetsov and B. L. Rozhdestvenskii, Z. Vycisl. Mat. Fiz. 1, 217 (1961).

An argument like that given in Sec. 4 now suggests approximating the Fourier transform of Ψ^k ($k \geq 1$) by

$$\exp(-\mu t\boldsymbol{\lambda}^2/k)\left\{h\left(\frac{\boldsymbol{\lambda}}{k}\right)\right\}^k(4\pi\mu t)^{-\frac{3}{2}(k-1)}k^{-\frac{3}{2}} \quad (5.5)$$

if Ψ is given by

$$\Psi(\mathbf{x}, t) = \int \exp(i\mathbf{x}\cdot\boldsymbol{\lambda} - \mu t\boldsymbol{\lambda}^2)h(\boldsymbol{\lambda})\,d\boldsymbol{\lambda} \quad (5.6)$$

where $\mathbf{x}$ and $\boldsymbol{\lambda}$ are 3-vectors, and h is bounded, real, and continuous. One can then proceed further and obtain estimates somewhat like (4.7) under further assumptions on h analogous to those made in Sec. 4. Instead of proceeding in this direction, we shall consider periodic solutions of system (5.1) in the manner of Sec. 2.

Let $\mathbf{v}$ be an irrotational solution of Eq. (5.1) with period 2π in each spatial variable and

$$\int_0^{2\pi} v^{(1)}(\mathbf{x}, t)\,dx_1 = \int_0^{2\pi} v^{(2)}(\mathbf{x}, t)\,dx_2 = \int_0^{2\pi} v^{(3)}(\mathbf{x}, t)\,dx_3 = 0. \quad (5.7)$$

Then the solution $\mathbf{v}$ is given by (5.2) in terms of a solution

$$\varphi = y_0 - \Psi = y_0 - \sum_{\mathbf{k}\neq\mathbf{0}} y_{\mathbf{k}} \exp(-\mathbf{k}^2\mu t)\exp(i\mathbf{k}\cdot\mathbf{x}) \quad (5.8)$$

of the heat equation. Notice that $\mathbf{k}$ is a wavenumber 3-vector of integers and $\mathbf{0}$ is the zero 3-vector. Assume, as in Sec. 2, that

$$\sum_{\mathbf{k}} |y_{\mathbf{k}}|\,\mathbf{k}^2 < y_0, \quad (5.9)$$

We can then get estimates for the vectors $\mathbf{z}_{\mathbf{k}}(t)$ and $\mathbf{z}'_{\mathbf{k}}(t)$, for large t, where the $\mathbf{z}_{\mathbf{k}}(t)$ are the Fourier coefficients in the harmonic analysis of $\mathbf{v}$:

$$\mathbf{v}(\mathbf{x}, t) = \sum \mathbf{z}_{\mathbf{k}}(t)e^{i\mathbf{k}\cdot\mathbf{x}} \quad (5.10)$$

Then

$$\mathbf{z}_{\mathbf{k}}(t) = 2\mu i\mathbf{k}\sum_{j=0}^{\infty}\frac{1}{(j+1)} \times \sum_{\substack{\mathbf{s}_1+\cdots+\mathbf{s}_{j+1}=\mathbf{k}\\ \mathbf{s}_i\neq\mathbf{0}}} e^{-(\mathbf{s}_1^2+\cdots+\mathbf{s}_{j+1}^2)\mu t}y_{\mathbf{s}_1}\cdots y_{\mathbf{s}_{j+1}}y_0^{-j-1}. \quad (5.11)$$

For large t, it is clear that the first-order contribution to $\mathbf{z}_{\mathbf{k}}(t)$ in (5.11) will come from vectors $\mathbf{s} \neq 0$ with the following properties. The component s_i of $\mathbf{s}$ will be zero if $k_i = 0$, while $s_i = 0, 1$ ($s_i = 0, -1$) if $k_i > 0$ ($k_i < 0$). Call this set of vectors $U(\mathbf{k})$. Thus, if all the components of $\mathbf{k}$ are positive, $U(\mathbf{k})$ consists of the vectors $(1, 0, 0)$, $(0, 1, 0)$, $(0, 0, 1)$, $(1, 1, 0)$, $(1, 0, 1)$, $(0, 1, 1)$, and $(1, 1, 1)$. If $\mathbf{k}$ has its first component negative and the others zero, $U(\mathbf{k})$ contains only $(-1, 0, 0)$. Let us set

$$A(\mathbf{k}) = \sum_{j\geq 0}\frac{1}{j+1}\sum_{\substack{\mathbf{s}_1+\cdots+\mathbf{s}_{j+1}=\mathbf{k}\\ \mathbf{s}_i\in U(\mathbf{k})}} y_{\mathbf{s}_1}\cdots y_{\mathbf{s}_{j+1}}y_0^{-j-1}. \quad (5.12)$$

Then, for fixed $\mathbf{k}$,

$$\mathbf{z}_{\mathbf{k}}(t) = 2\mu i\mathbf{k}A(\mathbf{k})\exp\left[-\sum_{i=1}^{3}|k_i|\,\mu t\right][1 + o(e^{-\mu t})] \quad (5.13)$$

and

$$\mathbf{z}'_{\mathbf{k}}(t) = -2\mu^2 i\mathbf{k}A(\mathbf{k})\left(\sum_{i=1}^{3}|k_i|\right) \times \exp\left(-\sum_{i=1}^{3}|k_i|\,\mu t\right)[1 + o(e^{-\mu t})], \quad (5.14)$$

as $t \to \infty$. Notice that the analog of the system (2.3) is

$$\mathbf{z}'_{\mathbf{k}}(t) + \sum_{\mathbf{k}'} i[\mathbf{k}'\cdot\mathbf{z}_{\mathbf{k}-\mathbf{k}'}(t)]\mathbf{z}_{\mathbf{k}'}(t) = -\mu\,|\mathbf{k}|^2\,\mathbf{z}_{\mathbf{k}}(t). \quad (5.15)$$

This is certainly satisfied by (5.11), for t sufficiently large. We shall just briefly remark on the analog of Eq. (2.18) for wavevectors $\mathbf{k}$ all of whose components are nonnegative. Given any two wavevectors $\mathbf{k}$ and $\mathbf{k}'$, $\mathbf{k} \geq \mathbf{k}'$ means that $\mathbf{k} - \mathbf{k}'$ is a vector all of whose components are nonnegative. The estimates (5.14) and (5.15) with error terms imply that

$$\mathbf{z}'_{\mathbf{k}}(t) + \sum_{\mathbf{k}\geq\mathbf{k}'\geq 0} i[\mathbf{k}'\cdot\mathbf{z}_{\mathbf{k}-\mathbf{k}'}(t)]\mathbf{z}_{\mathbf{k}'}(t) + \mu\,|\mathbf{k}|^2\,\mathbf{z}_{\mathbf{k}}(t) = O\left\{\exp\left[-\left(1 + \sum_{i=1}^{3}|k_i|\right)\mu t\right]\right\}, \quad (5.16)$$

if $\mathbf{k} \geq \mathbf{0}$, $\mathbf{k} \neq \mathbf{0}$ with $\mathbf{k}$ fixed as $t \to \infty$. The error estimate is small compared to each of the terms on the left of (5.16) which are of order of magnitude $\exp\{-\sum_{i=1}^{3}|k_i|\,\mu t\}$. One has a drift of spectral mass from wavevectors $\mathbf{k}'$, $0 \leq \mathbf{k}' \leq \mathbf{k}$ to $\mathbf{k}$ through the truncated convolution term as before. If $\mathbf{k} \leq 0$, the convolution term would be summed over $\mathbf{k}'$, $\mathbf{k} \leq \mathbf{k}' \leq 0$, with a similar modification for $\mathbf{k}$ belonging to some other octant.

6. ASYMPTOTIC DISTRIBUTION OF A CLASS OF RANDOM SOLUTIONS.

Let us now consider random solutions of the Burgers equation that are stationary in x. Our object is to get some information on the asymptotic distribution of the solution as $t \to \infty$. The case of periodic solutions satisfying condition (1.8) is rather uninteresting, since it is clear from (2.2) that

$$u(x, t) \cong -2\mu(y_1/y_0)e^{ix}e^{-\mu t} \quad (6.1)$$

as $t \to \infty$. The asymptotic distribution of $u(x, t)$ (assuming that $y_0 > \delta > 0$ with probability one) is determined by the joint distribution of y_0 and y_1.

Consider now a random solution of the Burgers equation given by

$$u = 2\mu\Psi_x(c - \Psi)^{-1}, \quad (6.2)$$

with c a positive constant and Ψ a strictly stationary solution of the heat equation (1.5) in x with

$$\max_x |\Psi_0(x)| < c, \quad E\Psi_0(x) \equiv 0, \quad (6.3)$$

where $\Psi_0(x) = \Psi(x, 0)$. Notice that (6.3) implies that $\max |\Psi(x, t)| < c$ for all $t > 0$. The stationarity of Ψ_0 in x implies that $u_0(x) = u(x, 0)$, $u(x, t)$, and $\Psi(x, t)$ are strictly stationary. Let the second-order spectral distribution function of Ψ_0 be $F(\lambda)$,

$$E\Psi_0(y + x)\Psi_0(y) = \int e^{ix\lambda}\, dF(\lambda). \quad (6.4)$$

Since $u_0(x) = 2\mu\Psi_{0x}(c - \Psi_{0x})^{-1}$, it is clear that at the very least we ought to have

$$\int \lambda^2\, dF(\lambda) < \infty, \quad (6.5)$$

in order that $u_0(x)$ be well defined at least in mean square. Let us assume that Ψ_0 is strongly (or uniformly) mixing in the following sense.[7,8] Let A, B be two events with A determined by conditions on the random variables $\Psi_0(x)$ with $x \le y$ and B determined by the random variables $\Psi_0(x)$ with $x \ge y + d$, $d > 0$. The process Ψ_0 is said to be strongly mixing if

$$|P(AB) - P(A)P(B)| < \alpha(d), \quad (6.6)$$

where $\alpha(d)$ is a function of d that tends to zero as $d \to \infty$ and that otherwise is independent of A, B, and y. This condition of strong mixing is a strong form of asymptotic independence as the events A and B are determined by random variables that are further and further removed spatially. Conditions sufficient for strong mixing are discussed in Refs. 7 and 8. Notice that strong mixing for Ψ_0 implies that u_0 is strongly mixing. Since

$$\Psi(x, t) = \int_{-\infty}^{\infty} (4\pi\mu t)^{-\frac{1}{2}} \exp\left\{-\frac{(x - y)^2}{4\mu t}\right\}\Psi_0(y)\, dy,$$

it is clear that $\max |\Psi(x, t)| \to 0$ as $t \to \infty$ and that

$$u(x, t) \cong 2\mu c^{-1}\Psi_x, \quad (6.7)$$

as $t \to \infty$. Now:

$$\Psi_x(x, t) = -\int (4\pi\mu t)^{-\frac{1}{2}} \exp\left\{-\frac{(x - y)^2}{4\mu t}\right\}\Psi_{0y}(y)\, dy. \quad (6.8)$$

Equation (6.8) indicates that Ψ_x is obtained from Ψ_{0x} by narrow band pass filtering about frequency zero and one knows that Ψ_{0x} is strongly mixing (since Ψ_0 is strongly mixing). Under appropriate conditions on second and fourth order moments (see Ref. 8), one would expect Ψ_x to be asymptotically normally distributed (when properly normalized) as $t \to \infty$. The assumption that Ψ_0 is strongly mixing implies that F is absolutely continuous with spectral density $f(\lambda)$. But then Ψ_{0x} has spectral density $\lambda^2 f(\lambda)$. One of the conditions required for asymptotic normality of Ψ_x is that the variance of $(4\pi\mu t)^{\frac{1}{2}}\Psi_x$ diverge as $t \to \infty$. But this variance

$$\sigma^2[(4\pi\mu t)^{\frac{1}{2}}\Psi_x] = 4\pi\mu t \int \lambda^2 e^{-2\mu t\lambda^2} f(\lambda)\, d\lambda = O(1). \quad (6.9)$$

We do not have divergence of (6.9) as $t \to \infty$ and so $u(x, t)$ as given by (6.7), is typically not asymptotically normal. The following objection might be brought against our argument. Even though u as given by (1.6) is stationary in x, the function φ might not be and it is this class of random solutions one ought to look at. Nonetheless, we shall see that random solutions of (5.1) stationary in $\mathbf{x}$ that are obtained from corresponding stationary solutions of the heat equation (5.3) by (5.2) are typically asymptotically normal under analogous conditions. This could be thought of as a difference between the one- and three-dimensional equations.

Let us consider a random solution of Eq. (5.1) given by

$$\mathbf{v} = 2\mu(c - \Psi)^{-1}\nabla\Psi \quad (6.10)$$

with c a positive constant and Ψ a strictly stationary solution of (5.1) in $\mathbf{x}$ with

$$\max_{\mathbf{x}} |\Psi_0(\mathbf{x})| < c, \quad E\Psi_0(\mathbf{x}) \equiv 0, \quad (6.11)$$

where $\Psi_0(\mathbf{x}) = \Psi(\mathbf{x}, 0)$. Let F be the second-order spectral distribution function of Ψ_0

$$E\Psi_0(\mathbf{y} + \mathbf{x})\Psi_0(\mathbf{y}) = \int \exp(i\mathbf{x}\cdot\boldsymbol{\lambda})\, dF(\boldsymbol{\lambda}). \quad (6.12)$$

One would have to require that

$$\int |\boldsymbol{\lambda}|^2\, dF(\boldsymbol{\lambda}) < \infty, \quad (6.13)$$

if $\mathbf{v}$ is to be well defined in mean square. The following definition is a possible generalization of the concept of strong mixing for a process $\Psi_0(\mathbf{x})$ with parameter $\mathbf{x}$ a point in 3-space. Consider any two events A, B determined by the random variables $\{\Psi_0(\mathbf{x}), \mathbf{x} \in I\}$ and $\{\Psi_0(\mathbf{x}'), \mathbf{x}' \in I'\}$, respectively. Let C_1, C_2 be the convex hulls of the set of points $\{\mathbf{x}, \mathbf{x} \in I\}$ and $\{\mathbf{x}', \mathbf{x}' \in I'\}$, respectively. Set d equal to the distance between C_1, C_2. We shall say that $\Psi_0(\mathbf{x})$ is strongly mixing if condition (6.6) is satisfied for all events A, B, where $\alpha(d)$ is a function of d that tends to zero as $d \to \infty$. If Ψ_0 is strongly mixing, then F is absolutely

[7] A. N. Kolmogorov and Y. A. Rozanov, Theory Probability Appl. (USSR) 5, 204 (1960).

[8] M. Rosenblatt, Quart. Appl. Math. 18, 387 (1961).

continuous with spectral density $f(\boldsymbol{\lambda})$. As $t \to \infty$ one can show that

$$\mathbf{v}(\mathbf{x}, t) \cong 2\mu c^{-1} \nabla \Psi \tag{6.14}$$

and

$$(4\pi\mu t)^{\frac{3}{2}} \nabla \Psi(\mathbf{x}, t) = -\int_{-\infty}^{\infty} \exp\left\{-\frac{|\mathbf{x}-\mathbf{y}|^2}{4\mu t}\right\} \nabla \Psi_0(\mathbf{y})\, d\mathbf{y}. \tag{6.15}$$

The spectral density of $(4\pi\mu t)^{\frac{3}{2}} \Psi_{x_i}(\mathbf{x}, t)$ is

$$\sigma^2[(4\pi\mu t)^{\frac{3}{2}} \Psi_{x_i}(\mathbf{x}, t)] = (4\pi\mu t)^3 \int \lambda_i^2 \exp(-2\mu t\, |\boldsymbol{\lambda}|^2) f(\boldsymbol{\lambda})\, d\boldsymbol{\lambda} \to \infty, \tag{6.16}$$

as $t \to \infty$ if $f(\boldsymbol{\lambda})$ is continuous and bounded away from zero at $\boldsymbol{\lambda} = 0$. Actually, less stringent conditions on $f(\boldsymbol{\lambda})$ at zero would suffice for (6.16). Again, it is clear from (6.15) that we are narrow band pass filtering about $\boldsymbol{\lambda} = 0$. But now because of (6.16), broad conditions on fourth-order moments assure asymptotic normality of (6.10).

ACKNOWLEDGMENTS

I should like to thank Professor Burgers, Professor Saffman, and Professor Whitham for their helpful comments.

R.A. Davis et al. (eds.), *Selected Works of Murray Rosenblatt*, Selected Works in Probability and Statistics, DOI 10.1007/978-1-4419-8339-8_25, © Springer Science+Business Media, LLC 2011

DENSITY ESTIMATES AND MARKOV SEQUENCES†

M. ROSENBLATT

1 INTRODUCTION

Estimates of the density function of a population based on a sample of independent observations have been considered in a number of papers [1, 6–7]. Questions of bias, variance and asymptotic distribution of the estimates have been dealt with at greatest length. Our object is to look at such estimates of the density function when the observations are dependent. The results will not be dealt with in the most general context or under very general conditions. To obtain results in a simple and readily understandable form, the observations are assumed to be sampled from a stationary Markov sequence with a fairly strong condition on the Markov transition operator. However, the extent to which some of the conditions can be obviously relaxed will be indicated.

It should be noted that the asymptotic results we obtain in the case of dependent observations are essentially the same as those in the case of independent observations. This initially is surprising because it is certainly not true when estimating the distribution function by means of the sample distribution function. The more complicated nature of asymptotic results for this problem in the case of dependence can be seen in Billingsley [2]. However, the happy fact that the results we obtain in estimating the density have the same character as in the case of independence is due to the local character of the estimates.

2 REMARKS ON THE CASE OF INDEPENDENT SAMPLING

We just briefly make a few remarks on estimation of the density function of a population with a sample of independent observations. More detailed and precise results can be found [1, 6–7]. Consider a population with continuous positive density function $f(x)$. Let the sample of independent observations be $X_1, X_2, \ldots, X_n$. Let the estimate of $f(x)$ be

$$\hat{f}_n(x) = n^{-1}b(n)^{-1} \sum_{j=1}^{n} h(b(n)^{-1}(x - X_j)), \tag{1}$$

† This research was supported by the Office of Naval Research.

where $h(x)$ is a given bounded continuous density function and $b(n)$ is a bandwidth parameter such that $b(n) \to 0$ and $nb(n) \to \infty$ as $n \to \infty$. The bias of the estimate is

$$b_n(x) = E\hat{f}_n(x) - f(x) = \int h(\alpha)\{f(x - b(n)\,\alpha) - f(x)\}\,d\alpha. \tag{2}$$

If f is continuously differentiable up to second order and the moments $m_i = \int \alpha^i h(\alpha)\,d\alpha, i = 1, 2,$ exist then

$$b_n(x) = f'(x)\,m_1 b(n) + f''(x)\,m_2 b(n)^2/2 + o(b(n)^2). \tag{3}$$

Thus $b_n(x) = 0(b(n)^2)$ if $m_1 = 0$. The variance

$$\sigma^2(\hat{f}_n(x)) \cong n^{-1}b(n)^{-1}f(x)\int h^2(\alpha)\,d\alpha \quad \text{as} \quad n \to \infty. \tag{4}$$

It is clear that the estimates are asymptotically normally distributed. Specifically $\sqrt{(nb(n))}[\hat{f}_n(x) - E\hat{f}_n(x)]$ is asymptotically normally distributed with mean zero and variance $f(x)\int h^2(\alpha)\,d\alpha$. It is also interesting to look at the covariance of estimates at different points x, y.

The covariance

$$\begin{aligned}
\operatorname{cov}(\hat{f}_n(x), \hat{f}_n(y)) &= n^{-1}b(n)^{-2}\operatorname{cov}(h(b(n)^{-1}(x - X)), h(b(n)^{-1}(y - X))) \\
&= n^{-1}b(n)^{-2}\left\{\int h(b(n^{-1}(x-\alpha))\,h(b(n)^{-1}(y-\alpha))f(\alpha)\,d\alpha\right. \\
&\quad \left. - \int h(b(n)^{-1}(x-\alpha))\,f(\alpha)\,d\alpha \int h(b(n)^{-1}(y-\alpha))f(\alpha)\,d\alpha\right\} \\
&= n^{-1}b(n)^{-1}\left\{\int h(\beta)\,h(b(n)^{-1}(y-x)+\beta)f(x-\beta\,b(n))\,d\beta\right. \\
&\quad \left. - b(n)\int h(\beta)f(x-b(n)\,\beta)\,d\beta \int h(\beta)\,f(y-b(n)\,\beta)\,d\beta\right\}.
\end{aligned} \tag{5}$$

This certainly implies that for fixed x, y with $x \neq y$

$$\sqrt{(nb(n))}[\hat{f}_n(x) - E\hat{f}_n(x)] \quad \text{and} \quad \sqrt{(nb(n))}[\hat{f}_n(y) - E\hat{f}_n(y)]$$

are asymptotically independent.

3 THE MARKOV ASSUMPTION

Let us now assume that $X_1, X_2, \ldots, X_n$ is a sequence of observations on a stationary Markov process with stationary continuous density function $f(x) > 0$ and continuous transition probability density function $f(y|x)$. Let $f_k(y|x)$ and $f_k(y,x)$ denote the conditional probability density of

X_{k+1} given X_1 and the joint probability density of X_{k+1} and X_1 respectively, $k = 1, 2, \ldots$. Clearly

$$f_k(y, x) = f_k(y|x) f(x) \tag{6}$$

and
$$f_{k+1}(y|x) = \int f_k(y|z) f(z|x)\, dz \quad (k = 1, 2, \ldots). \tag{7}$$

We use the same sort of estimate $\hat{f}_n(x)$ of the density function $f(x)$ as in the case of independent observations. The bias of the estimate in the case of a Markov sequence is the same as in the case of independent observations. Let us now look at the covariance of the estimate when one has dependence. Clearly

$$\operatorname{cov}[\hat{f}_n(x), \hat{f}_n(y)] = n^{-2} b(n)^{-2} \sum_{j=-n+1}^{n-1} (n - |j|) \operatorname{cov}[h(b(n)^{-1}(x - X_0)), h(b(n)^{-1} \times (y - X_j))]. \tag{8}$$

In §5 we will determine the asymptotic behavior of the covariance and of the distribution of the estimates $\hat{f}_n(x)$ themselves, given certain asymptotic conditions on the character of the dependence of the Markov sequence. Some of these conditions will be introduced and discussed in §4.

4 REMARKS ON THE TRANSITION PROBABILITY OPERATOR

In order to determine the asymptotic behaviour of covariance of density estimates at different points as well as the asymptotic distribution of estimates we shall have to impose certain conditions on the transition probability function of the Markov sequence. The conditions discussed here are undoubtedly too strong. Certainly weaker conditions that ensure the desired results can be determined. However, these conditions have a certain interest even outside the problem of density estimation and the discussion will be given in a broader context. Let T be the transition probability operator of a stationary Markov sequence. In our problem

$$(Th)(y) = \int h(x) f(x|y)\, dx, \tag{9}$$

where h can be thought of as a bounded function. The condition to be imposed on T corresponds roughly to what has been called 'geometric ergodicity' (see Kendall [5]) in some of the probability literature. Let $\|h\|_p$, $1 \leqslant p \leqslant \infty$, be the L^p norm of the function h with respect to the invariant measure μ ($d\mu = f(x)\,dx$ in our case of density estimation) so that

$$\|h\|_p = \begin{cases} \left[\int |h(x)|^p \mu(dx)\right]^{1/p} & \text{if} \quad 1 \leqslant p < \infty, \\ \operatorname*{ess\,sup}_x |h(x)| & \text{if} \quad p = \infty, \end{cases} \tag{10}$$

where ess sup is understood as referring to the measure μ. The L^p norm of the operator T^n, $n = 1, 2, \ldots$,

$$\|T^n\|_p = \sup_{h \neq 0} \frac{\|T^n h\|_p}{\|h\|_p} = 1 \tag{11}$$

for $1 \leqslant p \leqslant \infty$ since $T^n 1 = 1$. However, since we are interested in the rate at which $T^n h \to Eh = \int h(x)\,\mu(dx)$ as $n \to \infty$, the modified norm

$$\begin{aligned} |T^n|_p &= \sup_h \frac{\|T^n h - Eh\|_p}{\|h - Eh\|_p} \\ &= \sup_{h \perp 1} \frac{\|T^n h\|_p}{\|h\|_p} \end{aligned} \tag{12}$$

is introduced where by $h \perp 1$ one means that $\int h(x)\,\mu(dx) = 0$.

The transition probability operator T *is said to satisfy the condition* G_p ($1 \leqslant p \leqslant \infty$) *if there is some positive integer* n *such that*

$$|T^n|_p \leqslant \alpha < 1. \tag{13}$$

The reason for 'geometric ergodicity' is apparent since if (13) holds, it follows that

$$\lim_{m \to \infty} |T^m|_p / \alpha^{m/n} \leqslant 1. \tag{14}$$

Lemma. If T is a transition probability operator of a stationary Markov sequence satisfying condition G_p for some p, $1 < p < \infty$, then T satisfies the condition G_q for all q, $1 < q < \infty$.

Let U be such that $Uh = Th - Eh$. Notice that

$$\|h - Eh\|_q \leqslant \|h\|_q + |Eh| \leqslant 2\|h\|_q \tag{15}$$

and that

$$U^n h = U^{n-1}(Th - Eh) = T^{n-1}(Th - Eh). \tag{16}$$

This implies that

$$\|U^n\|_q \leqslant 2|T^n|_q \leqslant 2, \quad 1 \leqslant q \leqslant \infty. \tag{17}$$

However

$$|T^n|_q \leqslant \|U^n\|_q. \tag{18}$$

Since $\|U^n\|_1, \|U^n\|_\infty \leqslant 2$ the Riesz convexity theorem [4] can be applied to this situation. Suppose that for some p, $1 < p < \infty$ (for example $p = 2$), there is a positive integer n such that

$$|T^n|_p = \alpha < 1. \tag{19}$$

It then follows that $|T^{nk}|_p \leqslant \alpha^k$. Consider any other q with $1 < q < \infty$. For $q > p$ we have $1/q = (p/q)(1/p) + (1-p/q)$, while if $q < p$ then

$$1/q = \left(\frac{1-1/q}{1-1/p}\right) 1/p + \left(1 - \frac{1-1/q}{1-1/p}\right).$$

The Riesz convexity theorem tells us that if $q > p$

$$\log \|U^m\|_q \leqslant \frac{p}{q} \log \|U^m\|_p + \left(1 - \frac{p}{q}\right) \log \|U^m\|_\infty \tag{20}$$

and if $q < p$

$$\log \|U^m\|_q \leqslant \left(\frac{1-1/q}{1-1/p}\right) \log \|U^m\|_p + \left(1 - \frac{1-1/q}{1-1/p}\right) \log \|U^m\|_1. \tag{21}$$

Since $\|U^{nk}\|_p \leqslant 2\alpha^k$ and $\|U^m\|_\infty$, $\|U^m\|_1 \leqslant 2$ it is clear that for large enough k

$$\|U^{nk}\|_q \leqslant \beta < 1 \tag{22}$$

and therefore

$$|T^{nk}|_q \leqslant \beta < 1. \tag{23}$$

It is of some interest to consider the well known Doeblin condition and see how it relates to the conditions G_p. Let $P(x, A)$ be the transition probability function of the stationary Markov sequence defined for each real x and each Borel set A. Then

$$(Th)(x) = \int P(x, dy)\, h(y) \tag{24}$$

for bounded h and the n step transition probability function $P_n(x, A)$ is defined recursively via

$$P_{n+1}(x, A) = \int P(x, dy)\, P_n(y, A) \quad (n = 1, 2, \ldots). \tag{25}$$

The Doeblin condition D is said to be satisfied if there is a finite measure ϕ on the Borel sets with positive mass, an integer $n \geqslant 1$, and $\epsilon > 0$ such that

$$P_n(x, A) \leqslant 1 - \epsilon \tag{26}$$

for all x if $\phi(A) \leqslant \epsilon$. Doob (see [3]) considers the more stringent condition D_0 where one assumes that D is satisfied, there is only one ergodic set, and there are no cyclically moving sets. Under condition D_0, in Lemma 7.2 on p. 224 of Doob's book on stochastic processes, it is shown that there are constants γ and ρ, with $0 < \rho < 1$, such that if h is bounded

$$(|h| \leqslant M < \infty)$$

then

$$|(T^k h)(x) - Eh| \leqslant 2\gamma\rho^k. \tag{27}$$

However, it is easily seen that (27) is just the condition G_∞. By the Riesz convexity theorem, it follows that if G_∞ is satisfied then G_p holds for $1 < p \leqslant \infty$. Similarly, if G_1 is satisfied then G_p holds for $1 \leqslant p < \infty$.

Notice that the condition D_0 is satisfied if we have an ergodic aperiodic stationary Markov sequence with continuous stationary probability density $f(x)$ and transition probability density $p(y|x)$ satisfying

$$p(y|x) \leqslant K(1+y^2)^{-\alpha} \tag{28}$$

for some constant K and $\alpha > 1/2$.

However, it is easy to give a simple example which doesn't satisfy D or G_∞ but for which G_p, $1 < p < \infty$, holds. This example is provided by the aperiodic Gaussian stationary Markov sequences. For convenience, assume that the random variables of the sequence have mean zero and variance one. The invariant probability density is then

$$f(x) = \frac{1}{\sqrt{(2\pi)}} \exp(-x^2/2)$$

and the transition probability density is

$$p(y|x) = \frac{1}{\sqrt{(2\pi(1-\rho^2))}} \exp\left(-\frac{(y-\rho x)^2}{2(1-\rho^2)}\right) \tag{29}$$

with ρ a constant such that $|\rho| < 1$. Mehler's formula indicates that (29) can be written

$$p(y|x) = \sum_{\nu=0}^{\infty} \rho^\nu h_\nu(y)\, h_\nu(x) \frac{e^{-y^2/2}}{\sqrt{(2\pi)}}, \tag{30}$$

where the h_ν are the Hermite polynomials orthonormal with respect to the standard Gaussian density $f(x)$. By using (3) one can see that

$$|T^n|_2 = |\rho|^{2n} \tag{31}$$

so that condition G_2 is certainly satisfied. By the Lemma of this section it follows that G_p is satisfied for $1 < p < \infty$.

5 ASYMPTOTIC BEHAVIOR OF THE COVARIANCE AND DISTRIBUTION OF ESTIMATES

The asymptotic behavior of the covariance of a density estimate at two points x, y will first be determined before the asymptotic distribution is considered. Assume that h is a continuous density function with

$$h(u) = o(|u|^{-1}) \quad \text{as} \quad |u| \to \infty.$$

The density f is taken to be bounded and continuous. Consider the term corresponding to $j = 0$ in the sum (8).

Then
$$\operatorname{cov}\{h(b(n)^{-1}(x-X)), h(b(n)^{-1}(y-X))\}$$
$$= b(n)\int h(u)\,h(b(n)^{-1}(y-x)+u)f(x-b(n)\,u)\,du$$
$$-b(n)^2\int h(u)f(x-b(n)\,u)\,du\int h(u)f(y-b(n)\,u)\,du$$
$$\cong \begin{cases} b(n)f(x)\int h^2(u)\,du & \text{if} \quad x=y \\ 0(b(n)^2) & \text{if} \quad x \neq y \end{cases} \tag{32}$$

when $f(x) > 0$ as $n \to \infty$. The terms with $j \neq 0$ have the following asymptotic behavior
$$\operatorname{cov}\{h(b(n)^{-1}(x-X_0)), h(b(n)^{-1}(y-X_j))\}$$
$$= b(n^2)\int h(u)\,h(v)f_j(x-b(n)\,u, y-b(n)\,v)\,du\,dv$$
$$-b(n)^2\int h(u)f(x-b(n)\,u)\,du\int h(u)f(y-b(n)\,u)\,du$$
$$\cong b(n)^2\{f_j(x,y)-f(x)f(y)\} \tag{33}$$
as $n \to \infty$ if the joint density functions $f_j(\cdot\,,\,\cdot)$ are bounded continuous functions. We shall now get a bound on (33) under the assumption that the Markov sequence satisfies condition G_2. Let
$$g_x(\alpha) = h(b(n)^{-1}(x-\alpha)) - \int h(b(n)^{-1}(x-\alpha))f(\alpha)\,d\alpha. \tag{34}$$

Then
$$|\operatorname{cov}\{h(b(n)^{-1}(x-X_j)), h(b(n)^{-1}(y-X_0))\}|$$
$$= |E(g_y(X)\,(T^jg_x)\,(X))|$$
$$\leqslant \{E|g_y(X)|^2\,E|(T^jg_x)\,(X)|^2\}^{\frac{1}{2}}$$
$$\leqslant M\rho^j\{E|g_y(X)|^2\,E|g_x(X)|^2\}^{\frac{1}{2}}$$
$$\leqslant b(n)\,M'\rho^j\int h^2(u)\,du\,\sqrt{(f(x)f(y))} \tag{35}$$

for sufficiently large n with M, M' and ρ constants, $0 < \rho < 1$, if $f(x)$ and $f(y)$ are positive. The estimates (32), (33) and (35) indicate that
$$\lim_{n\to\infty} nb(n)\operatorname{cov}[\hat{f}_n(x), \hat{f}_n(y)]$$
$$= \begin{cases} f(x)\int h^2(u)\,du & \text{if} \quad x=y \\ 0 & \text{if} \quad x \neq y \end{cases} \tag{36}$$

under the conditions assumed.

We now consider the asymptotic distribution of

$$\sqrt{(nb(n))}\,[\hat{f}_n(x_j) - E\hat{f}_n(x_j)] \quad (j = 1, \ldots, m),$$

for any finite m-tuple of points x_j. These random variables are asymptotically jointly normal and independent with means zero and variances given by (36) as $n \to \infty$ under the assumptions made. So as to limit the notation the argument on asymptotic normality will be given for a single point, that is, $m = 1$. However, exactly the same argument and estimates can be used to obtain the multivariate result by applying them to any given linear combination

$$\sqrt{(nb(n))}\sum_{j=1}^{m} \alpha_j(\hat{f}_n(x_j) - E\hat{f}_n(x_j))$$

of the random variables. This is the standard reduction of a multivariate limit theorem to a univariate limit theorem. Consider

$$\sqrt{(nb(n))}\,[\hat{f}_n(x) - E\hat{f}_n(x)].$$

The argument for asymptotic normality will proceed by the classic device of proving such a limit theorem for dependent random variables, that is, by writing the sum as a sum of big blocks separated by small blocks. The contribution due to the small blocks is shown to be negligible and the big blocks approximately independent. A standard central limit theorem for independent random variables (the Liapounov theorem) is then used to get asymptotic normality. Now,

$$\begin{aligned}\sqrt{(nb(n))}\,&[\hat{f}_n(x) - E\hat{f}_n(x)]\\ &= \sum_{j=1}^{n} \frac{1}{\sqrt{(nb(n))}}\,[h(b(n)^{-1}(x-X_j)) - Eh(b(n)^{-1}(x-X_j))]\\ &= \sum_{l=1}^{k}(A_l + B_l) + H \qquad (37)\end{aligned}$$

with

$$A_l = \sum_{j=(l-1)(m+r)+1}^{lm+(l-1)r}, \quad B_l = \sum_{j=lm+(l-1)r+1}^{l(m+r)} \qquad (38)$$

and

$$H = \sum_{j=k(m+r)+1}^{n}. \qquad (39)$$

Here $m = m(n)$ and $r = r(n)$ are the summation ranges of the big and small blocks A_l and B_l respectively. They will be taken so that

$$m(n), r(n) \to \infty \quad \text{as} \quad n \to \infty \quad \text{but} \quad m(n) = o(n) \quad \text{and} \quad r(n) = o(m(n)).$$

Further $k = k(n) = [n/(m+r)]$ is the greatest integer less than or equal to $n/(m+r)$. The term H accounts for the additional few terms at the end not included in the big blocks or small blocks. Notice that $k(n) \to \infty$ as $n \to \infty$ since $m, r = o(n)$. Estimates like those used in (32), (33) and (35) imply that

$$E\left|\sum_{l=1}^{k} B_l + H\right|^2 \leqslant 2\left(\sum_{l=1}^{k-1} E|B_l|^2 + E|B_k + H|^2\right)$$

$$\leqslant C\left(\frac{kr(n)}{n} + \frac{m}{n}\right) = o(1), \tag{40}$$

with C a constant.

Thus $\sum_{l=1}^{k} B_l + H$ can be neglected. Consider the characteristic function of $\sum_{l=1}^{k} A_l$. We use the G_2 property of the Markov sequence to compare this characteristic function with the product of the characteristic functions of the individual A_l's and find that

$$\left|E\left\{\exp\left(it\sum_{l=1}^{k} A_l\right)\right\} - \prod_{l=1}^{k} E\{\exp(itA_l)\}\right|$$

$$\leqslant \sum_{j=2}^{k}\left|E\left\{\exp\left(it\sum_{l=1}^{j} A_l\right)\right\} - E\{\exp(itA_j)\}E\left\{\exp\left(it\sum_{l=1}^{j-1} A_l\right)\right\}\right|$$

$$\leqslant (k-1)M\rho^{r(n)}, \tag{41}$$

where M is a constant and $0 < \rho < 1$. The sequences $m(n)$, $r(n)$ will be chosen so that $k(n)\rho^{r(n)} \to 0$. However this means that we have asymptotic normality if we can show that $\sum_{l=1}^{k} A_l$ is asymptotically normal when the A_l's are treated as independent random variables with the same marginal distributions. This will be shown to be true by using the Liapounov theorem. For this we require an estimate of the fourth moment of the A_l's. Given that the joint density functions up to fourth order are continuous and bounded, estimates like those obtained in (32) and (33) show that

$$E|A_l|^4 \leqslant \sum_{j=1}^{k} M_j(mb(n))^j, \tag{42}$$

where the M_j's are constants. But then

$$\frac{kE|A_l|^4}{(E|\sum A_l|^2)^2} \leqslant \frac{M_1}{nb(n)} + M_2\frac{m}{n} + M_3\frac{m^2}{n}b(n) + M_4\frac{m^3}{n}b(n)^2. \tag{43}$$

If $nb(n) \to \infty$, $b(n) \to 0$, but $m(n) = o(n^{\frac{1}{3}})$ the expression (43) tends to zero as $n \to \infty$ and the Liapounov theorem is applicable. We can still choose $r(n) = o(m(n))$ so that (41) tends to zero as $n \to \infty$.

Theorem 1. *Let* $X_n, n = \ldots, -1, 0, 1, \ldots,$ *be a stationary Markov sequence with bounded continuous density function* $f(x)$ *and bounded transition probability density* $p(y|x)$ *continuous in the variables* (y, x). *Consider the estimate* $\hat{f}_n(x)$ *of the density* $f(x)$ *given by*

$$\hat{f}_n(x) = n^{-1}b(n)^{-1}\sum_{j=1}^{n} h(b(n)^{-1}(x-X_j)),$$

where h *is a given uniformly continuous density function. If* f *is continuously differentiable up to second order and the moments* $m_i = \int \alpha^i h(\alpha)\,d\alpha$, $i = 1, 2,$ *of* h *exist then the bias*

$$E\hat{f}_n(x) - f(x) = m_1 f'(x)\,b(n) + m_2 f''(x)\,b(n)^2 + o(b(n))^2,$$

where the bandwidth $b(n) \to 0$ *as* $n \to \infty$. *If the sequence* $\{X_n\}$ *satisfies condition* G_2 *and* $nb(n) \to \infty$ *as* $n \to \infty$ $(b(n) \to 0)$ *any* m-*tuple* (m *finite*) *of the normalized deviations*

$$\sqrt{(nb(n))}\,[\hat{f}_n(x_i) - E\hat{f}_n(x_i)] \quad (i = 1, \ldots, m)$$

(x_i *distinct*) *are asymptotically normal and independent with mean zero and variances*

$$f(x_i)\int h^2(\alpha)\,d\alpha \quad (i = 1, \ldots, m).$$

One could equally well consider estimation of the joint density function of a fixed number of random variables. Under conditions corresponding to those assumed in Theorem 1, one again finds that the asymptotic results are the same as those in the case of independent observations. Actually we shall explicitly note one such result under somewhat broader conditions of dependence. Now $X_n, n = \ldots, -1, 0, 1, \ldots,$ is assumed to be a stationary sequence of random variables. *Assume that all joint distributions of a finite number of distinct* X_j*'s are absolutely continuous with uniformly bounded continuous density functions.* Let $f(x, x')$ be the joint density function of X_j, X_{j+1}. A natural estimate of $f(x, x')$ is given by

$$f_n(x, x') = n^{-1}b(n)^{-2}\sum_{j=1}^{n-1} h(b(n)^{-1}(x-X_j), b(n)^{-1}(x'-X_{j+1})), \tag{44}$$

where $h(x, x')$ is a bounded continuous density function. Let $\mathscr{B}_n$ be the Borel field generated by X_k, $k \leqslant n$, and $\mathscr{F}_n$ be the Borel generated by X_k, $k \geqslant n$. We shall say that $\{X_m\}$ *satisfies the condition* S *if for every random variable* g *with* $Eg^2 < \infty$, $Eg = 0$, *that is* $\mathscr{F}_{n+k}$ *measurable it follows that*

$$E|E(g|\mathscr{B}_n)|^2 \leqslant a(k)\,E|g|^2, \tag{45}$$

where $a(k) = O(k^{-4-\epsilon})$ *for some* $\epsilon > 0$ *as* $k \to \infty$. By making estimates quite similar to those in the proof of the Theorem one can prove the following result.

Theorem 2. *Let* $\{X_m, m = \ldots, -1, 0, 1, \ldots\}$ *be stationary with uniformly bounded continuous joint density functions. Consider the estimate* $\hat{f}_n(x, x')$ *of the joint density* $f(x, x')$ *of* X_j, X_{j+1} *given by* (44). *If* f *is continuously differentiable up to second order and the moments,*

$$m_{i,j} = \int \alpha^i \beta^j h(\alpha, \beta)\, d\alpha\, d\beta \quad (i, j = 0, 1, 2),$$

when $i + j \leqslant 2$ *exist then the bias*

$$\begin{aligned} Ef_n(x, x') - f(x, x') &= [m_{1,0} f_x(x, x') + m_{0,1} f_{x'}(x, x')]\, b(n) \\ &\quad + \tfrac{1}{2}[m_{2,0} f_{x,x} + 2m_{1,1} f_{x,x'} + m_{0,2} f_{x',x'}]\, b(n)^2 + o(b(n)^2) \end{aligned} \tag{46}$$

as $b(n) \to 0$. *If* $\{X_m\}$ *satisfies the condition* S *and* $nb(n)^2 \to \infty$ *as*

$$n \to \infty \quad (b(n) \to 0)$$

then any finite m-tuple of the normalized deviations

$$\sqrt{(nb(n)^2)}\,[\hat{f}_n(x, x') - E\hat{f}_n(x, x')] \tag{47}$$

corresponding to distinct points (x, x') *are independent with mean zero and variances*

$$f(x, x') \int h^2(\alpha, \beta)\, d\alpha\, d\beta. \tag{48}$$

Regression estimates have been considered in [8] and [10]. Without going into details, one can show that under conditions similar to those specified in [8], the asymptotic behavior of these in the case of dependent observations is the same as the asymptotic behavior in the case of independent observations. In [9] estimation of the transition probability density for Markov sequences is discussed under condition D_0.

6 ACKNOWLEDGEMENT

I should like to thank D. Brillinger and M. Woodroofe for their helpful comments on this paper. A. Garsia suggested that the Riesz convexity theorem might be helpful in comparing the norms considered in § 4.

REFERENCES

[1] Bartlett, M. S. (1963). Statistical estimation of density functions. *Sankhyā*, Ser. A **25**, 245–54.

[2] Billingsley, P. (1968). *Convergence of probability measures*. New York: John Wiley.

[3] Doob, J. L. (1953). *Stochastic processes*. New York: John Wiley.

[4] Dunford, N. and Schwartz, J. L. (1958). *Linear Operators*, Part I. Interscience.

[5] Kendall, D. G. (1959). Unitary dilations of Markov transition operators, and corresponding integral representations for transition probability matrices. In *Probability and statistics* (ed. U. Grenander). New York: John Wiley.

[6] Parzen, E. (1962). On estimation of a probability density and mode, *Ann. Math. Statist.* **33**, 1065–76.

[7] Rosenblatt, M. (1956). Remarks on some nonparametric estimates of a density function, *Ann. Math. Statist.* **27**, 832–5.

[8] Rosenblatt, M. (1969). Conditional probability density and regression estimators. In *Multivariate Analysis*, vol. II. Academic Press.

[9] Roussas, G. (1967). Nonparametric estimation in Markov processes, *Technical Report No.* 110, Department of Statistics, University of Wisconsin, Madison, Wisconsin.

[10] Watson, G. S. (1964). Smooth regression analysis, *Sankhyā*, Ser. A **26**, 359–72.

DISCUSSION ON ROSENBLATT'S PAPER

MICHAEL WOODROOFE

I believe that Professor Rosenblatt's paper constitutes a significant contribution to the nonparametric literature and therefore feel quite honored to be a discussant of it. The paper not only contributes a major result by giving sufficient conditions for the asymptotic normality of density estimates from stationary Markov processes, but should also stimulate more research on the problem of density estimation from such processes. Indeed, it is to be hoped that Professor Rosenblatt's present paper on density estimation from dependent observations will stimulate as much research as did his first paper on density estimation from independent observations [4]. Among other things, I will indicate some directions in which I feel this research might profitably proceed.

In considering the asymptotic normality of the estimates, Professor Rosenblatt has solved a rather difficult problem and has left some less difficult results unstated. Since I feel these results to be important, I will take the liberty of stating some of them together with an indication of their proofs. Let X_n be a stationary process satisfying condition S with $a(k)$ summable. (This includes the case of a stationary Markov Process which satisfies G_2.) The argument of [3] then extends without difficulty to show that

$$M_n = \sup_x |\hat{f}_n(x) - f(x)| \to 0 \tag{1}$$

in probability as $n \to \infty$, provided that f is uniformly continuous, $\sqrt{[nb(n)]} \to \infty$, and the Fourier Transform of h is integrable. Thereafter, it follows, again as in [3], that the sample mode is a consistent estimate of the true mode, provided that the latter is unique. It would also be of interest to know the rate of convergence to zero of M_n and such related random variables as were studied in [6] and to determine conditions for and the exact rate of convergence to zero of the integrated mean square error.

$$I_n = \int_{-\infty}^{\infty} E[(f_n(x) - f(x))^2]\, dx.$$

To the best of my knowledge, the only other work which seriously considers density estimation from Markov processes is the unpublished report [5], in which consistency and asymptotic normality of kernel-type density estimates are obtained under the condition D_0. It is therefore of interest to consider Professor Rosenblatt's condition G_2 with some care. It is quite intuitively appealing in that it requires the dependence between segments of the process to decrease as the distance between them increases. I was therefore intrigued by the G_2 condition and hoped that it might be satisfied by a large enough class of stationary Markov processes to provide a reasonable setting for asymptotic inference from such processes. I still have this hope, but in the course of attempting to verify G_2 for linear processes of the form

$$X_k = \sum_{j=0}^{\infty} \beta^j Y_{k-j} \quad (k = 0, 1, 2, \ldots), \tag{2}$$

where the Y_i are independent and identically distributed and $|\beta| < 1$, I discovered an example which has shaken that hope slightly. If the Y_i have the distribution function

$$F_1(y) = 1-(\log y)^{-2}: x \geqslant e, \tag{3}$$

then (2) does not satisfy G_2. Indeed, by considering LaPlace transforms and applying the Tauberian theorems of [1], one may show that F, the stationary distribution function, has a slowly varying tail. Letting

$$g_n(x) = (1-F(n))^{-\frac{1}{2}} \quad \text{for} \quad x > n \quad \text{and} \quad g_n(x) = 0 \quad \text{for} \quad x \leqslant n,$$

it is then easily verified that

$$|T^k g_n|_2^2 = \|T^k g_n\|_2^2 - o(1) \geqslant \left[\frac{1-F(n\beta^{-k})}{1-F(n)}\right] - o(1) \to 1 \quad \text{as} \quad n \to \infty,$$

for each fixed k. Since $|g_n|_2 \leqslant \|g_n\|_2^2 = 1$ for all n, G_2 cannot be satisfied.

The distribution function in (3) is, of course, sufficiently pathological to be of little practical interest, but I still find this example disturbing. Since $X_k = \beta^k X_0 + Z$, where Z is independent of X_0, the process defined by (2) has the property that the dependence of X_k on X_0 decreases geometrically as $k \to \infty$, which property led me to believe that any process of the form (2) should be geometrically ergodic, that it should satisfy G_2. I believe the generality of the G_2 condition to be a question on which research could profitably be done. If it should turn out to be reasonably general, it might well simplify subsequent work on asymptotic inference from Markov processes.

Another area in which I feel research could profitably be done is that of density estimation from general linear processes of the form

$$X_k = \sum_{j=0}^{\infty} c_j Y_{k-j} \quad (k = 0, 1, 2, \ldots),$$

where the Y_i are as in (2). What conditions on the coefficients c_j and/or the distribution of Y_i are sufficient to insure the consistency and asymptotic normality of density estimates from such processes? These questions could be attacked either by verifying an appropriate version of condition S or from scratch, perhaps by using the techniques of [2], Part II. The former approach would be more desirable since conditions similar to S appear throughout the probability literature (e.g. in [2]), but the latter approach would probably be the easier.

REFERENCES

[1] Feller, William (1966). *An Introduction to probability theory and its applications*, **2**. New York: John Wiley.

[2] Ibragimov, I. A. (1962). Some limit theorems for stationary processes, *Theory Probab. Applic.* **7**, 349–82.

[3] Parzen, E. (1962). On estimation of a probability density and mode, *Ann. Math. Statist.* **33**, 1065–76.

[4] Rosenblatt, M. (1956). Remarks on some nonparametric estimates of a density, *Ann. Math. Statist.* **27**, 832–5.

[5] Roussas, G. (1967). Non-parametric estimation in Markov Processes, *Technical Report No.* 110, Department of Statistics, University of Wisconsin, Madison, Wisconsin.

[6] Woodroofe, M. (1967). On the maximum deviation of the sample density, *Ann. Math. Statist.* **38**, 475–81.

The Annals of Mathematical Statistics
1971, Vol. 42, No. 6, 1815–1842

CURVE ESTIMATES[1]

BY MURRAY ROSENBLATT

University of California, San Diego

1. Introduction. There is a large class of problems in which the estimation of curves arises naturally (see [15], [34]). It is curious that one of the earliest extensive investigations of this type involves the estimation of the spectral density function when sampling from a stationary sequence ([1], [17], [27], [33]). Even though the simple histogram has been used for years, it was only later that the simpler question of estimating a probability density function was dealt with at some length ([26], [25], [9]). Because the final character of the usual results obtained in both problem areas is quite similar, and the arguments are much more transparent in the case of the probability density function, we shall develop the results for the probability density function first. Later some corresponding results for spectra will be given. The similarities and differences in the two areas will be noted. Since the literature is rather extensive by now, any presentation of theory as given can only be a selection of topics and cannot claim to be exhaustive or perhaps even representative. There are a number of attractive open problems that one can suggest solutions to on heuristic grounds. A few of these problems will be examined. In most cases it is clear that one will not use the techniques to be proposed in estimating a density function unless there is a good deal of data (many observations), little a priori information about the density function available, but a great need to get additional information about the density function, even if it is fairly crude.

2. Estimating the probability density function by independent observations. Consider a population with absolutely continuous distribution function $F(x)$ and probability density function $f(x) = F'(x)$. A simple sequence of estimates is determined by the choice of an *integrable bounded weight function* $w(u)$ with

$$\int w(u)\,du = 1 \tag{1}$$

and a *sequence of bandwidths* $b(n) \downarrow 0$ as $n \to \infty$. Notice that this implies that $\int w^2(u)\,du < \infty$. An estimate $f_n(x)$ of $f(x)$

$$f_n(x) = \frac{1}{nb(n)} \sum_{j=1}^{n} w\left(\frac{x-X_j}{b(n)}\right) \tag{2}$$

is determined by a sample $X_1, \cdots, X_n$ of independent observations from the population. If w is chosen to be nonnegative, the estimate $f_n(x)$ itself will be a probability

Received February 11, 1971.

[1] Invited Wald Lectures given at the annual meeting of the Institute of Mathematical Statistics held at Laramie, Wyoming, August 25–27, 1970; research supported by Office of Naval Research Contract N00014–69–A–0200–6016, Task NR–042–217. Reproduction in whole or in part is permitted for any purpose of the United States Government.

R.A. Davis et al. (eds.), *Selected Works of Murray Rosenblatt*, Selected Works in Probability and Statistics, 255
DOI 10.1007/978-1-4419-8339-8_26, © Springer Science+Business Media, LLC 2011

density function. Usually the weight function w will satisfy some additional regularity conditions.

First consider the mean and variance of the estimate $f_n(x)$. The mean

$$Ef_n(x) = \frac{1}{b(n)}\int w\left(\frac{x-u}{b(n)}\right)f(u)\,du = \int w(v)f(x-b(n)v)\,dv \tag{3}$$

while the variance

$$\sigma^2[f_n(x)] = \frac{1}{n}\left[\frac{1}{b(n)}\int w^2(v)f(x-b(n)v)\,dv - \left(\int w(v)f(x-b(n)v)\,dv\right)^2\right]. \tag{4}$$

The mean square error can then be simply written as

$$\begin{aligned} E|f_n(x)-f(x)|^2 &= \sigma^2[f_n(x)] + |Ef_n(x)-f(x)|^2 \\ &= \frac{1}{n}\left[\frac{1}{b(n)}\int w^2(v)f(x-b(n)v)\,dv - \left(\int w(v)f(x-b(n)v)\,dv\right)^2\right] \\ &\quad + \left|\int w(v)\{f(x-b(n)v)-f(x)\}\,dv\right|^2. \end{aligned} \tag{5}$$

A simple bound shows that *if* $nb(n) \to \infty$ *as* $n \to \infty$ *and* f *is a bounded function that is continuous at* x, *then there is consistency in mean square at* x *as* $n \to \infty$. The bound is now given. Let $\sup_x f(x) \leqq M$. Then

$$\sigma^2[f_n(x)] \leqq M[nb(n)]^{-1}\int w^2(v)\,dv. \tag{6}$$

For each $\varepsilon > 0$ let $\delta(\varepsilon) = \sup_{|y|\leqq\varepsilon}|f(x+y)-f(x)|$. Then

$$|Ef_n(x)-f(x)| \leqq 2M\int_{b(n)|v|\geqq\varepsilon}|w(v)|\,dv + \delta(\varepsilon). \tag{7}$$

The result follows by first letting $n \to \infty$ and then $\varepsilon \to 0$.

R. H. Farrel [14] showed that one cannot get a uniformly consistent sequence of estimators of the density function at a point (say $x = 0$) for a plausible class of density functions. He specifically showed that if C_M is the class of density functions f on $(-\infty, \infty)$ with

(a) f continuously differentiable everywhere

(b) $\sup_{x\in R} f(x) \leqq M$,

then even with a sequential estimator δ_N of $f(0)$, *the supremum of the mean square error*

$$\sup_{f\in C_M} E|\delta_N - f(0)|^2 \geqq \tfrac{1}{16}$$

if $M \geqq 3$ *and* $\sup_{f\in C_M} E_f N < \infty$. However, if one requires a Hölder condition for f itself in the neighborhood of zero, a uniformly consistent sequence of nonsequential estimators of $f(0)$ is easy to obtain. Let $C_{M,\varepsilon}$ be the class of continuously differentiable density functions satisfying conditions (a) and (b) and the additional requirement

(c) $|f(y)-f(0)| \leqq M|y|^{\varepsilon}$ for $|y| \leqq 1$

with M, $\varepsilon > 0$ *independent* of f. The types of bounds obtained in (6) and (7) indicate that *the sequence* $f_n(0)$ *is uniformly consistent in mean square if* $n \to \infty$ and $nb(n) \to \infty$, *for* $f \in C_{M,\varepsilon}$.

A number of people have suggested using the integrated mean square error

$$E\int |f_n(x)-f(x)|^2\,dx = \int E|f_n(x)-f(x)|^2\,dx \tag{8}$$

as a global measure of how good the estimator $f_n(x)$ is over R. Notice that

$$\begin{aligned}\int_{-\infty}^{\infty} \sigma^2[f_n(y)]\,dy &\leqq \frac{1}{nb(n)}\int_{-\infty}^{\infty}\int_{-\infty}^{\infty} w^2(v)f(x-b(n)v)\,dv\,dx \\ &= \frac{1}{nb(n)}\int_{-\infty}^{\infty} w^2(v)\,dv.\end{aligned} \tag{9}$$

Further

$$\begin{aligned}\int_{-\infty}^{\infty}(Ef_n(x)-f(x))^2\,dx &= \int_{-\infty}^{\infty}\{\int_{-\infty}^{\infty} w(u)[f(x-b(n)u)-f(x)]\,du\}^2\,dx \\ &\leqq A\int_{-\infty}^{\infty}|w(u)|\int_{-\infty}^{\infty}\{f(x-b(n)u)-f(x)\}^2\,dx\,du\end{aligned} \tag{10}$$

where

$$A = \int |w(u)|\,du \tag{11}$$

if $f \in L^2$. Let

$$f(x) = \frac{1}{2\pi}\int_{-\infty}^{\infty} e^{-ix\lambda}\,\phi(\lambda)\,d\lambda. \tag{12}$$

Then

$$\int_{-\infty}^{\infty}\{f(x-b(n)u)-f(x)\}^2\,dx = \int_{-\infty}^{\infty}|\phi(\lambda)|^2|\exp[-ib(n)u\lambda]-1|^2\,d\lambda. \tag{13}$$

Since $|\exp[-b(n)u\lambda]-1| \leqq b(n)|u\lambda|$ it follows that (10) is bounded above by

$$Ab(n)^2\int_{-\infty}^{\infty}|w(u)|u^2\,du\int_{-\infty}^{\infty}|\phi(\lambda)|^2\lambda^2\,d\lambda.$$

We make use of (9) and (10) and note that *if* $f_n(x)$ *is a sequence of estimators of* $f(x)$ *derived from a weight function* w *with*

$$\int |w(u)|u^2\,du < \infty, \tag{14}$$

then the integrated mean square error of f_n *tends to zero uniformly as* n, $nb(n) \to \infty$ *for all* f *with* $\int |f(x)|^2dx$, $\int |f'(x)|^2dx \leqq k < \infty$. The condition $\int |f'(x)|^2dx \leqq k < \infty$ actually implies that $\int |f(x)|^2dx \leqq k+1$.

For convenience *assume that* f *and its first two derivatives are continuous and bounded.* Then simple approximations can be given locally for the variance and bias of $f_n(x)$ as $n \to \infty$, $nb(n) \to \infty$, if $\int |w(u)|u^2 du < \infty$. It is then immediately clear from (3) and (4) that

$$\sigma^2[f_n(x)] \cong \frac{f(x)}{nb(n)} \int w^2(v)\,dv \tag{15}$$

if $f(x) > 0$ and (assuming $\int w(u)u\,du = 0$) that

$$Ef_n(x) - f(x) = \tfrac{1}{2} b(n)^2 f''(x) \int w(u)u^2\,du + o(b(n))^2. \tag{16}$$

Notice that if $f(x) = 0$ with $f''(x) > 0$, then

$$\sigma^2[f_n(x)] \cong \frac{b(n)}{2n} f''(x) \int w^2(v)\,dv$$

as $n \to \infty$, $b(n) \to 0$.

With additional conditions on f (in terms of smoothness) and on w one can improve the rate at which the bias tends to zero. Such a discussion has been given in Bartlett [2] and we shall briefly indicate how this might be done. However, it will be clear that in order to bring this about weight functions which assume negative values must be used. In many situations such weight functions would not be plausible. *Suppose that f and its first four derivatives are continuous and bounded.* Further, let w be such that

$$\int w(u)u\,du = \int w(u)u^2\,du = \int w(u)u^3\,du = 0, \quad \int |w(u)|u^4\,du < \infty. \tag{17}$$

Then

$$Ef_n(x) - f(x) = \frac{1}{4!} b(n)^4 f^{(4)}(x) \int w(u)u^4\,du + o(b(n))^4. \tag{18}$$

A weight function satisfying the conditions (17) is given, for example, by

$$w(u) = \frac{3}{2}\left(1 - \frac{u^2}{3}\right) e^{-\frac{1}{2}u^2}.$$

There have been discussions in the past about the "optimal" shape of a window or weight function in spectral analysis (see [24]). The arguments usually given have been of an asymptotic character and it is a mistake to take them too literally from a finite sample point of view. But even asymptotic arguments if used and interpreted with care can yield meaningful ideas. Let us briefly consider a simple discussion due to Epanechnikov [13]. Under the assumptions leading to (17) and (18), the local estimate of mean square error obtained is

$$E|f_n(x) - f(x)|^2 = \frac{f(x)}{nb(n)} \int w^2(v)\,dv + \tfrac{1}{4}[b(n)]^4 [f''(x)]^2 \left(\int w(u)u^2\,du\right)^2 + o\left(\frac{1}{nb(n)} + (b(n))^2\right). \tag{19}$$

The rate at which the mean square error tends to zero as $n \to \infty$ is maximized (if $\int w(u)u^2 du \neq 0$) if we set

$$b(n) = Kn^{-1/5} \tag{20}$$

with K the constant

$$K = \left[\frac{4f(x) \int w^2(v)\, dv}{(f''(x) \int w(v)v^2\, dv)^2} \right]^{1/5}. \tag{21}$$

The mean square error then becomes

$$E|f_n(x) - f(x)|^2 \cong 2^{3/5}[f(x) \int w^2(v)\, dv]^{4/5} |f''(x) \int w(v)v^2\, dv|^{2/5} n^{-4/5} + o(n^{-4/5}) \tag{22}$$

and this suggests that the integrated mean square error $\int E|f_n(x) - f(x)|^2 dx$ will be no smaller than

$$2^{3/5} \int f(x)^{4/5} |f''(x)|^{2/5}\, dx\, [\int w^2(v)\, dv]^{4/5} (\int w(v)v^2\, dv)^{2/5} n^{-4/5}$$

to the first order as $n \to \infty$. However, this would require locally scaling $b(n)$ with K depending on x when estimating $f(x)$. If the scaling were to be global with K independent of x, we would expect

$$\int E|f_n(x) - f(x)|^2\, dx = \frac{1}{nb(n)} \int w^2(v)\, dv + \tfrac{1}{4} b^4(n) \int_{-\infty}^{\infty} |f''(x)|^2\, dx (\int w(u)u^2\, du)^2 + o\left(\frac{1}{nb(n)} + b^4(n)\right).$$

This suggests taking $b(n) = Kn^{-1/5}$ with

$$K = \frac{2^{2/5}[\int w^2(v)\, dv]^{1/5}}{(\int (f''(x))^2\, dx)^{1/5} \{\int w(u)u^2\, du\}^{2/5}}$$

leading to

$$\int E|f_n(x) - f(x)|^2\, dx = 2^{3/5} [\int w^2(v)\, dv]^{4/5} [\int |f''(x)|^2\, dx (\int w(v)v^2\, dv)^2]^{1/5} n^{-4/5} + o(n^{-4/5}). \tag{23}$$

The estimate (23) can be obtained rigorously under the assumptions that

(i) *f is bounded, twice continuously differentiable with $f, f'' \in L^2$*

(ii) *w is bounded nonnegative and symmetric with $\int w(u)u^2 du < \infty$.*

For then we can make use of the simple bound

$$\int [\int w(v) f(x - b(n)v)\, dv]^2\, dx = \int w(v)w(v') \{\int f(x - b(n)v) f(x - b(n)v')\, dx\}\, dv\, dv' \leqq [\int w(v)\, dv]^2 \int f(x)^2\, dx$$

and the estimate $f(x - b(n)u) - f(x) = -b(n)uf'(x) + \frac{1}{2} b(n)^2 u^2 f''(x - \theta b(n)u)$. The estimate implies that $\{\int w(u)[f(x - b(n)u) - f(x)]du\}^2 = \{\int w(u) \frac{1}{2} b(n)^2 u^2 f''(x - \theta b(n)u)\, du\}^2$. The expression for K and (23) can then be obtained with a small additional argument by using (4) and (10).

The type of discussion has been carried out several places some time ago. Epanechnikov's simple observation is that one simply ought to look for the weight function w minimizing

$$\int w^2(v)\,dv \tag{24}$$

subject to the restraints

$$\begin{aligned} &\text{(i)} \quad \int w(v)\,dv = 1 \\ &\text{(ii)} \quad w(v) = w(-v) \\ &\text{(iii)} \quad \int v^2 w(v)\,dv = 1. \end{aligned} \tag{25}$$

If δw represents a small deviation for an extremum subject to the restraints (i)–(iii), the variation of

$$\int_0^\infty w^2(v)\,dv + \lambda_1\{\int_0^\infty w(v)\,dv - \tfrac{1}{2}\} + \lambda_2\{\int_0^\infty w(v)v^2\,dv - \tfrac{1}{2}\}$$

should be zero (λ_1 and λ_2 are multipliers) and one is therefore led to

$$2\int_0^\infty w(v)\,\delta w(v)\,dv + \lambda_1 \int_0^\infty \delta w(v)\,dv + \lambda_2 \int_0^\infty \delta w(v) v^2\,dv = 0.$$

Thus

$$2w(v) + \lambda_1 + \lambda_2 v^2 = 0, \qquad w(v) = (-\lambda_1 - \lambda_2 v^2)/2.$$

The function $w(v)$ is zero at

$$v = \pm(-\lambda_1/\lambda_2)^{\frac{1}{2}}.$$

To have a function that is integrable and symmetric one ought to set

$$\begin{aligned} w(v) &= (-\lambda_1 - \lambda_2 v^2)/2 && \text{if } |v| \leqq (-\lambda_1/\lambda_2)^{\frac{1}{2}} \\ &= 0 && \text{otherwise.} \end{aligned}$$

The constants λ_1, λ_2 are determined by the conditions (i) and (iii). Then $\lambda_1 = -\frac{3}{4}5^{-\frac{1}{2}}$, $\lambda_2 = \frac{3}{4}t^{-\frac{3}{2}}$ so that the conjectured function is

$$\begin{aligned} w(v) &= \tfrac{3}{4}5^{-\frac{1}{2}}(1 - v^2/5) && \text{if } |v| \leqq 5^{\frac{1}{2}} \\ &= 0 && \text{otherwise.} \end{aligned} \tag{26}$$

This is a nonnegative weight function. Consider a variation $\delta w(v)$ about (26) which is such that $\int \delta w(v) dv = 0$, $\int v^2 \delta w(v) dv = 0$ and $\delta w(-v) = \delta w(v)$ with $\delta w(v) \geqq 0$ for $|v| \geqq 5^{\frac{1}{2}}$. Then

$$\begin{aligned} \int (w(v) + \delta w(v))^2\,dv &= \int (w(v) + \delta w(v))^2\,dv - \tfrac{3}{2}5^{-\frac{1}{2}}\{\int (w(v) + \delta w(v))\,dv - 1\} \\ &\quad + \tfrac{3}{2}5^{-\frac{3}{2}}\{\int (w(v) + \delta w(v))v^2\,dv - 1\} \\ &= \int w(v)^2\,dv + \int \{2w(v) - \tfrac{3}{2}5^{-\frac{1}{2}} - \tfrac{3}{2}5^{-\frac{3}{2}}v^2\}\,\delta w(v)\,dv \\ &\quad + \int (\delta w(v))^2\,dv \\ &\geqq \int w^2(v)\,dv + \int (\delta w(v))^2\,dv \end{aligned} \tag{27}$$

because the integrand in the second integral of line (27) is zero for $|v| \leqq 5^{\frac{1}{2}}$ and nonnegative for $|v| \geqq 5^{\frac{1}{2}}$. *Thus the function* $w(v)$ *given by* (26) *gives an absolute minimum for* (24) *under the restraints* (i) *to* (iii) *in the class of nonnegative weight functions*. Exactly this variational problem was solved by Lehmann and Hodges in a nonparametric investigation [21]. Condition (ii) could be replaced by (ii)′ $\int uw(u)du = 0$.

However, one should note that if weight functions w with negative values are allowed, expression (24) can be made as small as is desired even though the restraints (i)–(iii) are satisfied. We reproduce a small part of a table of Epanechnikov (this is the point of greatest interest in his paper) to show how insensitive (24) is to the shape of w when dealing with nonnegative weight functions satisfying (i)–(iii). The minimizing weight function (26) is referred to as w_0.

TABLE 1

w		$L = \int w^2(u)\, du$	$r = \int w^2(u)\, du / \int w_0{}^2(u)\, du$
w_0		$3.5^{-\frac{3}{2}}$	1
$1/6^{\frac{1}{2}} - \lvert y\rvert/6$	if $\lvert y\rvert \leqq 6$	$6^{\frac{1}{2}}/9$	1.015
0	if $\lvert y\rvert > 6$		
$(2\pi)^{-\frac{1}{2}}\, e^{-y^2/2}$		$2^{-1}\pi^{-\frac{1}{2}}$	1.051
$\frac{1}{2}\, 3^{-\frac{1}{2}}$	if $\lvert y\rvert \leqq 3^{\frac{1}{2}}$	$\frac{1}{2}\, 3^{-\frac{1}{2}}$	1.077
0	if $\lvert y\rvert > 3^{\frac{1}{2}}$		

From this table it is clear that one does almost as well with the Gaussian or even the rectangular weight function as one does with the optimal nonnegative weight function w_0. In computation, there would be clear advantages in dealing with a bandlimited weight function like w_0.

When $nb(n) \to \infty$ as $n \to \infty$ $(b(n) \to 0)$ the central limit theorem can be used to get a normal approximation to the probability distribution of $f_n(x)$. Some idea as to the error in the approximation is given by the Berry-Esseen theorem. Specifically, if the weight function w is bounded and nonnegative with f continuous, bounded and positive then

$$|P[\sigma(f_n(x))^{-1}[f_n(x) - Ef_n(x)] \leqq x] - \Phi(x)| \leqq K/(nb(n))^{\frac{1}{2}} \tag{28}$$

where K for sufficiently large n is bounded above by

$$9 \int w(u)^3\, du\, \{\int w(u)^2\, du\}^{-\frac{3}{2}} [f(x)]^{-\frac{1}{2}}. \tag{29}$$

If f is twice continuously differentiable with $f(x) = 0$ and $f''(x) > 0$, one would require $nb(n)^3 \to \infty$ and $\int w^2(v)v^2 dv < \infty$ for asymptotic normality of the estimate $f_n(x)$. It is of some interest to see whether one can get better and more detailed approximations for the distribution of such estimates when using a simple band-limited weight function $w(u)$ (say triangular or even uniform). Steepest descent methods (see Daniels [12]) may be useful. An alternative procedure would be to condition the number of summands in (2) that are nonzero, if the weight function is bandlimited. Approximations suggested by a paper of Kolmogorov [23] may then be helpful.

It is clear that corresponding results can be obtained for estimates of multi-dimensional probability density functions. Just a few remarks will be made in the case of two dimensional estimates. *Let $w(u, v)$ be a bounded integrable weight function with*

(30) $$\int w(u, v)\, du\, dv = 1.$$

Again let $b(n)$ be a linear bandwidth tending to zero as $n \to \infty$. The population is assumed to have an absolutely continuous two dimensional distribution function $F(x, y)$ with continuous probability density

$$f(x, y) = \frac{\partial}{\partial x}\frac{\partial}{\partial y} F(x, y).$$

The estimate $f_n(x, y)$ of $f(x, y)$

(31) $$f_n(x, y) = \frac{1}{nb(n)^2} \sum_{j=1}^{n} w\left(\frac{x - X_j}{b(n)}, \frac{y - Y_j}{b(n)}\right)$$

is computed from a sample $(X_1, Y_1), \cdots, (X_n, Y_n)$ of independent observations from the population. Assume that f and its partial derivatives up to second order are bounded and continuous. Further let

(32) $$\int |w(u, v)| u^2\, du\, dv, \int |w(u, v)| v^2\, du\, dv < \infty$$

with

(33) $$\int w(u, v) u\, du\, dv = \int w(u, v) v\, du\, dv = 0.$$

Then the variance and mean of $f_n(x, y)$ are given asymptotically by

(34) $$\sigma^2[f_n(x, y)] \cong n^{-1} b(n)^{-2} f(x, y) \int w^2(u, v)\, du\, dv$$

and

(35) $$Ef_n(x, y) = f(x, y) + \tfrac{1}{2} b(n)^2 [f_{xx}(x, y) m_{2,0} + 2f_{xy}(x, y) m_{1,1} + f_{yy}(x, y) m_{0,2}] + o(b(n)^2)$$

as $n \to \infty$, $b(n) \to 0$. Here the $m_{\alpha,\beta}$ are the moments

(36) $$\int u^\alpha v^\beta w(u, v)\, du\, dv$$

of the weight function w. The estimate is asymptotically normal if $nb(n)^2 \to \infty$ as $n \to \infty$. A more detailed discussion of multidimensional density estimates can be found in [7] and [13].

3. The process $[f_n(x) - Ef_n(x)]$. In this section the object is to study the process $[f_n(x) - Ef_n(x)]$, over an interval (say $x \in [0, 1]$), suitably normalized so as to obtain a nontrivial result as $n \to \infty$. The simple result obtained will be derived under unpleasant and completely impractical conditions. However, heuristically the result ought to hold under reasonable conditions. As we shall see, the type of result obtained suggests that limit theorems for sequences of random processes over finite intervals whose length is asymptotically unbounded may be of considerable interest. The asymptotic behavior of the process $[f_n(x) - Ef_n(x)]$ at a special set of points $x_j \in [0, 1]$, $j = 1, \cdots, n$, has been considered by Woodroofe [38]. To make it easier to follow the argument, the exposition is broken down into a number of stages. At the end the assumptions made in these stages are put together to get a coherent result. The argument is due to joint work with Peter Bickel.

(i) Consider the estimate $f_n(x)$ of $f(x)$ given by (2). Let us look at the covariance of $f_n(x)$, $f_n(y)$

$$\text{(37)}\quad \operatorname{Cov}(f_n(x), f_n(y)) = \frac{1}{nb(n)^2} \int w\left(\frac{x-u}{b(n)}\right) w\left(\frac{y-u}{b(n)}\right) f(u)\, du - \frac{1}{n} \int w(z) f(x - b(n)z)\, dz \int w(z) f(y - b(n)z)\, dz.$$

From remarks made in Section 2, it is clear that to get a nontrivial limiting distribution at a fixed point x as $n \to \infty$, one ought to look at

$$\text{(38)}\quad (nb(n))^{\frac{1}{2}} [f_n(x) - Ef_n(x)].$$

We shall be interested in (38) as a process with parameter x varying over a finite range, say $0 \leqq x \leqq 1$, for convenience. To get a reasonable limiting covariance function, the scale in x has to be renormalized so that we look at

$$\text{(39)}\quad (nb(n))^{\frac{1}{2}} [f_n(b(n)x) - Ef_n(b(n)x)]$$

with $0 \leqq x \leqq b(n)^{-1}$. Thus

$$\text{(40)}\quad nb(n) \operatorname{Cov}[f_n(x), f_n(x + \alpha b(n))] = \int w(z) w(z+\alpha) f(x - b(n)z)\, dz - b(n) \int w(z) f(x - b(n)z)\, dz \int w(z) f(x + (\alpha - z) b(n))\, dz$$

and (40) tends to

$$\text{(41)}\quad f(x) \int w(z) w(z+\alpha)\, dz$$

as $n \to \infty$ and $b(n) \to 0$ if f is bounded and continuous at x while w is integrable and bounded. For asymptotic normality of (39) as $n \to \infty$, we require that

$nb(n) \to \infty$. Let us now assume that f is continuous and bounded away from zero on [0, 1]. Then we could look at

$$(nb(n))^{\frac{1}{2}}[f(x)]^{-\frac{1}{2}}[f_n(x) - Ef_n(x)] \tag{42}$$

so as to get the limiting covariance function

$$r(\alpha) = \int w(z)w(z+\alpha)\,dz, \tag{43}$$

with the proper change of scale.

(ii) We now compare the process (42) with the process obtained by looking at (42) at the points $x_j = jc(n)$ where $j = 1, 2, \cdots, [b(n)^{-1}c(n)^{-1}]$ ($[y]$ denotes the greatest integer less than or equal to y here) and interpolating linearly over the range $0 \leqq x \leqq b(n)^{-1}$. To estimate the difference between these two processes it will be enough to look at

$$\begin{aligned}\sup_{0 \leqq x \leqq b(n)^{-1}} (nb(n))^{\frac{1}{2}} \big| f_n(b(n)x) - Ef_n(b(n)x) \\ - f_n(b(n)[x/c(n)]c(n)) + Ef_n(b(n)[x/c(n)]c(n))\big|.\end{aligned} \tag{44}$$

The expression (44) can be conveniently rewritten as

$$\begin{aligned}\sup_{0 \leqq x \leqq b(n)^{-1}} (nb(n))^{\frac{1}{2}} \Big| \int \{\exp[-itb(n)x] - \exp[-itb(n)[x/c(n)]c(n)]\} \\ \cdot h(b(n)t)\{\varphi_n(t) - \varphi(t)\}\,dt\Big|\end{aligned} \tag{45}$$

where

$$h(t) = \int e^{itu} w(u)\,du \tag{46}$$

and

$$\begin{aligned}\varphi_n(t) &= n^{-1} \textstyle\sum_{j=1}^{n} e^{itX_j}, \\ \varphi(t) &= E\exp(itX)\end{aligned} \tag{47}$$

if $h \in L$. Notice that the second moment

$$\begin{aligned} &E(\sup_{0 \leqq x \leqq b(n)^{-1}} (nb(n))^{\frac{1}{2}} |\cdots|)^2 \\ &\leqq nb(n) E\{\sup_{0 \leqq x \leqq b(n)^{-1}} \int |1 - \exp[itb(n)\{x - [x/c(n)]c(n)\}]|^2 |h(b(n)t)|\,dt \\ &\qquad \cdot \int |h(b(n)t)|\,|\varphi_n(t) - \varphi(t)|^2\,dt\}\end{aligned} \tag{48}$$

and that the right-hand side of inequality (48) is bounded by

$$nb(n)^3 c(n)^2 \int t^2 |h(b(n)t)|\,dt \int |h(b(n)t)| E|\varphi_n(t) - \varphi(t)|^2\,dt. \tag{49}$$

However,

$$E|\varphi_n(t) - \varphi(t)|^2 \leqq n^{-1}. \tag{50}$$

It follows from (49) and (50) that (44) tends to zero in probability as $n \to \infty$ if

$$c(n)^2/b(n) \to 0 \tag{51}$$

$(b(n) \to 0)$ and h, $t^2 h \in L$. Van Ryzin's paper [36] should be referred to for a related set of estimates.

(iii) Let ξ_i, $E\xi_i = 0$, $i = 1, \cdots, n$ be independent identically distributed random k vectors. Set

(52) $$\rho_i = E|\xi_{1i}|^3 / E^{\frac{3}{2}} \xi_{1i}^2, \qquad i = 1, \cdots, k$$

where ξ_{1i} is the ith component of the random variable ξ_1. $F_n(x)$ is to denote the distribution function of $n^{-\frac{1}{2}} \sum_{i=1}^n \xi_i$ and $G(x)$ the k-variate normal distribution function with means zero and the same covariance structure as $F_n(x)$. Sazanov's Theorem 3 in [31] then states that

(53) $$\sup_{x \in R^k} |F_n(x) - G(x)| \leqq C''(k)[\sum_{i=1}^k \rho_i n^{-\frac{1}{2}}]^{\frac{1}{3}}$$

where $C''(k) = Ck^2$ with C an absolute constant. In our context $k = [c(n)b(n)]^{-1}$ (the number of points $x_j = jc(n)$ in $0 \leqq x \leqq b(n)^{-1}$) and $\rho_i \leqq C'b(n)^{-\frac{1}{2}}$ for all i with C' an absolute constant. The estimate of the error on the right side of (53) becomes

$$C''c(n)^{-7/3} b(n)^{-5/2} n^{-1/6}$$

with C'' an absolute constant and if we take $c(n)$, $b(n) \to 0$ sufficiently slowly as $n \to \infty$, this estimate of the error will tend to zero.

(iv) The covariance function of the process (42) is

(54) $$\{f(x)f(y)\}^{-\frac{1}{2}} b(n)^{-1} \Big[\int w\left(\frac{x-u}{b(n)}\right) w\left(\frac{y-u}{b(n)}\right) f(u)\,du - b(n)^2 \int w(z) f(x - b(n)z)\,dz \int w(z) f(y - b(n)z)\,dz \Big].$$

Let ${}_0Y_n(x)$, $0 \leqq x \leqq 1$, be the Gaussian process with mean zero and covariance function (54). By using the Sazanov theorem, one can approximate the process (42) at the points $jc(n)b(n)$, $j = 1, \cdots, [c(n)b(n)]^{-1}$, in distribution by the Gaussian process ${}_0Y_n(x)$ at those same points. The type of argument used in (ii) shows that the supremum of the absolute difference between ${}_0Y_n(x)$ and the process obtained from ${}_0Y_n$ by linearly interpolating between the points $jc(n)b(n)$, $j = 1, \cdots, [c(n)b(n)]^{-1}$ tends to zero in probability as $n \to \infty$ if $c(n)^2/b(n) \to 0$ and $b(n) \to 0$. Let

(55) $${}_1Y_n(x) = {}_0Y_n(x) + [(b(n)/f(x))^{\frac{1}{2}} \int w(z) f(x - b(n)z)\,dz] U$$

where U is a normal random variable with mean zero and variance one independent of the process ${}_0Y_n$. The process ${}_1Y_n(x)$ can be written as

$${}_1Y_n(x) = (b(n))^{-\frac{1}{2}} \int w\left(\frac{x-u}{b(n)}\right) \left(\frac{f(u)}{f(x)}\right)^{\frac{1}{2}} dB(u)$$

where $B(u)$ is a Brownian motion process with $-\infty < u < \infty$. Here $B(u)$ is a Gaussian process with increments over non-overlapping intervals independent, $B(0) \equiv 0$, and

$$E|B(u) - B(v)|^2 = |u - v|, EB(u) \equiv 0,$$

for $-\infty < u, v < \infty$. Further, the process ${}_1Y_n(x)$ can be approximated in the sup norm over $0 \leqq x \leqq 1$ by

$$ {}_2Y_n(x) = (b(n))^{-\frac{1}{2}} \int w\left(\frac{x-u}{b(n)}\right) dB(u). $$

This can be seen by making use of an idea from a paper of Garsia, Rodemich, and Rumsey [16]. Let

$$ U(x) = {}_1Y_n(x) - {}_2Y_n(x). \tag{56} $$

We are interested in estimating $\sup_{0 \leqq x \leqq 1} |U(x)|$. $U(x)$ is a Gaussian process with mean zero and covariance function

$$ R(x, y) = \frac{1}{b(n)} \int w\left(\frac{x-u}{b(n)}\right) w\left(\frac{y-u}{b(n)}\right) \left[\left(\frac{f(u)}{f(x)}\right)^{\frac{1}{2}} - 1\right] \left[\left(\frac{f(u)}{f(y)}\right)^{\frac{1}{2}} - 1\right] du. \tag{57} $$

Let

$$ \Delta R(x, y) = R(x, x) + R(y, y) - 2R(x, y). $$

The following Lemma is an immediate corollary of Theorem 2.2 of [16].

LEMMA. *Let $U(x)$, $0 \leqq x \leqq 1$, be a Gaussian process with mean zero and covariance function $R(x, y)$. Consider $p(u)$ a nonnegative even function that is continuous and nonnegative for $u \geqq 0$ and such that*

$$ \int_0^1 \frac{dp(u)}{u} < \infty. $$

If

$$ \int_0^1 \int_0^1 \frac{\Delta R(x, y)}{p^2(x-y)} dx\, dy < \infty $$

then

$$ \sup_{0 \leqq x \leqq 1} |U(x)| \leqq |U(0)| + B^{\frac{1}{2}} \int_0^1 \frac{dp(u)}{u} $$

where B is a nonnegative random variable with

$$ EB \leqq c \int_0^1 \int_0^1 \frac{\Delta R(x, y)}{p^2(x-y)} dx\, dy $$

and c an absolute constant.

We shall now apply this Lemma to estimate $E[\sup_{0 \leqq x \leqq 1} |U(x)|]^2$ where $U(x)$ is the Gaussian process (56) with covariance (57). In this case

$$ \Delta R(x, y) = \frac{1}{b(n)} \int [a(x, u) - a(y, u)]^2 \, du $$

with

$$ a(x, u) = w\left(\frac{x-u}{b(n)}\right) \left[\left(\frac{f(u)}{f(x)}\right)^{\frac{1}{2}} - 1\right]. $$

We clearly have

$$|\Delta R(x,y)| \leqq \frac{2}{b(n)}\int \{a(x,u)^2+a(y,u)^2\}\,du.$$

Let f be continuously differentiable with f, f' bounded and f bounded away from zero on the interval [0, 1]. It then follows that

$$\frac{1}{b(n)}\int a(x,u)^2\,du \leqq K_1\int w(z)^2[f(x-b(n)z)-f(x)]^2\,dz$$
$$\leqq K_2 b(n)^2\int w(z)^2z^2\,dz \leqq K_3 b(n)^2$$

if $\int w(z)^2z^2dz < \infty$ with the K_i constants. The uniform bound $|\Delta R(x,y)| \leqq 4K_3b(n)^2$ for $\Delta R(x,y)$ is obtained. However, another bound will be required for $|\Delta R(x,y)|$ when x is close to y. Let $y-x = \alpha b(n)$ with $|\alpha| \leqq M < \infty$ where M is a constant. Then

$$|\Delta R(x,y)| \leqq 2\left[\int\left(w(z)-w\left(\frac{y-x}{b(n)}+z\right)\right)^2\left[\left(\frac{f(x-b(n)z)}{f(x)}\right)^{\frac{1}{2}}-1\right]^2dz\right.$$
$$\left.+\int w\left(\frac{y-x}{b(n)}+z\right)^2 f(x-b(n)z)\{f(x)^{-\frac{1}{2}}-f(y)^{-\frac{1}{2}}\}^2\,dz\right]$$
$$\leqq K_4\int w'(z)^2\,dz\,\alpha^2+K_5\alpha^2$$

if w' is continuous and bounded and in L^2. Thus $|\Delta R(x,y| \leqq K_6\alpha^2$ if $y-x = \alpha b(n)$ with $|\alpha| \leqq M < \infty$. Let us set $p'(u) = u^{-\frac{1}{2}}$, $0 < u$, in applying the lemma since $\int_0^1 dp(u)/u = \int_0^1 u^{-\frac{3}{2}}du$ is finite. Also

$$\int_0^1\int_0^1 \frac{\Delta R(x,y)}{p^2(x-y)}\,dx\,dy \leqq K\int_{b(n)^2\leqq|x|<2}\frac{b(n)^2}{x}\,dx+K'\int_{|x|\leqq b(n)^2}\frac{x^2}{b(n)^2|x|}\,dx$$
$$\leqq K''b(n)^2|\log b(n)|.$$

Notice that the Gaussian process $Y(x) = {}_2Y_n(xb(n))$ has the desired covariance function

$$r(\alpha) = \int w(z)w(z+\alpha)\,dz.$$

We should now like to apply the result on the asymptotic distribution of the maximum of a Gaussian process as given, for example, on pages 271–272 of Cramér and Leadbetter [11] to the process $Y(x)$ on the range $0 \leqq x \leqq b(n)^{-1}$. Assume that the covariance function $r(\alpha)$ of the process $Y(x)$ as given by (43) is an even function that is continuously differentiable up to fourth order and such that

$$r(\alpha) = O(|\alpha|^{-\varepsilon}) \tag{58}$$

as $|\alpha| \to \infty$ for some $\varepsilon > 0$. Let

$$\lambda_0 = r(0) \tag{59}$$
$$\lambda_2 = r''(0).$$

Then the result cited in [11] indicates that

$$(60) \qquad P\left[\max_{0 \leqq x \leqq T} \lambda_0^{-\frac{1}{2}} Y(x) \leqq (2\log T)^{\frac{1}{2}} + \frac{A+z}{(2\log T)^{\frac{1}{2}}}\right] \to e^{-e^{-z}}$$

as $T \to \infty$ where

$$(61) \qquad A = \log\{(-2\lambda_2/\lambda_0)^{\frac{1}{2}}/2\pi\}.$$

The estimates made in (ii), (iii) and (iv) (or obvious modifications of them) indicate that under appropriate conditions on f and w the distribution of the process (42) can be sufficiently well approximated by that of $Y(x)$ so as to obtain the following theorem.

THEOREM. *Let f be continuously differentiable and bounded away from zero on* $[0, 1]$. *Assume that the bounded, integrable weight function w used in the estimate f_n (see (2)) is twice continuously differentiable with $(1+v^4)|w''(v)|$ bounded and has a Fourier transform $h(t)$ with $(1+t^2)h(t) \in L$. Then if $n^{-1/24} = O(b(n))$ as $n \to \infty$, it follows that*

$$(62) \quad P\left[\max_{0 \leqq x \leqq 1} \left(\frac{nb(n)}{\alpha f(x)}\right)^{\frac{1}{2}} \{f_n(x) - Ef_n(x)\} \leqq \{2\log b(n)^{-1}\}^{\frac{1}{2}} + \frac{A+z}{(2\log b(n)^{-1})^{\frac{1}{2}}}\right] \to e^{-e^{-z}}$$

as $n \to \infty$, where

$$(63) \qquad \alpha = \int w^2(u)\,du$$

and

$$(64) \qquad A = \log\frac{B^{\frac{1}{2}}}{2\pi}, \qquad B = -\frac{2}{\alpha}\frac{d^2}{dt^2}\left(\int w(z)w(z+t)\,dz\right)\Big|_{t=0}.$$

The interest in the theorem given above is not in the detailed conditions. The result, in particular, must hold under much weaker requirements on $b(n)$. It is rather in its asymptotic nonparametric character and the ideas behind the proof. Notice that in the proof one approximates the process (42) in distribution with a corresponding Gaussian process.

The heuristics that suggest a result like that in the above Theorem also imply an asymptotic nonparametric character (under appropriate conditions) for

$$(65) \qquad \int_0^1 \frac{[f_n(x) - f(x)]^2}{f(x)}\,dx$$

when suitably normalized. There are the corresponding open questions for two-dimensional (and higher dimensional) density functions. For, example, let $f_n(x, y)$

be the estimate given in (31) of the density function $f(x, y)$. The marginal density estimates are

$$g_n(x) = \int f_n(x, y)\, dy \tag{66}$$
$$h_n(y) = \int f_n(x, y)\, dx.$$

A natural statistic to test independence over $[0, 1] \times [0, 1]$ is given by

$$\int_0^1 \int_0^1 \frac{|f_n(x, y) - g_n(x)h_n(y)|^2}{f_n(x, y)} dx\, dy \tag{67}$$

and its distribution should be investigated.

4. Probability density estimates for dependent sequences. It is also curious and pleasantly surprising that the asymptotic results on the behavior of probability density estimates obtained in Section 2 still hold even when sampling from a stationary process, if appropriate and rather reasonable conditions are satisfied. This is not true at all if the distribution function is estimated (see Billingsley [3, page 195] for a discussion) and the fact that it holds when estimating the probability density is due to the *local* character of the estimate.

Let $\{X_j, j = \cdots, -1, 0, 1, \cdots\}$ be a strictly stationary process. Assume that the instantaneous distribution function

$$F(x) = P[X_j \leqq x]$$

is absolutely continuous with continuous spectral density $f(x) = F'(x)$. We wish to estimate $f(x)$ by $f_n(x)$ as given in (2). The assumptions on the weight function w, the density function f, and the bandwidth $b(n)$ are the same as those in Section 2. The results on the bias of the estimate as given in (16) hold under the same conditions as in Section 2 since dependence has no effect on the bias. Let us now consider the covariance of the estimate. At this point the character of the dependence must be examined. Let the two dimensional distribution functions

$$F_j(x, y) = P[X_0 \leqq x, X_j \leqq y], \qquad j \neq 0, \tag{68}$$

be absolutely continuous with continuous density functions

$$f_j(x, y) = \frac{\partial}{\partial x}\frac{\partial}{\partial y} F_j(x, y). \tag{69}$$

The covariance of $f_n(x)$ and $f_n(y)$ is

$$\operatorname{Cov}[f_n(x), f_n(y)] = n^{-2}b(n)^{-2} \textstyle\sum_{j=-n}^{n} (n - |j|) \operatorname{Cov}[w(b(n)^{-1}(x - X_0)), \\ w(b(n)^{-1}(y - X_j))] \tag{70}$$

with

$$\operatorname{Cov}[w(b(n)^{-1}(x - X_0)), w(b(n)^{-1}(y - X_j))] \\ = b(n)^2 \iint w(u)w(v)\{f_j(x - b(n)u, y - b(n)v) - f(x - b(n)u)f(y - b(n)v)\}\, du\, dv \tag{71}$$

if $j \neq 0$ and

$$(72)\quad \operatorname{Cov}[w(b(n)^{-1}(x-X_0)), w(b(n)^{-1}(y-X_0))]$$
$$= b(n)\int w(u)w\left(u+\frac{y-x}{b(n)}\right)f(x-b(n)u)\,du$$
$$-b(n)^2\int w(u)f(x-b(n)u)\,du\int w(v)f(x-b(n)v)\,dv.$$

Thus, if

$$(73)\qquad \sum_{j\neq 0}|f_j(x,y)-f(x)f(y)| \leqq M < \infty$$

for all x and y, we have

$$74)\quad nb(n)[\operatorname{Cov}(f_n(b(n)x), f_n(b(n)y))] = \int w(u)w(u+y-x)f(x-b(n)u)\,du + O(b(n))$$

for all x and y as $n \to \infty$. The first term on the right of (74) is

$$f(x)\int w(u)w(u+y-x)\,du + O(b(n))$$

if f has a bounded continuous derivative. Additional conditions on the type of dependence are required to get a result on the asymptotic normality of the estimate $f_n(x)$. A standard result making use of the conditions will be described. The details of the proof can be found in [29]. Let $\mathscr{B}_n$ and $\mathscr{F}_m$ be the backward and forward Borel fields generated by the random variables $\{X_j; j \leqq n\}$ and $\{X_j; j \geqq m\}$ respectively. The stationary process $\{X_k\}$ satisfies the condition S if for every $\mathscr{F}_{n+k}$ measurable random variable $Y (k > 0)$ with $EY^2 < \infty$, $EY = 0$,

$$(75)\qquad E|E(Y\,|\,\mathscr{B}_n)|^2 \leqq a(k)EY^2$$

where

$$(76)\qquad a(k) = O(k^{-4-\varepsilon})$$

for some $\varepsilon > 0$ as $k \to \infty$. If $\{X_k\}$ satisfies condition S and the other requirements already specified, then

$$\{\int w^2(v)\,dv\}^{-\frac{1}{2}}(nb(n))^{\frac{1}{2}}[f_n(x) - Ef_n(x)]$$

is asymptotically normal with mean zero and variance one as $n \to \infty$, $nb(n) \to \infty$, $b(n) \to 0$. The paper of Roussas [30] discusses related questions.

The papers of Van Atta and Chen [34] and Frenkiel and Klebanoff [15] look at estimates of univariate and multivariate probability density functions without mentioning the estimation problem explicitly. They are interested in the probability density of the distribution of a component of the velocity in grid turbulence as well as the bivariate distribution of velocity components at the same point in space but with a time difference. They find that the univariate distribution appears to be normal but not the bivariate distribution.

A recent refinement in the "theory" of turbulence suggested by Kolmogorov is discussed in a paper of Yaglom [39]. The refinement suggests that a local spatial average of the energy dissipation might have a lognormal distribution. Currently, experimental results are being analyzed to estimate the probability density of such a local average of the energy dissipation to see whether it is reasonably approximated by a lognormal density.

5. Expansions in orthogonal functions. There are many ways of estimating a probability density function from a sample of independent observations on a population. Thus far in this paper, only estimates of the form (2) have been considered. Still another interesting class of estimates has been suggested by Van Ryzin [36]. An obvious class of estimates is also suggested by the notion of an orthogonal expansion. Just a few remarks will be made on this class but they will be enough to show that qualitatively results similar to those already obtained would be expected. However, it should be noted that there are still many open detailed questions.

Let $\{\varphi_j(x)\}$ be a complete orthonormal family of functions with respect to a given nonnegative weight function $w = w(x)$. Then if the probability density function f is in $L^2(w)$, the Fourier expansion

$$\sum_j c_j \varphi_j(x) \tag{77}$$

with

$$c_j = \int f(x)\overline{\varphi_j(x)}w(x)\,dx \tag{78}$$

will converge to f in $L^2(w)$. Let $F_n(x)$ be the sample distribution based on a sample of n independent observations $X_1, X_2, \cdots, X_n$ from the population with absolutely continuous distribution function $F(x)$ and continuous probability density $f(x) = F'(x)$. It is plausible to estimate c_j by

$$c_j{}^*(n) = \int \overline{\varphi_j(x)}w(x)\,dF_n(x) = n^{-1}\sum_{k=1}^{n} \overline{\varphi_j(X_k)}w(X_k). \tag{79}$$

Then

$$Ec_j{}^* = c_j \tag{80}$$

and

$$\begin{aligned} \operatorname{Cov}(c_j{}^*, c_k{}^*) &= n^{-1}\operatorname{Cov}(\varphi_j(X)w(X), \varphi_k(X)w(X)) \\ &= n^{-1}\{\textstyle\int \varphi_j(x)\overline{\varphi_k(x)}w^2(x)f(x)\,dx - c_j\bar{c}_k\}. \end{aligned} \tag{81}$$

In all the cases we shall think of, the functions φ_j will be continuous in the domain in which they are defined. Consider a sequence of weights $\{\alpha_j(n)\}$ that are zero except for a finite number of values of j. Let the estimate of $f(x)$ be

$$f_n(x) = \sum_j c_j{}^*(n)\alpha_j(n)\varphi_j(x). \tag{82}$$

The bias

$$Ef_n(x) - f(x) = \sum c_j(\alpha_j(n) - 1)\varphi_j(x) \tag{83}$$

and the covariance

$$(84)\quad \operatorname{Cov}(f_n(x), f_n(y)) = n^{-1}\sum_{j,k} \varphi_j(x)\overline{\varphi_k(y)}\alpha_j(n)\alpha_k(n) \cdot \{\textstyle\int \varphi_j(u)\overline{\varphi_k(u)}w^2(u)f(u)\,du - c_j\bar{c}_k\}.$$

If the Fourier expansion of f converges to f absolutely, one would obviously require that

$$(85)\quad \alpha_j(n) \to 1$$

as $n \to \infty$ for each fixed j. The integrated weighted mean square error has mean value

$$(86)\quad \begin{aligned} E\textstyle\int |f_n(x)-f(x)|^2 w(x)\,dx &= \textstyle\int \sigma^2[f_n(x)]w(x)\,dx + \int |Ef_n(x)-f(x)|^2\,dx \\ &= n^{-1}\textstyle\sum |\alpha_j(n)|^2\{\int |\alpha_j(u)|^2 w^2(u)f(u)\,du - |c_j|^2\} \\ &\quad + \textstyle\sum |c_j|^2|\alpha_j(n)-1|^2. \end{aligned}$$

It is sometimes convenient to rewrite $f_n(x)$ in the form

$$(87)\quad f_n(x) = n^{-1}\sum_{j=1}^n w(X_j)k_n(x, X_j)$$

where

$$(88)\quad k_n(x,u) = \sum_j \alpha_j(n)\varphi_j(x)\overline{\varphi_j(u)}$$

is a "generalized" kernel function (see G. S. Watson [37] for a related discussion). The variance of $f_n(x)$ can then also be written as

$$(89)\quad n^{-1}\{\textstyle\int w^2(u)|k_n(x,u)|^2 f(u)\,du - |\int w(u)k_n(x,u)f(u)\,du|^2\}.$$

The asymptotic behavior of the kernel function $k_n(x, u)$ is obviously of importance in determining the behavior of the variance (89) as $n \to \infty$ as well as the asymptotic distribution of $f_n(x)$.

We consider the case of the trigonometric functions

$$(90)\quad \varphi_j(x) = \frac{e^{ijx}}{(2\pi)^{\frac{1}{2}}}, \qquad -\pi \leqq x \leqq \pi,\ w(x) \equiv 1,$$

because it is simple and is amusingly like the case of spectral estimation. Here

$$(91)\quad k_n(x,u) = k_n(x-u) = \sum_j \alpha_j(u)\,e^{ij(x-u)}.$$

For economy in computation one would require $\alpha_j(n) = 0$ for $|j| \geqq n$. In fact, the usual situation will be that in which $\alpha_j(n) = 0$ for $|j| \geqq m(n)$ with $m(n) = o(n)$. Assume that

$$(92)\quad \textstyle\int k_n(x-u)\,du = \alpha_0(n) = 1.$$

For convenience, assume that $k_n(u)$ is a nonnegative weight function that behaves asymptotically like a δ function, that is, given any $\varepsilon > 0$

$$(93)\quad \textstyle\int_{|u| \geqq \varepsilon} k_n(u)\,du \to 0$$

as $n \to \infty$. Asymptotic unbiasedness of the estimate $f_n(x)$ follows since

$$Ef_n(x) = \int k_n(x, u) f(u)\, du \to f(x) \tag{94}$$

as $n \to \infty$ if *f is bounded and continuous*. The variance $\sigma^2[f_n(x)]$ is given asymptotically by

$$\sigma^2[f_n(x)] \cong \frac{f(x)}{n} \int k_n(u)^2\, du. \tag{95}$$

A similar argument shows that

$$E|k_n(x, X) - Ek_n(x, X)|^4 \cong f(x) \int k_n(u)^4\, du. \tag{96}$$

A sufficient condition for (87) to be asymptotically normal with mean $Ef_n(x)$ and variance (95) as $n \to \infty$ is that

$$\frac{\int k_n(u)^4\, du}{n(\int k_n(u)^2\, du)^2} \to 0 \tag{97}$$

by the Liapounov form of the central limit theorem. Suppose we look at the case of simple truncation.

$$\begin{aligned} \alpha_j(n) &= 1 \quad \text{if} \quad |j| \leqq m(n) \\ &= 0 \quad \text{if} \quad |j| > m(n) \end{aligned} \tag{98}$$

or the Fejer weights

$$\begin{aligned} \alpha_j(n) &= 1 - |j|/m \quad \text{if} \quad |j| \leqq m(n) \\ &= 0 \quad\quad\quad\quad \text{if} \quad |j| > m(n). \end{aligned} \tag{99}$$

Then

$$\begin{aligned} \int k_n(u)^2\, du &\cong c_1 m(n) \\ \int k_n(u)^4\, du &\cong c_2 m(n)^3 \end{aligned} \tag{100}$$

with $c_1, c_2 > 0$ so that (97) is satisfied if $m(n) = o(n)$ as $n \to \infty$.

Similar computations will now also be carried out for orthonormal polynomials with respect to a weight function. The conditions assumed will be somewhat restrictive but the results obtained are of interest because they suggest what ought to hold under much weaker conditions. Let $w(x)$ be a weight function on $-1 \leqq x \leqq 1$ with $p_j(x)$, $j = 0, 1, \cdots$, the orthonormal polynomials with respect to this weight function. We wish to consider estimating a continuous density function $f(x)$ on $-1 \leqq x \leqq 1$ by (87) with

$$k_n(x, u) = \textstyle\sum_j \alpha_j(n) p_j(x) p_j(u). \tag{101}$$

Let $x = \cos\theta$, $z = e^{i\theta}$ and

$$h(\theta) = w(\cos\theta)|\sin\theta|. \tag{102}$$

Assume that w is such that $h(\theta)$ *is a positive continuous function on the unit circle* $z = e^{i\theta}$, $-\theta < \theta \leqq \pi$ ($\theta = -\pi$ is identified with $\theta = \pi$), *and h satisfies a uniform Lipschitz condition.* Let $\phi_n(z)$, $n = 0, 1, 2, \cdots$, with $z = e^{i\theta}$, $-\pi < \theta \leqq \pi$, be the orthonormal trigonometric polynomials with weight function $h(\theta)$ generated from $1, z, z^2, \cdots$. The coefficient of z^n in $\phi_n(z)$ is κ_n. Then by [32, page 292]

$$p_n(x) = (2\pi)^{-\frac{1}{2}}\left\{1+\frac{\phi_{2n}(0)}{\kappa_{2n}}\right\}^{\frac{1}{2}}\{z^{-n}\phi_{2n}(z)+z^n\phi_{2n}(z^{-1})\} \tag{103}$$

for $n \geqq 1$. Also, if

$$g(h; r\,e^{i\theta}) = g(r\,e^{i\theta}) = \exp\left\{\frac{1}{2\pi}\int_{-\pi}^{\pi}\frac{e^{i\lambda}+r\,e^{i\theta}}{e^{i\lambda}-r\,e^{i\theta}}\log h(\lambda)\,d\lambda\right\} \tag{104}$$

then by [18, page 51]

$$\lim_{n\to\infty}\kappa_n = [g(o)]^{-1} \tag{105}$$

$$\lim_{n\to\infty}\phi_n(o) = 0.$$

Further, by [18, page 53] one knows that

$$\phi_n(z) = z^n/\overline{g(h;z)}+\frac{\log n}{n}O(1) \tag{106}$$

with the $O(1)$ uniform in n and z. *Assume that* $|\alpha_j(n)| \leqq 1$ with $\alpha_j(n) = 0$ for $|j| \geqq m(n)$ *where* $m(n) = O(n)$ *and*

$$\lim_{n\to\infty}\alpha_j(n) = 1$$

for each fixed j. Then, if the expansion of f in terms of the orthonormal polynomials p_n is absolutely convergent, the estimate $f_n(x)$ will be asymptotically unbiased. To avoid difficulties arising from possible singularities at $x = \pm 1$, we shall *assume that*

$$f(x)(1-x^2)^{-\frac{1}{2}} \tag{107}$$

is bounded on $-1 \leqq x \leqq 1$. The object is to again determine the asymptotic behavior of the variance of $f_n(x)$ and sufficient conditions for asymptotic normality of $f_n(x)$. From (103) and (106) it follows that

$$\begin{aligned} k_n(x,y) &= \sum\alpha_j(n)p_j(x)p_j(y) \\ &= \frac{1}{2\pi}\sum_j\alpha_j(n)\left(1+\frac{\phi_{2j}(o)}{\kappa_{2j}}\right)^{-1}\left\{z^j/\overline{g(h;z)}+z^{-j}/\overline{g(h,z^{-1})}\right. \\ &\quad\left.+\frac{\log j}{j}O(1)\right\}\left\{\bar{z}'^j/g(h;z')+z'^{-j}/g(h;z'^{-1})+\frac{\log j}{j}O(1)\right\} \end{aligned} \tag{108}$$

where $x = \cos\theta$, $z = e^{i\theta}$, $y = \cos\theta'$, $z' = e^{i\theta'}$. Let

$$c_n(\theta) = \sum_{j=1}^{n}\alpha_j(n)\,e^{ij\theta} \tag{109}$$

and assume that

$$|c_n(\theta)| \leqq M(\varepsilon) < \infty \qquad \text{for } |\theta| \geqq \varepsilon. \tag{110}$$

Then

$$\sigma^2[f_n(x)] = \frac{1}{n}\int w^2(u)|k_n(x,u)|^2 f(u)\,du \cong \frac{4}{n}\frac{f(x)}{(1-x^2)^{\frac{1}{2}}}\int_{-\pi}^{\pi}|c_n(\theta)|^2\,d\theta \tag{111}$$

as $n \to \infty$, if $|x \pm 1| \geqq \varepsilon > 0$ and

$$(\log m(n))^2 = o(\int |c_n(\theta)|^2\,d\theta)^{\frac{1}{2}}. \tag{112}$$

Further,

$$\int w^4(u)|k_n(x,u)|^4 f(u)\,du \leqq A\int_{-\pi}^{\pi}|c_n(\theta)|^4\,d\theta \tag{113}$$

with A a constant, as $n \to \infty$, if $|x \pm 1| \geqq \varepsilon > 0$ and

$$(\log m(n))^2 = o(\int |c_n(\theta)|^4 d)^{\frac{1}{4}}. \tag{114}$$

Thus, a sufficient condition for $f_n(x)$ to be asymptotically normally distributed with mean $Ef_n(x)$ and variance $\sigma^2[f_n(x)]$ as $n \to \infty$, given the requirements already specified, is that

$$\frac{\int |c_n(\theta)|^4\,d\theta}{n(\int |c_n(\theta)|^2\,d\theta)^2} \to 0 \tag{115}$$

as $n \to \infty$. The condition required on the weight function $w(u)$ is appropriate for the Chebyshev polynomials of the first kind but not for the other classical orthogonal polynomials. For just the Chebyshev polynomials of the first kind one would go through a direct and simple computation to get the results (111) and (115). The interest in deriving (111) and (115) under the conditions we have specified is that it suggests what ought to hold more generally for density estimates based on expansions in orthonormal polynomials on a finite interval. The case of classical orthonormal polynomials other than the Chebyshev polynomials of the first kind should be considered. Notice that asymptotic results like those obtained for expansions in the trigonometric functions can be derived for expansions in terms of the eigenfunctions of certain types of Sturm-Liouville problems by using the remarks made on page 117 of [18].

6. Spectra and stationary processes. There are many extended developments of the theory of stationary processes and discussions of spectral estimation [17], [33]. We shall briefly remark on some of the results that relate or are comparable to those already mentioned on estimation of the probability density.

Let $X(t)$, $t = \cdots, -1, 0, 1, \cdots$, be a discrete time parameter r-vector (column vector) valued stationary process with real-valued components. If second order moments exist, then

$$X(t) = \int_{-\pi}^{\pi} e^{it\lambda}\,dZ(\lambda) \tag{116}$$

where Z is an r-vector valued process (with complex-valued components) of orthogonal increments such that

$$E\,dZ(\lambda)\,dZ(\mu)' = \delta(\lambda+\mu)\,dF(\lambda), \qquad -\pi < \lambda, \mu < \pi. \tag{117}$$

In formula (117), δ is the Kronecker δ symbol

$$\begin{aligned}\delta(\lambda) &= 1 \quad \text{if} \quad \lambda = 0\\ &= 0 \quad \text{otherwise,}\end{aligned}$$

F is an $r \times r$ matrix-valued Hermitian non-decreasing function ($F(\lambda) - F(\mu)$ is positive semidefinite if $\lambda \geqq \mu$), and A' denotes the transpose of A. Since the components of $X(t)$ are real-valued

$$\overline{dZ(-\lambda)} = dZ(\lambda) \tag{118}$$

and $dF(\lambda) = dF(-\lambda)'$. In the case of a Gaussian process, the full probability structure is determined by the first and second order moments. It is convenient to assume that $EX(t) \equiv 0$.
The covariance matrices

$$r(t) = EX(\tau)X(t+\tau)' \tag{119}$$

are related to the second order spectral distribution function F by

$$r(t) = \textstyle\int_{-\pi}^{\pi} e^{it\lambda}\,dF(\lambda). \tag{120}$$

If the existence of all moments is assumed, we have

$$m_{a_1,\dots,a_k}(t_1,\cdots,t_k) = EX_{a_1}(t_1)\cdots X_{a_k}(t_k) = m_{a_1,\cdots a_k}(t_1+t,\cdots,t_k+t) \tag{121}$$

where $X_a(t)$ is the ath component of $X(t)$. Assume that the moments have Fourier representations.

$$\begin{aligned} m_{a_1,\dots,a_k}&(t_1,\cdots,t_k)\\ &= \textstyle\int_{-\pi}^{\pi}\cdots\int_{-\pi}^{\pi} \exp\left(i\sum_{j=1}^{k} t_j w_j\right) dG_{a_1,\cdots a_k}(w_1,\cdots,w_k)\,dG_{a_1,\cdots,a_k}(w_1,\cdots,w_k)\\ &= E\left(\textstyle\prod_1^k dZ_{a_j}(w_j)\right)\end{aligned} \tag{122}$$

with the functions G of bounded variation. The stationarity of the process $X(t)$ implies that

$$dG = 0 \quad \text{unless} \quad \textstyle\sum_1^k w_j = 0 \quad \text{modulo} \quad 2\pi. \tag{123}$$

The assumed representation (122) can also be written as

$$\begin{aligned} c_{a_1,\cdots,a_k}&(t_1,\cdots,t_k)\\ &= \textstyle\int_{-\pi}^{\pi}\cdots\int_{-\pi}^{\pi} \exp\left(i\sum_{j=1}^{k} t_j w_j\right) c\{dZ_{a_j}(w_j); j=1,\cdots,k\}\,dF_{a_1,\cdots,a_k}(w_1,\cdots,w_k)\\ &= c\{dZ_{a_j}(w_j); j=1,\cdots,k\}\end{aligned} \tag{124}$$

where c is the corresponding cumulant function. It is convenient to assume that the cumulants $c_{a_1,\cdots,a_k}(t_1,\cdots,t_k)$ are in L_1 as functions of their $k-1$ (or k) t arguments (see Condition I in (137)). This can be regarded, if assumed up to order k, as a curious mixing condition of kth order. This then implies that

$$dF_{a_1,\cdots,a_k}(w_1,\cdots,w_k)\eta(\textstyle\sum_1^k w_j) = f_{a_1,\cdots,a_k}(w_1,\cdots,w_k)\eta(\sum_1^k w_j)\,dw_1\cdots dw_k \tag{125}$$

($\eta(w) = 0$ if $w \neq 0$ modulo 2π and is equal to one otherwise) where the cumulant spectral density f is continuous and a function of only $k-1$ variables w since $\sum_1^k w_j = 0$ modulo 2π.

Let us first look at a real-valued stationary process $X(t)$ ($r = 1$) and consider estimating its spectral density of second order. Such an estimate is given by

$$f^{(N)}(\lambda) = \frac{1}{2\pi}\sum_{\nu=-N}^{N} w_\nu^{(N)} R_\nu^{(N)} e^{-i\nu\lambda} \tag{126}$$

with

$$w_\nu^{(N)} = \int_{-\pi}^{\pi} e^{i\nu\lambda}\, W_N(\lambda)\,d\lambda \tag{127}$$

and

$$R_\nu^{(N)} = \frac{1}{N}\sum_{t,\tau=1,\,t-\tau=\nu}^{N} X(t)X(\tau). \tag{128}$$

Formula (126) could also be written as

$$f^{(N)}(\lambda) = \int_{-\pi}^{\pi} W_N(\lambda-\alpha) I^{(N)}(\alpha)\,d\alpha \tag{129}$$

with

$$I^{(N)}(\alpha) = \frac{1}{2\pi N}\left|\sum_{t=1}^{N} X(t)e^{-it\lambda}\right|^2, \tag{130}$$

the periodogram, and it is convenient to assume that

$$W_N(u) = A_N B_N^{-1} W(B_N^{-1}u) \qquad \text{if } |u| \leqq \pi \tag{131}$$

with $B_N \downarrow 0$ as $N \to \infty$ and A_N a normalizing factor so that

$$\int_{-\pi}^{\pi} W_N(u)\,du = 1. \tag{132}$$

Assume W_N is periodically extended with period 2π. Then if W has a finite second moment, is nonnegative and symmetric about zero and $NB_N \to \infty$, we can show that

$$B_N^{-2}[Ef^{(N)}(\lambda) - f(\lambda)] = \int u^2 W(u)\,du\frac{f''(\lambda)}{2} + O(B_N^{-3}) \tag{133}$$

and

$$\begin{aligned}\operatorname{Cov}\{f^{(N)}(\lambda), f^{(N)}(\mu)\} &\\ = 2\pi N^{-1}\{&\textstyle\int_{-\pi}^{\pi} W_N(\lambda-\alpha)W_N(\mu+\alpha)f^2(\alpha)\,d\alpha \\ &+\textstyle\int_{-\pi}^{\pi} W_N(\lambda-\alpha)W_N(\mu-\alpha)f^2(\alpha)\,d\alpha + O(N^{-1})\}.\end{aligned} \tag{134}$$

By (131) and (134), it follows that

$$\lim_{N\to\infty} B_N N \operatorname{Cov}\{f^{(N)}(\lambda), f^{(N)}(\mu)\}$$

$$(135)\qquad \begin{aligned} &= 0 && \text{if } \lambda \neq \mu \\ &= 2\pi f^2(\lambda)\textstyle\int_{-\pi}^{\pi} W^2(u)\,du && \text{if } \lambda = \mu \neq 0, \pi \\ &= 4\pi f^2(\lambda)\textstyle\int_{-\pi}^{\pi} W^2(u)\,du && \text{if } \lambda = \mu = 0, \pi \end{aligned}$$

when $0 \leqq \lambda, \mu \leqq \pi$. If we look at the asymptotic mean square error of estimates, the analysis of Epanechnikov given in Section 2 still holds and one is led to the weight function $W(u)$ given in (26). The weight function $W(u)$ (26) is bandlimited but it leads to a sequence $w_v^{(N)}$ that is not bandlimited. But this need not be too frustrating since direct Fourier analysis via the fast Fourier transform (see [10]) may be preferable. One would directly compute the periodogram and then smooth it. The asymptotic results obtained in spectral estimation differ a bit from those obtained in estimating the probability density function. The basic range over which (135) holds is $0 < \lambda < \pi$ with a slightly different result at $\lambda = 0, \pi$. On the other hand, the results obtained in (15) for probability density estimation hold over the entire range of the independent variable x. Also, in the asymptotic formulas for the variance of the estimates, the constant $f(x)$ in (15) is replaced by $2\pi f^2(\lambda)$ in (135). Notice that if we rescale locally in the covariance result (134) for spectral estimation, let $\lambda = \mu + B_N a$ ($\lambda, \mu \neq 0, \pi$) and then normalize appropriately, as $N \to \infty$ one would expect a Gaussian process with covariance function

$$(136)\qquad \int W(b+a)W(b)\,db.$$

This is again very much like what was obtained in Section 3 when estimating a probability density function. It would be interesting to look at the functionals of a global character such as (65) mentioned in Section 3, when estimating the spectral density. The applications of second order spectral techniques are legion and the development of theory is due to many people—Bartlett [1], Tukey [33], Grenander and Rosenblatt [17], Parzen [24] and others.

The success of these second order techniques suggested that it might be worthwhile developing ways of estimating higher order spectra for situations in which there are nonlinear or nonGaussian effects. Tukey [33], Van Ness [34], Brillinger and Rosenblatt [4] developed a certain theoretical background. There have been a limited number of applications of which a few are mentioned here—the work of Hasselman, Munk and MacDonald [19] on bispectra of ocean waves, Haubrich [20] on earth noise, Brillinger and Rosenblatt [5] on sunspots, D. Cartwright [8] with a 4th order analysis of tides and surges, and Huber, Kleiner, Gasser and Dumermuth [22] on electroencephalographs. We shall just briefly mention some asymptotic results due to Brillinger and Rosenblatt[4]. Let $c'_{a_1,\dots,a_k}(v_1,\cdots,v_{k-1})$ and $f'_{a_1,\dots,a_k}(\lambda_1,\cdots,\lambda_{k-1})$ be the cumulant function $c_{a_1,\dots,a_k}$ and cumulant spectral density $f_{a_1,\dots,a_k}$ looked at as functions of $k-1$ variables. First assume

$$(137)\qquad \text{I}\quad \sum_{v_1,\dots,v_{k-1}=-\infty}^{\infty} \left|v_j c'_{a_1,\dots,a_k}(v_1,\cdots,v_{k-1})\right| < \infty$$

for $j = 1, \cdots, k-1$ and any k-tuple $a_1, \cdots, a_k$ with $k = 2, 3, \cdots$. Condition I implies that $f'_{a_1, \ldots, a_k}(\lambda_1, \cdots, \lambda_{k-1})$ has a bounded uniformly continuous gradient. Let W be a weight function with

$$W(-u_1, \cdots, -u_k) = W(u_1, u_2, \cdots, u_k) \tag{138}$$

and set

$$W_N(u_1, \cdots, u_k) = B_N^{-k+1} W(B_N^{-1}u_1, \cdots, B_N^{-1}u_k). \tag{139}$$

If

$$d_a^{(N)}(\lambda) = \textstyle\sum_{t=0}^{N-1} X_a(t) \exp(-i\lambda t), \tag{140}$$

then

$$I^{(N)}_{a_1, \ldots, a_k}(\lambda_1, \cdots, \lambda_k) = (2\pi)^{-k+1} N^{-1} \textstyle\prod_{j=1}^{k} d_{a_j}^{(N)}(\lambda_j), \tag{141}$$

with $\sum_1^k \lambda_j = 0$ modulo 2π, is a kth order analogue of the periodogram. Consider the estimate

$$\begin{aligned} (142)\quad & f^{(N)}_{a_1, \ldots, a_k}(\lambda_1, \cdots, \lambda_k) \\ &= (2\pi)^{k-1} N^{-k+1} \sum_{s_1=-\infty}^{\infty} \cdots \sum_{s_k=-\infty}^{\infty} W_N\left(\lambda_1 - \frac{2\pi s_1}{N}, \cdots, \lambda_k - \frac{2\pi s_k}{N}\right) \\ &\quad \cdot \Phi\left(\frac{2\pi s_1}{N}, \cdots, \frac{2\pi s_k}{N}\right) I^{(N)}_{a_1, \ldots, a_k}\left(\frac{2\pi s_1}{N}, \cdots, \frac{2\pi s_k}{N}\right) \end{aligned}$$

of $f_{a_1, \ldots, a_k}(\lambda_1, \cdots, \lambda_k)$ where it is understood that $\sum_1^k \lambda_j = 0$ modulo 2π. In (142) $\Phi(u_1, \cdots, u_k) = 1$ if $\sum_1^k u_k = 0$ modulo 2π but $\sum_{j \in J} u_j \neq 0$ on any nonvacuous proper subset J of $\{1, \cdots, k\}$. Make the additional assumption

$$\begin{aligned} (143)\quad \text{II}\quad & \left| W\left(u_1, \cdots, u_{k-1}, -\sum_1^{k-1} u_j\right)\right|, \left|\frac{\partial}{\partial u_i} W\left(u_1, \cdots, u_{k-1}, -\sum_1^{k-1} u_j\right)\right| \\ &\leqq A\left(1 + \left(\sum_1^{k-1} u_j^2\right)^{\frac{1}{2}}\right)^{-(k+\varepsilon-1)} \end{aligned}$$

for $i = 1, \cdots, k-1$ with $A, \varepsilon > 0$.

Under assumptions I and II

$$\begin{aligned} & \operatorname{Cov}\{f^{(N)}_{a_1, \ldots, a_k}(\lambda_1, \cdots, \lambda_k), f^{(N)}_{a_1', \cdots, a_k'}(\mu_1, \cdots, \mu_k)\} \\ (144)\quad &= 2\pi N^{-1} \textstyle\sum_P \int_{-\infty}^{\infty} \cdots \int W_N(\lambda_1 - \alpha_1, \cdots, \lambda_k - \alpha_k) \\ &\quad \cdot W_N(\mu_1 + \beta_1, \cdots, \mu_k + \beta_k) \textstyle\prod_1^k \eta(\alpha_j + \beta_{P(j)}) \eta(\sum_1^k \alpha_j) \\ &\quad \cdot \textstyle\prod_1^k f_{a_j, a'_{P(j)}}(\alpha_j)\, d\alpha_1 \cdots d\alpha_k\, d\beta_1 \cdots d\beta_k + O(B_N^{-k+2} N^{-1}) \end{aligned}$$

if $B_N^{k-1} N \to \infty$ as $B_N \to 0$ and $N \to \infty$. The summation in (144) is over all permutations P on the integers $1, \cdots, k$ and it is understood that

$$\textstyle\sum_1^k \lambda_j = \sum_1^k \mu_j = 0 \quad \text{modulo} \quad 2\pi.$$

Let $X(t)$ be a stationary process satisfying assumption I. Assume that ${}_jf^{(N)}$, $j = 1, \cdots, b$, are spectral estimates of order $k_1 \leqq \cdots \leqq k_b$ whose weight functions satisfy assumption II. Let the bandwidths $B_N^{(j)}$ be such that

$$B_N^{(j)} \to 0, \qquad (B_N^{(j)})^{k_j-1}N \to \infty \tag{145}$$

as $N \to \infty$ with $B_N^{(1)} \leqq \cdots \leqq B_N^{(b)}$. Bandwidths of estimates of the same order are assumed equal. Then, the estimates are asymptotically jointly normal as $N \to \infty$, with estimates of different order asymptotically independent, and estimates of the same order having

$$\begin{aligned}\lim_{N\to\infty} B_N^{k-1} N \operatorname{Cov}[f^{(N)}_{a_1,\cdots,a_k}(\lambda_1, \cdots, \lambda_k), f^{(N)}_{a_1',\cdots,a_k'}(\mu_1, \cdots, \mu_k)] \\ = 2\pi \textstyle\sum_P \eta_{\lambda_1-\mu_{P(1)}} \cdots \eta_{\lambda_k-\mu_{P(k)}} f_{a_1,a'_{P(1)}}(\lambda_1) \cdots f_{a_k,a'_{P(k)}}(\lambda_k) \\ \cdot \int_{-\infty}^{\infty} \cdots \int W(\tau_1, \cdots, \tau_k) W(\tau_{P(1)}, \cdots, \tau_{P(k)}) \delta(\textstyle\sum_1^k \tau_j)\, d\tau_1 \cdots d\tau_k.\end{aligned} \tag{146}$$

For a detailed discussion and interpretation of these results, [4] and [5] should be referred to. There are some marked differences between the case of spectral estimates and probability density estimates. In the higher order spectral case, asymptotic variances have products of 2nd order spectra in their principal term while in the case of probability density estimates, the multivariate density itself (see (34)) itself comes in. The stationarity and the real-valued character of the components of the time series imply that $\sum_1^k w_j = 0$ and $f(w_1, w_2, \cdots, w_k) = f(-w_1, -w_2, \cdots, -w_k)$ so that there are additional symmetries or restraints. This also leads to inhomogeneity on lower dimensional manifolds in w space unless one is careful. Related ideas are used by Brillinger [6] in a discussion of the spectral analysis of point processes.

Acknowledgment. I should like to thank colleagues who were kind enough to read the manuscript and send on their comments; and my thanks to Lillian Johnson for typing it.

REFERENCES

[1] Bartlett, M. S. (1950). Periodogram analysis and continuous spectra. *Biometrika* **37** 1–16.

[2] Bartlett, M. S. (1963). Statistical estimation of density functions. *Sankhyā, Ser. A* **25** 245–254.

[3] Billingsley, P. (1968). *Convergence of Probability Measures*. Wiley, New York.

[4] Brillinger, D. R. and Rosenblatt, M. (1967). Asymptotic theory of estimates of kth order spectra in *Advanced Seminar on Spectral Analysis of Time Series*, ed. B. Harris. 153–188.

[5] Brillinger, D. R. and Rosenblatt, M. (1967). Computation and interpretation of kth order spectra in *Advanced Seminar on Spectral Analysis of Time Series*, ed. B. Harris. 189–232.

[6] Brillinger, D. R. (1970). The spectral analysis of stationary interval functions. Proc. Sixth Berkeley Symp. *Math. Statist. Prob.* To appear.

[7] Cacoullos, T. (1966). Estimation of a multivariate density. *Ann. Inst. Statist. Math.* **18** 179–189.

[8] CARTWRIGHT, D. E. (1968). A unified analysis of tides and surges round North and East Britain. *Philos. Trans. Roy. Soc. London Ser. A* **263** 1–55.

[9] CENCOV, N. N. (1962). Evaluation of an unknown distribution density from observations. *Soviet Math.* **3** 1559–1562.

[10] COOLEY, J. W. and TUKEY, J. W. (1965). An algorithm for the machine computation of complex Fourier Series. *Math. Comp.* **19** 297–301.

[11] CRAMÉR, H. and LEADBETTER, M. R. (1967). *Stationary and Related Processes.* Wiley, New York.

[12] DANIELS, H. (1964). Saddlepoint approximations in statistics. *Ann. Math. Statist.* **3** 631–650.

[13] EPANECHNIKOV, V.A. (1969). Nonparametric estimates of a multivariate probability density. *Theor. Probability Appl.* **14** 153–158.

[14] FARREL, R. H. (1967). On the lack of a uniformly consistent sequence of estimators of a density function. *Ann. Math. Statist.* **38** 471–475.

[15] FRENKIEL, F. N. and KLEBANOFF, P. S. (1965). Two-dimensional probability distribution in a turbulent field. *Phys. Fluids* **8** 2291–2293.

[16] GARSIA, A. M., RODEMICH, E. and RUMSEY, H. Jr. (1970). A real variable lemma and the continuity of paths of some Gaussian processes. *Indiana Univ. Math. J.* **20** 565–578.

[17] GRENANDER, U. and ROSENBLATT, M. (1957). *Statistical Analysis of Stationary Time Series.* Wiley, New York.

[18] GRENANDER, U. and SZEGÖ, G. (1958). *Toeplitz Forms and Their Applications.* Univ. of California Press.

[19] HASSELMAN, K., MUNK, W. and MACDONALD, G. (1963). Bispectra of ocean waves in *Time Series Analysis*, ed. M. Rosenblatt. Wiley, New York.

[20] HAUBRICH, R A. (1965). Earth noise, 5 to 500 millicycles per second. *J. Geophys. Res.* **70** 1415–1427.

[21] HODGES, J. L. Jr. and LEHMANN, E. L. (1956). The efficiency of some nonparametric competitors of the t-test. *Ann. Math. Statist.* **27** 324–335.

[22] HUBER, P. J., KLEINER, B., GASSER, TH. and DUMERMUTH, G. (1971). Statistical methods for investigating phase relations in stationary stochastic processes. To appear in the *IEEE Trans. Audio and Electroacoustics.*

[23] KOLMOGOROV, A. (1963). On the approximation of distributions of sums of independent summands by infinitely divisible distributions. *Sankhyā Ser. A* **25** 159–174.

[24] PARZEN, E. (1961). Mathematical considerations in the estimation of spectra. *Technometrics* **3** 167–190.

[25] PARZEN, E. (1962). On the estimation of a probability density and mode. *Ann. Math. Statist.* **33** 1065–1076.

[26] ROSENBLATT, M. (1956). Remarks on some nonparametric estimates of a density function. *Ann. Math. Statist.* **27** 832–835.

[27] ROSENBLATT, M. (1959). Statistical analysis of stochastic processes with stationary residuals in *H. Cramér Volume*, ed. U. Grenander. 246–275.

[28] ROSENBLATT, M. (1969). Conditional probability density and regression estimates, in *Multivariate Analysis*, ed. Krishnaiah. 25–31.

[29] ROSENBLATT, M. (1970). Density estimates and Markov sequences, in *Nonparametric Techniques in Statistical Inference*, ed. M. Puri. 199–210.

[30] ROUSSAS, G. (1969). Nonparametric estimation in Markov processes. *Ann. Inst. Statist. Math.* **21** 73–87.

[31] SAZANOV, V. V. (1968). On the multidimensional central limit theorem. *Sankhyā Ser. A* **30** 181–204.

[32] SZEGÖ, G.(1959). *Orthogonal Polynomials.* American Mathematical Society, Providence.

[33] TUKEY, J. W. (1959). An introduction to the measurement of spectra, in *H. Cramér Volume*, ed. U. Grenander. 300–330.

[34] VAN ATTA, C. W. and CHEN, W. Y. (1968). Correlation measurements in grid turbulence using digital harmonic analysis. *J. Fluid Mech.* **34** 497–515.

[35] VAN NESS, J. (1966). Asymptotic normality of bispectral estimates. *Ann. Math. Statist.* **37** 1257–1272.

[36] VAN RYZIN, J. (1970). On a histogram method of density estimation. Technical Report No. 226 Dept. of Statistics, Univ. of Wisconsin.

[37] WATSON, G. S. (1969). Density estimation by orthogonal series. *Ann. Math. Statist.* **40** 1496–1498.

[38] WOODROOFE, M. (1967). On the maximum deviation of the sample density. *Ann. Math. Statist.* **38** 475–481.

[39] YAGLOM, A. M. (1966). The influence of fluctuations in energy dissipation on the shape of turbulence characteristics in the inertial interval. *Soviet Physics Dokl.* **11** 26–29.

The Annals of Statistics
1973, Vol. 1, No. 6, 1071-1095

ON SOME GLOBAL MEASURES OF THE DEVIATIONS OF DENSITY FUNCTION ESTIMATES

BY P. J. BICKEL[1] AND M. ROSENBLATT[2]

University of California, Berkeley; University of California, San Diego

We consider density estimates of the usual type generated by a weight function. Limt theorems are obtained for the maximum of the normalized deviation of the estimate from its expected value, and for quadratic norms of the same quantity. Using these results we study the behavior of tests of goodness-of-fit and confidence regions based on these statistics. In particular, we obtain a procedure which uniformly improves the chi-square goodness-of-fit test when the number of observations and cells is large and yet remains insensitive to the estimation of nuisance parameters. A new limit theorem for the maximum absolute value of a type of nonstationary Gaussian process is also proved.

1. Introduction. Let $X_1, X_2, \cdots, X_n$ be independent and identically distributed random variables with continuous density function $f(x)$. By now there are a goodly number of papers on estimation of the density function (see [13] for a bibliography). The class of estimates $f_n(x)$ that we consider are determined by a bounded integrable weight function w,

$$f_n(x) = \frac{1}{nb(n)} \sum_{j=1}^{n} w\left(\frac{x - X_j}{b(n)}\right) \tag{1.1}$$
$$= \int \frac{1}{b(n)} w\left(\frac{x - s}{b(n)}\right) dF_n(s) .$$

In formula (1.1), F_n is the sample distribution function. Also $b(n)$ is a bandwidth that tends to zero as $n \to \infty$ but is such that $n^{-1} = o(b(n))$.

The local properties of such estimates have been discussed extensively. Our object will be to get global measures of how good $f_n(x)$ is as an estimate of $f(x)$. In particular, the asymptotic distribution of the functionals

$$\max_{0 \le x \le 1} |f_n(x) - f(x)|/(f(x))^{\frac{1}{2}}$$

and

$$\int_0^1 \frac{[f_n(x) - f(x)]^2}{f(x)} dx$$

are evaluated under appropriate conditions as $n \to \infty$.

Received November 1971; revised March 1973.

[1] This work was performed while the author was a fellow of the John Simon Guggenheim Memorial Foundation and supported in part by the Office of Naval Research under Contract N00014-69-A0200-1038.

[2] This work was carried out while the author was a fellow of the John Simon Guggenheim Memorial Foundation and supported in part by the Office of Naval Research.

Key words and phrases. Density function estimates, weight function, bandwidth, global measure of deviation, asymptotic distribution, functionals of stochastic process, Gaussian process.

R.A. Davis et al. (eds.), *Selected Works of Murray Rosenblatt*, Selected Works in Probability and Statistics, DOI 10.1007/978-1-4419-8339-8_27, © Springer Science+Business Media, LLC 2011

We shall state two results, one concerned with absolute deviation of the estimate $f_n(x)$ from $f(x)$ and the other with integrated quadratic deviation. They will give some insight into the type of result that is obtained. However, in order to give the result on absolute deviation it is convenient to introduce at this point certain convenient assumptions which we shall refer to as A1, A2, A3, and A4.

A1. The weight function w also assigns mass one to the line and either (a) vanishes outside an interval $[-A, A]$ and is absolutely continuous on $[-A, A]$ with derivative w' or (b) is absolutely continuous on $(-\infty, \infty)$ with derivative w' such that $\int |w'(t)|^k \, dt < \infty$, $k = 1, 2$.

A2. The density f is continuous, positive and bounded.

A3. The function $f^{\frac{1}{2}}$ is absolutely continuous and its derivative $\frac{1}{2}f'/f^{\frac{1}{2}}$ is bounded in absolute value. Moreover,

$$\int_{[|z|\geqq 3]} |z|^{\frac{3}{2}}[\log\log|z|]^{\frac{1}{2}}[|w'(z)| + |w(z)|]\, dz < \infty .$$

A4. The second derivative f'' of f exists and is bounded. Moreover w is symmetric (about 0) and $z^2 w(z)$ is integrable.

We shall simply state a corollary of a main result on absolute deviations which is appealing because it is phrased in a form that is convenient if one wishes to set up a confidence band for the density function.

COROLLARY. *Let assumptions* A1—A4 *be satisfied with* $b(n) = n^{-\delta}$, $\frac{1}{5} < \delta < \frac{1}{2}$. *Then*

$$\lim_{n\to\infty} P\Big[f_n(x) - \Big(\frac{f_n(x)\lambda(w)}{nb(n)}\Big)^{\frac{1}{2}}\Big(\frac{z}{(2\delta\log n)^{\frac{1}{2}}} + d_n\Big)$$

$$\leqq f(x) \leqq f_n(x) + \Big(\frac{f_n(x)\lambda(w)}{nb(n)}\Big)^{\frac{1}{2}}\Big(\frac{z}{(2\delta\log n)^{\frac{1}{2}}} + d_n\Big) \text{ for all } 0 \leqq x \leqq 1\Big] \tag{1.2}$$

$$= e^{-2e^{-z}}$$

where

$$\lambda(w) = \int w^2(t)\, dt$$

and

$$d_n = (2\delta\log n)^{\frac{1}{2}} + \frac{1}{(2\delta\log n)^{\frac{1}{2}}}\Big\{\log\Big(\frac{K_1(w)}{\pi^{\frac{1}{2}}}\Big) + \tfrac{1}{2}[\log\delta + \log\log n]\Big\}$$

if (a) *of* A1 *holds and*

$$K_1(w) = \frac{w^2(A) + w^2(-A)}{2}\Big/\lambda(w) > 0 ,$$

and otherwise

$$d_n = (2\delta\log n)^{\frac{1}{2}} + \frac{1}{(2\delta\log n)^{\frac{1}{2}}}\Big[\log\frac{1}{\pi}\Big(\frac{K_2(w)}{2}\Big)^{\frac{1}{2}}\Big]$$

with

$$K_2(w) = \tfrac{1}{2}\int_{-\infty}^{\infty}[w'(t)]^2\, dt/\lambda(w) .$$

The following result for a quadratic functional is also of some interest. The function $a(x)$ used in the theorem is assumed to be a bounded piece-wise smooth integrable function.

THEOREM. *Let* A1—A4 *hold. Then if* $b(n) = o(n^{-\frac{2}{9}})$, $n^{-\frac{1}{4}}(\log n)^{\frac{1}{2}}(\log\log n)^{\frac{1}{4}} = o(b(n))$ *as* $n \to \infty$,

$$b(n)^{-\frac{1}{2}}[nb(n) \int [f_n(x) - f(x)]^2 a(x)\, dx - \int f(x)a(x)\, dx \int w^2(z)\, dz]$$

is asymptotically normally distributed with mean zero and variance

$$2 \int \left(\int w(x+y)w(x)\, dx\right)^2 dy \int a^2(x)f^2(x)\, dx$$

as $n \to \infty$.

The basic technique in obtaining the results is that of approximating the normalized and centered sample distribution function by an appropriate Brownian motion process on a convenient probability space by using a Skorohod-like imbedding due to Brillinger and Breiman. The details of this approximation and remarks on approximation of other functionals are given in Section 2. The asymptotic theory of the maximal deviation and that of quadratic deviations are developed in Sections 3 and 4 respectively. Some computations on the power of these procedures are also carried out. In particular, we show that a goodness-of-fit test based on a quadratic functional is strictly better than the χ^2 test. There is also an appendix on the asymptotic distribution of the maximal deviation for a type of nonstationary Gaussian process.

2. Approximations. As has been indicated in the introduction our technique is to consider the statistics of interest as functionals of certain stochastic processes on the interval [0, 1] and then to substitute Gaussian processes with the same covariance structure for the latter where possible.

It is convenient to introduce $Z_n^0(\cdot)$ given by

$$Z_n^0(t) = n^{\frac{1}{2}}(F_n^*(t) - t)\,, \qquad 0 \leqq t \leqq 1 \tag{2.1}$$

where $F_n^* = F_n(F^{-1})$ is the empirical distribution of $F(X_1), \cdots, F(X_n)$. This will be approximated by $Z^0(\cdot)$, the Brownian bridge, given by

$$Z^0(t) = Z(t) - tZ(1) \tag{2.2}$$

where Z is a standard Wiener process on [0, 1].

The process $[nb(n)f^{-1}(t)]^{\frac{1}{2}}(f_n(\cdot) - E(f_n(\cdot)))$ is central to our discussion. It can be written as

$$Y_n(t) = b^{-\frac{1}{2}}(n)f^{-\frac{1}{2}}(t) \int_{-\infty}^{\infty} w\left(\frac{t-s}{b(n)}\right) dZ_n^0(F(s))\,. \tag{2.3}$$

Approximations ${}_0Y_n$ and ${}_1Y_n$ to this process are obtained by substituting $Z^0(F(\cdot))$ and $Z(F(\cdot))$ respectively for the random measure in (2.3). The resulting processes are well defined, at least if $\int_{-\infty}^{\infty} w^2(t)\, dF(t) < \infty$.

Two other processes which also arise naturally are given by

$$ {}_2Y_n(t) = [b(n)f(t)]^{-\frac{1}{2}} \int w\left(\frac{t-s}{b(n)}\right)(f(s))^{\frac{1}{2}}\, dZ(s) \tag{2.4}$$

and

$$ {}_3Y_n(t) = [b(n)]^{-\frac{1}{2}} \int w\left(\frac{t-s}{b(n)}\right) dZ(s) \tag{2.5}$$

where Z is a two-sided Wiener process on $(-\infty, \infty)$ (dZ is Wiener measure). The process ${}_3Y_n$ is well defined if $\int w^2(t)\, dt < \infty$, and all the integrals with respect to $dZ^0(F(\cdot))$, $dZ(F(\cdot))$, $dZ(\cdot)$, $dZ^0(\cdot)$ are taken in the L^2 sense (cf. Doob [6] page 426). For convenience, suppose all our processes are realized as random elements taking their values in the space $D[0, 1]$ (cf. [3]). For $x \in D[0, 1]$ let $||x|| = \sup\{|x(t)| : 0 \leqq t \leqq 1\}$. Our approximations rest on the following theorem of Brillinger (1969). (A similar argument appeared simultaneously in Breiman (1969).)

THEOREM. *There exists a probability space (Ω, A, P) on which one can construct versions of Z_n^0 and Z such that*

$$||Z_n^0 - Z^0|| = O_p(n^{-\frac{1}{4}}(\log n)^{\frac{1}{2}}(\log\log n)^{\frac{1}{4}}) . \tag{2.6}$$

From this we can derive

PROPOSITION 2.1. *If the processes Z_n^0, Z^0 are constructed as above and* A1 *and* A2 *hold, then*

$$||Y_n - {}_0Y_n|| = O_p(b^{-\frac{1}{2}}(n)n^{-\frac{1}{4}}(\log n)^{\frac{1}{2}}(\log\log n)^{\frac{1}{4}}) . \tag{2.7}$$

PROOF. Write, using A1,

$$\begin{aligned} Y_n(q) = {} & [b(n)f(q)]^{-\frac{1}{2}}\{-w(A)Z_n^0(F(q - Ab(n))) \\ & + w(-A)Z_n^0(F(q + Ab(n)))\} \\ & + b^{-\frac{3}{2}}(n)f^{-\frac{1}{2}}(q) \int_{-\infty}^{\infty} Z_n^0(F(s))w'\left(\frac{q-s}{b(n)}\right) ds . \end{aligned} \tag{2.8}$$

(The first two terms inside the curly brackets are taken to be 0 in the event A1(b) holds but A1(a) does not.) A similar representation is valid for ${}_0Y_n$ and (2.7) follows.

PROPOSITION 2.2. *If* A2 *holds then*

$$||{}_0Y_n - {}_1Y_n|| = O_p(b^{\frac{1}{2}}(n)) . \tag{2.9}$$

If A2 *and* A3 *hold then*

$$||{}_2Y_n - {}_3Y_n|| = O_p(b^{\frac{1}{2}}(n)) . \tag{2.10}$$

PROOF. From the representation (2.2),

$$\begin{aligned} |{}_0Y_n(q) - {}_1Y_n(q)| = {} & |Z(1)|[b(n)f(q)]^{-\frac{1}{2}} \\ & \int w\left(\frac{q-s}{b(n)}\right) f(s)\, ds = |Z(1)|O(b^{\frac{1}{2}}(n)) . \end{aligned} \tag{2.11}$$

Applying (2.8) and its analogues, if A1(a) holds,

$$\begin{aligned}(2.12)\quad &|{}_2Y_n(q) - {}_3Y_n(q)| \\ &\leqq b^{-\frac{1}{2}}(n)\{[|Z(Ab(n)+q)||[f(Ab(n)+q)/f(q)]^{\frac{1}{2}} - 1| \\ &\quad + |Z(-Ab(n)+q)||[f(-Ab(n)+q)/f(q)]^{\frac{1}{2}} - 1|]\sup_t |w(t)| \\ &\quad + \int |Z(sb(n)+q)||[f(q+sb(n))/f(q)]^{\frac{1}{2}} - 1||w'(s)|\,ds \\ &\quad + \tfrac{1}{2}(b(n)) \int |Z(sb(n)+q)||f'(q+sb(n))|[f(q)f(q+sb(n))]^{-\frac{1}{2}}|w(s)|\,ds\} \\ &= O_p(b^{\frac{1}{2}}(n))\end{aligned}$$

by using A3 and the law of the iterated logarithm for the Wiener process. If A1(b) holds the first two terms vanish and the same argument applies.

To apply these propositions we make the elementary

REMARK. If $\{g_n\}$ is a sequence of functionals on $D[0, 1]$ satisfying Lipschitz conditions such that

$$(2.13)\qquad |g_n(x) - g_n(y)| \leqq M_n||x - y||$$

and A_n, B_n are stochastic processes realizable in D such that $||A_n - B_n|| = o_p(1/M_n)$, then $g_n(A_n)$ converges in law if and only if $g_n(B_n)$ does, and to the same limit.

We shall apply this proposition in the next two sections to the functionals

$$\text{I}\qquad (2|\log b(n)|)^{\frac{1}{2}}\left[\max\left\{\frac{|Y_n(t)|}{(\lambda(w))^{\frac{1}{2}}} : 0 \leqq t \leqq 1\right\} - B([b(n)]^{-1})\right]$$

where B is defined in Theorem A1 and,

$$\text{II}\qquad b^{-\frac{1}{2}}(n)\left[\int_{-\infty}^{\infty} Y_n^2(t)f(t)a(t)\,dt - \int_{-\infty}^{\infty} w^2(t)\,dt\right]$$

where a is an integrable weight function. Evidently, since ${}_1Y_n$ and ${}_2Y_n$ have the same joint laws, we can substitute ${}_3Y_n$ for Y_n in I if A1—A3 hold and

$$(2.14)\qquad o\left(\frac{b(n)}{|\log b(n)|}\right) = n^{-\frac{1}{2}} \log n(\log\log n)^{\frac{1}{2}}$$

and ${}_0Y_n$ can be substituted for Y_n in II if A1 and A2 hold and,

$$(2.15)\qquad o(b(n)) = n^{-\frac{1}{4}}(\log n)^{\frac{1}{2}}(\log\log n)^{\frac{1}{4}}\,.$$

Although we do not pursue this it is clear that the same technique can be applied to other functionals, e.g., a normalized version of the total time in [0, 1] spent by Y_n above a high level (cf. Berman (1971) [2]).

3. The maximum absolute deviation. The first measure of global deviation that we consider is $\tilde{M}_n = \max\{|Y_n(t)| : 0 \leqq t \leqq 1\}$. (There is no loss in considering [0, 1] rather than any other interval on which the density is bounded away from 0 and ∞.) The statistical interest of this functional is twofold as

(i) A convenient way of getting a confidence band for f.
(ii) A test statistic for the hypothesis $H: f = f_0$.

Under (ii) we shall also consider the possibility of testing composite hypotheses, for example, that f is Gaussian. The asymptotic theorem we need to discuss (i), and the behavior of (ii) under the null hypothesis is a consequence of our remarks in Section 2 and Theorem A1 of the appendix.

THEOREM 3.1. *Let w satisfy assumptions* A1—A3 *and*

$$b(n) = n^{-\delta}, \qquad 0 < \delta < \tfrac{1}{2}.$$

Then,

$$P\left[(2\delta \log n)^{\frac{1}{2}}\left(\frac{\tilde{M}_n}{(\lambda(w))^{\frac{1}{2}}} - d_n\right) < x\right] \to e^{-2e^{-x}}, \tag{3.1}$$

where

$$\lambda(w) = \int w^2(t)\, dt \tag{3.2}$$

and

$$d_n = (2\delta \log n)^{\frac{1}{2}} + \frac{1}{(2\delta \log n)^{\frac{1}{2}}}\left\{\log \frac{K_1(w)}{\pi^{\frac{1}{2}}} - \tfrac{1}{2}[\log \delta + \log\log n]\right\} \tag{3.3}$$

where

$$K_1(w) = \frac{w^2(A) + w^2(-A)}{2}\Big/\lambda(w),$$

if $K_1(w) > 0$, and otherwise

$$d_n = (2\delta \log n)^{\frac{1}{2}} + \frac{1}{(2\delta \log n)^{\frac{1}{2}}}\left[\log \frac{1}{\pi}\frac{K_2(w)}{2}\right]$$

where

$$K_2(w) = \tfrac{1}{2}\left[\int [w'(t)]^2\, dt\right]/\lambda(w).$$

REMARK 1. The natural weight function $w(t) = \frac{1}{2}$, $|t| \leqq 1$, $= 0$ otherwise, falls under the first case, while the "optimal" weight function of Epanechnikov (1969) $w(t) = 3/(4(5)^{\frac{1}{2}})(1 - (v^2/5))$ if $|v| \leqq 5^{\frac{1}{2}}$, $= 0$ otherwise, falls under the second.

REMARK 2. A similar result holds if one considers the maximum deviation (rather than absolute deviation) of a density function estimate as in Rosenblatt (1971). However, since one-sided deviations for density functions are unnatural the present result seems more interesting.

REMARK 3. The techniques of proof of this result may readily be adapted to prove limit theorems such as that of Woodroofe (1967) for the maximum deviation observed at an increasing finite number of points.

PROOF. It follows from Propositions 2.1 and 2.2 and the following remark that the limiting behavior of $(2\delta \log n)^{\frac{1}{2}}[(\tilde{M}_n/(\lambda(w))^{\frac{1}{2}}) - d_n]$ is the same as that of $(2 \log b(n))^{\frac{1}{2}}(\max\{|{}_2Y_n(t)|/(\lambda(w))^{\frac{1}{2}} : 0 \leqq t \leqq 1\} - d_n)$. By the similarity transform for the Wiener process, the law

$$L({}_3Y_n(t) : 0 \leqq t \leqq 1) = L\left(\int w\left(\frac{t}{b(n)} - s\right) dZ(s) : 0 \leqq t \leqq 1\right). \tag{3.4}$$

Since $1/(\lambda(w))^{\frac{1}{2}} \int w(t-s)\, dZ(s)$ is a stationary Gaussian process with mean 0 and covariance

$$r(t) = \frac{\int w(s+t)w(s)\, ds}{\lambda(w)}, \tag{3.5}$$

Theorem 3.1 follows from Corollary A.1 provided we show that r satisfies condition (v) and (vi) of Theorem A1 with $\alpha = 1, 2$. That (v) is satisfied is equivalent to Theorem B1. Moreover,

$$\int r^2(t)\, dt = \frac{1}{2\pi} \int |\hat{r}(t)|^2\, dt = \frac{1}{2\pi\lambda^2(w)} \int |\hat{w}(t)|^4\, dt \tag{3.6}$$

where $\hat{}$ denotes Fourier transformation. Since w is integrable and bounded $\hat{w}$ is square integrable and bounded and (vi) must hold.

APPLICATIONS. (i) To obtain a confidence band for f that is simple and explicit it is natural to consider δ such that $E(f_n)$ can be replaced by f. This is true if $\delta > \frac{1}{5}$ and A4 holds. Then,

$$\frac{1}{b(n)} \int w\left(\frac{t-s}{b(n)}\right) f(s)\, ds = f(t) + O(b^2(n)) \tag{3.7}$$

with 0 independent of t. If we now define $Y_n{}^*$ by replacing $E(f_n(t))$ with $f(t)$ we conclude that

$$||Y_n - Y_n{}^*|| = O([nb^5(n)]^{\frac{1}{2}})\,. \tag{3.8}$$

Using the usual approximations we conclude that $\max\{|Y_n{}^*(t)| : 0 \leqq t \leqq 1\}$ behaves like $\tilde{M}_n$ if A4 holds and $\delta > \frac{1}{5}$. In this case inverting as usual we obtain the confidence band

$$\begin{aligned} f &\leqq f_n + \left(\frac{f_n}{nb(n)}\right)^{\frac{1}{2}} c(\alpha)\left(1 + \frac{c^2(\alpha)}{4nb(n)f_n}\right)^{\frac{1}{2}} + \frac{c^2(\alpha)}{2nb(n)} \\ f &\geqq f_n - \left(\frac{f_n}{nb(n)}\right)^{\frac{1}{2}} c(\alpha)\left(1 + \frac{c^2(\alpha)}{4nb(n)f_n}\right)^{\frac{1}{2}} + \frac{c^2(\alpha)}{2nb(n)} \end{aligned} \tag{3.9}$$

where $c(\alpha)$ is given by (3.11). A simpler band is obtained if we further substitute f_n for f in the denominator of Y_n. The resulting process $Y_n{}^{**}$ (say) has

$$\begin{aligned} ||Y_n{}^* - Y_n{}^{**}|| &= O_p\left(\frac{||Y_n{}^*||^2}{(nb(n))^{\frac{1}{2}}}\, ||f_n{}^{-\frac{1}{2}}||\right) \\ &= O_p\left(\frac{\log n}{(nb(n))^{\frac{1}{2}}}\right) \end{aligned} \tag{3.10}$$

if A1—A4 hold and $\frac{1}{5} < \delta < \frac{1}{2}$. The approximate confidence band obtained by looking at the maximum of $|Y_n{}^{**}|$ is given in the introduction (1.2).

There is no choice of δ which asymptotically makes this simple band as thin as possible, i.e. one should choose δ as small as possible. This of course ignores the obvious—the speed with which bias disappears asymptotically depends on δ as does the speed of convergence to the asymptote. However, for fixed n there is an optimal $\delta(n)$ (depending on α) > 0 which for moderate n and small α may be the right thing to use if the choice of bandwidth is free.

(ii) To test $H: f = f_0$ it is natural to compute $\tilde{M}_n$ with $f = f_0$ and reject for large values of the statistic. According to the theorem to obtain approximate level α we should use as cutoff point,

$$c(\alpha) = -[\log|\log(1-\alpha)| - \log 2]\frac{(\lambda(w))^{\frac{1}{2}}}{(2\delta\log n)^{\frac{1}{2}}} + d_n(\lambda(w))^{\frac{1}{2}}. \tag{3.11}$$

Under some assumptions the same cutoff point may be used for testing composite hypotheses of the form $H: f = f_0(\cdot, \theta)$ where θ is an unknown vector parameter by using $\tilde{M}_n$ with an estimate $\hat{\theta}$ substituted for the unknown parameter θ. We need the following assumption.

A5. The estimate $\hat{\theta}$ is such that if $\theta = \theta_0$, for every θ_0,

$$\sup\{|\textstyle\int [f_0(t+sb(n), \hat{\theta}) - f_0(t+sb(n), \theta_0)]w(s)\,ds| : 0 \leqq t \leqq 1\} = o_p([nb(n)\log b(n)]^{-\frac{1}{2}}) \tag{3.12}$$

and

$$||f_0(\cdot, \theta_0) - f_0(\cdot, \hat{\theta})|| = o_p(|\log b(n)|^{-1}).$$

Typically for maximum likelihood and method of moments estimates

$$|\hat{\theta} - \theta_0| = O_p(n^{-\frac{1}{2}}). \tag{3.13}$$

If, moreover, $\theta = (\theta^{(1)}, \cdots, \theta^{(k)})$, $\partial f_0/\partial\theta^{(j)}$ is bounded for θ in a neighborhood of θ_0, all x, and $1 \leqq j \leqq k$, it is easy to see that A5 holds. To see that A5 is the needed assumption again introduce a process $\bar{Y}_n$ with $E_\theta(f_n)$ replaced by $E_{\hat{\theta}}(f_n)$ and $(f(\cdot, \theta))^{\frac{1}{2}}$ replaced in the denominator of Y_n by $(f(\cdot, \hat{\theta}))^{\frac{1}{2}}$. Then

$$||Y_n - \bar{Y}_n|| = o_p([\log b(n)]^{-\frac{1}{2}}) \tag{3.14}$$

and the result follows.

To make local power calculations on the test of the simple hypothesis described above we need to consider the behavior of $\tilde{M}_n$ (calculated under f_0) for a sequence of alternatives of the form,

$$g_n(x) = f_0(x) + \gamma_n\eta(x) + o(\gamma_n) \tag{3.15}$$

where g_n satisfy A2—A3 uniformly in n, $\gamma_n \downarrow 0$ at a suitable rate, and $o(\gamma_n)$ is uniform in x on $[0, 1]$. (Note that η must be continuous on $[0, 1]$.) Denote probabilities calculated under g_n by P_n. Our basic result is,

THEOREM 3.2. *Suppose that g_n are as above. Let w satisfy* A1—A3 *and define $\tilde{M}_n$ in terms of f_0. Let*

$$\gamma_n = n^{-\frac{1}{2}+\delta/2}[2\delta\log n]^{-\frac{1}{4}}.$$

Then,

$$P_n\left[(2\delta\log n)^{\frac{1}{2}}\left(\frac{\tilde{M}_n}{(\lambda(w))^{\frac{1}{2}}} - d_n\right) < x\right] \to \exp[-s(\eta)e^{-x}] \tag{3.16}$$

where

$$s(\eta) = \textstyle\int_0^1 \{\exp[\eta(t)/(f_0(t)\lambda(w))^{\frac{1}{2}}] + \exp[-\eta(t)/(f_0(t)\lambda(w))^{\frac{1}{2}}]\}\,dt. \tag{3.17}$$

This result follows from Theorem A1 quite readily.

One interesting consequence of this formula is that our test is asymptotically strictly unbiased for such alternatives. The reason is that $s(\eta) \geqq 2$ with $s(\eta) > 2$ unless $\eta = 0$ and the family of distributions $e^{-\theta e^{-x}}$ is an exponential family in θ.

Unfortunately these tests are asymptotically inadmissible (have Pitman efficiency 0) when compared to the test based on the quadratic functional of the next section based on the same w and $b(n)$. The reason is that alternatives there may be permitted to come in to f_0 at rate $n^{-\frac{1}{2}+\delta/4}$ rather than $n^{-\frac{1}{2}+\delta/2}$. However, this test for moderate sample sizes and some alternatives may well be preferable.

In analogy to the confidence band situation it would appear that maximum power is achieved by taking δ as small as possible. However, consideration of the approximation arguments suggests that $s(\eta_n)$ is a better measure of the "true shift" than $s(\eta)$ where,

$$\eta_n = (g_n - f_0)(2nb(n) \log b(n))^{\frac{1}{2}} . \tag{3.18}$$

Of course, $s(\eta_n)$ may well be maximized for $\delta > 0$. In all of these questions it would be desirable to have some small sample Monte Carlo explorations.

4. Quadratic functionals. We are interested in the behavior of the functional,

$$T_n = nb(n) \int_{-\infty}^{\infty} [f_n(x) - E(f_n(x))]^2 a(x)\, dx = \int_{-\infty}^{\infty} L_n^2(x) a(x)\, dx , \tag{4.1}$$

where $L_n = f^{\frac{1}{2}} Y_n$ and a is integrable. We have already remarked that if A1 and A2 hold and (say) $b_n = n^{-\delta}$, $\delta < \frac{1}{4}$, then.

$$|T_n - \int {}_0L_n^2(x) a(x)\, dx| = o(b^{\frac{1}{2}}(n)) . \tag{4.2}$$

Moreover, if a is bounded as well as integrable and w and f are bounded, we can replace ${}_0L_n$ by ${}_1L_n = f^{\frac{1}{2}}{}_1Y_n$ and hence by ${}_2L_n = f^{\frac{1}{2}}{}_2Y_n$. To see this note that,

$$\begin{aligned} &|\int ({}_1L_n^2(x) - {}_0L_n^2(x)) a(x)\, dx| \\ &= \left|\int \frac{1}{b(n)} \left\{\left(Z(1) \int w\left(\frac{t-s}{b(n)}\right) f(s)\, ds\right)^2 \right.\right. \\ &\qquad \left.\left. - 2Z(1) \int w\left(\frac{t-s}{b(n)}\right) dZ(F(s)) \int w\left(\frac{t-s}{b(n)}\right) f(s)\, ds\right\} a(t)\, dt\right| \\ &\leqq Z(1)^2 b(n) \sup_x |f(x)| \int |a(t)\, dt| \\ &\qquad + 2|Z(1)| b(n)| \int (\int w(y) c(s + b(n)y) a(s + b(n)y)\, dy)\, dZ(F(s))| \end{aligned} \tag{4.3}$$

where

$$c(t) = \int w(y) f(t - b(n)y)\, dy .$$

But,

$$\begin{aligned} &E(\int (\int w(y) c(s + b(n)y) a(s + b(n)y)\, dy)\, dZ(F(s)))^2 \\ &\qquad = \int (\int w(y) c(s + b(n)y) a(s + b(n)y)\, dy)^2\, dF(s) \end{aligned} \tag{4.4}$$

is bounded.

By (4.3) and (4.4),

$$|T_n - \int {}_2L_n^2(x) a(x)\, dx| = O_p(b(n)) . \tag{4.5}$$

(The infinite range poses no problem since we are approximating L_n rather than the normalized Y_n.)

The following lemma lets us determine the characteristic function of a quadratic functional

$$Z = \int Y(x)^2 a(x)\, dx \tag{4.6}$$

of a Gaussian process $Y(x)$ under appropriate conditions.

LEMMA 4.1. *Let $Y(x)$, $EY(x) \equiv 0$, be a Gaussian process with bounded, uniformly continuous covariance function $r(x, y)$. If $a(x)$ is a piecewise smooth integrable function, the quadratic functional* (4.6) *has characteristic function formally given by*

$$E(e^{itZ}) = \exp\{\textstyle\sum_{k=1}^{\infty} 2^{k-1}(it)^k c_k/k\} \tag{4.7}$$

with

$$c_k = \int \cdots \int r(x_1, x_2) r(x_2, x_3) \cdots r(x_k, x_1) a(x_1) a(x_2) \cdots a(x_k)\, dx_1 \cdots dx_k\,.$$

The representation (4.6) is valid for $|t| < 1/2M$ where $M = \|r\| \int |a(t)|\, dt$. The quantities $(k-1)!\, 2^{k-1} c_k$ are of course the cumulants of (4.6).

The lemma is obtained by considering the form

$$\textstyle\sum_{j=1}^{n} \bar{\mathbf{Y}}_j^2 a_j \tag{4.8}$$

in jointly Gaussian random variables $\bar{\mathbf{Y}}_j$, $E\bar{\mathbf{Y}}_j \equiv 0$ with the a_j's constants. Let R be the covariance matrix of the $\bar{\mathbf{Y}}_j$'s with A the diagonal matrix with diagonal entries a_j. The characteristic function of (4.8) is then

$$|1 - 2itRA|^{-\frac{1}{2}} = \textstyle\prod_{j=1}^{n} (1 - 2\lambda_j it)^{-\frac{1}{2}} = \exp\{\sum_{k=1}^{\infty} 2^{k-1}(it)^k \operatorname{tr}(RA)^k/k\}\,,$$

at least if $|t| < 1/2 \operatorname{tr}(RA)$.

Here $\operatorname{tr}(M)$ denotes the trace of M, $|M|$ its determinant and $\lambda_1, \cdots, \lambda_n$ are the eigenvalues of RA. Lemma 4.1 is then obtained by going through an appropriate limiting operation.

The covariance function of the Gaussian process ${}_2L_n(x)$ can be written

$$\begin{aligned} r(x, y) &= \int w(z) w(\alpha + z) f(x - b(n)z)\, dz \\ &= f(x) \int w(z) w(\alpha + z)\, dz + O(b(n)) \end{aligned} \tag{4.9}$$

where

$$\alpha = (y - x)/(b(n))$$

and $O(b(n))$ is independent of x if f is bounded and has a uniformly bounded derivative and $w^2(z)(1 + |z|)$ is integrable. Then

$$E(\textstyle\int {}_2L_n(x)^2 a(x)\, dx) = \int f(x) a(x)\, dx \int w(z)^2\, dz + O(b(n))\,. \tag{4.10}$$

Similarly if a is bounded as well as integrable and w is bounded and f is as above, the variance of $\int {}_2L_n^2(x) a(x)\, dx$ is $2b(n) \int [w * \bar{w}(u)]^2\, du \int a^2(x) f^2(x)\, dx$ to first order as $n \to \infty$, where $\bar{w}(t) = w(-t)$ and $*$ denotes convolution. A similar argument shows that under the same conditions the kth cumulant of $\int {}_2L_n^2(x) a(x)\, dx$ equals to first order $(k-1)!\, 2^{k-1} b^{k-1}(n) [w * \bar{w}]^{(k)}(0) \int a^k(x) f^k(x)\, dx$ as $n \to \infty$ where the superscript (k) indicates that $w * \bar{w}$ is convoluted with itself k times. As a result we have the following theorem which actually holds under the weaker assumptions indicated above.

THEOREM 4.1. *Let* A1—A3 *hold and suppose that* a *is integrable piecewise continuous and bounded. Suppose moreover that* (2.16) *holds. Then* $b^{-\frac{1}{2}}(n)(T_n - (\int f(x)a(x)\,dx) \int w^2(z)\,dz)$ *is asymptotically normally distributed with mean* 0 *and variance* $2(w * \bar{w})^{(2)}(0) \int a^2(x)f^2(x)\,dx$ *as* $n \to \infty$.

A particular case of interest for the application of the theorem is that in which as in Section 3, $a(x)$ vanishes off an interval, say [0, 1], and one sets $a(x) = f(x)^{-1}$ on [0, 1]. In this case under A1—A3, T_n is asymptotically Gaussian with mean $\int w^2(z)\,dz$ and variance $2b(n)(w * \bar{w})^{(2)}(0)$.

The statistic

$$\tilde{T}_n = nb(n) \int [f_n(x) - f(x)]^2 a(x)\,dx \tag{4.11}$$

is probably of greater interest than that considered in Theorem 4.1. However, let us expand $\tilde{T}_n$ in the form

$$\begin{aligned} nb(n)\{&\int [f_n(x) - Ef_n(x)]^2 a(x)\,dx \\ &+ 2 \int [f_n(x) - Ef_n(x)][Ef_n(x) - f(x)]a(x)\,dx \\ &+ \int [Ef_n(x) - f(x)]^2 a(x)\,dx\}. \end{aligned} \tag{4.12}$$

Let w be positive and symmetric about zero with

$$c = \int w(u)u^2\,du < \infty\,. \tag{4.13}$$

Then if $n^{-1} = O(b(n))$, $b(n) \to 0$ as $n \to \infty$, A1 holds and f has a continuous bounded second derivative, the second term of (4.12) may, by the usual approximation arguments, be shown to be asymptotically normal with mean zero and variance

$$n^{-1}b(n)^4c^2 \int f''(x)^2a(x)^2f(x)\,dx \tag{4.14}$$

to the first order. Also, under the same conditions, the last term of (4.12) can be shown to be

$$b(n)^4c^2 \int f''(x)^2a(x)\,dx \tag{4.15}$$

to the first order. Then $[b(n)]^{-\frac{1}{2}}[\tilde{T}_n - T_n] = o_p(1)$ if and only if $b(n) = o(n^{-\frac{2}{9}})$. (The term (4.14) is then negligible.) The theorem quoted in the introduction follows.

APPLICATIONS. An explicit confidence band is hard to obtain from Theorem 4.1 and the theorem of the introduction. However we can test $H: f = f_0$ at (approximate) level α by calculating T_n for $f = f_0$ and rejecting when $T_n \geqq d(\alpha)$ where by Theorem 4.1

$$\begin{aligned} d(\alpha) = &[\int f_0(x)a(x)\,dx][\int w^2(z)\,dz] \\ &+ b^{\frac{1}{2}}(n)\Phi^{-1}(1-\alpha)/[2(w * \bar{w})^{(2)}(0) \int a^2(x)f_0^2(x)\,dx]^{\frac{1}{2}}\,. \end{aligned} \tag{4.16}$$

As in Section 3 it is easy to see that in testing $H: f = f_0(\cdot, \theta)$ where θ is an unknown vector parameter we may use T_n with f replaced by $f_0(\cdot, \hat{\theta})$ and $d(\alpha)$ with f_0 replaced by $f_0(\cdot, \hat{\theta})$, provided that A6 below holds.

A6. For each θ_0, $(\partial^2 f(x, \theta)/\partial\theta^{(i)}\partial\theta^{(j)})$ is bounded in absolute value for all θ in a neighborhood of θ_0 and all x, i, j. Moreover, if θ_0 is true,

$$|\hat{\theta} - \theta_0| = o_p([nb(n)]^{-\frac{1}{2}}) . \tag{4.17}$$

To see this, taking $k = 1$ for simplicity, expand as in (4.12) and note that it suffices to show that

$$\int [f_n(x) - E_{\theta_0}(f_n(x))][E_{\theta_0}(f_n(x)) - E_{\hat{\theta}}(f_n(x))]a(x)\, dx = o_p([nb^{\frac{1}{2}}(n)]^{-1}) \tag{4.18}$$

and

$$\int [E_{\theta_0}(f_n(x)) - E_{\hat{\theta}}(f_n(x))]^2 a(x)\, dx = o_p([nb^{\frac{1}{2}}(n)]^{-1}) . \tag{4.19}$$

Taylor expanding the integral in (4.18) about θ_0 we obtain a first term

$$(\hat{\theta} - \theta_0) \int [f_n(x) - E_{\theta_0}(f_n(x))]\left[\int \frac{\partial f(x + b(n)z, \theta)}{\partial\theta}\bigg|_{\theta=\theta_0} w(z)\, dz\right] a(x)\, dx$$

which is $O_p(|\hat{\theta} - \theta_0|n^{-\frac{1}{2}})$, and a second term which is $O_p([nb(n)]^{-\frac{1}{2}}(\hat{\theta} - \theta_0)^2)$, and (4.18) follows. A similar argument yields (4.19).

To make local power calculations we again suppose g_n is as in (3.15) with g_n satisfying A2—A3 uniformly in n and $o(\gamma_n)$ uniform in x and η is bounded.

THEOREM 4.2. *Let g_n be as above, w satisfy* A1—A4, *a be integrable piecewise continuous and bounded, $b(n) = n^{-\delta}$, $\delta < \frac{1}{4}$, $\gamma_n = n^{-\frac{1}{2}+\delta/4}$. Define T_n in terms of f_0. Then,*

$$b^{-\frac{1}{2}}(n)(T_n - [\int f_0(x)a(x)\, dx] \int w^2(z)\, dz) \tag{4.20}$$

is asymptotically normally distributed with mean $\int \eta^2(x)a(x)\, dx$ and variance

$$2(w * \bar{w})^{(2)}(0) \int a^2(x)f_0^2(x)\, dx .$$

The proof is straightforward. As in Section 3 it follows that the test which rejects when T_n is $\geq d(\alpha)$ is locally strictly unbiased if $a(x) > 0$ for all x.

Also as before the asymptotics lead to choosing δ as large as possible and again this conclusion is shaken if one uses the better approximation to the asymptotic mean, $\int \eta_n^2(x)a(x)\, dx$ where

$$\eta_n(x) = \int w(z)[g_n(x + b(n)z) - f_0(x + b(n)z)]\, dz . \tag{4.21}$$

It is also clear that for fixed δ we can let $\lambda_n \to 0$ more quickly than for the sup functional and still get power. Thus the Pitman efficiency of the T_n test to the $\tilde{M}_n$ test for the same δ is ∞.

Suppose that f_0 is the uniform density on $[0, 1]$ an effect we can always achieve by applying the probability integral transformation to our observations before making the test. Let $a(x) = 1$ on $[0, 1]$ and 0 otherwise, w be the uniform density on $[-\frac{1}{2}, \frac{1}{2}]$. Neglecting fringe effects we may then write

$$T_n = \int_0^1 \frac{(N[t - \frac{1}{2}b(n), t + \frac{1}{2}b(n)] - b(n))^2}{nb(n)} dt \tag{4.22}$$

where $N[x, y]$ is the number of observations falling in the interval $[x, y]$. A related statistic for testing uniformity on the circle was considered by Watson in [15]. This is, of course, very similar to the χ^2 statistic for the problem based on the cells $[0, b(n)]$, $[b(n), 2b(n)]$, $\cdots$, $[(K-1)\frac{1}{2}b(n), (K+1)\frac{1}{2}b(n)]$ given by,

$$\chi_n^2 = \sum_{k=1}^{*K} \frac{(N[\frac{1}{2}kb(n) - \frac{1}{2}b(n), \frac{1}{2}kb(n) + \frac{1}{2}b(n)] - nb(n))^2}{nb(n)} \tag{4.23}$$

where $(K+1)\frac{1}{2}b(n) \leqq 1 < (K+2)\frac{1}{2}b(n)$ and $\sum^*$ is a sum over odd index.

Now we can write,

$$\chi_n^2/K = nb(n) \int_0^1 (f_n(t) - E(f_n(t)))^2 \, dA_n(t) \tag{4.24}$$

where A_n places mass $1/K$ at each of the points $\frac{1}{2}b(n), \cdots, K\frac{1}{2}b(n)$. It is easy to see that the arguments leading to Theorem 4.2 apply to functionals of this type also and that under the conditions of that theorem, if $b(n) = n^{-\delta}$, $\delta < \frac{1}{2}$, χ_n^2/K is asymptotically normal with the natural parameters $E(\chi_n^2/K)$ and $\operatorname{Var}(\chi_n^2/K)$.

This result is, of course, known. A rigorous proof under milder conditions but using a different method may be found in Steck (1957). Now

$$\begin{aligned} E\left(\frac{\chi_n^2}{K}\right) &= 1 + \frac{1}{K} \sum_{j=1}^{*K} nb(n) \left(1 - \frac{1}{b(n)} \int_{(j-1)b(n)/2}^{(j+1)b(n)/2} g_n(x)\, dx\right)^2 \\ &= 1 + nb(n)\gamma_n^2 \frac{1}{K} \sum_{j=1}^{*K} \left[\frac{1}{b(n)} \int_{(j-1)b(n)/2}^{(j+1)b(n)/2} \eta(x)\, dx\right]^2 + o(nb(n)\gamma_n^2) \end{aligned} \tag{4.25}$$

$$\operatorname{Var}\left(\frac{\chi_n^2}{K}\right) = \frac{1}{K} \operatorname{Var}\left(\frac{N^2[0, b(n)]}{nb(n)}\right) + o\left(\frac{1}{K}\right) = \frac{2}{K} + o\left(\frac{1}{K}\right). \tag{4.26}$$

Thus if we take $\gamma_n = n^{-\frac{1}{2}+\delta/4}$ as in Theorem 4.2, under g_n the statistics

$$W_n = b^{-\frac{1}{2}}(n)\left(\frac{\chi_n^2}{K} - 1\right) \tag{4.27}$$

have a limiting Gaussian distribution with mean $\int_0^1 \eta^2(x)\, dx$ and variance 2. Under the same circumstances the asymptotic mean of $b^{-\frac{1}{2}}(n)(T_n - 1)$ with T_n given by (4.22) is also $\int_0^1 \eta^2(x)\, dx$ while its asymptotic variance is,

$$2w^{(4)}(0) = 2\int_{-1}^{1} (1 - |t|)^2 \, dt = \tfrac{4}{3}\,. \tag{4.28}$$

The Pitman efficiency of the tests based on T_n to those based on W_n is thus by the usual calculations,

$$e(T_n, W_n) = (\tfrac{3}{2})^{\frac{1}{2}-\delta} \tag{4.29}$$

and thus at least $(\frac{3}{2})^{\frac{1}{2}} = 1.217$ on the range $\delta > 0$. For the Mann–Wald (1942) prescription $\delta = \frac{2}{5}$ we get an efficiency of 1.292.

Although as we have seen these asymptotic calculations are to be taken with a grain of salt we feel that the procedure T_n has promise as a competitor to the χ^2 test, at least for moderate sample sizes.

Acknowledgment. We are grateful to S. Berman and J. Pickands, III for providing us with preprints of some of the papers cited in the list of references.

5. Appendix A. On the extrema of some nonstationary Gaussian processes. Let $Y_T(\cdot)$ be a sequence of separable Gaussian processes with mean $\mu_T(\cdot)$ such that $Y_T(\cdot) - \mu_T(\cdot)$ is stationary. Let $r(\cdot)$ be the covariance function of Y_T,

$$M_T = \max\{Y_T(t) : 0 \leqq t \leqq T\}, \qquad m_T = \min\{Y_T(t) : 0 \leqq t \leqq T\}.$$

Let $b_T(t) = \mu_T(t)(2\log T)^{\frac{1}{2}}$.

THEOREM A1. *Suppose that,*

(i) $b_T(t)$ *is uniformly bounded in* t *and* T *on* $[0, T]$ *as* $T \to \infty$.
(ii) $b_T(t) \to b(t)$ *uniformly on* $[0, T]$ *as* $T \to \infty$.
(iii) $T^{-1}\lambda[t : b(t) \leqq x, 0 \leqq t \leqq T] \to \eta(x)$ *the* cdf *of a probability measure as* $T \to \infty$. *(λ as usual denotes Lebesgue measure.)*
(iv) $b(\cdot)$ *is uniformly continuous on* R.
(v) $r(t) = 1 - C|t|^\alpha + o(|t|^\alpha)$, $0 < \alpha \leqq 2$, *as* $t \to \infty$.
(vi) $\int_0^\infty r^2(t)\, dt < \infty$.

Let

$$B(t) = (2\log t)^{\frac{1}{2}} + \frac{1}{(2\log t)^{\frac{1}{2}}} \times \left\{\left(\frac{1}{\alpha} - \frac{1}{2}\right)\log\log t + \log (2\pi)^{-\frac{1}{2}}(C^{1/\alpha} H_\alpha 2^{(2-\alpha)/2\alpha})\right\}$$

where

$$H_\alpha = \lim_{T\to\infty} \frac{1}{T} \int_0^\infty e^s P[\sup_{0\leq t\leq T} Y(t) > s]\, ds$$

and Y *is a Gaussian process with,*

$$E(Y(t)) = -|t|^\alpha, \qquad \operatorname{Cov}(Y(t_1), Y(t_2)) = |t_1|^\alpha + |t_2|^\alpha - |t_1 - t_2|^\alpha. \tag{A.1}$$

Then,

$$U_T = (2\log T)^{\frac{1}{2}}(M_T - B(T)) \quad \textit{and} \quad V_T = -(2\log T)^{\frac{1}{2}}(m_T + B(T))$$

are asymptotically independent with,

$$P[U_T < z] \to e^{-\lambda_1 e^{-z}}, \qquad P[V_T < z] \to e^{-\lambda_2 e^{-z}}; \tag{A.2}$$

where,

$$\lambda_1 = \int e^z\, d\eta(z), \qquad \lambda_2 = \int e^{-z}\, d\eta(z). \tag{A.3}$$

An immediate consequence of Theorem A1 is,

COROLLARY A1. *If $\tilde{M}_T = \max\{|Y_T(t)| : 0 \leqq t \leqq T\}$ then under the conditions of the theorem,*

$$P[(2 \log T)^{\frac{1}{2}}(\tilde{M}_T - B(T)) < x] \to \exp[-(\lambda_1 + \lambda_2)e^{-x}]. \tag{A.4}$$

Note. $\lambda_1 + \lambda_2 \geqq 2$ with strict inequality unless η concentrates at 0.

COROLLARY A2. *Let $Y_0(t) - \mu(t)$ be a stationary mean 0 Gaussian process with covariance function $r(t)$ satisfying the conditions of the theorem. Suppose that $b(t) = (2 \log (t + 2))^{\frac{1}{2}}\mu(t)$ is a bounded uniformly continuous function of t and that $b(\cdot)$ satisfies condition* (iii) *of the theorem. Then,*

$$P[(2 \log T)^{\frac{1}{2}}(\max\{Y_0(s) : 0 \leqq s \leqq T\} - B(T)) < x] \to e^{-\lambda_1 e^{-x}}. \tag{A.5}$$

Similar assertions hold about the independence of maximum and minimum and the asymptotic distribution of the minimum.

This corollary may be viewed as complementing Theorem 4.1 of Qualls and Watanabe (1971) which deals with the extrema of a mean 0 process whose covariance function is asymptotically locally approximated by that of a stationary process while we deal with a process which is stationary when centered and asymptotically stationary.

The constants H_1 and H_2 are the only ones known explicitly. They are given by $H_1 = 1$, $H_2 = \pi^{-\frac{1}{2}}$ (cf. [11]).

PROOF OF COROLLARY A2. Define,

$$\begin{aligned} Y_T(t) &= Y_0(t) \qquad \text{on} \quad [\varepsilon(T), T] \\ &= Y_0(t) + ([\log(t+2)/\log(T+2)]^{\frac{1}{2}} - 1)\mu(t) \quad \text{otherwise} \end{aligned} \tag{A.6}$$

where $\varepsilon(T) = o(T)$, $\log \varepsilon(T) \sim \log T$. Evidently, $(2 \log T)^{\frac{1}{2}}E(Y_T(t)) \to b(t)$ uniformly and

$$\begin{aligned} &\Big| P\Big[\max\{Y_T(s) : 0 \leqq s \leqq T\} < \frac{x}{(2 \log T)^{\frac{1}{2}}} + B(T)\Big] \\ &\quad - P\Big[\max\{Y_0(s) : 0 \leqq s \leqq T\} < \frac{x}{(2 \log T)^{\frac{1}{2}}} + B(T)\Big]\Big| \\ &\leqq 2P\Big[\max\{Y_0(s) - E(Y_0(s)) : 0 \leqq s \leqq \varepsilon(T)\} \geqq \frac{x}{(2 \log T)^{\frac{1}{2}}} \\ &\qquad - K + B(T)\Big] \end{aligned} \tag{A.7}$$

where $K = \max\{\mu(t) : 0 \leqq t \leqq \varepsilon(t)\}$. Since $B(\varepsilon(T)) - B(T) \to -\infty$ the term on the right of (A.7) tends to 0 by the theorem. □

PROOF OF THEOREM A1. The theorem is argued much as Theorem 3.1 of Pickands (1969). We refer the reader to this paper and Berman (1971) for the details of the argument.

LEMMA A1. *Let* $\psi(x) = \phi(x)/x$ *where* ϕ *is the standard normal density. Let* $C = 1$, $x = x(T) = B(T) + z_1/(2\log T)^{\frac{1}{2}}$. *Then for* $a > 0$,

$$\begin{aligned} P[\max\{Y_T(t+akx^{-2/\alpha}), 0 \leqq k \leqq n\} > x] \\ = \psi(x)e^{b(t)}H_\alpha(n,\alpha) + o(\psi(x)) \end{aligned} \tag{A.8}$$

$$\begin{aligned} P[\min\{Y_T(t+kax^{-2/\alpha}), 0 \leqq k \leqq n\} < -x] \\ = \psi(x)e^{-b(t)}H_\alpha(n,a) + o(\psi(x)) \end{aligned}$$

as $T \to \infty$ *uniformly in* $0 \leqq t \leqq T$ *where*

$$H_\alpha(n,a) = \textstyle\int_{-\infty}^{\infty} e^s P[\max\{Y(ka): 0 \leqq k \leqq n\} > s]\,ds\,. \tag{A.9}$$

Moreover, if $y = y(T) = B(T) + z_2/(2\log T)^{\frac{1}{2}}$ *then*

$$\begin{aligned} P[\max\{Y_T(t+kax^{-2/\alpha}): 0 \leqq k \leqq n\} > x, \\ \min\{Y_T(t+kax^{-2/\alpha}): 0 \leqq k \leqq n\} < -y] \\ = o(\psi(x)) = o(\psi(y))\,, \end{aligned} \tag{A.10}$$

uniformly in $0 \leqq t \leqq T$. (Throughout, k may take on integer values only.)

PROOF. As in [11] consider the "local" process

$$\tilde{Y}_T(s) = x(Y_T(t+sx^{-2/\alpha}) - \mu_T(t) - x)\,. \tag{A.11}$$

$$\begin{aligned} P[\max\{Y_T(t+akx^{-2/\alpha}): 0 \leqq k \leqq n\} < x] \\ = \textstyle\int_{-\infty}^{\infty} \gamma(z)P[\max\{\tilde{Y}_T(ka): 0 \leqq k \leqq n\} > -x\mu_T(t) \mid \tilde{Y}_T(0) = z]\,dz \end{aligned} \tag{A.12}$$

where γ is the density of $\tilde{Y}_T(0)$,

$$\gamma(z) = \frac{1}{x}\phi\left(x + \frac{z}{x}\right) = \psi(x)\exp[-z - z^2/2x^2]\,. \tag{A.13}$$

It is easy to see using (ii) and (iv) that the finite dimensional conditional distributions of $\tilde{Y}_T(s)$ given $\tilde{Y}_T(0) = z$ converge uniformly in t to those of the process $Y(s) + z$ where Y is given by (A.1). Arguing as in [11] the first part of (A.8) follows since $x\mu_T(t) \to b(t)$ uniformly as required. By considering $-Y_T$ we obtain the second part. To prove (A.10) let A be the event whose probability is being estimated. Then,

$$\begin{aligned} P\left(A, Y_T(t) > x - \frac{1}{x^{\frac{1}{2}}} + \mu_T(t)\right) \\ \leqq \textstyle\int_{-x^{\frac{1}{2}}}^{\infty} \gamma(z)P[\min\{\tilde{Y}_T(ka): 0 \leqq k \leqq n\} - z \\ \leqq -z - x(y + x + \mu_T(t)) \mid \tilde{Y}_T(0) = z]\,dz \\ \leqq \psi(x)\textstyle\int_{-\infty}^{x^{\frac{1}{2}}} e^z P[\min\{\tilde{Y}_T(ka) + z: 0 \leqq k \leqq n\} \\ < z - x(y + x + \mu_T(t)) \mid \tilde{Y}_T(0) = -z]\,dz \\ \leqq \psi(x)\textstyle\sum_{k=0}^{n}\{\int_{-\infty}^{0} P[\tilde{Y}_T(ka) + z < z \\ - x(y + x + \mu_T(t)) \mid \tilde{Y}_T(0) = -z]\,dz \\ + x^{\frac{1}{2}}\exp x^{\frac{1}{2}} \max\{P[\tilde{Y}_T(ka) + z \\ < x^{\frac{1}{2}} - x(y + x + \mu_T(t)) \mid \tilde{Y}_T(0) = -z]: 0 \leqq z \leqq x^{\frac{1}{2}}\}\}]\,. \end{aligned} \tag{A.14}$$

Applying the usual estimate $\Phi(z) \leqq \psi(|z|)$ for $z \leqq 0$ we conclude that the left-hand side of (A.14) is $o(\psi(x))$. Similarly,

$$P\left(A, Y_T(t) - \mu_T(t) < -y + \frac{1}{y^{\frac{1}{2}}}\right) = o(\psi(y)) . \tag{A.15}$$

Finally,

$$\begin{aligned} &P\left(A, -y + \frac{1}{y^{\frac{1}{2}}} \leqq Y_T(t) - \mu_T(t) \leqq x - \frac{1}{x^{\frac{1}{2}}}\right) \\ &\leqq \textstyle\int_{-\infty}^{-x^{\frac{1}{2}}} \gamma(z) P[\max\{\tilde{Y}_T(ka): 0 \leqq k \leqq n\} > -x\mu_T(t) \mid \tilde{Y}(0) = z]\, dz \\ &\leqq \psi(x) \textstyle\int_A^\infty e^z P[\max\{\tilde{Y}_T(ka): 0 \leqq k \leqq n\} > z]\, dz \end{aligned} \tag{A.16}$$

for every $A < \infty$.

The final statement of the lemma follows.

LEMMA A2. *The assertion of Lemma* A1 *remains valid if* $a = 1$, k *is permitted to range over all values in* $[0, n]$ *and* $H_\alpha(n, a)$ *is replaced by*

$$\bar{H}_\alpha(n) = \cdot \textstyle\int_{-\infty}^{\infty} e^t P[\max\{Y(s): 0 \leqq s \leqq n\} > t]\, dt \tag{A.17}$$

PROOF. We prove the analogue of (A.8); the other assertions follow similarly. We need to check that uniformly in T,

(a) The conditional distributions of the continuous processes $\tilde{Y}_T(t) - z$ given $\tilde{Y}_T(0) = z$ converge weakly (in the sense of Prohorov) to that of $Y(\cdot)$,

(b) $$P[\max\{\tilde{Y}_T(k): 0 \leqq k \leqq n\} > x\mu_T(t) \mid \tilde{Y}_T(0) = z] \leqq g(z)$$

where $\int e^{-z} g(z)\, dz < \infty$.

To see that (a) holds it suffices to note that,

$$\operatorname{Var}[(\tilde{Y}_T(s_1) - \tilde{Y}_T(s_2)) \mid \tilde{Y}_T(0) = z] \leqq C |s_1 - s_2|^\alpha \tag{A.18}$$

and then apply Billingsley [3] page 95. To see that (b) is valid use the estimate of Fernique (1970) given below on the tails of $\max\{|\tilde{Y}_T(k)|: 0 \leqq k \leqq n\}$.

LEMMA. *Let* $Z(\cdot)$ *be a Gaussian process on* $(0, 1)$. *Let a be such that* $P[\|Z\| \leqq a] \geqq \frac{3}{4}$, $P[\|Z\| \geqq a] \geqq \frac{1}{4}$. *Then, for* $z \geqq a$

$$P[\|Z\| > z] \leqq \exp\left\{-\frac{z^2}{24a^2} \log 3\right\} .$$

LEMMA A3. *Fix* $t > 0$ *such that* $\inf\{s^{-\alpha}(1 - r(s)): 0 \leqq s \leqq t\} \geqq A(t) > 0$. *Define x and y as before. Let,*

$$H_\alpha(a) = \lim_{n\to\infty} \frac{H_\alpha(n, a)}{n} . \tag{A.19}$$

$$0 < H_\alpha = \lim_{a\to 0} \frac{H_\alpha(a)}{a} = \lim_{n\to\infty} \frac{\bar{H}_\alpha(n)}{n} . \tag{A.20}$$

(See the note at the end of the lemma.)

Then,

$$P\left[\max\left\{Y_T(v+kax^{-2/\alpha}) : 0 \leqq k \leqq \left[\frac{x^{2/\alpha}}{a}t\right]\right\} > x\right] = x^{2/\alpha}\psi(x)\frac{H_\alpha(a)}{a}\int_v^{v+t}\exp b(s)\,ds + o(x^{2/\alpha}\psi(x)), \tag{A.21}$$

$$P[\max\{Y_T(v+s) : 0 \leqq s \leqq t\} > x] = x^{2/\alpha}\psi(x)\left[\int_v^{v+t}\exp b(s)\,ds\right]H_\alpha + o(x^{2/\alpha}\psi(x)), \tag{A.22}$$

uniformly in $0 \leqq v \leqq T$. *Similar assertions hold for* $P[\min\{Y_T(v+s) : 0 \leqq s \leqq t\} < -x]$ *with* $-b$ *replacing* b. *Finally,*

$$P[\max\{Y_T(v+s) : 0 \leqq s \leqq t\} > x, \min\{Y_T(v+s) : 0 \leqq s \leqq t\} < -y] = o(x^{2/\alpha}\psi(x)). \tag{A.23}$$

Note. The existence of the limit in (A.19) was first proved in [11]. An incorrect proof of (A.20) was also given. Subsequently, a correct proof was communicated to the author by J. Pickands and another is included in [12]. We provide yet a third in Appendix B.

PROOF. We prove (A.22); (A.21) is argued similarly. Begin by bounding the left-hand side of (A.22) from above by,

$$\sum_{k=0}^{M} P[\max\{Y_T(v+knx^{-2/\alpha}+s) : 0 \leqq s \leqq nx^{-2/\alpha}\} > x] \tag{A.24}$$

where $M = [tx^{2/\alpha}/n]$. By Lemma A2 the expression above is asymptotic to

$$\frac{t\bar{H}_\alpha(n)}{n}x^{2/\alpha}\psi(x)\left[\frac{1}{M+1}\sum_{k=0}^{M}\exp b(v+knx^{-2/\alpha})\right] = \frac{\bar{H}_\alpha(n)}{n}x^{2/\alpha}\psi(x)\left[\int_v^{v+t}\exp b(s)\,ds + o(1)\right] \tag{A.25}$$

since b is assumed uniformly continuous and bounded. On the other hand we can bound from below by the left-hand side of (A.21) which in turn is bounded from below by,

$$\sum_{r=0}^{M_a} P(A_r) - \sum_{0\leqq r\neq s\leqq M_a} P(A_r A_s) \tag{A.26}$$

where $A_r = [\{\max\{Y_T(v+kax^{-2/\alpha}), rn \leqq k < (r+1)n\} > x]$, $M_a = [x^{-2/\alpha}t/na]$. If we apply Lemma A.1 to the first term on the right of (A.26) we obtain that,

$$\sum_{r=0}^{M_a} P(A_r) \sim \frac{H_\alpha(n,a)}{na}x^{2/\alpha}\psi(x)\left[\int_v^{v+t}e^{b(s)}\,ds\right]. \tag{A.27}$$

Finally,

$$P(A_r A_s) \leqq P(C_r C_s) \tag{A.28}$$

where

$$C_r = \left[\max\{Y_T(kax^{-2/\alpha}+v) - \mu_T(kax^{-2/\alpha}) : rn \leqq k < (r+1)n\} > x - \frac{K}{(2\log T)^{\frac{1}{2}}}\right] \tag{A.29}$$

where $K = \sup\{(2\log T)^{\frac{1}{2}}|\mu_T(t)| : 0 \leqq t \leqq T\}$. Now applying Lemma 2.3 of [11] and arguing as in Lemma 2.5 of the same paper we see that,

$$\sum P(C_r C_s) = o(x^{2/\alpha}\psi(x)) . \tag{A.30}$$

Applying (A.20) we see that (A.22) follows. To prove (A.23) it suffices to show that,

$$\begin{aligned} &P\{[\max\{Y_T(v+s): 0 \leqq s \leqq t\} > x] \\ &\quad \cup [\min\{Y_T(v+s): 0 \leqq s \leqq t\} < -y]\} \\ &\quad = x^{2/\alpha}\psi(x)H_\alpha \textstyle\int_v^{v+t} [\exp b(\xi)]\, d\xi \\ &\qquad + y^{2/\alpha}\psi(y)H_\alpha \textstyle\int_v^{v+t} [\exp -b(\xi)]\, d\xi + o(x^{2/\alpha}\psi(x)) . \end{aligned} \tag{A.31}$$

But we can bound the expression on the left of (A.31) from above by

$$P[\max\{Y_T(v+s): 0 \leqq s \leqq t\} > x] + P[\min\{Y_T(v+s): 0 \leqq s \leqq t\} < -y]$$

and from below as in (A.26) where we add $A_{M_a+1}, \cdots, A_{2M_a+1}$ with $A_{M_a+j} = \{\min\{Y_T(v + kax^{-2/\alpha}): (j-1)n \leqq k < jn\}\} < -y\}$. Now by (A.10)

$$\frac{1}{n} \textstyle\sum_{j=0}^{M_a} P(A_j A_{M_a+j+1}) = o(x^{2/\alpha}\psi(x)) . \tag{A.32}$$

Finally, again arguing as for the previous case,

$$\frac{1}{n} \textstyle\sum_{0 \leqq j \neq k \leqq M_a} P(A_j A_k) , \quad \frac{1}{n} \textstyle\sum_{1 \leqq j \neq k \leqq M_a+1} P(A_{j+M_a} A_{k+M_a}) \quad \text{and} \tag{A.33}$$

$$\frac{1}{n} \textstyle\sum_{0 \leqq j \neq k \leqq M_a} P(A_j A_{M_a+k+1}) \quad \text{are all} \quad o(x^{2/\alpha}\psi(x)) . \ \square$$

The rest of the proof goes much as in Berman [1]. Neglecting fringe effects break the interval $[0, T]$ up into $2N$ intervals of which half, $W_1, \cdots, W_N$ are of length t and the others $V_1, \cdots, V_N$ of length ε so that V_i follows W_i which follows V_{i-1}, $i = 2, \cdots, N$. Of course, $N \sim T/(t+\varepsilon)$. Define x and y as in Lemma A1 and note that,

$$x^{2/\alpha}\psi(x)H_\alpha \sim \frac{1}{T} e^{-z_1} . \tag{A.34}$$

Then, by Lemma A3,

$$\begin{aligned} P[\max\{Y_T(\tau): \tau \in \textstyle\bigcup_{j=1}^N V_j\} \geqq x] &\leqq \textstyle\sum_{j=1}^N P[\max\{Y_T(\tau): \tau \in V_j\} \geqq x] \\ &\sim [\textstyle\sum_{j=1}^N \int_{V_j} \exp b(s)\, ds] \frac{e^{-z_1}}{T} \\ &= \varepsilon O\left(\frac{N}{T}\right) = \varepsilon O(1) \end{aligned} \tag{A.35}$$

where the O term is independent of ε and the V_j. A similar assertion holds for $\min\{Y_T(\tau): \tau \in \bigcup_{j=1}^N V_j\}$ and hence we need only show that,

$$\begin{aligned} \lim_{\varepsilon\to 0} \varliminf_{T\to\infty} P[&\max\{Y_T(\tau): \tau \in \textstyle\bigcup_{j=1}^N W_j\} \leqq x, \\ &\min\{Y_T(\tau): \tau \in \textstyle\bigcup_{j=1}^N W_j\} \geqq -y] \\ &\qquad = \exp -\{\lambda_1 e^{-z_1} + \lambda_2 e^{-z_2}\} , \end{aligned} \tag{A.36}$$

where λ_1, λ_2 are defined in (A.3) and the bars above and below the limit sign indicate lim sup and lim inf respectively. Next choose $a > 0$. If $W_j = [a_j, a_j + t)$, $j = 1, \cdots, N$.

$$\begin{aligned} &\Big| P[\max\{Y_T(\tau) : \tau \in \textstyle\bigcup_{j=1}^N W_j\} \leqq x] \\ &\quad - P\Big[Y_T(a_j + kax^{-2/\alpha}) \leqq x : 0 \leqq k \leqq \Big[\frac{tx^{2/\alpha}}{a}\Big], 1 \leqq j \leqq N\Big]\Big| \\ &\leqq \textstyle\sum_{j=1}^N \Big| P[\max\{Y_T(\tau) : \tau \in W_j\} \leqq x] \\ &\quad - P\Big[\max\Big\{Y_T(a_j + kax^{-2/\alpha}) : 0 \leqq k \leqq \Big[\frac{tx^{2/\alpha}}{a}\Big]\Big\} \leqq x\Big]\Big| \\ &\sim [\textstyle\sum_{j=1}^N \int_{W_j} \exp b(s)\, ds] x^{2/\alpha} \psi(x) \Big[H_\alpha - \frac{H_\alpha(a)}{a}\Big] e^{-z_1}, \end{aligned} \tag{A.37}$$

by Lemma A3.

A similar argument holds for $P[\min\{Y_T(\tau) : \tau \in \bigcup_{j=1}^N W_j\} \geqq -y]$ and by simple probability manipulations it follows that to prove the theorem we need only show,

$$\begin{aligned} &\lim_{a\to 0} \lim_{\varepsilon\to 0} \overline{\underline{\lim}}_T\, P\Big[-y \leqq Y_T(a_j + kax^{-2/\alpha}) \leqq x : 1 \leqq j \leqq N, \\ &\qquad 0 \leqq k \leqq \Big[\frac{tx^{2/\alpha}}{a}\Big]\Big] \\ &\qquad\qquad = \exp\{-[\lambda_1 e^{-z_1} + \lambda_2 e^{-z_2}]\}. \end{aligned} \tag{A.38}$$

Now in view of Lemma A3 it is easy to show that,

$$\begin{aligned} &\underline{\lim}_T \textstyle\sum_{j=1}^N \Big(1 - P\Big[-y \leqq Y_T(a_j + kax^{-2/\alpha}) \leqq x : 0 \leqq k \leqq \Big[\frac{tx^{2/\alpha}}{a}\Big]\Big]\Big) \\ &\quad = \frac{H_\alpha(a)}{aH_\alpha} \underline{\lim} \frac{1}{T} \textstyle\sum_{j=1}^N \int_{W_j} \{\exp[b(s) - z_1] + \exp -[b(s) + z_2]\}\, ds. \end{aligned} \tag{A.39}$$

Since, by the boundedness of b, $T^{-1}[\sum_{j=1}^N \int_{W_j} \exp b(s)\, ds - \int_0^T \exp b(s)\, ds] = O(\varepsilon)$ uniformly in T it follows from (A.39) and (A.20) that

$$\begin{aligned} &\lim_{a\to 0} \lim_{\varepsilon\to 0} \underline{\lim}_T \textstyle\sum_{j=1}^N \Big(1 - P\Big[-y \leqq Y_T(a_j + kax^{-2/\alpha}) \\ &\qquad \leqq x : 0 \leqq k \leqq \Big[\frac{tx^{2/\alpha}}{a}\Big]\Big]\Big) \\ &\qquad\qquad = \lambda_1 e^{-z_1} + \lambda_2 e^{-z_2}. \end{aligned} \tag{A.40}$$

Let E_j, $j = 1, \cdots, N$ be the events whose probabilities are being summed in (A.40). The assertion (A.38) corresponds to a limiting statement about $P(E_1 \cdots E_N)$. If the E_j were independent assertion (A.38) would follow readily from (A.40). Let $\tilde{P}$ be the measure which makes the vectors $(Y_T(a_1), Y_T(a_1 + ax^{-2/\alpha}), \cdots, Y_T(a_1 + ax^{-2/\alpha}[tx^{2/\alpha}/a]))$, $(Y_T(a_2), \cdots, Y_T(a_2 + ax^{-2/\alpha}[tx^{2/\alpha}/a]))$, $\cdots$, $(Y_T(a_N), \cdots, Y_T(a_N + ax^{-2/\alpha}[tx^{2/\alpha}/a])$ independent and otherwise agrees with P.

To conclude the proof of the theorem we need to show that.

$$\lim_{\varepsilon\to 0} \overline{\lim}_T |(P - \tilde{P})(E_1 \cdots E_N)| = 0 . \tag{A.41}$$

To do this apply the following modification of Lemma 4.1 of [1].

LEMMA A4. *Let*

$$\phi(x, y, p) = \frac{1}{2\pi(1-p^2)^{\frac{1}{2}}} \exp - \frac{(x^2 - 2pxy + y^2)}{2(1-p^2)} . \tag{A.42}$$

Let $\Sigma_1 = |r_{ij}|$, $\Sigma_2 = |s_{ij}|$ *be* $k \times k$ *nonnegative semi-definite matrices with* $r_{ii} = s_{ii} = 1$ *for all* i. *Let* $\mathbf{X} = (X_1, \cdots, X_k)$ *be a mean* 0 *Gaussian vector with covariance matrix* Σ_1 *or* Σ_2. *Let* $u_1, \cdots, u_k$ *be nonnegative numbers and* $u = \min_j u_j$. *Then,*

$$\begin{aligned}|P_{\Sigma_1}[X_j \leqq u_j, 1 \leqq j \leqq k] - P_{\Sigma_2}[X_j \leqq u_j, 1 \leqq j \leqq k]| \\ \leqq 4 \textstyle\sum_{i,j} |\int_{s_{ij}}^{r_{ij}} \phi(u, u; \lambda)\, d\lambda| .\end{aligned} \tag{A.43}$$

PROOF. By the usual argument (see [1] page 931) the left-hand side of (A.43) is bounded by, $4 \sum_{i,j} |\int_{s_{ij}}^{r_{ij}} \phi(u_i, u_j; \lambda)\, d\lambda|$. But, by an elementary inequality

$$x^2 - 2pxy + y^2 \geqq \frac{(1-p)}{2} (x+y)^2 . \tag{A.44}$$

Thus,

$$\phi(u_i, u_j, \lambda) \leqq \phi\left(\frac{u_i + u_j}{2}, \frac{u_i + u_j}{2}, \lambda\right) \leqq \phi(u, u, \lambda) . \qquad \square \tag{A.45}$$

Take $X_1 = Y_T(a_1) - \mu_T(a_1)$, $X_2 = -Y_T(a_1) + \mu_T(a_1)$ etc., $k = 2N[tx^{2/\alpha}/a]$, $|r_{ij}|$ corresponding to the distribution of $\mathbf{X}$ under P, $|s_{ij}|$ corresponding to $\tilde{P}$, $u_1 = x - \mu_T(a_1)$, $u_2 = y + \mu_T(a_1)$ etc. Evidently,

$$u = (2 \log T)^{\frac{1}{2}} + O((\log T)^{-\frac{1}{2}}) . \tag{A.46}$$

It is clear now that we can apply to the bound of (A.43) exactly the same analysis as that given by Berman on pages 933–936 of [1] to arrive at the conclusion of the theorem.

Note. By applying the more refined analysis of Pickands [11] pages 64–72 we can show that the conclusion of the theorem also holds if (vi) is replaced by,

$$\lim_{t\to\infty} r(t) \log t = 0 . \tag{A.47}$$

Unfortunately, the analysis of Berman appears to only yield the conclusion under the stronger

$$r(t)[\log t]^{2/\alpha} \to 0 . \tag{A.48}$$

We do not enter into this further since (vi) is what we need for Theorems 1.1 and 1.2.

5. Appendix B. Miscellanea.

THEOREM B1. *Let* w *be an absolutely continuous square integrable function with* *a square integrable derivative w'. Let,*

$$r(t) = \int w(t+s)w(s)\,ds \,. \tag{B.1}$$

Then r is twice differentiable and

$$r''(t) = -\int w'(t+s)w'(s)\,ds \,. \tag{B.2}$$

PROOF. We first show that

$$r'(t) = \int w'(t+s)w(s)\,ds = \int w(s-t)w'(s)\,ds \,. \tag{B.3}$$

Let $\hat{w}$, $\hat{w}'$ be the Fourier transforms of w, w'. Then by Parseval,

$$\frac{r(t+h)-r(t)}{h} = \frac{1}{2\pi}\int \frac{(e^{-i(t+h)u} - e^{-itu})}{h} |\hat{w}(u)|^2\,du \,. \tag{B.4}$$

Applying the dominated convergence theorem we obtain the existence of r' given by

$$r'(t) = -\frac{i}{2\pi}\int e^{-itu}u|\hat{w}(u)|^2\,du = \int w'(t+s)w(s)\,ds \,.$$

Similarly

$$\begin{aligned} \frac{r'(t+h)-r'(t)}{h} &= \int \frac{w(s-t-h)-w(s-t)}{h} w'(s)\,ds \\ &= \frac{i}{2\pi}\int \left(\frac{e^{i(t+h)u} - e^{itu}}{h}\right) u|\hat{w}(u)|^2\,du \\ &\to -\frac{1}{2\pi}\int e^{itu}u^2|\hat{w}(u)|^2\,du = -\int w'(s-t)w'(s)\,ds \,. \end{aligned} \tag{B.5}$$

The theorem follows. Note that $r'(0) = 0$ from (B4) since $|\hat{w}|$ is symmetric.

THEOREM B2. *Let w be absolutely continuous on $[-A, A]$ and 0 otherwise. Then r has left and right derivatives at 0 and*

$$r_+'(0) = -r_-'(0) = -\tfrac{1}{2}(w^2(A) + w^2(-A)) \,. \tag{B.6}$$

PROOF. Write, for $h > 0$,

$$\begin{aligned} &\int \frac{w(s+h)-w(s)}{h} w(s)\,ds \\ &= \int_{-A}^{A-h}\left[\frac{1}{h}\int_s^{s+h} w'(z)\,dz\right] w(s)\,ds - \frac{1}{h}\int_{A-h}^{A} w^2(s)\,ds \\ &\to \int_{-A}^{A} w'(s)w(s)\,ds - w^2(A) = -\tfrac{1}{2}(w^2(A) + w^2(-A)) \end{aligned} \tag{B.7}$$

by arguing as in Theorem A1 and using Lebesgue's theorem. Since $r(-t) = r(t)$ the result follows.

THEOREM B3. (Pickands) *If $H_\alpha(n, a)$, $\bar{H}_\alpha(n)$ are defined as in* (A.17), (A.19) *then* (A.20) *holds.*

PROOF. Suppose first that $0 < \alpha < 2$. Let for $\gamma > 0$,

$$\bar{H}_\alpha(n, \gamma) = \int_{-\infty}^{\infty} e^s[\max_{0\le t\le n} Y(t) > s+\gamma]\,ds = e^{-\gamma}\bar{H}_\alpha(n) \,. \tag{B8}$$

Then

$$\frac{1}{n}|H_\alpha(n, a) - \bar{H}_\alpha(na, \gamma)|$$

$$\text{(B.9)} \quad \begin{aligned} &\leqq \frac{1}{n}[\textstyle\int_{-\infty}^{\infty} e^s P[\max_{0\leqq t\leqq na} Y(t) > s + \gamma, \max_{0\leqq k\leqq n} Y(ka) \leqq s]\, ds \\ &\qquad + \textstyle\int_{-\infty}^{\infty} e^s P[s < \max_{0\leqq t\leqq na} Y(t) \leqq s + \gamma]\, ds] \\ &\leqq \frac{1}{n}\textstyle\sum_{k=0}^{n-1} \int_{-\infty}^{\infty} e^s P[Y(ka) \leqq s, \max_{ka\leqq t\leqq (k+1)a} Y(t) > s + \gamma]\, ds \\ &\qquad + \frac{1}{n}[\bar{H}_\alpha(na) - \bar{H}_\alpha(na, \gamma)]\,. \end{aligned}$$

If the summands on the right of the first term of (B.9) are denoted by $A(k, \gamma, a)$ then,

$$\text{(B.10)} \quad \begin{aligned} A(k, \gamma, a) = \textstyle\int_{-\infty}^{\infty} e^s \int_{-\infty}^{s} \tau(z, ka) \\ \times P[\max_{0\leqq t\leqq a} Y(t + ka) > s + \gamma \mid Y(ka) = z]\, dz\, ds \end{aligned}$$

where $\tau(z, ka)$ is the density of $Y(ka)$. After some manipulation we obtain

$$\text{(B.11)} \quad \begin{aligned} A(k, \gamma, a) = \textstyle\int_{-\infty}^{\infty} \phi(w) \int_0^\infty e^s P[\max_{0\leqq t\leqq a}(Y(t + ka) - Y(ka)) > s + \gamma \mid Y(ka) \\ = w + (ka)^\alpha]\, ds\, dw\,. \end{aligned}$$

As $k \to \infty$, the finite dimensional conditional distributions of $Y(t + ka) - Y(ka)$ given $Y(ka) = w + (ka)^\alpha$ tend for each w to those of $Y(t)$, $0 \leqq t \leqq a$. Arguing as in Lemma A1 we conclude that,

$$\text{(B.12)} \quad \lim_k A(k, \gamma, \alpha) = A(\gamma, \alpha) = \textstyle\int_0^\infty e^s P[\max_{0\leqq t\leqq a} Y(t) > s + \gamma]\, ds\,.$$

Let $Y^*(t) = Y(t) + |t|^\alpha$. Then,

$$\text{(B.13)} \quad \begin{aligned} A(\gamma, \alpha) &\leqq \textstyle\int_0^\infty e^s P[\max_{0\leqq t\leqq a} Y^*(t) > s + \gamma]\, ds \\ &= \textstyle\int_0^\infty e^s P[\max_{0\leqq t\leqq 1} Y^*(t) > (s + \gamma)a^{-\alpha/2}]\, ds \\ &= a^{\alpha/2} e^{-\gamma} \textstyle\int_{\gamma a^{-\alpha/2}}^{\infty} e^{wa^{\alpha/2}} P[\max_{0\leqq t\leqq 1} Y^*(t) > w]\, dw\,. \end{aligned}$$

Applying Fernique's estimate the right-hand side of (B.13) is $O(\exp -a^{-\alpha/2})$ for every $\gamma > 0$. We conclude that,

$$\text{(B.14)} \quad \begin{aligned} \limsup_a \limsup_n \frac{1}{na}|H_\alpha(n, a) - \bar{H}_\alpha(na, \gamma)| \\ \leqq (1 - e^{-\gamma}) \limsup_a \limsup_n \frac{\bar{H}_\alpha(na)}{na} \end{aligned}$$

for every $\gamma > 0$. Since,

$$\text{(B.15)} \quad \begin{aligned} &P[\max_{0\leqq t\leqq n} Y(t) > s] \\ &\qquad \leqq \textstyle\sum_{k=0}^{n-1} P[\max_{k\leqq t\leqq k+1} Y(t) > s] \\ &\qquad \leqq \textstyle\sum_{k=0}^{n-1} \{P[Y(k) \leqq s, \max_{k\leqq t\leqq k+1} Y(t) > s] + P[Y(k) > s]\}\,, \end{aligned}$$

it is easy to see that,

$$\sup_{x \geq 1} \frac{\bar{H}_\alpha(x)}{x} < \infty . \tag{B.16}$$

Hence,

$$\lim_a \limsup_n \frac{1}{na} |H_\alpha(n, a) - \bar{H}_\alpha(na)| = 0 . \tag{B.17}$$

But from the argument of Lemma A3 it is clear that for every $a > 0$,

$$\limsup_n \frac{H_\alpha(n, a)}{na} \leq \liminf_n \frac{\bar{H}_\alpha(na)}{na} . \tag{B.18}$$

The theorem follows for $0 < \alpha < 2$. For $\alpha = 2$ we can use the representation $Y(t) = 2^{\frac{1}{2}} tZ - t^2$ where Z is a standard normal deviate. Evidently,

$$\begin{aligned} \max_{0 \leq s \leq na/2^{\frac{1}{2}}} Y(s) &= \frac{Z^2}{2} \quad \text{if} \quad 0 \leq Z < \frac{na}{2^{\frac{1}{2}}} \\ &= naZ - \frac{n^2a^2}{2} \quad \text{otherwise} . \end{aligned} \tag{B.19}$$

It follows that,

$$\begin{aligned} &\frac{1}{na} \left| \bar{H}_2\left(\frac{na}{2^{\frac{1}{2}}}\right) - H_2\left(n, \frac{a}{2^{\frac{1}{2}}}\right) \right| \\ &\quad \leq \frac{1}{na} \int_0^{n^2a^2} e^{s/2} P[s^{\frac{1}{2}} < Z < (s + a^2)^{\frac{1}{2}}] \, ds \sim 2(1 - e^{-a^2/2}) \end{aligned} \tag{B.20}$$

by standard arguments. The theorem now follows generally.

REFERENCES

[1] Berman, S. N. (1971). Asymptotic independence of the numbers of high and low level crossings of stationary Gaussian processes. *Ann. Math. Statist.* **42** 927–946.

[2] Berman, S. N. (1971). Maxima and high level excursions of stationary Gaussian processes. *Trans. Amer. Math. Soc.* To appear.

[3] Billingsley, P. (1968). *Convergence of Probability Measures.* Wiley, New York.

[4] Breiman, L. (1969). *Probability.* Addison-Wesley, Reading.

[5] Brillinger, D. (1969). An asymptotic representation of the sample distribution function. *Bull. Amer. Math. Soc.* **75** 545–547.

[6] Doob, J. L. (1953). *Stochastic Processes.* Wiley, New York.

[7] Epanechnikov, V. A. (1969). Nonparametric estimates of a multivariate probability density. *Theor. Probability Appl.* **14** 153–158.

[8] Fernique, X. (1970). Intégrabilité des vecteurs gaussiens. *C.R. Acad. Sci. Paris* **270** A 1698–99.

[9] Hajek, J. and Sidak, Z. (1967). *Theory of Rank Tests.* Academic Press, New York.

[10] Mann, H. B. and Wald, A. (1942). On the choice of the number of class intervals in the application of the χ^2 test. *Ann. Math. Statist.* **13** 306–317.

[11] Pickands, J. S. III (1960). Upcrossing probabilities for Gaussian processes. *Trans. Amer. Math. Soc.* **145** 51–73.

[12] Qualls, C. and Watanabe, H. (1971). Asymptotic properties of Gaussian processes. Tech. Report No. 736, Univ. of North Carolina, Chapel Hill.

[13] ROSENBLATT M. (1971). Curve estimates. *Ann. Math. Statist.* **42** 1815–1842.

[14] STECK, G. P. (1957). Limit theorems for conditional distributions. *Univ. California Publ. Statist.* **2** 237–284.

[15] WATSON, G. S. (1967). Some problems in the statistics of directions. *Bull. I.S.I.* **17** 374–385.

[16] WOODROOFE, M. (1967). On the maximum deviation of the sample density. *Ann. Math. Statist.* **38** 475–481.

DEPARTMENT OF STATISTICS
UNIVERSITY OF CALIFORNIA
BERKELEY, CALIFORNIA 94720

DEPARTMENT OF MATHEMATICS
UNIVERSITY OF CALIFORNIA
LA JOLLA, CALIFORNIA 92037

Comp. & Maths with Appls. Vol. 1. pp. 223–235. Pergamon Press, 1975. Printed in Great Britain

ASYMPTOTIC BEHAVIOR OF A SPLINE ESTIMATE OF A DENSITY FUNCTION

KEH-SHIN LII and M. ROSENBLATT
University of California, San Diego, La Jolla, California 92037, U.S.A.

Communicated by B. R. Aging

(*Received December* 1974)

Abstract—A class of cubic spline estimators of probability density functions over a finite interval are considered in this paper. The precise asymptotic behavior of the bias and covariance of such estimators is obtained in the interior of the interval. The estimators are shown to be asymptotically normally distributed. The properties of these estimators are compared with those of kernel estimators. The kernel and spline estimators are compared in some Monte Carlo simulations as well as in the analysis of some data obtained in turbulent wind flow.

1. INTRODUCTION

Particular examples of spline functions have been used for years but it was I. J. Shoenberg who singled out their particular structure for special study and named them in the middle of 1940's [6]. From 1960 on they have attracted wide attention. After more than ten years in the hands of analysts, the idea of splines has been generalized and applied in many fields such as interpolation, approximation and differential equations. In 1970, Boneva, Kendall and Stefanov proposed "histosplines" in a density estimation [2]. This appears to have been the first time spline-like techniques have been proposed in density estimation.

Here we propose a rather traditional appraoch to density estimation via splines. We restrict our attention to the equal-spaced cubic spline for its simplicity. It can be generalized to quintic splines but the analysis of these would require more labor. In Section 2, a statement of the results on bias, covariance and asymptotic distribution is given. The derivation of these results is given in Section 3. And in Section 4 we try to compare spline estimation with kernel type estimation both theoretically and numerically in terms of a number of examples. The results obtained are new and give detailed information about the fine structure of the spline estimates due to "binning".

2. STATEMENT OF THE RESULTS ON BIAS, COVARIANCE AND ASYMPTOTIC DISTRIBUTION

Let $f(x)$ be a continuous density function on $[0, 1]$. Suppose that $X_1, X_2, \ldots, X_n$ are independent, identically distributed random variables with density f. Consider

$$y_k = F_n\left(\frac{k}{N}\right), \; k = 0, 1, \ldots, N = \frac{1}{h} \tag{1}$$

where $F_n(x)$ is the sample distribution function and $h = 1/N$ is the bin size. *Let $s_n(x)$ be the cubic spline interpolator of F_n with knots at the points $x_j = j/N$, $j = 0, 1, \ldots, N$ with* $f(0) = s_n'(0) = y_0'$, $f(1) = s_n'(1) = y_N'$. Notice that $s_n(x)$ also depends on N. The derivative of the spline interpolator

$$s_n'(x) = -M_{i-1}\frac{(x-x_i)^2}{2h} + M_i\frac{(x-x_{i-1})^2}{2h} - \frac{h}{6}(M_i - M_{i-1}) + \frac{1}{h}(y_i - y_{i-1}) \tag{2}$$

when $x \in [x_{i-1}, x_i]$ where

$$M_i = s_n''(x_i) \tag{3}$$

(see [1] p. 10), and $s_n'(x)$ is used as the estimator of $f(x)$. At this point we shall give statements of a number of the results on bias and covariance. The proofs given later are for the boundary conditions for the spline $f(0) = s_n'(0)$, $f(1) = s_n'(1)$ cited above. However, simple modifications of

R.A. Davis et al. (eds.), *Selected Works of Murray Rosenblatt*, Selected Works in Probability and Statistics, DOI 10.1007/978-1-4419-8339-8_28, © Springer Science+Business Media, LLC 2011

the proof show that the results are still valid with any other conventional boundary conditions for cubic splines, for example, if

$$s_n''(0) = 0 = s_n''(1)$$

or if one has periodic boundary conditions for the cubic spline (see "Asymptotics and representations of cubic splines" an ONR report of Rosenblatt).

The first result concerns bias and is a direct consequence of a result of Rosenblatt[4]. We introduce the following notation for convenience

$$\sigma = \sqrt{3} - 2$$

$$r = \frac{1}{h}(x - x_{i-1}). \tag{4}$$

Theorem 1: *Let* $f \in C^3[0, 1]$ (continuously differentiable up to third order). *Then the bias*

$$b_n(x) = Es_n'(x) - f(x) = \frac{f^{(3)}(x)}{4!}\left\{\frac{1}{h}[(x_i - x)^4 - (x_{i-1} - x)^4] - h[(x_i - x)^2 - (x_{i-1} - x)^2] + o(h^3)\right\} \tag{5}$$

$$= \frac{f^{(3)}(x)}{4!}h^3\{(1-r)^4 - r^4 - (1-r)^2 + r^2 + o(1)\}$$

if $0 < x < 1$ *is fixed and* $x \in [x_{i-1}, x_i]$ (that is, $x_{i-1} = [Nx]/N$ where $[y]$ is the greatest integer less than or equal to y) as $N \to \infty$. The following result on asymptotic behavior of the variance is obtained.

Theorem 2. Let f be continuous on [0, 1]. *The variance of the spline estimator* $s_n'(x)$ *of* $f(x)$ *is then given by*

$$\frac{f(x)}{nh}A(r) + O\left(\frac{h}{n}\right). \tag{6}$$

if $0 < x < 1$ *is fixed and* $nh \to \infty$, $h \to 0$.
Here

$$A(r) = 1 - \frac{3(1-\sigma)}{2+\sigma}\left(2r^2 - 2r + \frac{1}{3}\right) + \frac{9}{4}\left(\frac{1-\sigma}{2+\sigma}\right)^2\left[\left(2r^2 - 2r + \frac{1}{3}\right)^2 + \left[\left(r^2 - \frac{1}{3}\right) + \sigma\left(\frac{1}{3} - (1-r)^2\right)\right]^2\frac{1}{1-\sigma^2} + \left[\left(r^2 - \frac{1}{3}\right) + \frac{1}{\sigma}\left(\frac{1}{3} - (1-r)^2\right)\right]^2\frac{\sigma^2}{1-\sigma^2}\right] \tag{7}$$

Notice that $A'(0) = A'(1) = 0$.

In Fig. 1, the function $(1-r)^4 - r^4 - (1-r)^2 + r^2$ that describes the asymptotic behavior of the bias is given over the range $0 \le r \le 1$. The function $A(r)$ that determines the asymptotic behavior of the variance is represented in Fig. 2.

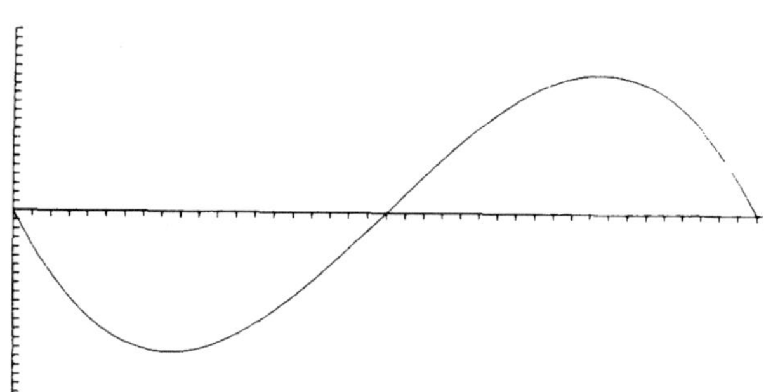

Fig. 1. Bias and the function $(1-r)^4 - r^4 - (1-r)^2 + r^2$.

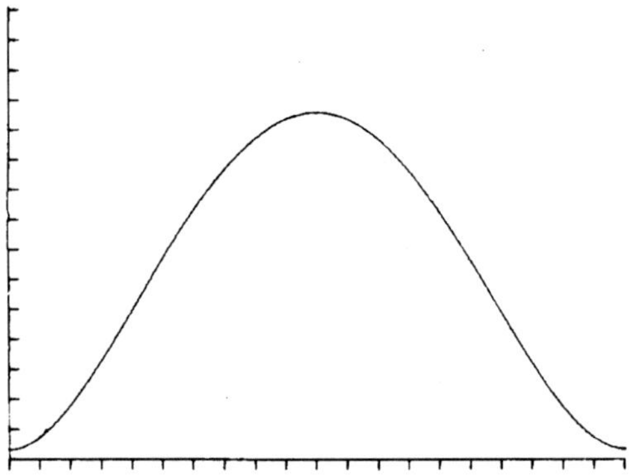

Fig. 2. Variance and the function $A(r)$.

One can show that $s_n'(x)$, $s_n'(y)$ are asymptotically uncorrelated if x, y are fixed and not equal when $nh \to \infty$, $h \to 0$. However, a more detailed result on the behavior of the covariance is given in the following theorem.

Theorem 3. Consider x, y fixed with $0 < x, y < 1$. Let $x_{i-1} = [Nx]/N$, $x_{j-1} = [Ny]/N$ *and* $d = i - j$. Set $r_1 = h^{-1}(x - x_{i-1})$, $r_2 = h^{-1}(y - x_{j-1})$.
Then

$$\begin{aligned}\operatorname{cov}[s_n'(x), s_n'(y)] = \frac{1}{nh}\frac{3}{4}f(x)\Big\{\Big[\Big(r_1^2 - \frac{1}{3}\Big)\Big(r_2^2 - \frac{1}{3}\Big) \\ + \Big(\frac{1}{3} - (1-r_1)^2\Big)\Big(\frac{1}{3} - (1-r_2)^2\Big)\Big] \cdot \Big[6|d|\sigma^{|d|} - \frac{12\sigma^{|d|+1}}{1-\sigma^2}\Big] \\ + \Big(r_1^2 - \frac{1}{3}\Big)\Big(\frac{1}{3} - (1-r_2)^2\Big)\Big[6|d+1|\sigma^{|d+1|} - \frac{12\sigma^{|d+1|+1}}{1-\sigma^2}\Big] \\ + \Big(r_2^2 - \frac{1}{3}\Big)\Big(\frac{1}{3} - (1-r_1)^2\Big)\Big[6|d-1|\sigma^{|d-1|} - \frac{12\sigma^{|d-1|+1}}{1-\sigma^2}\Big]\Big\} \\ + \frac{f(y)}{nh}\frac{\sqrt{3}}{2}\Big[\Big(r_1^2 - \frac{1}{3}\Big)(\sigma^{|d+1|} - \sigma^{|d|}) + \Big(\frac{1}{3} - (1-r_1)^2\Big)(\sigma^{|d|} - \sigma^{|d-1|})\Big] \\ + \frac{f(x)}{nh}\frac{\sqrt{3}}{2}\Big[\Big(r_2^2 - \frac{1}{3}\Big)(\sigma^{|d-1|} - \sigma^{|d|}) + \Big(\frac{1}{3} - (1-r_2)^2\Big)(\sigma^{|d|} - \sigma^{|d+1|})\Big] \\ + \delta_{d,0}\frac{1}{nh}f(x) + o\Big(\frac{1}{n}\Big) \\ \text{as} \quad nh \to \infty, h \to 0.\end{aligned} \tag{8}$$

Theorem 4. Given a finite number of fixed distinct points $0 < x_1, x_2, \ldots, x_m < 1$, *the estimates* $s_n'(x_1), \ldots, s_n'(x_m)$ *are jointly asymptotically normal and independent with bias given by* (4) *and variance given by* (8) *if* $nh \to \infty$, $h \to 0$.

3. PROOFS OF THE RESULTS ON ASYMPTOTIC BEHAVIOR OF THE ESTIMATES

We shall first give the proof of theorem 2 which describes the asymptotic behavior of the variance of $s_n'(x)$. The computations involved in the proof of Theorem 3 (involving the covariance) are similar but somewhat more involved.

Proof of Theorem 2: We first note that

$$M_i = s_n''(x_i) = \sum_{j=0}^{N} A_{i,j}^{-1} d_j \tag{9}$$

where

$$d_j = \frac{3}{h^2}(y_{j+1} - 2y_j + y_{j-1}),\ j = 1, \dots, N-1$$

$$d_0 = \frac{6}{h}\left(\frac{y_1 - y_0}{h} - y_0'\right) \tag{10}$$

$$d_N = \frac{6}{h}\left(y_N' - \frac{y_N - y_{N-1}}{h}\right)$$

and $A_{i,j}^{-1}$ is the (i, j) entry of A^{-1} with the

$$A = \begin{bmatrix} 2 & 1 & & & & \\ \frac{1}{2} & 2 & \frac{1}{2} & & & \\ & \frac{1}{2} & 2 & \frac{1}{2} & & 0 \\ & & \ddots & \ddots & \ddots & \\ & 0 & & \frac{1}{2} & 2 & \frac{1}{2} \\ & & & & 1 & 2 \end{bmatrix} \tag{11}$$

(see [1] p. 11). Notationally let

$$\Delta F_{n,k-1} = y_k - y_{k-1}. \tag{12}$$

Then

$$M_i = \sum_{j=0}^{N} A_{i,j}^{-1} \frac{3}{h^2}(\Delta F_{n,j} - \Delta F_{n,j-1}) \tag{13}$$

where it is understood that

$$y_{-1} = y_1 - 2hy_0'$$
$$y_{N+1} = y_{N-1} + 2hy_N'. \tag{14}$$

To further simplify notation let

$$a_{i,j} = \frac{h}{2}\left(r^2 - \frac{1}{3}\right) A_{i,j}^{-1} + \frac{h}{2}\left(\frac{1}{3} - (1-r)^2\right) A_{i-1,j}^{-1}. \tag{15}$$

Then using (12), (13) and (15) we find that if $x \in [x_{i-1}, x_i]$

$$\begin{aligned}
\operatorname{var}(s_n'(x)) &= \operatorname{var}\left(\frac{3}{h^2}\sum_{j=0}^{N} a_{i,j}(\Delta F_{n,j} - \Delta F_{n,j-1}) + \frac{1}{h}\Delta F_{n,i-1}\right) \\
&= \frac{9}{h^4}\left[\sum_{j=0}^{N} a_{i,j}^2 \{\operatorname{var}(\Delta F_{n,j}) + \operatorname{var}(\Delta F_{n,j-1})\}\right. \\
&\quad \left. + 2\sum_{0 \le j < k \le N} a_{i,j}a_{ik}\{\operatorname{cov}(\Delta F_{n,j}, \Delta F_{n,k}) + \operatorname{cov}(\Delta F_{n,j-1}, \Delta F_{n,k-1})\}\right] \\
&\quad + \frac{1}{h^2}\operatorname{var}(\Delta F_{n,i-1}) - \frac{18}{h^4}\left[\sum_{j=0}^{N-1} a_{i,j}a_{i,j+1}\operatorname{var}(\Delta F_{n,j})\right. \\
&\quad + 2\sum_{0 \le j < k \le N-1}(a_{i,j}a_{i,k+1} + a_{i,j+1}a_{i,k})\operatorname{cov}(\Delta F_{n,j}, \Delta F_{n,k}) \\
&\quad \left. + \sum_{j=0}^{N} a_{i,0}a_{i,j}\operatorname{cov}(\Delta F_{n,-1}, \Delta F_{n,j}) + \sum_{j=0}^{N} a_{i,N}a_{i,j}\operatorname{cov}(\Delta F_{n,N}, \Delta F_{n,j-1})\right] \\
&\quad + \frac{6}{h^3}\left[\sum_{\substack{j=0 \\ j \ne i-1}}^{N} a_{i,j}\operatorname{cov}(\Delta F_{n,j}, \Delta F_{n,i-1}) - \sum_{\substack{j=0 \\ j \ne i}}^{N} a_{i,j}\operatorname{cov}(\Delta F_{n,j-1}, \Delta F_{n,i-1})\right. \\
&\quad \left. + (a_{i,i-1} - a_{i,i})\operatorname{var}(\Delta F_{n,i-1})\right].
\end{aligned} \tag{16}$$

However, only the terms in the ungainly expression above involving variances make a major contribution asymptotically as one can see from the following remarks on the sample distribution. Since

$$\operatorname{cov}(F_n(x), F_n(x')) = \frac{1}{n}[F(\min(x, x')) - F(x)F(x')] \tag{17}$$

we have

$$\operatorname{cov}(\Delta F_{n,i}, \Delta F_{n,j}) = -\frac{1}{n}\Delta F\left(\frac{i}{N}\right)\Delta F\left(\frac{j}{N}\right) = O\left(\frac{h^2}{n}\right) \quad \text{if} \quad i \neq j$$

and

$$\operatorname{var}(\Delta F_{n,i}) = \frac{1}{n}\Delta F\left(\frac{i}{N}\right) + O\left(\frac{h^2}{n}\right). \tag{19}$$

Here it is to be understood that $F(x) = \int_0^x f(u)du$ and $\Delta F(i/N) = F(i/N) - F((i-1)/N)$. Using (16), this implies that

$$\operatorname{var}(s_n'(x)) = \frac{9}{h^4}\sum_{j=0}^{N}(a_{i,j} - a_{i,j+1})^2 \frac{1}{n}\Delta F\left(\frac{j}{N}\right) + \frac{1}{h^2}\frac{1}{n}\Delta F\left(\frac{i-1}{N}\right)$$
$$+ \frac{6}{h^3}(a_{i,i-1} - a_{i,i})\frac{1}{n}\Delta F\left(\frac{i-1}{N}\right) + O\left(\frac{h^2}{n}\right). \tag{20}$$

The elements of A^{-1} (see [1], p. 38) are given by

$$A_{i,j}^{-1} = \frac{\sigma^{j-i}(1+\sigma^{2i})(1+\sigma^{2N-2j})}{(2+\sigma)(1-\sigma^{2N})} \quad \text{if} \quad 0 < i \leq j < N$$
$$A_{i,N}^{-1} = \frac{\sigma^{N-i}(1+\sigma^{2i})}{(2+\sigma)(1-\sigma^{2N})} \quad \text{if} \quad 0 < i \leq N$$
$$A_{0,j}^{-1} = \frac{2\sigma^{j}(1+\sigma^{2N-2j})}{(2+\sigma)(1-\sigma^{2N})} \quad \text{if} \quad 0 < j < N$$
$$A_{0,N}^{-1} = \frac{2\sigma^N}{(2+\sigma)(1-\sigma^{2N})}$$
$$A_{0,0}^{-1} = \frac{1}{2}\frac{2-\sigma^{2N-1}(1+\sigma)^2}{(2+\sigma)(1-\sigma^{2N})} \tag{21}$$

with $\sigma = \sqrt{3} - 2$, $A_{i,j}^{-1} = A_{j,i}^{-1}$, $0 < i,j < N$ and $A_{i,j}^{-1} = A_{N-i,N-j}^{-1}$. For all i,j, $|A_{i,j}^{-1}| \leq 0{\cdot}3^{|i-j|}$ and for any given $\epsilon > 0$ and

$$\epsilon < \frac{i}{N}, \frac{j}{N} < 1 - \epsilon \tag{22}$$

it follows that $A_{i,j}^{-1}$ is to the first order

$$\sigma^{|j-i|}/(2+\sigma) \tag{23}$$

where

$$\sum_i \frac{\sigma^{|i-j|}}{2+\sigma} = \frac{1}{3}. \tag{24}$$

Using (23), for i,j in the range (22)

$$a_{i,j} - a_{i,j+1} \tag{25}$$

$$\simeq \begin{cases} \dfrac{h}{2}\left(r^2-\dfrac{1}{3}\right)\dfrac{\sigma(1-\sigma^{-1})}{2+\sigma}+\dfrac{h}{2}\left(\dfrac{1}{3}-(1-r)^2\right)\dfrac{1-\sigma}{2+\sigma} & \text{if } \ j=i-1 \\ \dfrac{h}{2}\left(r^2-\dfrac{1}{3}\right)\dfrac{\sigma^{|j-i|}(1-\sigma)}{2+\sigma}+\dfrac{h}{2}\left(\dfrac{1}{3}-(1-r)^2\right)\dfrac{\sigma^{|j-i+1|}(1-\sigma)}{2+\sigma} & \text{if } \ j\geq i \\ \dfrac{h}{2}\left(r^2-\dfrac{1}{3}\right)\dfrac{\sigma^{|j-i|}(1-\sigma^{-1})}{2+\sigma}+\dfrac{h}{2}\left(\dfrac{1}{3}-(1-r)^2\right)\dfrac{\sigma^{|j-i+1|}(1-\sigma^{-1})}{2+\sigma} & \text{if } \ j\leq i-2. \end{cases}$$

Since

$$\Delta F\left(\frac{i-1}{N}\right)=f(x)h+O(h^2)$$

we have

$$\begin{aligned} \operatorname{var}(s_n'(x)) &= \frac{9}{h^4}\frac{h^2}{4}\frac{(1-\sigma)^2}{(2+\sigma)^2}\left[\frac{1}{3}-(1-r)^2-\left(r^2-\frac{1}{3}\right)\right]^2\frac{f(x)h}{n} \\ &+\frac{9}{h^4}\frac{h^2}{4}\frac{(1-\sigma)^2}{(2+\sigma)^2}\left[\left(r^2-\frac{1}{3}\right)+\sigma\left(\frac{1}{3}-(1-r)^2\right)\right]^2\sum_{j\geq i}^{N}\sigma^{2|j-i|}\frac{1}{n}\Delta F\left(\frac{j}{N}\right) \\ &+\frac{9}{h^4}\frac{h^2}{4}\frac{(1-\sigma^{-1})^2}{(2+\sigma)^2}\left[\left(r^2-\frac{1}{3}\right)+\frac{1}{\sigma}\left(\frac{1}{3}-(1-r)^2\right)\right]^2\sum_{j=0}^{i-2}\sigma^{2|j-i|}\frac{1}{n}\Delta F\left(\frac{j}{N}\right) \\ &+\frac{1}{h^2}\frac{hf(x)}{n}+\frac{6}{h^3}\frac{h}{2}\frac{1-\sigma}{2+\sigma}\left[\frac{1}{3}-(1-r)^2-\left(r^2-\frac{1}{3}\right)\right]\frac{hf(x)}{n}+O\left(\frac{h}{n}\right). \end{aligned} \tag{26}$$

Since σ^{2k} damps out exponentially and f is continuous, by using (24) in the summations in formula (26) we obtain the final result (6).

Proof of Theorem 3: In looking at the asymptotic covariance of $s_n'(x)$, $s_n'(y)$, consider $x\in[x_{i-1},x_i]$, $y\in[x_{k-1},x_k]$ with $i-k=d$ with d fixed as $nh\to\infty$, $h\to 0$. This is reasonable, since we shall see, $s_n'(x)$, $s_n'(y)$ are asymptotically uncorrelated as $nh\to\infty$, $h\to 0$, if x, y are fixed with $x\neq y$. Using arguments similar to those given in Theorem 2 one has

$$\begin{aligned} \operatorname{cov}[s_n'(x),\ s_n'(y)] &= \frac{9}{h^4}\sum_j (a_{i,j}-a_{i,j+1})\,(a_{k,j}-a_{k,j+1})\operatorname{var}(\Delta F_{n,j}) \\ &+\frac{1}{h^2}\operatorname{cov}(\Delta F_{n,i-1},\Delta F_{n,k-1})+\frac{3}{h^2}\{(a_{i,k-1}-a_{i,k})\operatorname{var}(\Delta F_{n,k-1}) \\ &+(a_{k,i-1}-a_{k,i})\operatorname{var}(\Delta F_{n,i-1})\}+O\left(\frac{h}{n}\right). \end{aligned} \tag{27}$$

By (25) it is seen that

$$\begin{aligned} (a_{i,j}-a_{i,j+1})\,(a_{k,j}-a_{k,j+1}) &\simeq \frac{h^2}{4(2+\sigma)^2}\left\{\left(r_1^2-\frac{1}{3}\right)\left(r_2^2-\frac{1}{3}\right)(\sigma^{|j-i|}-\sigma^{|j-i+1|})\,(\sigma^{|j-k|}-\sigma^{|j+1-k|})\right. \\ &+\left(\frac{1}{3}-(1-r_1)^2\right)\left(\frac{1}{3}-(1-r_2)^2\right)(\sigma^{|j+1-i|}-\sigma^{|j+2-i|})(\sigma^{|j+1-k|}-\sigma^{|j+2-k|}) \\ &+\left(r_1^2-\frac{1}{3}\right)\left(\frac{1}{3}-(1-r_2)^2\right)(\sigma^{|j-i|}-\sigma^{|j+1-i|})\,(\sigma^{|j+1-k|}-\sigma^{|j+2-k|}) \\ &\left.+\left(r_2^2-\frac{1}{3}\right)\left(\frac{1}{3}-(1-r_1)^2\right)(\sigma^{|j+1-i|}-\sigma^{|j+2-i|})\,(\sigma^{|j-k|}-\sigma^{|j+1-k|})\right\}. \end{aligned} \tag{28}$$

Let

$$A_j(i,k)=(\sigma^{|j-i|}-\sigma^{|j+1-i|})\,(\sigma^{|j-k|}-\sigma^{|j+1-k|}). \tag{29}$$

Then

$$B(d)=\sum_j A_j(i,k)=B(-d)=|d|\sigma^{|d|}(1-\sigma^{-1})\,(1-\sigma)+2\sigma^{|d|}(1-\sigma)^2/(1-\sigma^2). \tag{30}$$

We similarly have

$$\sum_j A_j(i-1,k-1) = \sum_j A_j(i,k) = B(d)$$

$$\sum_j A_j(i,k-1) = B(d+1)$$

and

$$\sum_j A_j(i-1,k) = B(d-1).$$

By using the fact that

$$a_{i,k-1} - a_{i,k} \simeq \frac{h}{2(2+\sigma)}\left\{\left(r_1^2 - \frac{1}{3}\right)(\sigma^{|k-1-i|} - \sigma^{|k-i|}) + \left(\frac{1}{3} - (1-r_1)^2\right)(\sigma^{|k-i|} - \sigma^{|k-i+1|})\right\}$$

and previous estimates the conclusion of Theorem 3 follows.

Let us now establish the asymptotic normality of the estimates $s'_n(x)$.

Proof of Theorem 4: For convenience let

$$I_k(u,v) = \begin{cases} 1 & \text{if } u < X_k \le v \\ & \text{otherwise.} \end{cases} \tag{31}$$

Then

$$\begin{aligned} s'_n(x) &= \frac{1}{h}(y_i - y_{i-1}) + \sum_{j=0}^{N} \frac{3a_{i,j}}{h^2}[y_{j+1} - 2y_j + y_{j-1}] \\ &= \frac{1}{nh}\sum_{k=1}^{n}\left\{I_k\left(\frac{i-1}{N},\frac{i}{N}\right) + \sum_{j=0}^{N}\frac{3}{h}a_{ij}\left[I_k\left(\frac{j}{N},\frac{j+1}{N}\right) - I_k\left(\frac{j-1}{N},\frac{j}{N}\right)\right]\right\}. \end{aligned}$$

The terms involving y_{-1}, y_{N+1} are negligible asymptotically compared to the others and that will be made use of in the following discussion.

Let

$$Z_k = \frac{1}{nh}\left\{I_k\left(\frac{i-1}{N},\frac{i}{N}\right) + \sum_j \frac{3}{h}a_{i,j}\left[I_k\left(\frac{j}{N},\frac{j+1}{N}\right) - I_k\left(\frac{j-1}{N},\frac{j}{N}\right)\right]\right\}.$$

Then

$$s'_n(x) \simeq \sum_{k=1}^{n} Z_k \tag{32}$$

with the Z_k's independent and identically distributed. Now

$$E|Z_k - EZ_k|^3 \le E|Z_k|^3 + 3EZ_k^2|EZ_k| + 4|EZ_k|^3 \tag{33}$$

and

$$Z_k^2 \le \left(\frac{1}{nh}\right)^2\left\{I_k\left(\frac{i-1}{N},\frac{i}{N}\right) + 24\,I_k\left(\frac{i-1}{N},\frac{i}{N}\right) + 36\sum_{l=0}^{N}\frac{1}{h^2}(a_{i,l}^2 + a_{i,l+1}^2)I_k\left(\frac{l-1}{N},\frac{l}{N}\right)\right\}$$

since

$$\left|\frac{1}{h}a_{i,j}\right| \le 2$$

(see formula (15) and $A_{i,j}^{-1}$ as explicitly given in [1] p. 39).

Therefore

$$EZ_k^2 \le \left(\frac{1}{nh}\right)^2 0(h) = O\left(\frac{1}{n^2h}\right) \tag{34}$$

and

$$|Z_k|^3 \le Z_k^2 O\left(\frac{1}{nh}\right)$$

since $|Z_k|$ is bounded by a constant multiple of $1/nh$. Hence

$$E|Z_k|^3 = O\left(\frac{1}{n^3h^2}\right)$$

and this together with (33), (34) imply

$$\sum_{k=1}^{n} E|Z_k - EZ_k|^3 = O\left(\frac{1}{n^2h^2}\right)$$

Since $\sigma^2[s_n'(x)] \approx C/nh$ for some constant C,

$$\frac{\sum_{k=1}^{n} E|Z_k - EZ_k|^3}{\sigma[s_n'(x)]^3} = o(1)$$

and an application of Liapounov's central limit theorem gives us the desired result for $s_n'(x)$. A trivial modification of this argument yields the result for asymptotic joint normality and independence of $s_n'(x_1), \ldots, s_n'(x_m)$.

4. COMPARISON OF KERNEL AND SPLINE ESTIMATES

Let us briefly recall some properties of kernel estimates. Assume that w is a bounded integrable weight function with integral one and that $X_1, \ldots, X_n$ are independent, identically distributed random variables with common density $f(x)$. A kernel estimate based on w with bandwidth $h(n)$ is given by

$$f_n(x) = \frac{1}{nh(n)} \sum_{j=1}^{n} w\left(\frac{x - X_j}{h(n)}\right) = \int \frac{1}{h(n)} w\left(\frac{x-u}{h(n)}\right) dF_n(u) \tag{35}$$

where F_n is the sample distribution function. If f is bounded and continuous the variance

$$\sigma^2(f_n(x)) \simeq \frac{f(x)}{nh(n)} \int w(u)^2 \, du \tag{36}$$

as $nh(n) \to \infty$ and $h(n) \to 0$. Further, if f is twice continuously differentiable with bounded derivatives and $\int u^2 |w(u)| \, dn < \infty$ then the bias

$$b_n(x) = Ef_n(x) - f(x) = \frac{h(n)^2}{2} f''(x) \int u^2 w(u) \, du + o[h(n)^2]. \tag{37}$$

Also if $f(x) > 0$, $f_n(x) - f(x)$ is asymptotically normally distributed with mean zero and variance given by (36) (see [5] for details).

Notice that in terms of the order of magnitude of the variance (and covariance) spline and kernel estimates have the same behavior as can be seen from formulas (6) and (36). The bias of the spline (4) appears to be of smaller magnitude than that of the kernel estimate. But we can think of formula (37) being characteristic for non-negative weight functions w. If we allow weight functions w for which the first three moments are zero (w must then take on some negative values), we will get an asymptotic behavior of the bias, if f is smooth enough, as least as good as that of the spline estimate. Notice that splines may take on negative values occasionally (when data is sparse).

However, one should carefully notice that the spline estimate has a detailed fine structure [see formulas (4), (6) and (8)] even asymptotically. From that point of view its asymptotic structure is more complicated than that of the kernel estimate. Let us be explicit in the simplest case of the bias of the spline. In formula (4) of Theorem 1, the function

$$(1-r)^4 - r^4 - (1-r)^2 + r^2$$

where

$$r = \frac{1}{h}(x - x_{i-1}), \quad 0 \le r \le 1,$$

for $x \in [x_{i-1}, x_i]$ appears as the fine structure factor. Clearly, in formulas (6) and (8) of Theorems 2 and 3 the fine structure factors are more complicated.

Often one takes the weight function w of a kernel estimate to be bandlimited, say to $(-\frac{1}{2}, \frac{1}{2})$, and then the estimates $f_n(x)$, $f_n(y)$ are asymptotically independent if separated by more than $h(n)$. However, one needs a somewhat broader separation in the case of a spline estimate for this to happen as can be seen from formula (8) and Table 1. This, of course, reflects the fact that the spline estimate smooths the data in a more global manner than does such a kernel estimate (also see Boneva, Kendall and Stefanov [2]).

A number of Monte Carlo simulations were carried out using P. Lewis' package for generating pseudo-random numbers [3]. In each of these simulations 401 random numbers were generated. We give examples of generation from

1. uniform distribution
2. normal distribution
3. two superimposed triangular distributions.

In each of these we generate a kernel estimate of the density function based on Epanechnikov's weight function

$$w(u) = \begin{cases} 3(1-x^2)/4 & \text{if } |x| \le 1 \\ 0 & \text{otherwise} \end{cases}$$

and a spline estimate of the density function. These are given in Figs. 3–6. Notice that generally in these graphs the fluctuations of the kernel estimate and the spline estimate in magnitude are the same. However, the spline estimate is smoother over a range larger than the bandwidth. Spline

Table 1. The covariance function $D = i - j$, $R_1 = (x - x_i)/h$, $R_2 = (y - x_j)/h$.

D= 0	R2= 0.1667	R2= 0.3333	R2= 0.5000	R2= 0.6667	R2= 0.8333
R1= 0.1667	1.039637E 00	1.124466E 00	1.101175E 00	9.697651E-01	7.302354E-01
R1= 0.3333	1.124466E 00	1.310470E 00	1.346688E 00	1.233120E 00	9.697651E-01
R1= 0.5000	1.101175E 00	1.346688E 00	1.428525E 00	1.346688E 00	1.101175E 00
R1= 0.6667	9.697651E-01	1.233120E 00	1.346688E 00	1.310470E 00	1.124466E 00
R1= 0.8333	7.302354E-01	9.697651E-01	1.101175E 00	1.124466E 00	1.039637E 00
D= 1	**R2= 0.1667**	**R2= 0.3333**	**R2= 0.5000**	**R2= 0.6667**	**R2= 0.8333**
R1= 0.1667	3.031611E-02	1.266032E-01	2.563839E-01	4.196582E-01	6.164262E-01
R1= 0.3333	-2.230762E-01	-1.873926E-01	-6.837567E-02	1.339746E-01	4.196582E-01
R1= 0.5000	-3.775907E-01	-3.853630E-01	-2.822913E-01	-6.837567E-02	2.563839E-01
R1= 0.6667	-4.332275E-01	-4.673079E-01	-3.853630E-01	-1.873926E-01	1.266032E-01
R1= 0.8333	-3.899864E-01	-4.332275E-01	-3.775907E-01	-2.230762E-01	3.031611E-02
D= 2	**R2= 0.1667**	**R2= 0.3333**	**R2= 0.5000**	**R2= 0.6667**	**R2= 0.8333**
R1= 0.1667	-7.842418E-02	-9.872056E-02	-1.042741E-01	-9.508468E-02	-7.115242E-02
R1= 0.3333	4.679060E-02	3.824557E-02	1.175135E-02	-3.269207E-02	-9.508468E-02
R1= 0.5000	1.277767E-01	1.277767E-01	8.910162E-02	1.175135E-02	-1.042741E-01
R1= 0.6667	1.645343E-01	1.698730E-01	1.277767E-01	3.824557E-02	-9.872056E-02
R1= 0.8333	1.570631E-01	1.645343E-01	1.277767E-01	4.679060E-02	-7.842418E-02
D= 3	**R2= 0.1667**	**R2= 0.3333**	**R2= 0.5000**	**R2= 0.6667**	**R2= 0.8333**
R1= 0.1667	3.985079E-02	4.381449E-02	3.747277E-02	2.082562E-02	-6.126941E-03
R1= 0.3333	-9.058856E-03	-7.041555E-03	-1.388375E-03	7.900685E-03	2.082562E-02
R1= 0.5000	-4.136556E-02	-4.080754E-02	-2.748181E-02	-1.388375E-03	3.747277E-02
R1= 0.6667	-5.706931E-02	-5.748346E-02	-4.080754E-02	-7.041555E-03	4.381449E-02
R1= 0.8333	-5.617012E-02	-5.706931E-02	-4.136556E-02	-9.058856E-03	3.985079E-02
D= 4	**R2= 0.1667**	**R2= 0.3333**	**R2= 0.5000**	**R2= 0.6667**	**R2= 0.8333**
R1= 0.1667	-1.572537E-02	-1.639230E-02	-1.259505E-02	-4.333645E-03	8.391932E-03
R1= 0.3333	1.495212E-03	1.027650E-03	-9.968083E-05	-1.886779E-03	-4.333645E-03
R1= 0.5000	1.299378E-02	1.269474E-02	8.330249E-03	-9.968083E-05	-1.259505E-02
R1= 0.6667	1.877032E-02	1.860896E-02	1.269474E-02	1.027650E-03	-1.639230E-02
R1= 0.8333	1.882485E-02	1.877032E-02	1.299378E-02	1.495212E-03	-1.572537E-02
D= 5	**R2= 0.1667**	**R2= 0.3333**	**R2= 0.5000**	**R2= 0.6667**	**R2= 0.8333**
R1= 0.1667	5.566044E-03	5.638867E-03	4.059246E-03	8.271808E-04	-4.057328E-03
R1= 0.3333	-1.508853E-04	-4.515488E-05	1.530996E-04	4.438782E-04	8.271808E-04
R1= 0.5000	-3.993431E-03	-3.873239E-03	-2.491062E-03	1.530996E-04	4.059246E-03
R1= 0.6667	-5.961594E-03	-5.845385E-03	-3.873239E-03	-4.515488E-05	5.638867E-03
R1= 0.8333	-6.055373E-03	-5.961594E-03	-3.993431E-03	-1.508853E-04	5.566044E-03

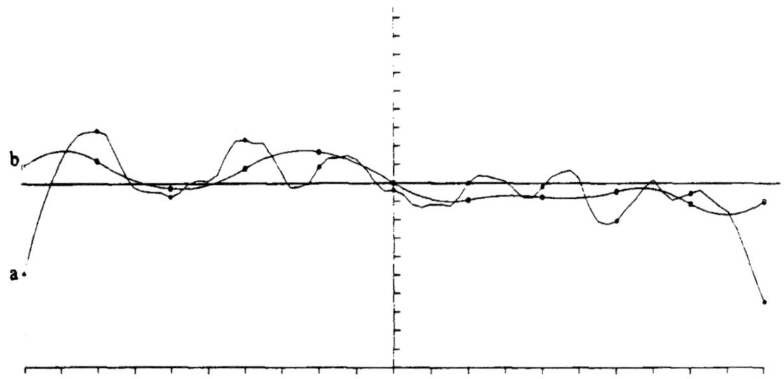

Fig. 3. Monte Carlo simulation from a uniform distribution. Kernel estimate (a) with bin size 1/10. Spline estimate (b) with cell size 1/5. y_0', y_N' estimated by two cell width.

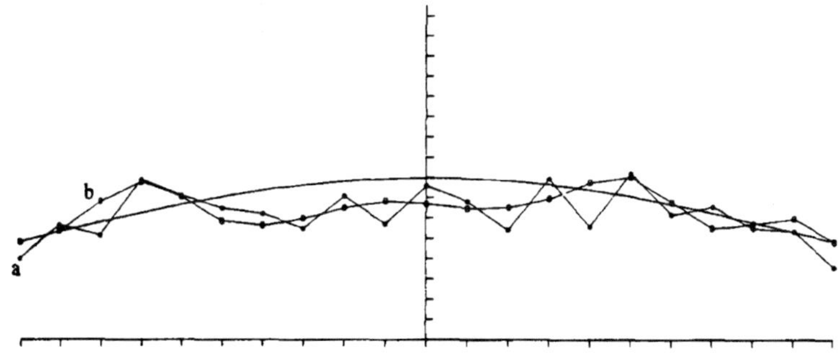

Fig. 4. Monte Carlo simulation from a standard normal on (−1, 1). Kernel and spline estimate as in Fig. 3 except that y_0' and y_N' are specified.

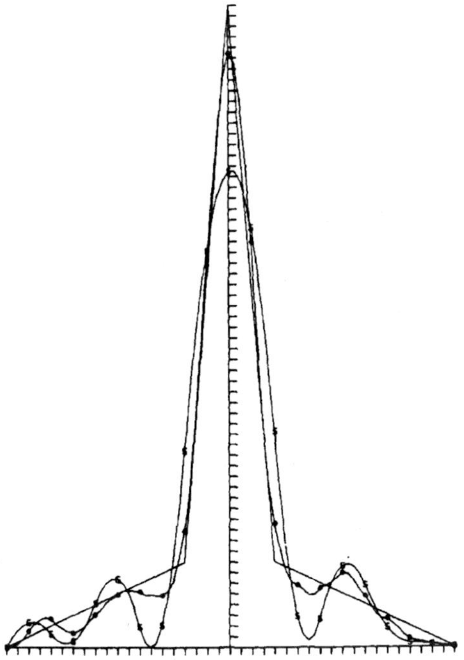

Fig. 5. Monte Carlo simulation from half the sum of two triangular densities with base length 2 and 0·4, respectively. Spline S and kernel □ estimates of the density function.

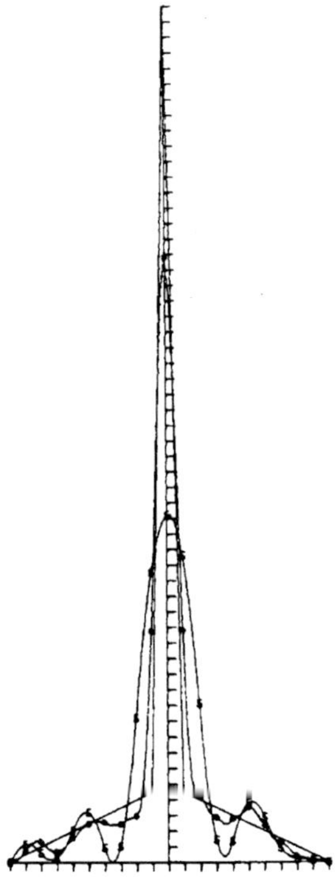

Fig. 6. Monte Carlo simulation from half the sum of two triangular densities with base length 2 and 0·2 respectively.

estimates do seem to work very effectively in smoothing data that is already fairly regular. However, splines seem to have a marked disadvantage relative to a bandlimited kernel in following rapidly changing data as is indicated in Figs. 5 and 6 of the estimates of a superimposed triangular. Notice the undershooting lobes in the spline estimate on either side of the peak in the density function. This might be explained by the character of the covariance function of the spline (8).

We have also analyzed some data that was very kindly given to us by Dr. Wyngaard. The data was already considered in an article of Tennekes and Wyngaard [7]. It consists of readings of the derivative of a turbulent wind velocity (in a fixed direction) taken on the Kansas plains and sampled 3200 times per second for roughly an hour. We estimated the probability density of the wind velocity derivative using kernel and spline estimates and in this way it was different from that given in [7]. The data was already binned. The graph of estimates are given for the left tail of the density of $-(\partial u/\partial t)$ (u the velocity and t the time) in three sections, Figs. 7–9. Notice that even where the spline is a good estimate there appears to be a series of small oscillations which may be due to the fine structure. It is curious that here the kernel estimate (based on a triangular-like weight function) appears to be smoother than the spline. Also, as we get far out into the tail (and closer to zero) one has the impression that the spline can show much greater instability than the kernel estimate.

We give a more detailed description of the turbulence data. The original data is formed into a histogram by combining 12 horizontal units into one bin. The left tail contains data from the -31st bin to the -337th bin. The horizontal scale has the standard deviation $\sigma = 114{\cdot}65193175$ *units or* $\sigma/12$ bins. The vertical scale is the usual one after taking a logarithmic transformation of the data. If the observation or the fitted value is less than or equal to zero, we set it equal to 9·1. The histogram is denoted by +. The kernel type fit is the peicewise linear curve denoted by X. The weight function used here is

$$w(x) = \begin{cases} \frac{1}{2} \text{ if } |x| \leq \frac{1}{2} \\ \frac{1}{4} \quad \text{if } \frac{1}{2} \leq |x| \leq \frac{3}{2} \\ 0 \text{ otherwise} \end{cases}$$

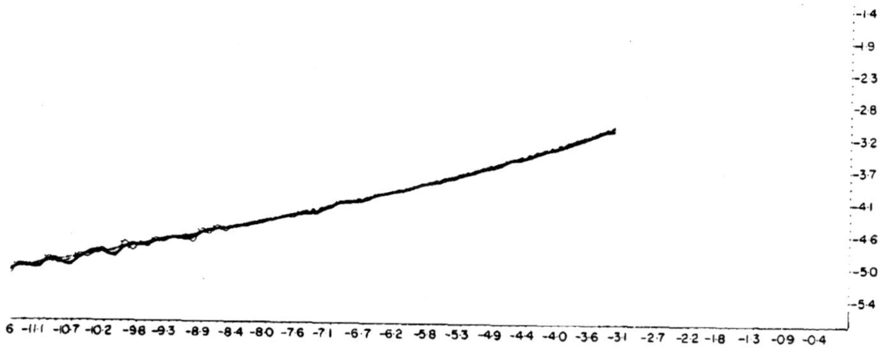

Fig. 7.

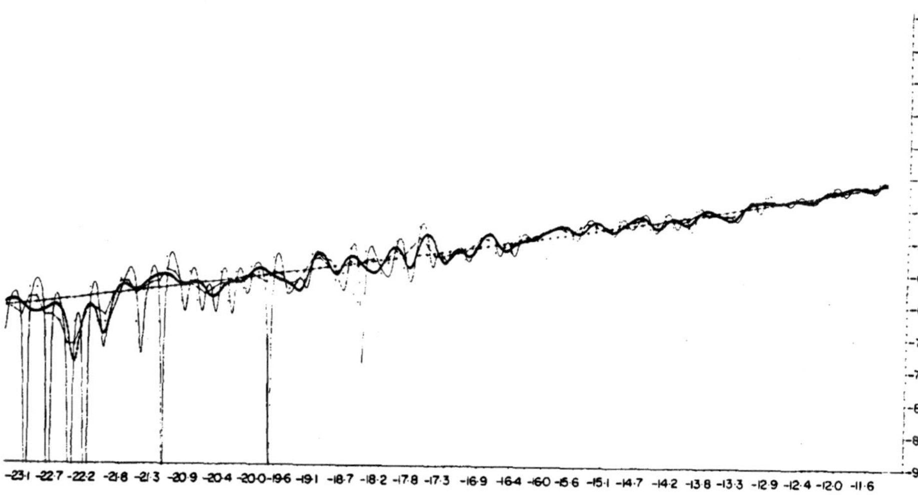

Fig. 8.

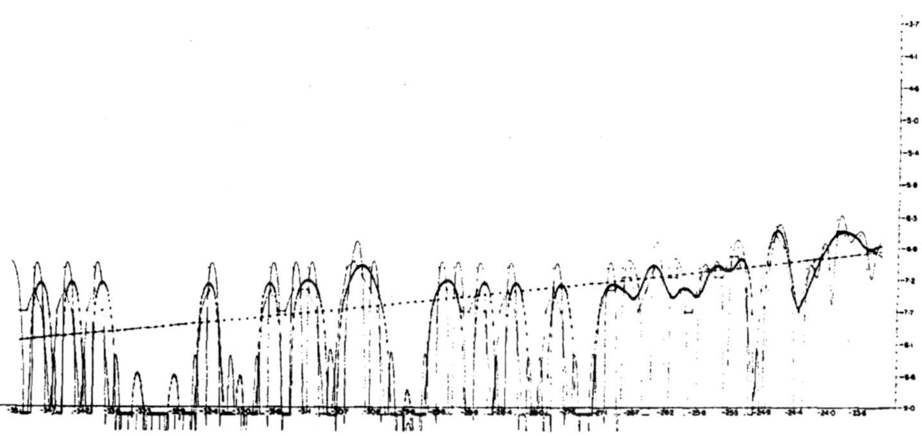

Fig. 9.

Figs. 7–9. Estimation of left tail of the probability density of turbulent wind velocity derivative. Turbulent Reynold's number ≅ 8000. Histogram, kernel and two spline estimates.

and the bandwidth is 3 bins. The spline fit with cell-width equal to one bin is the oscillatory curve without any symbol on it. The boundary condition is of Boneva–Kendall–Stefanov type, that is $y''_0 = y''_N = 0$. The spline fit with cell-width equal to 2 bins is the thick curve denoted by 5. We also try to fit the data with an exponential function $f(x) = A\ e^{-B|x|^C}$ by a least squares technique using the data from -32nd bin to -192nd bin. The procedure is as follows. Given

$$\log f(x) = \log A - B|x|^C$$

$$\frac{d}{dx}\log f(x) = -BC|x|^{C-1}$$

minimize $\Sigma[-a \log f(x_i) + |x_i|\,(d/dx)(\log f(x_i)) + \beta]^2$ and take α as the estimate of C, β as the estimate of $C \log A$, and B as estimated by

$$\frac{\Sigma |x_i|^\alpha . (\log A - \log f(x_i))}{\Sigma(|x_i|^\alpha)^2}$$

where we approximate $(d/dx)(\log f(x_i))$ by $[\log f(x_{i+1}) - \log f(x_{i-1})]/2$. The smooth curve denoted by 8 is the fitted curve $\log_{10}(A\ e^{-B|x|^C})$. The estimated values of A, B and C are $A = 0{\cdot}74$, $B = 4{\cdot}2$, $C = 0{\cdot}41$.

Kolmogorov and Obukhov suggested that the rate of energy dissipation in high Reynold's number turbulence would have a log normal distribution and this was later predicted in a statistical model of Yaglom [8]. The interest in estimating the probability density of the velocity derivative is in part motivated by a desire to see to what extent this prediction is consistent with experimental data.

Acknowledgement—Research supported in part by the office of Naval Research and the National Science Foundation.

REFERENCES

1. J. Ahlberg, E. Nilson and J. Walsh, *The Theory of Splines and their Applications*. Academic Press New York (1967).
2. L. Boneva D. Kendall and I. Stefanov, Spline transformations, *J.. R. statist. Soc.* Ser. B, **33**, 1–70 (1971).
3. G. P. Learmonth and P. A. W. Lewis, Statistical tests of some widely used and recently proposed uniform random number generators. Research report NPS55LW73111A, Naval Postgraduate School, Monterey, California, 1973.
4. M. Rosenblatt, A note on the local behavior of the derivative of a cubic spline interpolator. To appear in J. Approx. Th.
5. M. Rosenblatt Curve Estimates *Ann. math. Statist.* **42**, 1815–1842 (1971).
6. I. J. Schoenberg, Contributions to the problem of approximation of equidistant data by analytic functions, *Q. appl. Math.* **4**, 45–99 112–141 (1946).
7. H. Tennekes and J. C. Wyngaard, The intermittent small-scale structure of turbulence: data processing hazards. *J. Fluid Mech.* **55**, part 1 93–103 (1972).
8. A. M. Yaglom, Influence of fluctuations in dissipation rates in turbulence characteristics in the inertial subrange. *Dokl. Akad. Nauk. SSSR* **166**, 49 (1966).

J. Appl. Prob. **13**, 723–732 (1976)
Printed in Israel

FRACTIONAL INTEGRALS OF STATIONARY PROCESSES AND THE CENTRAL LIMIT THEOREM

M. ROSENBLATT, *University of California, San Diego*

Abstract

A class of limit theorems involving asymptotic normality is derived for stationary processes whose spectral density has a singular behavior near frequency zero. Generally these processes have 'long-range dependence' but are generated from strongly mixing processes by a fractional integral or derivative transformation. Some related remarks are made about random solutions of the Burgers equation.

FRACTIONAL INTEGRAL; STATIONARY PROCESS; CENTRAL LIMIT THEOREM; SINGULAR SPECTRUM; BURGERS EQUATION

1. Introduction

Let $X(t)$, $-\infty < t < \infty$, be a strictly stationary real-valued separable random process with $EX(t) \equiv 0$ and $EX(t)^4 < \infty$. Our interest is in limit theorems for

$$A(T) = \int X(t)g(t/T)dt \tag{1}$$

as $T \to \infty$, where g is a real-valued integrable function. The standard central limit theorem is for

$$g(t) = \begin{cases} 1 & \text{if } 0 < t < 1, \\ 0 & \text{otherwise,} \end{cases} \tag{2}$$

and is often proved under a number of possible strong mixing conditions (see [4] and [5]). Suppose that the process $X(t)$ has an absolutely continuous spectral distribution function F with spectral density f. The usual assumption on the variance of (1) for g given by (2) is that it is proportional to T as $T \to \infty$. In order to arrive at this rate of growth it is usual to assume that

$$\lim_{|\lambda| \to 0} f(\lambda) = c > 0. \tag{3}$$

However, there are a number of special results in which asymptotic normality has been obtained where the spectral density has a behavior like

Received 2 February 1976.
Research carried out while the author was visiting the Australian National University and supported in part by the Office of Naval Research.

R.A. Davis et al. (eds.), *Selected Works of Murray Rosenblatt*, Selected Works in Probability and Statistics, 321
DOI 10.1007/978-1-4419-8339-8_29, © Springer Science+Business Media, LLC 2011

$$f(\lambda) \simeq |\lambda|^{-\alpha}, \qquad \alpha < 1$$

as $|\lambda| \to 0$ (see [1], [3] and [7]). If α is not an even integer, such processes are not strongly mixing (see [2]). Our object is to obtain a broader class of such limit theorems in which the basic behavior is laid out in a way that may be more apparent. The argument is based on the assumption that $X(t)$ can be identified as a fractional integral of a strongly mixing process and g as a fractional derivative of a function.

We shall want to specify conditions on the process $X(t)$ in terms of an auxiliary stationary process $Y(t)$. The conditions on $Y(t)$ will be those used in the theorem of [5]. Let $\mathscr{B}_t(Y)$ be the Borel field generated by the random variables $Y(\tau)$, $\tau \leqq t$, and $\mathscr{F}_t(Y)$ the Borel field generated by the random variables $Y(\tau)$, $\tau \geqq t$. We shall say that the process Y is *strongly mixing* if

$$\sup_{\substack{B \in \mathscr{B}_t \\ F \in \mathscr{F}_\tau}} |P(B \cap F) - P(B)P(F)| = g(\tau - t) \downarrow 0 \tag{4}$$

as $\tau - t \to \infty$. The process Y will be said to satisfy Assumption A if the following conditions are fulfilled.

Assumption A.

(a) Y is a stationary separable process with $EY(t) \equiv 0$, $EY(t)^4 < \infty$, and $E|Y(t) - Y(\tau)|^4 \to 0$ as $t - \tau \to 0$.

(b) Y is strongly mixing.

(c) The covariance function $R(t)$ of Y, $R(t_2 - t_1) = E[Y(t_2)Y(t_1)]$, is absolutely integrable with the spectral density $f(\lambda)$ of Y,

$$f(\lambda) = \frac{1}{2\pi} \int e^{-i\lambda t} R(t)\, dt,$$

positive everywhere.

(d) The fourth-order cumulant function of Y,

$$\begin{aligned} Q(t_2 - t_1, t_3 - t_1, t_4 - t_1) = {} & E[Y(t_1)Y(t_2)Y(t_3)Y(t_4)] \\ & - R(t_2 - t_1)R(t_4 - t_3) - R(t_3 - t_1)R(t_4 - t_2) \\ & - R(t_4 - t_1)R(t_3 - t_2), \end{aligned}$$

is absolutely integrable.

Notice that under Assumption A the spectral density f is continuous as is the fourth-order cumulant spectral density,

$$q(\lambda_1, \lambda_2, \lambda_3) = \iiint \exp(-i\{t_1\lambda_1 + t_2\lambda_2 + t_3\lambda_3\}) Q(t_1, t_2, t_3)\, dt_1 dt_2 dt_3.$$

The notion of a fractional integral (or derivative), when meaningful, will be of some interest. Because of the concern with stationarity and time translation, the Weyl definition of a fractional integral seems to be most appropriate (see [8], Vol. 2, p. 133). Let us call

$$(5)\qquad (I_\alpha Y)(t) = \frac{1}{\Gamma(\alpha)} \int_{-\infty}^{t} (t-\tau)^{\alpha-1} Y(\tau)d\tau$$

$$= \underset{L\to\infty}{\text{l.i.m.}} \frac{1}{\Gamma(\alpha)} \int_{-L}^{t} (t-\tau)^{\alpha-1} Y(\tau)d\tau$$

the αth fractional integral, $0<\alpha<1$, of Y if it is well-defined. Our definition is understood to be given by the limit in mean square on the right of Equation (5). One can define

$$(I_1 Y)(t) = \underset{\alpha\uparrow 1}{\text{l.i.m.}} (I_\alpha Y)(t).$$

An alternative equivalent definition is given by

$$(I_1 Y)(t) = \underset{L\to\infty}{\text{l.i.m.}} \frac{1}{L} \int_0^L \int_{-s}^{t} Y(\tau)d\tau\, ds.$$

Then, recursively, one can define $I_\beta Y$ for all $\beta>0$ for which it exists by

$$(6)\qquad I_{1+\beta} Y = I_1(I_\beta Y), \qquad \beta>0.$$

The fractional derivative $D^\alpha Y$, $0<\alpha<1$, is given by

$$(7)\qquad D^\alpha Y = D^1(I_{1-\alpha} Y)$$

where by D^1 one understands the derivative in mean square. The derivative D^α for all $\alpha>0$ would then be given recursively when it exists by

$$D^{1+\alpha} Y = D^1(D^\alpha Y), \qquad \alpha>0.$$

It is then clear that one could just as well make the identification

$$I_\alpha = D^{-\alpha}$$

for all α. These definitions could be extended slightly to that of fractional integrals (derivatives) centred at frequency μ. Thus the αth fractional integral $I_{\alpha,\mu}$ centred at frequency μ would be given by

$$(8)\qquad (I_{\alpha,\mu} Y)(t) = \frac{1}{\Gamma(\alpha)} \int_{-\infty}^{t} (t-\tau)^{\alpha-1} e^{i\mu\tau} Y(\tau)d\tau$$

for $0<\alpha<1$ with the obvious modification given for other values of α as in the case of an ordinary fractional integral (or derivative).

Lemma 1. *Let Y be a separable weakly stationary process* $(EY(t)\equiv 0)$

$$Y(t) = \int_{-\infty}^{\infty} e^{it\lambda} dZ_Y(\lambda) \tag{9}$$

$$EdZ_Y(\lambda)dZ_Y(\mu) = \delta(\lambda + \mu)dF_Y(\lambda)$$

with corresponding process Z_Y of orthogonal increments and spectral distribution function F_Y. The αth ($0 \leqq \alpha$) fractional integral (derivative) $I_{\alpha,\mu}Y$ ($D^{\alpha,\mu}Y$) centred at μ exists as a weakly stationary process if and only if

$$\int |\lambda - \mu|^{-2\alpha} dF_Y(\lambda) < \infty \tag{10}$$

$$\left(\int |\lambda - \mu|^{2\alpha} dF_Y(\lambda) < \infty\right)$$

and is given by

$$(I_{\alpha,\mu}Y)(t) = \int (i(\lambda - \mu))^{-\alpha} e^{it(\lambda-\mu)} dZ_Y(\lambda)$$

$$\left((D^{\alpha,\mu}Y)(t) = \int (i(\lambda - \mu))^{\alpha} e^{it(\lambda-\mu)} dZ_Y(\lambda)\right). \tag{11}$$

It will be enough to go through the argument for $I_{\alpha,\mu}Y$ with $0 < \alpha < 1$. Then

$$(I_{\alpha,\mu}Y)(t) = \underset{L\to\infty}{\text{l.i.m.}} \frac{1}{\Gamma(\alpha)} \int_{-L}^{t} (t-\tau)^{\alpha-1} e^{-i\mu\tau} Y(\tau)d\tau$$

$$= \int_{-\infty}^{\infty} \frac{1}{\Gamma(\alpha)} \int_{-\infty}^{t} e^{i\tau(\lambda-\mu)}(t-\tau)^{\alpha-1} d\tau\, dZ_Y(\lambda).$$

But

$$\int_{-\infty}^{t} e^{i\tau(\lambda-\mu)}(t-\tau)^{\alpha-1} d\tau = \Gamma(\alpha) e^{it(\lambda-\mu)}(i(\lambda-\mu))^{-\alpha}.$$

The following result is a useful formal analogue of integration by parts.

Lemma 2. Let $Y(t)$ be a separable weakly stationary process, $EY(t) \equiv 0$, and $g(t)$ a bounded integrable function. Then if $(I_\alpha Y)(t)$ exists, $0 < \alpha < 1$, and if

$$g_T(t) = g(t/T)$$

it follows that

$$\int (I_\alpha Y)(t) g_T(t)dt = \int Y(t)(\tilde{I}_\alpha g_T)(t)dt \tag{12}$$

where

$$(\tilde{I}_\alpha g_T)(t) = \int_{-t}^{\infty} (t+\tau)^{\alpha-1} g\left(\frac{\tau}{T}\right) d\tau\, \Gamma(\alpha)^{-1}$$

$$= T \int_{-(t/T)}^{\infty} (t + Tu)^{\alpha-1} g(u) du\, \Gamma(\alpha)^{-1}. \tag{13}$$

Further, if $\alpha < \frac{1}{2}$

$$\int |\tilde{I}_\alpha g_T(t)|^2 dt = \frac{T^{1+2\alpha}}{2\pi} \int |\hat{g}(\lambda)|^2 |\lambda|^{-2\alpha} d\lambda \tag{14}$$

where $\hat{g}$ *is the Fourier transform of* g. *Also*

$$|\tilde{I}_\alpha g_T(t)| \leqq KT^\alpha \tag{15}$$

for all t *with* K *an absolute constant.*

The equality (12) follows straightforwardly on using the Parseval relation twice. One also notes that $(\tilde{I}_\alpha g_T)(t)$ is what one obtains on applying the operator I_α to $g(-t/T)$. The inequality (15) is obtained by using the inequality

$$\begin{aligned}
|\tilde{I}_\alpha g_T(t)| &\leqq \int_{-(t/T)}^{\infty} h(u) \frac{du}{\Gamma(\alpha)} \\
&= \int_{(t/T)}^{(t/T)+1} h(u) \frac{du}{\Gamma(\alpha)} + \int_{-(t/T)+1}^{\infty} h(u) \frac{du}{\Gamma(\alpha)} \\
&\leqq K' \int_{-(t/T)}^{-(t/T)+1} T(t+Tu)^{\alpha-1} du + T^\alpha \int |g(u)|\, du \\
&\leqq KT^\alpha.
\end{aligned}$$

A corresponding result can be given for fractional derivatives.

Lemma 3. *Let* $Y(t)$ *be a separable weakly stationary process,* $EY(t) \equiv 0$, *and* $g(t)$ *a bounded integrable function that satisfies a Lipschitz condition of fixed order* β, $0 < \beta < 1$, *uniformly over* $(-\infty, \infty)$. *Then if* $(D^\alpha Y)(t)$ *exists,* $0 < \alpha < \beta$, *and if* $g_T(t) = g(t/T)$ *it follows that*

$$\int (D^\alpha Y)(t) g_T(t) dt = \int Y(t)(\tilde{D}^\alpha g_T)(t) dt \tag{16}$$

where

$$(\tilde{D}^\alpha g_T)(t) = \int_{-t}^{\infty} (t+\tau)^{-\alpha} g\left(\frac{\tau}{T}\right) d\tau \Gamma(1-\alpha)^{-1}. \tag{17}$$

Moreover, if $\hat{g}(\lambda)|\lambda|^\alpha \in L^2$

$$\int |\tilde{D}^\alpha g_T(t)|^2 dt = \frac{T^{1-2\alpha}}{2\pi} \int |\hat{g}(\lambda)|^2 |\lambda|^{2\alpha} d\lambda. \tag{18}$$

Further,

$$|\tilde{D}^\alpha g_T(t)| \leqq KT^{-\alpha} \tag{19}$$

for all t *with* K *an absolute constant.*

Relation (16) follows by the Parseval relation as in Lemma 2. Now

$$\Gamma(1-\alpha)\tilde{D}^{\alpha} g_T(t) = \frac{d}{dt}\int_{-t}^{\infty} (t+\tau)^{-\alpha} g_T(\tau)d\tau$$

$$= \frac{d}{dt}\int_0^{\infty} u^{-\alpha}\frac{d}{du}\{G_T(u-t) - G_T(-t)\}du$$

where G_T is the indefinite integral of g_T. On integrating by parts we obtain

$$\Gamma(1-\alpha)\tilde{D}^{\alpha} g_T(t) = \frac{d}{dt}\int_0^{\infty} \alpha u^{-1-\alpha}\{G_T(u-t) - G_T(-t)\}du$$

$$= \int_0^{\infty} \alpha u^{-1-\alpha}\{g_T(u-t) - g_T(-t)\}du.$$

The expression

$$(20) \quad \int_0^{\infty} u^{-1-\alpha}\{g_T(u-t) - g_T(-t)\}du = T^{-\alpha}\int_0^{\infty} v^{-1-\alpha}\left\{g\left(v-\frac{t}{T}\right) - g\left(-\frac{t}{T}\right)\right\}dv.$$

Notice that the integral on the right-hand side of (20) is bounded in absolute value by

$$K\int_0^1 v^{-1+\beta-\alpha}dv + \int_{-\infty}^{\infty} |g(v)|\,dv + \frac{1}{\alpha}\left|g\left(-\frac{t}{T}\right)\right|.$$

The assumption that $g \in \text{Lip}(\beta)$ has been used. K is a constant.

The range within which the result of Lemma 3 holds can be broadened somewhat. Let us say that $g \in \Lambda_\beta$, $\beta > 0$, if it is $[\beta]$ times continuously differentiable ($[\beta]$ is the greatest integer less than or equal to β) and its derivative $D^{[\beta]}g$ satisfies a Lipschitz condition of order $\beta - [\beta]$ uniformly over $(-\infty, \infty)$. The result of Lemma 3 then holds for g a bounded integrable function in Λ_β, $\beta > 0$, and $[\beta] \leqq \alpha < \beta$.

It is clear that results corresponding to Lemmas 2 and 3 for the more general context of fractional integrals (derivatives) centred at μ hold and can be derived by the same type of argument.

2. A central limit theorem

We can now obtain a central limit theorem for the sequence $A(T)$ as $T \to \infty$ under appropriate conditions by applying the main theorem of [5].

Theorem. Let $X(t)$ be a separable strictly stationary process that is the αth fractional integral for some α, $|\alpha| < \frac{1}{2}$ of a process $Y(t)$ satisfying Assumption A. *Then the spectral density $h(\lambda)$ of the process $X(t)$ is such that*

$$(21) \qquad \lim_{\lambda \to 0} h(\lambda)|\lambda|^{2\alpha} = f(0) > 0$$

where $f(0)$ is the value of the spectral density of the process $Y(t)$ at zero. Consider the sequence of random variables

$$A(T) = \int X(t)g\left(\frac{t}{T}\right) dt \tag{22}$$

as $T \to \infty$, where g is a bounded integrable function. If $\alpha < 0$ we also assume that g satisfies a Lipschitz condition of order β, $0 < |\alpha| < \beta \leqq 1$, uniformly over $(-\infty, \infty)$. Then

$$A(T)/\sigma(T) \tag{23}$$

is asymptotically normally distributed with mean zero and variance $2\pi f(0)$ as $T \to \infty$, where

$$\sigma^2(T) = \frac{T^{1+2\alpha}}{2\pi} \int |\hat{g}(\lambda)|^2 |\lambda|^{-2\alpha} d\lambda. \tag{24}$$

The remark on the spectral density $h(\lambda)$ of $X(t)$ near zero follows immediately from the fact that $X(t)$ is the αth fractional integral of the process $Y(t)$. By Lemmas 2 and 3 $A(T)$ can be written in the form

$$\int Y(t)\omega_T(t)dt$$

with

$$\omega_T(t) = (\tilde{I}_\alpha g_T)(t)$$

if $0 \leqq \alpha < \frac{1}{2}$ and

$$\omega_T(t) = \tilde{D}^{-\alpha} g_T(t)$$

if $-\frac{1}{2} < \alpha \leqq 0$. Then

$$\sigma^2(T) = W(T) = \int \omega_T(t)^2 dt = \frac{T^{1+2\alpha}}{2\pi} \int |\hat{g}(\lambda)|^2 |\lambda|^{-2\alpha} d\lambda.$$

Further,

$$\lim_{T\to\infty} W(T)^{-1} \int \omega_T(t+h)\omega_T(t)dt = 1$$

for all h. The bounds obtained on $\tilde{I}_\alpha g_T(t)$, $\tilde{D}^{-\alpha} g_T(t)$ in Lemmas 2 and 3 together with Assumption A imply that the conditions for the validity of the basic theorem of [5] are satisfied. Here $M(\lambda)$ is a jump function with total mass one at $\lambda = 0$.

Notice that if $Y(t)$ satisfies Assumption A then the two derived processes (with $|\alpha| < \frac{1}{2}$)

$$(25)\qquad \begin{aligned}(C_{\alpha,\mu}Y)(t) &= \frac{1}{\Gamma(\alpha)}\int_{-\infty}^{t} (t-\tau)^{\alpha-1}\cos\mu\tau\, Y(\tau)d\tau\\ (S_{\alpha,\mu}Y)(t) &= \frac{1}{\Gamma(\alpha)}\int_{-\infty}^{t} (t-\tau)^{\alpha-1}\sin\mu\tau\, Y(\tau)d\tau\end{aligned}$$

are jointly stationary. Of course, these processes are just the real and imaginary parts of $(I_{\alpha,\mu}Y)(t)$. One can also remark that the processes $(C_{\alpha,\mu}Y)(t)$ and $(S_{\alpha,\mu}Y)(t)$ have spectral densities which behave like

$$K\,|\,\lambda \pm \mu\,|^{-2\alpha}$$

in the neighborhood of $\pm\mu$ respectively with K a constant. The following result can be regarded as a corollary of the theorem since it is obtained by essentially the same type of argument.

Corollary. Consider the processes

$$(26)\qquad (C_{\alpha_j,\mu_j}Y)(t), \qquad (S_{\alpha_j,\mu_j}Y)(t)$$

$|\alpha_j| < \frac{1}{2}$, $0 < \mu_1 < \mu_2 < \cdots < \mu_m$, $j = 1, \cdots, m$ *derived from a process* $Y(t)$ *satisfying Assumption* A. *Let* $g_j(t)$, $j = 1, \cdots, m$, *be bounded integrable functions. If* $\alpha_j < 0$ *we also assume that* $g_j(t)$ *satisfies a Lipschitz condition of order* β_j, $0 < |\alpha_j| < \beta_j \leqq 1$. *Let*

$$(27)\qquad \begin{aligned}A_j(T) &= \int (C_{\alpha_j,\mu_j}Y)(t)g_j\left(\frac{t}{T}\right)dt\\ B_j(T) &= \int (S_{\alpha_j,\mu_j}Y)(t)g_j\left(\frac{t}{T}\right)dt.\end{aligned}$$

Then

$$(28)\qquad A_j(T)/\sigma_j(T), \qquad B_j(T)/\sigma_j(T)$$

$j = 1, \cdots, m$ *are asymptotically normal and independent with mean zero and variance* $\pi f(\mu_j)$, *where*

$$(29)\qquad \sigma_j^2(T) = \frac{T^{1+2\alpha_j}}{2\pi}\int |\hat{g}_j(\lambda)|^2|\lambda|^{-2\alpha_j}d\lambda.$$

In the context of the theorem, one could consider

$$A(aT)/\sigma(T), \qquad a > 0,$$

as a random process and it is clear that on heuristic grounds one would expect to have weak convergence as $T \to \infty$ to a Gaussian random process $Z(a)$ with mean zero and covariance function

$$\mathrm{Cov}(Z(a), Z(b)) = 2\pi f(0)\{|a|^{1+2\alpha} + |b|^{1+2\alpha} - |a-b|^{1+2\alpha}\},$$

$a, b > 0$. This type of result has been obtained by Taqqu in [7] for partial sums of instantaneous functions of Gaussian stationary sequences under appropriate conditions. His result corresponds to $g(t)$ given by (2).

3. Random solutions of Burgers equation

In Section 6 of [6], a class of random solutions to the Burgers equation was considered. Some doubt was expressed as to the asymptotic normality of the solutions in the one-dimensional case. The question is related to some of the issues brought up in this paper. We shall remark again about the reasons for the doubt in [6], but shall show that under appropriate conditions one does in fact get asymptotic normality.

The Burgers equation for $u = u(t,x)$ is

$$u_t + uu_x = \mu u_{xx}, \qquad \mu > 0. \tag{30}$$

Consider a random solution of (30), $u = 2\mu\Psi_x(c - \Psi)^{-1}$, with c a positive constant and $\Psi(t, x)$ a solution of the heat equation $\phi_t = \mu\phi_{xx}$, strictly stationary in x with

$$\max_x |\Psi_0(x)| < c, \qquad E\Psi_0(x) \equiv 0 \tag{31}$$

where $\Psi_0(x) = \Psi(x, 0)$. Let the spectral distribution function of Ψ_0 be $F(\lambda)$. Since $u_0(x) = u(0, x) = 2\mu\Psi_{0x}(c - \Psi_o)^{-1}$, we require that $\int \lambda^2 dF(\lambda) < \infty$. Notice that (31) implies that $\max_x |\Psi(x, t)| < c$ for all t. We also assume that Ψ_0 satisfies Assumption A and so is in particular strongly mixing. These conditions could be weakened somewhat but they are kept so as to be consistent with [6]. We are interested in the asymptotic distribution of $u(x, t)$ for fixed x as $t \to \infty$ when it is appropriately centred and normalized. Since

$$\Psi(x, t) = \int_{-\infty}^{\infty} (4\pi\mu t)^{-\frac{1}{2}} \exp\left\{-\frac{(x-y)^2}{4\mu t}\right\} \Psi_0(y)dy$$

it follows that $\max_x |\Psi(x, t)| \to 0$ as $t \to \infty$ and $u(x, t) \simeq 2\mu c^{-1}\Psi_x$. But

$$\Psi_x(x, t) = -\int (4\pi\mu t)^{-\frac{1}{2}} \exp\left\{-\frac{(x-y)^2}{4\mu t}\right\} \Psi_{0y}(y)dy. \tag{32}$$

Notice that $\Psi_{0y}(y)$ is a strongly mixing process with spectral density $\lambda^2 f(\lambda)$, $f(\lambda) = F'(\lambda)$. In looking at (32) one is narrow band-pass filtering the process $\Psi_{0y}(y)$ and this is a strongly mixing process whose spectral density has a high order zero at $\lambda = 0$. Under such circumstances the usual form of central limit theorem need not hold, and this led to possible doubts about the asymptotic normality of (32). We give the following simple example. Consider a homogeneous process $\xi(t)$, $-\infty < t < \infty$, with independent increments. In particular, one could take $\xi(t)$ to be a Poisson process. Set $X(t) = \xi(t) - 2\xi(t-1) + \xi(t-2)$. The process $X(t)$ is then a strongly mixing stationary process with covariance function

$$R(t) = \begin{cases} 2 - 3|t| & \text{if } |t| \leqq 1, \\ -2 + |t| & \text{if } 1 \leqq |t| \leqq 2, \\ 0 & \text{otherwise.} \end{cases}$$

Notice that

$$\int_0^T X(u)du = \int_{T-1}^T \{\xi(u) - \xi(u-1)\}du - \int_0^1 \{\xi(u-1) - \xi(u-2)\}du, \tag{33}$$

so that the sequence (33) has a non-normal limiting distribution as $T \to \infty$ if one does not normalize at all.

Returning to (32), it is clear that one can write it as

$$\Psi_x(x,t) = \int \Psi_0(y)(4\pi\mu t)^{-\frac{1}{2}} \frac{\partial}{\partial y} \exp\left\{-\frac{(x-y)^2}{4\mu t}\right\} dy.$$

Without loss of generality set $x = 0$. In this form one can see that the theorem of [5] is applicable and so one does have asymptotic normality. The key is that we are dealing with an expression of the form

$$\int g\left(\frac{t}{T}\right) \frac{d}{dt} X(t)dt, \tag{34}$$

with $X(t)$ a process satisfying Assumption A. If g is absolutely continuous with $\lim_{|t|\to\infty} g'(t) = 0$ and g, g' absolutely integrable, (34) can be rewritten as $-(1/T)\int g'(t/T)X(t)dt$ and the expression is asymptotically normal as $T \to \infty$. In the case of (33), the function g is not differentiable.

The global asymptotic behavior of the whole process (32) can be determined as $T \to \infty$. However, that would be better pursued elsewhere.

References

[1] Davydov, Yu. A. (1970) The invariance principle for stationary processes. *Theor. Prob. Appl.* **15**, 487–498.

[2] Helson, H. and Sarason, D. (1967) Past and future. *Math. Scand.* **21**, 5–16.

[3] Mandelbrot, B. B. (1975) Limit theorems on the self-normalized range for weakly and strongly dependent processes. *Z. Wahrscheinlichkeitsth.* **31**, 271–285.

[4] McLeish, D. L. (1975) Invariance principles for dependent variables. *Z. Wahrscheinlichkeitsth.* **32**, 165–178.

[5] Rosenblatt, M. (1961) Some comments on narrow band-pass filters. *Quart. Appl. Math.* **18**, 387–393.

[6] Rosenblatt, M. (1968) Remarks on the Burgers equation. *J. Math. Phys.* **9**, 1129–1136.

[7] Taqqu, M. (1975) Weak convergence to fractional Brownian motion and to the Rosenblatt process. *Z. Wahrscheinlichkeitsth.* **31**, 287–302.

[8] Zygmund, A. (1968) *Trigonometric Series.* Cambridge University Press.

J. Appl. Prob. 17, 265–270 (1980)
Printed in Israel
0021–9002/80/010265–06 $00.85

LINEAR PROCESSES AND BISPECTRA

M. ROSENBLATT,* *University of California, San Diego*

Abstract

A linear process is generated by applying a linear filter to independent, identically distributed random variables. Only the modulus of the frequency response function can be estimated if only the linear process is observed and if the independent identically distributed random variables are Gaussian. In this case a number of distinct but related problems coalesce and the usual discussion of these problems assumes, for example, in the case of a moving average that the zeros of the polynomial given by the filter have modulus greater than one. However, if the independent identically distributed random variables are non-Gaussian, these problems become distinct and one can estimate the transfer function under appropriate conditions except for a possible linear phase shift by using higher-order spectral estimates.

LINEAR PROCESS; BISPECTRUM; NON-GAUSSIAN; FREQUENCY RESPONSE FUNCTION; MODULUS; PHASE; AUTOREGRESSIVE MOVING AVERAGE MODEL; LINEAR AND NON-LINEAR PREDICTION; ESTIMATION

Let v_t, $t = \cdots, -1, 0, 1, \cdots$, be independent identically distributed random variables with mean zero, $Ev_t \equiv 0$, and variance one, $Ev_t^2 \equiv 1$. Consider a sequence of real constants $\{\alpha_j\}$ with

$$\sum_{j=-\infty}^{\infty} \alpha_j^2 < \infty.$$

The process

$$x_t = \sum_{j=-\infty}^{\infty} \alpha_j v_{t-j} \tag{1}$$

is the linear process generated by $\{\alpha_j\}$ and $\{v_t\}$. The frequency response function is

$$\alpha(e^{-i\lambda}) = \sum_j \alpha_j e^{-ij\lambda} \tag{2}$$

Received 5 December 1978; revision received 23 February 1979.
* Postal address: Department of Mathematics, University of California, San Diego, La Jolla, CA 92093, U.S.A.
Research partially supported by the Office of Naval Research Contract N00014–75–C–0428.

R.A. Davis et al. (eds.), *Selected Works of Murray Rosenblatt*, Selected Works in Probability and Statistics, DOI 10.1007/978-1-4419-8339-8_30, © Springer Science+Business Media, LLC 2011

and the spectral density of $\{x_t\}$ is

(3) $$f(\lambda) = \frac{1}{2\pi} |\alpha(e^{-i\lambda})|^2.$$

We shall actually assume that $\Sigma|\alpha_j| < \infty$ (and often more) so that $\alpha(e^{-i\lambda})$ and $f(\lambda)$ are continuous. Of course, if both x_t and v_t, $t = 1, \cdots, T$, are observed $\alpha(e^{-i\lambda})$ can be estimated by estimating the cross-spectral density of $\{x_t\}$ and $\{v_t\}$.

However, if only $\{x_t\}$ is observed a number of distinct but related problems arise. If $\{v_t\}$ (and hence $\{x_t\}$) is Gaussian, then only $f(\lambda)$ or equivalently the modulus of $|\alpha(e^{-i\lambda})|$ can be estimated. It is then usual to assume that if $\{x_t\}$ is an autoregressive moving average process

(4) $$\sum_{j=0}^{p} b_j x_{t-j} = \sum_{k=0}^{q} a_k v_{t-k}, \qquad (b_0, q_0 \neq 0)$$

that the roots of the polynomials

(5) $$A(z) = \sum_{k=0}^{q} a_k z^k$$
$$B(z) = \sum_{j=0}^{p} b_j z^j$$

are distinct and all have modulus greater than one (see [1] and [4]). It is clear that

(6) $$\alpha(e^{-i\lambda}) = \frac{A(e^{-i\lambda})}{B(e^{-i\lambda})}.$$

In the Gaussian case any zero z_j can be replaced by $\bar{z}_j^{-1}$ without changing $f(\lambda)$ so that there are generally 2^{p+q} possible specifications of the zeros without changing the structure of $\{x_t\}$. However, the specification in which the roots all have modulus greater than one corresponds to the prediction problem in the following sense. Then $\{x_t\}$ can be represented as a one-sided moving average in terms of $\{v_t\}$:

(7) $$x_t = \sum_{j=0}^{\infty} \beta_j v_{t-j}, \qquad \sum|\beta_j| < \infty.$$

v_t can be represented as a one-sided moving average in terms of x_t

(8) $$v_t = \sum_{j=0}^{\infty} \gamma_j x_{t-j}, \qquad \sum|\gamma_j| < \infty.$$

Thus v_t is independent of the past of $\{x_t\}$, that is, $x_{t-1}, x_{t-2}, \cdots$, and the best one-step predictor of x_t in terms of $x_{t-1}, x_{t-2}, \cdots$ is

(9) $$x_t^* = -\sum_{j=1}^{p} \frac{b_j}{b_0} x_{t-j} + \sum_{k=1}^{q} \frac{a_k}{b_0} v_{t-k}$$

with prediction error

(10) $$x_t - x_t^* = \frac{a_0}{b_0} v_t, \qquad E\,|x_t - x_t^*|^2 = \left|\frac{a_0}{b_0}\right|^2.$$

It is also true that if one regards an autoregressive scheme

$$\sum_{j=0}^{p} b_j x_{t-j} = v_t$$

as a system characterized by x_t being driven by v_t, then the condition on roots of $B(z)$ having modulus greater than one is equivalent to 'stability' of the system (see [2], [5]). Of course, if one is just interested in estimating the spectral density

(11) $$f(\lambda) = \frac{1}{2\pi} \left| \frac{A(e^{-i\lambda})}{B(e^{-i\lambda})} \right|^2,$$

of an autoregressive moving average scheme, the roots taken with modulus greater than one may be as convenient a set of parameters for such a finite parameter scheme as any others.

The prediction problem for a Gaussian sequence has the same character for forward prediction as for backward prediction. This is decidedly not the case for linear processes with non-Gaussian v_t. The simplest case is that of the stationary autoregressive sequence

(12) $$x_t = \frac{1}{2} x_{t-1} + v_t$$

with

$$v_t = \begin{cases} \frac{1}{2} \text{ with probability } \frac{1}{2} \\ 0 \text{ with probability } \frac{1}{2} \end{cases}$$

and v_t independent of the past of the x_t process. Here the best predictor of x_t in terms of the past (in the sense of minimizing mean square error of prediction) is

$$x_t^* = \frac{1}{2} x_{t-1} + \frac{1}{4}.$$

The fact that v_t, x_t do not have mean zero here should not trouble one. It just accounts for the constant $\frac{1}{4}$ in the expression for the best predictor. Notice that the marginal distribution of x_t is uniform $[0, 1]$. The best predictor of x_t given the future is

$$\bar{x}_t^* = 2x_{t+1} \text{ modulo one} = x_t.$$

Thus the backward predictor is non-linear and has variance of prediction error zero (perfect prediction) while the forward predictor is linear and has positive variance of prediction error. The process (12) is purely non-deterministic going forwards in time and purely deterministic going backwards in time. Of course, this simple example is a bit extreme. If one has an autoregressive moving average process $\{x_t\}$ with the v_t's non-Gaussian and by chance the zeros of $A(z)$, $B(z)$ of modulus greater than one, then it is still true that the best predictor of x_t in terms of the past is given by (9) with prediction error (10). But if the zeros of $A(z)$, $B(z)$ do not all have modulus greater than one, then the best predictor of x_t given the past will no longer have the form (9) and will generally be non-linear. Also, as we shall see, the function (6) and hence the zeros (as contrasted with the Gaussian case) will generally be identifiable.

A simple example is now given of a situation in which there is interest in estimating $\alpha(e^{-i\lambda})$ or as much of it as one can manage. Suppose $\{v_t\}$ is independent, identically distributed non-Gaussian. It is passed through the linear filter with coefficients $\{\alpha_j\}$ by (1) to get $\{x_t\}$. The experimenter observes only $\{x_t\}$ (not $\{v_t\}$) and wishes to estimate $\alpha(e^{-i\lambda})$. A particular case is that of the moving average

$$x_t = 6v_t - 5v_{t-1} + v_{t-2} \tag{13}$$

where the roots of $A(z)$ are 2 and 3. The moving average

$$x_t = 3v_t - 7v_{t-1} + 2v_{t-2} \tag{14}$$

has a polynomial $A(z)$ with roots at $\frac{1}{2}$ and 3. Both schemes (13) and (14) have the same spectral density but if the independent, identically distributed v_t's are exponential, the marginal distributions of x_t for these schemes are different.

Assume that one has a linear process where the independent, identically distributed non-Gaussian variables v_t have finite moment $m_k = Ev_t^k (k > 2)$ and corresponding cumulant γ_k. The kth-order cumulant spectral density is then

$$\begin{aligned}\frac{1}{(2\pi)^{k-1}} \sum_{j_1,\cdots,j_{k-1}} \operatorname{cum}(x_t, x_{t+j_1}, \cdots, x_{t+j_{k-1}}) \exp\left(-\sum_{s=1}^{k-1} i j_s \lambda_s\right) \\ = \frac{1}{(2\pi)^{k-1}} \alpha(e^{-i\lambda_1})\alpha(e^{-i\lambda_2}) \cdots \alpha(e^{-i\lambda_{k-1}})\alpha(e^{i(\lambda_1+\lambda_2+\cdots+\lambda_{k-1})})\gamma_k.\end{aligned} \tag{15}$$

We shall now indicate circumstances under which $\alpha(e^{-i\lambda})$ can be identified except for a linear phase shift.

Theorem. Let $\{x_t\}$ be a linear process generated by independent, identically distributed random variables $\{v_t\}$ with finite third cumulant $\gamma = \gamma(v) \neq 0$. Further let the coefficients α_j be such that

$$\sum_j |j| |\alpha_j| < \infty \tag{16}$$

and $\alpha(e^{-i\lambda}) \neq 0$ for any λ. Then $\alpha(e^{-i\lambda})$ is identifiable except for a linear phase shift if one observes the process $\{x_t\}$. The conclusion of the theorem is still valid if the $\{v_t\}$ have finite kth $(k > 2)$ cumulant $\gamma_k = \gamma_k(v) \neq 0$.

Obviously $|\alpha(e^{-i\lambda})|$ is identifiable since $|\alpha(e^{-i\lambda})| = (2\pi)^{\frac{1}{2}}[f(\lambda)]^{\frac{1}{2}}$. We thus have only to deal with $\arg\{\alpha(e^{-i\lambda})\}$. Let

$$h(\lambda) = \arg\{\alpha(e^{-i\lambda})\} - \arg\{\alpha(1)\}.$$

Since the α_j's are real, $h(-\lambda) = -h(\lambda)$. The bispectrum of the x process is

$$b(\lambda, \mu) = \frac{1}{(2\pi)^2}\alpha(e^{-i\lambda})\alpha(e^{-i\mu})\alpha(e^{i(\lambda+\mu)})\gamma.$$

$(\alpha(1)/|\alpha(1)|)\gamma$ can be determined since

$$\frac{\alpha(1)}{|\alpha(1)|}\gamma = (2\pi)^{\frac{1}{2}}\{b(0,0)\}/\{f(0)\}^{\frac{3}{2}}.$$

Also

$$h(\lambda) + h(\mu) - h(\lambda+\mu)$$

can be determined since

$$\begin{aligned} &h(\lambda) + h(\mu) - h(\lambda+\mu) \\ &= \frac{1}{i}\log[(2\pi)^{\frac{1}{2}}\gamma^{-1}b(\lambda,\mu)\{f(\lambda)f(\mu)f(\lambda+\mu)\}^{-\frac{1}{2}}] - \arg\{\alpha(1)\}. \end{aligned}$$

Notice that

$$h'(0) - h'(\lambda) = \lim_{\Delta \downarrow 0}\frac{1}{\Delta}\{h(\lambda) + h(\Delta) - h(\lambda+\Delta)\}.$$

Set $c = h'(0)$. Then

$$h(\lambda) = \int_0^\lambda \{h'(u) - h'(0)\}du + c\lambda.$$

Thus

$$\alpha(e^{-i\lambda}) = u(\lambda)e^{ic\lambda}$$

and by the simple argument given above $\alpha(e^{-i\lambda})$ is identifiable except for sign and for the linear phase shift $c\lambda$. The shift $c\lambda$ is clearly not identifiable under our assumptions unless one makes additional specifications. Notice that if $c = n$ (integer), this just corresponds to a time shift of n units.

The argument given can be immediately elaborated to take care of finite kth cumulant $\gamma_k \neq 0$ $(k > 3)$ if $\gamma_3 = 0$.

The kth cumulant spectrum of the process is given by (15) and a simple adaptation of the proof of the theorem yields the result.

Given that $x_1, x_2, \cdots, x_N$ is observed one can easily suggest a consistent estimate of $\alpha(e^{-i\lambda})$ up to the unidentifiable phase shift $c\lambda$. The estimate is suggested by the approximation

$$\int_0^\lambda \{h'(u) - h'(0)\}du \cong \sum_{j=1}^{[m\lambda]} \left\{h'\left(\frac{j}{m}\right) - h'(0)\right\}\frac{1}{m}$$

$$\cong -\sum_{j=1}^{[m\lambda]} \left\{h\left(\frac{j}{m}\right) + h\left(\frac{1}{m}\right) - h\left(\frac{j+1}{m}\right)\right\}$$

as $m \to \infty$. Let $f_N(\lambda)$, $b_N(\lambda, \mu)$ be estimates of the spectral density and bispectral density with symmetric weight functions (see [3] for a discussion of such estimates). Let β_N be the linear bandwidth of the estimates so that $\beta_N \downarrow 0$, $N\beta_N^2 \to \infty$. Now $f_N(\lambda)^{\frac{1}{2}}(2\pi)^{\frac{1}{2}}$ is a consistent estimate of $|\alpha(e^{-i\lambda})|$. Further

$$(2\pi)^{\frac{1}{2}}\{b_N(0,0)\}\{f_N(0)\}^{-3/2}$$

is a consistent estimate of $(\alpha(1)/|\alpha(1)|)\gamma$. This implies that

$$(2\pi)^{\frac{1}{2}} f_N(\lambda)^{\frac{1}{2}} \prod_{j=1}^{[m\lambda]} \left(\left[b_N\left(\frac{1}{m}, \frac{j}{m}\right)\left\{f_N\left(\frac{1}{m}\right) f_N\left(\frac{j}{m}\right) f_N\left(\frac{j+1}{m}\right)\right\}^{-\frac{1}{2}}\right]^{-1} b_N(0,0) f_N(0)^{-\frac{3}{2}}\right)$$

will be a consistent estimate of $\alpha(e^{-i\lambda})$ up to sign and the linear phase shift $c\lambda$ if $m\beta_N^2 \to 0$, $m(N\beta_N^2)^{-1} \to 0$ as $m, N \to \infty$, $\beta_N \to 0$. Of course, this is a complicated and unwieldy estimate.

Notice that when dealing with the autoregressive moving average scheme (4), the specification $a_0, b_0 \neq 0$ implies that the phase shift $c\lambda \equiv 0$.

References

[1] Anderson, T. W. (1977) Estimation for autoregressive moving average models in the time and frequency domain, *Ann. Statist.* **5**, 842–865.

[2] Åström, K. (1970) *Introduction to Stochastic Control Theory*. Academic Press, New York.

[3] Brillinger, D. R. and Rosenblatt, M. (1967) Asymptotic theory of estimates of kth order spectra. In *Spectral Analysis of Time Series*, ed. B. Harris, Wiley, New York, 153–188.

[4] Hannan, E. J. (1970) *Multiple Time Series*. Wiley, New York.

[5] Whittle, P. (1963) *Prediction and Regulation*. English Universities Press, London.

Z. Wahrscheinlichkeitstheorie verw. Gebiete
55, 123–132 (1981)

Zeitschrift für
Wahrscheinlichkeitstheorie
und verwandte Gebiete

Limit Theorems for Fourier Transforms of Functionals of Gaussian Sequences

M. Rosenblatt*

Dept. of Mathematics, University of California, San Diego
La Jolla, California 92093, USA

Summary. Limit theorems with a non-Gaussian (in fact nonstable) limiting distribution have been obtained under suitable conditions for partial sums of instantaneous nonlinear functions of stationary Gaussian sequences with long range dependence. Analogous limit theorems are here obtained for finite Fourier transforms of instantaneous nonlinear functions of stationary Gaussian sequences with long range dependence.

Introduction

A number of authors (see [2–7]) have considered non-central limit theorems for partial sums derived from nonlinear functionals of Gaussian sequences. One considers a stationary Gaussian sequence X_n, $n=\ldots,-1,0,1,\ldots$, $EX_n=0$, $EX_n^2=1$. Let the correlation function of $\{X_n\}$ be

$$r(n)=EX_0X_n. \tag{1}$$

A real function $H(x)$ is considered with

$$\int_{-\infty}^{\infty} H(x)\exp\left(-\frac{x^2}{2}\right)dx=0 \tag{2}$$

and

$$\int_{-\infty}^{\infty} H(x)^2\exp\left(-\frac{x^2}{2}\right)dx<\infty. \tag{3}$$

The derived sequence $H(X_n)$, $n=\ldots,-1,0,1,\ldots$ is defined and the limiting behavior of the sequence

$$Y_n^N=\frac{1}{A_N}\sum_{j=N(n-1)}^{Nn-1} H(X_j), \qquad \begin{array}{l} n=\ldots,-1,0,1,\ldots \\ N=1,2,\ldots \end{array} \tag{4}$$

* Research supported in part by the Office of Naval Research Contract N00014-75C-0428.

0044-3719/81/0055/0123/$02.00

is determined as $N\to\infty$, where A_N is a suitable positive norming factor. Under appropriate conditions of long-range dependence on the sequence $\{r(n)\}$ and other conditions on the function H, nonnormal (non-central) limiting distributions for the sequence $\{Y_n^N, n=\ldots,-1,0,1,\ldots\}$ have been obtained as $N\to\infty$.

The object of this note is to obtain analogous limit theorems for finite Fourier transforms

$$Y_n^N(\beta)=\frac{1}{A_N}\sum_{j=N(n-1)}^{Nn-1} H(X_j)\,e^{-ij\beta}. \tag{5}$$

We could deal with the Fourier transform in complex form or equivalently the real and imaginary parts under appropriate conditions as $N\to\infty$. The behavior of the Fourier transform is of interest because it is a basic ingredient in the construction of spectral estimates [1]. Various of the ideas used in the paper of Dobrushin and Major [3] are helpful in deriving such a result.

We shall assume that the covariance

$$r(n)=|n|^{-\alpha}L(|n|)\sum_{j=0}^{m} s_{\lambda_j}\cos n\lambda_j, \qquad s_{\lambda_j}>0 \tag{6}$$

for some positive integer m with $0=\lambda_0<\lambda_1<\lambda_2<\ldots<\lambda_m$ and $L(t)$, $0<t<\infty$, a slowly varying function, i.e.,

$$\lim_{s\to\infty}\frac{L(st)}{L(s)}=1 \tag{7}$$

for every $t\in(0,\infty)$. Let $H_j(x)$ be the j-th Hermite polynomial with leading coefficient 1. The function $H(x)$ is then expanded in terms of the Hermite polynomials

$$H(x)=\sum_{j=1}^{\infty} c_j H_j(x) \tag{8}$$

where

$$\sum_{j=1}^{\infty} c_j^2\, j!<\infty. \tag{9}$$

For convenience, let B_n^N denote

$$B_n^N=\{j\,|\,j\in Z,\ nN\leqq j<(n+1)N\}. \tag{10}$$

Also let Λ_k be the set of frequencies obtained by taking sums of any k elements (with repetition allowed) out of the set $\{0, \pm\lambda_1, \pm\lambda_2, \ldots, \pm\lambda_m\}$.

We also introduce the following complex-valued Gaussian random measures. W_0 is the spectral measure of the white noise process so that

$$W_0(\Delta)=\overline{W_0(-\Delta)} \tag{11}$$

$$E|W_0(\Delta)|^2=\frac{1}{2\pi}|\Delta| \tag{11'}$$

for any interval Δ. Also, for any disjoint intervals $\Delta_1, \ldots, \Delta_j$ on the positive axis $W_0(\Delta_1), \ldots, W_0(\Delta_j)$ are independent. Further $\operatorname{Re} W_0(\Delta)$, $\operatorname{Im} W_0(\Delta)$ are independent Gaussian variables with mean zero and equal variances if Δ is an interval on the positive axis. W_μ for $\mu>0$ is a Gaussian random measure with the same properties as W_0 with the following exception. We no longer have (11). Also $W_\mu(\Delta_1), \ldots, W_\mu(\Delta_j)$ are independent for *any* disjoint intervals $\Delta_1, \ldots, \Delta_j$ on the real axis. Further the random measure $W_{-\mu}$, $\mu>0$, is specified so that

$$W_{-\mu}(\Delta)=\overline{W_\mu(-\Delta)} \tag{12}$$

for any interval Δ. Such measures W_μ are introduced for $\mu=\pm\lambda_1, \ldots, \pm\lambda_m$ and it is assumed that $W_{\lambda_1}, \ldots, W_{\lambda_m}$ are independent.

Asymptotic Distribution of Fourier Transforms

We state our result below.

Theorem. *Let* (6) *hold with* $\alpha<\frac{1}{k}$ *where k is the smallest index in the series* (8) *for which* $c_k \neq 0$. *Set*

$$A_N=N^{1-\frac{k\alpha}{2}} L(N)^{\frac{k}{2}}. \tag{13}$$

Then the finite dimensional distributions of

$$Y_n^N(\beta), \qquad n=\ldots,-1,0,1,\ldots,\beta\in\Lambda_k,$$

tend to those of

$$Y_n^*(\beta), \qquad n=\ldots,-1,0,1,\ldots,\beta\in\Lambda_k,$$

given by

$$\begin{aligned} Y_n^*(\beta)=D^{-k/2} c_k \int & e^{in(x_1+\ldots+x_k)} \frac{e^{i(x_1+\ldots+x_k)}-1}{i(x_1+\ldots+x_k)} \\ & \cdot |x_1|^{\frac{\alpha-1}{2}} \ldots |x_k|^{\frac{\alpha-1}{2}} \\ & \cdot \sum_{\mu_1+\ldots+\mu_k\equiv\beta \bmod 2\pi}{}' \left(s_{\mu_1}\frac{1+\delta(\mu_1)}{2} \ldots s_{\mu_k}\frac{1+\delta(\mu_k)}{2}\right)^{\frac{1}{2}} \\ & \cdot W_{\mu_1}(dx_1)\ldots W_{\mu_k}(dx_k) \end{aligned} \tag{14}$$

where

$$D=\int_{-\infty}^{\infty} \exp(ix)|x|^{\alpha-1}\,dx=2\Gamma(\alpha)\cos\left(\frac{\alpha\pi}{2}\right) \tag{15}$$

and Σ' *denotes a sum over k-tuples of* μ *values where the* μ_j*'s can only range over* $0, \pm\lambda_1, \ldots, \pm\lambda_m$.

The multiple Wiener integral has to be modified so as to take account of the fact that $W_0(-dx)=\overline{W_0(dx)}$ and $W_{-\mu}(-dx)=\overline{W_\mu(dx)}$. See [2] for a discussion of such questions.

For convenience let

$$\lambda_{-j}=\lambda_j$$

and

$$s_{\lambda_{-j}}=s_{\lambda_j}, \quad j=1,\ldots,m.$$

Lemma 1. *Let $a>0$ be any real number satisfying*

$$a<\tfrac{1}{2}\min_{i\neq j}|\lambda_i-\lambda_j|.$$

Then there is an infinitely differentiable symmetric function f satisfying

$$\begin{aligned} 0\leqq f(x)\leqq 1 &\quad \text{for all } x \\ f(x)\equiv 1 &\quad \text{for all } |x|<a/2 \\ f(x)\equiv 0 &\quad \text{for all } |x|>a. \end{aligned} \tag{16}$$

Let

$$X_n=\int_{-\infty}^{\infty} e^{inx}\,dz(x),\ {}_jX_n=\int_{-\infty}^{\infty} e^{inx}f(x-\lambda_j)\,dz(x)=\int_{-\infty}^{\infty} e^{inx}f_j(x)\,dz(x),$$

$$j=-m,\ldots,m, \tag{17}$$

and

$$\begin{aligned} {}_\phi X_n=X_n-\sum_{j=-m}^{m} {}_jX_n &= \int_{-\pi}^{\pi} e^{inx}f_\phi(x)\,dz(x) \\ &= \int_{-\pi}^{\pi} e^{inx}\left(1-\sum_{j=-m}^{m} f_s(x)\right)dz(x). \end{aligned} \tag{18}$$

Let $r_j(n)$ be the covariance sequence of ${}_jX_n$, $j=-m,\ldots,m,\phi$. Then

$$\begin{aligned} r_j(n) &= \int e^{inx}|f_j(x)|^2\,dG(x) \\ &= \tfrac{1}{2}(1+\delta_{\lambda_j})\,|n|^{-\alpha}L(|n|)\,s_{\lambda_j}e^{in\lambda_j}(1+o(1)) \end{aligned} \tag{19}$$

if $j=-m,\ldots,m$ and

$$r_\phi(n)=\int_{-\pi}^{\pi} e^{inx}|f_\phi(x)|^2\,dG(x)=o(|n|^{-\alpha}L(|n|)). \tag{20}$$

Here G is the spectral measure of the stationary Gaussian sequence $\{X_n\}$.

It is easy to construct a function f having the desired properties. Consider

$$g(x)=\begin{cases} 0 & \text{if } x\leqq 0 \text{ or } x\geqq 1 \\ \exp\left\{-\dfrac{1}{x}-\dfrac{1}{1-x}\right\} & \text{if } 0<x<1. \end{cases}$$

Let

$$A=\int_0^1 g(u)\,du$$

and

$$h(x)=A^{-1}\int_0^x g(u)\,du.$$

Then we can take

$$f(x)=h\left(\frac{x}{2a}-2\right)h\left(2-\frac{x}{2a}\right).$$

Consider now

$$r_j(n)=\int_{-\infty}^{\infty} e^{inx}\,h_j(x)\,dG(x) \tag{21}$$

with

$$h_j(x)=|f_j(x)|^2. \tag{22}$$

Let

$$h_j(x)=\sum c_{j,n}\,e^{inx}.$$

Then

$$\begin{aligned}\int e^{inx}\,h_j(x)\,dG(x)&=\sum_k c_{j,k}\,r(k+n)\\ &=\sum_{p=-m}^{m}\tfrac{1}{2}s_{\lambda_p}(1+\delta_{\lambda_p})\sum_k c_{j,k}\,|n+k|^{-\alpha}L(|n+k|)\\ &\quad\cdot e^{i(n+k)\lambda_p}.\end{aligned}$$

However if $j=-m,\ldots,m$

$$\sum_k c_{j,k}\,e^{ik\lambda_j}=h_j(\lambda_j)=|f_j(\lambda_j)|^2=1$$

$$\sum_k c_{j,k}\,e^{ik\lambda_p}=h_j(\lambda_p)=|f_j(\lambda_p)|^2=0 \quad \text{for } p\neq j.$$

Since h_j is infinitely differentiable

$$c_{j,k}=O(|k|^{-\beta})$$

where $\beta>0$ can be chosen arbitrarily large. Thus for $j=-m,\ldots,m$

$$\begin{aligned}&\sum_k c_{j,k}\,|n+k|^{-\alpha}L(|n+k|)\,e^{i(n+k)\lambda_j}\\ &\quad=\sum_{|k|<\sqrt{n}}+O(n^{-2\alpha})\\ &\quad=|n|^{-\alpha}L(|n|)\,e^{in\lambda_j}\Big(\sum_{|k|<\sqrt{n}} c_{j,k}\,e^{ik\lambda_j}\Big)(1+o(1))\\ &\quad=|n|^{-\alpha}L(|n|)(1+o(1))\end{aligned}$$

and by a similar argument

$$\sum_k c_{j,k}\,|n+k|^{-\alpha}L(|n+k|)\,e^{i(n+k)\lambda_p}=o(|n|^{-\alpha}L(|n|))$$

if $p\neq j$. In the same way, one can show that

$$\sum c_{\phi,k}\,|n+k|^{-\alpha}L(|n+k|)\,e^{i(n+k)\lambda_p}=o(|n|^{-\alpha}L(|n|))$$

for all $p=-m,\ldots,m$. The lemma follows from these observations.

Proof of the Theorem. Let I denote the index set $\{-m,\dots,0,\dots,m,\phi\}$. Then

$$X_n=\int_{-\pi}^{\pi} e^{inx}\,dZ_G(x)=\sum_{j\in I}{}_jX_n$$

where G is the spectral measure of the Gaussian sequence $\{X_n\}$ and Z_G is the corresponding random spectral measure of the process. Also

$${}_jX_n=\int_{-\pi}^{\pi} e^{inx}\,dZ_{G_j}(x) \tag{23}$$

with

$$dZ_{G_j}(x)=f_j(x)\,dZ_G(x) \tag{24}$$

and

$$G_j(\lambda)=\int_{-\pi}^{\pi} |f_j(u)|^2\,dG(u). \tag{25}$$

Let us consider the case $H(x)=H_k(x)$. Now

$$\begin{aligned} H_k(X_n)=H_k\Big(\sum_{j\in I}{}_jX_n\Big)&=\int e^{in(x_1+\dots+x_k)}Z_G(dx_1)\dots Z_G(dx_k)\\ &=\sum_{j_1,\dots,j_k\in I}\int e^{in(x_1+\dots+x_k)}Z_{G_{j_1}}(dx_1)\dots Z_{G_{j_k}}(dx_k). \end{aligned} \tag{26}$$

Then

$$\begin{aligned} Y_n^N(\beta)&=\frac{1}{A_N}\sum_{j=N(n-1)}^{Nn-1} H_k(X_j)\,e^{-ij\beta}\\ &=\sum_{j_1,\dots,j_k\in I}\frac{1}{A_N}\int e^{iN(n-1)(x_1+\dots+x_k-\beta)}\\ &\quad\cdot k_N(x_1+\dots+x_k-\beta)\,Z_{G_{j_1}}(dx_1)\dots Z_{G_{j_k}}(dx_k) \end{aligned} \tag{27}$$

where

$$\begin{aligned} k_N(x_1+\dots+x_k)&=\sum_{j\in B_0^N} e^{ij(x_1+\dots+x_k)}\\ &=\frac{e^{iN(x_1+\dots+x_k)}-1}{e^{i(x_1+\dots+x_k)}-1}. \end{aligned}$$

We wish to first show that the variance of a term of (27) with one of the subscripts $j_i=\phi$ or else with $j_1,\dots,j_k\in\{-m,\dots,m\}$ but $\lambda_{j_1}+\dots+\lambda_{j_k}\neq\beta \bmod 2\pi$ tends to zero as $N\to\infty$. For simplicity in notation the computation is carried out for distinct $j_1,\dots,j_k$. The variance is then

$$\begin{aligned} &\frac{1}{A_N^2}\int |k_N(x_1+\dots+x_k-\beta)|^2\,dG_{j_1}(x_1)\dots dG_{j_k}(x_k)\\ &=\frac{1}{N^{2-k\alpha}L(N)^k}\sum_{p\in B_0^N}\sum_{q\in B_0^N} r_{j_1}(p-q)\dots r_{j_k}(p-q)\,e^{-i(p-q)\beta}\\ &=\frac{1}{N^{2-k\alpha}L(N)^k}\sum_{p\in \bar B_N}(N-|p|)\,r_{j_1}(p)\dots r_{j_k}(p)\,e^{-ip\beta} \end{aligned}$$

where

$$\tilde{B}_N = \{p \mid -N < p < N\}.$$

If one of the subscripts $j_i = \phi$, the estimates (19) and (20) together with an estimation like that given in the proof of Lemma 1 in [3] imply that (28) tends to zero as $N \to \infty$.

The following lemma is helpful in showing that (28) tends to zero as $N \to \infty$ if $\lambda_{j_1} + \dots + \lambda_{j_k} \not\equiv \beta \bmod 2\pi$.

Lemma 2. *Let* $\alpha > 0$. *If* $L(t)$ *is a slowly varying function there exist* $L_1(t)$ *and* $L_2(t)$ *such that* $L(t) = L_1(t) + L_2(t)$

$$L_2(t) = o(L_1(t)) \tag{29}$$

and

$$n^{-\alpha} L_1(n) \tag{30}$$

is monotone decreasing.

This lemma follows from Karamata's theorem (for Karamata's theorem refer to the book of Ibragimov and Linnik).

By Lemma 2 we can estimate (28) when $\lambda_{j_1} + \dots + \lambda_{j_k} \not\equiv \beta \bmod 2\pi$ by

$$\frac{1}{N^{2-k\alpha} L(N)^k} \sum_p p^{-k\alpha} L_1(p)\, e^{ip(\lambda_{j_1} + \dots + \lambda_{j_k} - \beta)} \tag{31}$$

where L_1 is a slowly varying function such that $n^{-k\alpha} L_1(n)$ is monotone decreasing. The infinite sum in (31) is convergent since $\lambda_{j_1} + \dots + \lambda_{j_k} \not\equiv \beta \bmod 2\pi$ and so the whole expression (31) tends to zero as $N \to \infty$. The case in which several subscripts are the same can be carried out similarly but in a more tedious manner. All these terms can therefore be neglected as $N \to \infty$. We now have to consider the terms for which $\lambda_{j_1} + \dots + \lambda_{j_k} \equiv \beta \bmod 2\pi$. The asymptotic behavior of one such term will be determined as $N \to \infty$ but the argument given when trivially elaborated can be applied to the linear combination of any finite number of such terms. Thus the joint asymptotic distribution of the terms can be determined as $N \to \infty$. Let us consider the term

$$\frac{1}{A_N} \int e^{iN(n-1)(x_1 + \dots + x_k - \beta)}\, k_N(x_1 + \dots + x_k - \beta)\, Z_{G_{j_1}}(dx_1) \dots Z_{G_{j_k}}(dx_k) \tag{32}$$

of the sum (27) with $\lambda_{j_1} + \dots + \lambda_{j_k} \equiv \beta \bmod 2\pi$. Let

$$Z_{N,j}(A) = \frac{N^{\alpha/2}}{L(N)^{\frac{1}{2}}} Z_{G_j}(\lambda_j + N^{-1} A) \tag{33}$$

and

$$G_{N,j}(A) = \frac{N^{\alpha}}{L(N)} G_j(\lambda_j + N^{-1} A) \tag{34}$$

with A a Borel set. Our object is to show that (32) converges in distribution to that of

$$D^{-k/2} \int e^{i(n-1)(x_1 + \dots + x_k)} \frac{e^{i(x_1 + \dots + x_k)} - 1}{i(x_1 + \dots + x_k)} \cdot s_{\lambda_1} \frac{1 + \delta(\lambda_1)}{2} \dots s_{\lambda_k} \frac{1 + \delta(\lambda_k)}{2} W_{\lambda_1}(dx_1) \dots W_{\lambda_k}(dx_k). \tag{35}$$

Of course, the finite dimensional distributions of (5) will converge in distribution to those of (14) and as remarked earlier a simple but notationally tedious elaboration of our argument will yield that result in the case $H(x)=H_k(x)$.

Let

$$K_N(u)=\frac{1}{N}k_N\left(\frac{1}{N}u\right).$$

Then expression (32) can be rewritten as

$$\int e^{i(n-1)(x_1+\ldots+x_k)}K_N(x_1+\ldots+x_k)Z_{N,j_1}(dx_1)\ldots Z_{N,j_k}(dx_k). \tag{36}$$

The measure $G_{N,j}(\cdot)$ corresponding to $Z_{N,j}(\cdot)$ can be shown to converge locally weakly as $N\to\infty$ to

$$s_{\lambda_j}\left(\frac{1+\delta(\lambda_j)}{2}\right){}_0G(\cdot)$$

where ${}_0G$ has density $D^{-1}|x|^{\alpha-1}$. Set

$$\varphi_{N,\bar\lambda}(t_1,\ldots,t_k)$$
$$=\int e^{i\frac{1}{N}(u_1x_1+\ldots+u_kx_k)}|K_N(x_1+\ldots+x_k)|^2 G_{N,j_1}(dx_1)\ldots G_{N,j_k}(dx_k) \tag{37}$$

where $u_p=[t_pN]$, $p=1,\ldots,k$ and $\bar\lambda=(\lambda_{j_1},\ldots,\lambda_{j_k})$. Then

$$\varphi_{N,\bar\lambda}(t_1,\ldots,t_k)$$
$$=\frac{1}{N^{2-k\alpha}L(N)^k}\sum_{p\in B_0^N}\sum_{q\in B_0^N}r(p-q+u_1)\ldots r(p-q+u_k)$$
$$\cdot e^{-i(p-q)\beta}e^{-i(u_1\lambda_{j_1}+\ldots+u_k\lambda_{j_k})}$$
$$=\frac{1}{N^{2-k\alpha}L(N)^k}\sum_{p\in\bar B^N}(N-|p|)r(p+u_1)\ldots$$
$$\cdot r(p+u_k)e^{-ip\beta}e^{-i(u_1\lambda_{j_1}+\ldots+u_k\lambda_{j_k})}. \tag{38}$$

Just as in the proof of Lemma 1 of [3] one can show that

$$\lim_{N\to\infty}\varphi_{N,\bar\lambda}(t_1,\ldots,t_k)$$
$$=\int_{-1}^{1}(1-|x|)\frac{1}{|x_1+t_1|^\alpha}\cdots\frac{1}{|x+t_k|^\alpha}dx$$
$$\cdot s_{\lambda_{j_1}}\ldots s_{\lambda_{j_k}}\frac{1+\delta(\lambda_{j_1})}{2}\cdots\frac{1+\delta(\lambda_{j_k})}{2}$$
$$=\varphi_{\bar\lambda}(t_1,\ldots,t_k). \tag{39}$$

with the limit function $\varphi_{\bar\lambda}$ a continuous function. Also the function $\varphi_{N,\bar\lambda}$ of (38) is the Fourier transform ($u=(u_1,\ldots,u_k), u_p=[t_pN], x=(x_1,\ldots,x_k)$)

$$\varphi_{N,\bar\lambda}(t)=\int e^{iu\cdot x/N}\eta_{N,\bar\lambda}(dx)$$

of a finite measure $\eta_{N,\bar\lambda}$ on R^k with support on $[-N\pi,N\pi]^k$. Since $\varphi_{N,\bar\lambda}(t)$ tends to a limit function $\varphi_{\bar{\lambda}}$ that is continuous, it follows that the sequence of measures $\eta_{N,\bar{\lambda}}$ tends to a finite measure $\eta_{\bar{\lambda}}$ and $\varphi_{\bar{\lambda}}$ is the Fourier transform of $\eta_{\bar{\lambda}}$ (see Lemma 2 of [3]).

We consider special functions h taking on a finite number of values of the following character. Consider sets $A_1, A_2, \ldots, A_s$, $s=1,2,\ldots$, Borel sets (of finite ${}_0G$ mass) such that $A_{-i}=-A_i$ and $A_{-s}, \ldots, A_{-1}$, $A_1, \ldots, A_s$ are disjoint. Let $\hat{h}(i_1, \ldots, i_k)$ be complex numbers and

$$h(x_1, \ldots, x_k)=\hat{h}(i_1, \ldots, i_k) \quad \text{if } x_1 \in A_{i_1}, \ldots, x_k \in A_{i_k}$$

where $i_1, \ldots, i_k$ take on the values $\pm 1, \ldots, \pm s$ but with $i_j \neq i_{j'}$ if $j \neq j'$ and $h(x_1, \ldots, x_k)=0$ for all other $x_1, \ldots, x_k$. One can show that

$$\int h(x_1, \ldots, x_k) Z_{N,j_1}(dx_1) \ldots Z_{N,j_k}(dx_k) \tag{40}$$

asymptotically as $N \to \infty$ has the same distribution as

$$\int h(x_1, \ldots, x_k) W_{\lambda_{j_1}}(dx_1) \ldots W_{\lambda_{j_k}}(dx_k).$$

The integral (40) is a polynomial in the random variables $Z_{N,j_i}(B)$ with the B's Borel sets. Now the joint distribution of the random variables $Z_{N,j_i}(B)$ tends to the joint distribution of the random variables $W_{\lambda_{j_i}}(B)$ and so we have the desired limit behavior. Also $K_N(x_1+\ldots+x_k)$ tends to

$$K_0(x_1+\ldots+x_k)=\frac{e^{i(x_1+\ldots+x_k)}-1}{i(x_1+\ldots+x_k)}$$

uniformly on every finite rectangle in R^k. Equations (37) and (39) imply that

$$\lim_{A \to \infty} \int_{R^k-[-A,A]^k} |K_N(x_1+\ldots+x_k)|^2 G_{N,j_1}(dx_1) \ldots G_{N,j_k}(dx_k)=0 \tag{41}$$

uniformly for $N=0,1,2,\ldots$. An adaptation of Lemma 3 of [3] then implies that expression (36) tends in distribution to expression (35) as $N \to \infty$. For relation (41) implies that for any $\varepsilon>0$ one can find a function of type h (depending on ε) such that

$$E\left|\int [e^{i(n-1)(x_1+\ldots+x_k)} K_N(x_1+\ldots+x_k)-h(x_1, \ldots, x_k)] \cdot Z_{N,j_1}(dx_1) \ldots Z_{N,j_k}(dx_k)\right|^2<\varepsilon$$

for $N>N(\varepsilon)$ and

$$E\left|\int [e^{i(n-1)(x_1+\ldots+x_k)} K_N(x_1+\ldots+x_k)-h(x_1, \ldots, x_k)] \cdot W_{\lambda_{j_1}}(dx_1) \ldots W_{\lambda_{j_k}}(dx_k)\right|^2<\varepsilon.$$

The same argument as that given on p. 36 of [3] implies the validity of the theorem for general H.

The theorem was derived for stationary sequences. However, it is clear that a corresponding result could have been obtained for random fields under conditions comparable to those given in [3]. The normalization in the theorem is of the form $N^\gamma L(N)$ with $\gamma>\frac{1}{2}$ and $L(N)$ slowly varying. The greatest interest is most likely in the case of a non-Gaussian limiting process. However, one can obtain nonnormal limiting distributions with a normalization having exponent

$\gamma<\frac{1}{2}$ by considering processes somewhat like those analyzed in [6] and [7]. A parallel argument carried out in the case of continuous time parameter suggests that the limiting distributions obtained are self-similar.

Additional Remarks

Taqqu noted orally that one can determine non-Gaussian self-similar processes whose second order properties are the same as the Wiener process by using the techniques of [6] or [7]. This means that the exponent in the self-similarity is 1/2 as it is for the Wiener process.

In all of the results discussed in [3] and [4] there is a smallest index in a Hermite expansion that plays a basic role. Given the results of [4] it is easy and of interest to concoct special examples in which one needs much more than one index to characterize a self-similar distribution. Such examples are particular cases of a general class of such processes. Let ${}_kW$, $k=1,2,\ldots$, be independent processes each having the same probability structure as W_0. Let

$$\begin{aligned} U_n(k)=\int &\exp[in(x_1+\ldots+x_k)]\,K_0(x_1,\ldots,x_k) \\ &\cdot|x_1+\ldots+x_k|^{\frac{1}{4}}\,|x_1|^{3/(4k)-\frac{1}{2}}\ldots|x_k|^{3/(4k)-\frac{1}{2}} \\ &\cdot{}_kW(dx_1)\ldots{}_kW(dx_k). \end{aligned} \tag{42}$$

The discussion in [4] indicates that each of the process $\{U_n(k), n=\ldots,-1,0,1,\ldots\}$, $k=1,2,\ldots$ is self-similar with exponent 1/2. Let

$$U_n=\sum_{k=1}^{\infty} U_n(k)\,g_k \tag{43}$$

where the g_k are assumed to approach zero sufficiently fast as $k\to\infty$ so that (43) converges in mean square. The resulting process $\{U_n\}$ is still self-similar with exponent 1/2 but involves polynomial forms in the W processes of all powers.

Acknowledgement. I wish to thank a referee who suggested using Lemmas 1 and 2.

References

1. Brillinger, D.: Time series: data analysis and theory. New York: Holt, Rinehart and Winston 1975
2. Dobrushin, R.L.: Gaussian and their subordinated self-similar random generalized fields. Ann. Probab. **7**, 1–28 (1979)
3. Dobrushin, R.L., Major, P.: Non-central limit theorems for nonlinear functionals of Gaussian fields. Z. Wahrscheinlichkeitstheorie verw. Gebiete **50**, 27–52 (1979)
4. Major, P.: Limit theorems for nonlinear functionals of Gaussian sequences. [To be published in Z. Wahrscheinlichkeitstheorie verw. Gebiete]
5. Rosenblatt, M.: Independence and dependence, Proc. 4th Sympos. Math. Statist. Probab. 431–443. Univ. California (1961)
6. Rosenblatt, M.: Some limit theorems for partial sums of quadratic forms in stationary Gaussian variables. Z. Wahrscheinlichkeitstheorie verw. Gebiete, **49**, 125–132 (1979)
7. Taqqu, M.S.: Weak convergence to fractional Brownian motion and to the Rosenblatt process. Z. Wahrscheinlichkeitstheorie verw. Gebiete **31**, 287–302 (1975)

Received March 6, 1980; in revised form September 20, 1980

The Annals of Statistics
1982, Vol. 10, No. 4, 1195–1208

DECONVOLUTION AND ESTIMATION OF TRANSFER FUNCTION PHASE AND COEFFICIENTS FOR NONGAUSSIAN LINEAR PROCESSES[1]

BY K. S. LII AND M. ROSENBLATT

University of California, Riverside and San Diego

NonGaussian linear processes are considered. It is shown that the phase of the transfer function can be estimated under broad conditions. This is not true of Gaussian linear processes and in this sense Gaussian linear processes are atypical. The asymptotic behavior of a phase estimate is determined. The phase estimates make use of bispectral estimates. These ideas are applied to a problem of deconvolution which is effective even when the transfer function is not minimum phase. A number of computational illustrations are given.

1. Introduction. Most of the literature on finite parameter time series models is either centered on Gaussian models or the results are motivated by what one can do in the case of Gaussian models. Here we deal with stationary nonGaussian linear processes and show that under broad conditions, aspects of the structure that are completely nonidentifiable in the Gaussian case can be resolved in the nonGaussian case. Assume that the random variables v_t, $t = \cdots, -1, 0, 1, \cdots$ are independent and identically distributed with mean zero, $Ev_t \equiv 0$, and variance one $Ev_t^2 \equiv 1$. Let $\{\alpha_j\}$ be a sequence of real constants with

$$\sum_{j=-\infty}^{\infty} \alpha_j^2 < \infty.$$

Consider the linear process generated by $\{\alpha_j\}$ and $\{v_t\}$

$$x_t = \sum_{j=-\infty}^{\infty} \alpha_j v_{t-j}. \tag{1}$$

Let $\alpha(z) = \sum_j \alpha_j z^j$ be the z-transform corresponding to the process $\{x_t\}$. Then

$$\alpha(e^{-i\lambda}) = \sum_j \alpha_j e^{-ij\lambda}$$

is called the frequency response function or transfer function. We are concerned with the estimation of $\alpha(e^{-i\lambda})$ on the basis of observations only on the process $\{x_t\}$. Further, the linear nonGaussian process serves as a model for types of seismic exploration. The weights α_j can be regarded as the signature of a disturbance passing through a medium and the random weights v_j as reflectivity of slabs in a layered medium. It has been observed that in many geophysical contexts the observed data (the x_t's) are distinctly nonGaussian and a basic object is to deconvolve, estimating the α_j's and v_j's in the process. A discussion of related questions in the geophysical context can be found in Donoho (1981), Godfrey and Rocca (1981), and Wiggins (1978).

The spectral density of $\{x_t\}$ is

$$f(\lambda) = \frac{1}{2\pi} |\alpha(e^{-i\lambda})|^2.$$

In the Gaussian case (when $\{x_t\}$ is normally distributed) the full probability structure of $\{x_t\}$ is determined by $f(\lambda)$ or equivalently by the modulus of $\alpha(e^{-i\lambda})$, $|\alpha(e^{-i\lambda})|$. The phase

Received November 1981; revised May 1982.
[1] Research supported in part by Office of Naval Research Contract N00014-81-K-0003.
AMS 1980 *subject classification.* Primary 62M15; secondary 62G99.
Key words and phrases. Asymptotics, bispectrum, deconvolution, linear process, non-Gaussian, phase.

R.A. Davis et al. (eds.), *Selected Works of Murray Rosenblatt*, Selected Works in Probability and Statistics, DOI 10.1007/978-1-4419-8339-8_32, © Springer Science+Business Media, LLC 2011

information in $\alpha(e^{-i\lambda})$ is not identifiable in the Gaussian case.

If $\alpha(z)$ is a rational function

$$\alpha(z) = \frac{A(z)}{B(z)}$$

with $A(z)$, $B(z)$ polynomials

$$A(z) = \sum_{k=0}^{q} a_k z^k, \qquad a_0 \neq 0, \qquad B(z) = \sum_{k=0}^{p} b_k z^k, \qquad b_0 = 1,$$

the process $\{x_t\}$ is a finite parameter autoregressive moving average process; that is,

$$\sum_{j=0}^{p} b_j x_{t-j} = \sum_{k=0}^{q} a_k v_{t-k}. \tag{2}$$

If $\{x_t\}$ is a Gaussian process satisfying (2), then any real root $z_j \neq 0$ of $A(z)$ or $B(z)$ can be replaced by its inverse and pairs of nonzero conjugate roots by their paired conjugated inverses $\bar{z}^{-1}$ without changing the probability structure of $\{x_t\}$ if the process is appropriately rescaled. This follows since $|e^{i\lambda} - z_0| = |z_0|\,|e^{-i\lambda} - \bar{z}_0^{-1}|$. With real distinct roots there are 2^{p+q} ways of specifying the roots without changing the structure of $\{x_t\}$. To ensure unique determination of the coefficients a_k and b_j of (2) (since there is a different specification of these coefficients corresponding to each of the possible root specifications) in the Gaussian case, it is the custom to assume that all the roots of $A(z)$ and $B(z)$ are outside the unit circle $|z| \leq 1$ in the complex plane.

However, for a nonGaussian stationary process satisfying (2) (in which case the independent v_t's are nonGaussian) the different 2^{p+q} specifications of roots mentioned above generally correspond to different probability structures and different stationary processes. As a simple example, consider the moving average

$$x_t = 6v_t - 5v_{t-1} + v_{t-2}$$

with the roots of $A(z)$ being 2 and 3, and the moving average

$$y_t = 3v_t - 7v_{t-1} + 2v_{t-2}$$

having a polynomial $A(z)$ with roots ½ and 3. Both $\{x_t\}$ and $\{y_t\}$ have the same spectral density but if the independent random sequence $\{v_t\}$ is exponentially distributed, the marginal distributions of the $\{x_t\}$ and $\{y_t\}$ sequences are different. In the problem of deconvolution where one wishes to recover the process $\{v_t\}$ (assumed nonGaussian which is most often the case in applications) in some sense, the proper specification of nonzero roots (which are inside and which are outside) becomes crucial (see Rosenblatt, 1974). There is a discussion concerning the distribution of roots as related to prediction problems in Rosenblatt (1980). The assumption that all roots are outside the unit disc is often called the minimum phase condition.

There are results on the estimation of the coefficients a_j and b_k of (2) (corresponding to roots assumed outside the unit circle) in Box and Jenkins (1976). In the Gaussian case these are essentially equivalent asymptotically to maximum likelihood procedures. In the nonGaussian case the computations are carried out as if the process were Gaussian. One has a least squares but not a maximum likelihood solution in the nonGaussian case. The coefficients estimated are those corresponding to roots outside the unit circle even though the actual structure of the process may not be one with all the roots outside the unit circle. Thus one will typically not be able to resolve the actual structure using these procedures in the nonGaussian case. Of course, if one knows the actual nonGaussian distribution of the v_t, one can use the maximum likelihood estimate or an asymptotically equivalent procedure to estimate the coefficients in (2) even if the roots z_j are not all outside the unit circle (see Basawa, Feigen and Heyde, 1975). Higher order spectral methods discussed in the next section do not require this knowledge. Our discussion follows that of Rosenblatt (1980).

2. Higher order spectral method. We prove the following lemma.

LEMMA 1. *Let $\{x_t\}$ be a nonGaussian linear process (see (1)) with the independent random variables $\{v_t\}$ having all moments finite. Assume that*

$$\sum |j|\,|a_j| < \infty$$

and $\alpha(e^{-i\lambda}) \neq 0$ for all λ. Then the function $\alpha(e^{-i\lambda})$ is identifiable on the basis of observations on $\{x_t\}$ alone up to the integer a in a factor $e^{ia\lambda}$ and the sign of $\alpha(1) = \sum \alpha_k$. Actually it is enough to have moments of order $k > 2$ finite with cumulant $\gamma_k \neq 0$.

Since the v_t's are assumed nonGaussian with all moments finite, there must be a cumulant of v_t, $\gamma_k \neq 0$ of smallest subscript $k > 2$. The kth order cumulant spectral density of the process $\{x_t\}$ is given by

$$\begin{aligned} b_k(\lambda_1, \cdots, \lambda_{k-1}) &= \frac{1}{(2\pi)^{k-1}} \sum_{j_1,\cdots,j_{k-1}} \operatorname{cum}(x_t, x_{t+j_1}, \cdots, x_{t+j_{k-1}}) \exp(-\textstyle\sum_{s=1}^{k-1} ij_s\lambda_s) \\ &= \frac{\gamma_k}{(2\pi)^{k-1}} \alpha(e^{-i\lambda_1}) \cdots \alpha(e^{-i\lambda_{k-1}})\alpha(e^{i(\lambda_1+\cdots+\lambda_{k-1})}). \end{aligned} \tag{3}$$

Let

$$h(\lambda) = \arg\left\{\alpha(e^{-i\lambda}) \frac{\alpha(1)}{|\alpha(1)|}\right\}.$$

Then

$$\left\{\frac{\alpha(1)}{|\alpha(1)|}\right\}^k \gamma_k = (2\pi)^{k/2-1} b_k(0, \cdots, 0)/\{f(0)\}^{k/2},$$

and

$$h(\lambda_1) + \cdots + h(\lambda_{k-1}) - h(\lambda_1 + \cdots + \lambda_{k-1}) = \arg\left[\left\{\frac{\alpha(1)}{|\alpha(1)|}\right\}^k \gamma_k^{-1} b_k(\lambda_1, \cdots, \lambda_{k-1})\right] \tag{4}$$

since $h(-\lambda) = -h(\lambda)$. Further

$$h'(0) - h'(\lambda) = \lim_{\Delta\to 0} \frac{1}{(k-2)\Delta} \{h(\lambda) + (k-2)h(\Delta) - h(\lambda + (k-2)\Delta)\}. \tag{5}$$

Now

$$h(\lambda) = \int_0^\lambda \{h'(u) - h'(0)\}du + c\lambda = h_1(\lambda) + c\lambda$$

where $c = h'(0)$. In particular

$$h(\pi) = h_1(\pi) + c\pi.$$

Since the α_j's are real we must have $h(\pi) = a\pi$ for some integer a. Set $h_1(\pi)/\pi = \delta$. Then $h(\pi) = a\pi = (\delta + c)\pi$, so that

$$c = a - \delta.$$

The integer a cannot be determined without further assumptions since it corresponds to reindexing or subscripting the v_t's. The sign of $\alpha(1)$ is also intrinsically undecidable since one can multiply α_j's and v_t's by (-1) without changing the observed process $\{x_t\}$. Thus, under the conditions specified above for a nonGaussian linear process $\{x_t\}$, $\alpha(e^{-i\lambda})$ is identifiable up to the integer a and the sign on the basis of observations on $\{x_t\}$ only and is given by

$$\alpha(e^{-i\lambda}) = |2\pi f(\lambda)|^{1/2} \exp\{ih(\lambda)\}$$

with

$$h(\lambda) = \int_0^{\lambda} \{h'(u) - h'(0)\} du + c\lambda = h_1(\lambda) - \frac{h_1(\pi)}{\pi}\lambda + a\lambda . \tag{6}$$

Notice that $h_1(\lambda)$ can actually be computed. In Section 3 we will consider an estimate of the identifiable aspects of $\alpha(e^{-i\lambda})$ suggested by Lemma 1 and the convergence properties of this estimate. Comments on the computations using spectral methods follow in Section 4. A few remarks on other computational methods are made in Section 5. Examples of estimation and deconvolution are given in Section 6. It should be noted that in some seismic investigations, a can be estimated through knowledge of speed of wave (e.g. sound) propagation through the medium.

3. Phase estimation and convergence of estimates. There are many discussions concerned with the estimation of the second order spectral density $f(\lambda)$ (see Anderson 1971, or Jenkins and Watts, 1968). We will concentrate on the estimation of $h(\lambda)$. For simplicity of discussion we will assume that the third order cumulant γ_3 of v_t is nonzero. The program in the higher order case can be carried out in a similar manner. Equation (5) becomes

$$h'(0) - h'(\lambda) = \lim_{\Delta\to 0} \frac{1}{\Delta} \{h(\lambda) + h(\Delta) - h(\lambda + \Delta)\}$$

when $k = 3$. From (4) we find that up to a sign

$$h(\lambda) + h(\Delta) - h(\lambda + \Delta) = \arg\{b_3(\lambda, \Delta)\}.$$

From this point on we will drop the subscript and understand that we are dealing with the bispectral density $b(\lambda, \mu)$.

The following approximation is useful in deriving the theorem of this section.

LEMMA 2. *Let the assumptions of Lemma* 1 *be satisfied. Assume that* ${}_nb(\lambda, \mu)$ *is a consistent estimate of* $b(\lambda, \mu)$ *(as* $n \to \infty$*) based on a sample of size n. If one considers*

$$\theta_n(\lambda, \mu) = \arctan(Im_n b(\lambda, \mu)/Re_n b(\lambda, \mu))$$

as an estimate of

$$\theta(\lambda, \mu) = \arg b(\lambda, \mu)$$

then

$$\begin{aligned}\theta_n(\lambda, \mu) - \theta(\lambda, \mu) = &-\frac{Im\, b(\lambda, \mu)}{|b(\lambda, \mu)|^2}\{Re_n b(\lambda, \mu) - Re\, b(\lambda, \mu)\} \\ &+ \frac{Re\, b(\lambda, \mu)}{|b(\lambda, \mu)|^2}\{Im_n b(\lambda, \mu) - Im\, b(\lambda, \mu)\} + o_p({}_nb(\lambda, \mu) - b(\lambda, \mu)).\end{aligned} \tag{7}$$

We note that for a complex number

$$z = x + iy = re^{i\theta}$$

with $r = |z|$ and $\theta = \arctan(y/x)$ a principal value determination, one has

$$\frac{\partial\theta}{\partial y} = \frac{x}{r^2}, \quad \frac{\partial\theta}{\partial x} = -\frac{y}{r^2}, \quad \frac{\partial^2\theta}{\partial x^2} = \frac{2xy}{r^4}, \quad \frac{\theta^2\theta}{\partial y^2} = -\frac{2xy}{r^4}$$

and

$$\frac{\partial^2\theta}{\partial x\partial y} = \frac{1}{r^2} - \frac{2x^2}{r^4} = -\frac{1}{r^2} + \frac{2y^2}{r^2}.$$

The relation (7) follows on using these observations in a Taylor expansion of the arctan function.

Let us consider estimating

$$h_1(\lambda) - \frac{h_1(\pi)}{\pi}\lambda.$$

Set $\Delta = \Delta(n)$, $k\Delta = \lambda$, and let $\Delta = \Delta(n) \to 0$ as $n \to \infty$. Assume for convenience that $b(0, 0)$ is positive. A simple modification indicated later takes care of the case in which $b(0, 0)$ is negative. Now

$$h_1(\lambda) = h(\lambda) - h'(0)\lambda \cong h(k\Delta) - \frac{h(\Delta)}{\Delta} k\Delta$$

$$= -\textstyle\sum_{j=1}^{k-1} \{h(j\Delta) + h(\Delta) - h((j+1)\Delta)\} = -\sum_{j=1}^{k-1} \arg b(j\Delta, \Delta).$$

This suggests

$$H_n(\lambda) = -\textstyle\sum_{j=1}^{k-1} \arg_n b(j\Delta, \Delta)$$

as an appropriate estimate of $h_1(\lambda)$.

THEOREM. *Assume that $b(0, 0)$ is positive and that $EX_t^6 < \infty$. Consider*

$$H_n(\lambda) = -\textstyle\sum_{j=1}^{k-1} \arg_n b(j\Delta, \Delta), \tag{8}$$

$k\Delta = \lambda$, as an estimate of

$$h_1(\lambda) = h(\lambda) - h'(0)\lambda,$$

where it is understood that the bispectral estimates are weighted averages of 3rd order periodogram values. If $b(\lambda, \mu) \in \mathscr{C}^2$ and the weight function is symmetric and bandlimited with bandwidth Δ, then

$$H_n(\lambda) - h_1(\lambda) = R_n(\lambda) + o_p(H_n(\lambda) - h_1(\lambda))$$

with

$$ER_n(\lambda) \sim \int_0^\lambda \frac{1}{2}\frac{\operatorname{Im} b(u, 0)}{|b(u, 0)|^2}\{AD_u^2 \operatorname{Re} b(u, 0) + 2BD_uD_v \operatorname{Re} b(u, 0) + CD_v^2 \operatorname{Re} b(u, 0)\}\, du\Delta$$

$$- \int_0^\lambda \frac{1}{2}\frac{\operatorname{Re} b(u, 0)}{|b(u, 0)|^2}\{AD_u^2 \operatorname{Im} b(u, 0) + 2BD_uD_v \operatorname{Im} b(u, 0) + CD_v^2 \operatorname{Im} b(u, 0)\}\, du\Delta + o(\Delta)$$

and

$$\operatorname{Cov}\{R_n(\lambda), R_n(\mu)\} \cong \frac{2\pi^2}{\Delta^3 n \gamma_3^2}\min(\lambda, \mu)\int w^2(u, v)\, du\, dv \tag{9}$$

for $\Delta(n) \to 0$, $\Delta^2 n \to \infty$ as $n \to \infty$. Here A, B, and C are the second moments of the weight function w of the bispectral estimates

$$A = \int u^2 w(u, v)\, du\, dv, \qquad B = \int uvw(u, v)\, du\, dv, \qquad C = \int v^2\, w(u, v)\, du\, dv$$

and D_u and D_v are partial derivatives with respect to u and v respectively.

It has been shown in Brillinger and Rosenblatt (1967) under the assumptions of the theorem that

$$E\{{}_nb(\lambda, \mu)\} - b(\lambda, \mu) \sim \frac{1}{2}\int\int (uD_\lambda + vD_\mu)^2 b(\lambda, \mu) w(u, v)\, du\, dv\Delta^2 + o(\Delta^2) \tag{10}$$

and

$$\text{(11)} \qquad \operatorname{Var}\{{}_nb(\lambda, \mu)\} \sim \frac{f(\lambda)f(\mu)f(\lambda+\mu)}{\Delta^2 n}\int w^2(u, v)\,du\,dv, \qquad 0 \neq \lambda \neq \mu \neq \pi,$$

if $\Delta^2 n \to \infty$ as $n \to \infty$, $\Delta(n) \to 0$. Further estimates ${}_nb(\lambda, \mu)$, ${}_nb(\lambda', \mu')$, $0 \le \mu \le \lambda$, $0 \le \mu' \le \lambda'$, are asymptotically uncorrelated as $n \to \infty$ if (λ, μ) and (λ', μ') are distinct. Using (7) and (8) we can write

$$H_n(\lambda) - h_1(\lambda) = R_n(\lambda) + o_p(H_n(\lambda) - h_1(\lambda)),$$

$$\text{(12)} \qquad R_n(\lambda) = \sum \left[\frac{\operatorname{Im} b(j\Delta, \Delta)}{|b(j\Delta, \Delta)|^2} \{\operatorname{Re}_n b(j\Delta, \Delta) - \operatorname{Re} b(j\Delta, \Delta)\} - \frac{\operatorname{Re} b(j\Delta, \Delta)}{|b(j\Delta, \Delta)|^2} \{\operatorname{Im}_n b(j\Delta, \Delta) - \operatorname{Im} b(j\Delta, \Delta)\} \right]$$

and show by employing (10) that the approximation for ER_n given in the statement of the theorem is valid. Further, by using (11) it follows that (9) holds.

COROLLARY. *Assume that all moments of $\{x_t\}$ exist and that the assumptions of the theorem are satisfied. Then $EH_n(\lambda) \to h_1(\lambda)$ and $H_n(\lambda)$ is asymptotically normal with variance given by* (9). *The mean square error of $R_n(\lambda)$ is of order*

$$C_1\Delta^2 + \frac{C_2}{\Delta^3 n}.$$

The optimal rate of convergence is $n^{-2/5}$ when $\Delta(n) \sim n^{-1/5}$. If $b(\lambda, \mu) \in \mathscr{C}^3$ and the weight function is bandlimited with first and second moments zero, the mean square error of $R_n(\lambda)$ is of the order

$$C_1\Delta^4 + \frac{C_2}{\Delta^3 n}.$$

The optimal rate of convergence is then $n^{-4/7}$ with $\Delta(n) \sim n^{-1/7}$.

The asymptotic normality with variance given by (9) follows on using (12).

Generally we will estimate $h_1(\lambda)$ and hence $h(\lambda)$ for a whole range of λ values. The sign of $b(0, 0)$ may not be positive. We estimate it by noting the real part of ${}_nb(\Delta, \Delta)$. If it is negative we multiply all ${}_nb(j\Delta, \Delta)$ with a minus sign. The estimate $H_n(\lambda)$ is then given by

$$H_n(\lambda) = -\sum_{j=1}^{k-1} \arg\{-{}_nb(j\Delta, \Delta)\}.$$

4. Computations using spectral methods. We remark on the computational aspect of phase estimation of $\alpha(e^{-i\lambda})$ and give a few illustrative examples to indicate its effectiveness.

Given a sample $\{x_t\}$ of size $n = kN$, we center and normalize it so that it has mean zero and variance one. Break up the sample into k disjoint subsections of equal length N so that the variance of the bispectral estimate from each section is not too large. Then choose a grid of points $\lambda_j = j\Delta$ in $(0, 2\pi)$, $j = 1, \cdots, M$, $\Delta = 2\pi L/N$ for an appropriate integer L. Though the symmetry condition $h(\lambda) = -h(-\lambda)$ implies that one need only deal with λ in $(0, \pi)$, there may be some advantage in considering $\lambda \in (0, 2\pi)$. We will comment on this point later on. Form the bispectral estimate ${}_Nb(j\Delta, \Delta)$ of the type discussed above with a weight function of bandwidth Δ from each subsection. Average the estimates from the different subsections so as to arrive at a final estimate ${}_nb(j\Delta, \Delta)$. A detailed discussion of this kind of algorithm can be found in Lii and Helland (1981). Compute $\theta_n(j) = \arg\{{}_nb(j\Delta, \Delta)\} + 2k\pi$ where the integer k is chosen to ensure continuity of $H_n(\ell\Delta) = H_n(\lambda_\ell) = -\sum_{j=1}^{\ell-1} \theta_n(j)$, $\ell = 2, \cdots, M+1$ (neighboring values are as close to each other as possible). Since the upper index is $\ell - 1$ we start with $\ell = 2$. Since $h(0) = 0$ one sets $H_n(0) = 0$ and estimates $H_n(\Delta) = H_n(\lambda_1)$ by an interpolation between 0 and $H_n(\lambda_2)$, $\lambda_2 = 2\Delta$. $H_n(\pi)$ is also computed by an interpolation procedure. This amounts to a complete procedure for estimating $h(\lambda)$.

Since

$$\alpha_k = \frac{1}{2\pi} \int_0^{2\pi} \alpha(e^{-i\lambda}) e^{ik\lambda} \, d\lambda$$

an estimate $\hat{\alpha}_k$ of α_k is given by

$$\text{(13)} \qquad \begin{aligned} \hat{\alpha}_k &= \frac{1}{2\pi} \int_0^{2\pi} \hat{\alpha}(e^{-i\lambda}) e^{ik\lambda} \, d\lambda \\ &\cong \frac{1}{(M+2)} \sum_{j=0}^{M+1} \sqrt{2\pi f_n(\lambda_j)} \exp\left\{ i\left(H_n(\lambda_j) - \frac{H_n(\pi)}{\pi} \lambda_j + k\lambda_j \right) \right\} \end{aligned}$$

and this computation can be carried out by using the fast Fourier transform.

The α_k's are real numbers and so the $\hat{\alpha}_k$'s may or may not be real. If the symmetric property of $f(\lambda)$ and $h(\lambda)$ is used and the integration is carried out from $-\pi$ to π almost real $\hat{\alpha}_k$'s will be obtained. The imaginary part of the α_k's will only be the size of rounding errors. In practice there is no indication of how good or bad the estimates are apart from asymptotic results. In actual practice $\{j\Delta\}_{j=0}^{M+1}$ may not be symmetric about π. If the estimates $H_n(\lambda_j)$ are reasonably good the estimated $\hat{\alpha}_j$'s from (13) should still be almost real. The size of the imaginary part reflects the level of noise. When the estimates $H_n(\lambda_j)$ are not good the imaginary part of the $\hat{\alpha}_k$'s becomes comparable to (or larger than) its real part. This can serve as a direct indication of the quality of the estimation.

If the linear process is one-sided with a finite number of parameters one has a moving average of order q

$$x_t = \sum_{j=0}^{q} \alpha_j v_{t-j}, \qquad \alpha_0 \neq 0.$$

We could estimate $\alpha(z) = \sum_{j=0}^{q} \alpha_j z^j$ by $\hat{\alpha}(z) = \sum_{j=0}^{q} \hat{\alpha}_j z^j$. In deconvolution we try to recover the process $\{v_t\}$, $v_t = \dfrac{1}{\alpha(B)} x_t$ (B is the backward shift operator so that $B^j x_t = x_{t-j}$) by computing the approximation $\hat{v}_t = \dfrac{1}{\hat{\alpha}(B)} x_t$. If all the roots of $\alpha(z)$ (and $\hat{\alpha}(z)$) are outside the unit circle (the frequency function is minimum delay) then $\hat{\alpha}^{-1}(z)$ has a one-sided expansion $\sum_{j=0}^{\infty} \alpha_j' B^j$. In the computation, the series is truncated after a certain number of terms. If some of the roots of $\hat{\alpha}(z)$ have modulus less than one we can still expand $\hat{\alpha}^{-1}(B)$ with a Laurent series expansion. Once the roots of $\hat{\alpha}(z)$ are computed, one can easily get the Laurent series expansion of $\hat{\alpha}^{-1}(B)$ by partial fractions as described in Rosenblatt (1974) or Henrici (1974).

Another way to find the inverse weights in deconvolution is to use a least squares criterion as described in Wiggins (1978). Another general method of deconvolution will be mentioned in the section on computation. A discussion of related questions for autoregressive schemes can be found in Huzii (1981).

5. Other possible computational methods. We briefly discuss two other possible methods of estimating the coefficients of a nonGaussian moving average process of order q

$$x_t = \sum_{j=0}^{q} \alpha_j v_{t-j}. \tag{14}$$

As noted earlier, second order moments will not allow us to determine the location of the roots of

$$\alpha(z) = \sum_{j=0}^{q} \alpha_j z^j. \tag{15}$$

Higher order moments will be used in the first method which makes use of a *least squares procedure*. Assume $Ev_t \equiv 0$, $Ev_t^3 \equiv \gamma \neq 0$. Consider

$$c_k = Ex_t x_{t+k}^2 = \gamma \sum_j \alpha_j \alpha_{j+k}^2, \qquad k = -q, -q+1, \cdots, q. \tag{16}$$

Estimate c_k by

$$\hat{c}_k = \frac{1}{n} \sum_{k=1}^n x_t x_{t+k}^2$$

and solve the extremal problem

$$\min_{a_j} \sum_{k=-q}^q (\hat{c}_k - \gamma \sum_l a_l a_{l+k}^2)^2. \tag{17}$$

There are $q+2$ unknowns $a_0, \cdots, a_q$ and γ in (17). Due to the homogeneity of the α_j's we have to normalize the problem appropriately; all the α_j's can be multiplied by a constant $c \neq 0$ and γ can be divided by c^3 without changing (17). There are a number of ways of carrying out such a normalization. One could set $Ev_t^3 = \gamma = 1$. Alternatively $a_0 = 1$ could be the normalization condition. Some comments on the asymptotic distribution of the $\hat{c}_k$'s are given in Appendix 2.

The second method is a searching procedure. One uses a typical second order method to estimate the roots of $\alpha(z)$, $r_j, j = 1, \cdots, q$, assuming all the roots have modulus greater than one. An accurate estimate of the distribution of roots is obtained by taking the conjugated inverse of an appropriate number of the r_j's. Suppose all of the r_j's are real and distinct. Then there are 2^q possible sets of roots that give the same second order structure. Each of these sets yields a distinct set of $\hat{\alpha}_j$'s which in turn lead to a distinct set of the c_k's. Choose the set of $\hat{\alpha}_j$'s which determine the set of c_k's minimizing

$$\sum_{k=-q}^q (\hat{c}_k - c_k)^2$$

among all the possible sets of $\{\hat{\alpha}_j\}$. If some of the roots r_j are complex, the inverse complex conjugates are taken in pairs. If there are multiple roots, the solution of roots in terms of coefficients is unstable. Some comments on this question are made in Appendix 1. The initial set of coefficient estimates corresponding to roots all outside the unit circle can be obtained by the method described in Box and Jenkins (1976). Alternatively, one could try to obtain the roots directly by solving for the roots of the polynomial

$$p(z) = z^q g(z)$$

where

$$g(z) = \alpha(z)\alpha(z^{-1}) = \sum_{j=-q}^q \beta_j z^j$$

with

$$\beta_{|j|} = Ex_t x_{t+|j|} = \sum_l \alpha_l \alpha_{l+|j|}.$$

We estimate β_j by

$$\hat{\beta}_j = \frac{1}{n} \sum_{t=1}^{n-|j|} x_t x_{t+|j|}.$$

The roots of $p(z)$ with modulus greater than one are the initial set of roots.

6. Examples. We will consider a few simple examples generated by Monte Carlo simulation to illustrate the computation and to give a qualitative feeling of the effectiveness of the theory. Details and possible "fine tuning" of the computational method will be considered elsewhere.

We generate $x_t = v_t + \alpha_1 v_{t-1} + \alpha_2 v_{t-2}$, $t = 1, \cdots, 640$ where $v_t = v_t' - 1$ and v_t''s are independent exponentially distributed random deviates with mean one obtained from the GGEXN subroutine in the International Mathematical and Statistical Library (IMSL).

Then

$$Ev_t = 0, \qquad \mathrm{var}(v_t) = Ev_t^2 = 1, \qquad Ev_t^3 = 2.$$

We partition $\{x_t\}_{t=1}^{640}$ into five sections, each of which has 128 points. Compute the bispectrum estimate ${}_{128}b^{(i)}(j\Delta, \Delta), j = 1, \cdots, 13; i = 1, \cdots, 5$ by the algorithm described in Lii and Helland (1981). Here we set $\Delta = 18\pi/128 = 0.442$. Our final bispectrum estimate is

$${}_{640}b(j\Delta, \Delta) = 1/5 \textstyle\sum_{i=1}^{5} {}_{128}b^{(i)}(j\Delta, \Delta).$$

Compute

$$\theta_n(j) = \arg\{{}_n\hat{b}(j\Delta, \Delta)\} = \arctan\{\mathrm{Im}\, {}_n\hat{b}(j\Delta, \Delta)/\mathrm{Re}\, {}_n\hat{b}(j\Delta, \Delta)\}$$

by taking the principal value as well as

$$-H_n(j\Delta) \equiv -H_n(\lambda_j) = \textstyle\sum_{i=1}^{j-1} \theta_n(i), \qquad j = 2, \cdots, 14.$$

Let

$$H_n(0) \equiv H_n(\lambda_0) = 0,$$

$$H_n(\Delta) \equiv H_n(\lambda_1) = \tfrac{1}{2} H_n(\lambda_2),$$

$$H_n(\pi) = \tfrac{1}{2}\{H_n(7\Delta) + H_n(8\Delta)\} = \tfrac{1}{2}\{H_n(3.093) + H_n(3.534)\}$$

and

$$\hat{c} = -H_n(\pi)/\pi = -\,\delta.$$

Recall that $\hat{c}$ is an estimate of

$$c = \lim_{\Delta \to 0} \frac{h(\Delta)}{\Delta} = h'(0)$$

up to an integer. From formula (13) we compute the α_k's. A standard smoothed periodogram with uniform weights and bandwidth Δ was used to compute $f_n(\lambda)$ as an estimate of the spectrum $f(\lambda)$ of $\{x_t\}$; $f_n(0)$ is obtained by a linear extrapolation. These examples are as follows:

MODEL: $x_t = v_t + \alpha_1 v_{t-1} + \alpha_2 v_{t-2}$ with four cases specified in Table 1.

Here and from this point on all $\hat{\alpha}_k$'s are adjusted by sign and index shift.

Table 2 gives the details of the computations for Case 2. Figure 1 compares the theoretical h with the estimated h.

As an indication of discretization error, notice that if we use the exact $h(\lambda)$ and $|\alpha(e^{-i\lambda})|$ instead of estimated ones, we get

$$\hat{\alpha}_0 = 0.9136, \qquad \hat{\alpha}_1 = -2.247, \qquad \hat{\alpha}_3 = 0.5977.$$

When we increase the number of parameters to four, we observe qualitatively the same type of result as in the three parameter case. But generally speaking, the instability is increased. We give one example. The model is

$$x_t = v_t - 4.25v_{t-1} + 4.75v_{t-2} - 0.938v_{t-3}$$

with roots ⅔, ⅖, 4 and we get

$$\hat{\alpha}_0 = 1.106, \qquad \hat{\alpha}_1 = -3.572, \qquad \hat{\alpha}_2 = 4.293, \qquad \hat{\alpha}_3 = -0.9762.$$

These simple examples indicate that one can estimate the unknown coefficients reasonably well and one is able to discriminate between different models even though they have the same spectral structure.

TABLE 1
Coefficients, roots and estimated coefficients for four cases

Case	Coefficients			Roots		Estimated coefficients		
	α_0	α_1	α_2	r_1	r_2	$\hat{\alpha}_0$	$\hat{\alpha}_1$	$\hat{\alpha}_2$
1	1.0	−0.833	0.167	2.0	3.0	0.9593	−0.5816	0.1158
2	1.0	−2.333	0.667	0.5	3.0	0.7164	−2.175	0.7605
3	1.0	−3.50	1.50	2.0	0.333	0.7561	−3.334	1.778
4	1.0	−5.0	6.0	0.5	0.333	0.9603	−3.904	4.966

TABLE 2
Computations for Case 2

λ	Length $\lvert\alpha(e^{-i\lambda})\rvert$	Est. Length	Argument $h(\lambda)$	Argument by Sum $-\mathrm{H}_n(\lambda)$	Argument at J $\theta_n(\lambda)$	Adjusted $-H_n(\lambda)+\hat{c}\lambda$	$-H_n(\lambda)+(\hat{c}-1)\lambda$
0	0.6667	0.473	0.0000	0	0	0	0
0.442	0.8389	0.894	−0.6125	−0.2713	0	0.0904	−0.352
0.884	1.3018	1.315	−1.0828	−0.5425	−0.5425	0.1808	−0.703
1.325	1.9553	1.888	−1.4915	−0.4851	0.0574	−0.0575	−1.382
1.767	2.6785	2.748	−1.8895	−0.6904	−0.2053	−0.0331	−1.80
2.209	3.3369	3.348	−2.2893	−0.6481	0.0423	−0.2563	−2.465
2.651	3.8064	3.520	−2.6920	−1.0097	−0.3615	−0.0756	−2.727
3.093	3.9980	3.809	−3.0966	−1.0884	−0.0787	−0.1778	−3.27
3.534	3.8752	3.951	−3.5014	−1.4842	−0.3958	0.0371	−3.497
3.976	3.4614	3.297	−3.9047	−1.6797	−0.1956	0.0518	−3.924
4.418	2.8351	2.802	−4.3051	−1.5538	0.1259	−0.2550	−4.693
4.860	2.1146	2.168	−4.7031	−1.8428	−0.2890	−0.1469	−5.007
5.301	1.4344	1.429	−5.1070	−1.8785	−0.0357	−0.2921	−5.593
5.743	0.9199	0.936	−5.5558	−2.3478	−0.4693	−0.0037	−5.747
6.185	0.6755	0.443	−6.1366	−2.6185	−0.2706	0.0861	−6.099

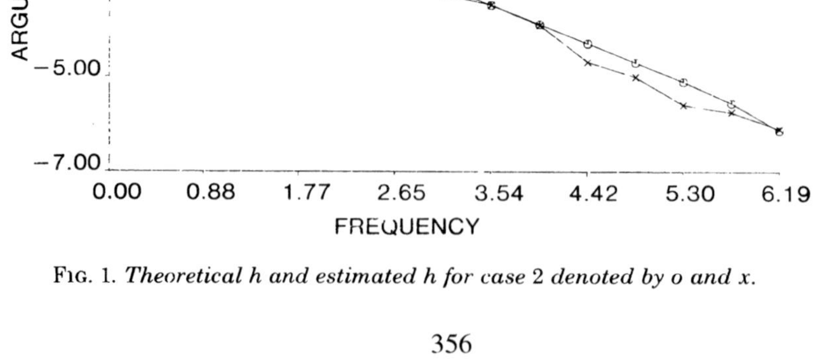

FIG. 1. *Theoretical h and estimated h for case 2 denoted by o and x.*

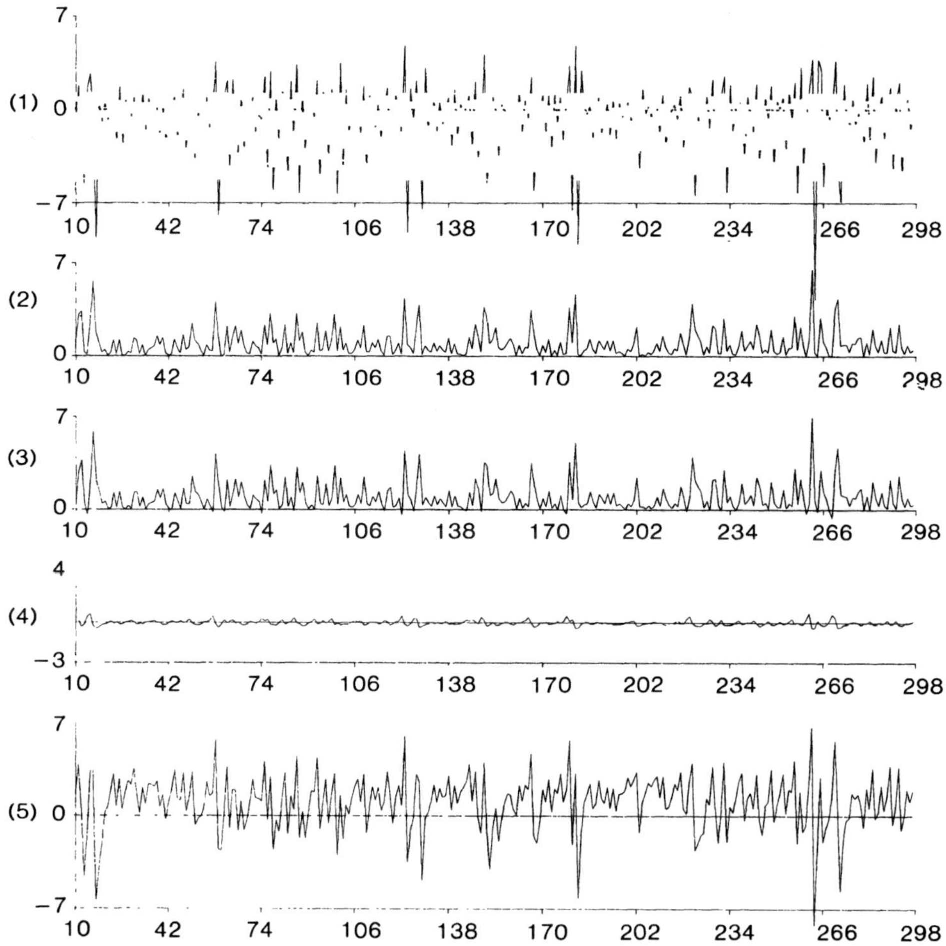

FIG. 2. *Deconvolution for case* 2. (1) *Moving average process* x_t, $t = 10$ to $t = 298$, (2) exponential v_t *sequence that generates* x_t, (3) *estimated* $\hat{v}_t$ *as deconvolved by our procedure,* (4) $v_t - \hat{v}_t$, (5) *deconvolution of* x_t *using minimum phase assumption.*

To illustrate our deconvolution procedure we consider the moving average

$$x_t = v_t - 2.333v_{t-1} + 0.667v_{t-2}$$

with the v_t's independent exponential variables with mean one. The v_t's are generated as pseudo-random variates. The estimated model (using our bispectral techniques) is

$$x_t = 0.7164v_t - 2.175v_{t-1} + 0.7605v_{t-2}.$$

Let $\hat{\alpha}(B) = 0.7164 - 2.175B - 0.7605B^2$. Set

$$\hat{v}_t = \sum_{j=-9}^{9} \hat{a}_j x_{t-j} \cong \frac{1}{\hat{\alpha}(B)} x_t. \tag{18}$$

The first line of Figure 2 is the moving average process x_t as generated for $t = 10$ to $t = 298$. The second line is the independent exponential v_t sequence that generates x_t for the same t range. Line 3 shows the estimated $\hat{v}_t$ as deconvolved by our procedure. The fourth line is the difference $v_t - \hat{v}_t$. The last line is a deconvolution of x_t using the assumption of minimum phase. The sample mean of $v_t - \hat{v}_t$ is .00032 and the sample variance .030. Notice how our deconvolution has generally preserved the shape of the original v_t sequence while the minimum phase deconvolution has no resemblance to the original v_t sequence.

A general way to find the deconvolution weights can be described as follows. We have an estimate

$$\hat{\alpha}(e^{-i\lambda}) = \sqrt{2\pi \hat{f}_n(\lambda)}\, \exp\{i(H_n(\lambda) + \hat{c}\lambda)\}.$$

We compute $b(e^{-i\lambda}) = \hat{\alpha}^{-1}(e^{-i\lambda})$. Then

$$\hat{\omega}_k = \frac{1}{2\pi}\int_0^{2\pi} b(e^{-i\lambda})e^{ik\lambda}\, d\lambda$$

give the coefficients of the series expansion of $\hat{\alpha}^{-1}(e^{-i\lambda})$ which are the weights desired in deconvolution. This method is general. We do not require knowledge of the order of the moving average process and there is no need to compute the roots from the estimated coefficients and how the roots are distributed is irrelevant.

APPENDIX 1

In this appendix we consider the relationship between the coefficients of a polynomial and the roots of the polynomial, at least locally. The polynomial is

$$\sum_{j=0}^{p} a_j z^j = \prod_{j=1}^{p} (z - z_j)$$

where the roots are $z_j, j = 1, \cdots, p$, and the coefficients $a_j, j = 0, 1, \cdots, p$, with $a_p = 1$. It is well known that

$$a_{p-1} = -\sum_j z_j, \qquad a_{p-2} = \sum_{j\neq k} z_j z_k,$$
$$a_{p-3} = -\sum_{j\neq k\neq \ell} z_j z_k z_\ell, \cdots, a_0 = (-1)^p z_1 \cdots z_p.$$

Let us consider the relationship between the differentials of the coefficients a_j and the differentials of the roots z_k. Now

$$\frac{\partial q_{p-1}}{\partial z_\ell} = -1, \qquad \frac{\partial a_{p-2}}{\partial z_\ell} = \sum_{j\neq \ell} z_j = -a_{p-1} - z_\ell,$$
$$\frac{\partial a_{p-3}}{\partial z_\ell} = -\sum_{j\neq k\neq \ell} z_j z_k = -a_{p-2} - 2a_{p-1}z_\ell - 2z_\ell^2,$$

$\ell = 1, \cdots, p$. Thus

$$da_{p-1} = -\sum_j dz_j, \qquad da_{p-2} = -a_{p-1}\sum_j dz_j - \sum_j z_j\, dz_j,$$
$$da_{p-3} = -a_{p-2}\sum_j dz_j - 2a_{p-1}\sum_j z_j\, dz_j - 2\sum_j z_j^2\, dz_j, \cdots$$

and this can be written in matrix notation as

$$\begin{pmatrix} da_{p-1} \\ da_{p-2} \\ da_{p-3} \\ \vdots \end{pmatrix} = UV \begin{pmatrix} dz_1 \\ dz_2 \\ dz_3 \\ \vdots \end{pmatrix}$$

where

$$U = \begin{pmatrix} -1 & 0 & 0 & 0 \\ -a_{p-1} & -1 & 0 & 0 \\ & -2a_{p-1} & -2 & 0 \\ & \cdots & & \end{pmatrix}$$

and

$$V = \begin{pmatrix} 1 & 1 & 1 & \cdots & 1 \\ z_1 & z_2 & z_3 & \cdots & z_n \\ z_1^2 & z_2^2 & z_3^2 & \cdots & z_n^2 \\ & & \cdots & & \end{pmatrix}.$$

U is a triangular nonsingular matrix and V is the Vander Monde matrix. V is nonsingular as long as the roots z_i are distinct.

APPENDIX 2

The object is to remark on some aspects of the asymptotic behavior of estimates of the c_k's in the context of a general linear process. Let x_t be a nonGaussian linear process (1) with

$$E\,v_t \equiv 0, \qquad E\,v_t^2 \equiv 1, \qquad E\,v_t^3 \equiv \gamma_3$$
$$\operatorname{cum}(v_t^4) = \gamma_4, \qquad \operatorname{cum}(v_t^6) = \gamma_6.$$

Set

$$y_t = x_t^2.$$

For convenience we introduce

$$g_u = \operatorname{Cov}(x_t, y_{t-u}), \qquad r_u = \operatorname{Cov}(x_t, x_{t-u}), \qquad h_u = \operatorname{Cov}(y_t, y_{t-u}).$$

Consider the estimates

$$\hat{g}_a = \frac{1}{N} \sum_{t=1}^{N} x_t y_{t+a}$$

of $Ex_t y_{t+a}$. Then

$$\operatorname{Cov}(\hat{g}_a, \hat{g}_b) = \frac{1}{N^2} \sum_{t,\tau=1}^{N} \operatorname{Cov}(x_t y_{t+a}, x_\tau y_{\tau+b})$$

where

$$\operatorname{Cov}(x_t y_{t+a}, x_\tau y_{\tau+b}) = r_{t-\tau} c_{t-\tau+a-b} + g_{t-\tau-b} g_{t-\tau+a} + \operatorname{cum}(x_t, x_\tau, y_{\tau+b}) \operatorname{cum}(y_{t+a})$$
$$+ \operatorname{cum}(x_t, x_\tau, y_{t+a}) \operatorname{cum}(y_{t+b}) + \operatorname{cum}(x_t, y_{t+a}, x_\tau, y_{\tau+b}).$$

Now

$$r_u = \sum_j \alpha_j \alpha_{j-u}, \qquad g_u = \gamma_3 \sum_j \alpha_j \alpha_{j-u}^2, \qquad h_u = \gamma_4 \sum \alpha_j^2 \alpha_{j-u}^2 + 2\,(\textstyle\sum_j \alpha_j \alpha_{j-u})^2.$$

Further

$$\operatorname{cum}(y_{t+a}) = \sum_j \alpha_j^2,$$
$$\operatorname{cum}(x_t, x_\tau, y_{\tau+b}) = \gamma_4 \sum_j \alpha_j \alpha_{j+b}^2 \alpha_{j+(t-\tau)} = k(b, t-\tau),$$
$$\operatorname{cum}(x_t, x_\tau, y_{t+a}) = \gamma_4 \sum_j \alpha_j \alpha_{j+a}^2 \alpha_{j+(\tau-t)} = k(a, \tau-t),$$
$$\operatorname{cum}(x_t, y_{t+a}, x_\tau, y_{\tau+b}) = \gamma_6 \sum_u \alpha_u \alpha_{u+(\tau-t)} \alpha_{u+a}^2 \alpha_{u+b+(\tau-t)}^2$$
$$+ 4\gamma_4 (\textstyle\sum_u \alpha_u \alpha_{u+(\tau-t)} \alpha_{u+a} \alpha_{u+b+(\tau-t)})(\sum_u \alpha_u \alpha_{u+(\tau-t)+b-a})$$
$$+ 2\gamma_3^2 (\textstyle\sum_u \alpha_u \alpha_{u+(\tau-t)} a_{u+a})(\sum \alpha_u^2 \alpha_{u+(t-\tau)+a-b})$$
$$+ 2\gamma_3^2 (\textstyle\sum_n \alpha_u \alpha_{u+(t-\tau)} \alpha_{u+b})(\sum \alpha_u^2 \alpha_{u+(\tau-t)+b-2})$$
$$= k(a, b, t-\tau).$$

It is then clear that

$$\lim_{N\to\infty} N \operatorname{Cov}(\hat{g}_a, \hat{g}_b) = \sum_s r_s c_{s+a-b} + \sum_s g_{s-b} g_{s+a} + r_0 \sum_s \{k(b, s) + k(a, s)\} + \sum_s k(a, b, s).$$

Under the assumption that $\sum |\alpha_j| < \infty$, with a truncation argument like that employed in Anderson (1970), one can show that

$$\sqrt{N}\,(\hat{g}_a - g_a)$$

are asymptotically normally distributed with covariance structure given by the preceding formula.

Acknowledgment. Research of the second author was carried out in part in the University of Manchester Institute of Science and Technology.

REFERENCES

ANDERSON, T. W. (1971). *The Spectral Analysis of Time Series.* Wiley, New York.

BASAWA, I. V., FEIGEN, P. D. and HEYDE, C. C. (1975). Asymptotic properties of maximum likelihood estimates for stochastic processes. *Sankhya, Ser. A* **38** 259–270.

BOX, G. E. P. and JENKINS, G. M. (1976). *Time Series Analysis, Forecasting and Control.* Holden-Day, San Francisco.

BRILLINGER, D. and ROSENBLATT, M. (1967). Asymptotic theory of estimates of kth order spectra. In *Spectral Analysis of Time Series* (ed. B. Harris) 153–188. Wiley, New York.

DONOHO, D. (1981) On minimum entropy deconvolution. In *Applied Time Series Analysis II* (ed. D. F. Findley) Academic, New York. 565–608.

GODFREY, R. and ROCCA, F. (1981) Zero memory nonlinear deconvolution. *Geophys. Prospect.* **29** 189–228.

HUZII, M. (1981) Estimation of coefficients of an autoregressive process by using a higher order moment. *J. Time Series Analysis* **2(2),** 87–93.

HENRICI, P. (1974). *Applied and Computational Complex Analysis,* vol. **1.** Wiley, New York.

JENKINS, G. M. and WATTS, D. G. (1968). *Spectral Analysis and Its Applications.* Holden-Day, San Francisco.

LII, K. S. and HELLAND, K. N. (1981). Cross-bispectrum computation and variance estimation. *ACM Trans. Math. Software* **7** No. 3.

ROSENBLATT, M. (1974). *Random Processes.* Springer, Berlin.

ROSENBLATT, M. (1980). Linear processes and bispectra. *J. Appl. Probab.* **17** 265–270.

WIGGINS, R. A. (1978). Minimum entropy deconvolution. *Geoexploration* **17.**

DEPARTMENT OF STATISTICS
UNIVERSITY OF CALIFORNIA
RIVERSIDE, CALIFORNIA 92521

DEPARTMENT OF MATHEMATICS
UNIVERSITY OF CALIFORNIA, SAN DIEGO
LA JOLLA, CALIFORNIA 92093

The Annals of Probability
1984, Vol. 12, No. 4, 1167–1180

ASYMPTOTIC NORMALITY, STRONG MIXING AND SPECTRAL DENSITY ESTIMATES[1]

BY M. ROSENBLATT

University of California, San Diego

Asymptotic normality is proven for spectral density estimates assuming strong mixing and a limited number of moment conditions for the process analyzed. The result holds for a large class of processes that are not linear and does not require the existence of all moments.

1. Introduction.

Strong mixing as a sufficient condition (usually together with some auxiliary moment conditions) for classical limit theorems was initially proposed and discussed in the period from the mid 50's to the early 60's (see Rosenblatt 1956, 1961; Kolmogorov and Rozanov, 1960; and Ibragimov, 1962). There had already been some investigation of the asymptotic properties of spectral density estimates. But from that time to the present day, most results on the asymptotic normality of spectral density estimates assume either a very special structure for the process (a linear process in Anderson, 1971, and Hannan, 1970) or else existence of all moments (as in Brillinger, 1975). Approximability by multilinear schemes in independent, identically distributed random variables has been considered in Rosenblatt, 1959. Here we shall show that a direct application of a central limit theorem using strong mixing together with a limited number of moment conditions will imply asymptotic normality for a large class of spectral density estimates.

2. Strong mixing and a triangular central limit theorem.

Let $X = \{X_n\}$ be a strictly stationary process. Let $\mathscr{B}_n = \mathscr{B}(X_k, k \le n)$ be the σ-field generated by the random variables. X_k, $k \le n$ and $\mathscr{F}_m = \mathscr{B}(X_k, k \ge m)$ the σ-field generated by X_k, $k \ge m$. The process $X = \{X_n\}$ is said to be *strongly mixing* if

$$\sup_{B \in \mathscr{B}_0, F \in \mathscr{F}_n} | P(BF) - P(B)P(F) | = \alpha(n) \to 0 \tag{2.1}$$

as $n \to \infty$. The following central limit theorem can be established by a big block-small block argument just as in Rosenblatt (1959, 1961). This result will be assumed. It will be shown that the result can be applied to the case of spectral density estimates.

THEOREM 1. *Let* $\{Y_j^{(n)}, j = \cdots, -1, 0, 1, \cdots\}$, $EY_j^{(n)} \equiv 0$, $n = 1, 2, \cdots$ *be a sequence of strictly stationary processes defined on the probability space of the*

Received March 1983; revised December 1983.

[1] Research supported in part by Office of Naval Research Contract N00014-81-K-003 and National Science Foundation Grant DMS 83-12106.

AMS 1980 *subject classifications*. Primary 60F05; secondary 62M15.

Key words and phrases. Asymptotic normality, strong mixing, spectral density estimates, cumulants, nonlinear functions of Gaussian processes.

R.A. Davis et al. (eds.), *Selected Works of Murray Rosenblatt*, Selected Works in Probability and Statistics, 361
DOI 10.1007/978-1-4419-8339-8_33, © Springer Science+Business Media, LLC 2011

strongly mixing stationary process $X = \{X_n\}$. Let $Y_j^{(n)}$ be measurable with respect to $\mathscr{F}_{j-c(n)} \cap \mathscr{B}_{j+c(n)}$ where $c(n) = o(n)$, $c(n) \uparrow \infty$ as $n \to \infty$. Set

$$(2.2) \qquad h_n(b-a) = E\,|\textstyle\sum_{j=a}^{b} Y_j^{(n)}|^2.$$

Assume that for any two sequences $s(n)$, $m(n)$ with $c(n) = o(m(n))$, $m(n) \le n$ and $s(n)/m(n) \to 0$ one has

$$(2.3) \qquad \frac{h_n(m(n))}{h_n(s(n))} \to \infty.$$

Let $F_{n,m}(x)$ be the distribution function of

$$\sum_{k=1}^{m} \frac{Y_k^{(n)}}{\{(n/m)h_n(m)\}^{1/2}}$$

with $m = m(n)$ and $c(n) = o(m(n))$, $m(n) = o(n)$. Assume that

$$(2.4) \qquad \frac{n}{m(n)} \int_{|x|>\eta} x^2\, dF_{n,m(n)}(x) \to 0$$

as $n \to \infty$ for each $\eta > 0$. There are then sequences $k(n)$, $p(n) \to \infty$ as $n \to \infty$ with $k(n)p(n) \cong n$ such that

$$(2.5) \qquad \textstyle\sum_{j=1}^{n} Y_j(n)/\sqrt{k(n)h_n(p(n))}$$

is asymptotically normally distributed with mean zero and variance one. Also if $k(n)h_n(p(n)) \cong h_n(n)$ the normalization in (2.5) can be replaced by $\sqrt{h_n(n)}$.

Condition (2.4) is like a Lindeberg condition. A corresponding result with this condition replaced by a stronger Liapounov condition is valid.

COROLLARY 1. *The conclusion of Theorem 1 still holds if all assumptions are the same except for condition (2.4) and condition (2.4) is replaced by the following condition (2.4)′. Let*

$$(2.4)' \qquad \{h_n(m)\}^{-(2+\delta)/2} E\,|\textstyle\sum_{k=1}^{m} Y_k^{(n)}|^{2+\delta} = O(1)$$

for $m = m(n)$ and some $\delta > 0$.

Because conditions in terms of cumulants are used a brief discussion relating to them is given. Let $Z = (Z_j, j = 1, \cdots, k)$ be a random k-vector with $t = (t_j, j = 1, \cdots, k)$ a k-vector of coefficients. The characteristic function of the random vector Z is

$$\varphi(t) = E \exp(it \cdot Z).$$

Let $m = (m_1, \cdots, m_k)$, $|m| = \sum m_i$. If moments up to order s exist there is a Taylor expansion

$$\varphi(t) = \sum_{|m| \le s} \frac{i^{|m|}\mu^{(m)}}{m!} t^m + o(|t|^s)$$

with

$$t^m = \prod_{j=1}^k t_j^{m_j},\ m! = \prod_{j=1}^k m_j!,$$

and the moments

$$\mu^{(m)} = \mu^{(m_1,\cdots,m_k)} = E(\prod_{j=1}^k Z_j^{m_j}).$$

There is a corresponding Taylor expansion for $\log \varphi(t)$

$$\log \varphi(t) = \sum_{|m|\le s} \frac{i^{|m|}c^{(m)}}{m!} t^m + o(|t|^s)$$

with the coefficients $c^{(m)}$ the cumulants of Z

$$c^{(m)} = c^{(m_1,\cdots,m_k)} = \mathrm{cum}(Z_1^{m_1}, \cdots, Z_k^{m_k}).$$

There are identities linking the moments $\mu^{(m)}$, $|m| \le s$, up to order s with the cumulants $c^{(m)}$, $|m| \le s$, up to order s. In fact, the moments up to order s exist if and only if the cumulants up to order s exist.

At times one will deal with random variables $W_1, \cdots, W_J$ given in terms of doubly indexed random variables

$$U_{k,\ell},\quad \ell = 1, \cdots, \ell_k,\quad k = 1, \cdots, J$$

by

$$W_k = \prod_{\ell=1}^{\ell_k} U_{k,\ell}.$$

The object is to carry out the computation of the cumulant

$$\mathrm{cum}(W_1, \cdots, W_J)$$

in terms of the cumulants of the $U_{k,\ell}$. Such a computation can be carried out in the following manner. Lay out the following array of the $U_{k,\ell}$'s in J rows

$$\begin{matrix} U_{1,1} & U_{1,2} & \cdots & U_{1,\ell_1} \\ U_{2,1} & U_{2,2} & \cdots & U_{2,\ell_2} \\ \vdots & \vdots & & \vdots \\ U_{J,1} & U_{J,2} & \cdots & U_{J,\ell_J}. \end{matrix}$$

A partition of this array into sets $\nu_1, \cdots, \nu_r$ is called decomposable if the union of the elements in a nonvacuous proper subcollection of sets of the partition yields exactly the elements in a nonvacuous proper subset of the rows of the table. A partition is indecomposable if it is not decomposable. One can show that

$$\mathrm{cum}(W_1, \cdots, W_J) = \sum \prod_{s=1}^r \mathrm{cum}(U_{j,\ell}, (j, \ell) \in \nu_s)$$

where the sum on the right is extended over all indecomposable partitions of the table given above (see Leonov and Shiryaev, 1959).

Let us note that if kth order moments of the process $X = \{X_n\}$ are finite, that the cumulants

$$\mathrm{cum}(X_j, X_{j+s_1}, \cdots, X_{j+s_{k-1}}) = r^{(k)}_{s_1,\cdots,s_{k-1}}$$

are well-defined. Summability conditions on cumulants such as

$$\sum_{s_1,\cdots,s_{k-1}} |r^{(k)}_{s_1,\cdots,s_{k-1}}| < \infty \tag{2.6}$$

are convenient to make at times. In Brillinger (1975) and Brillinger and Rosenblatt (1967) assumptions of the form

$$\sum_{s_1,\cdots,s_{k-1}} (1 + |s_j|)\, |r^{(k)}_{s_1,\cdots,s_{k-1}}| < \infty,$$

$j = 1, \cdots, k-1$, are made for all $k \geq 2$. We shall only have to assume a limited number of the conditions (2.6) together with strong mixing.

Most central limit theorems for dependent processes making use of mixing or mixingale conditions make an assumption on the rate of decay of the mixing coefficient (see Hall and Heyde, 1980). Notice that this is not the case for the limit theorems in this paper. The conditions are basically summability conditions on cumulants. In fact, we shall later give examples of processes that satisfy the assumptions of the theorems but still have a slow rate of decay of the mixing coefficient.

3. Spectral density estimates. The assumption (2.6) for $k = 2$ is just summability of the covariances

$$\sum_k |r_k| < \infty, \; r_k = \operatorname{cov}(x_j, x_{j+k}).$$

The spectral density $f(\lambda)$ of the process is then continuous with

$$f(\lambda) = 1/2\pi \sum_k r_k e^{-ik\lambda}.$$

Covariance estimates $r_k^{(n)}$ in terms of a sequence of observations, $X_1, \cdots, X_n$, $EX_k \equiv 0$, on X are given by

$$r_k^{(n)} = 1/n \sum_{j=1}^{n-k} X_j X_{j+k}, \quad 0 \leq k \leq n,$$

with $r_{-k}^{(n)} = r_k^{(n)}$. A large class of spectral density estimates $f_n(\lambda)$ are of the form

$$f_n(\lambda) = 1/2\pi \sum_{k=-n+1}^{n-1} r_k^{(n)} w_k^{(n)} \cos k\lambda$$

with

$$w_k^{(n)} = a(kb_n), \quad a(0) = 1,$$

with $a(x)$ *continuous at zero*, bounded and satisfying

$$a(x) = a(-x).$$

Here $b_n \to 0$ with $n^{-1} = o(b_n)$ as $n \to \infty$. The following proposition amounts to a simple remark on the asymptotic unbiasedness of a class of estimates of the form specified above.

PROPOSITION. *Let* $X = (X_k)$ *be a weakly stationary process with* $EX_k \equiv 0$, $\sum |r_k| < \infty$. *Assume that*

$$f_n(\lambda) = 1/2\pi \sum_{k=-n+1}^{n-1} r_k^{(n)} a(kb_n) \cos k\lambda.$$

Let $a(x)$ *be bounded, continuous at zero with* $a(0) = 1$ *and symmetric about zero.*

Then

$$Ef_n(\lambda) - f(\lambda) \to 0$$

as $b_n \to 0$, $n \to \infty$.

Notice that

$$\begin{aligned} Ef_n(\lambda) - f(\lambda) &= \frac{1}{2\pi} \sum_{|k| \le n} r_k \left(\frac{n - |k|}{n}\right)[a(kb_n) - 1] \cos k\lambda \\ &\quad - \frac{1}{2\pi} \sum_{|k| \le n} \frac{|k|}{n} r_k \cos k\lambda - \frac{1}{2\pi} \sum_{|k| > n} r_k \cos k\lambda. \end{aligned} \tag{3.1}$$

The absolute summability of r_k implies that the second and third sums on the right of (3.1) tend to zero as $n \to \infty$. The assumptions on $a(\cdot)$ together with the summability of r_k imply that the first sum on the right of (3.1) tends to zero as $b_n \to \infty$ and $n \to \infty$.

The following lemma will be useful in deriving some results on the asymptotic distribution of spectral density estimates.

LEMMA. *Let*

$$g(u) = 1/2\pi \sum g_k e^{-iku}, \quad \sum |g_k| < \infty.$$

Further assume that $a(\cdot)$ *is piecewise continuous, continuous at* 0 *with* $a(0) = 1$, *symmetric and such that* $a(x) = O(|x|^{-1/2-\varepsilon})$ *for some* $\varepsilon > 0$ *as* $|x| \to \infty$. *Then if*

$$W_n(u) = 1/2\pi \sum w_k^{(n)} e^{-iku}$$

with $w_k^{(n)} = a(kb_n)$ *it follows that*

$$b_n \int_{-\pi}^{\pi} W_n^2(u + \lambda) g(u)\, du \to g(\lambda) \int W^2(u)\, du \tag{3.2}$$

as $b_n \to 0$ *where*

$$W(u) = \frac{1}{2\pi} \int a(u) e^{-iu\alpha}\, du$$

while

$$\int_{-\pi}^{\pi} W_n(u + \lambda) W_n(u + \mu) g(u)\, du = o(b_n^{-1}) \tag{3.3}$$

as $b_n \to 0$ *if* $\lambda \ne \mu$, $|\lambda - \mu| < 2\pi$.

Notice that

$$b_n \int_{-\pi}^{\pi} W_n(u + \lambda) W_n(u + \mu) g(u)\, du$$

$$= \frac{1}{2\pi} \sum_{\ell} g_{\ell} e^{-i\ell u} \frac{1}{2\pi} \sum_k b_n a(kb_n) a((k - \ell) b_n) e^{ik(\mu - \lambda)}.$$

Clearly

$$|\sum_k b_n a(kb_n)a((k-\ell)b_n)| \le \sum b_n |a(kb_n)|^2 \tag{3.4}$$

and the right-hand side of (3.4) tends to

$$\int |a(u)|^2\,du$$

as $b_n \to 0$. Also, if $\mu = \lambda$

$$\sum_k b_n a(kb_n)a((k-\ell)b_n) \to \int |a(u)|^2\,du$$

as $b_n \to 0$ for each fixed ℓ. This yields the limiting relationship (3.2). Let us now consider the case in which $\lambda \neq \mu$, $|\lambda-\mu| < 2\pi$. Relation (3.3) will follow if we can show that

$$\sum_k b_n a(kb_n)a((k-\ell)b_n)e^{ik(\mu-\lambda)} \to 0 \tag{3.5}$$

as $b_n \to 0$ for each fixed ℓ. Given $\varepsilon > 0$ there is an $L(\varepsilon)$ such that

$$\sum_{b_n|k|>L(\varepsilon)} b_n |a(kb_n)|^2 < \varepsilon.$$

The function $a(x)$ can be uniformly approximated arbitrarily well by step functions over the interval $|x| \le L(\varepsilon)$. It is therefore enough to show that (3.5) holds for a step function $a(x)$ of finite support. In fact, this will follow if one can prove (3.5) for $a(x)$ the indicator function of a finite interval, say the interval (α_1, α_2), $\alpha_1 < \alpha_2$. Then

$$\begin{aligned}\sum_k a(kb_n)a((k-\ell)b_n)e^{ik(\mu-\lambda)} &= \sum_{\alpha_1<kb_n<\alpha_2} e^{ik(\mu-\lambda)}b_n + O(b_n)\\ &= e^{i\alpha_1 c_n(\mu-\lambda)} \sum_{0<kb_n<\alpha_2-\alpha_1} e^{ik(\mu-\lambda)}b_n + O(b_n)\\ &= b_n e^{i\alpha_1 c_n(\mu-\lambda)} \frac{e^{i(\alpha_2-\alpha_1)c_n(\mu-\lambda)}-1}{e^{i(\mu-\lambda)}-1} + O(b_n)\end{aligned}$$

with $c_n = b_n^{-1}$.

Let

$$\eta(\lambda) = \begin{cases} 1 & \text{if} \quad \lambda = k\pi, \quad k \text{ integer} \\ 0 & \text{otherwise.}\end{cases}$$

We shall prove the following theorem.

THEOREM 2. *Let $X = \{X_n\}$ be a strictly stationary strongly mixing process with $EX_j \equiv 0$. Assume that the cumulant functions (2.6) of order two and four are summable. Further, let the spectral density estimate $f_n(\lambda)$ have weights $w_k^{(n)}$ defined in terms of a function $a(\cdot)$ that is piecewise continuous, continuous at zero with $a(0) = 1$, symmetric about zero and is such that $xa(x)$ is bounded. Let*

$$Y_u^{(n)}(\lambda) = \sum_{k=-c(n)}^{c(n)} X_u X_{u+k} w_k^{(n)} \cos k\lambda$$

with $w_k^{(n)} = a(kb(n))$ *and* $c(n) = \alpha b_n^{-1}$ *for all sufficiently large fixed* α. *Set*

$$Z_n(\lambda) = \sum_{u=1}^{m} \frac{Y_u^{(n)}(\lambda)}{\{nb_n^{-1}\}^{1/2}}$$

with $m = m(n)$ *and* $c(n) = o(m(n))$, $m(n) = o(n)$. *Consider the distribution function* $F_{n,m}(x)$ *of* $Z_n(\lambda)$ *and assume that*

$$\frac{n}{m(n)} \int_{|x|>\eta} x^2 \, dF_{n,m(n)}(x) \to 0 \tag{3.6}$$

as $n \to \infty$, $b_n \to 0$, $nb_n \to \infty$ *for each* $\eta > 0$. *Then* $f_n(\lambda) - Ef_n(\lambda)$ *is asymptotically normally distributed with mean zero and variance*

$$\frac{2\pi(1 + \eta(\lambda))}{nb_n} f^2(\lambda) \int W^2(\alpha)\, d\alpha. \tag{3.6$'$}$$

Let us first consider the case in which $a(\cdot)$ has finite support. We first consider replacing $f_n(\lambda)$ by

$$\tilde{f}_n(\lambda) = \frac{1}{2\pi} \sum_{k=-c(n)}^{c(n)} \frac{1}{n} \sum_{j=1}^{n} x_j x_{j+k} w_k^{(n)} \cos k\lambda.$$

Here the support of the function $a(\cdot)$ has been taken as the interval $[-1, 1]$ to simplify notation. The argument goes through in exactly the same way for any other interval. Notice that

$$2\pi n \tilde{f}_n(\lambda) = \sum_{u=1}^{n} Y_u^{(n)} \tag{3.7}$$

where

$$Y_u^{(n)} = \sum_{k=-c(n)}^{c(n)} X_u X_{u+k} w_k^{(n)} \cos k\lambda.$$

Thus, the expression (3.7) is of the type discussed in Theorem 1. Now,

$$\sigma^2(f_n(\lambda) - \tilde{f}_n(\lambda)) \le \frac{1}{\pi^2 n^2} \sigma^2\Big[\sum_{k=1}^{c(n)} \sum_{j=n-k}^{n} X_j X_{j+k} w_k^{(n)} \cos k\lambda\Big] \tag{3.8}$$

and the right-hand side of (3.8) is less than

$$\begin{aligned}
&\frac{1}{\pi^2 n^2} \sum_{k,k'=1}^{c(n)} \sum_{j=n-k}^{n} \sum_{j'=n-k'}^{n} \\
&\qquad \cdot \{|r_{j-j'} r_{j-j'+k-k'}| + |r_{j-j'-k'} r_{j-j'+k}| + |r_{k,j'-j,j'+k'-j}^{(4)}|\}\, |w_k^{(n)} w_{k'}^{(n)}| \\
&\le \frac{1}{\pi^2 n^2} \sum_{k,k'=1}^{c(n)} \sum_s \min(k', k) \{|r_s|\,|r_{s+k'-k}| + |r_{s-k'} r_{s+k}| + |r_{k,s,s+k'}^{(4)}|\} \\
&\qquad \cdot |w_k^{(n)}|\,|w_{k'}^{(n)}|.
\end{aligned}$$

The assumptions on the weights $w_k^{(n)}$ and the function $a(\cdot)$ imply that the weights are bounded and

$$|k|\,|w_k^{(n)}| \le Lb_n^{-1}$$

for all k and some constant L. This together with the summability of (2.6) for

$k = 2, 4$ imply that

$$\sigma^2(f_n(\lambda) - \tilde{f}_n(\lambda)) = O((nb_n)^{-2}) = o((nb_n)^{-1}).$$

As we shall see this is of smaller order of magnitude than $\sigma^2(\tilde{f}_n(\lambda))$. Let us now consider

$$\begin{aligned} \sigma^2\left(\textstyle\sum_{u=1}^m Y_u^{(n)}\right) &= h_n(m(n)) \\ &= \textstyle\sum_{u,u'=1}^m \sum_{k,k'=-c(n)}^{c(n)} \{r_{u-u'}r_{u+k-u'-k'} + r_{u'-u+k'}r_{u'-u-k} + r^{(4)}_{k,u'-u,u'-u+k'}\} \\ &\qquad \cdot \cos k\lambda \cos k'\lambda w_k^{(n)} w_{k'}^{(n)} \\ &= (1) + (2) + (3) \end{aligned} \tag{3.9}$$

where $m = m(n) \le n$ with $b_n^{-1} = o(m(n))$ as $n \to \infty$. The first term on the right of (3.9) is

$$(1) = \int \int_{-\pi}^{\pi} \frac{\sin^2(m/2)(\alpha - \beta)}{\sin^2(1/2)(\alpha - \beta)} f(\alpha) f(\beta) \left| \textstyle\sum_{v=-c(n)}^{c(n)} w_v^{(n)} \cos v\lambda\, e^{iv\beta} \right|^2 d\alpha\, d\beta.$$

The sequence of weight functions $W_n(u)$ have the property

$$\max_{|u| \le Am^{-1}} \left| \frac{\int_{-\pi}^{\pi} W_n(v) W_n(v + u)\, dv}{\int_{-\pi}^{\pi} W_n^2(u)\, du} - 1 \right| \to 0 \tag{3.10}$$

as $m = m(n) \to \infty$ for each $A > 0$ since $c(n) = b_n^{-1} = o(m(n))$. An argument like that given on page 176 of Rosenblatt (1974) implies that

$$(1) = 2\pi m(1 + o(1)) \int_{-\pi}^{\pi} f^2(\beta) \left| \frac{1}{2} W_n(\beta + \lambda) + \frac{1}{2} W_n(\beta - \lambda) \right|^2 d\beta \tag{3.11}$$

holds. The lemma then yields

$$(1) = \pi m(1 + \eta(\lambda) + o(1)) f^2(\lambda) \int_{-\pi}^{\pi} W_n^2(v)\, dv. \tag{3.12}$$

The second term on the right of (3.9) is

$$\begin{aligned} (2) = \int \int_{-\pi}^{\pi} \frac{\sin^2 (m/2)(\alpha - \beta)}{\sin^2 \frac{1}{2}(\alpha - \beta)} f(\alpha) f(\beta) \left(\frac{1}{2} W_n(\alpha + \lambda) + \frac{1}{2} W_n(\alpha - \lambda)\right) \\ \cdot \left(\frac{1}{2} W_n(\beta + \lambda) + \frac{1}{2} W_n(\beta - \lambda)\right) d\alpha\, d\beta. \end{aligned}$$

Property (3.10) of the family of weight functions as well as the continuity of f imply that relation (3.11) holds for (2) as well as (1). This implies that we get estimate (3.12) valid for (2) as well as (1). A direct estimate shows that

$$|(3)| \le m \textstyle\sum_{k,s,k'} |r^{(4)}_{k,s,k'}|$$

and this is of smaller order than (3.12) under the assumptions made. Notice that

$$\int_{-\pi}^{\pi} W_n(u)^2\, du = (1 + o(1))b_n^{-1} \int W^2(u)\, du.$$

The estimates of (1), (2) and (3) imply that $\sigma^2(\tilde{f}_n(\lambda))$ and therefore $\sigma^2(f_n(\lambda))$ have the same asymptotic behavior as (3.1) as $n \to \infty$. This is obtained by setting $m(n) = n$. Also notice that if $m(n) = o(n)$ and $k(n)m(n) = n$, that then

$$k(n)h_n(m(n)) \cong h_n(n)$$

because of the summability of the cumulant functions for $k = 2$ and $k = 4$.

Assumption (3.6) and Theorem 1 now imply the conclusion of this theorem in the case of a function $a(\cdot)$ with finite support.

We now indicate how the argument can be modified so as to take care of a function $a(\cdot)$ that is not of finite support. One can replace $f_n(\lambda)$ by

$$\hat{f}_n(\lambda) = 1/2\pi \sum_{|k|\le c(n)} r_k^{(n)} w_k^{(n)} \cos k\lambda$$

with error term

$$g_n(\lambda) = 1/2\pi \sum_{c(n)<|k|\le n} r_k^{(n)} w_k^{(n)} \cos k\lambda$$

and

$$c(n) = Lb_n^{-1}$$

where L is fixed but sufficiently large. The asymptotic normality of $\hat{f}_n(\lambda)$ has already been determined by the argument given since it corresponds to a function $a(\cdot)$ with finite support. We just have to estimate the variance of the error $g_n(\lambda)$ and show that it is sufficiently small if L is large. That will establish the asymptotic normality of $f_n(\lambda)$ for $a(\cdot)$ not of bounded support. Now

$$\sigma^2(g_n(\lambda)) = \frac{1}{4\pi^2 n^2} \sum_{|k|,|k'|\ge c(n)} \sum_{u,u'=1}^{n} \{r_{u-u'} r_{u-u'+k-k'} + r_{u-u'-k'} r_{u-u'+k} + r^{(4)}_{k,u'-u,u'-u+k'}\} w_k^{(n)} w_{k'}^{(n)} \cos k\lambda \cos k'\lambda$$

and this is bounded by

$$\frac{1}{4\pi^2 n} \sum_{|k|,|k'|\ge c(n)} \sum_s \{|r_s|\,|r_{s+k-k'}| + |r_{s-k'} r_{s+k}| + |r^{(4)}_{k,s,s+k'}|\}\, |w_k^{(n)}|\, |w_{k'}^{(n)}|. \tag{3.13}$$

Notice that the terms $|w_k^{(n)}| = |a(kb_n)|$ are bounded by an absolute constant. The sum in (3.13) involving fourth cumulants is $O(1/n)$. The other terms in (3.13) are less than or equal to

$$\frac{1}{4\pi^2 n} \sum_{|k|,|k'|\ge c(n)} 2\,|r_{k-k'}|^{(2^*)} |w_k^{(n)}|\, |w_{k'}^{(n)}| \tag{3.14}$$

where $|r_k|^{(2^*)}$ represents the convolution of the sequence $|r_k|$ with itself. Now

$$|kw_k^{(n)}| = |ka(kb_n)| \le Mb_n^{-1} \tag{3.15}$$

with M an absolute constant. This implies that the sum (3.14) is bounded by

$$\frac{1}{n} \sum_s |r_s|^{(2^*)} M^2 b_n^{-2} \sum_{|k| \ge c(n)} \{|k| (|k| + |s|)\}^{-1} \le \frac{M^2}{n} L^{-1} b_n b_n^{-2} \sum_s |r_s|^{(2^*)}$$

because of (3.15). The variance $\sigma^2(g_n(\lambda))$ is bounded by an absolute constant multiplied by $L^{-1}(b_n n)^{-1}$ and this is small compared to $(b_n n)^{-1}$ if L is large.

Joint asymptotic normality and asymptotic independence of spectral density estimates at distinct frequencies λ, $0 \le \lambda \le \pi$, is dealt with in the following corollary.

COROLLARY 2. *Let all the assumptions of Theorem 2 be satisfied with the exception of* (3.6). *This is modified as follows. Consider*

$$\sum_{j=1}^{s} \alpha_j Z_n(\lambda_j) \tag{3.16}$$

with the λ_j *distinct frequencies in* $[0, \pi]$. *Let* $F_{n,m(n)}(x)$ *be the distribution function of* (3.16). *Let*

$$\frac{n}{m(n)} \int_{|x|>\eta} x^2 \, dF_{n,m(n)}(x) \to 0 \tag{3.17}$$

as $n \to \infty$, $b_n \to 0$, $nb_n \to \infty$ *for all values of the* α_j*'s and each* $\eta > 0$. *It then follows that* $(nb_n)^{1/2}\{f_n(\lambda_j) - Ef_n(\lambda_j)\}$ *are jointly asymptotically independent and normally distributed with variances*

$$2\pi(1 + \eta(\lambda_j)) f^2(\lambda_j) \int W^2(\alpha) \, d\alpha, \; j = 1, \cdots, s.$$

The conditions of Corollary 2 imply that every linear combination

$$(nb_n)^{1/2} \sum_{j=1}^{s} \alpha_j \{f_n(\lambda_j) - Ef_n(\lambda_j)\}$$

is asymptotically normal. That implies that the spectral estimates $(nb_n)^{1/2}\{f_n(\lambda_j) - Ef_n(\lambda_j)\}$, $j = 1, \cdots, s$, are jointly asymptotically normal. The asymptotic orthogonality and hence asymptotic independence follows from (3.3) of the lemma.

The conditions of the form (3.17) are sometimes difficult to verify directly. The following corollary replaces these by stronger cumulant conditions that are sometimes easier to implement and interpret.

COROLLARY 3. *Let the assumptions of the corollary be satisfied except for* (3.17). *These are replaced by summability of the cumulant functions up to order eight. The conclusion of the Corollary then holds.*

It is enough to show this in the case of a single spectral estimate. The argument for several estimates is the same except for the additional notation.

In order to apply Corollary 2 in this situation we shall verify condition (3.17) by showing that

$$\sigma^{-4} \left(\sum_{u=1}^{m} Y_u^{(n)}\right) E \left| \sum_{u=1}^{m} (Y_u^{(n)} - EY_u^{(n)}) \right|^2 = O(1)$$

where $Y_u^{(n)} = Y_u^{(n)}(\lambda)$, for $m = m(n)$ as $n \to \infty$. Now

$$E\,|\textstyle\sum_{u=1}^m (Y_u^{(n)} - EY_u^{(n)})|^4 = \sigma^4(\sum_{u=1}^m Y_u^{(n)}) + \mathrm{cum}_4(\sum_{u=1}^m Y_u^{(n)}).$$

Because of the multilinear character of the cumulant

$$\mathrm{cum}_4(\textstyle\sum_{u=1}^m Y_u^{(n)}) = \sum_{u_1,u_2,u_3,u_4=1}^m \mathrm{cum}_4(Y_{u_1}^{(n)}, Y_{u_2}^{(n)}, Y_{u_3}^{(n)}, Y_{u_4}^{(n)})$$

and

$$\mathrm{cum}_4(Y_{u_1}^{(n)}, Y_{u_2}^{(n)}, Y_{u_3}^{(n)}, Y_{u_4}^{(n)})$$

$$= \left(\textstyle\prod_{i=1}^4 w_{k_i} \cos k_i\lambda\right) \sum_\nu \mathrm{cum}(X_s, s \in \nu_1) \cdots \mathrm{cum}(X_s, s \in \nu_p)$$

where the summation over $\nu = \nu_1 \cup \cdots \cup \nu_p$ is over all indecomposable partitions (see the discussion in Section 2) of the following table

$$\begin{array}{c} x_{u_1}x_{u_1+k_1} \\ x_{u_2}x_{u_2+k_2} \\ x_{u_3}x_{u_3+k_3} \\ x_{u_4}x_{u_4+k_4}. \end{array}$$

One of the several indecomposable partitions consisting entirely of pairs leads to the sum

$$\begin{aligned} &\textstyle\sum_{k_i} \sum_{u_i} r_{u_1-u_2} r_{u_3-u_4} r_{u_3-u_1+k_3-k_1} r_{u_4-u_2+k_4-k_2} \\ &\qquad w_{k_1}^{(n)} \cos k_1\lambda w_{k_2}^{(n)} \cos k_2\lambda\, w_{k_3}^{(n)} \cos k_3\lambda\, w_{k_4}^{(n)} \cos k_4\lambda. \end{aligned} \tag{3.18}$$

Introduce $a = u_1 - u_2$, $b = u_3 - u_4$, $\alpha = k_3 - k_1$, $\beta = k_4 - k_2$. Expression (3.18) is bounded by

$$\textstyle\sum_{a,b,u_1,u_3,k_1,k_2,\alpha,\beta} |r_a|\; |r_b|\; |r_{u_3-u_1+\alpha}|\; |r_{u_3-u_1-b+\alpha+\beta}|\; |w_{k_1}^{(n)}|\; |w_{k_1+\alpha}^{(n)}|\; |w_{k_2}^{(n)}|\; |w_{k_2+\beta}^{(n)}|.$$

If one first sums over k_1, k_2 the bound

$$b_n^{-2} \textstyle\sum_{a,b,u_1,u_3,\alpha,\beta} |r_a|\; |r_b|\; |r_{u_3-u_1+\alpha}|\; |r_{u_3-u_1-b+\alpha+\beta}|$$

is obtained. A sum over α, β yields the bound

$$b_n^{-2} \textstyle\sum_{a,b,u_1,u_3} |r_a|\; |r_b|.$$

The sum over a, b and then u_1, u_3 gives us the final bound

$$b_n^{-2} m^2.$$

All the other indecomposable partitions consisting entirely of pairs lead to the same bound. Indecomposable partitions which do not consist entirely of pairs yield smaller bounds. An example is given by the partition leading to the sum

$$\textstyle\sum_{k_i} \sum_{u_i} r_{u_2-u_1,u_3-u_1,u_4-u_1}^{(4)} r_{k_2-k_1+u_2-u_1} r_{k_4-k_3+u_4-u_3} \prod_{i=1}^4 (w_{k_i}^{(n)} \cos k_i\lambda). \tag{3.19}$$

Introduce $u_2 - u_1 = a$, $u_3 - u_1 = b$, $u_4 - u_1 = c$, $\alpha = k_2 - k_1$, $\beta = k_4 - k_3$. Expression (3.19) is then bounded by

$$\textstyle\sum_{u_1,a,b,c,\alpha,\beta,k_1,k_2} |r_{a,b,c}^{(4)}|\; |r_{\alpha+a}|\; |r_{\beta+b}|\; |w_{k_1}^{(n)}|\; |w_{k_1+\alpha}^{(n)}|\; |w_{k_2}^{(n)}|\; |w_{k_2+\beta}^{(n)}|.$$

Summing first over k_1, k_2 leads to the bound

$$b_n^{-2} \sum_{u_1,a,b,c,\alpha,\beta} | r_{a,b,c}^{(4)} | \; | r_{\alpha+a} | \; | r_{\beta+b} | .$$

Then sum over α, β and then over a, b, c to obtain

$$b_n^{-2} \sum_{u_1} 1 = m b_n^{-2}.$$

These estimates imply that (3.17) is valid so that Corollary 2 can be applied to obtain the asymptotic normality of such spectral density estimates.

A little reflection shows one that Corollary 3 is still valid if one replaces strict stationarity by stationarity up to eighth order.

If one has a vector-valued strongly mixing strictly stationary process, comparable conditions lead to corresponding results on the asymptotic behavior of cross-spectral estimates (see Brillinger, 1975, and Rosenblatt, 1959). These ideas can be adapted to get results for higher order spectral estimates.

4. Additional remarks. Kolmogorov and Rozanov (1960) have shown a sufficient condition for a Gaussian stationary process to be strongly mixing is that it have a continuous positive spectral density function. In their book, Ibragimov and Rozanov (1978) give an example of a strongly mixing Gaussian process with an unbounded spectral density function. This example is constructed by making use of a result of Helson and Sarason (1967). Non-Gaussian strongly mixing processes can be constructed by taking instantaneous or bandlimited functions of Gaussian strongly mixing processes.

Let $\varphi(z)$ be the standard Gaussian density function and

$$H_v(z) = (-1)^v \frac{(d^v/dz^v)\varphi(z)}{\varphi(z)},$$

$v = 0, \cdots$, the Hermite polynomials. Let $Y_1, \cdots, Y_m$ be jointly Gaussian random variables with means zero, variances one, and covariance matrix

$$\rho = (\rho_{ij};\ i, j = 1, \cdots, m).$$

Slepian (1972) has shown that the moment

$$\begin{aligned} E\{H_{s_1}(Y_1) \cdots H_{s_m}(Y_m)\} &= s_1! \cdots s_m! \sum{}' \frac{\rho^v}{v!} \\ &= s_1! \cdots s_m! \sum{}' \prod_{i<j} \frac{\rho_{ij}^{v_{ij}}}{v_{ij}!} \end{aligned} \tag{4.1}$$

where the primed summation is over all $m \times m$ symmetric matrices $v = (v_{ij};\ i, j = 1, \cdots, m)$ with nonnegative integer entries such that

$$s_k = \sum_{j \neq k} v_{jk}, \quad k = 1, \cdots, m.$$

This result leads directly to the following result.

COROLLARY 4. *Let $X = (X_k)$ be a strictly stationary Gaussian sequence with mean zero and summable covariance function. Assume that $\psi(x)$ is a function with* *a Fourier-Hermite expansion that has nonnegative coefficients. Further let*

$$E\,|\,\psi(x)\,|^{2k} < \infty$$

for some positive integer k. *Then the cumulant sequences up to order* $2k$ *are summable.*

It is easy to generate functions ψ with nonnegative Fourier-Hermite coefficients such that all moments

$$E\,|\,\psi(x)\,|^{2m}$$

are not finite. Consider the identity

$$e^{tx}e^{-t^2/2} = \sum_{k=0}^{\infty} H_k(x)t^k/k!.$$

For any positive c

$$\psi(x) = \left(\frac{2\pi}{1+c}\right)^{1/2} \exp\left\{\frac{x^2}{2(1+c)}\right\} = \int_{-\infty}^{\infty} e^{tx}e^{-(1+c)t^2/2}\,dt$$

$$= \sum_{k=0}^{\infty} \frac{H_k(x)}{k!} \int_{-\infty}^{\infty} t^k e^{-ct^2/2}\,dt = \sum_{k=0}^{\infty} \frac{H_{2k}(x)}{2^k k!}\, c^{-k-1/2}.$$

With c sufficiently large the $(2k)$th moment of $\psi(x)$ will be finite but not higher order moments. If $X = (X_n)$ is a stationary Gaussian sequence with summable covariance sequence and positive spectral density, it will be strongly mixing. The non-Gaussian process $\psi(X_n)$ will be strongly mixing and have cumulant sequences summable up to order $2k$. If $k = 4$ the assumptions of Corollary 3 will be satisfied.

We now show how one can construct examples satisfying the assumptions of Corollary 3 but have a slow rate of decay of the mixing coefficient. A result of Ibragimov and Rozanov (1978, page 181) states that the mixing coefficient of a stationary Gaussian sequence

$$\alpha(n) = O(n^{-\beta})$$

for a fixed β, $0 < \beta < 1$, if and only if the process has a spectral density of the form

$$f(\lambda) = |\,P(e^{i\lambda})\,|^2 w(\lambda)$$

where P is a polynomial with zeros on $|\,z\,| = 1$ and w is strictly positive and satisfies a Hölder condition of order β. One can show that if $\{Y_n\}$, $EY_n \equiv 0$, is a stationary Gaussian sequence with covariances

$$r_n = |\,n\,|^{-1-\beta}$$

for $|\,n\,| \geq 1$, its spectral density is of the form

$$f(\lambda) = c\,|\,\lambda\,|^{\beta} + g(\lambda)$$

where $c > 0$ is a constant and $g(\lambda)$ an infinitely differentiable function for $|\,\lambda\,| < \pi$ (see Zygmund, 1968, vol. 2, page 135). Notice that for β in this range

$$\sum |\,r_n\,| < \infty$$

but not

$$\sum |n| \, |r_n| < \infty.$$

By choosing β sufficiently close to zero, we can effect as slow an inverse polynomial rate of decay of $\alpha(n)$ as may be desired. Consider the process

$$X_n = Y_n^{2k-1}, \quad n = 1, 2, \cdots,$$

for k some integer greater than two. By (4.1) it is clear that all cumulant functions are absolutely summable. Corollary 3 is therefore applicable to the process $\{X_n\}$. However, $\{X_n\}$ has exactly the same mixing coefficient sequence as the process $\{Y_n\}$.

One can ask whether absolute summability of cumulants up to fourth order is enough to get the conclusion of Corollary 3. The examples constructed by Herrndorf (1983), indicating that moment conditions up to second order are not enough for a standard version of a central limit theorem, suggest that one may need more than summability of cumulants up to fourth order in the context of spectral density estimates.

REFERENCES

ANDERSON, T. W. (1971). *Statistical Analysis of Time Series.* Wiley, New York.

BRILLINGER, D. R. and ROSENBLATT, M. (1967). Asymptotic theory of kth order spectra. *Spectral Analysis of Time Series* (ed. B. Harris), 153–188. Wiley, New York.

BRILLINGER, D. R. (1975). *Time Series, Data Analysis and Theory.* Holt, New York.

HALL, P. and HEYDE, C. C. (1980). *Martingale Limit Theory and Its Application.* Academic, New York.

HANNAN, E. J. (1970). *Multiple Time Series.* Wiley, New York.

HELSON, H. and SARASON, D. (1967). Past and future. *Math. Scand.* **21** 5–16.

HERRNDORF, N. (1983). Stationary strongly mixing sequences not satisfying the central limit theorem. *Ann. Probab.* **11** 809–813.

IBRAGIMOV, I. A. (1962). Some limit theorems for stationary processes. *Theory Probab. Appl.* **7** 349–382.

IBRAGIMOV, I. A. and ROZANOV, Y. A. (1978). *Gaussian Random Processes.* Springer, Berlin.

KOLMOGOROV, A. and ROZANOV, Y. A. (1960). On a strong mixing condition for stationary Gaussian processes. *Theory Probab. Appl.* **5** 204–208.

LEONOV, V. P. and SHIRYAEV, A. N. (1959). On a method of calculation of semi-invariants. *Theory Probab. Appl.* **4** 319–329.

ROSENBLATT, M. (1956). A central limit theorem and a strong mixing condition. *Proc. Nat. Acad. Sci. U.S.A.* **42** 43–47.

ROSENBLATT, M. (1959). Statistical analysis of stochastic processes with stationary residuals. In *Probability and Statistics* (ed. U. Grenander) 246–275. Wiley, New York.

ROSENBLATT, M. (1961). Some comments on narrow band-pass filters. *Quart. Appl. Math.* **18** 387–393.

ROSENBLATT, M. (1974). *Random Processes.* Springer, Berlin.

SLEPIAN, D. (1972). On the symmetrized Kronecker power of a matrix and extensions of Mehler's formula for Hermite polynomials. *SIAM J. Math. Anal.* **3** 606–616.

MATHEMATICS DEPARTMENT
UNIVERSITY OF CALIFORNIA, SAN DIEGO
LA JOLLA, CALIFORNIA 92093

Proc. Natl. Acad. Sci. USA
Vol. 83, pp. 199–200, January 1986
Applied Mathematical Sciences

Deconvolution of non-Gaussian linear processes with vanishing spectral values

(non-minimum phase/cumulant spectra/filter/noise)

KEH-SHIN LII* AND MURRAY ROSENBLATT†

*Department of Statistics, University of California, Riverside, CA 92521; and †Department of Mathematics, University of California at San Diego, La Jolla, CA 92093

Contributed by Murray Rosenblatt, September 9, 1985

ABSTRACT We consider the problem of estimating the filter generating a non-Gaussian linear process and the deconvolution of that process when the spectral density of the process has zeros. Without using a minimum phase assumption we show that often if there are only finitely many zeros there are procedures to effect such an estimation and deconvolution.

Linear processes $\mathbf{X}_t$ are generated by an independent identically distributed input sequence η_t convolved with a linear filter α_t

$$\mathbf{X}_t = \sum_j \alpha_j \eta_{t-j}. \quad [1]$$

This model is of considerable interest in geoexploration (1–4). Given observations on $\mathbf{X}_t$ only, a main interest is to estimate the filter coefficients α_t and the input sequence η_t by deconvolving $\mathbf{X}_t$.

Let $\alpha(z) = \Sigma\alpha_j z^j$ be the z-transform of the filter α_t. Then the spectral density of $\{\mathbf{X}_t\}$ is $f(\lambda) = \sigma_\eta^2|\alpha(e^{-i\lambda})|^2/2\pi$, which is completely determined by the second-order properties of $\{\mathbf{X}_t\}$ and does not contain the phase information of $\alpha(z)$. If η_t has a Gaussian distribution then the process $\{\mathbf{X}_t\}$ is completely determined by second-order properties and one can only determine $\alpha(z)$ up to $|\alpha(z)|$. There is the indeterminancy of the phase of $\alpha(z)$. Often this is resolved by making the ad hoc assumption that $\alpha(z)$ is minimum phase. If $\alpha(z)$ is polynomial this means the roots are outside the unit disc.

Often, η_t appears to have a non-Gaussian distribution and then one need not assume $\alpha(z)$ minimum phase. By using information provided by the higher-order (greater than 2) cumulants, the phase of $\alpha(z)$ can be estimated. Results using higher-order cumulant spectral techniques have been given by Rosenblatt (5) and by Lii and Rosenblatt (3, 6). A main result is summarized below.

LEMMA. *Consider a non-Gaussian linear process* X_t *given in Eq.* 1 *with* η_t *having all their moments finite. Let* $\Sigma|j\alpha_j| < \infty$ *and assume* $\alpha(e^{-i\lambda}) \neq 0$ *for all* λ. *The transfer function* $\alpha(e^{-i\lambda})$ *can then be identified in terms of observations only on* $\{\mathrm{X}_t\}$ *up to an indeterminate integer* a *in a factor* $e^{ia\lambda}$ *that corresponds to a shift of index of* X_t *and an indeterminate sign of* $\alpha(1) = \Sigma\alpha_k$. *For this result it is sufficient to have some finite moment of order* $k > 2$ *with cumulant* $\gamma_k \neq 0$.

The basic idea is to obtain the modulus of $\alpha(e^{-i\lambda})$ using the spectral density $f(\lambda)$ of $\{\mathbf{X}_t\}$ [since $|\alpha(e^{-i\lambda})| = |2\pi f(\lambda)|^{1/2}$] assuming $E\eta_t \equiv 0$ and $E\eta_t^2 \equiv 1$. There are a number of consistent estimates of $f(\lambda)$ discussed by Priestley (7). The phase $h(\lambda) = \arg\{\alpha(e^{-i\lambda})\alpha(1)/|\alpha(1)|\}$ of $\alpha(e^{-i\lambda})$ is obtained by making use of the relation

$$h(\lambda_1) + \cdots + h(\lambda_{k-1}) - h(\lambda_1 + \cdots + \lambda_{k-1}) = \arg\left[\left\{\frac{\alpha(1)}{|\alpha(1)|}\right\}^k \gamma_k^{-1} b_k(\lambda_1, \ldots, \lambda_{k-1})\right], \quad [2]$$

which is derived from the kth-order cumulant spectral density of the process $\{\mathbf{X}_t\}$

$$b_k(\lambda_1, \ldots, \lambda_{k-1}) = \frac{\gamma_k}{(2\pi)^{k-1}} \alpha(e^{-i\lambda_1}) \cdots \alpha(e^{-i\lambda_{k-1}})\alpha(e^{i(\lambda_1+\cdots+\lambda_{k-1})}). \quad [3]$$

Given an estimate $\hat{b}_k$ of b_k one can obtain an estimate $\hat{h}(\lambda)$ of $h(\lambda)$. $\hat{\alpha}(e^{-i\lambda}) = |\hat{\alpha}(e^{-i\lambda})|e^{i\hat{h}(\lambda)}$ is an estimate of $\alpha(e^{-i\lambda})$. One can deconvolve $\mathbf{X}_t$ and obtain estimates of b_k by using

$$\hat{\eta}_t = \hat{\alpha}^{-1} * \mathbf{X}_t. \quad [4]$$

THE RESULTS

In many contexts the spectral density $f(\lambda)$ does have zeros and it is of some interest to see whether one can carry out deconvolution and in the case of this model recover the original non-Gaussian white noise sequence η_n. We have already remarked that, under the assumption of the lemma, *Condition A* is valid.

Condition A. The function $\alpha(e^{-i\lambda})$ can be consistently estimated up to a factor $e^{ia\lambda}$ with a an integer and the sign $\pm$ 1.

In the results we mention, the important details of how to estimate $f(\lambda)$ and $\alpha(e^{-i\lambda})$ will not be dealt with in any detail here. Some of the aspects of estimation of $\alpha(e^{-i\lambda})$ can be found in the papers by Lii and Rosenblatt (3, 6).

One should note that *Condition A* will be satisfied under the following assumption concerning $\alpha(e^{-i\lambda})$.

PROPOSITION. *If* $\alpha(e^{-i\lambda})$ *is the restriction to the unit circle of a function* $\alpha(z) \not\equiv 0$ *analytic in an annulus about the unit circle,* Condition A *is satisfied. Such a function can have at most a finite number of zeros on* $|z| = 1$.

Under this assumption the procedure suggested via equations 2 and 3 of the paper for the estimation of $\alpha(e^{-i\lambda})$ will still be asymptotically valid (see additional remark at end of paper).

THEOREM 1. *Under* Condition A *the process* X_n *can be deconvolved so as to obtain the process* η_n *up to sign and a time shift by making use of the filter corresponding to*

$$\{f_n(\lambda) + a_n\}^{-1/2} \exp\{-i \arg \alpha_n(e^{-i\lambda})\}, \quad [5]$$

where $f_n(\lambda)$ (≥ 0), $\alpha_n(e^{i\lambda})$ *are appropriate consistent estimates of* $f(\lambda)$, $\alpha(e^{-i\lambda})$ *based on a sample size* n *and* a_n *an appropriate sequence of positive numbers with limit zero.*

The publication costs of this article were defrayed in part by page charge payment. This article must therefore be hereby marked "*advertisement*" in accordance with 18 U.S.C. §1734 solely to indicate this fact.

R.A. Davis et al. (eds.), *Selected Works of Murray Rosenblatt*, Selected Works in Probability and Statistics, 375
DOI 10.1007/978-1-4419-8339-8_34, © Springer Science+Business Media, LLC 2011

We shall also consider the case in which the process x_n has added to it Gaussian white noise ξ_t with small variance $a > 0$. We thus have the observed process

$$\xi_n = x_n + \xi_n = \int_{-\pi}^{\pi} e^{in\lambda} \alpha(e^{-i\lambda}) dz_\eta(\lambda) + \int_{-\pi}^{\pi} e^{in\lambda} dz_\xi(\lambda), \quad [6]$$

with $z_\eta(\lambda)$, $z_\xi(\lambda)$ random spectral processes corresponding to the processes η and ξ.

THEOREM 2. *Under* Condition A *and* $\Sigma|ja_j| < \infty$, *the process* X_n *can be deconvolved so as to obtain the process* η_n *up to sign and a time shift by applying the filter corresponding to Eq.* **5** *to the process* ξ_n *and letting* $n \to \infty$.

Details, particularly those concerned with estimation and derivation of results, will be provided later.

This research was partially supported by Office of Naval Research Contract N00014-81-K0003 and National Science Foundation Grant DMS83-12106.

1. Wiggins, R. (1978) *Geoexploration* **17,** 21–35.
2. Jurkevics, A. & Wiggins, R. (1984) *Geophysics* **49,** 2109–2116.
3. Lii, K. S. & Rosenblatt, M. (1982) *Ann. Stat.* **10,** 1195–1208.
4. Matsuoka, T. & Ulrych, T. J. (1984) *Proc. IEEE* **72,** 1403–1411.
5. Rosenblatt, M. (1980) *J. Appl. Probab.* **17,** 265–270.
6. Lii, K. S. & Rosenblatt, M. (1985) *Multivariate Analysis* (Elsevier, New York), Vol. 6, pp. 395–410.
7. Priestley, M. B. (1981) *Spectral Analysis and Time Series* (Academic, New York).

J. Appl. Prob. **24**, 328–338 (1987)
Printed in Israel

SCALE RENORMALIZATION AND RANDOM SOLUTIONS OF THE BURGERS EQUATION

M. ROSENBLATT,* *University of California, San Diego*

Abstract

Solutions of the Burgers equation with a stationary (spatially) stochastic initial condition are considered. A class of limit laws for the solution which correspond to a scale renormalization is considered.

STATIONARY INITIAL CONDITION; SHORT-RANGE DEPENDENCE; LONG-RANGE DEPENDENCE

Introduction

The Burgers equation

$$u_t + uu_x = \mu u_{xx}, \qquad \mu > 0, \tag{1}$$

$u = u(x, t)$ is a simple example of a non-linear partial differential equation that has been suggested as a simple model for turbulence (see Burgers (1948)) as well as for shock waves when $\mu \downarrow 0$ (see Whitham (1974)). Spatially stochastic solutions of the Navier–Stokes equation have been used in discussions of turbulence (see Batchelor (1953)) and Vishik et al., (1979)). We shall consider solutions of the Burgers equation with spatially stationary stochastic initial conditions.

Both Hopf (1950) and Cole (1951) have shown how the discussion of the Burgers equation can be reduced to that of the diffusion equation by a simple non-linear transformation. We shall give Hopf's formulation in the following theorem.

Theorem 0 (Hopf). *Let* $u(x, t)$ *be a solution of Equation* (1) *with continuous*

Received 25 November 1985; revision received 22 April 1986.
* Postal address: Department of Mathematics, University of California, San Diego, La Jolla, CA 92093, USA.
This research is supported in part by Office of Naval Research Contract N00014-81-K-003 and National Science Foundation Grant DMS 83-12106.

R.A. Davis et al. (eds.), *Selected Works of Murray Rosenblatt*, Selected Works in Probability and Statistics, DOI 10.1007/978-1-4419-8339-8_35, © Springer Science+Business Media, LLC 2011

initial condition $u_0(x) = u(x, 0)$. *If*

$$\int_0^x u_0(\xi)d\xi = o(x^2) \tag{2}$$

as $|x| \to \infty$, *then*

$$u(x, t) = \frac{\int_{-\infty}^{\infty} \frac{x-y}{t} \exp\left[-\frac{1}{2\mu} F(x, y, t)\right] dy}{\int_{-\infty}^{\infty} \exp\left[-\frac{1}{2\mu} F(x, y, t)\right] dy}$$

with

$$F(x, y, t) = \frac{(x-y)^2}{2t} + \int_0^y u_0(\xi)d\xi.$$

Notice that if we set

$$\varphi_0(y) = \exp\left[-\frac{1}{2\mu}\int_0^y u_0(y)dy\right]$$

and

$$\varphi(x, t) = (4\pi\mu t)^{-1/2} \int_{-\infty}^{\infty} \varphi_0(y) \exp\left[-\frac{(x-y)^2}{4\mu t}\right] dy$$

then

$$u(x, t) = -2\mu \frac{\partial}{\partial x} \log \varphi(x, t). \tag{3}$$

Random solutions. We shall be interested in the case in which $u_0(x)$ is a continuous and stationary process in x. To use Hopf's representation of the solution and its consequences, we shall require that the stationary process $u_0(\cdot)$ satisfy (2). For the moment assume that this is the case. Simple conditions on the process $u_0(\cdot)$ that imply (2) will be given a bit later. Our interest is in the asymptotics of

$$\int_0^R u(x, t)dx = -2\mu \log \frac{\varphi(R, t)}{\varphi(0, t)} \tag{4}$$

as $R \to \infty$ with $t > 0$ fixed. Notice that

$$\varphi(R, t) = (4\pi\mu t)^{-1/2} \int_{-\infty}^{\infty} \varphi_0(R + v) \exp\left[-\frac{v^2}{4\mu t}\right] dv$$

with

$$\varphi_0(R+v) = \exp\left[-\frac{1}{2\mu}\int_0^R u_0(y)dy\right]\exp\left[-\frac{1}{2\mu}\int_R^{R+v} u_0(y)dy\right]$$

so that

$$\log \varphi(R,t) = -\frac{1}{2\mu}\int_0^R u_0(y)dy$$

$$(5)\qquad + \log\left\{(4\pi\mu t)^{-1/2}\int_{-\infty}^{\infty}\exp\left[-\frac{1}{2\mu}\int_R^{R+v} u_0(y)dy - \frac{v^2}{4\mu t}\right]dv\right\}.$$

The following simple remark can be made.

Proposition 1. *If* $u_0(x)$ *is a continuous stationary process in* x *that satisfies* (2), *then the process* $u(x,t)$ *is stationary in* x *for each fixed* $t>0$.

The proposition can be seen to be valid in the following way. If (2) is satisfied by $u_0(\cdot)$, the expression

$$(6)\qquad (4\pi\mu t)^{-1/2}\int_{-\infty}^{\infty}\exp\left[-\frac{1}{2\mu}\int_x^{x+v} u_0(y)dy - \frac{v^2}{4\mu t}\right]dv$$

is well defined and determines a process that is jointly stationary in x with $u_0(\cdot)$. On differentiating $\log\varphi(x,t)$ with respect to x and making use of (4) and (5), one sees that $u(x,t)$ is stationary in x.

As an immediate consequence of the ergodic theorem we have the following remark.

Proposition 2. *If* u_0 *is a stationary process with*

$$E\,|u_0(x)| < \infty$$

then

$$(7)\qquad \int_0^x u_0(y)dy = O(x)$$

as $|x|\to\infty$ *and consequently* (2) *is satisfied.*

Under the assumptions of this proposition

$$\frac{1}{x}\int_0^x u_0(y)dy \to E(u_0(0)\,|\,\mathscr{I})$$

with probability 1 as $|x|\to\infty$ where $\mathscr{I}$ is the σ-field of invariant sets for the process $u_0(\cdot)$. However, this immediately yields (7).

Let us now consider stationary processes $u_0(\cdot)$ with finite second moment

$Eu_0(x)^2$. For most such processes the variance

$$\sigma^2(R) = \operatorname{var}\left(\int_0^R u_0(y)dy\right)$$

will diverge as $R \to \infty$ though it is of course easy to construct processes for which this variance remains bounded as $R \to \infty$. Nonetheless, the most interesting processes are those for which the variance diverges and we shall assume that this is the case. Consider the scale renormalization

$$\sigma(R)^{-1} \int_0^{Ry} u(\eta, t)d\eta$$

as $R \to \infty$. Given a process $z(x)$ consider

$$h(R)^{-1} \int_0^{Ry} \{z(x) - c\}dx = T_R(y)$$

where $h(R) \to \infty$ as $R \to \infty$. We shall say that $T_R(y)$ converges weakly (to some process $W(y)$) if for any given continuous functional ψ on $C[0, 1]$ the distribution of

$$\psi(T_R(y), 0 \leqq y \leqq 1)$$

tends to the distribution of $\psi(W(y), 0 \leqq y \leqq 1)$. We then have the following theorem.

Theorem 1. *Let $u_0(\cdot)$ be a continuous stationary process with finite second moment such that $\sigma(R)$ diverges as $R \to \infty$. Assume that $u_0(\cdot)$ is the initial condition for the Burgers equation and $u(x, t)$ the solution of the Burgers equation. If the processes*

$$\sigma(R)^{-1} \int_0^{Ry} \{u_0(x) - Eu_0(x)\}dx, \tag{8}$$

$$\sigma(R)^{-1} \int_0^{Ry} \{u(x, t) - Eu_0(x)\}dx \tag{9}$$

both converge weakly as $R \to \infty$, they must converge weakly to the same random process.

The theorem follows from the stationarity of (6) in x for each $t > 0$, together with relations (4) and (5).

Theorem 1 is of interest because it implies that the low frequency behavior of the spatial spectrum of $u(x, t)$ in the immediate neighborhood of 0 is the same independently of $t > 0$ and is the same as that of $u_0(x)$. Remarks suggesting something like this have been made in the case of homogeneous turbulence (see Batchelor (1953), Section 3.3).

We shall mention a number of conditions on the process $u_0(\cdot)$ (the initial condition) that allow one to determine the asymptotic behavior of (8) and (9) as $R \to \infty$ (but with $\mu > 0$ and t fixed). Let

$$\mathscr{B}_x = \mathscr{B}\{u_0(w), w \leqq x\},$$
$$\mathscr{F}_x = \mathscr{B}\{u_0(w), w \geqq x\},$$

be the σ-fields of events generated by $u_0(w)$, $w \leqq x$, and $u_0(w)$, $w \geqq x$, respectively. Thus, $\mathscr{B}_x$ and $\mathscr{F}_x$ correspond to the spatial information to the left of x, and the spatial information to the right of x for the process $u_0(\cdot)$. Let $\alpha(d)$, $d > 0$ be the coefficient

$$\alpha(d) = \sup_{\substack{A \in \mathscr{B}_x \\ B \in \mathscr{F}_{x+d}}} |P(A \cap B) - P(A)P(B)|.$$

Notice that $\alpha(d)$ is independent of x because of the assumption that $u_0(x)$ is stationary in x. The process $u_0(\cdot)$ is said to be strongly mixing if

$$\alpha(d) \to 0$$

as $d \to \infty$. The strong mixing property is one of the many non-equivalent ways of formulating a concept of short-range dependence. Let $u_0(\cdot)$ be stationary with

$$(10) \qquad \begin{aligned} &Eu_0(x) \equiv 0, \\ &E\,|u_0(x)|^{2+\delta} < \infty \end{aligned}$$

with $0 < \delta \leqq 1$, and strongly mixing with

$$\alpha(d) = O(d^{-(1+\varepsilon)(1+2/\delta)})$$

for some $0 < \varepsilon \leqq 1$. If $\sigma(R) \to \infty$ as $R \to \infty$, these conditions are enough to guarantee that (8) and (9) converge weakly to a Brownian motion process $B(y)$, $0 \leqq y$. This and even stronger results are directly implied by Theorem 1 and remarks in Dehling and Philipp (1982).

A more restricted but less complicated result can be formulated in terms of the concept of an asymptotically uncorrelated process. If $\mathscr{A}$ is a σ-field, let $L^2(\mathscr{A})$ be the collection of random variables measurable with respect to $\mathscr{A}$ that have finite second moment. Let

$$\beta(d) = \sup_{\substack{f \in L^2(\mathscr{B}_x) \\ g \in L^2(\mathscr{F}_{x+d})}} |\operatorname{corr}(f, g)|$$

with corr (f, g) the correlation of f and g. The process $u_0(\cdot)$ is said to be asymptotically uncorrelated if $\beta(d) \to 0$ as $d \to \infty$. The asymptotically uncorrelated condition is an even stronger formulation of a concept of short-range dependence than is that of strong mixing.

Theorem 2. *Let $u_0(\cdot)$ be a stationary process that is asymptotically uncorrelated, satisfies* (10), *and is such that $\sigma(R)\to\infty$ as $R\to\infty$. It then follows that* (8) *converges weakly to a Brownian motion process $B(y)$ as $R\to\infty$.*

Theorem 2 follows directly from our Theorem 1 and a result of Ibragimov (1975) concerning asymptotic normality for asymptotically uncorrelated processes. It is easy to see how to modify these results if the mean of $u_0(\cdot)$ is non-zero.

Let us now consider a special class of processes $u_0(\cdot)$ as initial condition which imply a weak limiting process different from the Brownian motion. These processes all exhibit a long-range dependence as contrasted with those processes dealt with earlier. This long-range dependence shows itself in terms of a singular mass accumulation in the neighborhood of 0 for the spectrum of the process $u_0(\cdot)$. All the processes $u_0(\cdot)$ we exhibit are obtained by an instantaneous function of a Gaussian process. Let $G(x)$ be a stationary Gaussian process with mean zero and covariance function

$$r(x)=EG(0)G(x)\cong x^{-D}L(x) \tag{11}$$

as $x\to\infty$, $L(\cdot)$ slowly varying and $0<D<1$ For convenience, assume that $r(0)=1$. Let $h(\cdot)$ be a function with

$$\begin{aligned} &Eh(G(x))\equiv 0,\\ &Eh(G(x))^2<\infty. \end{aligned} \tag{12}$$

Our object is to take the initial condition for the Burgers equation $u(x,0)=u_0(x)$ a process of the form

$$u_0(x)=h(G(x)). \tag{13}$$

Consider the Fourier–Hermite expansion of $h(\cdot)$

$$h(u)=\sum_{q=1}^{\infty}\frac{J(q)}{q!}h_j(u)$$

and let m be the first index with non-zero coefficient $J(m)$. The Hermite polynomials $h_j(u)$ have leading coefficient 1. We shall be interested in the range

$$0<D<\frac{1}{m}.$$

The process $u_0(\cdot)$ given by (13) satisfied the conditions of Theorem 1 and so the asymptotic behavior of (8) will be the same as that of (9). Now

$$\sigma^2(R) = E\left(\int_0^R h(G(\dot{x}))dx\right)^2$$
$$\cong \left(\frac{J(m)}{m!}\right)^2 \frac{2}{(1-mD)(2-mD)} R^{2-mD} L^m(R)$$

as $R \to \infty$. Let $W(\cdot)$ be the complex Gaussian process with

$$EW(\Delta) \equiv 0$$
$$W(\Delta) = \overline{W(-\Delta)}$$

for Δ an interval,

$$W\left(\bigcup_{i=1}^{n} \Delta_i\right) = \sum_{i=1}^{n} W(\Delta_i)$$

for disjoint intervals Δ_i, and covariance

$$E(W(\Delta_1)\overline{W(\Delta_2)}) = |\Delta_1 \cap \Delta_2|$$

($|\Delta|$ is the length of Δ). Let the symbol $\int_m'$ denote integration over all of m-dimensional Euclidean space except for $\lambda_i = \pm \lambda_j$, $i \neq j$, $i, j = 1, \cdots, m$. We introduce the process

$$Z_m(x) = C(m, D) \int_m' \frac{\exp(i(\lambda_1 + \cdots + \lambda_m)x) - 1}{i(\lambda_1 + \cdots + \lambda_m)}$$
$$\times |\lambda_1|^{(D-1)/2} \cdots |\lambda_m|^{(D-1)/2} W(d\lambda_1) \cdots W(d\lambda_m)$$

with

$$C(m, D) = \left\{\frac{(1-mD)(2-mD)}{2(m!)(2\Gamma(D)\cos D\pi/2)^m}\right\}^{1/2}.$$

Notice that the process is Gaussian only in the case $m = 1$.

Theorem 3. *Let* $u_0(\cdot)$ *be the process* (13) *satisfying* (11) *and* (12). *Then*

$$\sigma(R)^{-1} \int_0^{Ry} u(x, t)dx$$

will converge weakly to the process

$$Z_m(y)$$

as $R \to \infty$.

Theorem 3 follows directly from our Theorem 1 and results of Taqqu (1981) (his Theorem 3.2) and Dobrushin and Major (1979) (their Theorem 1). The limiting process $Z_m(y)$ is Gaussian when $m = 1$ and is then a fractional Brownian motion.

A counterexample for a growth conditon. The growth condition (2) has played an important part in our discussion. It is of some interest to give a simple example of a stationary process $u_0(\cdot)$ that does not satisfy the condition. The process will understandably not have moments in view of Proposition 2. It will be defined in terms of a stable symmetric process $\xi(\cdot)$ with coefficient α and with stationary independent increments. Let $F(y)$ be the distribution function of $\xi(1)$. It is well known that

$$1 - F(y) \cong \frac{C}{y^\alpha}$$

as $y \to \infty$ *with some constant* $c > 0$ and $0 < \alpha < 2$. The characteristic function of $\xi(1)$ is $\exp[-|u|^\alpha]$. Since the characteristic function of $\xi(\tau)$, $\tau > 0$, is $\exp[-\tau|u|^\alpha]$ it follows that the distribution function $\xi(\tau)$ is $F(\tau^{-1/2}y)$ with

$$1 - F(\tau^{-1/2}y) \cong \frac{c\tau}{y^\alpha} \tag{14}$$

as $y \to \infty$. Let $g(t)$ be an increasing continuous non-negative function defined for $t > 0$. Now $g(t)$ is called an upper function for $\xi(\cdot)$ if

$$P\left\{\limsup_{t\to\infty} \xi(t)/g(t) < 1\right\} = 1$$

and a lower function for $\xi(\cdot)$ if

$$\mathrm{P}\left\{\limsup_{t\to\infty} \xi(t)/g(t) > 1\right\} = 1.$$

There are sufficient conditions for $g(t)$ to be an upper or lower function for $\xi(\cdot)$ (see Gikhman and Skorokhod (1969)). A sufficient condition for $g(\cdot)$ to be an upper function for $\xi(\cdot)$ is

$$\int_1^\infty \frac{1}{t} P\{\xi(t) > g(t)\}dt < \infty. \tag{15}$$

A sufficient condition for $g(\cdot)$ to be a lower function for $\xi(\cdot)$ is that

$$\sum_{k=1}^{\infty} P\{\xi(a^k) > g(a^k)\} \tag{16}$$

diverge for all $a > 1$. By using (14) and (15) it is clear that

$$g(t) = t^{1/\alpha}(\log t)^{1/\alpha+\varepsilon}$$

or

$$g(t) = t^{1/\alpha}(\log t)^{1/\alpha}(\log\ \log t)^{1/\alpha+\varepsilon}$$

are upper functions for $\xi(\cdot)$ for any $\varepsilon > 0$.

Similarly (14) and (16) imply that

$$g(t) = t^{1/\alpha}(\log \log t)^{1/\alpha}$$

or

$$g(t) = t^{1/\alpha}(\log t)^{1/\alpha}(\log\ \log t)^{1/\alpha}$$

are lower functions for $\xi(\cdot)$. We shall now generate a stationary process $u_0(\cdot)$ with continuous sample functions from the stable process $\xi(\cdot)$ with independent stationary increments as follows. Let

$$u_0(x) = \int_x^{x+1} [\xi(w+1) - \xi(w)]dw. \tag{17}$$

Now

$$\begin{aligned} \int_0^x u_0(y)dy &= \int_0^x \int_y^{y+1} [\xi(w+1) - \xi(w)]dw\, dy \\ &= \int_0^1 y[\xi(y+1) - \xi(y)]dy + \int_{x-1}^x (x-y)[\xi(y+1) - \xi(y)]dy \\ &\quad + \int_1^{x-1} [\xi(y+1) - \xi(y)]dy \end{aligned} \tag{18}$$

while

$$\int_1^{x-1} [\xi(y+1) - \xi(y)]dy = \int_{x-1}^x \xi(y)dy - \int_1^2 \xi(y)dy. \tag{19}$$

From (18) and (19) it follows that the functions $g(x)$ mentioned above as upper and lower functions for $\xi(x)$ as $x \to \infty$ are also upper and lower functions for

$$\int_0^x u_0(y)dy$$

with $u_0(\cdot)$ given by (17). Notice that this implies that (17) satisfies the growth condition (2) if $\alpha > \frac{1}{2}$ but does not if $\alpha \leqq \frac{1}{2}$.

Varying μ. In the discussion of asymptotics in the earlier part of the paper one considered letting $R \to \infty$ in (4) but with μ and t fixed. It's clearly also of interest to look at the case in which $R \to \infty$ and $\mu \downarrow 0$ but with t fixed or at least bounded. A simple estimate will show us that the asymptotics obtained earlier are still valid in such a circumstance if one can assume that for the initial condition one has the finite bound

$$|u_0(x)| \leqq c$$

for all x with probability 1. For then

$$\int_x^{x+v} u_0(y)dy + \frac{v^2}{2t} \geqq \begin{cases} -cv + \dfrac{v^2}{2t} & \text{if } v \geqq 0 \\ cv + \dfrac{v^2}{2t} & \text{if } v \leqq 0. \end{cases} \tag{20}$$

This implies that an upper bound for (6) is given by

$$\frac{1}{\sqrt{4\pi\mu t}} \exp\left(\frac{c^2 t}{4\mu}\right)\left[\int_0^\infty \exp\left(-\frac{(v-ct)^2}{4\mu t}\right)dv + \int_{-\infty}^0 \exp\left(-\frac{(v+ct)^2}{4\mu t}\right)dv\right]$$
$$\leqq 2\exp\left(\frac{c^2 t}{4\mu}\right).$$

(21)

A lower bound for (6) can be obtained by noting that

$$\int_x^{x+v} u_0(y)dy + \frac{v^2}{2t} \leqq \begin{cases} cv + \dfrac{v^2}{2t} & \text{if } v \geqq 0 \\ -cv + \dfrac{v^2}{2t} & \text{if } v \leqq 0. \end{cases} \tag{20$'$}$$

The lower bound for (6) is then

$$\frac{1}{\sqrt{4\pi\mu t}} \exp\left(\frac{c^2 t}{4\mu}\right)\left[\int_0^\infty \exp\left(-\frac{(v+ct)^2}{4\mu t}dv\right) + \int_{-\infty}^0 \exp\left(-\frac{(v-ct)^2}{4\mu t}\right)dv\right]$$

$$= \exp\left(\frac{c^2 t}{4\mu}\right) 2\left[1 - \Phi\left(\frac{ct}{\sqrt{2\mu t}}\right)\right] \tag{22}$$

$$\leqq \begin{cases} \dfrac{1}{c}\left(\dfrac{\mu}{\pi t}\right)^{1/2} & \text{if } c\left(\dfrac{t}{2\mu}\right)^{1/2} > 1 \\ \exp\left(\dfrac{c^2 t}{4\mu}\right)[1 - \Phi(1)] & \text{if } c\left(\dfrac{t}{2\mu}\right)^{1/2} \leqq 1. \end{cases}$$

The upper and lower bounds (21) and (22) for (6) together with the relations (4) and (5) imply that for a bounded initial condition $u_0(\cdot)$ the results already obtained for $R \to \infty$ but with μ and t fixed are still valid when $R \to \infty$, $\mu \downarrow 0$ and t is bounded.

Acknowledgment

I should like to thank S. Karlin and the Mathematics Department at Stanford University for their hospitality during a sabbatical when this paper was written.

References

BATCHELOR, G. (1953) *The Theory of Homogeneous Turbulence*. Cambridge University Press.

BURGERS, J. M. (1948) A mathematical model illustrating the theory of turbulence. *Adv. Appl. Mech.* **1**, 171–199.

COLE, J. D. (1951) On a quasilinear parabolic equation occurring in aerodynamics. *Q. Appl. Math.* **9**, 225–236.

DEHLING, H. AND PHILIPP, W. (1982) Almost sure invariance principles for weakly dependent vector-valued random variables. *Ann. Prob.* **10**, 689–701.

DOBRUSHIN, R. AND MAJOR, P. (1979) Non-central limit theorems for non-linear functionals of Gaussian fields. *Z. Wahrscheinlichkeitsth.* **50**, 27–52.

GIKHMAN, I. AND SKOROKHOD, A. V. (1969) *Introduction to the Theory of Random Processes*. Saunders, Philadelphia, PA.

HOPF, E. (1950) The partial differential equation $u_t + uu_x = \mu u_{xx}$. *Comm. Pure Appl. Math.* **3**, 201–230.

IBRAGIMOV, I. (1975) A note on the central limit theorem for dependent random variables. *Theory Prob. Appl.* **20**, 135–141.

TAQQU, M. (1981) Self-similar processes and related ultraviolet and infrared catastrophes. In *Random Fields*, ed. J. Fritz, J. Lebowitz, and D. Szasz, North-Holland, Amsterdam, 1057–1096.

VISHIK, M., KOMECH, A. AND FURSIKOV, A. (1979) Some mathematical problems of statistical hydrodynamics. *Russian Math. Surveys* **34**, 149–234.

WHITHAM, G. B. (1974) *Linear and Nonlinear Waves*. Wiley, New York.

Biometrika (1988), 75, 3, pp. 477–84
Printed in Great Britain

On frequency estimation

BY JOHN A. RICE AND MURRAY ROSENBLATT
Department of Mathematics, University of California, San Diego, La Jolla, California 92093, U.S.A.

SUMMARY

This paper discusses a least-squares procedure and the use of the periodogram for isolating a discrete harmonic of a time series. It is shown that the usual asymptotics on estimation of frequency, amplitude and phase of such a harmonic have to be used with great caution from a moderate sample perspective. Computational issues are discussed and some illustrations are provided. Bolt & Brillinger (1979) make use of these asymptotic results.

Some key words: Frequency estimation; Harmonic analysis; Periodogram; Time series.

1. INTRODUCTION

We consider a time series model of the form $X_t = a\cos(\omega t+\phi)+\varepsilon_t$, where ε_t is a stationary noise sequence, and one is interested in estimating the amplitude, frequency and phase of the harmonic component. The asymptotic theory of the least-squares estimates of these parameters has a long history. Whittle (1951, 1953) obtained some of the earliest results. More recent results are by Hasan (1982), Hannan (1973) and Walker (1971), who formalize and extend Whittle's results. In these works it is shown that the asymptotic variance of the frequency estimate is of order n^{-3} and that the asymptotic variances of the other two components are of the more usual order n^{-1}. These results extend when there are several harmonic components.

The rate for the estimate of ω seems almost unbelievably good, and our work was motivated by a desire to see how reliable the asymptotic theory is. In brief, we find that the product of the amplitude and the sample size, n, must be quite large in order for the asymptotic theory to be meaningful. If this product is not large, the frequency estimate is much more variable than indicated by the asymptotic theory and the amplitude estimate is severely biased. In applications in which the amplitude is small, giving rise to a small peak in the periodogram, these results suggest that naive application of the asymptotic theory to gauge resolution can be quite misleading. Section 2 of this paper is devoted to a review and examination of the asymptotic theory.

We are also concerned with computational issues arising from the least-squares problem. This problem is nonlinear in the parameters, so that some sort of iterative search must be employed. Typically, search methods start from an initial guess and then proceed by a sequence of modified Newton–Raphson steps. For this nonlinear least-squares problem, it turns out that there are many local minima with a separation in frequency about n^{-1} which makes the stationary point to which the iterative scheme converges extremely sensitive to the starting values, and this problem gets worse as the sample size increases. Furthermore, it follows from the results of § 2 that the estimate of the amplitude is very biased unless the frequency is resolved with order $o(n^{-1})$ so that failure to converge to the global minimum may give a very poor estimate of amplitude. The problem becomes especially severe if one is estimating the parameters of several harmonic components simultaneously since then the iteration is taking place in a high-dimensional space with many local minima. Walker (1971) suggests finding the maximum of the periodogram and then computing the amplitude and phase estimates. However, it turns out that this approach has some drawbacks. A bias can arise for moderate sample sizes that is appreciable compared to the standard deviation suggested by asymptotic theory. These issues are further discussed in § 3. There we also report the results of some modest simulations which support the qualms about asymptotic theory raised in § 2.

The bulk of the paper deals with a general context where the procedures are suggested by the likelihood function computed for Gaussian residuals. A discussion of the likelihood function with residuals a Gaussian stationary sequence is given in the Appendix.

2. Asymptotic theory

We initially assume that

$$X_t = A_0 \cos(\omega_0 t) + B_0 \sin(\omega_0 t) + \varepsilon_t,$$

where ε_t is a strictly stationary sequence with a continuous spectral density. If ε_t is Gaussian, the log likelihood of $X_1, \ldots, X_n$ is

$$-\tfrac{1}{2}\log|R_n| - \tfrac{1}{2}\sum_{t,\tau=1}^{n}\{X_t - A_0\cos(\omega_0 t) - B_0\sin(\omega_0 t)\}r_{t,\tau}^{(-1)}\{X_\tau - A_0\cos(\omega_0\tau) - B_0\sin(\omega_0\tau)\},$$

where R_n is the covariance matrix of the X's and $r_{t,\tau}^{(-1)}$ is the (t, τ) element of the inverse of the Toeplitz matrix R_n. Under a number of assumptions on the spectral density $f(\lambda)$ of the residuals ε_t, one can give a heuristic argument showing that the part of the log likelihood above depending on ω_0, after fitting A_0 and B_0, can be approximated by a constant multiplied by

$$\frac{1}{f(\omega_0)}\frac{1}{n}\left|\sum_{t=1}^{n} X_t e^{it\omega_0}\right|^2.$$

We shall give a brief heuristic discussion of this in the Appendix. In the main body of the paper we shall restrict ourselves to the case in which the ε_t are independent and identically distributed $N(0, v)$. In this case the log likelihood is

$$-\tfrac{1}{2}n\log(2\pi v) - \tfrac{1}{2}S_n(A, B, \omega)/v,$$

with

$$S_n(A, B, \omega) = \sum_{t=1}^{n}\{X_t - A\cos(\omega t) - B\sin(\omega t)\}^2.$$

The maximum likelihood estimates of A, B, ω are determined by minimizing the sum of squares S_n. The maximum likelihood estimate of v is given by the minimum sum of squares divided by n.

Suppose that one replaces $S_n(A, B, \omega)$ by

$$U_n(A, B, \omega) = \sum_{t=1}^{n} X_t^2 - 2\sum_{t=1}^{n} X_t\{A\cos(\omega t) + B\sin(\omega t)\} + \tfrac{1}{2}n(A^2 + B^2).$$

The difference between $S_n(A, B, \omega)$ and $U_n(A, B, \omega)$ is $O(1)$ as $n \to \infty$ if $\omega \neq 0, \pi$ and this suggests that if U_n is minimized instead of S_n the difference will be small for large

n if the true value ω_0 of ω is not zero or π. The estimates A, B, ω obtained by minimizing U_n instead of S_n are simpler and are given by

$$\hat{A}_n = \frac{2}{n}\sum_{t=1}^{n} X_t \cos(\hat{\omega}t), \quad \hat{B}_n = \frac{2}{n}\sum_{t=1}^{n} X_t \sin(\hat{\omega}t),$$

with $\hat{\omega}_n$ such that

$$I_n(\hat{\omega}_n) = \max_{0\leq\omega\leq\pi} \{I_n(\omega)\}.$$

Following the notation of Walker (1971), the periodogram $I_n(\omega)$ is given by

$$\tfrac{1}{2}nI_n(\omega) = \left|\sum_{t=1}^{n} e^{i\omega t}\{A_0\cos(\omega_0 t) + B_0\sin(\omega_0 t) + \varepsilon_t\}\right|^2. \tag{1}$$

At first the analysis will be carried out for A_0 and B_0 assumed constant. Later a corresponding computation will be considered where A_0 and B_0 may depend on the sample size n. Walker (1971) considers estimating the unknown frequency ω_0 first. In the asymptotic theory it is usual to assume that ω_0 is estimated with an error $o(n^{-1})$. We shall finally see what happens to the estimates of the unknown parameters if ω_0 is not estimated with this precision.

First consider A_0, B_0 to be independent of n. Let $D_0 = \frac{1}{2}(A_0 - iB_0)$ and

$$M_n(u) = \sum_{t=1}^{n} e^{iut} = \begin{cases} \exp\{\frac{1}{2}i(n+1)u\}\sin(\frac{1}{2}nu)/\sin(\frac{1}{2}u) & (0<u<2\pi), \\ n & (u=0). \end{cases}$$

Equation (1) can thus be rewritten

$$\tfrac{1}{2}nI_n(\omega) = \left|\sum_{t=1}^{n}\varepsilon_t e^{i\omega t}\right|^2 + 2\Re\left[\sum_{t=1}^{n}\varepsilon_t e^{-i\omega t}\{D_0M_n(\omega+\omega_0) + \bar{D}_0M_n(\omega-\omega_0)\}\right]$$
$$+ |D_0M_n(\omega+\omega_0) + \bar{D}_0M_n(\omega-\omega_0)|^2. \tag{2}$$

Let $0<\omega_0<\pi$. Following Walker (1971) it can be shown that when $\omega=\omega_0$ expression (2) is given to the first order by

$$|\bar{D}_0M_n(0)|^2 = \tfrac{1}{4}n^2(A_0^2 + B_0^2)$$

which arises from the last term on the right of (2). It is clear that

$$\max_{0\leq\omega\leq\pi}\{|M_n(\omega+\omega_0)|\} = O(1), \quad \max_{0\leq\omega\leq\pi}\{|M_n(\omega-\omega_0))|\} = n.$$

Under assumptions of short range dependence on the sequence $\{\varepsilon_t\}$ (Brillinger, 1981; Turkman & Walker, 1984) one can show that

$$\max_{0\leq\omega\leq\pi}\left(\left|\sum_{t=1}^{n}\varepsilon_t e^{i\omega t}\right|^2\right) = O(n\log n)^{\frac{1}{2}}. \tag{3}$$

It then follows that

$$\max_{0\leq\omega\leq\pi}\{|I_n(\omega) - \tfrac{1}{2}n^{-1}(A_0^2+B_0^2)|M_n(\omega-\omega_0)|^2|\} = O(n\log n)^{\frac{1}{2}}. \tag{4}$$

The argument given by Walker (1971) shows that if $\delta>0$ is chosen small enough so that $\{\sin(\frac{1}{2}\delta)/(\frac{1}{2}\delta)\}^2 > 1/\pi^2$ then

$$K(n,\delta) = \max_{|\omega-\omega_0|\geq n^{-1}\delta}\{I_n(\omega)\}$$
$$\leq \tfrac{1}{2}n^{-1}(A_0^2+B_0^2)\sin^2(\tfrac{1}{2}\delta)/\sin^2(\tfrac{1}{2}n^{-1}\delta) + O(n\log n)^{\frac{1}{2}},$$

using estimate (3). Since

$$\lim_{n\to\infty} \mathrm{pr}\{K(n,\delta) < I_n(\omega_0\} = 1$$

it follows that

$$\lim_{n\to\infty} \mathrm{pr}\,(n|\hat{\omega}_n - \omega_0| < \delta) = 1,$$

where $\hat{\omega}_n$ is the location of the absolute maximum of $I_n(\omega)$ on $0 \leq \omega \leq \pi$. This itself suggests that if one wishes effectively to estimate the maximum one would have to sample the periodogram deterministically by a grid mesh with the mesh length $o(n^{-1})$ as $n \to \infty$. Because of the many local maxima referred to in § 1, it is clear that the use of a nonlinear least-squares procedure implemented by a recursive local computation has great disadvantages.

Walker (1971) goes on to show that $\hat{A}_n$, $\hat{B}_n$, $\hat{\omega}_n$ are asymptotically normally distributed. This result is established by a linearization. It is part of the point of the present paper to note that this linearization only holds in a very narrow range.

To examine the situation in which the amplitude of the harmonic component is small, let us now consider the analysis again where A_0 and B_0 can depend on n. Estimate (3) implies again that the first term on the right-hand side of (2) is $O(n \log n)^{\frac{1}{2}}$. We shall estimate the remaining terms on the right-hand side of (2) over the range $|\omega - \omega_0| > h$, where h decreases to zero as $n \to \infty$. The second term on the right-hand side of (2) is of order of magnitude

$$O(n \log n)^{\frac{1}{2}}|D_0|\{1 + \min\,(nh, 1)/h\}$$

and the last term is

$$O\{|D_0| \min\,(nh, 1)/h\}^2.$$

These estimates imply that if $|D_0|^2 \gg (\log n)/n$ asymptotically as $n \to \infty$ one will be able to estimate ω_0 with an error $o(n^{-1})$.

Suppose that one had obtained an estimate ω_n of ω_0 by searching along a grid mesh of mesh length h which was not $o(n^{-1})$ or that an iterative procedure had converged to a local maximum which was not within $o(n^{-1})$ away from ω_0. One would still be tempted to use the estimates $\hat{A}_n$, $\hat{B}_n$ of the coefficients A_0 and B_0

$$\hat{A}_n + i\hat{B}_n = \frac{2}{n}\sum_{t=1}^{n} X_t\, e^{i\omega_n t} = \frac{2}{n}\sum_{t=1}^{n} (D_0\, e^{i\omega_0 t} + \bar{D}_0\, e^{-i\omega_0 t} + \varepsilon_t)\, e^{i\omega_n t}.$$

Then

$$\hat{A}_n - A_0 + i(\hat{B}_n - B_0) = \frac{2}{n}\left[D_0 M_n(\omega_n + \omega_0) + \bar{D}_n\{M_n(\omega_n - \omega_0) - n\} + \sum_{t=1}^{n} \varepsilon_t\, e^{i\omega_n t}\right].$$

Assuming that $\omega_n - \omega_0 = o(1)$, it then follows that

$$\hat{A}_n - A_0 + i(\hat{B}_n - B_0) = \frac{2}{n}\,\bar{D}_0\{M_n(\omega_n - \omega_0) - n\} + O\{n^{-\frac{1}{2}}(\log n)^{\frac{1}{2}}\}.$$

If $\omega_n - \omega_0 \sim cn^{-\beta}$, for $0 < \beta < 1$, $c \neq 0$, then

$$\hat{A}_n - A_0 + i(\hat{B}_n - B_0) = -2D_0 + O\{n^{-1+\beta} + n^{-\frac{1}{2}}(\log n)^{-\frac{1}{2}}\}$$

and this suggests that instead of using the conjugate of $\hat{D}_n$, $\hat{D}_n^c = \hat{A}_n + i\hat{B}_n$ as the appropriate estimate one ought to use $-\hat{D}_n^c$. However, if $\omega_n - \omega_0 \sim cn^{-1}$, one would obtain

$$\hat{D}_n^c - \bar{D}_0 = 2\bar{D}_0\{\exp(\tfrac{1}{2}ic)\sin(\tfrac{1}{2}c)/(\tfrac{1}{2}c) - 1\} + o(1).$$

Since one could not determine how close ω_n is to ω_0 such relations are meaningless in obtaining an estimate of $\bar{D}_0$. We thus see that, if ω_n is not within $o(n^{-1})$ of ω_0, the estimates of A_0 and B_0 are inconsistent. Note that the fast Fourier transform allows us only to compute the periodogram at the points $2\pi j/n$ which is $O(n^{-1})$ and not $o(n^{-1})$.

3. Computational considerations

In this section we consider some computational aspects and the results of some modest simulations for the simple model $x_t = a_0 \cos(\omega_0 t + \phi_0) + \varepsilon_t$, where the noise sequence ε_t is Gaussian with mean 0 and variance 1. We will see that certain difficulties arise already in fitting this simple model.

As mentioned in the introduction, computation of the least-squares estimates is complicated by the presence of many local minima. Figure 1 illustrates this problem. It shows minus the residual sum of squares for one realization of the model above with $n = 100$, $a_0 = 1$, $\omega_0 = \frac{1}{2}$, and $\phi_0 = 0$ as a function of the variables a, ϕ, ω, for $a = 1$, $-\pi \leqslant \phi \leqslant \pi$ and $0{\cdot}3 \leqslant \omega \leqslant 0{\cdot}7$. A Newton-type algorithm iterating in a three-dimensional space runs a risk of converging to a local extrema. The situation is not as bad as it could be since the ridges corresponding to local extrema in frequency make a small angle with the ϕ axis and since the estimate of a is asymptotically uncorrelated with the estimates of frequency and phase. Nevertheless, it is clear that real problems can arise, especially if one were fitting two or more peaks simultaneously. The initial values provided to the search algorithm are thus critical; in particular initial values suggested by complex demodulation may be unsuitable because of the low pass filter employed in this process. We have found it necessary to use a large number of random starting points in our computations.

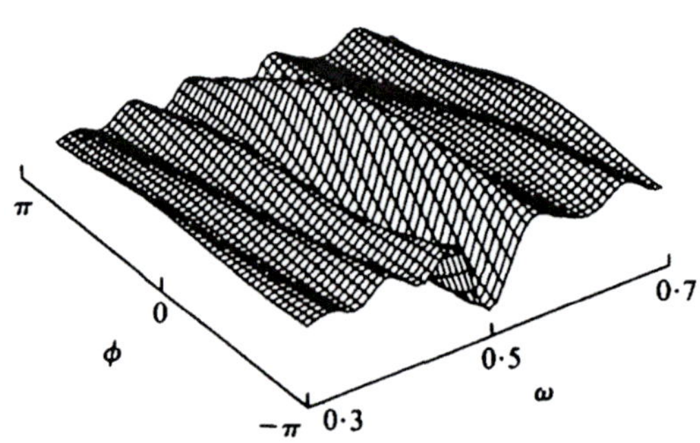

Fig. 1. Minus residual sum of squares; $n = 100$, $a_0 = 1$, $\omega_0 = \frac{1}{2}$, $\phi_0 = 0$.

A direct search of the periodogram at a fine grid of points substantially finer than that given by the frequencies $2\pi j/n$ used by the fast Fourier transform is thus appealing, but unfortunately has its drawbacks as well. In the previous section it was shown that the contribution to the expected value of the periodogram by the harmonic component is proportional to $|M_n(\omega + \omega_0) + M_n(\omega - \omega_0)|^2$, and that for large n this expression is dominated by the second factor. However, for small n, the first factor can make a nonnegligible contribution. For example, for $n=100$ and $\omega_0=0{\cdot}5$, the maximum of this expression occurs at $\omega=0{\cdot}5013$ rather than at $0{\cdot}5$. According to our simulations, if $a=8$, the bias of $\hat{\omega}$ is consequently more than twice its standard error, and similarly for the estimate of phase. While this is noticeable, it is not as large as that caused by using a local rather than the global minimum of the sum of squares. As we will see below, the least-squares estimate does not appear to be similarly affected.

We now turn to a discussion of some simulations. For the simple model above we took $n=100$ and varied a as $0{\cdot}5, 1, 2, 4, 8$. For each such model, 100 realizations were constructed, and the least-squares estimates were calculated by a modified Levenberg-Marquardt algorithm, SNLSE from MINPACK. Because of the many local minima alluded to above we found it necessary to use many starting points to locate the global minimum. Thus 50 starting values for the algorithm were generated on each run as $a+a(U_1-\frac{1}{2})$, $\omega+8\pi(U_2-\frac{1}{2})n$, and $2\pi U_3$, where the U_i were pseudo random uniform variates on [0, 1]. The choice of 50 may have been unduly conservative, but the only penalty can have been in computer time. The proper choice of a range for the random starting points is more problematical. The fast Fourier transform computes the periodogram at frequencies which are multiples of $2\pi/n$, and such a periodogram would often be the starting point of an investigation of periodicities; we thus generate the starting values for $\hat{\omega}$ in an interval of width $8\pi/n$ around ω_0 assuming that a crude guess is obtained from the periodogram. We assume that some knowledge, but not much, about amplitude is obtained from the periodogram, and that there is no information about phase.

To compare the results to the asymptotic theory, we standardized the least-squares estimates by subtracting the true values and dividing by the asymptotic standard deviations. Table 1 gives the means and standard deviations of the standardized variables for each model. Table 1 has been supplemented by graphical analysis of the frequency distributions. Estimates of amplitude are positively biased and the bias becomes smaller with increasing amplitude.

Table 1. *Least-squares estimates of a, ω, ϕ minus true value divided by asymptotic standard error. Mean and standard deviation of resulting standardized variable*

	$a=\frac{1}{2}$	$a=1$	$a=2$	$a=4$	$a=8$
a	0·94, 1·03	0·45, 1·30	0·31, 1·33	0·24, 1·34	0·20, 1·34
ω	−0·71, 6·98	0·02, 1·26	0·05, 1·25	0·06, 1·25	0·07, 1·25
ϕ	0·05, 1·34	−0·05, 1·31	−0·08, 1·28	−0·10, 1·28	−0·10, 1·28

When $a=\frac{1}{2}$ the frequency is not well resolved, and the asymptotics fail to give a reasonable indication of variability. Direct examination of periodograms shows that in this case the periodogram frequently has a local maximum near the true value, but larger maxima at other locations. In this case the contribution of the noise sequence to the expected value of the periodogram, the first term in (2), is 2 and the contribution from the harmonic component, the last term in (2), is $\frac{1}{2}na^2=12{\cdot}5$. At $\omega=0{\cdot}5$ the contribution to the variance of the periodogram from the product of the noise sequence and the harmonic component, the middle term in (2), is about $(7{\cdot}07)^2$, which is large enough to occasionally result in a false dominant peak. For larger amplitudes, the frequency resolution is in rough accord with the asymptotics, although there are occasional outlying values. Apart from some inflation of the standard deviation, the distribution of the phase estimates is about as predicted by the asymptotic theory.

Appendix

An approximation relating to the likelihood function for a Gaussian process

We consider a Gaussian stationary process and note that if the spectral density $f(\lambda)$ is continuously differentiable and bounded away from zero the following approximations appear to be appropriate. The basic approximation is that of replacing $r_{t,\tau}^{(-1)}$ $(t, \tau = 1, \ldots, n)$ by

$$\int_{-\pi}^{\pi} e^{i(t-\tau)\lambda} \frac{1}{4\pi^2} f^{-1}(\lambda)\, d\lambda. \tag{A1}$$

Aside from the term $-\frac{1}{2}\log|R_n|$ the log likelihood can be written as

$$\sum_{t,\tau=1}^{n} X_t X_\tau r_{t,\tau}^{(-1)} - 2\sum_{t,\tau=1}^{n} X_t\{A\cos(\omega\tau) + B\sin(\omega\tau)\} r_{t,\tau}^{(-1)} + A^2 \sum_{t,\tau=1}^{n} \cos(\omega t)\cos(\omega\tau) r_{t,\tau}^{(-1)}$$

$$+ B^2 \sum_{t,\tau=1}^{n} \sin(\omega t)\sin(\omega\tau) r_{t,\tau}^{(-1)} + 2AB \sum_{t,\tau=1}^{n} \cos(\omega t)\sin(\omega\tau) r_{t,\tau}^{(-1)}.$$

The third term, on using the approximation (A1) mentioned above, is to first order

$$\frac{A^2}{4\pi^2}\int_{-\pi}^{\pi} \frac{1}{f(\lambda)} \left|\sum_{t=1}^{n} \cos(\omega t)\, e^{it\lambda}\right|^2 d\lambda$$

and if $0 < \omega < \pi$ this is approximated by $\frac{1}{4} n\pi^{-1} A^2/f(\omega)$. The same argument shows that the fourth term is approximated by $\frac{1}{4} n\pi^{-1} B^2/f(\omega)$. The fifth term can be shown to be $o(n)$ if $0 < \omega < \pi$, since it is to the first order

$$\frac{AB}{4\pi^2}\int_{-\pi}^{\pi} \frac{1}{f(\lambda)} \left\{\sum_{t=1}^{n} \cos(\omega t)\, e^{it\lambda}\right\}\left\{\sum_{\tau=1}^{n} \sin(\omega\tau)\, e^{-i\tau\lambda}\right\} d\lambda.$$

If we insert these approximations and minimize the resulting expressions as a function of A, B, ω the estimates of A, B, ω resulting are

$$\hat{A} = \frac{4\pi f(\hat{\omega})}{n} \sum_{t,\tau=1}^{n} X_t r_{t,\tau}^{(-1)} \cos(\hat{\omega}\tau), \quad \hat{B} = \frac{4\pi f(\hat{\omega})}{n} \sum_{t,\tau=1}^{n} X_t r_{t,\tau}^{(-1)} \sin(\hat{\omega}\tau),$$

where $\hat{\omega}$ maximizes

$$\frac{4\pi f(\omega)}{n}\left[\left\{\sum_{t,\tau=1}^{n} X_t r_{t,\tau}^{(-1)} \cos(\omega\tau)\right\}^2 + \left\{\sum_{t,\tau=1}^{n} X_t r_{t,\tau}^{(-1)} \sin(\omega\tau)\right\}^2\right]. \tag{A2}$$

Again using the approximation (A1), it would appear to be the case that (A2) can be approximated by

$$\frac{2}{nf(\hat{\omega})}\left|\sum_{t=1}^{n} X_t e^{-i\hat{\omega}t}\right|^2.$$

From the same approximation, the estimates of A_0 and B_0 are given approximately by

$$\hat{A} = \frac{2}{n}\sum_{t=1}^{n} X_t \cos(\hat{\omega}t), \quad \hat{B} = \frac{2}{n}\sum_{t=1}^{n} X_t \sin(\hat{\omega}t).$$

Acknowledgement

This research is supported in part by the Office of Naval Research and National Science Foundation of the U.S.A.

REFERENCES

BOLT, B. A. & BRILLINGER, D. R. (1979). Estimation of uncertainties in eigenspectral estimates from decaying geophysical time series. *Geophysics. J. R. Astr. Soc.* **59**, 593-603.

BRILLINGER, D. R. (1981). *Time Series: Data Analysis and Theory.* Oakland, CA: Holden-Day.

HANNAN, E. J. (1973). The estimation of frequency. *J. Appl. Prob.* **10**, 510-9.

HASAN, T. (1982). Nonlinear time series regression for a class of amplitude modulated cosinusoids. *J. Time Series Anal.* **3**, 109-22.

TURKMAN, K. & WALKER, A. M. (1984). On the asymptotic distributions of maxima of trigonometric polynomials with random coefficients. *Adv. Appl. Prob.* **16**, 819-42.

WALKER, A. M. (1971). On the estimation of a harmonic component in a time series with stationary independent residuals. *Biometrika* **58**, 21-36.

WHITTLE, P. (1951). *Hypothesis Testing in Time Series Analysis.* Uppsala: Almqvist and Wiksells Boktryckeri.

WHITTLE, P. (1953). The simultaneous estimation of a time series' harmonic components and covariance structure. *Trab. Estad.* **3**, 43-57.

[*Received August* 1987. *Revised January* 1988]

JOURNAL OF MULTIVARIATE ANALYSIS 36, 175–198 (1991)

Maximum Likelihood Estimation for Noncausal Autoregressive Processes

F. JAY BREIDT* AND RICHARD A. DAVIS*

Colorado State University

KEH-SHIN LII†

University of California, Riverside

AND

MURRAY ROSENBLATT†

University of California, San Diego

Communicated by the Editors

We discuss a maximum likelihood procedure for estimating parameters in possibly noncausal autoregressive processes driven by i.i.d. non-Gaussian noise. Under appropriate conditions, estimates of the parameters that are solutions to the likelihood equations exist and are asymptotically normal. The estimation procedure is illustrated with a simulation study for AR(2) processes.

1. INTRODUCTION

In this paper we discuss maximum likelihood estimation for possibly noncausal autoregressive (AR) processes. We assume that $\{X_t\}$ satisfies the difference equations

$$X_t - \phi_1 X_{t-1} - \cdots - \phi_p X_{t-p} = Z_t,$$

where $\{Z_t\}$ is an independent and identically distributed (i.i.d.) sequence of random variables with mean zero, variance σ^2, and common probability

Received October 31, 1989; revised June 26, 1990.

AMS 1980 subject classifications: primary, 62M09, secondary, 62E20, 60F05, 62F10.

Key words and phrases: maximum likelihood estimates, asymptotic normality, autoregressive process, nonminimum phase, noncausal.

* Research supported by NSF Grant 8802559.

† Research supported by ONR Contract N00014-85-k-0468 and Grant N00014-90-J-1371 and NSF Grant DMS 83-12106.

683/36/2-3

0047-259X/91 $3.00

R.A. Davis et al. (eds.), *Selected Works of Murray Rosenblatt*, Selected Works in Probability and Statistics, DOI 10.1007/978-1-4419-8339-8_37, © Springer Science+Business Media, LLC 2011

density function f_σ, σ a scale parameter. A unique stationary solution to these difference equations exists provided the autoregressive polynomial $\phi(z) = 1 - \phi_1 z - \cdots - \phi_p z^p$ has no roots on the unit circle ($\phi(z) \neq 0$ for $|z| = 1$). This solution is said to be *causal* (or *minimum phase*) if $\phi(z)$ has no roots inside the unit circle ($\phi(z) \neq 0$ for $|z| \leqslant 1$), since then X_t can be expressed as a function of only the present and past of the noise process, $Z_t, Z_{t-1}, \ldots$. The solution is said to be *noncausal* (or *nonminimum phase*) if $\phi(z)$ has any roots inside the unit circle. More specifically, we will say that $\{X_t\}$ is *purely noncausal* if $\phi(z)$ has all of its roots inside the unit circle ($\phi(z) \neq 0$ for $|z| \geqslant 1$); in this case X_t is a function of only the future of the noise process, $Z_{t+1}, Z_{t+2}, \ldots$. Finally, if $\phi(z)$ has roots both inside and outside the unit circle, we will say that the solution $\{X_t\}$ is *mixed* in the sense that X_t is then a function of both the future and the past of the noise process. These ideas are made precise in Section 2.

Now if $\{X_t\}$ is a noncausal AR(p) driven by the i.i.d. sequence $\{Z_t\}$ with mean zero and variance σ^2, then $\{X_t\}$ can be reexpressed as a causal (or purely noncausal) AR(p) driven by a new white noise sequence $\{\tilde{Z}_t\}$ with mean zero and variance $\tilde{\sigma}^2$ (Brockwell and Davis [3, p. 125]). (Note that in the non-Gaussian case, $\{\tilde{Z}_t\}$ is uncorrelated, but not independent (Breidt and Davis [2].) In any of these representations, the second-order structure of $\{X_t\}$—namely its autocovariance function—remains unchanged. Thus any estimation method based solely on the second-order properties of the system will be unable to distinguish among causal and noncausal models. In particular, moment estimation techniques such as Yule–Walker estimation will always yield causal models.

Nonidentifiability of causal and noncausal models also appears in Gaussian maximum likelihood estimation. Classically in time series analysis, estimation of the parameters has been carried out for causal models using a Gaussian likelihood (Rosenblatt [10], Brockwell and Davis [3]). Since in the Gaussian case the probabilistic structure of $\{X_t\}$ is wholly determined by its autocovariance function, causal and noncausal systems cannot be distinguished, and so it is conventional to restrict the parameter space to the causal region. On the other hand, for $\{Z_t\}$ non-Gaussian, causal and noncausal models are identifiable from the likelihood function.

We propose a maximum likelihood procedure for estimating σ and the parameters of the autoregressive polynomial in possibly noncausal AR(p) processes driven by i.i.d. noise with mean zero, variance σ^2, and common probability density function f_σ, σ a scale parameter. We show that under appropriate conditions estimates of these parameters which are solutions to the likelihood equations exist and are asymptotically normal, and we derive the form of the asymptotic covariance matrix. Inherent in the estimation procedure is the identification of the order of causality of the model—whether the process $\{X_t\}$ will be modeled as causal, purely noncausal, or mixed. We demonstrate the effectiveness of this procedure with a simulation study for AR(2) processes.

A natural criticism is that the practicality of this estimation procedure is limited by the assumption that the probability density function of the noise process $\{Z_t\}$ is known to within the value of a scale parameter. Of course, this criticism applies equally to much of classical parametric inference. It seems that a reasonable first step is to consider the case of f_σ known. Consideration of the case of f_σ unknown is the next step, which we are currently taking. It involves developing methods of obtaining reasonably efficient initial estimates of the order of causality and of the parameters in the model. Such estimates might then be used in an adaptive estimation procedure, like those described by Beran [1] and Kreiss [6]. One approach to obtaining these estimates is based on the use of higher order cumulant spectra (Lii and Rosenblatt [8], Nikias and Raghuveer [9]). Preliminary simulation results indicate that maximizing an appropriate non-Gaussian likelihood, such as the likelihood obtained when Z_1 has a Laplace density, may be a more efficient approach; these results are discussed briefly in Section 4. This next step in the problem, however, is not our focus here.

2. Approximating the Likelihood

In this section we derive an approximation to the likelihood of a possibly noncausal AR(p) process and calculate the asymptotic covariance matrix of the partial derivatives of the likelihood.

Let $\{X_t\}$ be the mean zero AR(p) process satisfying the difference equations

$$X_t - \phi_1 X_{t-1} - \cdots - \phi_p X_{t-p} = Z_t, \tag{2.1}$$

where $\phi(z) = 1 - \phi_1 z - \cdots - \phi_p z^p \neq 0$ for $|z| = 1$, $\phi_p \neq 0$, and $\{Z_t\}$ is an i.i.d. sequence of random variables with mean zero, variance σ^2, and common probability density function f_σ, σ a scale parameter. Specifically, we assume $f_\sigma(x) = \sigma^{-1} f(x/\sigma)$ for some probability density function (pdf) $f(x)$. It is well known (see, for example, Brockwell and Davis [3, p. 88]) that there exists a unique stationary solution to (2.1) given by the two-sided moving average

$$X_t = \sum_{j=-\infty}^{\infty} \psi_j Z_{t-j},$$

where ψ_j is the coefficient of z^j in the Laurent series expansion of $1/\phi(z)$, viz.,

$$\phi(z)^{-1} = \sum_{j=-\infty}^{\infty} \psi_j z^j \tag{2.2}$$

which exists in some annulus $d<|z|<d^{-1}$, $d<1$. If $\phi(z)\neq 0$ for $|z|\leqslant 1$, then $\psi_j=0$ for $j<0$ and we call $\{X_t\}$ *causal*, since it now is a causal function of $\{Z_t\}$, i.e.,

$$X_t=\sum_{j=0}^{\infty}\psi_j Z_{t-j}.$$

On the other hand, if $\phi(z)\neq 0$ for $|z|\geqslant 1$ then

$$X_t=\sum_{j=0}^{\infty}\psi_{-j} Z_{t+j}$$

is a function of the future values of $\{Z_t\}$. We call such a process *purely noncausal*. In this case, the coefficients ψ_j satisfy

$$(1-\phi_1 z-\cdots-\phi_p z^p)(\psi_0+\psi_{-1}z^{-1}+\cdots)=1$$

which implies

$$\psi_0=\psi_{-1}=\cdots=\psi_{1-p}=0, \qquad \psi_{-p}=-\phi_p^{-1} \tag{2.3}$$

so that X_t is independent of Z_s, $s\leqslant t+p-1$.

In the causal and purely noncausal cases, it is rather straightforward to approximate the likelihood by the conditional likelihood (conditional on the first p observations in the causal case and the last p observations in the noncausal case). However, in the *mixed case*, when $\phi(z)$ has zeros lying inside and outside the unit circle, approximating the likelihood is more difficult since X_t now depends on both the future and past values of $\{Z_t\}$. To handle the mixed case, we first reparameterize the model by decomposing the autoregressive polynomial, $\phi(z)$, into its causal and purely noncausal components, and then analyze the corresponding AR processes that arise from this decomposition.

Factor the autoregressive polynomial as

$$\phi(z)=1-\phi_1 z-\cdots-\phi_p z^p=\phi^{\dagger}(z)\,\phi^{*}(z), \tag{2.4}$$

where

$$\phi^{\dagger}(z)=1-\theta_1 z-\cdots-\theta_r z^r\neq 0 \qquad \text{for } |z|\leqslant 1,$$

$$\phi^{*}(z)=1-\theta_{r+1} z-\cdots-\theta_p z^s\neq 0 \qquad \text{for } |z|\geqslant 1,$$

and $r, s\geqslant 0$, with $r+s=p$. In other words, if $m_1, ..., m_r, m_{r+1}, ..., m_p$ are the p zeros of $\phi(z)$ with $|m_i|>1$, $i=1, ..., r$, and $|m_i|<1$, $i=r+1, ..., p$, then

$$\phi^{\dagger}(z)=\prod_{i=1}^{r}(1-m_i^{-1}z) \qquad \text{and} \qquad \phi^{*}(z)=\prod_{i=r+1}^{p}(1-m_i^{-1}z) \tag{2.5}$$

and the ϕ_j's can be determined from the θ_j's through the equations

$$\phi_j = \begin{cases} \theta_j - \sum_{i=1}^{j} \theta_{j-i}\theta_{r+i}, & j=1, \dots, r, \\ -\sum_{i=j-r}^{j} \theta_{j-i}\theta_{r+i}, & j=r+1, \dots, p, \end{cases} \tag{2.6}$$

where we set $\theta_0 = -1$ and $\theta_j = 0$ whenever $j \notin \{0, \dots, p\}$.

Now define the causal and purely noncausal AR processes by

$$U_t = \phi^*(B)X_t \qquad \text{and} \qquad V_t = \phi^\dagger(B)X_t,$$

respectively, where B is the backwards shift operator ($B^k X_t = X_{t-k}$, $k = 0, \pm 1, \dots$). Since $\phi^\dagger(B)\,\phi^*(B)X_t = Z_t$,

$$\phi^\dagger(B)U_t = Z_t \qquad \text{and} \qquad \phi^*(B)V_t = Z_t,$$

and hence

$$U_t = \sum_{j=0}^{\infty} \alpha_j Z_{t-j}, \qquad V_t = \sum_{j=s}^{\infty} \beta_j Z_{t+j}, \tag{2.7}$$

where

$$\phi^\dagger(z)^{-1} = \sum_{j=0}^{\infty} \alpha_j z^j \qquad \text{and} \qquad \phi^*(z)^{-1} = \sum_{j=s}^{\infty} \beta_j z^{-j}. \tag{2.8}$$

Since U_t is independent of V_{t-s+1}, the pdf of the random vector $(U_1, \dots, U_n, V_{n-s+1}, \dots, V_n)'$ is

$$h_U(U_1, \dots, U_r) \prod_{t=r+1}^{n} f_\sigma(U_t - \theta_1 U_{t-1} - \cdots - \theta_r U_{t-r})\, h_V(V_{n-s+1}, \dots, V_n),$$

where h_U and h_V are the joint pdf's of $(U_1, \dots, U_r)'$ and $(V_{n-s+1}, \dots, V_n)'$, respectively. The joint pdf of $(U_1, \dots, U_s, X_1, \dots, X_n)'$ is obtained via the transformation

$$\begin{bmatrix} U_1 \\ \vdots \\ U_s \\ U_{s+1} \\ \vdots \\ U_n \\ V_{n-s+1} \\ \vdots \\ V_n \end{bmatrix} = \begin{bmatrix} U_1 \\ \vdots \\ U_s \\ X_{s+1} - \theta_{r+1}X_s - \cdots - \theta_p X_1 \\ \vdots \\ X_n - \theta_{r+1}X_{n-1} - \cdots - \theta_p X_{n-s} \\ X_{n-s+1} - \theta_1 X_{n-s} - \cdots - \theta_r X_{n-s+1-r} \\ \vdots \\ X_n - \theta_1 X_{n-1} - \cdots - \theta_r X_{n-r} \end{bmatrix} = T \begin{bmatrix} U_1 \\ \vdots \\ U_s \\ X_1 \\ \vdots \\ X_n \end{bmatrix},$$

where T is an $(n+s) \times (n+s)$ matrix. The joint pdf of $(U_1, \dots, U_s, X_1, \dots, X_n)'$ is then

$$h_U(U_1, \ldots, U_r)\left(\prod_{t=r+1}^{n} f_\sigma(X_t - \phi_1 X_{t-1} - \cdots - \phi_p X_{t-p})\right)$$
$$\times h_V(\phi^\dagger(B)X_{n-s+1}, \ldots, \phi^\dagger(B)X_n)\,|\det T|,$$

where U_j is replaced by $\phi^*(B)X_j$ for $s < j \leqslant r$. From the form of the transformation, it follows for $s > 0$ that $\ln|\det T| \sim \ln|\theta_p|^{n-p}$ which suggests approximating the log-likelihood by

$$L(\theta_1, \ldots, \theta_{p,\sigma}) = \sum_{t=p+1}^{n} (\ln f_\sigma(U_t - \theta_1 U_{t-1} - \cdots - \theta_r U_{t-r}) + \ln|\theta_p|)$$
$$= \sum_{t=p+1}^{n} g_t(\boldsymbol{\theta}), \tag{2.9}$$

where

$$g_t(\boldsymbol{\theta}) = \ln f_\sigma(U_t - \theta_1 U_{t-1} - \cdots - \theta_r U_{t-r}) + \ln|\theta_p|$$
$$= \ln f_\sigma(V_t - \theta_{r+1} V_{t-1} - \cdots - \theta_p V_{t-s}) + \ln|\theta_p|$$

and $\boldsymbol{\theta} = (\theta_1, \ldots, \theta_{p+1})'$ denotes the parameter vector with $\theta_{p+1} = \sigma$. Note that in the causal case ($s = 0$), the $\ln|\theta_p|$ term does not appear in (2.9).

In order to calculate the asymptotic covariance matrix of the partials of L, we make the following assumptions on f:

A1. $f(x) > 0$ for all x.

A2. $f \in C^2$.

A3. $f' \in L^1$ with $\int f'(x)\,dx = f(x)|_{-\infty}^{\infty} = 0$.

A4. $\int x f'(x)\,dx = x f(x)|_{-\infty}^{\infty} - \int f(x)\,dx = -1$.

A5. $\int f''(x)\,dx = f'(x)|_{-\infty}^{\infty} = 0$.

A6. $\int x f''(x)\,dx = x f'(x)|_{-\infty}^{\infty} - \int f'(x)\,dx = 0$.

A7. $\int x^2 f''(x)\,dx = 2x f'(x)|_{-\infty}^{\infty} - 2\int x f'(x)\,dx = 2$.

A8. $\int (1 + x^2)(f'(x))^2 / f(x)\,dx < \infty$.

Evaluating the partials of g_t at the true values, we obtain

$$\frac{\partial g_t}{\partial \theta_i} = \begin{cases} -U_{t-i}\dfrac{f_\sigma'(Z_t)}{f_\sigma(Z_t)}, & i = 1, \ldots, r \\ -V_{t+r-i}\dfrac{f_\sigma'(Z_t)}{f_\sigma(Z_t)}, & i = r+1, \ldots, p-1 \\ -V_{t-s}\dfrac{f_\sigma'(Z_t)}{f_\sigma(Z_t)} + \dfrac{1}{\theta_p}, & i = p \\ -\sigma^{-1}\left(Z_t \dfrac{f_\sigma'(Z_t)}{f_\sigma(Z_t)} + 1\right), & i = p+1. \end{cases} \tag{2.10}$$

The assumptions on f imply

$$EZ_s \frac{f'_\sigma(Z_t)}{f_\sigma(Z_t)} = \begin{cases} 0, & \text{if } s \neq t, \\ -1, & \text{if } s = t, \end{cases} \tag{2.11}$$

and hence, using the representation of (2.7) and applying (2.3) to the β_j sequence, we have

$$E \frac{\partial g_t}{\partial \theta_i} = 0, \qquad i = 1, ..., p+1. \tag{2.12}$$

Next, we determine the limiting covariance matrix of $(n-p)^{-1/2} \sum_{t=p+1}^{n} (\partial g_t / \partial \boldsymbol{\theta})$. From (2.11) and the assumptions on f, we have

$$\operatorname{Cov}\left(Z_{t-i} \frac{f'_\sigma(Z_t)}{f_\sigma(Z_t)}, Z_{k-j} \frac{f'_\sigma(Z_k)}{f_\sigma(Z_k)}\right) = \begin{cases} \sigma^2 \tilde{J}, & \text{if } t = k, i = j = 0, \\ \sigma^2 \tilde{I}, & \text{if } t = k, i = j \neq 0, \\ 1, & \text{if } t \neq k, i = t-k, j = k-t, \\ 0, & \text{otherwise}, \end{cases} \tag{2.13}$$

where

$$\tilde{I} = \sigma^{-2} \int (f'(x))^2 / f(x)\, dx, \qquad \tilde{J} = \sigma^{-2} \left(\int x^2 (f'(x))^2 / f(x)\, dx - 1 \right).$$

Let $\gamma_U(\cdot)$ and $\gamma_V(\cdot)$ denote the autocovariance functions of $\{U_t\}$ and $\{V_t\}$. Then from the representations of U_t and V_t given in (2.7) together with (2.13), we obtain

$$\operatorname{Cov}\left(U_{t-i} \frac{f'_\sigma(Z_t)}{f_\sigma(Z_t)}, U_{k-j} \frac{f'_\sigma(Z_k)}{f_\sigma(Z_k)}\right) = \begin{cases} \gamma_U(i-j)\tilde{I}, & k = t, 1 \leqslant i \leqslant j \leqslant r, \\ 0, & k \neq t, 1 \leqslant i \leqslant j \leqslant r, \end{cases}$$

$$\operatorname{Cov}\left(V_{t+r-i} \frac{f'_\sigma(Z_t)}{f_\sigma(Z_t)}, V_{k+r-j} \frac{f'_\sigma(Z_k)}{f_\sigma(Z_k)}\right)$$

$$= \begin{cases} \gamma_V(i-j)\tilde{I}, & k = t, r < i \leqslant j \leqslant p, i \neq p, \\ \gamma_V(0)\tilde{I} + \beta_s^2 \sigma^2 (\tilde{J} - \tilde{I}), & k = t, i = j = p, \\ 0, & k \neq t, r \leqslant i \leqslant j \leqslant p, \end{cases}$$

$$\operatorname{Cov}\left(U_{t-i} \frac{f'_\sigma(Z_t)}{f_\sigma(Z_t)}, V_{k+r-j} \frac{f'_\sigma(Z_k)}{f_\sigma(Z_k)}\right)$$

$$= \sum_{b=s}^{\infty} \sum_{a=0}^{\infty} \alpha_a \beta_b \operatorname{Cov}\left(Z_{t-i-a} \frac{f'_\sigma(Z_t)}{f_\sigma(Z_t)}, Z_{k+r-j+b} \frac{f'_\sigma(Z_k)}{f_\sigma(Z_k)}\right)$$

$$= \begin{cases} \alpha_{t-k-i} \beta_{t-k+j-r}, & \text{if } t > k, 1 \leqslant i \leqslant r < j \leqslant p, \\ 0, & \text{if } t \leqslant k, 1 \leqslant i \leqslant r < j \leqslant p, \end{cases}$$

and

$$\operatorname{Cov}\left(\frac{\partial g_t}{\partial \theta_i}, \frac{\partial g_k}{\partial \theta_{p+1}}\right) = \begin{cases} 0, & \text{if } i=1, \ldots, p-1 \text{ or } t \neq k, \\ -\theta_p^{-1}\sigma \mathcal{J}, & \text{if } i=p,\, t=k, \\ \mathcal{J}, & \text{if } i=p+1,\, t=k. \end{cases}$$

Also, for $1 \leqslant i \leqslant r$ and $r < j \leqslant p$,

$$\begin{aligned} &(n-p)\operatorname{Cov}\left(\frac{1}{n-p}\sum_{t=p+1}^{n}\frac{\partial g_t}{\partial \theta_i}, \frac{1}{n-p}\sum_{k=p+1}^{n}\frac{\partial g_k}{\partial \theta_j}\right) \\ &= \frac{1}{n-p}\sum_{t=p+1}^{n}\sum_{k=p+1}^{n}\operatorname{Cov}\left(\frac{\partial g_t}{\partial \theta_i}, \frac{\partial g_k}{\partial \theta_j}\right) \\ &= \frac{1}{n-p}\sum_{k=p+1}^{n-1}\sum_{t=k+1}^{n}\alpha_{t-k-i}\beta_{t-k+j-r} \\ &= \frac{1}{n-p}\sum_{k=p+1}^{n-1}\sum_{t=0}^{n-k-i}\alpha_t\beta_{t+i+j-r} \\ &\to \sum_{t=0}^{\infty}\alpha_t\beta_{t+i+j-r}, \end{aligned}$$

the convergence holding due to the geometric decay of $\{\alpha_t\}$ and $\{\beta_t\}$. Combining the preceding results, we conclude that

$$(n-p)^{-1}\operatorname{Cov}\left(\sum_{t=p+1}^{n}\frac{\partial g_t}{\partial \boldsymbol{\theta}}, \left(\sum_{k=p+1}^{n}\frac{\partial g_k}{\partial \boldsymbol{\theta}}\right)'\right) \to \Sigma, \tag{2.14}$$

where, if $s > 0$, Σ is the symmetric matrix with (i, j) element $(i \leqslant j)$,

$$\sigma_{ij} = \begin{cases} \mathcal{I}\gamma_U(i-j), & 1 \leqslant i \leqslant j \leqslant r, \\ \mathcal{I}\gamma_V(i-j), & r < i \leqslant j \leqslant p,\, i \neq p, \\ \mathcal{I}\gamma_V(0) + \beta_s^2\sigma^2(\mathcal{J} - \mathcal{I}), & i=j=p, \\ \sum_{k=0}^{\infty}\alpha_k\beta_{k+i+j-r}, & 1 \leqslant i \leqslant r < j \leqslant p, \\ -\theta_p^{-1}\sigma\mathcal{J}, & i=p,\, j=p+1, \\ \mathcal{J}, & i=j=p+1, \\ 0, & \text{otherwise.} \end{cases} \tag{2.15}$$

If $s = 0$ then

$$\sigma_{ij} = \begin{cases} \mathcal{I}\gamma_U(i-j), & 1 \leqslant i \leqslant j \leqslant p, \\ \mathcal{J}, & i=j=p+1, \\ 0, & \text{otherwise.} \end{cases}$$

Remark. In view of (2.14), we call Σ the information matrix. Note that in the $s>0$ case, only the four elements in the southeast corner of Σ depend on the scale parameter σ. Of course, in the causal case ($r=p$, $s=0$), $\sigma_{p+1,p+1}$ is the only entry which depends on σ.

PROPOSITION 1. *Let f be a non-normal pdf satisfying the conditions* A1–A8. *Then the matrix Σ is positive definite.*

Proof. First partition Σ as

$$\Sigma=\begin{bmatrix} A & B \\ B' & C \end{bmatrix},$$

where A is $r\times r$, C is $(s+1)\times(s+1)$, and B is $r\times(s+1)$. Next consider the random vectors, $\mathbf{R}=(R_1,...,R_r)'$ and $\mathbf{S}=(S_1,...,S_{s+1})'$ defined by

$$R_t=U_{-t}\frac{f_\sigma'(Z_0)}{f_\sigma(Z_0)}=\sum_{k=0}^{\infty}\alpha_k Z_{-t-k}\frac{f_\sigma'(Z_0)}{f_\sigma(Z_0)}, \qquad \text{for} \quad t=1,...,r,$$

$$S_t=\sum_{j=s}^{\infty}\beta_j Z_{t-j}\frac{f_\sigma'(Z_0)}{f_\sigma(Z_0)}+\begin{cases}0, & 1\leqslant t<s,\\ -\theta_p^{-1}, & t=s,\end{cases} \qquad \text{for} \quad t=1,...,s,$$

and

$$S_{s+1}=\sigma^{-1}\left(Z_0\frac{f_\sigma'(Z_0)}{f_\sigma(Z_0)}+1\right).$$

It is readily verified that the covariance matrices, Σ_{RR} and Σ_{SS}, of $\mathbf{R}$ and $\mathbf{S}$ are equal to A and C, respectively. Also, from (2.13), we have for $j\leqslant s$,

$$\begin{aligned}\operatorname{Cov}(R_i,S_j)&=\sum_{a=0}^{\infty}\sum_{b=s}^{\infty}\alpha_a\beta_b E\left(Z_{-i-a}Z_{j-b}\left(\frac{f_\sigma'(Z_0)}{f_\sigma(Z_0)}\right)^2\right)\\ &=\sum_{a=0}^{\infty}\beta_{a+i+j}\sigma^2\tilde{I}\end{aligned}$$

and

$$\operatorname{Cov}(R_i,S_{s+1})=0,$$

whence

$$\Sigma_{RS}=\operatorname{Cov}(\mathbf{R},\mathbf{S})=\sigma^2\tilde{I}B.$$

Moreover,

$$\sigma^2\tilde{I}>1, \tag{2.16}$$

since, by the Cauchy–Schwarz inequality and the assumptions on f,

$$1 = \left| E\left[-Z_0 \frac{f'_\sigma(Z_0)}{f_\sigma(Z_0)} \right] \right|^2 \leqslant \sigma^2 \tilde{I}$$

with equality holding if and only if f is normal.

The matrices A and C are positive definite since there is no linear dependence within the vectors $\mathbf{R}$ and $\mathbf{S}$. Thus, to prove positive definiteness of Σ, it suffices to show that

$$C - B'A^{-1}B$$

is positive definite. The matrix $C - (\sigma^2 \tilde{I})^2 B'A^{-1}B$, being the covariance matrix of $\mathbf{S} - \Sigma_{SR}\Sigma_{RR}^{-1}\mathbf{R}$, is non-negative definite and hence for a nonzero vector $\mathbf{a} \in \mathbb{R}^{s+1}$ with $B\mathbf{a} \neq \mathbf{0}$, we have, by (2.16),

$$\mathbf{a}'(C - B'A^{-1}B)\mathbf{a} > \mathbf{a}'(C - (\sigma^2\tilde{I})^2 B'A^{-1}B)\mathbf{a} \geqslant 0.$$

On the other hand, if $B\mathbf{a} = \mathbf{0}$, then

$$\mathbf{a}'(C - B'A^{-1}B)\mathbf{a} = \mathbf{a}'C\mathbf{a} > 0$$

by the positive definiteness of C. This concludes the proof. ∎

PROPOSITION 2. *If f satisfies the assumptions* A1–A8, *then*

$$(n-p)^{-1/2} \sum_{t=p+1}^{n} \frac{\partial g_t}{\partial \boldsymbol{\theta}} \xrightarrow{d} N(\mathbf{0}, \Sigma),$$

where Σ is the matrix given in (2.15).

Proof. By the Cramér–Wold device, it suffices to show that

$$(n-p)^{-1/2} \sum_{t=p+1}^{n} \mathbf{a}' \frac{\partial g_t}{\partial \boldsymbol{\theta}} \xrightarrow{d} N(0, \mathbf{a}'\Sigma\mathbf{a}) \tag{2.17}$$

for all $\mathbf{a} \in \mathbb{R}^{p+1}$. Define the sequence of $(p+1)$ dimensional random vectors $\{\mathbf{Y}_{tm}, t = 0, \pm 1, ...\}$ to be the partials defined in (2.10), but with the U_t's and V_t's replaced by the sums in (2.7) truncated at a large positive integer m. In addition, let Σ_m be the matrix corresponding to Σ, obtained by truncating the U_t's and V_t's. Then the stationary sequence $\{\mathbf{Y}_{tm}, t = 0, \pm 1, ...\}$ is $m + p$ dependent and it follows easily that (2.14) holds with $\partial g_t / \partial \boldsymbol{\theta}$ replaced by $\mathbf{Y}_{tm}$ and limit covariance matrix given by Σ_m. Applying a standard central limit theorem for finite dependent stationary sequences (for example, Theorem 6.4.2 in [3], we obtain

$$(n-p)^{-1/2} \sum_{t=p+1}^{n} \mathbf{a}'\mathbf{Y}_{tm} \xrightarrow{d} N(0, \mathbf{a}'\Sigma_m\mathbf{a}).$$

Now $\Sigma_m \to \Sigma$ as $m \to \infty$ and since

$$\lim_{m\to\infty} \lim_{n\to\infty} \operatorname{Var}\left((n-p)^{-1/2} \sum_{t=p+1}^{n} \left(\mathbf{a}'\mathbf{Y}_{tm} - \mathbf{a}'\frac{\partial g_t}{\partial \boldsymbol{\theta}}\right)\right) = 0,$$

the convergence in (2.17) is immediate from Proposition 6.3.9 in [3]. ∎

3. Asymptotic Normality

In this section, we show that there exists a sequence of solutions, $\hat{\boldsymbol{\theta}}_n$, to the *likelihood equations*,

$$\frac{\partial L(\boldsymbol{\theta})}{\partial \theta_j} = 0, \qquad j = 1, ..., p+1, \tag{3.1}$$

L given in (2.9), which is consistent and asymptotically efficient in the sense that

$$n^{1/2}(\hat{\boldsymbol{\theta}}_n - \boldsymbol{\theta}) \xrightarrow{d} N(\mathbf{0}, \Sigma^{-1}),$$

where Σ is the Fisher information matrix computed in (2.15). We temporarily assume that L does not depend on s, the number of zeros of the autoregressive polynomial which lie inside the unit circle. In addition to the assumptions A1–A8 of Section 2, we assume that

$$h'(x) := \frac{d}{dx} h(x) = \frac{d}{dx}\left(\frac{f'(x)}{f(x)}\right) = h_1(x) - h_2(x), \tag{3.2}$$

where h_1 and h_2 are nondecreasing functions with

$$h_i(x) = O(|x|^k) \qquad \text{as} \qquad |x| \to \infty$$

for some $k \geqslant 0$ satisfying

$$E\,|Z_0|^{2+k} < \infty.$$

In particular, this condition implies

$$E\,|Z_0|^j\,|h'(\sigma_0^{-1}Z_0)| \leqslant E\,|Z_0|^j\,|h_1(\sigma_0^{-1}Z_0)| + E\,|Z_0|^j\,h_2(\sigma_0^{-1}Z_0)| < \infty$$

for $j = 0, 1, 2$.

To establish the existence of a consistent sequence of estimators, $\hat{\boldsymbol{\theta}}_n$, satisfying (3.1), we follow the argument given on p. 430 of Lehmann [7]. Since we are assuming here that s is fixed, the parameter space for the model is

$$\begin{aligned}\Omega_s = \{\boldsymbol{\theta} \in \mathbb{R}^{p+1} &: 1 - \theta_1 z - \cdots - \theta_r z^r \neq 0 \qquad \text{for} \quad |z| \leqslant 1,\\ &1 - \theta_{r+1} z - \cdots - \theta_p z^s \neq 0 \qquad \text{for} \quad |z| \geqslant 1,\\ &\theta_r \neq 0, \theta_p \neq 0, \text{and } \theta_{p+1} = \sigma > 0\}.\end{aligned}$$

Let $\boldsymbol{\theta}_0 = (\theta_{01}, ..., \theta_{0p}, \sigma_0)' \in \Omega_s$ be the true parameter value and let Q_ε denote the closed ball of radius ε centered at $\boldsymbol{\theta}_0$, i.e., $Q_\varepsilon =$

$\{\boldsymbol{\theta} \in \mathbb{R}^{p+1}: |\boldsymbol{\theta} - \boldsymbol{\theta}_0| \leqslant \varepsilon\}$, where $|\cdot|$ represents the max norm on $\mathbb{R}^{p+1}$. The parameter space Ω_s is open and for ε small, there exists a $d<1$ such that for all $\boldsymbol{\theta} \in Q_\varepsilon$

$$\phi^\dagger(z) = 1 - \theta_1 z - \cdots - \theta_r z^r \neq 0 \qquad \text{for} \quad |z| < d^{-1},$$

$$\phi^*(z) = 1 - \theta_{r+1} z - \cdots - \theta_p z^s \neq 0 \qquad \text{for} \quad |z| > d,$$

and

$$\phi(z) = \phi^\dagger(z)\, \phi^*(z) = 1 - \phi_1 z - \cdots - \phi_p z^p \neq 0 \qquad \text{for} \quad d \leqslant |z| \leqslant d^{-1}.$$

It follows from these relations and (2.5) and (2.6) that there exists a constant $C>0$ such that

$$\begin{aligned} \sup_{\boldsymbol{\theta} \in Q_\varepsilon} |\phi_j - \phi_{0j}| &\leqslant C\varepsilon, & j &= 1, ..., p, \\ \sup_{\boldsymbol{\theta} \in Q_\varepsilon} |\psi_j| &\leqslant C d^{|j|}, & j &= 0, \pm 1, ..., \\ \sup_{\boldsymbol{\theta} \in Q_\varepsilon} |\psi_j - \psi_{0j}| &\leqslant C\varepsilon d^{|j|}, & j &= 0, \pm 1, ..., \end{aligned} \tag{3.3}$$

where the $\{\psi_j\}$ and $\{\psi_{0j}\}$ are the coefficients in the power series (2.2) with parameter $\boldsymbol{\theta}$ and $\boldsymbol{\theta}_0$, respectively.

Expanding $L(\boldsymbol{\theta})$ in a neighborhood of $\boldsymbol{\theta}_0$, we have

$$\begin{aligned} &\frac{1}{n-p}(L(\boldsymbol{\theta}) - L(\boldsymbol{\theta}_0)) \\ &= \frac{1}{n-p} \sum_{j=1}^{p+1} A_j(\boldsymbol{\theta}_0)(\theta_j - \theta_{0j}) \\ &\quad + \frac{1}{2(n-p)} \sum_{j=1}^{p+1} \sum_{k=1}^{p+1} B_{jk}(\boldsymbol{\theta}_0)(\theta_j - \theta_{0j})(\theta_k - \theta_{0k}) \\ &\quad + \frac{1}{2(n-p)} \sum_{j=1}^{p+1} \sum_{k=1}^{p+1} (B_{jk}(\boldsymbol{\theta}^*) - B_{jk}(\boldsymbol{\theta}_0))(\theta_j - \theta_{0j})(\theta_k - \theta_{0k}) \\ &= S_1 + S_2 + S_3, \end{aligned} \tag{3.4}$$

where

$$A_j(\boldsymbol{\theta}) = \sum_{t=p+1}^{n} \frac{\partial g_t(\boldsymbol{\theta})}{\partial \theta_j},$$

$$B_{jk}(\boldsymbol{\theta}) = \sum_{t=p+1}^{n} \frac{\partial^2 g_t(\boldsymbol{\theta})}{\partial \theta_j\, \partial \theta_k},$$

and $\boldsymbol{\theta}^*$ is between $\boldsymbol{\theta}_0$ and $\boldsymbol{\theta}$. By the ergodic theorem and (2.12),

$$S_1 = \sum_{j=1}^{p+1} \frac{1}{n-p} \sum_{t=p+1}^{n} \frac{\partial g_t(\boldsymbol{\theta}_0)}{\partial \theta_j}(\theta_j - \theta_{0j}) \to \sum_{j=1}^{p+1} E\frac{\partial g_1(\boldsymbol{\theta}_0)}{\partial \theta_j}(\theta_j - \theta_{0j}) = 0 \qquad \text{a.s.} \tag{3.5}$$

To analyze the terms involving mixed partials, we begin with some preliminaries. In what follows, a tilde over a random variable will indicate dependence of that random variable on the parameter $\boldsymbol{\theta}$, while a tildeless random variable will depend only on the true model ($\boldsymbol{\theta}=\boldsymbol{\theta}_0$). For example,

$$\begin{aligned}
\tilde{U}_t &= X_t-\theta_{r+1}X_{t-1}-\cdots-\theta_p X_{t-s}, & U_t &= X_t-\theta_{0,r+1}X_{t-1}-\cdots-\theta_{0p}X_{t-s},\\
\tilde{Z}_t &= \tilde{U}_t-\theta_1\tilde{U}_{t-1}-\cdots-\theta_r\tilde{U}_{t-r}, & Z_t &= U_t-\theta_{01}U_{t-1}-\cdots-\theta_{0r}U_{t-r},\\
&= X_t-\phi_1 X_{t-1}-\cdots-\phi_p X_{t-p}, & &= X_t-\phi_{01}X_{t-1}-\cdots-\phi_{0p}X_{t-p},\\
\tilde{X}_t &= \sum_{j=-\infty}^{\infty}\psi_j Z_{t-j}, & X_t &= \sum_{j=-\infty}^{\infty}\psi_{0j}Z_{t-j}.
\end{aligned}\tag{3.6}$$

It follows easily from (3.3) and (3.6) that

$$\begin{aligned}
&\sup_{\boldsymbol{\theta}\in Q_\varepsilon}|\tilde{U}_t-U_t|\leqslant\varepsilon(|X_{t-1}|+\cdots+|X_{t-s}|)\\
&\sup_{\boldsymbol{\theta}\in Q_\varepsilon}|\tilde{V}_t-V_t|\leqslant\varepsilon(|X_{t-1}|+\cdots+|X_{t-r}|)\\
&\sup_{\boldsymbol{\theta}\in Q_\varepsilon}|\tilde{X}_t-X_t|\leqslant\sup_{\boldsymbol{\theta}\in Q_\varepsilon}\sum_{j=-\infty}^{\infty}|\psi_j-\psi_{0j}|\,|Z_{t-j}|\leqslant C\varepsilon\sum_{j=-\infty}^{\infty}d^{|j|}|Z_{t-j}|.
\end{aligned}\tag{3.7}$$

Also, writing $\tilde{Z}_t=\tilde{Z}_t-Z_t+Z_t$, we have for all $\boldsymbol{\theta}\in Q_\varepsilon$

$$Z_t-C\varepsilon\left(\sum_{i=1}^{p}|X_{t-i}|\right)\leqslant\tilde{Z}_t\leqslant Z_t+C\varepsilon\left(\sum_{i=1}^{p}|X_{t-i}|\right).\tag{3.8}$$

The mixed partials of $g_t(\boldsymbol{\theta})$ are calculated to be

$$\frac{\partial^2 g_t(\boldsymbol{\theta})}{\partial\theta_j\theta_k}=
\sigma^{-2}\begin{cases}
\tilde{U}_{t-j}\tilde{U}_{t-k}h'(\sigma^{-1}\tilde{Z}_t), & 1\leqslant j\leqslant k\leqslant r,\\
\sigma X_{t-j-k+r}h(\sigma^{-1}\tilde{Z}_t)+\tilde{U}_{t-j}\tilde{V}_{t-k+r}h'(\sigma^{-1}\tilde{Z}_t), & 1\leqslant j\leqslant r<k\leqslant p,\\
\tilde{U}_{t-j}h(\sigma^{-1}\tilde{Z}_t)+\tilde{U}_{t-j}\sigma^{-1}\tilde{Z}_t h'(\sigma^{-1}\tilde{Z}_t), & 1\leqslant j\leqslant r,\ k=p+1,\\
\tilde{V}_{t-j+r}\tilde{V}_{t-k+r}h'(\sigma^{-1}\tilde{Z}_t), & r<j\leqslant k\leqslant p,\ j<p,\\
\tilde{V}_{t-s}^2 h'(\sigma^{-1}\tilde{Z}_t)-\sigma^2\theta_p^{-2}, & j=k=p,\\
\tilde{V}_{t-j+r}h(\sigma^{-1}\tilde{Z}_t)+\tilde{V}_{t-j+r}\sigma^{-1}\tilde{Z}_t h'(\sigma^{-1}\tilde{Z}_t), & r<j\leqslant p,\ k=p+1,\\
\sigma^{-2}\tilde{Z}_t^2 h'(\sigma^{-1}\tilde{Z}_t)+2\sigma^{-1}\tilde{Z}_t h(\sigma^{-1}\tilde{Z}_t)+1, & j=k=p+1.
\end{cases}$$

One verifies that

$$E\frac{\partial^2 g_t(\boldsymbol{\theta}_0)}{\partial\theta_j\,\partial\theta_k}=-\sigma_{jk},$$

where σ_{jk} is given in (2.15), and hence by the ergodic theorem,

$$S_2 = \frac{1}{2} \sum_{j=1}^{p+1} \sum_{k=1}^{p+1} \frac{1}{n-p} B_{jk}(\boldsymbol{\theta}_0)(\theta_j - \theta_{0j})(\theta_k - \theta_{0k})$$

$$\to -\frac{1}{2} (\boldsymbol{\theta} - \boldsymbol{\theta}_0)' \Sigma(\boldsymbol{\theta} - \boldsymbol{\theta}_0) \qquad \text{a.s.} \tag{3.9}$$

As for S_3, we show

$$\limsup_{n \to \infty} \sup_{\boldsymbol{\theta} \in Q_\varepsilon} \frac{1}{n-p} |B_{jk}(\boldsymbol{\theta}) - B_{jk}(\boldsymbol{\theta}_0)| \to 0 \qquad \text{a.s.} \tag{3.10}$$

as $\varepsilon \to 0$ for $j, k = 1, ..., p+1$. For $1 \leqslant j \leqslant k \leqslant r$, we have

$$\frac{1}{n-p} |B_{jk}(\boldsymbol{\theta}) - B_{jk}(\boldsymbol{\theta}_0)|$$

$$\leqslant \frac{1}{n-p} \sum_{t=p+1}^{n} |\sigma^{-2} \tilde{U}_{t-j} \tilde{U}_{t-k} h'(\sigma^{-1} \tilde{Z}_t) - \sigma_0^{-2} U_{t-j} U_{t-k} h'(\sigma_0^{-1} Z_t)|$$

$$\leqslant \frac{|\sigma^2 - \sigma_0^2|}{\sigma^2 \sigma_0^2} \frac{1}{n-p} \sum_{t=p+1}^{n} |U_{t-j} U_{t-k} h'(\sigma_0^{-1} Z_t)|$$

$$+ \sigma^{-2} \frac{1}{n-p} \sum_{t=p+1}^{n} |(U_{t-j} - \tilde{U}_{t-j})\, U_{t-k} h'(\sigma_0^{-1} Z_t)|$$

$$+ \sigma^{-2} \frac{1}{n-p} \sum_{t=p+1}^{n} |(U_{t-k} - \tilde{U}_{t-k})\, \tilde{U}_{t-j} h'(\sigma_0^{-1} Z_t)|$$

$$+ \sigma^{-2} \frac{1}{n-p} \sum_{t=p+1}^{n} |\tilde{U}_{t-j} \tilde{U}_{t-k} (h'(\sigma_0^{-1} Z_t) - h'(\sigma^{-1} \tilde{Z}_t))|$$

$$= T_1 + T_2 + T_3 + T_4.$$

For $\varepsilon < \sigma_0/2$, we have by the ergodic theorem

$$\sup_{\theta \in Q_\varepsilon} T_1 \leqslant \varepsilon 10 \sigma_0^{-3} \frac{1}{n-p} \sum_{t=p+1}^{n} |U_{t-j} U_{t-k} h'(\sigma_0^{-1} Z_t)|$$

$$\to \varepsilon 10 \sigma_0^{-3} E|U_{-j} U_{-k} h'(\sigma_0^{-1} Z_0)| \qquad \text{a.s.}$$

$$= \varepsilon 10 \sigma_0^{-3} E|U_{-j} U_{-k}|\, E|h'(\sigma_0^{-1} Z_0)| \to 0$$

as $\varepsilon \to 0$. Using the bounds in (3.7) and applying the ergodic theorem once again, we obtain

$$\limsup_{n \to \infty} \sup_{\boldsymbol{\theta} \in Q_\varepsilon} T_2 \leqslant 4 \sigma_0^{-2} \varepsilon E|(|X_{-j-1}| + \cdots + |X_{-j-s}|)\, U_{-k} h'(\sigma_0^{-1} Z_0))| \quad \text{a.s.}$$

$$\to 0 \qquad \text{as} \quad \varepsilon \to 0,$$

and, by a similar argument,

$$\limsup_{n\to\infty} \sup_{\boldsymbol{\theta}\in Q_\varepsilon} T_3 \to 0 \qquad \text{a.s. as } \varepsilon \to 0.$$

As for T_4, set $W_t = \sum_{i=1}^{p} |X_{t-i}|$ and define for $i = 1, 2$,

$$Y_{t,i}^{\varepsilon} = \begin{cases} h_i\left(\dfrac{Z_t + C\varepsilon W_t}{\sigma_0 - \varepsilon}\right) - h_i\left(\dfrac{Z_t - C\varepsilon W_t}{\sigma_0 + \varepsilon}\right), & \text{if } Z_t - C\varepsilon W_t > 0, \\ h_i\left(\dfrac{Z_t + C\varepsilon W_t}{\sigma_0 + \varepsilon}\right) - h_i\left(\dfrac{Z_t - C\varepsilon W_t}{\sigma_0 - \varepsilon}\right), & \text{if } Z_t + C\varepsilon W_t < 0, \\ h_i\left(\dfrac{Z_t + C\varepsilon W_t}{\sigma_0 - \varepsilon}\right) - h_i\left(\dfrac{Z_t - C\varepsilon W_t}{\sigma_0 - \varepsilon}\right), & \text{otherwise,} \end{cases}$$

where h_1 and h_2 are the nondecreasing functions specified in (3.2). Then by (3.7), (3.8), and the ergodic theorem,

$$\limsup_{n\to\infty} \sup_{\boldsymbol{\theta}\in Q_\varepsilon} T_4 \leqslant 4\sigma_0^{-2} E[(|U_{-j}| + \varepsilon W_{-j})(|U_{-k}| + \varepsilon W_{-k})(Y_{0,1}^{\varepsilon} + Y_{0,2}^{\varepsilon})] \quad \text{a.s.} \tag{3.11}$$

Using the inequality

$$\begin{aligned} \left| h_i\left(\frac{Z_0 \pm C\varepsilon W_0}{\sigma_0 \pm \varepsilon}\right)\right| &\leqslant A_1 + A_2 \left|\frac{Z_0 \pm C\varepsilon W_0}{\sigma_0 \pm \varepsilon}\right| \\ &\leqslant A_1 + A_3\left(|Z_0|^k + \sum_{i=1}^{p} |X_{-i}|^k\right) \\ &\leqslant A_1 + A_3\left(|Z_0|^k + \sum_{i=1}^{p} \sum_{i_1} \cdots \sum_{i_k} |\psi_{i_1} \cdots \psi_{i_k}|\, |Z_{-i-i_1} \cdots Z_{-i-i_k}|\right), \end{aligned}$$

where A_1, A_2, and A_3 are constants, the expectation in (3.11) is finite, from which it follows by the assumptions on h_i and dominated convergence that the limit of the right hand side in (3.11) is 0 as $\varepsilon \to 0$. This proves (3.10), at least for the $1 \leqslant j \leqslant k \leqslant r$ case. For the other cases, the arguments follow the same ideas as used above and hence are omitted.

Combining the results in (3.5), (3.9), (3.10), and Proposition 1, we conclude that for ε small

$$\sup(S_1 + S_2 + S_3) < 0 \qquad \text{a.s.}$$

as $n \to \infty$, where the sup is taken over $\boldsymbol{\theta}$ on the boundary of Q_ε. Consequently, for n large,

$$L(\boldsymbol{\theta}) < L(\boldsymbol{\theta}_0) \qquad \text{a.s.}$$

for all $\boldsymbol{\theta}$ on the boundary of Q_ε and so $L(\boldsymbol{\theta})$ must have a local maximum on the interior of Q_ε. Such a local maximum must satisfy the likelihood equations. Now as discussed in Lehmann [7], a sequence of local maxima can be chosen, independent of ε, so as to converge a.s. to $\boldsymbol{\theta}_0$.

Having established the existence of a consistent sequence of estimators, $\hat{\boldsymbol{\theta}}_n$, satisfying the likelihood equations (3.1), asymptotic normality of $\hat{\boldsymbol{\theta}}_n$ is practically immediate from Proposition 2. To see this, a Taylor series expansion of $\partial L(\boldsymbol{\theta})/\partial\boldsymbol{\theta}$ about $\boldsymbol{\theta}_0$ gives

$$0 = n^{-1/2}\frac{\partial L(\hat{\boldsymbol{\theta}}_n)}{\partial\boldsymbol{\theta}} = n^{-1/2}\sum_{t=p+1}^{n}\frac{\partial g_t(\boldsymbol{\theta}_0)}{\partial\boldsymbol{\theta}} + n^{-1}B(\boldsymbol{\theta}^*)\, n^{1/2}(\hat{\boldsymbol{\theta}}_n - \boldsymbol{\theta}_0),$$

where $B(\boldsymbol{\theta})$ is the $(p+1)\times(p+1)$ matrix with entries $B_{jk}(\boldsymbol{\theta})$ and $\boldsymbol{\theta}^*$ is on the line segment joining $\boldsymbol{\theta}_0$ and $\hat{\boldsymbol{\theta}}_n$. Since $\boldsymbol{\theta}^* \to \boldsymbol{\theta}_0$ a.s., it follows from (3.10) and the ergodic theorem that

$$n^{-1}B(\boldsymbol{\theta}^*) = n^{-1}B(\boldsymbol{\theta}_0) + n^{-1}(B(\boldsymbol{\theta}^*) - B(\boldsymbol{\theta}_0)) \to -\Sigma$$

a.s. and hence, from Proposition 2, that

$$n^{1/2}(\hat{\boldsymbol{\theta}}_n - \boldsymbol{\theta}_0) \xrightarrow{d} N(\mathbf{0}, \Sigma^{-1}).$$

We record the preceding results in the following theorem.

THEOREM 3. *Let $\{X_t\}$ be the mean zero* AR(p) *process satisfying*

$$X_t - \phi_1 X_{t-1} - \cdots - \phi_p X_{t-p} = Z_t,$$

where the autoregressive polynomial $\phi(z)$ has the factorization given in (2.4) *and $\{Z_t\}$ is an i.i.d. sequence of random variables with mean zero, variance σ^2 and common pdf, $\sigma^{-1}f(x/\sigma)$. Further suppose that f is a non-normal pdf which satisfies conditions* A1–A8 *of Section* 2 *and* (3.2). *Then there exists a sequence of solutions, $\hat{\boldsymbol{\theta}}_n$, to the likelihood equations* (3.1) *which is asymptotically normal with mean $\boldsymbol{\theta}_0$ and asymptotic covariance matrix $n^{-1}\Sigma^{-1}$, where Σ is specified in* (2.15).

From this theorem, it is also possible to get the limit distribution for the estimates of the original autoregressive parameters $\phi_1, ..., \phi_p$. In this case the resulting estimate, $\hat{\boldsymbol{\phi}}_n$, is computed by replacing the θ_j's in (2.6) by their estimated values. A standard argument shows that

$$n^{1/2}(\hat{\boldsymbol{\phi}}_n - \boldsymbol{\phi}) \xrightarrow{d} N(\mathbf{0}, R\Sigma_p^{-1}R'), \tag{3.12}$$

where Σ_p^{-1} is the $p\times p$ northwest submatrix of Σ^{-1} and

$$R = \left[\frac{\partial\phi_j}{\partial\theta_k}\right]_{j,k=1}^{p}.$$

R is quite trivial to compute from (2.6).

Theorem 3 and (3.12) remain valid if the likelihood function $L(\boldsymbol{\theta})$ in (2.9) is allowed to depend on s, the number of zeros of $\phi(z)$ inside the unit circle. The proof is the same as in the s fixed case, since the key observation is that if $|\boldsymbol{\phi}-\boldsymbol{\phi}_0|<\varepsilon$, ε small, then $\phi(z)$ and $\phi_0(z)$ have the same number of zeros inside the unit circle.

If there happens to be more than one solution to the likelihood equations with s either known or unknown, Theorem 3 does not indicate which of the local maxima to choose as the estimator. The obvious candidate is, of course, the maximum likelihood estimator, $\tilde{s}$, $\tilde{\boldsymbol{\theta}}_n$, found by maximizing $L(\boldsymbol{\theta})$ with respect to s and $\boldsymbol{\theta}$. Subject to mild restrictions, $\tilde{s}$, $\tilde{\boldsymbol{\theta}}_n$ will be consistent, from which the asymptotic normality of $\tilde{\boldsymbol{\theta}}_n$ ensues. An argument for this assertion is as follows.

It will be convenient here to work with the parameters $\boldsymbol{\phi}$ and σ rather than $\boldsymbol{\theta}$ so that the log-likelihood becomes

$$L(\boldsymbol{\phi}, \sigma)=\sum_{t=p+1}^{n} g_t(\phi, \sigma),$$

where

$$g_t(\boldsymbol{\phi}, \sigma)=\ln(f_\sigma(X_t-\phi_1 X_{t-1}-\cdots-\phi_p X_{t-p})|\theta_p|)$$

and from (2.5),

$$\theta_p^{-1}=\begin{cases}\prod_{i=r+1}^{p} m_i, & \text{if } s\neq 0,\\ 1, & \text{if } s=0.\end{cases}$$

Restrict the parameter space,

$$\Omega := \{(\boldsymbol{\phi}', \sigma)' \in \mathbb{R}^{p+1}: \phi_p\neq 0, \phi(z)\neq 0 \text{ for } |z|=1, \text{and } \sigma>0\},$$

to any compact subset, Ω_c, containing the true parameter $\boldsymbol{\phi}_0$, σ_0. Now, using the same method of argument given for the consistency part of Theorem 3, it can be shown that with probability one,

$$\begin{aligned}\frac{1}{n-p} L(\boldsymbol{\phi}, \sigma) &\to E g_{p+1}(\boldsymbol{\phi}, \sigma)\\ &= E\ln(f_\sigma(X_{p+1}-\phi_1 X_p-\cdots-\phi_p X_1)|\theta_p|)\end{aligned}$$

uniformly on Ω_c. Provided the limit, $Eg_{p+1}(\boldsymbol{\phi}, \sigma)$, has a unique maximum at $\boldsymbol{\phi}_0$, σ_0, then, by a standard compactness argument, any maximum of $(1/(n-p))\, L(\boldsymbol{\phi}, \sigma)$ must converge to the maximum of the limit and, therefore, the mle is consistent.

4. Simulation Methods and Results

The standard recursion for simulating n observations from a causal AR(r) process $\{U_t\}$ is to begin far back in the past of the process, say at $t=-k$, to set $U_{-k-1}=U_{-k-2}=\cdots=U_{-k-r}=0$, and then to compute recursively

$$U_t=\phi_1 U_{t-1}+\phi_2 U_{t-2}+\cdots+\phi_r U_{t-r}+Z_t$$

for $t=-k, \quad -k+1, ..., 1, 2, ..., n$. Purely noncausal processes can be generated similarly by reversing the time scale; the recursion begins far forward in the future of the process, say at $t=n+m$, and computes U_t for $t=n+m,\ n+m-1, ..., n, n-1, ..., 1$.

Of course, neither of these recursions will generate a series of mixed causality, since they force exclusive dependence on the past or on the future of the i.i.d. sequence $\{Z_t\}$. To simulate time series of mixed causality, we use the factorization (2.4) and a two-stage recursive procedure. We first simulate $n+m$ observations from the causal AR(r)$\{U_t\}$; as above, we begin at $t=-k$, far back in the past of the process. Then, since $\phi^*(B)X_t=U_t$, we have that

$$X_t=\theta_p^{-1}(X_{t+s}-\theta_{r+1}X_{t+s-1}-\cdots-\theta_{p-1}X_{t+1}-U_{t+s}). \tag{4.1}$$

We set $X_{n+m+s}=X_{n+m+s-1}=\cdots=X_{n+m+1}=0$, compute (4.1) recursively for $t=n+m,\ n+m-1, ..., n, n-1, ..., 1$, and then retain the last n computed values as the simulated time series $\{X_t\}$, $t=1, 2, ..., n$.

Using the above recursions, we simulated causal and mixed AR(2) processes driven by non-Gaussian noise, emphasizing the mixed case. Noise distributions included the Laplace (two-sided exponential) distribution and the Student's t distribution; the underlying random number generator is discussed in Kahaner, Moler, and Nash [5]. We approximated the log-likelihood of each simulated series as in (2.9) and, allowing $L(\boldsymbol{\theta})$ to depend on s, the number of zeros of $1-\phi_1 z-\phi_2 z^2$ inside the unit circle, maximized this function using a non-linear optimizer described by Dennis and Schnabel [4].

In the Laplace case, the mle of σ can be expressed in closed form as a function of the data and of the parameters in $\phi(z)$. The approximate log-likelihood can thus be reduced to a function of ϕ_1, ϕ_2, and s, or in the parameterization of (2.5), as a function of m_1^{-1}, m_2^{-1}, and s. This parameterization is used in the surface and contour plots of Figs. 4.1 and 4.2, which show the reduced log-likelihood for 100 observations from the simulated AR(2) process

$$\begin{aligned} Z_t &= (1-m_1^{-1}B)(1-m_2^{-1}B)X_t \\ &= (1+0.9B)(1-1.1B)X_t, \end{aligned}$$

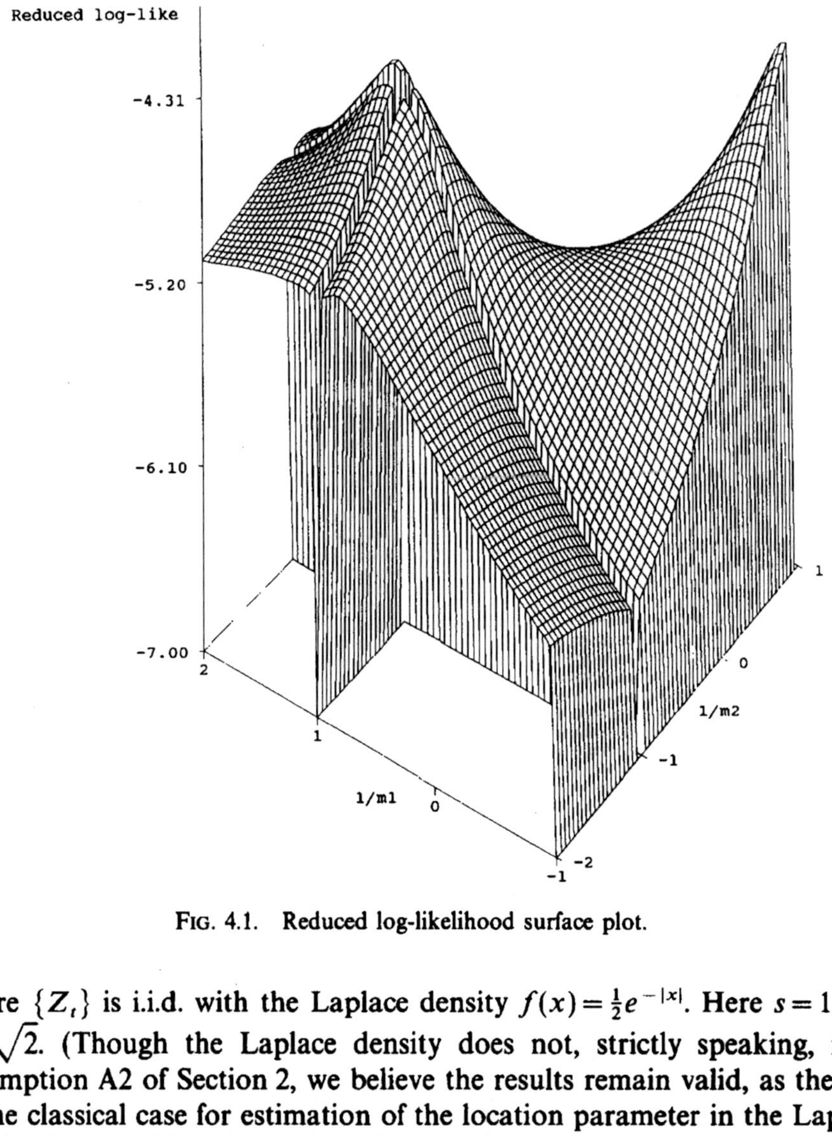

FIG. 4.1. Reduced log-likelihood surface plot.

where $\{Z_t\}$ is i.i.d. with the Laplace density $f(x)=\frac{1}{2}e^{-|x|}$. Here $s=1$ and $\sigma=\sqrt{2}$. (Though the Laplace density does not, strictly speaking, meet assumption A2 of Section 2, we believe the results remain valid, as they do in the classical case for estimation of the location parameter in the Laplace density [7, p. 419].) These plots show sections of the causal ($s=0$), mixed ($s=1$), and purely noncausal ($s=2$) regions. Note that the surface is unbounded along the lines $m_1^{-1}=\pm 1$ and $m_2^{-1}=\pm 1$, since these correspond to roots of $\phi(z)$ on the unit circle. Note also that because m_1^{-1} and m_2^{-1} commute in the reduced log-likelihood function, the surface is symmetric about the line $m_1^{-1}=m_2^{-1}$. In this example, the estimation procedure chose $\tilde{s}=1$, $\tilde{m}_1^{-1}=-0.9270$, $\tilde{m}_2^{-1}=1.1075$, and $\tilde{\sigma}=1.1746$.

For the following simulation study, we applied the maximum likelihood estimation procedure to each of 1000 time series of length n and recorded the number of times R that the procedure chose the correct order of causality ($\tilde{s}=s$). To allow comparison with the asymptotic theory, we com-

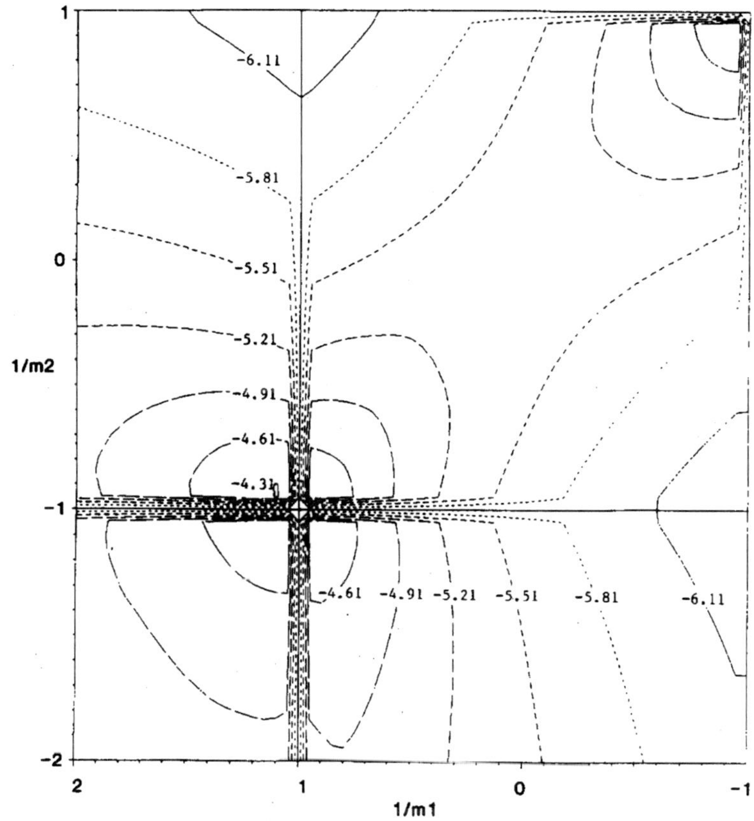

FIG. 4.2. Reduced log-likelihood contour plot.

puted sample means and standard deviations only for those R estimates. Results for the mixed case appear in Tables I–IV, while an example for the causal case and a few other numerical examples of interest appear below.

The tabled results for the mixed case show that the estimation procedure does an excellent job of choosing the order of causality of the model even when the roots m_1 and m_2 of $\phi(z)$ are near the unit circle (Tables I and II).

TABLE I

$\{Z_t\} \sim$ i.i.d. Laplace, $m_1^{-1} = -0.9$, $m_2^{-1} = 1.1$, $\sigma = \sqrt{2}$

		$\tilde{m}_1^{-1}$			$\tilde{m}_2^{-1}$			$\tilde{\sigma}$		
n	R	Mean	SD	ASD	Mean	SD	ASD	Mean	SD	ASD
50	613	−0.8671	0.0671	0.0436	1.1395	0.0830	0.0504	1.4309	0.2064	0.2103
100	805	−0.8935	0.0364	0.0309	1.1141	0.0505	0.0357	1.4096	0.1581	0.1487
200	935	−0.8887	0.0284	0.0218	1.1080	0.0285	0.0252	1.4174	0.1029	0.1051
400	1000	−0.8961	0.0176	0.0154	1.1026	0.0219	0.0178	1.4121	0.0674	0.0743

TABLE II

$\{Z_t\} \sim$ i.i.d. $t(4)$, $m_1^{-1} = -0.9$, $m_2^{-1} = 1.1$, $\sigma = \sqrt{2}$

		$\tilde{m}_1^{-1}$			$\tilde{m}_2^{-1}$			$\tilde{\sigma}$		
n	R	Mean	SD	ASD	Mean	SD	ASD	Mean	SD	ASD
50	684	−0.8752	0.0736	0.0517	1.1414	0.0845	0.0598	1.4730	0.2424	0.2029
100	791	−0.8810	0.0423	0.0366	1.1225	0.0550	0.0423	1.5077	0.1702	0.1435
200	925	−0.8913	0.0299	0.0259	1.1079	0.0351	0.0299	1.5073	0.1066	0.1014
400	980	−0.8973	0.0183	0.0183	1.1042	0.0221	0.0211	1.5089	0.0762	0.0717

When roots are near the unit circle—the boundary for causality—models with high likelihoods occur in the causal, mixed, and purely noncausal regions, as illustrated in Figs. 4.1 and 4.2, and the estimation procedure must choose among them. Not surprisingly, as the roots move away from the unit circle (Tables III and IV), the success rate for determining the order of causality is even higher.

For a causal example, we simulated $n = 200$ observations from the series

$$\begin{aligned} Z_t &= (1 - m_1^{-1}B)(1 - m_2^{-1}B)X_t \\ &= (1 - .8B)(1 + .8B)X_t, \end{aligned}$$

where $\{Z_t\}$ was i.i.d. $t(4)$ ($\sigma = \sqrt{2}$). Sample means and standard deviations (SDs) computed for the R estimates which fell in the correct region, as well as asymptotic standard deviations (ASDs) computed as in Theorem 3 and (2.15), are recorded here:

		$\tilde{m}_1^{-1}$			$\tilde{m}_2^{-1}$			$\tilde{\sigma}$		
n	R	Mean	SD	ASD	Mean	SD	ASD	Mean	SD	ASD
200	882	0.7971	0.0421	0.0455	−0.7930	0.0404	0.0455	1.5014	0.1188	0.0939

TABLE III

$\{Z_t\} \sim$ i.i.d. Laplace, $m_1^{-1} = -0.5$, $m_2^{-1} = 1.3$, $\sigma = \sqrt{2}$

		$\tilde{m}_1^{-1}$			$\tilde{m}_2^{-1}$			$\tilde{\sigma}$		
n	R	Mean	SD	ASD	Mean	SD	ASD	Mean	SD	ASD
50	753	−0.4898	0.1102	0.0884	1.3642	0.2102	0.1102	1.4656	0.2710	0.2332
100	921	−0.4936	0.0694	0.0625	1.2972	0.0931	0.0779	1.3968	0.1558	0.1649
200	994	−0.5023	0.0472	0.0442	1.3189	0.0696	0.0551	1.4292	0.1283	0.1166
400	1000	−0.4965	0.0384	0.0312	1.3065	0.0489	0.0390	1.4161	0.0949	0.0824

TABLE IV

$\{Z_t\} \sim$ i.i.d. $t(4)$, $m_1^{-1} = -0.5$, $m_2^{-1} = 1.3$, $\sigma = \sqrt{2}$

		$\tilde{m}_1^{-1}$			$\tilde{m}_2^{-1}$			$\tilde{\sigma}$		
n	R	Mean	SD	ASD	Mean	SD	ASD	Mean	SD	ASD
50	687	−0.4895	0.1071	0.1067	1.3548	0.2212	0.1331	1.5326	0.3084	0.2371
100	855	−0.4926	0.0805	0.0755	1.3321	0.1163	0.0941	1.5098	0.1965	0.1677
200	957	−0.4907	0.0550	0.0534	1.3113	0.0736	0.0665	1.5091	0.1355	0.1186
400	1000	−0.5012	0.0395	0.0377	1.3022	0.0507	0.0471	1.4998	0.0904	0.0838

In addition to computing sample means and standard deviations, we computed sample correlation matrices and compared them to the asymptotic theory. The following example is for $n = 200$ observations from the mixed AR(2)

$$\begin{aligned} Z_t &= (1 - m_1^{-1}B)(1 - m_2^{-1}B)X_t \\ &= (1 + .5B)(1 - 1.3B)X_t, \end{aligned}$$

where $\{Z_t\}$ was i.i.d. $t(4)$ $(\sigma = \sqrt{2})$. In this case, the asymptotic theory indicates that estimates of m_2^{-1} and σ should be quite highly correlated. Estimates of m_1^{-1} should be weakly correlated with the estimates of the other parameters. Here are the sample and asymptotic standard deviation-correlation matrices:

	Sample				Asymptotic		
	$\tilde{m}_1^{-1}$	$\tilde{m}_2^{-1}$	$\tilde{\sigma}$		$\tilde{m}_1^{-1}$	$\tilde{m}_2^{-1}$	$\tilde{\sigma}$
$\tilde{m}_1^{-1}$	0.055	0.283	0.208	$\tilde{m}_1^{-1}$	0.053	0.280	0.171
$\tilde{m}_2^{-1}$		0.074	0.636	$\tilde{m}_2^{-1}$		0.067	0.611
$\tilde{\sigma}$			0.136	$\tilde{\sigma}$			0.119

More results from this simulation are included in Table IV.

We also loked at normal probability plots for the optimized parameters to check the asymptotic normality of the estimates. In the mixed case, the asymptotic normal approximation to the distribution of the estimates of m_1^{-1}, the inverse of the root of $\phi(z)$ outside the unit circle, is good even for n as small as 50. The distributions of the estimates of m_2^{-1} and σ are, however, quite skewed for the smaller samples ($n = 50$ and 100); this skewness diminishes as n increases.

An example with the parameter set $m_1^{-1} = -0.5$ and $m_2^{-1} = 1.1$ and with $\{Z_t\}$ i.i.d. $t(10)$ $(\sigma = \sqrt{5/2})$ illustrates the identifiability problem we encounter as the noise distribution approaches a Gaussian distribution. As the degrees of freedom ν in the $t(\nu)$ distribution are increased, $t(\nu)$ approaches the standard Gaussian distribution. Since in the Gaussian case the parameters m_1^{-1}, m_2^{-1}, and σ are not identifiable [3, pp. 123–125], processes driven by near-Gaussian noise will be similarly troublesome. In this example, the estimation procedure is still largely successful in choosing the order of causality, with a 70.5% success rate. These results should be compared with the results for the $t(4)$ in Table IV.

		$\tilde{m}_1^{-1}$			$\tilde{m}_2^{-1}$			$\tilde{\sigma}$		
n	R	Mean	SD	ASD	Mean	SD	ASD	Mean	SD	ASD
400	705	−0.5010	0.0452	0.0455	1.3106	0.0624	0.0567	1.2858	0.0799	0.0664

For this final example, we used approximate log-likelihoods corresponding to AR(2) series with Laplace error distributions to estimate the parameters in models for simulated AR(2) series with $t(4)$ error distributions. (These simulated series were the same as those corresponding to $n = 200$ in Table II, so the results below are directly comparable.) This procedure is analogous to estimating parameters in general ARMA(p, q) models by maximizing the Gaussian likelihood even when the process is known to be non-Gaussian, a standard estimation procedure in time series. The resulting estimators are, for noncausal models, the analogue of least absolute deviation estimators. In this example, maximizing the Laplace likelihood yields remarkably good estimates. Note that the asymptotic standard deviation (ASD) recorded here is for an AR(2) driven by $t(4)$ noise.

		$\tilde{m}_1^{-1}$			$\tilde{m}_2^{-1}$		
n	R	Mean	SD	ASD	Mean	SD	ASD
200	837	−0.8915	0.0331	0.0259	1.1090	0.0388	0.0299

This example suggests that maximizing the Laplace likelihood can give reasonably efficient estimates of the parameters in the model. It appears that these estimates might in fact be $\sqrt{n}$-consistent, in which case they could be used as initial values in an adaptive estimation procedure like those advocated by Beran [1] and Kreiss [6].

In Tables I–IV the sample mean and standard deviation (SD) are calculated from the R estimates out of the 1000 replications for which the estimation procedure chose the correct order of causality ($\tilde{s} = s$). The asymptotic standard deviation (ASD) as computed from Theorem 3 and (2.15) is also recorded.

Note added in proof. It should be noted that this paper contains detailed derivations of the results of the paper "Nonminimum Phase non-Gaussian Autoregressive Processes" which appeared in the *Proc. Natl. Acad. Sci. USA* **87**, 179–181.

REFERENCES

[1] BERAN, R. (1976). Adaptive estimates for autoregressive processes. *Ann. Inst. Statist. Math.* **28** 77–89.

[2] BREIDT, F. J., AND DAVIS, R. A. (1990). Time-reversibility, identifiability, and independence of innovations for stationary time series. Preprint.

[3] BROCKWELL, P. J., AND DAVIS, R. A. (1987). *Time Series: Theory and Methods.* Springer-Verlag, New York.

[4] DENNIS, J. E., AND SCHNABEL, R. B. (1983). *Numerical Methods for Unconstrained Optimization and Nonlinear Equations.* Prentice–Hall, New York.

[5] KAHANER, D., MOLER, C., AND NASH, S. (1988). *Numerical Methods and Software.* Prentice–Hall, New York.

[6] KREISS, J. (1987). On adaptive estimation in stationary ARMA processes. *Ann. Statist.* **15** 112–133.

[7] LEHMANN, E. L. (1983). *Theory of Point Estimation.* Wiley, New York.

[8] LII, K. S., AND ROSENBLATT, M. (1982). Deconvolution and estimation of transfer function phase and coefficients for non-Gaussian linear processes. *Ann. Statist.* **10,** 1195–1208.

[9] NIKIAS, C. L., AND RAGHUVEER, M. R. (1987). Bispectrum estimation: A digital signal processing framework. *Proc. IEEE* **75** 869–891.

[10] ROSENBLATT, M. (1985). *Stationary Sequences and Random Fields.* Birkhäuser, Boston.

The Annals of Statistics
2002, Vol. 30, No. 1, 258–297

SPECTRAL ANALYSIS FOR HARMONIZABLE PROCESSES

BY KEH-SHIN LII AND MURRAY ROSENBLATT

University of California, Riverside and University of California, San Diego

Spectral estimation of nonstationary but harmonizable processes is considered. Given a single realization of the process, periodogram-like and consistent estimators are proposed for spectral mass estimation when the spectral support of the process consists of lines. Such a process can arise in signals of a moving source from array data or multipath signals with Doppler stretch from a single receiver. Such processes also include periodically correlated (or cyclostationary) and almost periodically correlated processes as special cases. We give detailed analysis on aliasing, bias and covariances of various estimators. It is shown that dividing a single long realization of the process into nonoverlapping subsections and then averaging periodogram-like estimates formed from each subsection will not yield meaningful results if one is estimating spectral mass with support on lines with slope not equal to 1. If the slope of a spectral support line is irrational, then spectral masses do not fold on top of each other in estimation even if the data are equally spaced. Simulation examples are given to illustrate various theoretical results.

1. Introduction. Spectral analysis of stationary processes has a long history with interest in both theory and applications. Wide-ranging applications in various practical problems in engineering, economics, science and medicine are well documented [see Yaglom (1987)]. These developments have been closely coupled with the stationarity assumption. In recent years there has been growing interest in the spectral analysis of various specific models of nonstationarity. A class of models of this type that have generated considerable interest are the periodically correlated (or cyclostationary, periodically stationary) or almost periodically correlated processes [Alekseev (1988), Hurd (1989), Gardner (1991), Gerr and Allen (1994a, 1994b), Dandawate and Giannakis (1994) and Leskow and Weron (1992)]. The stationary and almost periodically correlated processes are proper subsets of the class of harmonizable processes. A harmonizable continuous-time-parameter process $X(t)$, $EX(t) \equiv 0$, can be represented as a Fourier–Stieltjes integral [Loève (1963)]

$$X(t) = \int_{-\infty}^{\infty} e^{it\lambda}\, dZ(\lambda) \tag{1.1}$$

Received December 1999; revised August 2001.

AMS 2000 subject classifications. Primary, 62M15; secondary, 62G07, 60G12, 62F12.

Key words and phrases. Spectral density function, harmonizable processes, estimation, bias, covariance, aliasing, cyclostationary, consistency, Doppler, multipath.

DOI 10.1007/978-1-4419-8339-8_38,

of a process $Z(\lambda)$ and the covariance function

$$
\begin{aligned}
r_{t,\tau} &= \operatorname{cov}(X(t), X(\tau)) \\
&= \iint_{-\infty}^{\infty} e^{it\lambda - i\tau\mu} \operatorname{cov}(dZ(\lambda), dZ(\mu)) \qquad (1.2)\\
&= \iint_{-\infty}^{\infty} e^{it\lambda - i\tau\mu} dH(\lambda, \mu),
\end{aligned}
$$

with the spectral function

$$
\begin{aligned}
&H(\lambda, \mu) = \operatorname{cov}(Z(\lambda), Z(\mu)), \qquad (1.3)\\
&\int_{-\infty}^{\infty}\int |dH(u, v)| < \infty
\end{aligned}
$$

of bounded variation. A readable discussion of the properties of harmonizable processes with many references is given in Yaglom (1987).

In the case of stationary processes, one has a representation of the form (1.1) with $Z(\lambda)$ a process of orthogonal increments, $E(dZ(\lambda)\overline{dZ(\mu)}) = \delta_{\lambda,\mu} dF(\lambda)$ and F a bounded nondecreasing function. Major interest focuses on the estimation of the spectral density $f(\lambda) = F'(\lambda)$ when F is differentiable. The concept of harmonizability in the sense of Loève is a natural generalization that includes a large class of nonstationary processes. The spectral mass may not concentrate on the diagonal line $\lambda = \mu$ because the process $Z(\lambda)$ in (1.1) may not have orthogonal increments. A natural question concerns what one can estimate consistently about the spectral function $H(\lambda, \mu)$ of a harmonizable process from a *single* realization $X(t)$, $|t| \leq T$, as $T \to \infty$.

If the assumption of stationarity is dropped, there are many different nonstationary models that can be considered [see Dahlhaus (1997) for some references and examples]. The basic model treated here differs from that considered in recent papers where nonstationary processes bounded in mean square are considered. Much of the research in these papers [see Dahlhaus (1997), Mallat, Papanicolaou and Zhang (1998) and Neumann and von Sachs (1997)] is based on the concept of a locally stationary process, that is, one that can be approximated locally by a stationary process. If a process has spectral mass with support on lines that are not very close to the diagonal line $\lambda = \mu$, then locally stationary processes are not suitable to model such a process. The approximate methods are based on the asymptotics suggested by the local approximation of the process. Here there is no notion of such an approximation—the fixed given process is analyzed. To be specific, consider a process $X(t)$ with periodic covariance (period $T > 0$) $\operatorname{cov}(X(t), X(\tau)) = R(t, \tau) = \operatorname{cov}(X(t + T), X(\tau + T))$. All discrete time parameter processes of this type are harmonizable but they are not locally stationary if they are periodic but not stationary.

It should be noted that there are continuous-time-parameter periodic processes that are not harmonizable. Gladyshev (1963) gives the example $R(t, \tau) = g(t)g(\tau)$

with g continuous periodic with period T but having Fourier coefficients that are not absolutely summable. If the Fourier coefficients are absolutely summable, the process with this covariance function is, of course, harmonizable. Let the number τ be a translation number of the complex continuous function f belonging to $\varepsilon > 0$ if $\sup_{-\infty<x<\infty} |f(x+\tau) - f(x)| \le \varepsilon$. If for any $\varepsilon > 0$ there is an $l(\varepsilon) > 0$ such that any interval of length $l(\varepsilon)$ contains a translation number of f belonging to ε, then f is almost periodic in the sense of Bohr ($f \in B$) [Bohr (1951)]. Any finite trigonometric polynomial $\sum_{n=1}^{m} \alpha_n \exp(i\lambda_n x)$ with the λ's real is in B. Furthermore, B is an algebra closed under uniform convergence. A process $X(t)$ is said to have an almost periodic covariance function $B(t, \tau)$ if $B(t+s, \tau+s)$ is almost periodic as a function of s for each t, τ. Most processes with almost periodic covariance functions that are not stationary are harmonizable but not locally stationary. The class of processes we consider is qualitatively broader than the processes with almost periodic covariance function and the nonstationary subclass is not locally stationary in general.

In this paper we consider the estimation of spectra of harmonizable processes with spectral mass concentrated on lines on the basis of a time-ordered sequence of observations. Estimation of spectra on lines with slopes not parallel to the diagonal has been of certain interest in signal processing [see Allen and Hobbs (1992)]. Some preliminary results were announced in Lii and Rosenblatt (1998). Further remarks on where such problems arise are given after (2.7), where specific models are discussed. In Section 2 we lay out some basic properties relevant to the estimation problem and motivate the consideration of harmonizable processes with spectral mass concentrated on lines. In Section 3 a periodogram-like estimator is proposed and its bias is examined. The covariance properties of this estimator are given in Section 4. A consistent estimator is considered in Section 5. Remarks are made about the multivariate case. Section 6 has simulated examples and remarks on estimation and computation. Section 7 has most of the proofs. We note that, while the basic form of the estimator is a smoothing of a periodogram-like form, the analysis is quite different from that of the usual periodogram-based estimators. It is shown that averaging periodograms of nonoverlapping subsections of a long section of data will not produce a meaningful result when the slope of the line of the spectral support is not 1 in contrast to what happens in the case of stationary or almost periodically correlated processes. Results obtained include periodically correlated or almost periodically correlated processes as special cases for which much more detailed analysis and results are presented here than in the existing literature. In particular, the effects of aliasing are given. Extensive applications of these processes are well documented in Gardner (1994).

2. Preliminaries. Consider a zero-mean harmonizable process with the representation (1.1). If the process is real, then $r_{t,\tau} = r_{\tau,t}$ and is real. This implies that

$$dH(u, v) = dH(-v, -u). \tag{2.1}$$

This is symmetry with respect to $u = -v$. Also,

$$dH(u,v) = \overline{dH(-u,-v)} = \overline{dH(v,u)}.$$

This is complex conjugate symmetry with respect to the point $(0,0)$ and the line $u = v$. From this we see that information of $H(u,v)$ in the quadrant given by $\{(u,v) \mid v \geq |u|\}$ determines $H(u,v)$ everywhere. If the process $X_t \equiv X(t)$ is stationary, the spectral mass is concentrated on the diagonal line $dH(u,v) = \delta_{v-u,0} \times dF(u)$. If the process X_t is periodically or almost periodically correlated, then the spectral mass is concentrated on at most countably many lines which are parallel to the diagonal [Gladyshev (1963) and Dehay and Hurd (1993)]

$$dH(u,v) = \sum_j \delta_{u-v-w_j,0}\, dF_j(u). \tag{2.2}$$

In this paper *we assume the continuous-time harmonizable process $X(t)$ has all its spectral mass concentrated on a finite number of lines of the form $u = \alpha v + w$, $\alpha > 0$.* We will discuss cases in which there is spectral mass on a set of two-dimensional positive measure, on a curve which is not a straight line, as well as when $\alpha \leq 0$. We are interested in estimation of the spectral density $f_{\alpha,w}(\mu)$ on a line $\lambda = \alpha\mu + w$:

$$dH(\lambda,\mu)|_{\lambda=\alpha\mu+w} = \delta_{\lambda-\alpha\mu-w,0}\, dF_{\alpha,w}(\mu) = \delta_{\lambda-\alpha\mu-w,0}\, f_{\alpha,w}(\mu)\, d\mu. \tag{2.3}$$

The special case with $\alpha = 1$ was considered in Hurd (1989) and Dandawate and Giannakis (1994). Because of the symmetries in the case of a real-valued process, if $u = av + b$ is a line of spectral support, then so are $u = a^{-1}v - a^{-1}b$, $u = av - b$ and $u = a^{-1}v + a^{-1}b$ with

$$f_{a,b}(v) = a\bar{f}_{a^{-1},-a^{-1}b}(av+b) = \bar{f}_{a,-b}(-v) = af_{a^{-1},a^{-1}b}(-av-b).$$

Sampling at the integers n, m, we have

$$X(n) \equiv X_n = \int_{-\infty}^{\infty} e^{in\lambda}\, dZ(\lambda) = \int_{-\pi}^{\pi} e^{in\lambda} \sum_{k=-\infty}^{\infty} dZ(\lambda + 2\pi k),$$

with

$$r_{n,m} \equiv \operatorname{cov}(X_n, X_m) = EX_n\bar{X}_m = \iint_{-\pi}^{\pi} e^{in\lambda - im\mu}\, d\tilde{H}(\lambda,\mu)$$

and

$$d\tilde{H}(\lambda,\mu) = \sum_{k,j} dH(\lambda + 2\pi k, \mu + 2\pi j). \tag{2.4}$$

From this it is seen that an aliasing problem (in the sense of spectral mass folding onto $[-\pi,\pi]^2$) occurs if the harmonizable process is broadband.

It is well known that one generally does not have consistent estimates of spectral mass for a harmonizable process when $\tilde{H}$ is absolutely continuous with a spectral density $\tilde{h}$, $d\tilde{H}(\lambda, \mu) = \tilde{h}(\lambda, \mu)\, d\lambda\, d\mu$ with $\tilde{h}(\lambda, \mu) \neq 0$ on a set of two-dimensional Lebesgue measure, if the sample is $X_{-n}, \dots, X_n$, letting $n \to \infty$. The simplest example is given by $X_0 \sim N(0, 1)$ with $X_k = 0$ for $k \neq 0$, where $dH(\lambda, \mu) = C\, d\lambda\, d\mu$ for some constant C. The most one could possibly hope for is to estimate spectral mass whose support has two-dimensional Lebesgue measure 0. The simplest such support is a finite collection of curves and, in particular, lines. This motivates the consideration of estimation of spectral mass concentrated on lines given a single realization.

A simple example of a harmonizable process with spectral mass on lines is given by

$$X_t = Y_t + \sum_{s=1}^{k} \beta_s Y_{\alpha_s t}, \tag{2.5}$$

with Y_t stationary and β_s and α_s real and positive numbers, respectively. In such case the spectral support lines are $\lambda = \mu$, $\lambda = \alpha_s \mu$, $\lambda = \alpha_s^{-1} \mu$ and $\lambda = \alpha_s \alpha_{s'}^{-1} \mu$ for all s, $s' = 1, \dots, k$. In Section 6 we give a slightly more complicated example where it is shown that the spectral support lines have the given form.

A small variant of this model is given by $X_t = \sum_{s=1}^{k} \beta_s Y_{\alpha_s (t - \tau_s)}$. Such processes can be encountered in contexts when signals are of a multipath character and contain time delays τ_s and Doppler stretches α_s due to different propagation speed along different paths for a single receiver. For a multiple-receiver example, consider an acoustic signal that emanates from a moving point source and propagates through an acoustic medium, before being received at two spatially separated sensors. If, for instance, the source is in motion while the sensors are stationary, then the actual waveform observed at the two sensor locations can be modeled by [Ferguson (1999), page 261].

$$x_1(t) = s(\beta_1 t) + v_1(t),$$
$$x_2(t) = \sigma s\bigl(\beta_2(t - d)\bigr) + v_2(t),$$

where $v_1(t)$ and $v_2(t)$ represent additive noise observed in the absence of signal, d is the time delay, σ is the relative attenuation and β_1, β_2 are the respective receiver time scales that are introduced to account for the differential Doppler effect. In the present case, the relative motion between the source and the sensors results in the radial velocity component of the source at any given time being different for each receiver. An example of aircraft transits overhead illustrating this model is given in Ferguson (1999), page 261. Alternatively, the received waveforms can be modeled by

$$x(t) = s(t) + n_x(t),$$
$$y(t) = \sigma_1 s\bigl(\beta(t - \tau)\bigr) + n_y(t),$$

where $\beta = \beta_1/\beta_2$ [Ferguson (1999), (16)]. It is also demonstrated in Ferguson that if the Doppler effect is ignored then the estimated delay will not be correct.

Equation (16) of Ferguson (1999) is a version of a special case of the model we dealt with (see the remarks at the end of Section 5 on multivariate harmonizable processes). Additional examples can be found in Jin, Wong and Luo (1995), page 904, for a wideband communication example and in Munk, Worcester and Wunsch (1995), page 204, for one in ocean acoustic tomography. A related example can be found in Chiu (1986) for seismic applications.

Throughout the paper we shall assume that $X(t)$ is a continuous-time-parameter harmonizable process with:

ASSUMPTION 1. All spectral support is on a finite number of lines given by $\mathcal{L} = \{(a_i, b_i) | u = a_i v + b_i,\ i = 1, \dots, K\}$ with positive slopes $a_i > 0$.

ASSUMPTION 2. The spectral mass on the line $u = av + b$ is given by a continuously differentiable spectral density $f_{a,b}(v)$, $(a, b) \in \mathcal{L}$. The spectral densities $f_{a,b}(v)$ and their first derivatives are continuous and bounded in absolute value by a function $g(v)$ that is a monotonic decreasing function of $|v|$ that decreases to 0 as $|v| \to \infty$ and that is integrable as a function of v.

The following simple remark indicates how a harmonizable process with line spectra may differ from a stationary or almost periodically correlated process in terms of aliasing. The aliasing then can have a more complicated character. We say that a line $u = av + b$ is a line of spectral support if the spectral density $f_{a,b}(v)$ on the line is nonzero on a set of positive one-dimensional Lebesgue measure.

PROPOSITION 2.1. *Let $X(t)$ be a continuous-time-parameter process continuous in mean square satisfying Assumptions* 1 *and* 2. *The discretely observed process X_n then has a countably infinite number of lines of support in $(-\pi, \pi]^2$ if and only if one of the lines of spectral support of $X(t)$ has irrational slope a and its spectral density is nonzero at an infinite number of points v_j, with $|v_j| \to \infty$. Furthermore, if there is a line of spectral support with irrational slope with spectral density nonzero at all v, $|v| > s$ for some $s > 0$, then the discretely observed process has a countably dense set of lines of support in $(-\pi, \pi]^2$.*

Before we prove the proposition, we set the following convention which will be followed in the paper. When we refer to $w = z \bmod 2\pi$, it is understood that $-\pi < w \le \pi$. Let $\{u\}$ be the integer l such that $-1/2 < u - l \le 1/2$. Our version of $z \bmod 2\pi$ is then $(z \bmod 2\pi) = z - \{z/(2\pi)\}2\pi$.

PROOF OF PROPOSITION 2.1. The proposition follows from the following simple remarks. Let $u = av + b$ be a line with nonzero spectral density for the continuous-time process. The aliased line segments of this line, in the discretely sampled case, in $(-\pi, \pi]^2$ all have slope a. Let us look at the intersections of these line segments with $v = v_0 \in (-k\pi, k\pi]$, $-\pi < u \le \pi$, for integer k. The points

$(a(v_0+2\pi n)+b,\ v_0+2\pi n)$, n integer, on $u=av+b$ are mapped onto the points $((2\pi na+av_0+b) \bmod 2\pi,\ v_0)$ of $v=v_0$, $-\pi<u\le\pi$. If a is rational there are at most a finite number of values $(2\pi na+av_0+b) \bmod 2\pi$ for n integer. Notice that if $a=1$ there are at most two such aliased line segments. If a is irrational there are an infinite number of values $(2\pi na+av_0+b) \bmod 2\pi$ for n integer and they are dense on $(-k\pi,k\pi]$. Since v_0 is an arbitrary point in $(-k\pi,k\pi]$, the aliased line segments in $(-\pi,\pi]^2$ are dense. Now the proposition follows using Assumptions 1 and 2. □

We note that the previous argument also implies that spectral mass with its support on a line with irrational slope never folds back on its own aliased line segments. Equally spaced sampling of a continuous-time process does produce aliasing. In the stationary case the "Nyquist" frequency is the frequency such that spectral mass beyond that frequency folds back on spectral mass of a lower frequency. In the harmonizable case aliasing does not produce a Nyquist frequency in the conventional sense for a spectral density whose spectral line support has irrational slope. Spectral masses of two different frequencies do not overlap due to aliasing in the case of a spectral density with support line having irrational slope. In this sense there is no overlap "aliasing problem" for spectral density estimation if the support line has irrational slope. The effect of aliasing on estimation in almost periodically correlated processes has not appeared in the literature and is given as a special case of our analysis in Sections 3–5.

3. A periodogram and its bias. In the classical case of a stationary process, an estimate proposed by Daniell (1946) was obtained by dividing a sequence of observations $X_0,\dots,X_{N-1}$ into m disjoint sections of length n, $N=mn$, computing a periodogram for each section and averaging the resulting periodograms. On letting $m,n\to\infty$ as $N\to\infty$, a consistent estimate is obtained under appropriate conditions. If a harmonizable process has mass on a line of slope $a\ne 1$, one object of our discussion is to show that a Daniell-like estimate cannot be used to estimate the spectral mass on that line.

To show the effect relative to a Daniell-like estimate in sampling from the continuous-time harmonizable process $X(t)$, $-\infty<t<\infty$, let us consider observations X_{t+lJ} with $t=0,\dots,n-1$ with $J,n\ge 0$ integers and l an index value. Let

$$F_{l,n}(\lambda)=\sum_{t=0}^{n-1}X_{t+lJ}e^{-i(t+lJ)\lambda}, \tag{3.1}$$

with corresponding periodogram

$$I_{l,n}(\lambda,\mu)=\frac{1}{2\pi n}F_{l,n}(\lambda)\overline{F_{l,n}(\mu)}. \tag{3.2}$$

Notice that, for $J=n$, $F_{l,n}(\lambda)$ uses nonoverlapping subsections of $X(t)$ for different integers l. When $J<n$, we have overlapping sections [Zurbenko (1986),

page 214]. Given a sample $X_0, \ldots, X_{N-1}$, $N = mn$, a Daniell-like estimate is, when $J = n$, the average of periodograms of nonoverlapping subsections of the data,

$$D_{m,n}(\lambda, \mu) = \frac{1}{m} \sum_{l=0}^{m-1} I_{l,n}(\lambda, \mu). \tag{3.3}$$

Another concern is to show the effect of aliasing that is the consequence of discretely sampling a continuous-time-parameter process. In the case of a stationary process, one tries to correct the aliasing by either narrowband filtering or discretely sampling more frequently. Similar methods can be used for the processes we deal with, but it should be noted that the effects of aliasing are more complicated than they are for a stationary process.

Given $(\lambda, \mu) \in (-\pi, \pi]^2$, let (α, w) be a fixed pair with $\alpha > 0$ such that $\lambda = \alpha\mu + w$. Then set

$$\begin{aligned} y &\equiv y(k; a, b; \lambda, \mu) \\ &\equiv \big((2\pi ka + (a - \alpha)\mu + b - w) \bmod 2\pi\big) \\ &= \big((2\pi ka + a\mu + b - \lambda) \bmod 2\pi\big) \end{aligned} \tag{3.4}$$

for $(a, b) \in \mathcal{L}$ with k an integer. The meaning of some aspects of y will be remarked on later. One can then obtain the following theorem.

THEOREM 3.1. *Assume Assumptions* 1 *and* 2 *hold. Then*

$$\begin{aligned} EI_{l,n}(\lambda, \mu) = \sum_{(a,b)\in\mathcal{L}} \ \sum_{|y(k;a,b;\lambda,\mu)|\le\nu} & e^{ilJy(k;a,b;\lambda,\mu)} \\ & \times f_{a,b}(\mu + 2\pi k) G\big(n, l, J, y(k; a, b; \lambda, \mu)\big) \\ + O\left(\frac{\log n}{n}\right) & \end{aligned} \tag{3.5}$$

uniformly in (λ, μ) *with a fixed* $\nu \in (0, \pi]$ *and*

$$\begin{aligned} G &= G\big(n, l, J, y(k; a, b; \lambda, \mu)\big) \\ &= \frac{1}{a} \int_{-\infty}^{\infty} 1_{[0,a]}(t) 1_{[-lJ(a-1)/n, 1-lJ(a-1)/n]}(t) e^{inyt/a}\, dt \\ &= \begin{cases} \dfrac{1}{iny}\left[\exp\left(\dfrac{iny}{a}\left[\min\left(a, 1 - \dfrac{lJ(a-1)}{n}\right)\right]\right)\right. \\ \qquad \left. - \exp\left(\dfrac{iny}{a} \max\left(0, -\dfrac{lJ(a-1)}{n}\right)\right)\right], \\ \qquad \text{if } (0, a) \cap \left(-\dfrac{lJ(a-1)}{n}, 1 - \dfrac{lJ(a-1)}{n}\right) \neq \varnothing, \\ 0, \qquad \text{otherwise,} \end{cases} \end{aligned} \tag{3.6}$$

where $\varnothing$ *is the empty set.*

We note that for an error term like that in (3.5) we need ν fixed and cannot let $\nu \downarrow 0$.

The theorem gives a detailed estimate for the expectation of this type of periodogram. It is worthwhile making a few remarks about the theorem. First, notice that if $(a,b) \in \mathcal{L}$ and $y(k;a,b;\lambda,\mu) \neq 0$ for a fixed k the contribution to the expectation from that term in the double sum on the right-hand side of (3.5) [see (3.6)] is $O(1/n)$. For an $(a,b) \in \mathcal{L}$, the condition

$$y = ((2\pi ka - \lambda + (a\mu + b)) \bmod 2\pi) = 0$$

for some integer k means that the line of spectral support $u = av + b$ or an aliased version of it passes through the point (λ, μ). From (3.4) we also see that $y = 0$ implies $2\pi ka + a\mu + b - (\alpha\mu + w) = 0 \bmod 2\pi$. Notice that if $(\alpha, w) \in \mathcal{L}$ with $(a,b) = (\alpha, w)$ and $k = 0$ then $y = 0$ is satisfied. This means that if (λ, μ) is a point on a spectral support line then there is a contribution to (3.5) in the summation. If $(\alpha, w) \notin \mathcal{L}$ then $y = 0$ in (3.4) implies there exists $(a,b) \in \mathcal{L}$ such that at least one of its aliased lines (i.e., at least one k) passes through the point (λ, μ). So for a given point (λ, μ) if it is not on any spectral support lines or their aliased versions then $EI_{l,n}(\lambda, \mu) = O(\log n/n) + G(M)$, where

$$G(M) = \sum_{(a,b)\in\mathcal{L}} \sum_{|y|\le\nu \text{ and } |k|>M} e^{ilJy} f_{a,b}(\mu + 2\pi k) G(n,l,J,y)$$

for any fixed integer M from (3.4) and (3.5). However, $|G(M)| \to 0$ as $M \to \infty$ by Assumptions 1 and 2. The rate at which $G(M) \to 0$ depends on the rate of $g(v) \to 0$ in Assumption 2. As a result we have:

COROLLARY 3.1. *Under Assumptions* 1 *and* 2, *if* $(\lambda, \mu) \in [-\pi, \pi]^2$ *is not on any lines given by* $(a,b) \in \mathcal{L}$ *or their aliased lines, then* $EI_{l,n}(\lambda, \mu) \to 0$ *as* $n \to \infty$.

For the term with $a = 1$, we have

$$G = \frac{1}{iny}[e^{iny} - 1].$$

Then, if $y \neq 0$, $y = ((a\mu + b - \lambda) \bmod 2\pi)$ is fixed and independent of k and $G = O(1/n)$ whatever the values of l and J. If $a \neq 1$, then $G \to 0$ if $lJ/n \to \infty$ as $n \to \infty$. Also, if $y(k;a,b;\lambda,\mu) = 0$, then, for $a > 0$,

$$G = \frac{1}{a} \|[0,a] \cap [-lJ(a-1)/n,\ 1 - lJ(a-1)/n]\|, \tag{3.7}$$

with $\|\cdot\|$ denoting length. Notice that, if $y = 0$, then $G \neq 0$ if and only if either

$$a \le 1 \quad \text{and} \quad -\frac{1}{1-a} < \frac{lJ}{n} < \frac{a}{1-a}$$

or

$$a \geq 1 \quad \text{and} \quad -\frac{a}{a-1} < \frac{lJ}{n} < \frac{1}{a-1}.$$

To guarantee that $G \neq 0$ for arbitrary $a > 0$, we need $-1 \leq lJ/n \leq 0$. Also notice that, from (3.7),

$$G = \frac{1}{a}\min(a, 1)$$

if (i) $a = 1$ or $l = 0$ for any J or (ii) $l \in [-1, 0]$ and $J = n$. This suggests that, for $J = n$ and $y = 0$, any $l \in [-1, 0]$ can be used in (3.1) and in these cases G terms in (3.5) have a simple form. However, for computational purposes, the following three forms of a finite Fourier transform, which correspond to the cases when $J = n$ and $l = 0$, -1 and $-\frac{1}{2}$, respectively, in (3.1), can be conveniently used, for any $a > 0$,

$$\text{(3.8)} \qquad \begin{aligned} &\text{(a)} \quad F_n(\lambda) = \sum_{t=0}^{n} X_t e^{-it\lambda}, \\ &\text{(b)} \quad F_n(\lambda) = \sum_{t=-n}^{0} X_t e^{-it\lambda}, \\ &\text{(c)} \quad F_n(\lambda) = \sum_{t=-n/2}^{n/2} X_t e^{it\lambda}, \end{aligned}$$

with a corresponding periodogram

$$\text{(3.9)} \qquad I_n(\lambda, \mu) = \frac{1}{2\pi(n+1)} F_n(\lambda)\overline{F_n(\mu)}.$$

We will use the form (c) given by (3.8) later and in such cases we take n to be even.

COROLLARY 3.2. *Assume Assumptions* 1 *and* 2 *hold and* $J = n$ *with* $l \in [-1, 0]$. *Then, as* $n \to \infty$,

$$\text{(3.10)} \qquad EI_{l,n}(\lambda, \mu) \to \sum_{(a,b)\in\mathcal{L}} \sum_{y(k;a,b;\lambda,\mu)=0} f_{a,b}(\mu + 2\pi k)\frac{1}{a}\min(a, 1).$$

Notice that if there is no aliasing the right-hand side of (3.10) becomes

$$\sum_{(a,b)\in\mathcal{L}} \sum_{y(k;a,b;\lambda,\mu)=0} f_{a,b}(\mu)\frac{1}{a}\min(a, 1).$$

Since we assume only a finite number of lines of spectral support, there are at most a finite number of points of intersection of these lines. Aside from these points, the mean value tends to

$$f_{\alpha,w}(\mu)\frac{1}{\alpha}\min(\alpha, 1)$$

if $\lambda = \alpha\mu + w$ is a line of spectral support. In the case of a Daniell-like estimate (3.3) (note $J = n$), where l takes on integer index values, if $a \neq 1$ then, except for a fixed number of terms, $|l(a-1)| > 1$ and so the corresponding G [see (3.7)] value is 0. Alternatively, we saw earlier that, to guarantee $G \neq 0$ for arbitrary $a > 0$, we needed $l \in [-1, 0]$ when $J = n$. Consequently, the mean of the Daniell estimate (3.3) which averages over periodograms of nonoverlapping subsections of the data will generally not tend to $f_{\alpha,w}(\cdot)$ as $m, n \to \infty$ even when $\alpha < 1$.

If a process is almost periodically correlated, the spectral mass is located on lines of slope 1 and $y(k; 1, b; \mu + w, \mu) = (b - w) \bmod 2\pi$. The following corollary holds.

COROLLARY 3.3. *If the process is almost periodically correlated, under Assumptions* 1 *and* 2 *one has*

$$\begin{aligned} EI_{l,n}(\lambda,\mu) = \sum_{(1,b)\in\mathcal{L}} \sum_{|y(k;1,b;\mu+w,\mu)|\le v} & e^{ilJy(k;1,b;\mu+w,\mu)} \\ & \times f_{1,b}(\mu + 2\pi k)\frac{1}{iny}[e^{iny} - 1] \qquad (3.11) \\ & + O\left(\frac{\log n}{n}\right). \end{aligned}$$

As $n \to \infty$, *the expected value tends to*

$$\sum_{(1,b)\in\mathcal{L}} \sum_{y(k;1,b;\mu+w,\mu)=0} f_{1,b}(\mu + 2\pi k). \qquad (3.12)$$

Notice that $y(k; 1, b; \mu + w, \mu) = (b - w) \bmod 2\pi$ from (3.4).

From the previous discussions we see that the expectations of the periodogram from an equally spaced discrete sample [see (3.1)] asymptotically converge to the "aliased" version of the spectral density function on the support lines. This is analogous to the stationary case where the expectation of the periodogram converges asymptotically to the aliased version of the spectral density function. The difference is that the aliasing is more complicated for the harmonizable processes than for the stationary processes. Bias and consistency properties of an estimator based on discrete samples are properties relative to these aliased spectral densities. We also note here that when the spectral support line has slope $a < 0$, previous and subsequent discussions and results still hold if we use the symmetrized version (c) in (3.8) with very minor modifications in the derivations. We have degenerate cases when the slope of the spectral support line is $a = 0$ which is symmetric to the line with slope $a^{-1} = \infty$. It is also plausible, based on the discussions in this section, that if the spectral mass is concentrated on a smooth curve, similar results on the expectations and covariances should hold. These details are not pursued here.

4. Periodogram covariance. Our objective is to now give the asymptotic behavior of the covariance of the periodogram. Given fixed (λ, μ), (λ', μ'), assume that the fixed pair (α, w), (α', w') with $\alpha > 0$, $\alpha' > 0$ is such that $\lambda = \alpha\mu + w$ and $\lambda' = \alpha'\mu' + w'$. Let $(a, b), (a', b') \in \mathcal{L}$. To state the result in a convenient form, the following expressions are introduced:

$$y(1) = \big(y(k; a, b; \lambda, \lambda')\big) = \big((2\pi ka + a\lambda' + b - \lambda) \bmod 2\pi\big), \tag{4.1}$$

$$y(2) = \big(y(k'; a', b'; -\mu, -\mu')\big) = \big((2\pi k'a' - a'\mu' + \mu + b') \bmod 2\pi\big), \tag{4.2}$$

$$y(3) = \big(y(k; a, b; \lambda, -\mu')\big) = \big((2\pi ka - a\mu' + b - \lambda) \bmod 2\pi\big), \tag{4.3}$$

$$y(4) = \big(y(k'; a', b'; -\mu, \lambda')\big) = \big((2\pi k'a' + a'\lambda' + \mu + b') \bmod 2\pi\big). \tag{4.4}$$

This additional notation is introduced so that formulas later on do not become too cluttered. Noticc that in (4.1)–(4.4) two lines $\lambda = \alpha\mu + w$, $\lambda' = \alpha'\mu' + w'$ are dealt with. The notation in this case is introduced to take care of the covariance computation. In formula (3.4) the bias of a point on a single line was involved.

The covariance properties of the periodogram are given in the following theorem.

THEOREM 4.1. *Let $X(t)$ be a normal harmonizable process satisfying Assumptions* 1 *and* 2. *Then*

$$\begin{aligned}
&\operatorname{cov}\big(I_{l,n}(\lambda, \mu), I_{l,n}(\lambda', \mu')\big)\\
&\quad= \Bigg[\sum_{(a,b)\in\mathcal{L}} \sum_{|y(1)|\le\nu} f_{a,b}(\lambda' + 2k\pi) e^{ilJy(1)} G(n, l, J, y(1)) + O\Big(\frac{\log n}{n}\Big)\Bigg]\\
&\qquad\times \Bigg[\sum_{(a',b')\in\mathcal{L}} \sum_{|y(2)|\le\nu} f_{a',b'}(-\mu' + 2k'\pi) e^{ilJ(2)} G(n, l, J, y(2)) + O\Big(\frac{\log n}{n}\Big)\Bigg]\\
&\qquad+ \Bigg[\sum_{(a,b)\in\mathcal{L}} \sum_{|y(3)|\le\nu} f_{a,b}(-\mu' + 2\pi k) e^{ilJy(3)} G(n, l, J, y(3)) + O\Big(\frac{\log n}{n}\Big)\Bigg]\\
&\qquad\times \Bigg[\sum_{(a',b')\in\mathcal{L}} \sum_{|y(4)|\le\nu} f_{a',b'}(\lambda' + 2\pi k') e^{ilJy(4)} G(n, l, J, y(4)) + O\Big(\frac{\log n}{n}\Big)\Bigg],
\end{aligned} \tag{4.5}$$

with $-\pi < \lambda, \lambda', \mu, \mu' \le \pi$, ν *given in Theorem* 3.1 *and* $G(\cdot)$ *as given in* (3.6).

Theorem 4.1 will be useful in deriving the covariance properties of locally smoothed estimates of the spectral density on lines of spectral support.

COROLLARY 4.1. *The result of Theorem* 4.1 *holds for non-Gaussian harmonizable processes with finite fourth-order moments if the fourth-order cumulants satisfy* $\frac{1}{n}\sup_t \sum_{\tau,t',\tau'} |\text{cum}(X_t, X_\tau, X_{t'}, X_{\tau'})| \to 0$ *as* $n \to \infty$. This condition is satisfied if $\sup_t \sum_{\tau,t',\tau'} |\text{cum}(X_t, X_\tau, X_{t'}, X_{\tau'})| < \infty$.

5. Consistent estimates. The estimates that we consider are smoothed versions of the periodogram and for that reason we introduce the following additional assumption.

ASSUMPTION 3. Let $K(\eta)$ be a continuous nonnegative symmetric weight function of finite support with $\int K(x)\,dx = 1$. Also, $K_n(\eta) = b_n^{-1} K(b_n^{-1}\eta)$, where $b_n \downarrow 0$ and $nb_n \to \infty$ as $n \to \infty$.

The periodogram has the doubly periodic property

$$I_n(\lambda + 2\pi k, \mu + 2\pi j) = I_n(\lambda, \mu) \tag{5.1}$$

for all λ, μ and all integers k, j. This suggests that

$$\hat{f}_{\alpha,w}(\eta) = \int_{-\pi^+}^{\pi^+} I_n(\alpha\mu + w, \mu) K_n(\mu - \eta)\,d\mu, \tag{5.2}$$

where $\pi^+ = \pi + \delta$ ($\delta > 0$) is a plausible estimate of $f_{\alpha,w}(\eta)$ with $-\pi \le \eta$, $\alpha\eta + w \le \pi$.

For convenience in exposition, form (c) of (3.8) of the finite Fourier transform will be used in (3.9). Let us then note that we have, from (3.5) and (4.5), Theorem 5.1.

THEOREM 5.1. *Under Assumptions* 1 *and* 2 *and* $I_n(\lambda, \mu)$ *given by* (3.9) *and* (3.8)(c), *using* (3.4),

$$\begin{aligned} EI_n(\lambda, \mu) = & \sum_{(a,b)\in\mathcal{L}} \sum_{|y|\le\nu} f_{a,b}(\mu + 2\pi k) \sin\Big(\frac{n+1}{2a} y \min(a, 1)\Big) \Big/ \Big(\frac{n+1}{2} y\Big) \\ & + O\Big(\frac{\log n}{n}\Big) \end{aligned} \tag{5.3}$$

uniformly in (λ, μ) with y *given in* (3.4) *with the sum over* k *through the expression* y *and*

$$\begin{aligned}
&\operatorname{cov}\left(I_n(\alpha\mu + w, \mu), I_n(\alpha'\mu' + w', \mu')\right)\\
&\quad = \Bigg[\sum_{(a,b)\in\mathcal{L}}\ \sum_{|y(1)|\le\nu} f_{a,b}(\lambda' + 2k\pi)\operatorname{sinc}(\alpha, y(1)) + O\left(\frac{\log n}{n}\right)\Bigg]\\
&\qquad \times \Bigg[\sum_{(a',b')\in\mathcal{L}}\ \sum_{|y(2)|\le\nu} f_{a',b'}(-\mu' + 2k'\pi)\operatorname{sinc}(\alpha', y(2)) + O\left(\frac{\log n}{n}\right)\Bigg]\\
&\qquad + \Bigg[\sum_{(a,b)\in\mathcal{L}}\ \sum_{|y(3)|\le\nu} f_{a,b}(-\mu' + 2\pi k)\operatorname{sinc}(\alpha, y(3)) + O\left(\frac{\log n}{n}\right)\Bigg]\\
&\qquad \times \Bigg[\sum_{(a',b')\in\mathcal{L}}\ \sum_{|y(4)|\le\nu} f_{a',b'}(\lambda' + 2\pi k')\operatorname{sinc}(\alpha', y(4)) + O\left(\frac{\log n}{n}\right)\Bigg]
\end{aligned} \tag{5.4}$$

for $-\pi < \lambda, \lambda', \mu, \mu' \le \pi$, *where* $y(1), \ldots, y(4)$ *are given in* (4.1)–(4.4) *and*

$$\begin{aligned}
\operatorname{sinc}(a, y) &= \sin\left(\frac{n+1}{2a} y \min(a, 1)\right) \Big/ \left(\frac{n+1}{2} y\right)\\
&= \operatorname{sinc}\left(\frac{n+1}{2} y l(a)\right) l(a),
\end{aligned} \tag{5.5}$$

with

$$l(a) = \frac{\min(a, 1)}{a}, \qquad \operatorname{sinc}(x) = \frac{\sin x}{x}.$$

Then, by (5.2), (5.3) and (5.5),

$$\begin{aligned}
E\hat{f}_{\alpha,w}(\eta) = &\sum_{(a,b)\in\mathcal{L}}\ \sum_{|y|\le\nu} \int f_{a,b}(\mu + 2\pi k)\operatorname{sinc}(a, y) K_n(\mu - \eta)\, d\mu\\
&+ O\left(\frac{\log n}{n}\right),
\end{aligned} \tag{5.6}$$

where y is given in (3.4). We then have the following result.

THEOREM 5.2. *Under Assumptions* 1–3 *when* $b_n \downarrow 0$, $nb_n \to \infty$ *with* $\xi = \alpha\eta + w$,

$$E\hat{f}_{\alpha,w}(\eta) = o(b_n) + O\left(\frac{\log n}{nb_n}\right)$$

if α *is not a slope of a spectrum support line. However,*

$$\begin{aligned}
E\hat{f}_{\alpha,w}(\eta) = l(\alpha) &\sum_{k,(\alpha,b)\in\mathcal{L}} f_{\alpha,b}(\eta + 2\pi k)\frac{\sin(((n+1)/2)l(\alpha)y')}{((n+1)/2)y'l(\alpha)}\\
&+ o(b_n) + O\left(\frac{\log n}{n}\right)
\end{aligned} \tag{5.7}$$

if α *is a slope of a spectrum support line with* $y' = ((2\pi k\alpha + b - w) \bmod 2\pi)$.

Equation (5.7) says, asymptotically, if (ξ, η) is not on a spectral support line or, alternatively, if it is on a spectral support line but the smoothing in the estimate is not on a spectral support line, then $E\hat{f}_{\alpha,w}(\eta) \to 0$. If (ξ, η) is on a spectral support line and the smoothing is on the spectral support line given by (α, w) which passes through (ξ, η) (there could be more than one support line passing through the same point), then $E\hat{f}_{\alpha,w}(\eta)$ converges to $l(\alpha)$ times the sum of all spectral masses on spectral support lines which have the same slope α and which are aliased onto (ξ, η) through the condition $y' = 0$.

In the almost periodically correlated case, a simpler estimate leads to the following proposition.

PROPOSITION 5.1. *If the process $X(t)$ is almost periodically correlated, then under Assumptions* 1–3, $b_n \downarrow 0, nb_n \to \infty$,

$$E\hat{f}_{1,w}(\eta) = \sum_{(1,b)\in\mathcal{L}} \sum_k \frac{\sin((n+1)/2)((b-w) \bmod 2\pi))}{((n+1)/2)((b-w) \bmod 2\pi)} \times \int f_{1,b}(\mu + 2\pi k) K_n(\mu - \eta)\, d\mu + O\left(\frac{\log n}{n}\right).$$

If the functions $f_{1,b}$ are continuously differentiable up to second order, then

$$\int f_{1,b}(\mu + 2\pi k) K_n(\mu - \eta)\, d\mu = f_{1,b}(\eta + 2\pi k) + \frac{b_n^2}{2} f''_{1,b}(\eta + 2\pi k) \int z^2 K(z)\, dz + o(b_n^2)$$

and

$$\begin{aligned} E\hat{f}_{1,w}(\eta) = {} & \sum_{(1,b)\in\mathcal{L}} \sum_k \operatorname{sinc}\left(\frac{n+1}{2}((b-w) \bmod 2\pi)\right) f_{1,b}(\eta + 2\pi k) \\ & + \frac{b_n^2}{2} \int z^2 K(z)\, dz \sum_{(1,b)\in\mathcal{L}} \sum_k \operatorname{sinc}\left(\frac{n+1}{2}((b-w) \bmod 2\pi)\right) \\ & \qquad \times f''_{1,b}(\eta + 2\pi k) \\ & + o(b_n^2) + O\left(\frac{\log n}{n}\right). \end{aligned}$$

We note that if $((b-w) \bmod 2\pi) \neq 0$, then the first term on the right-hand side of the last equation is $O(1/n)$. If $((b-w) \bmod 2\pi) = 0$, then the sinc function is 1. We now consider the asymptotic behavior of the covariance of spectral estimates.

THEOREM 5.3. *Let $(\alpha, w), (\alpha', w')$ be fixed with $\alpha > 0$, $\alpha' > 0$. Also, let $(a, b), (a', b')$ range over the finite set $\mathcal{L}$. If $-\pi \le \xi, \xi', \eta, \eta' \le \pi$ with $\xi = \alpha\eta + w$, $\xi' = \alpha'\eta' + w'$, then*

$$
\begin{aligned}
&\operatorname{cov}\left(\hat{f}_{\alpha,w}(\eta), \hat{f}_{\alpha',w'}(\eta')\right) \\
&= \frac{2\pi}{(n+1)b_n}\Bigg\{ \sum_{\substack{k,k',a,a'\\ b,b'}} \Bigg[\delta_{a\alpha'-a'\alpha,0}\,\delta_{(y(k;a,b;\xi,\xi')),0}\,\delta_{(y(k';a',b';-\eta,-\eta')),0} \\
&\qquad \times f_{a,b}(a'\eta' + w' + 2\pi k) f_{a',b'}(-\eta' + 2\pi k') \\
&\qquad \times \int K(a'x)K(x)\,dx \min\left(\alpha'\frac{\min(a,1)}{\min(a',1)}, 1\right)\frac{\min(a',1)}{a\alpha'} \\
&\qquad + \delta_{a-a'\alpha'\alpha,0}\,\delta_{(y(k';a',b';-\eta,\xi')),0}\,\delta_{(y(k;a,b;\xi,-\eta')),0} \\
&\qquad \times f_{a,b}(-\eta' + 2\pi k) f_{a',b'}(a'\eta' + w' + 2\pi k') \\
&\qquad \times \int K(-a'\alpha' x)K(x)\,dx \min\left(\frac{1}{\alpha'}\frac{\min(a,1)}{\min(a',1)}, 1\right) \\
&\qquad \times \frac{\min(a',1)}{a'\alpha}\Bigg]\Bigg\} \\
&\quad + o\left(\frac{1}{nb_n}\right) + O\left(\frac{\log n}{n}\right) + O\left(\frac{\log^2 n}{n^2 b_n^2}\right)
\end{aligned} \tag{5.8}
$$

if $n \to \infty$, $b_n \downarrow 0$, $nb_n \to \infty$.

In Theorem 5.3 the covariance properties of the smoothed spectral estimates are given. Thus Theorems 5.2 and 5.3 show that the smoothed spectral estimates give consistent estimates of the spectral mass on the lines of concentration if there is no aliasing. The possible effects of aliasing can be read off from Theorem 5.2. Also consider formula (5.8) for the case of the variance when there is no aliasing. Then, except for a finite number of points, the expression on the right-hand side is

$$
\frac{2\pi}{(n+1)b_n} f_{1,0}(\eta + w) f_{1,0}(-\eta) \int K(x)^2\,dx,
$$

the contribution from the diagonal $a = 1$, $b = 0$. The additional terms that can arise at this finite number of points correspond to the doubling at $0, \pi$ in the case of spectral estimates for a stationary process.

If the process is almost periodically correlated, all the spectral lines have slope 1, $a = a' = \alpha = \alpha' = 1$. The following corollary is an immediate consequence of the theorem.

COROLLARY 5.1. *If the process is almost periodically correlated, then, under the conditions of Theorem* 5.3,

$$
\begin{aligned}
&\operatorname{cov}(\hat{f}_{1,w}(\eta), \hat{f}_{1,w'}(\eta')) \\
&\quad = \frac{2\pi}{(n+1)b_n} \sum_{\substack{k',k \\ b,b'}} \Big[\delta_{((\eta'-\eta+w'-w+b) \bmod 2\pi),\,0}\,\delta_{((\eta-\eta'+b') \bmod 2\pi),\,0} \\
&\qquad\qquad \times f_{1,b}(\eta'+w'+2\pi k) f_{1,b'}(-\eta'+2\pi k') \int K(x)^2\,dx \\
&\qquad\qquad + \delta_{((-\eta'+b-\eta-w) \bmod 2\pi),0}\,\delta_{((\eta'+w'+\eta+b') \bmod 2\pi),0} \\
&\qquad\qquad \times f_{1,b}(-\eta'+2\pi k) f_{1,b'}(\eta'+w'+2\pi k') \\
&\qquad\qquad\qquad \times \int K(-x)K(x)\,dx \Big] \\
&\quad + o\Big(\frac{1}{nb_n}\Big) + O\Big(\frac{\log n}{n}\Big) + O\Big(\frac{\log^2 n}{n^2 b_n^2}\Big).
\end{aligned}
$$

Now we discuss briefly the estimation of the cross-spectrum of harmonizable processes.

DEFINITION. Stochastic processes $\{X^{(j)}(t) \mid j = 1, \ldots, J\}$ are jointly harmonizable if, for $l, j = 1, \ldots, J$,

$$
X_t^{(j)} \equiv X^{(j)}(t) = \int e^{it\lambda}\,dZ_j(\lambda)
$$

and

$$
\begin{aligned}
\operatorname{cov}(X^{(l)}(t), X^{(j)}(\tau)) &= \int\int_{-\infty}^{\infty} e^{it\lambda - i\tau\mu} \operatorname{cov}(dZ_l(\lambda), dZ_j(\mu)) \\
&= \int\int_{-\infty}^{\infty} e^{it\lambda - i\tau\mu}\,dH_{l,j}(\lambda, \mu),
\end{aligned}
$$

with

$$
\int\int_{-\infty}^{\infty} |dH_{l,j}(\lambda, \mu)|\,d\lambda\,d\mu < \infty.
$$

If cross-spectral mass is concentrated on lines in the set

$$
\mathcal{L} = \{(a, b) \mid \lambda = a\mu + b\}
$$

and spectral densities exist, then

$$
dH_{i,j}(\lambda, \mu) = \sum_{(a,b)\in\mathcal{L}} \delta_{\lambda - a\mu - b,0}\, f_{a,b}^{(i,j)}(\mu)\,d\mu.
$$

An obvious periodogram is

$$I_{l,n}^{(i,j)}(\lambda,\mu)=\frac{1}{2\pi n}F_{l,n}^{(i)}(\lambda)\overline{F_{l,n}^{(j)}(\mu)},$$

with

$$F_{l,n}^{(k)}(\lambda)=\sum_{t=0}^{n-1}X_{t+lJ}^{(k)}e^{-i(t+lJ)\lambda}.$$

Under assumptions similar to Assumptions 1–3 analogs of the previous results in the univariate case hold for the multivariate case.

6. Simulation examples. In this section we give an example of a harmonizable process with spectral support on lines and carry out spectral density estimation. Let

$$Y_t=\int e^{it\lambda}\,dZ(\lambda) \tag{6.1}$$

be a stationary process with spectral density $f(\cdot)$ and spectral distribution $F(\cdot)$. Let

$$X_t=Y_t+\cos(wt)Y_{\alpha t}, \tag{6.2}$$

with $w,\alpha\in(0,\infty)$. This is the multipath problem of (2.5) with $k=1$ where we have made the example more complicated by including a "carrier" $\cos(wt)$ for the Doppler stretched signal. Then

$$\begin{aligned}X_t&=\int e^{it\lambda}\,dZ(\lambda)+\frac{1}{2}\int(e^{iwt}+e^{-iwt})e^{i\alpha t\lambda}\,dZ(\lambda)\\&=\int e^{it\lambda}\,dZ(\lambda)+\frac{1}{2}\int e^{it\lambda}\Big[dZ\Big(\frac{\lambda-w}{\alpha}\Big)+dZ\Big(\frac{\lambda+w}{\alpha}\Big)\Big]\\&=\int e^{it\lambda}\,dZ_1(\lambda),\end{aligned}$$

with $dZ_1(\lambda)=dZ(\lambda)+\frac{1}{2}dZ(\frac{1}{\alpha}(\lambda-w))+\frac{1}{2}dZ(\frac{1}{\alpha}(\lambda+w))$. Then the spectrum of X_t is given by $dH(\lambda,\mu)\equiv E\,dZ_1(\lambda)\overline{dZ_1(\mu)}$. Using the orthogonal increment property of the $Z(\lambda)$ process and the fact that $E\,dZ(\lambda)\overline{dZ(\lambda)}=f(\lambda)\,d\lambda$, we have

$$E\,dZ(\lambda)\overline{dZ(\mu)}=f(\mu)\delta_{\lambda,\mu}\,d\mu,$$

$$\begin{aligned}&E\frac{1}{2}dZ\Big(\frac{1}{\alpha}(\lambda-w)\Big)\frac{1}{2}\overline{dZ\Big(\frac{1}{\alpha}(\mu-w)\Big)}\\&\quad=\frac{1}{4}E\Big\{\lim_{h\to 0}\frac{1}{h}\Big|Z\Big(\frac{1}{\alpha}(\mu+h-w)\Big)-Z\Big(\frac{1}{\alpha}(\mu-w)\Big)\Big|^2\delta_{\lambda,\mu}\,d\mu\Big\}\end{aligned}$$

$$= \frac{1}{4} \lim_{h\to 0} \frac{1}{h} \left[F\left(\frac{1}{\alpha}(\mu - w) + \frac{h}{\alpha}\right) - F\left(\frac{1}{\alpha}(\mu - w)\right) \right] \delta_{\lambda,\mu}\, d\mu$$

$$= \frac{1}{4\alpha} f\left(\frac{1}{\alpha}(\mu - w)\right) \delta_{\lambda,\mu}\, d\mu,$$

$$E \frac{1}{2} dZ\left(\frac{1}{\alpha}(\lambda - w)\right) \overline{dZ(\mu)} = \frac{1}{2} E\, dZ(\mu)\, \overline{dZ(\mu)} \delta_{(1/\alpha)(\lambda - w),\mu}\, d\mu$$

$$= \frac{1}{2} f(\mu) \delta_{\lambda,\alpha\mu + w}\, d\mu,$$

$$E\, dZ(\lambda) \overline{\frac{1}{2} dZ\left(\frac{1}{\alpha}(\mu - w)\right)}$$

$$= \frac{1}{2} E \left| dZ\left(\frac{1}{\alpha}(\mu - w)\right)\right|^2 \delta_{\lambda,(1/\alpha)(\mu - w)}\, d\mu$$

$$= \frac{1}{2\alpha} f\left(\frac{1}{\alpha}(\mu - w)\right) \delta_{\lambda,(1/\alpha)(\mu - w)}\, d\mu,$$

with corresponding expressions for all other possible terms. Therefore

$$\begin{aligned} dH(\lambda,\mu) = {} & \left[f(\mu) + \frac{1}{4\alpha} f\left(\frac{1}{\alpha}(\mu - w)\right) + \frac{1}{4\alpha} f\left(\frac{1}{\alpha}(\mu + w)\right)\right] \delta_{\lambda,\mu}\, d\mu \\ & + \frac{1}{2} f(\mu)\delta_{\lambda,\alpha\mu+w}\, d\mu + \frac{1}{2} f(\mu)\delta_{\lambda,\alpha\mu - w}\, d\mu \\ & + \frac{1}{2\alpha} f\left(\frac{1}{\alpha}(\mu - w)\right)\delta_{\lambda,(1/\alpha)(\mu - w)}\, d\mu \\ & + \frac{1}{2\alpha} f\left(\frac{1}{\alpha}(\mu + w)\right)\delta_{\lambda,(1/\alpha)(\mu + w)}\, d\mu \\ & + \frac{1}{4\alpha} f\left(\frac{1}{\alpha}(\mu - w)\right)\delta_{\lambda,\mu - 2w}\, d\mu \\ & + \frac{1}{4\alpha} f\left(\frac{1}{\alpha}(\mu + w)\right)\delta_{\lambda,\mu + 2w}\, d\mu. \end{aligned} \tag{6.3}$$

The spectral mass of X_t is supported on the seven lines given by

$$\begin{aligned} & \lambda = \mu, \quad \lambda = \alpha\mu + w, \quad \lambda = \alpha\mu - w, \quad \lambda = \frac{1}{\alpha}(\mu - w), \\ & \lambda = \frac{1}{\alpha}(\mu + w), \quad \lambda = \mu - 2w \quad \text{and} \quad \lambda = \mu + 2w, \end{aligned} \tag{6.4}$$

which we shall refer to as lines (1)–(7), respectively. These lines are shown in Figure 1 for $(\alpha, w) = (0.5, 0.5)$.

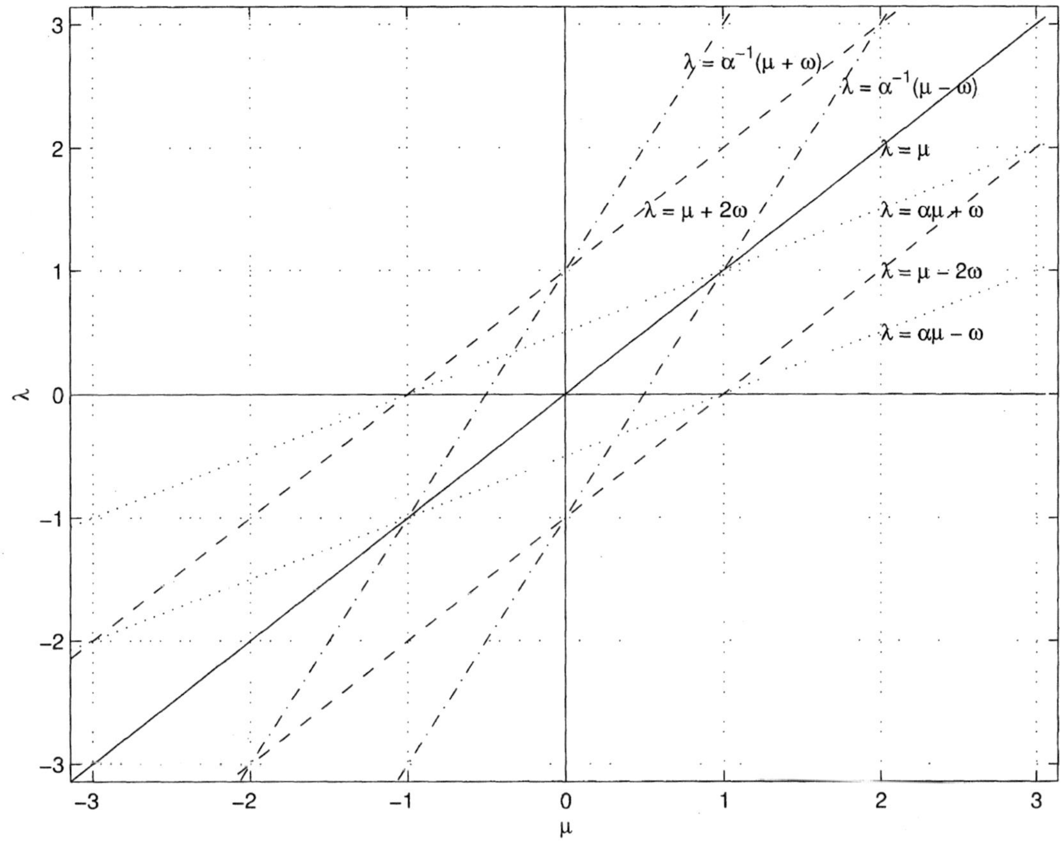

FIG. 1. *Spectral support lines for* $y_t = x_t + \cos(\omega t)x_{\alpha t}$ *with* $(\alpha, \omega) = (0.5, 0.5)$.

As an example, consider

$$Y_t = \varepsilon_t + 0.5\varepsilon_{t-1}, \tag{6.5}$$

with ε_t i.i.d. $N(0, 1)$.

Five thousand independent Y_t with $t = -128, \ldots, 128$ are generated. Shannon interpolation is used to obtain $Y_{\alpha t}$, $t = -128, \ldots, 128$, for $\alpha = 0.5$. Using (6.2), X_t, $t = -128, \ldots, 128$, is obtained. Periodograms are formed from (3.9) on each of the seven lines given by (6.4). Five thousand independent such periodograms are averaged and for each line both real and imaginary parts of the corresponding periodogram are plotted against the corresponding theoretical values given by (6.3). These plots are shown in Figure 2 in order starting from term 1. Relative to term 1, the sum of the three terms which gives the theoretical spectral mass on the diagonal line [see line 1 from (6.3)] is plotted as a dashed curve. The averaged periodogram of X_t is plotted with it as the solid line and we see that it is roughly unbiased with some variability. The theoretical spectral mass and its estimate are almost superimposed. The theoretical spectral mass is also plotted as the dashed curve slightly below its estimate with an offset. The two spikes located at ± 1 are

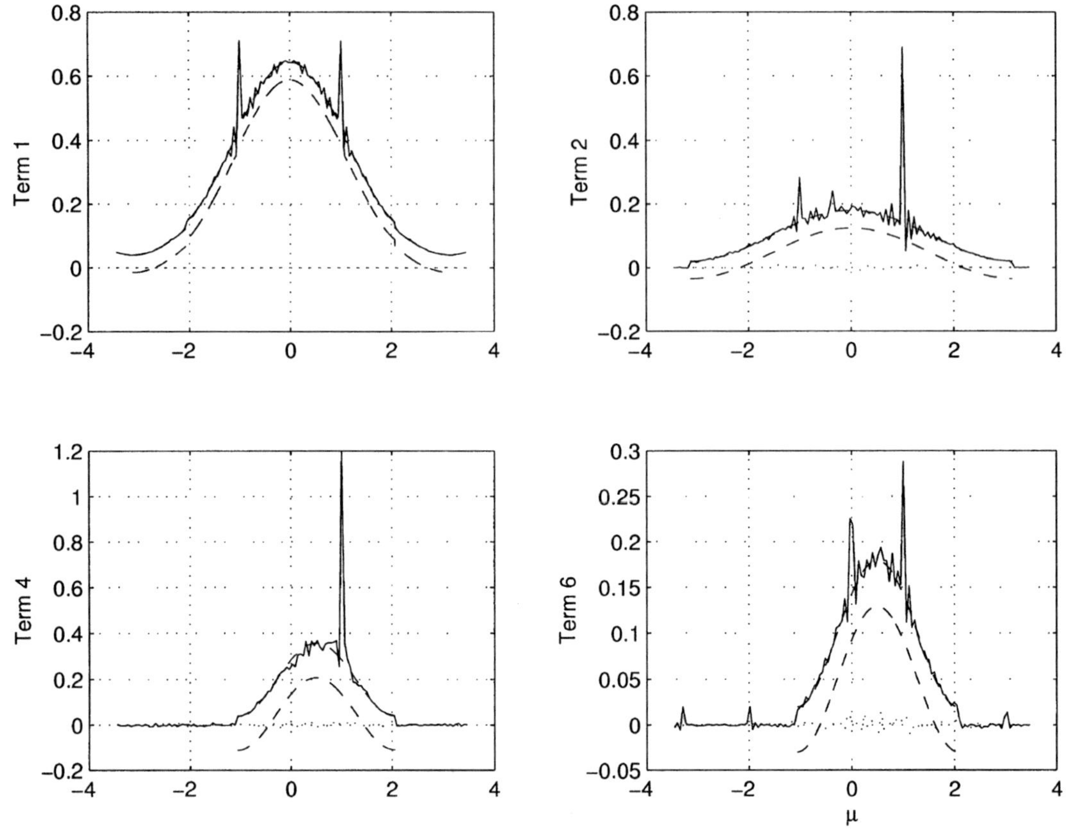

FIG. 2. *Averaged periodogram spectral estimates on different lines*; $(\alpha, \omega) = (0.5, 0.5)$; *sample size*, 257, *each averaged* 5000 *times*; —, *estimated real part*; $\cdots$, *estimated imaginary part*; - - -, *theoretical real part*; *theoretical imaginary parts are all* 0 *for this example and are not plotted. Theoretical real parts are almost superimposed on their estimates so they are also plotted with an arbitrary offset.*

due to the intersection of the diagonal line with two other spectral support lines. Lines 2 and 4 intersect line 1 at $+1$ while lines 3 and 5 intersect line 1 at -1 (see Figure 1). There are extra spectral masses contributing to the diagonal lines at these intersection points. Similar remarks can be made for all other graphs where there are spikes. These graphs are consistent with the properties of the periodogram given by (3.10) or (5.4). Since the theoretical spectral functions are almost the same as their estimates, we also plot a shift of the real part of the theoretical spectral functions to better indicate their character. Terms 2, 4 and 6 correspond to lines 2, 4 and 6 from (6.3). Terms 3, 5 and 7 are not plotted since they are symmetric versions of terms 2, 4 and 6, respectively. Overall, we see that the procedure is effective in estimating the spectral mass supported on lines. One way to remove the spikes is to use the average of neighboring spectral mass on the line. From the graphs in Figure 2, we also see the effects of the band-limitness of Y_t [see (6.5)] in the sudden drop-offs.

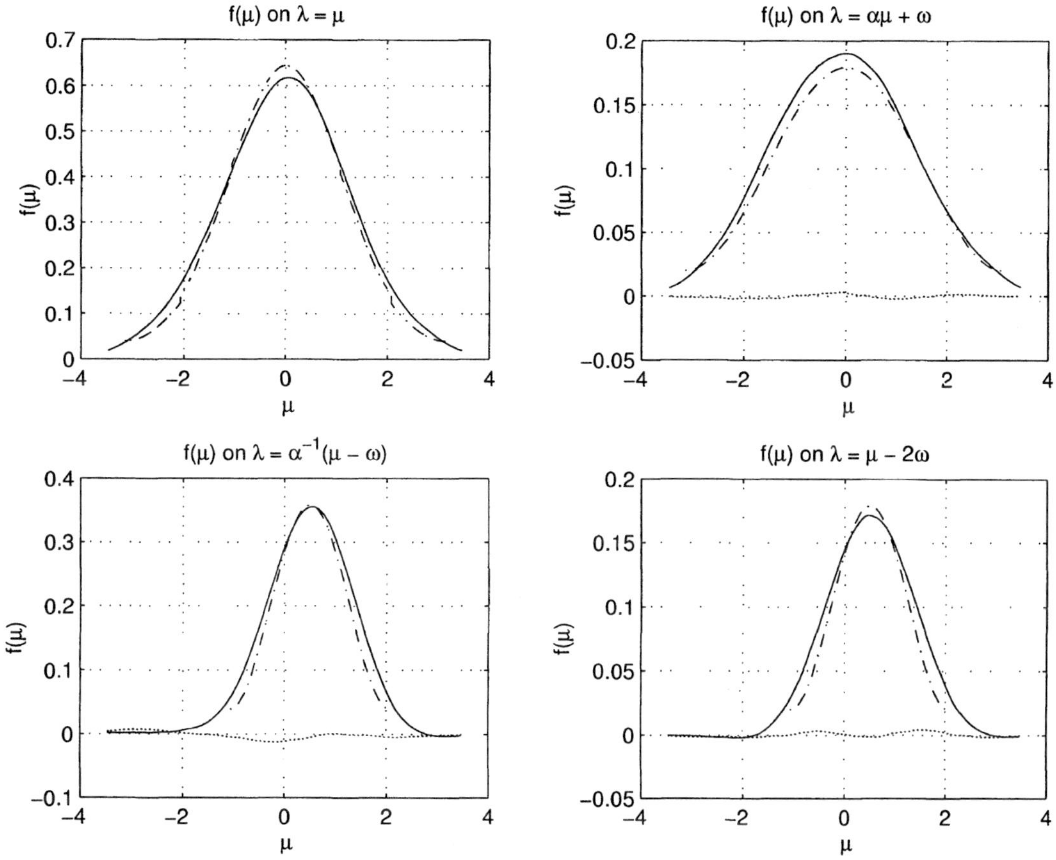

FIG. 3. *Spectral estimates on different lines*; $(\alpha, \omega) = (0.5, 0.5)$; *sample size*, 16385; —, *estimated real part*; $\cdots$, *estimated imaginary part*; - · -, *theoretical real part*; *theoretical imaginary parts are all* 0 *for this example and are not plotted.*

In reality, we do not have independent realizations with which to average the periodograms. Using the same model, we generated *one* realization of X_t for $t = -2^{13}, \ldots, 2^{13}$. Then a smoothed periodogram of the form in (5.2) is used to estimate the spectral mass on each of the support lines. Results are given in Figure 3.

Symmetry properties given in (2.5) make some graphs redundant. Consequently, some graphs are not plotted. We see that the estimates are generally very close to the theoretical ones. These computations confirm the results of Theorems 5.2 and 5.3. Spikes at the intersection of spectral support lines (see Figure 2) are removed. The values at such intersection points are determined by the direction of smoothing. If the lines of spectral support are not known a priori, the results of Theorems 5.2 and 5.3 imply that one can search for these using a grid of points, by computing smoothed periodograms in a grid of directions at each of these points and recording the largest magnitude among these directions at each point. Set an appropriate threshold level to remove all those points whose spectral magnitude is below the threshold. Then by connecting the remaining points one obtains information about the locations of the spectral support lines. Alternatively, one can search for (α, w) in that if certain statistics from the spectral estimate along the line represented by (α, w) are significantly different from 0 then (α, w) represents a line of spectral support. Spectral mass on these lines of spectral support can be estimated. The spectral function $f(\cdot)$ of the unobserved signal $Y(t)$ can be estimated from estimates of off-diagonal spectral lines in (6.3). These are obviously computationally intensive and are not pursued here.

7. Proofs. We need the following lemmas in the derivations of the results and they are used often.

LEMMA 7.1. *Let*

$$D_n(x) = \sum_{t=0}^{n-1} e^{itx} = \frac{e^{inx} - 1}{e^{ix} - 1} \tag{7.1}$$

and

$$\tilde{D}_n(x) = \frac{e^{inx} - 1}{ix}.$$

Then $D'_n(x) = D_n(x) - \tilde{D}_n(x)$ *is uniformly bounded in* $[-2\pi + c, 2\pi - c]$ *for a fixed* $c > 0$. *In particular,* $D'_n(x)$ *is uniformly bounded in* $[-\pi, \pi]$ *with*

$$D'_n(x) = (e^{inx} - 1)\frac{ix - e^{ix} + 1}{(e^{ix} - 1)ix}.$$

The derivation is straightforward.

LEMMA 7.2. *For any real* a, $|a| > 0$, $|y| \le \pi$,

$$\int_{-\pi}^{\pi} D_n(ax + y) D_n(x)\, dx = \int_{-\pi}^{\pi} \tilde{D}_n(ax + y)\tilde{D}_n(x)\, dx + O(\log n).$$

PROOF. Since $D_n(x) = D'_n(x) + \tilde{D}_n(x)$,

$$\begin{aligned}
&\int_{-\pi}^{\pi} D_n(ax + y) D_n(x)\, dx - \int_{-\pi}^{\pi} \tilde{D}_n(ax + y)\tilde{D}_n(x)\, dx \\
&\quad = \int_{-\pi}^{\pi} D'_n(ax + y) D'_n(x)\, dx + \int_{-\pi}^{\pi} D'_n(ax + y)\tilde{D}_n(x)\, dx \\
&\qquad + \int_{-\pi}^{\pi} \tilde{D}_n(ax + y) D'_n(x)\, dx = (1) + (2) + (3).
\end{aligned}$$

We first note that, for a fixed $M > 0$,

$$\int_{-M}^{M} |\tilde{D}_n(x)|\,dx = \int_{|x|\le 1/n^2} |\tilde{D}_n(x)|\,dx + \int_{M\ge|x|\ge 1/n^2} |\tilde{D}_n(x)|\,dx$$
$$\le \frac{2}{n} + O(\log n) + O(\log M) = O(\log n).$$

Using Lemma 7.1, for a constant C°,

$$|(1)| \le C^\circ \int_{-\pi}^{\pi} |D_n'(ax+y)|\,dx = \frac{C^\circ}{|a|}\int_{-a\pi+y}^{a\pi+y} |D_n'(x)|\,dx.$$

For a fixed $d \in [d', \pi)$ with $d' > 0$, by Lemma 7.1,

$$\int_{-d}^{d} |D_n'(x)|\,dx = O(1).$$

For $k \neq 0$,

$$\int_{2k\pi-d}^{2k\pi+d} |D_n'(x)|\,dx = \int_{2k\pi-d}^{2\pi k+d} \left|\frac{e^{inx}-1}{e^{ix}-1}\right|\left|\frac{ix-e^{ix}+1}{ix}\right| dx$$
$$\le \left(1+\frac{2}{\pi}\right)\int_{2k\pi-d}^{2k\pi+d} \left|\frac{\sin\frac{n}{2}x}{\sin\frac{1}{2}x}\right| dx$$
$$= \left(1+\frac{2}{\pi}\right)\int_{-d}^{d} \left|\frac{\sin\frac{n}{2}x}{\sin\frac{1}{2}x}\right| dx \le O(\log n).$$

It is clear that $\int_A |D_n'(x)|\,dx = O(1)$ for a closed interval A not containing $2\pi k$ for integer k. Therefore $|(1)| = O(\log n)$. A similar argument applied to $\tilde{D}_n(ax+y)$ shows that $|(3)| = O(\log n)$. Finally,

$$|(2)| = \int_{|x|\le d} |D_n'(ax+y)\tilde{D}_n(x)|\,dx + \int_{\pi\ge|x|\ge d} |D_n'(ax+y)\tilde{D}_n(x)|\,dx = (4)+(5).$$

For fixed a, we can choose a d such that $|ax+y| \le 2\pi - c$ for a fixed $c > 0$ for all $|x| \le d$ and therefore $D_n'(ax+y)$ is uniformly bounded by Lemma 7.1. Hence

$$(4) \le C^\circ \int_{|x|\le d} |\tilde{D}_n(x)|\,dx = O(\log n).$$

For the same d, $|\tilde{D}_n(x)| \le 2d^{-1}$ for $\pi \ge |x| \ge d$. Hence

$$(5) \le 2d^{-1} \int_{\pi\ge|x|\ge d} |D_n'(ax+y)|\,dx = O(\log n).$$

This completes the proof of Lemma 7.2. □

PROOF OF THEOREM 3.1. Given $\{X_t\}_{t=-N}^{N}$, $\lambda' = \alpha\mu' + w$. Let [using (3.1) and (3.2)]

$$G_0 \equiv EI_{l,n}(\lambda',\mu') = \frac{1}{2\pi n}\iint_{-\infty}^{\infty}\sum_{t,\tau=0}^{n-1} e^{it(\lambda-\lambda')-i\tau(\mu-\mu')} \times e^{ilJ(\lambda-\mu)}dH(\lambda,\mu)e^{-ilJ(\lambda'-\mu')} \tag{7.2}$$

$$= e^{-ilJ(\lambda'-\mu')}\sum_{(a,b)\in\mathcal{L}}\sum_k G_1(k,a,b),$$

where

$$G_1(k,a,b) \equiv G_1(k,a,b,\lambda',\mu';n,l,J) \equiv G_1$$
$$\equiv \frac{1}{2\pi n}\int_{\mu'-\pi+2k\pi}^{\mu'+\pi+2k\pi} D_n(\lambda-\lambda')D_n(-\mu+\mu')e^{ilJ(\lambda-\mu)} \times f_{a,b}(\mu)\delta_{\lambda,a\mu+b}\,d\mu.$$

Let $x = \mu - \mu' - 2\pi k$, $\mu = x + \mu' + 2\pi k$. Then, using $\lambda = a\mu + b$ and $\lambda' = \alpha\mu' + w$,

$$G_1 = \frac{1}{2\pi n}\int_{-\pi}^{\pi} D_n\big(ax + (a-\alpha)\mu' + b - w + 2\pi ka\big)D_n(-x) \times e^{ilJ(a-1)x} f_{a,b}(x+\mu'+2\pi k)\,dx\, e^{ilJ[b+(a-1)(\mu'+2\pi k)]}$$

since

$$-\mu+\mu' = -x-2\pi k, \qquad D_n(-\mu+\mu') = D_n(-x-2\pi k) = D_n(-x),$$
$$\lambda-\mu = (a-1)x + b + (a-1)(\mu'+2\pi k).$$

Notice that $e^{-ilJ(\lambda'-\mu')}e^{ilJ[b+(a-1)(\mu'+2\pi k)]} = \exp\{ilJ(2\pi ka + (a-\alpha)\mu' + b - w)\}$. Then, using (3.4),

$$G_0 = \sum_{(a,b)\in\mathcal{L}}\sum_k e^{ilJ\cdot y(k,a,b;\lambda,\mu)}G_2, \tag{7.3}$$

with

$$G_2 = \frac{1}{2\pi n}\int_{-\pi}^{\pi} D_n(ax+y)D_n(-x)e^{ilJ(a-1)x} f_{a,b}(x+\mu'+2\pi k)\,dx.$$

Because of the continuous differentiability of $f_{a,b}$, we have

$$f_{a,b}(x+\mu'+2\pi k) = f_{a,b}(\mu'+2\pi k) + x f'_{a,b}(x^*+2\pi k),$$

with x^* between μ' and $\mu'+x$. Then

$$G_2 = G_3 + G_4, \tag{7.4}$$

with

$$(7.5)\qquad \begin{aligned} G_4 &= \frac{1}{2\pi n}\int_{-\pi}^{\pi} D_n(ax+y)D_n(-x)xe^{ilJ(a-1)x} f'_{a,b}(x^*+2\pi k)\,dx,\\ |G_4| &\le g(2\pi k)\cdot O\Big(\frac{\log n}{n}\Big) \end{aligned}$$

by Assumption 2 and

$$(7.6)\qquad \begin{aligned} G_3 &= f_{a,b}(\mu'+2\pi k)\frac{1}{2\pi n}\int_{-\pi}^{\pi} D_n(ax+y)D_n(-x)e^{ilJ(a-1)x}\,dx\\ &= f_{a,b}(\mu'+2\pi k)\Big[\frac{1}{2\pi n}\int_{-\pi}^{\pi} \tilde{D}_n(ax+y)\tilde{D}_n(-x)e^{ilJ(a-1)x}\,dx\\ &\qquad + O\Big(\frac{\log n}{n}\Big)\Big] \end{aligned}$$

using Lemma 7.2. Consider

$$(7.7)\qquad G_5 = \frac{1}{2\pi n}\int_{-\pi}^{\pi} \frac{e^{in(ax+y)}-1}{i(ax+y)}\frac{e^{-inx}-1}{-ix}e^{ilJ(a-1)x}\,dx.$$

Let $nx = z$. Then

$$(7.8)\qquad \begin{aligned} G_5 &= \frac{1}{2\pi n}\int_{-n\pi}^{n\pi} \frac{e^{i(az+ny)}-1}{i(az+ny)/n}\frac{e^{-iz}-1}{-iz/n}e^{ilJ(a-1)z/n}\,dz/n\\ &= \frac{1}{2\pi}\int_{-\infty}^{\infty} \frac{e^{i(az+ny)}-1}{i(az+ny)}\frac{e^{-iz}-1}{-iz}e^{ilJ(a-1)z/n}\,dz\\ &\quad - \frac{1}{2\pi}\int_{|z|>n\pi} \frac{e^{i(az+ny)}-1}{i(az+ny)}\frac{e^{-iz}-1}{-iz}e^{ilJ(a-1)z/n}\,dz\\ &= G_6 - G_7. \end{aligned}$$

To evaluate the second term G_7, we first note that $|y| \le \pi$ and

$$\begin{aligned} &\int_{|z|\ge n\pi} \Big|\frac{e^{i(az+ny)}-1}{i(az+ny)}\frac{e^{-iz}-1}{-iz}\Big|\,dz\\ &\le \frac{1}{|a|}\Big[\int_{z\in A_1\cap A} + \int_{z\in A_1^c\cap A_2} + \int_{|z|\ge n\pi+|n\pi/a|}\Big]\Big|\frac{e^{i(az+ny)}-1}{i(z+ny/a)}\frac{e^{-iz}-1}{-iz}\Big|\,dz, \end{aligned}$$

where $A_1 = [-ny/a - c, -ny/a + c]$ with fixed $c > 0$ and $A = \{z|\ |z| \ge n\pi\}$, $A_2 = \{z \mid n\pi \le |z| \le n\pi + |n\pi/a|\}$. For the first integral, $z \in A_1 \cap A$ and $|(e^{i(az+ny)}-1)/(i(z+ny/a))| \le |a|$, $|(e^{-iz}-1)/(-iz)| \le 1/n$. Therefore

$$\int_{z\in A_1\cap A} \Big|\frac{e^{i(az+ny)}-1}{i(z+ny/a)}\frac{e^{-iz}-1}{-iz}\,dz\Big| \le \frac{2|a|c}{n} = O\Big(\frac{1}{n}\Big).$$

The second integral is bounded by

$$\frac{1}{n\pi}\int_{z\in A_1^c\cap A_2}\left|\frac{1}{z+ny/a}\right|dz = O\left(\frac{\log n}{n}\right).$$

The third integral is bounded by

$$C\int_{|z|>n\pi}\frac{1}{z^2}\,dz = O\left(\frac{1}{n}\right).$$

From these we see

$$G_7 = \frac{1}{2\pi}\int_{|z|>n\pi}\frac{e^{i(az+ny)}-1}{i(az+ny)}\,\frac{e^{-iz}-1}{-iz}\,e^{ilJ(a-1)z/n}\,dz = O\left(\frac{\log n}{n}\right). \tag{7.9}$$

To evaluate G_6, we note that

$$\frac{e^{i(az+ny)}-1}{i(az+ny)} = \int_{-\infty}^{\infty}\frac{1}{a}1_{[0,a]}(t)\;e^{inyt/a}e^{izt}\,dt, \tag{7.10}$$

$$\left[\frac{e^{-iz}-1}{-iz}e^{ilJ(a-1)z/n}\right]^* = \int_{-\infty}^{\infty}1_{[-lJ(a-1)/n,1-lJ(a-1)/n]}(t)e^{izt}\,dt \tag{7.11}$$

and, using the Parseval relation, we have

$$\begin{aligned} G_6 &= \frac{1}{a}\int_{-\infty}^{\infty}1_{(0,a]}(t)1_{[-lJ(a-1)/n,1-lJ(a-1)/n]}(t)e^{inyt/a}\,dt \\ &= \begin{cases} \dfrac{1}{iny}\Bigg[\exp\left(i\dfrac{ny}{a}\left[\min(a,1-\dfrac{lJ(a-1)}{n})\right]\right) \\ \qquad - \exp\left(i\dfrac{ny}{a}\left[\max(0,\dfrac{-lJ(a-1)}{n})\right]\right)\Bigg] \\ \qquad\qquad \text{if } [0,a]\cap\left[\dfrac{-lJ(a-1)}{n},1-\dfrac{lJ(a-1)}{n}\right]\neq\varnothing, \\ 0, \qquad \text{otherwise.} \end{cases} \end{aligned} \tag{7.12}$$

Successively using (7.2)–(7.9) and (7.12), we obtain Theorem 3.1. □

PROOF OF THEOREM 4.1. Let $d_n = 1/(2\pi n)$. Then

$$\begin{aligned} &\operatorname{cov}\left(I_{l,n}(\lambda,\mu), I_{l,n}(\lambda',\mu')\right) \\ &\quad = d_n^2\sum_{t,\tau,t',\tau'=0}^{n}\left[r_{t+lJ,t'+lJ}r_{\tau+lJ,\tau'+lJ} + r_{t+lJ,\tau'+lJ}r_{\tau+lJ,t'+lJ}\right] \\ &\qquad\qquad \times\exp\{i[-(t+lJ)\lambda+(\tau+lJ)\mu+(t'+lJ)\lambda'-(\tau'+lJ)\mu']\} \\ &\quad = d_n^2\sum_{t,\tau,t',\tau'=0}^{n}\Bigg[E\int e^{i(t+lJ)u}\,dZ(u)\int e^{-i(t'+lJ)v}\,\overline{dZ(v)} \\ &\qquad\qquad \times E\int e^{i(\tau+lJ)r}\,dZ(r)\int e^{-i(\tau'+lJ)s}\,\overline{dZ(s)} \end{aligned}$$

$$+ E \int e^{i(t+lJ)u}\, dZ(u) \int e^{-i(\tau'+lJ)v}\, \overline{dZ(v)}$$

$$\times E \int e^{i(\tau+lJ)r}\, dZ(r) \int e^{-i(t'+lJ)s}\, \overline{dZ(s)}\Big]$$

$$\times \exp\{i[-(t+lJ)\lambda + (\tau+lJ)\mu + (t'+lJ)\lambda' - (\tau'+lJ)\mu']\}$$

$$= d_n^2 \iiiint \sum_{t,\tau,t',\tau'=0}^{n} \Big[\exp\{i[t(u-\lambda) + \tau(r+\mu) + t'(\lambda'-v)$$

$$- \tau'(\mu'+s)]\}\exp\{ilJ[u-v+r-s]\}$$

$$+ \exp\{i[t(u-\lambda) + \tau(r+\mu) + t'(\lambda'-s)$$

$$- \tau'(\mu'+v)]\}$$

$$\times \exp\{ilJ[u-v+r-s]\}\Big]$$

$$\times \exp\{ilJ[-\lambda+\mu+\lambda'-\mu']\}\, dH(u,v)\, dH(r,s)$$

$$= d_n^2 \iiiint \Big[D_n(u-\lambda)D_n(\lambda'-v)D_n(\mu+r)D_n(-\mu'-s)$$

$$\times \exp\{ilJ[u-v+r-s]\}$$

$$+ D_n(u-\lambda)D_n(-\mu'-v)D_n(\mu+r)D_n(\lambda'-s)$$

$$\times \exp\{ilJ(u-v+r-s)\}\Big]$$

$$\times \exp\{ilJ(-\lambda+\mu+\lambda'-\mu')\}\, dH(u,v)\, dH(r,s)$$

$$= c_1 + c_2,$$

where

$$c_1 = \Big\{\Big[d_n \sum_{(a,b)\in\mathcal{L}} \int D_n(av+b-\alpha\mu-w)D_n(\alpha'\mu'+w'-v) f_{a,b}(v)$$

$$\times \exp\{ilJ(av+b-v)\}\, dv\Big]\exp\{-ilJ(\lambda-\lambda')\}\Big\}$$

$$\times \Big\{\Big[d_n \sum_{(a',b')\in\mathcal{L}} \int D_n(\mu+a's+b')D_n(-\mu'-s) f_{a',b'}(s)$$

$$\times \exp\{ilJ(a's+b'-s)\}\, ds\Big]\exp\{-ilJ(\mu'-\mu)\}\Big\},$$

$$c_2 = \left\{\left[d_n \sum_{(a,b)\in\mathcal{L}} \int D_n(av+b-\alpha\mu-w)D_n(-\mu'-v)f_{a,b}(v) \times \exp\{ilJ(av+b-v)\}\,dv\right]\exp\{-ilJ(\lambda-\lambda')\}\right\}$$

$$\times\left\{\left[d_n \sum_{(a,b)\in\mathcal{L}} \int D_n(\mu+a's+b')D_n(\alpha'\mu'+w'-s)f_{a',b'}(s) \times \exp\{ilJ(a's+b-s)\}\,ds\right]\exp\{-ilJ(\mu'-\mu)\}\right\},$$

with $u = av+b$, $r = a's+b'$, $\lambda = \alpha\mu + w$, $\lambda' = \alpha'\mu' + w'$. Now each term in the outer brackets has exactly the same form as the expectation of the periodogram [see (7.2)]. Therefore the covariance of the periodogram has the form of the product of the expectations of the periodogram. □

PROOF OF THEOREM 5.2. We now take the symmetrized version. From (5.7), (5.3), (5.5) and (3.4),

$$E\hat{f}_{\alpha,w}(\eta) = \sum_{(a,b)\in\mathcal{L}}\sum_k \int f_{a,b}(\mu+2\pi k)\,\mathrm{sinc}(a,y)K_n(\mu-\eta)\,d\mu + O\left(\frac{\log n}{n}\right). \tag{7.13}$$

Let

$$\begin{aligned}E_0 &\equiv \int_{-\infty}^{\infty} f_{a,b}(\mu+2\pi k)\,\mathrm{sinc}(a,y)K_n(\mu-\eta)\,d\mu\\ &= \int_{-\infty}^{\infty} f_{a,b}(\eta+b_nz+2\pi k)\\ &\qquad\times\mathrm{sinc}\left\{\frac{n+1}{2}l(a)[[2\pi ka+(a-\alpha)\eta+b-w\\ &\qquad\qquad+(a-\alpha)b_nz]\bmod 2\pi]\right\}l(a)K(z)\,dz.\end{aligned} \tag{7.14}$$

By Assumption 2, we have $f(\eta+b_nz+2\pi k) = f(\eta+2\pi k)+b_nzf'_{a,b}(\eta^*_{k,z}+2\pi k)$ for $\eta^*_{k,z}$ between $\eta+2\pi k$ and $\eta+2\pi k+b_nz$. Then $E_0 = E_1+E_2$ with

$$\begin{aligned}E_1 &= l(a)f_{a,b}(\eta+2\pi k)E_3,\\ E_2 &= b_n\int zf'_{a,b}(\eta^*_{k,z}+2\pi k)\\ &\qquad\times\mathrm{sinc}\left\{\frac{n+1}{2}l(a)[[y_\eta+(a-\alpha)b_nz]\bmod 2\pi]\right\}l(a)K(z)\,dz,\\ E_3 &= \int\mathrm{sinc}\left\{\frac{n+1}{2}l(a)[[y_\eta+(a-\alpha)b_nz]\bmod 2\pi]\right\}K(z)\,dz\end{aligned} \tag{7.15}$$

and $y_\eta \equiv 2\pi ka + (a-\alpha)\eta + b - w$. We evaluate E_2 first and then E_3.

$$E_2 = b_n l(a) f'_{a,b}(\eta + 2\pi k) \int_{-c}^{c} zK(z) \times \operatorname{sinc}\left\{\frac{n+1}{2} l(a)[[y_\eta + (a-\alpha)b_n z] \bmod 2\pi]\right\} dz + o(b_n)$$

by the continuity of f' with $[-c, c]$ the support of K for a finite c. For $a = \alpha$,

$$E_2 = b_n l(a) f'_{a,b}(\eta + 2\pi k) \operatorname{sinc}\left\{\frac{n+1}{2} l(a)(y_\eta \bmod 2\pi)\right\} \int_{-c}^{c} zK(z)\, dz + o(b_n)$$
$$= o(b_n)$$

by Assumption 3. For $a \neq \alpha$, $E_2 = o(b_n) + E'_2$ with

$$E'_2 = b_n l(a) f'_{a,b}(\eta + 2\pi k) \int_{-c}^{c} zK(z) \times \operatorname{sinc}\left\{\frac{n+1}{2} l(n)[[y_\eta + (a-\alpha)b_n z] \bmod 2\pi]\right\} dz$$
$$= b_n l(a) f'_{a,b}(\eta + 2\pi k) \int_{-c}^{c} zK(z) \operatorname{sinc}\left\{\frac{n+1}{2} l(a)[(y_\eta)_{2\pi} + (a-\alpha)b_n z]\right\} dz,$$

where $(y_\eta)_{2\pi} = (y_\eta \bmod 2\pi)$. Therefore

$$(7.16)\qquad \begin{aligned} |E'_2| &\le b_n C^\circ \int_{-c}^{c} \left| \operatorname{sinc}\left\{\frac{n+1}{2} l(a)[(y_\eta)_{2\pi} + (a-\alpha)b_n z]\right\}\right| dz \\ &= C^\circ \int_{(y_\eta)_{2\pi}-(a-\alpha)b_n c}^{(y_\eta)_{2\pi}+(a-\alpha)b_n c} \left| \operatorname{sinc}\left\{\frac{n+1}{2} l(a)x\right\}\right| \frac{dx}{(a-\alpha)} \\ &\le C'\left[\int_{|x|\le 1/n} + \int_{C''>|x|>1/n}\right] \left| \operatorname{sinc}\left\{\frac{n+1}{2} l(a)x\right\}\right| dx \\ &= O\left(\frac{1}{n}\right) + O\left(\frac{\log n}{n}\right), \end{aligned}$$

where C°, C' are constants.

To evaluate E_3, we see that, by a similar argument, for $a \neq \alpha$, $|E_3| = O(\log n/nb_n)$.

If $a = \alpha$, then

$$E_3 = \int_{-c}^{c} \operatorname{sinc}\left\{\frac{n+1}{2} l(a)(y_\eta \bmod 2\pi)\right\} K(z)\, dz$$
$$= \operatorname{sinc}\left\{\frac{n+1}{2} l(a)(y_\eta \bmod 2\pi)\right\}.$$

Substituting these results on E_3, E_2, E_1 into E_0 using (7.13)–(7.16), we have the results. □

PROOF OF THEOREM 5.3. We evaluate the covariance

$$\begin{aligned} C_4 &\equiv \operatorname{cov}(\hat{f}_{\alpha,w}(\eta), \hat{f}_{\alpha',w'}(\eta')) \\ &= \iint_{-\pi^+}^{\pi^+} K_n(\mu-\eta)K_n(\mu'-\eta') \\ &\qquad\qquad \times \operatorname{cov}(I_n(\alpha\mu+w,\mu), I_n(\alpha'\mu'+w',\mu'))\,d\mu\,d\mu' \\ &= C_5 + C_6 + O\Big(\frac{\log n}{n}\Big), \end{aligned} \tag{7.17}$$

with

$$\begin{aligned} C_5 = \iint_{-\pi^+}^{\pi^+} & K_n(\mu-\eta)K_n(\mu'-\eta')\Sigma^1\Sigma^2 f_{a,b}(\lambda'+2\pi k) f_{a',b'}(-\mu'+2\pi k') \\ & \times \operatorname{sinc}(a, y(1))\operatorname{sinc}(a', y(2))\,d\mu\,d\mu', \end{aligned} \tag{7.18}$$

$$\begin{aligned} C_6 = \iint_{\pi^+}^{\pi^+} & K_n(\mu-\eta)K_n(\mu'-\eta')\Sigma^3\Sigma^4 f_{a,b}(-\mu'+2\pi k) f_{a',b'}(\lambda'+2\pi k') \\ & \times \operatorname{sinc}(a, y(3))\operatorname{sinc}(a', y(4))\,d\mu\,d\mu' \end{aligned} \tag{7.19}$$

using (5.5). The sum Σ^1 is over $|y(1)| < \nu$ and $|y(2)| \le \nu$, Σ^2 and Σ^4 are over $(a,b) \in \mathcal{L}$ and $(a',b') \in \mathcal{L}$, while Σ^3 is over $|y(3)| < \nu$ and $|y(4)| < \nu$.

Now consider a typical summand of (7.18),

$$\begin{aligned} C_7 \equiv \iint & K_n(\mu-\eta)K_n(\mu'-\eta') f_{a,b}(\lambda'+2\pi k) f_{a',b'}(-\mu'+2\pi k') \\ & \times \operatorname{sinc}(a, y(1))\operatorname{sinc}(a', y(2))\,d\mu\,d\mu'. \end{aligned} \tag{7.20}$$

For fixed k, k', $(a,b) \in \mathcal{L}$, $(a',b') \in \mathcal{L}$, we evaluate C_7 in the following cases:

(i) If $a\alpha' \neq a'\alpha$, because of the form of $y(1)$ and $y(2)$, we introduce the change of variables

$$a\alpha'\mu' - \alpha\mu = x, \qquad a'\mu' - \mu = x',$$

with the Jacobian

$$J = \begin{vmatrix} a\alpha' & -\alpha \\ a' & -1 \end{vmatrix} = -a\alpha' + a'\alpha \neq 0$$

and

$$\mu = J^{-1}(a\alpha' x' - a'x), \qquad \mu' = J^{-1}(\alpha x' - x).$$

We have

$$C_7 = \iint_\Omega K_n(J^{-1}(a\alpha' x' - a'x) - \eta) K_n(J^{-1}(\alpha x' - x) - \eta')$$
$$\times f_{a,b}(\alpha' J^{-1}(\alpha x' - x) + w' + 2\pi k) f_{a',b'}(-J^{-1}(\alpha x' - x) + 2\pi k')$$
$$\times \operatorname{sinc}(a, (2\pi ka + x + aw' + b - w) \bmod 2\pi)$$
$$\times \operatorname{sinc}(a', (2\pi k'a' - x' + b') \bmod 2\pi) J^{-1}\, dx\, dx',$$

where Ω is a bounded region. Using Assumption 3 and the fact that K is bounded, there is a constant c such that

$$\begin{aligned} |C_7| &\le \frac{c}{b_n^2} g(x_1 + 2\pi k) g(x_2 + 2\pi k') \\ &\quad \times \int_\Omega |\operatorname{sinc}(a', (2\pi ka + x + aw' + b - w) \bmod 2\pi) \\ &\qquad \times \operatorname{sinc}(a', (2\pi k'a' - x' + b') \bmod 2\pi)|\, dx\, dx' \\ &= O\Big(\frac{\log^2 n}{n^2 b_n^2}\Big) g(x_1 + 2\pi k) g(x_2 + 2\pi k'), \end{aligned} \tag{7.21}$$

with x_1, x_2 some fixed values. Using Assumptions 1 and 2, the contribution to C_5 is shown to be $O(\log^2 n/(n^2 b_n^2))$.

(ii) If $a\alpha' = a'\alpha$, then, by a change of variable,

$$x = a\alpha'\mu' - \alpha\mu, \qquad \mu = \frac{1}{\alpha}(a\alpha'\mu' - x) = a'\mu' - \frac{1}{\alpha}x,$$

$$C_7 = -\frac{1}{\alpha} \int_{-\pi^+}^{\pi^+} \int_A K_n\Big(-\frac{1}{\alpha}x + a'\mu' - \eta\Big) K_n(\mu' - \eta') f_{a,b}(\alpha'\mu' + w' + 2\pi k)$$
$$\times f_{a',b'}(-\mu' + 2\pi k') \operatorname{sinc}(a, (x + y_{k,1})_{2\pi})$$
$$\times \operatorname{sinc}\Big(a', \Big(\frac{-1}{\alpha}(x - \alpha y_{k',2})\Big)_{2\pi}\Big) I_{A(\mu')}(x)\, dx\, d\mu',$$

where

$$y_{k,1} = 2\pi ka + aw' + b - w, \qquad y_{k',2} = 2\pi k'a' + b',$$

with

$$A = [-(a\alpha' + \alpha)\pi^+, (a\alpha' + \alpha)\pi^+], \qquad A(\mu') = [a\alpha'\mu' - \alpha\pi^+, a\alpha'\mu' + \alpha\pi^+].$$

Let $x' = (\mu' - \eta')b_n^{-1}$, $\mu' = \eta' + b_n x'$. Then

$$C_7 = -\frac{1}{\alpha b_n} \int_{A_1} \int_A K\Big(a'x' + \frac{1}{b_n}\Big(a'\eta' - \eta - \frac{1}{\alpha}x\Big)\Big) K(x')$$
$$\times f_{a,b}(\alpha'\eta' + \alpha' b_n x' + w' + 2\pi k) f_{a',b'}(-\eta' - b_n x' + 2\pi k')$$

$$\times \operatorname{sinc}(a, (x + y_{k,1})_{2\pi}) \operatorname{sinc}\left(a', -\left(\frac{1}{\alpha}(x - \alpha y_{k',2})\right)_{2\pi}\right)$$

$$\times I_{A(\eta' + b_n x')}(x)\, dx\, dx',$$

with $A_1 = [(-\pi^+ - \eta')b_n^{-1}, (\pi^+ - \eta')b_n^{-1}]$. Let $z = x + y_{k,1}$, $x = z - y_{k,1}$ [we note that since $(x + y_{k,1})_{2\pi} = (x + (y_{k,1})_{2\pi})_{2\pi}$ we can replace $y_{k,1}$ by $(y_{k,1})_{2\pi}$ in the following]

$$C_7 = \frac{-1}{\alpha b_n} \int_{A_1} \int_{A'} K\left(a'x' + \frac{1}{b_n}\left(a'\eta' - \eta + \frac{1}{\alpha} y_{k,1} - \frac{1}{\alpha} z\right)\right) K(x')$$

$$\times f_{a,b}(\alpha'\eta' + w' + 2\pi k + b_n \alpha' x') f_{a',b'}(-\eta' + 2\pi k' - b_n x')$$

$$\times \operatorname{sinc}(a, z_{2\pi}) \operatorname{sinc}\left(a', -\left(\frac{1}{\alpha}(z - y_{k,1} - \alpha y_{k',2})\right)_{2\pi}\right)$$

$$\times I_{A(\eta' + b_n x')}(z - y_{k,1})\, dz\, dx',$$

where $A' = A + y_{k,1}$.

Let $\mathcal{L}' = \{l \mid 2\pi l \in A', l \text{ integers}\}$ which is a finite set.

For each $l \in \mathcal{L}'$, let $A^l = [-\pi + 2\pi l, \pi + 2\pi l] \cap A'$. Then

$$C_7 = \frac{-1}{\alpha b_n} \sum_{l \in \mathcal{L}'} \int_{A_1} \int_{A^l} K\left(a'x' + \frac{1}{b_n}\left(a'\eta' - \eta + \frac{1}{\alpha} y_{k,1} - \frac{1}{\alpha} z\right)\right) K(x')$$

$$\times f_{a,b}(\alpha'\eta' + \alpha' b_n x' + w' + 2\pi k) f_{a',b'}(-\eta' - b_n x' + 2\pi k')$$

$$\times \operatorname{sinc}(a, z_{2\pi}) \operatorname{sinc}\left(a', -\left(\frac{1}{\alpha}(z - y_{k,1} - \alpha y_{k',2})\right)_{2\pi}\right)$$

$$\times I_{A(\eta' + b_n x')}(z - y_{k,1})\, dz\, dx'.$$

Letting $z' = z - 2\pi l$,

$$C_7 = \sum_{l \in \mathcal{L}'} \frac{-1}{\alpha b_n} G_l, \tag{7.22}$$

with

$$G_l = \int_{A_1} \int_{A^l - 2\pi l} K\left(a'x' + \frac{1}{b_n}\left(a'\eta' - \eta + \frac{1}{\alpha} y_{k,1} - \frac{1}{\alpha} 2\pi l - \frac{1}{\alpha} z'\right)\right) K(x')$$

$$\times f_{a,b}(\alpha'\eta' + w' + 2\pi k + b_n \alpha' x') f_{a',b'}(-\eta' + 2\pi k' - b_n x')$$

$$\times \operatorname{sinc}(a, z') \operatorname{sinc}\left(a', -\left(\frac{1}{\alpha}(z' + 2\pi l - y_{k,1} - \alpha y_{k',2})\right)_{2\pi}\right)$$

$$\times I_{A(\eta' + b_n x')}(z' + 2\pi l - y_{k,1})\, dz'\, dx'.$$

We note that $A_c^l \equiv A^l - 2\pi l$ is centered on 0 with length at most 2π. For each $l \in \mathcal{L}'$, rewriting this last expression with $q_l \equiv a'\eta' - \eta + \frac{1}{\alpha}y_{k,1} - \frac{1}{\alpha}2\pi l \neq 0$ and $J(l, z', x') = I_{A(\eta'+b_n x')}(z' + 2\pi l - y_{k,1})$, we have

$$\begin{aligned} G_l = \int_{A_1}\int_{A_c^l} & K\left(a'x' + \frac{1}{b_n}\left(q_l - \frac{1}{\alpha}z'\right)\right)K(x') \\ & \times f_{a,b}(\alpha'\eta' + w' + 2\pi k + b_n\alpha'x')f_{a',b'}(-\eta' + 2\pi k' - b_n x') \\ & \times \operatorname{sinc}(a, z')\operatorname{sinc}\left(a', -\left(\frac{1}{\alpha}(z' + 2\pi l - y_{k,1} - \alpha y_{k',2})\right)_{2\pi}\right) \\ & \times J(l, z', x')\,dz'\,dx'. \end{aligned}$$

Let e_n be a sequence such that

$$e_n = o(b_n), \qquad n^{-1} = o(e_n), \qquad n^{-1} = o(b_n), \qquad c_0, c_1 \text{ constants.}$$

Then there exist constants c_0, x_1 and x_2 such that

$$\begin{aligned} |G_l| \le {} & c_0 g(x_1 + 2\pi k)g(x_2 + 2\pi k') \\ & \times \int_{A_1}\left[\int_{|z'|\le c_1 e_n} + \int_{\pi \ge |z'| > c_1 e_n}\right]|G(z', x')|\,dz', \end{aligned} \tag{7.23}$$

where $G(z', x') = G'(z', x')G''(z')$, $G'(z', x') = K(a'x' + (1/b_n)(q_l - (1/\alpha)z'))$ $K(x')$,

$$G''(z') = \operatorname{sinc}(a, z')\operatorname{sinc}\left(a', -\left(\frac{1}{\alpha}(z' + 2\pi l - y_{k,1} - \alpha y_{k',2})\right)_{2\pi}\right).$$

We see that the first double integral in (7.23) is $\int_{|z'|\le c_1 e_n}(\int_{A_1}|G'(z', x')|\,dx')$ $\times|G''(z')|\,dz'$ and $\int_{A_1}|G'(z', x')|\,dx' = o(1)$ uniformly in z' since $e_n = o(b_n)$ and $(1/b_n)(q_l - \frac{1}{\alpha}z') \to \infty$. (We can also use the fact that x' has finite range or K has finite support.) Note that $\int_{|z'|\le ce_n}|G''(z')|\,dz' = O(\frac{1}{n})$ by the Schwarz inequality. The second double integral in (7.23) is $\int_{c_1 e_n \le |z'| \le \pi}(|G''(z')|\int_{A_1}|G'(x', x')|\,dx')$ $\times\,dz'$. But

$$\int_{A_1}|G'(z', x')|\,dx' \le |a'|^{-1/2}\int K^2(y)\,dy$$

by the Schwarz inequality and

$$\begin{aligned} & \int_{c_1 e_n \le |z'| \le \pi}|G''(z')|\,dz' \\ & = O\left(\frac{1}{n^2}\right)\int_{c_1 e_n \le |z'| \le \pi}\frac{\sin(((n+1)/2)z'l(a))}{z'} \end{aligned}$$

$$\times \frac{\sin\{((n+1/2)[-(1/\alpha)(z'+2\pi l-y_{k,1}-\alpha y_{k',2})]_{2\pi} l(a')\}}{(-(1/\alpha)(z'+2\pi l-y_{k,1}-\alpha y_{k',2}))_{2\pi}}\,dz'$$

$$= o\left(\frac{1}{n}\right).$$

Therefore, when $q_l \neq 0$, $|G_l| = o(1/n)g(x_1+2\pi k)g(x_2+2\pi k')$. In this case, by Assumptions 1 and 2, the contribution from C_7 to C_5 is shown to be $o(1/nb_n)$.

If $q_l = a'\eta' - \eta + \frac{1}{\alpha}y_{k,1} - \frac{1}{\alpha}2\pi l = 0$,

$$\begin{aligned} G_l = \int_{A_1}\int_{A_c^l} & K\left(a'x' - \frac{1}{\alpha}z'/b_n\right)K(x') \\ & \times f_{a,b}(\alpha'\eta' + w' + 2\pi k + b_n\alpha' x') f_{a',b'}(-\eta' + 2\pi k' - b_n x') \\ & \times J(l,z',x')G''(z')\,dz'\,dx' \\ = \Bigg[\int_{A_1}\int_{\{|z'|\le c_1 e_n\}\cap A_c^l} & + \int_{A_1}\int_{\{|z'|\ge c_1 e_n\}\cap A_c^l}\Bigg] G'''(z',x')G''(z') \\ & \times J(l,z',x')\,dz'\,dx', \end{aligned} \tag{7.24}$$

where

$$\begin{aligned} G'''(z',x') = K\left(a'x' - \frac{1}{\alpha}z'/b_n\right)K(x') f_{a,b}(\alpha'\eta' + w' + 2\pi k + b_n\alpha' x') \\ \times f_{a',b'}(-\eta' + 2\pi k' - b_n x'). \end{aligned}$$

Using the same argument as before, we see that the second double integral in (7.24) is $o(1/n)g(x_1+2\pi k)g(x_2+2\pi k')$. Its total contribution to C_5 is $o(1/nb_n)$. The first double integral can be approximated by $G_{l,1}$ given below using the fact that $|z'| \le c_1 e_n$, $e_n = o(b_n)$ and Lemma 7.3 (at the end of this section) with $y = x'$ and $z = z'$, noting that G'' and J are bounded.

$$\begin{aligned} G_{l,1} \equiv \int_{\{|z'|\le c_1 e_n\}\cap A_c^l}\int_{A_1} & K(a'x')K(x') f_{a,b}(a'\eta' + w' + 2\pi k + b_n\alpha' x') \\ & \times f_{a',b'}(-\eta' + 2\pi k' - b_n x') I_{A(\eta'+b_n x')}(z' + 2\pi l - y_{k,1}) G''(z')\,dx'\,dz'. \end{aligned}$$

The weight function K has finite support A_K and by Assumption 2 we have

$$\begin{aligned} G_{l,1} = \int_{A_K} & K(a'x')K(x')\,dx' f_{a,b}(a'\eta' + w' + 2\pi k) f_{a',b'}(-\eta' + 2\pi k') \\ & \times \int_{\{|z'|\le c_1 e_n\}\cap A_c^l} \operatorname{sinc}(a,z')\operatorname{sinc}\left(a', -\left(\frac{1}{\alpha}(z' + 2\pi l - y_{k,1} - \alpha y_{k',2})\right)_{2\pi}\right) \\ & \times I_{A(\eta'+b_n x')}(z' + 2\pi l - y_{k,1})\,dz'\,dx' \end{aligned}$$

$$+ O(b_n)g(a'\eta'^* + w' + 2\pi k)g(-\eta'^* + 2\pi k') \int_{\{|z'| \le c_1 e_n\} \cap A_c^l} |G''(z')|\, dz'$$

$$\equiv G'_{l,1} + G''_{l,1}.$$

It is clear that $|G''_{l,1}| \le O(b_n)g(a'\eta'^* + w' + 2\pi k)g(-\eta'^* + 2\pi k')O(\frac{1}{n})$ and its total contribution to C_5 is $O(\frac{1}{n})$. To evaluate $G'_{l,1}$, let

$$A_3 \equiv [a\alpha'\eta' - \alpha\pi^+ - 2\pi l + y_{k,1}, a\alpha'\eta' + \alpha\pi^+ - 2\pi l + y_{k,1}].$$

Note that

$$I_{A(\eta'+b_n x')}(z' + 2\pi l - y_{k,1}) = I_{A'_3}(z'),$$

with

$$A'_3 = \big[a\alpha'(\eta' + b_n x') - \alpha\pi^+ - 2\pi l + y_{k,1}, a\alpha'(\eta' + b_n x') + \alpha\pi^+ - 2\pi l + y_{k,2}\big].$$

The condition $q_l = 0$ means $a'\alpha\eta' - \alpha\eta + y_{k,1} - 2\pi l = 0$. This means $a\alpha'\eta' - \alpha\eta + y_{k,1} - 2\pi l = 0$ by the condition $a'\alpha = a\alpha'$. Now $\eta \in [-\pi, \pi]$, $\pi^+ > \pi$, x' is bounded. Therefore $0 \in A_3^0$, the interior of A_3, and then $I_{A_3}(z') = 1$ when $|z'| \le c_1 e_n$ for large n. Also $q_l = 0$ implies $0 \in A_c^l$, so, for large n, the inner integral of $G'_{l,1}$ is

$$G^4 \equiv \int_{|z'| \le c_1 e_n} \operatorname{sinc}(a, z') \operatorname{sinc}\left(a', -\left(\frac{1}{\alpha}(z' + 2\pi l - y_{k,1} - \alpha y_{k',2})\right)_{2\pi}\right) dz'.$$

If $q'_l \equiv (\frac{1}{\alpha} y_{k,1} + y_{k',2} - \frac{1}{\alpha} 2\pi l)_{2\pi} \ne 0$, then $G^4 = o(\frac{1}{n})$. If $q'_l = 0$, then, for large n,

$$G^4 = \int_{|z'| \le c_1 e_n} \operatorname{sinc}(a, z') \operatorname{sinc}\left(a', -\frac{1}{\alpha} z'\right) dz'$$

$$= -\alpha \int_{|x| \le c_1 e_n / \alpha} \operatorname{sinc}(a, -\alpha x) \operatorname{sinc}(a', x)\, dx.$$

Following an argument like that leading to (7.7)–(7.12) and using the Parseval identity, we have

$$\int_{|x| \le c_1 e_n/\alpha} \operatorname{sinc}(a, -\alpha x) \operatorname{sinc}(a', x)\, dx = \frac{2\pi}{n+1} l(\alpha, \alpha', a, a') + o\left(\frac{1}{n}\right)$$

with

$$l(\alpha, \alpha', a, a') = \min\left(\alpha' \frac{\min(a, 1)}{\min(a', 1)}, 1\right) \frac{\min(a', 1)}{a\alpha'}.$$

In this case,

$$G_{l,1} = f_{a,b}(a'\eta' + w' + 2\pi k) f_{a',b'}(-\eta' + 2\pi k') \int_{A_K} K(a'x') K(x')\, dx'$$

$$\times \left[\frac{-\alpha 2\pi}{n+1} l(\alpha, \alpha', a, a') + o\left(\frac{1}{n}\right)\right].$$

In summary, under the assumption $a\alpha' = a'\alpha$,

$$
\begin{aligned}
C_7 = {}& \frac{1}{b_n} \sum_{l \in \mathcal{L}''} f_{a,b}(a'\eta' + w' + 2\pi k) f_{a',b'}(-\eta' + 2\pi k') \\
&\times \left[\int K(a'x') K(x')\, dx' \frac{2\pi}{n+1} l(\alpha, \alpha', a, a') + o\left(\frac{1}{n}\right) \right],
\end{aligned}
\tag{7.25}
$$

where, for each k, k', a, b, a', b', $\mathcal{L}'' = \mathcal{L}' \cap \mathcal{L}_1' \cap \mathcal{L}_2'$ with

$$\mathcal{L}' = \{ l \mid 2\pi l \in A' \equiv [-(a\alpha' + \alpha)\pi^+, (a\alpha + \alpha)\pi^+] + y_{k,1} \},$$

$$y_{k,1} = 2\pi k a + a w' + b - w,$$

$$\mathcal{L}_1' = \left\{ l \,\middle|\, q_l \equiv a'\eta' - \eta + \frac{1}{\alpha}(y_{k,1} - 2\pi l) = 0 \right\},$$

$$\mathcal{L}_2' = \left\{ l \,\middle|\, q_l' = \left(\frac{1}{\alpha}(y_{k,1} - 2\pi l) + y_{k',2} \right)_{2\pi} = 0 \right\}$$

and

$$y_{k',2} = 2\pi k' a' + b'.$$

However, we note that if there is an $l \in \mathcal{L}_1'$ then $l \in \mathcal{L}'$ and the condition $q_l = 0$ is

$$
\begin{aligned}
0 &= \alpha a'\eta' - \alpha\eta + 2\pi k a + a w' + b - w - 2\pi l \\
&= 2\pi k a + b + a(\alpha'\eta' + w') - (\alpha\eta + w) - 2\pi l
\end{aligned}
$$

by the condition $a\alpha' = \alpha a'$. This is the same as $y(k, a, b; \xi, \xi') = 0 \bmod 2\pi$. Substituting $q_l = 0$ into $q_l' = 0$ in $\mathcal{L}_2'$, we have $(-a'\eta' + \eta + y_{k',2})_{2\pi} = 0$. This is the same as $y(k', a', b', -\eta, -\eta') = 0 \bmod 2\pi$. Therefore the sum over $l \in \mathcal{L}''$ in (7.25) can be replaced by the restriction

$$\delta_{a\alpha' - a'\alpha, 0}\, \delta_{(y(k; a, b; \xi, \xi')), 0}\, \delta_{(y(k'; a', b'; -\eta, -\eta')), 0}.$$

Substituting the modified (7.25) into (7.20) and (7.18), we have C_5 in (7.17) and the first term in (5.8). By a similar argument for C_6 in (7.19), we obtain the second term in (5.8). This completes the proof. □

LEMMA 7.3. *If $K(x) \in L_\infty \cap L_1$ and*

$$K(x) \to 0 \quad \text{as } |x| \to \infty,$$

then, for $e \neq 0$,

$$\int (K(ey + z) - K(ey)) K(y) g(y, z)\, dy \to 0 \quad \text{as } z \to 0$$

for a bounded function g.

PROOF. Let $|g| \leq c$. Then

$$\int (K(ey+z) - K(ey))g(y,z)K(y)\,dy$$

$$= \left[\int_{|y|\geq L} + \int_{|y|\leq L}\right](K(y)g(y,z)(K(ey+z) - K(ey)))\,dy.$$

Given any $\varepsilon > 0$, since $K(x) \in L_1$ and $K(x) \to 0$ as $|x| \to \infty$ there exists an $L > 0$ such that

$$\left|\int_{|y|\geq L} g(y,z)K(y)(K(ey+z) - K(ey))\,dy\right| < \varepsilon/2.$$

By Lusin's theorem, there exists a continuous function $K_{c,\varepsilon}(y)$ such that

$$\int_{|y|\leq L} |K(ey) - K_{c,\varepsilon}(ey)|^2\,dy \leq \frac{\varepsilon^2}{36M^2},$$

where $M^2 = c^2 \int K^2(y)\,dy$. Then

$$\left|\int_{|y|\leq L} g(y,z)K(y)(K(ey) - K(ey+z))\right|dy$$

$$\leq \left[\int_{|y|\leq L} g^2(y,z)K^2(y)\,dy \int_{|y|\leq L} (K(ey) - K(ey+z))^2\,dy\right]^{1/2}$$

$$= M\left\{\int_{|y|\leq L} [K(ey) - K_{c,\varepsilon}(ey) + K_{c,\varepsilon}(ey) - K_{c,\varepsilon}(ey+z) \right.$$

$$\left. + K_{c,\varepsilon}(ey+z) - K(ey+z)]^2\,dy\right\}^{1/2}$$

$$\leq \sqrt{3}M\left\{\int_{|y|\leq L} [K(ey) - K_{c,\varepsilon}(ey)]^2\,dy\right.$$

$$+ \int_{|y|\leq L} [K_{c,\varepsilon}(ey) - K_{c,\varepsilon}(ey+z)]^2\,dy$$

$$\left. + \int_{|y|\leq L} [K_{c,\varepsilon}(ey+z) - K(ey+z)]^2\right\}^{1/2}$$

$$\leq \sqrt{3}M\left[\frac{\varepsilon^2}{36M^2} + \frac{\varepsilon^2}{36M^2} + \frac{\varepsilon^2}{36M^2}\right]^{1/2} = \frac{\varepsilon}{2}$$

since

$$\int_{|y|\leq L} (K_{c,\varepsilon}(ey) - K_{c,\varepsilon}(ey+z))^2\,dy \leq \frac{\varepsilon^2}{36M^2}$$

for z sufficiently small by the continuity of $K_{c,\varepsilon}(u)$ for $|u| \leq 2L$.

Therefore, for any $\varepsilon > 0$, one can choose z sufficiently small such that

$$\left| \int g(y,z)K(y)\bigl(K(ey) - K(ey+z)\bigr)\,dy \right| \le \varepsilon$$

and the lemma is proved. □

REFERENCES

ALEKSEEV, V. G. (1988). Estimating the spectral densities of a Gaussian periodically correlated stochastic process. *Problems Inform. Transmission* **24** 109–115.

ALLEN, J. and HOBBS, S. (1992). Detecting target motion by frequency-plane smoothing. In *Asilomar Conference on Signals, Systems and Computers* (A. Singh, ed.) 1042–1047.

BOHR, H. A. (1951). *Almost Periodic Functions*. Chelsea, New York.

CHIU, S.-T. (1986). Statistical estimation of the parameters of a moving source from array data. *Ann. Statist.* **14** 559–578.

DAHLHAUS, R. (1997). Fitting time series models to nonstationary processes. *Ann. Statist.* **25** 1–37.

DANDAWATE, A. V. and GIANNAKIS, G. B. (1994). Nonparametric polyspectral estimators for kth-order (almost) cyclostationary processes. *IEEE Trans. Inform. Theory* **40** 67–84.

DANIELL, P. J. (1946). Discussion of paper by M. S. Bartlett. *J. Roy. Statist. Soc.* (*Suppl.*) **8** 88–90.

DEHAY, D. and HURD, H. L. (1993). Representation and estimation for periodically and almost periodically correlated random processes. In *Cyclostationarity in Communications and Signal Processing* (W. A. Gardner, ed.). IEEE Press, New York.

FERGUSON, B. G. (1999). Time-delay estimation techniques applied to the acoustic detection of jet aircraft transits. *J. Acoust. Soc. Amer.* **106** 255–264.

GARDNER, W. A. (1991). Exploitation of spectral redundancy in cyclostationary signals. *IEEE Signal Processing Magazine* **8** 14–36.

GARDNER, W. A. (1994). *Cyclostationarity in Communications and Signal Processing*. IEEE Press, New York.

GERR, N. and ALLEN, J. (1994a). The generalized spectrum and spectral coherence of a harmonizable time series. *Digital Signal Processing* **4** 222–238.

GERR, N. and ALLEN, J. (1994b). Time-delay estimation for harmonizable signals. *Digital Signal Processing* **4** 49–62.

GLADYSHEV, E. G. (1963). Periodically and almost periodically correlated random processes with continuous time parameters. *Theory Probab. Appl.* **8** 173–177.

HURD, H. (1989). Nonparametric time series analysis for periodically correlated processes. *IEEE Trans. Inform. Theory* **35** 350–359.

HURD, H. and GERR, N. (1991). Graphical methods for determining the presence of periodic correlation. *J. Time Ser. Anal.* **12** 337–350.

JIN, Q., WONG, K. M. and LUO, Z. Q. (1995). The estimation of time delay and Doppler stretch of wideband signals. *IEEE Trans. Signal Processing* **43** 904–916.

LESKOW, J. and WERON, A. (1992). Ergodic behavior and estimation for periodically correlated processes. *Statist. Probab. Lett.* **15** 299–304.

LII, K. S. and ROSENBLATT, M. (1998). Line spectral analysis for harmonizable processes. *Proc. Nat. Acad. Sci. USA* **95** 4800–4803.

LOÈVE, M. (1963). *Probability Theory*, 3rd ed. Van Nostrand, Princeton, NJ.

MALLAT, S., PAPANICOLAOU, G. and ZHANG, Z. (1998). Adaptive covariance estimation of locally stationary processes. *Ann. Statist.* **26** 1–47.

MUNK, W., WORCESTER, P. and WUNSCH, C. (1995). *Ocean Acoustic Tomography*. Cambridge Univ. Press.

NEUMANN, M. H. and VON SACHS, R. (1997). Wavelet thresholding in anisotropic function classes and application to adaptive estimation of evolutionary spectra. *Ann. Statist.* **25** 38–76.

YAGLOM, A. M. (1987). *Correlation Theory of Stationary and Related Random Functions* **1**, **2**. Springer, New York.

ZURBENKO, I. G. (1986). *The Spectral Analysis of Time Series*. North-Holland, New York.

DEPARTMENT OF STATISTICS
UNIVERSITY OF CALIFORNIA, RIVERSIDE
RIVERSIDE, CALIFORNIA 92521
E-MAIL: ksl@gauss.ucr.edu

DEPARTMENT OF MATHEMATICS
UNIVERSITY OF CALIFORNIA, SAN DIEGO
LA JOLLA, CALIFORNIA 92093
E-MAIL: mrosenblatt@ucsd.edu

The Annals of Statistics
2003, Vol. 31, No. 5, 1693

CORRECTION

SPECTRAL ANALYSIS FOR HARMONIZABLE PROCESSES

BY KEH-SHIN LII AND MURRAY ROSENBLATT

The Annals of Statistics (2002) **30** 258–297

In the statement of Theorem 5.3, page 273, $f_{a,b}(a'\eta' + \omega' + 2\pi k)$ should be replaced by $f_{a,b}(\alpha'\eta' + \omega' + 2\pi k)$ and $f_{a',b'}(a'\eta' + \omega' + 2\pi k')$ by $f_{a',b'}(\alpha'\eta' + \omega' + 2\pi k')$. The theorem holds in the non-Gaussian case if

$$\sup_t \sum_{\tau,t',\tau'} |\text{cum}(X_t, X_\tau, X_{t'}, X_{\tau'})| < \infty.$$

On the same page $f_{1,0}(\eta + \omega)$ should be replaced by $f_{1,0}(\alpha\eta + \omega)\ell(\alpha)$.

From page 292 on, wherever $a'\eta' + \omega'$ occurs it should be replaced by $\alpha'\eta' + \omega'$.

DEPARTMENT OF STATISTICS
UNIVERSITY OF CALIFORNIA, RIVERSIDE
RIVERSIDE, CALIFORNIA 92521
E-MAIL: ksl@gauss.ucr.edu

DEPARTMENT OF MATHEMATICS
UNIVERSITY OF CALIFORNIA, SAN DIEGO
LA JOLLA, CALIFORNIA 92093
E-MAIL: mrosenblatt@ucsd.edu

Received December 2002.

The Annals of Statistics
2006, Vol. 34, No. 3, 1115–1139
DOI: 10.1214/009053606000000218

ESTIMATION FOR ALMOST PERIODIC PROCESSES

BY KEH-SHIN LII AND MURRAY ROSENBLATT

University of California, Riverside and University of California, San Diego

Processes with almost periodic covariance functions have spectral mass on lines parallel to the diagonal in the two-dimensional spectral plane. Methods have been given for estimation of spectral mass on the lines of spectral concentration if the locations of the lines are known. Here methods for estimating the intercepts of the lines of spectral concentration in the Gaussian case are given under appropriate conditions. The methods determine rates of convergence sufficiently fast as the sample size $n \to \infty$ so that the spectral estimation on the estimated lines can then proceed effectively. This task involves bounding the maximum of an interesting class of non-Gaussian possibly nonstationary processes.

1. Introduction. The main objective of this paper is to present a constructive method to determine the lines of support of spectra or, equivalently, the frequencies or periods of Gaussian harmonizable processes with almost periodic covariances under appropriate conditions. The processes considered provide an interesting set of random processes that are generally not transient and not stationary and for which Fourier methods of analysis are helpful and meaningful. The lines of support of spectra are lines parallel to the diagonal in the two-dimensional spectral plane. A number of papers have discussed spectral estimation in this context but they all assume knowledge of the lines of support of the spectra. The novelty of this paper is in the presentation of *constructive methods for estimating* the lines of support under appropriate conditions with a rate of convergence good enough to imply bias and covariance of spectral estimation using the estimated support lines with the same asymptotic properties as if the actual lines of support were known precisely.

Almost periodically correlated processes have been considered by Alekseev [1], Hurd [10], Gardner [5], Gladyshev [9], Gerr and Allen [7, 8], Dandawate and Giannakis [3], Leskow and Weron [14] and Lii and Rosenblatt [15]. Extensive application of periodically correlated (or cyclostationary) processes is noted in [6] and references therein. For the sake of simplicity and economy, we limit ourselves to a discrete time parameter process $\{X_n, n = \dots, -1, 0, 1, \dots\}$ with mean $EX_n \equiv 0$. The process $\{X_n\}$ with covariance function $r_{n,m} = EX_n\overline{X}_m$ is said to be

Received February 2004; revised May 2005.

AMS 2000 subject classifications. Primary 62M15, 62M10; secondary 62G05, 62M99.

Key words and phrases. Almost periodic covariance, spectral estimation, periodogram, process maximum, Gaussian and non-Gaussian process, frequency detection and estimation.

R.A. Davis et al. (eds.), *Selected Works of Murray Rosenblatt*, Selected Works in Probability and Statistics, DOI 10.1007/978-1-4419-8339-8_39, © Springer Science+Business Media, LLC 2011

periodically correlated with period T if

$$r_{n+T,m+T} = r_{n,m}. \tag{1.1}$$

We shall call $\{X_n\}$ a process with *almost periodically correlated covariance function* if for each (s,τ), there are functions $\alpha_j(s,\tau)$ and real $\lambda_j(s,\tau)$ with $\sum_j |\alpha_j(s,\tau)| < \infty$ such that, for each s, τ, $r_{s+t,\tau+t} = \sum_j \alpha_j(s,\tau)e^{i\lambda_j(s,\tau)t}$ with the convergence uniform in s, τ. We shall call a sequence $\{a(n)\}$ almost periodic if there are sequences $\{\alpha_j\}$, real $\{\beta_j\}$ such that $a(n) = \sum_j \alpha_j e^{i\beta_j n}$ with $\sum_j |\alpha_j| < \infty$.

The process $\{X_n\}$ is harmonizable (in the sense of Loève) if it has a Fourier representation in mean square,

$$X_n = \int_{-\pi}^{\pi} e^{in\lambda}\, dZ(\lambda), \tag{1.2}$$

with the random spectral function having a covariance

$$\begin{gathered}\operatorname{cov}\big(Z(\lambda), Z(\mu)\big) = F(\lambda,\mu),\\ r(n,m) = \int_{-\pi}^{\pi}\int_{-\pi}^{\pi} e^{in\lambda - im\mu}\, dF(\lambda,\mu), \qquad \int_{-\pi}^{\pi}\int_{-\pi}^{\pi} |dF(\lambda,\mu)| < \infty\end{gathered} \tag{1.3}$$

[16, 20]. Almost periodicity of the covariance function of a harmonizable process implies that F has mass (complex) on at most a countable number of diagonal lines

$$\lambda = \mu + b, \qquad b = b_j, \qquad j = \dots, -1, 0, 1, \dots.$$

If the process is real-valued (as we assume to be the case) and $\lambda = \mu + b$ is a line of spectral support, then so is $\lambda = \mu - b$.

If the spectra on the lines of support of F are absolutely continuous with respect to Lebesgue measure with spectral densities $f_b(\mu)$, the assumption that $\{X_n\}$ is real-valued implies that

$$f_b(\mu) = \bar{f}_{-b}(\mu + b) = \bar{f}_{-b}(-\mu) = f_b(-\mu - b). \tag{1.4}$$

Also,

$$|f_b(\mu)|^2 \le f_0(\mu) f_0(\mu + b). \tag{1.5}$$

Let us call $\mathcal{L}$ the collection of b's corresponding to lines with nontrivial spectral support. In [3], the spectra $f_b(\mu)$ are estimated under appropriate conditions assuming that the elements of $\mathcal{L}$ are known.

We consider estimation of $b \in \mathcal{L}$ under the assumption that $\mathcal{L}$ is finite and $\int_{-\pi}^{\pi} f_b(\lambda)\, d\lambda \neq 0$ and see whether the final estimates of $f_b(\mu)$ based on these estimated elements of $\mathcal{L}$ produce an effect.

At this point, a few additional remarks are made on related papers. Tian [18] considers time domain estimation of the integral period of a sequence with periodically correlated covariance function observed at integral time. Hurd and Gerr [11] and Gerr and Allen [7] use the coherence between spectral components to test for a specific period of a periodically correlated covariance function against the null hypothesis of stationarity. They used the integrated coherence along diagonal spectral support lines. Lund et al. [17] use an averaged squared coherence statistic to test existence of periodic correlation against stationarity. The case of a continuous parameter process $\{X(\cdot)\}$ is considered in [4] and [12].

Before the main discussion we describe a simple but interesting class of processes with almost periodic covariance function. Additional examples can be found in [6] and [20]. Set

$$X_n = \sum_{j=1}^{m} \cos(w_j n) Y^{(j)}_{n-\tau_j}, \qquad 0 \le w_j \le \pi, \tag{1.6}$$

where the processes $\{Y_n^{(j)}\}$ are jointly stationary with means zero and the cross-spectral density of $Y^{(j)}(\cdot)$ and $Y^{(k)}(\cdot)$ is $f_{jk}(\cdot)$. This commonly used model in communication is a mixture of amplitude modulated (AM) signals with carrier frequency w_j modulated by the signal $Y_n^{(j)}$ with possible relative delay τ_j. If the stationary processes $\{Y_n^{(j)}\}$ have the spectral representation

$$Y_n^{(j)} = \int_{-\pi}^{\pi} e^{in\lambda}\, dZ_j(\lambda), \tag{1.7}$$

the process $\{X_n\}$ is given by $X_n = \int_{-\pi}^{\pi} e^{in\lambda}\, dZ(\lambda)$ with

$$dZ(\eta) = \frac{1}{2}\sum_j [e^{-i\tau_j(\eta-w_j)}\, dZ_j(\eta - w_j) + e^{-i\tau_j(\eta+w_j)}\, dZ_j(\eta + w_j)]. \tag{1.8}$$

Since the vector of processes $\{Z_j(\lambda)\}$ has orthogonal increments, the spectral mass of the process $\{X_n\}$ is given by

$$E\, dZ(\lambda)\overline{dZ(\mu)} = \begin{cases} \frac{1}{4}\sum_j \{f_{jj}(\lambda - w_j) + f_{jj}(\lambda + w_j)\}\, d\lambda, & \text{on } \lambda = \mu, \\ \frac{1}{4}\left\{\sum_{j,k} e^{-i(\tau_j-\tau_k)(\lambda-w_j)} f_{jk}(\lambda - w_j) \right. \\ \qquad \left. + \sum_{j,k}{}' e^{-i(\tau_j-\tau_k)(\lambda+w_j)} f_{jk}(\lambda + w_j)\right\} d\lambda, \\ \qquad \text{where the first sum is over } (j,k) \text{ such that} \\ \qquad \lambda - \mu = w_j \pm w_k = \eta \neq 0 \\ \qquad \text{and the second primed sum over } (j,k) \text{ such that} \\ \qquad \lambda - \mu = -w_j \pm w_k = \eta \neq 0, \\ 0, \quad \text{otherwise.} \end{cases} \tag{1.9}$$

The estimation method proposed in this paper can be used to estimate the frequencies w_j, the spectral densities of the $Y^{(j)}$'s and the τ_j's under certain assumptions. We note that the remarks for almost periodic sequences imply that a moving average process

$$X_n = \sum_{j=0}^{m} a_j(n)\xi_{n-j}, \qquad E\xi_n \equiv 0,$$

with the sequences (in n) $\{a_j(n)\}$, $j = 0, 1, \ldots, m$, almost periodic and the process $\{\xi_n\}$ a white noise process, is a process with almost periodic covariance function. That is also true of the autoregressive scheme $X_n - a_n X_{n-1} = \xi_n$ if $\{a_n = \sum_j \alpha_j e^{i\beta_j n}\}$ is an almost periodic sequence with $\sum_j |\alpha_j| < 1$ (see also [13]).

We note that in meteorological, astronomical and engineering applications questions of periodic or almost periodic covariances are often of interest (see [17] and [20] for examples). The organization of the rest of the paper is as follows. In the next section notation and assumptions are given. The main result on the estimation of intercepts of almost periodic spectral support lines using an integrated periodogram [see (2.6)] is presented after Corollary 4. The estimation method is based on Theorem 4. Theorems 5 and 7 show that the asymptotic bias and variance of spectral estimates on the estimated lines are the same as those when the support lines are known. Section 3 contains a simulated example to demonstrate the effectiveness of the methods. The Appendix contains most of the proofs of the results.

2. Results. In this section we introduce notation and assumptions, and state the main results. The principal thread of the derivation will be indicated. However, the detailed proofs of the results will be given in the Appendix. By

$$W = Z \bmod 2\pi, \tag{2.1}$$

it is to be understood that $-\pi < W \le \pi$ and if $\{u\}$ is the integer ℓ with $-1/2 < u - \ell \le 1/2$, then $(Z \bmod 2\pi) = Z - \{Z/(2\pi)\}2\pi$. Consider the finite Fourier transform

$$F_n(\lambda) = \sum_{t=-n/2}^{n/2} X_t e^{-it\lambda} \tag{2.2}$$

and the periodogram

$$I_n(\lambda, \mu) = \frac{1}{2\pi(n+1)} F_n(\lambda)\overline{F_n(\mu)}. \tag{2.3}$$

Here we assume that n is even and t is summed over integers. Let $K(\eta)$ be a continuous nonnegative symmetric weight function with finite support and such that $\int_{-\infty}^{\infty} K(\eta)\,d\eta = 1$. Set $K_n(\eta) = b_n^{-1}K(b_n^{-1}\eta)$ with $b_n \downarrow 0$ and $nb_n \to \infty$ as $n \to \infty$. A plausible estimate of $f_w(\eta)$ is given by

$$\hat{f}_w(\eta) = \int_{-\pi}^{\pi} I_n(\mu + w, \mu)K_n(\mu - \eta)\,d\mu. \tag{2.4}$$

ASSUMPTION 1. $\{X_k\}$, $EX_k = 0$, is a harmonizable Gaussian sequence with almost periodic covariance function. The set $\mathcal{L}$ of b's with nontrivial support is finite, and the $f_b(\mu)$, $b \in \mathcal{L}$, are continuously differentiable on the curve $\lambda = (b + \mu) \bmod 2\pi$ on the torus of points $(\lambda, \mu) \in [-\pi, \pi]^2$, with $(-\pi, \mu)$ identified with (π, μ) and $(\lambda, -\pi)$ with (λ, π).

Now

$$
\begin{aligned}
r_{n,m} &= EX_nX_m = \sum_{b\in\mathcal{L}} \int_{-\pi}^{\pi} e^{in(\mu+b)} e^{-im\mu} f_b(\mu)\, d\mu = \sum_{b\in\mathcal{L}} e^{inb} c_b(n-m), \\
c_b(n) &:= \int_{-\pi}^{\pi} e^{in\lambda} f_b(\lambda)\, d\lambda.
\end{aligned} \tag{2.5}
$$

ASSUMPTION 2. There is a function $c(j) \geq 0$ on the integers with $|c_b(j)| \leq c(j)$ for all $b \in \mathcal{L}$ and $\sum_j c(j) < \infty$. Unless otherwise noted, the summation over j is on the integers.

ASSUMPTION 3. The $f_b(\mu)$, $b \in \mathcal{L}$, are such that $\int_{-\pi}^{\pi} f_b(\mu)\, d\mu \neq 0$.

In general, it is possible that there may be spectral lines b with $\int_{-\pi}^{\pi} f_b(\mu)\, d\mu = 0$ (see [11], page 347). In that case let $\mathcal{L}$ consist only of those b with $\int_{-\pi}^{\pi} f_b(\mu)\, d\mu \neq 0$ and our arguments will be valid for the restricted $\mathcal{L}$. *The local extrema of the absolute value of the integrated periodogram*

$$
U_n(b) := \int_{-\pi}^{\pi} I_n(b+\mu, \mu)\, d\mu = \frac{1}{n+1} \sum_{k=-n/2}^{n/2} X_k^2 e^{-ibk} \tag{2.6}
$$

will be used to estimate the $b \in \mathcal{L}$ under Assumption 3. The fact that (2.6) is a periodogram computed for X_k^2 is due to the uniform weighting in integrating $I_n(b+\mu, \mu)$. With nonuniform weighting, one would get more complicated quadratic forms that might be useful if $\int_{-\pi}^{\pi} f_b(\mu)\, d\mu = 0$.

Theorems 1 and 2 give estimates for $EI_n(\lambda, \mu)$ and $\operatorname{cov}(I_n(\mu + w, \mu), I_n(\mu' + w', \mu'))$ and it is *crucial* that they are given with error terms uniform in the w's and μ's. Let

$$
y(b, w) = (b - w) \bmod 2\pi. \tag{2.7}
$$

It will also be convenient to make use of the following expressions for $b, b' \in \mathcal{L}$ when $\lambda = \mu + w$, $\lambda' = \mu' + w'$:

$$
\begin{aligned}
y(1) &= (\lambda' + b - \lambda) \bmod 2\pi, \\
y(2) &= (-\mu' + \mu + b') \bmod 2\pi, \\
y(3) &= (-\mu' + b - \lambda) \bmod 2\pi, \\
y(4) &= (\lambda' + \mu + b') \bmod 2\pi.
\end{aligned} \tag{2.8}
$$

Also let

$$\operatorname{sinc}(y) = \sin\left(\frac{n+1}{2}y\right) \Big/ \left(\frac{n+1}{2}y\right). \tag{2.9}$$

THEOREM 1. *If* $\lambda = \mu + w$,

$$EI_n(\lambda, \mu) = \sum_{b \in \mathcal{L}} f_b(\mu) \frac{\sin(((n+1)/2)y(b, w))}{((n+1)/2)y(b, w)} + O\left(\frac{\log n}{n}\right) \tag{2.10}$$

with uniformity in w, μ *when Assumption* 1 *is satisfied.*

THEOREM 2. *Assumption* 1 *and* $\lambda = \mu + w$, $\lambda' = \mu' + w'$ *imply that*

$$\begin{aligned} &\operatorname{cov}\big(I_n(\mu + w, \mu), I_n(\mu' + w', \mu')\big) \\ &\quad = \left\{\sum_{b \in \mathcal{L}} f_b(\mu' + w') \operatorname{sinc}(y(1)) + O\left(\frac{\log n}{n}\right)\right\} \\ &\qquad \times \left\{\sum_{b' \in \mathcal{L}} f_{b'}(-\mu') \operatorname{sinc}(y(2)) + O\left(\frac{\log n}{n}\right)\right\} \\ &\qquad + \left\{\sum_{b \in \mathcal{L}} f_b(-\mu') \operatorname{sinc}(y(3)) + O\left(\frac{\log n}{n}\right)\right\} \\ &\qquad \times \left\{\sum_{b' \in \mathcal{L}} f_{b'}(\mu' + w') \operatorname{sinc}(y(4)) + O\left(\frac{\log n}{n}\right)\right\}. \end{aligned} \tag{2.11}$$

Theorem 2 almost immediately implies Corollary 1.

COROLLARY 1. *Assumption* 1 *implies that*

$$\begin{aligned} &\operatorname{cov}(\hat{f}_w(\eta), \hat{f}_{w'}(\eta')) \\ &\quad = \sum_{b, b' \in \mathcal{L}} \int_{-\pi}^{\pi} \int_{-\pi}^{\pi} f_b(\mu' + w') f_{b'}(-\mu') \operatorname{sinc}(y(1)) \operatorname{sinc}(y(2)) K_n(\mu - \eta) \\ &\qquad \times K_n(\mu' - \eta')\, d\mu\, d\mu' \\ &\qquad + \sum_{b, b' \in \mathcal{L}} \int_{-\pi}^{\pi} \int_{-\pi}^{\pi} f_b(-\mu') f_{b'}(\mu' + w') \operatorname{sinc}(y(3)) \operatorname{sinc}(y(4)) \\ &\qquad \times K_n(\mu - \eta) K_n(\mu' - \eta')\, d\mu\, d\mu' \\ &\qquad + O\left(\frac{\log n}{n}\right). \end{aligned} \tag{2.12}$$

Further,

$$\operatorname{cov}(\hat{f}_w(\eta), \hat{f}_{w'}(\eta')) = O\left(\frac{1}{nb_n}\right) + O\left(\frac{\log n}{n}\right). \tag{2.13}$$

Bounds on $\sup_w |U_n(w) - EU_n(w)|$ will be obtained by making use of Theorem 3. Let

$$V_n(w) = (n+1)\{U_n(w) - EU_n(w)\}. \tag{2.14}$$

Cumulants of $V_n(w)$ will be estimated and an argument akin to Brillinger [2] will be used to gauge $\sup_w |V_n(w)|$.

THEOREM 3. *Let $\{X_k\}$ be a Gaussian sequence with almost periodic covariance function satisfying Assumptions* 1 *and* 2. *Then*

$$\limsup_{n\to\infty} \sup_w |V_n(w)|/(n \log n)^{1/2} \leq 2^{5/2} q\left(\sum_j c(j)\right), \tag{2.15}$$

with probability 1 *where q is the number of elements in $\mathcal{L}$.*

The arguments used to prove Theorem 3 will imply the following two corollaries.

COROLLARY 2. *Let $c > 0$. Then under the conditions of Theorem* 3, *there is a constant $C > 0$ such that*

$$P\left[\sup_w |U_n(w) - EU_n(w)| > c\right] \leq \exp[-Cn^{1/2}(\log n)^{1/2}]. \tag{2.16}$$

COROLLARY 3. *If $c > 0$ is fixed but large,*

$$P\left[\sup_w |U_n(w) - EU_n(w)| > 2cn^{-1/2}(\log n)^{1/2}\right] \leq \exp\left\{-\frac{c}{2}\log n\right\}. \tag{2.17}$$

An additional corollary indicates that $\sup_w |V_n(w)|$ diverges as $n \to \infty$.

COROLLARY 4. *Assume that $f_0(\cdot)$ is not identically zero. Then under Assumptions* 1–3, *there are a countable number of distinct w_i in the neighborhood of zero such that, for each finite subcollection $\mathfrak{C}$,*

$$\frac{1}{\sqrt{n+1}} V_n(w_i), \qquad w_i \in \mathfrak{C} \tag{2.18}$$

are jointly asymptotically normal and independent with variances each greater than a constant $\zeta > 0$. Further,

$$1 = o\left[\sup_w |V_n(w)|\right] \tag{2.19}$$

in probability as $n \to \infty$.

We now describe the estimation procedure. *Fix a given level $\delta > 0$. Consider the number of $b \in \mathcal{L}$ with*

$$\left| \int_{-\pi}^{\pi} f_b(\mu)\, d\mu \right| > \delta. \tag{2.20}$$

The following procedure will yield estimates $w_n(b)$ of $b \in \mathcal{L}$ satisfying (2.20) *which are such that $n|w_n(b) - b| < a_n = n^{-1/5}$ except for a set of probability less than $O(\log n/n^2)$ if n is sufficiently large (i.e., as $n \to \infty$). Divide the range (it can be taken as $-\pi \le w \le \pi$) of w into blocks of length $1/n$ and compute the global maximum of $|\int_{-\pi}^{\pi} I_n(\mu + w, \mu)\, d\mu|$ in each block. Throw out the blocks with maxima that fall less than or equal to $\delta + (\frac{\log n}{n})^{1/4}$. Also, if there are two maxima located within $1/n$ of each other, throw out the block with the smaller maximum. As $n \to \infty$, the locations $w_n(b)$ at which the remaining maxima occur satisfying* (2.20) *are such that $n|w_n(b) - b| < a_n = n^{-1/5}$ except for a set of probability less than $O(\log n/n^2)$ as $n \to \infty$.*

We sketch the argument for this. Using (2.10), it is seen that

$$EU_n(w) = \sum_{b \in \mathcal{L}} \frac{\sin((n+1)/2(b-w))}{(n+1)/2(b-w)} \int_{-\pi}^{\pi} f_b(\mu)\, d\mu + O\left(\frac{\log n}{n}\right).$$

The maximum of

$$\left| \frac{\sin((n+1)/2)(b-w)}{((n+1)/2)(b-w)} \right| \left| \int_{-\pi}^{\pi} f_b(\mu)\, d\mu \right| \tag{2.21}$$

is clearly at $w = b$ given Assumption 1. The contribution from the other terms in $|EU_n(w)|$ with $b' \ne b, b' \in \mathcal{L}$ is asymptotically $O(\frac{1}{n})$ in a neighborhood $\{w : |b - w| < \min_{b' \ne b, b' \in \mathcal{L}} \frac{1}{2}|b' - b|\}$ of $b \in \mathcal{L}$. First note that, uniformly in w,

$$\big| |U_n(w)| - |EU_n(w)| \big| \le C_1 \left(\frac{\log n}{n}\right)^{1/2} \tag{2.22}$$

except on a set of probability less than $O(n^{-2})$ by Corollary 3. We shall show that the local maximum $w_n(b)$ converging to b must be such that $|n(w_n(b) - b)| < n^{-1/5}$ except for a set of probability less than $O(\log n/n^2)$. First note that

$$\left| \left| \int_{-\pi}^{\pi} I_n(\mu + w, \mu)\, d\mu \right| - \left| \int_{-\pi}^{\pi} f_b(\mu)\, d\mu \frac{\sin((n+1)/2)(b-w)}{((n+1)/2)(b-w)} \right| \right|$$
$$\le C_2 \left(\frac{\log n}{n}\right)^{1/2}$$

in a fixed nontrivial interval $I(b, \varepsilon) = \{w : |b - w| < \varepsilon\}, 0 < \varepsilon < \frac{1}{2}$, containing b but excluding the other points $b' \in \mathcal{L}$ except for a set of probability $O(\log n/n^2)$. The value of $|\int I_n(\mu + w, \mu)\, d\mu|$ at $w = b$ is at least

$$\text{①} := \left| \int_{-\pi}^{\pi} f_b(\mu)\, d\mu \right| - C_2 \left(\frac{\log n}{n}\right)^{1/2}.$$

However, the value outside $|n(w-b)| < n^{-1/5}$, but inside $I(b,\varepsilon)$, of $|\int I_n(\mu + w, \mu)\, d\mu|$, is at most

$$②:= \left|\int_{-\pi}^{\pi} f_b(\mu)\, d\mu\right| \frac{\sin n^{-1/5}}{n^{-1/5}} + C_2\left(\frac{\log n}{n}\right)^{1/2},$$

and since

$$\frac{\sin n^{-1/5}}{n^{-1/5}} = 1 - \frac{n^{-2/5}}{3!} + O(n^{-4/5}),$$

①is larger than ②for n sufficiently large. Consequently, we have the following theorem.

THEOREM 4. *Let $\delta > 0$ and Assumptions* 1–3 *hold. There are then estimates $w_n(b)$ of values $b \in \mathcal{L}$ with $|\int_{-\pi}^{\pi} f_b(\mu)\, d\mu| > \delta$ such that $n|w_n(b) - b| < n^{-1/5}$ except for a set of probability $O(\log n/n^2)$ if n is sufficiently large.*

A similar but more detailed argument yields a better estimate for $|w_n(b) - b|$.

COROLLARY 5. *Under the assumptions of Theorem* 4, $(n+1)(b - w_n(b)) = O(n^{-1/4}(\log n)^{1/4})$ *and* $[(n+1)(b - w_n(b))]^2 = O(\sup_w |U_n(w) - EU_n(w)|)$ *except for a set of probability $O(\log n/n^2)$.*

The following results show that if one uses these estimates $w_n(b)$ of b to estimate the spectral densities at b, the bias and variance are asymptotically as good as for the estimates with $b \in \mathcal{L}$ given.

THEOREM 5. *If $|\int_{-\pi}^{\pi} f_b(\mu)\, d\mu| > \delta$ and Assumptions* 1–3 *hold,*

$$E\hat{f}_{w_n(b)}(\eta) = E\left[\frac{\sin((n+1)/2)y(b, w_n(b))}{((n+1)/2)y(b, w_n(b))} \int_{-\pi}^{\pi} f_b(\mu) K_n(\mu - \eta)\, d\mu\right] + O\left(\frac{\log n}{n}\right)$$

and the right-hand side is

$$f_b(\mu) + \frac{b_n^2}{2} f_b''(\mu) \int_{-\infty}^{\infty} x^2 K(x)\, dx + O(b_n^2) + O(n^{-1/2}(\log n)^{1/2}) + o\left(\frac{1}{nb_n}\right).$$

Corollaries 2 and 3 are used in the derivation of this result.

THEOREM 6. *Let $\delta > 0$ and Assumptions* 1–3 *hold. If*

$$\left|\int_{-\pi}^{\pi} f_b(\mu)\, d\mu\right| > \delta, \qquad \left|\int_{-\pi}^{\pi} f_{b'}(\mu)\, d\mu\right| > \delta$$

for $b, b' \in \mathcal{L}$, then

$$\operatorname{cov}(\hat{f}_{w_n(b)}(\eta), \hat{f}_{w_n(b')}(\eta')) = O\left(\frac{1}{nb_n}\right) + O\left(\frac{\log n}{n}\right)$$

as $n \to \infty$, $b_n \downarrow 0$, $nb_n \to \infty$.

THEOREM 7. *If $|\int_{-\pi}^{\pi} f_b(\mu)\,d\mu| > \delta$ and Assumptions 1–3 hold with $f_0(\cdot)$ not identically zero, setting $\gamma(x) = 1$ if $x = 0 \bmod 2\pi$ and 0 otherwise, then*

$$\begin{aligned}\operatorname{var}(\hat{f}_{w_n(b)}(\eta)) &= \frac{2\pi}{nb_n}\Bigg[f_0(\eta + b) f_0(-\eta) \\ &\quad + \sum_{b' \in \mathcal{L}} \gamma(-2\eta + b' - b)|f_{b'}(-\eta)|^2\Bigg] \int_{-\infty}^{\infty} K^2(x)\,dx + O\left(\frac{\log n}{n}\right).\end{aligned}$$

A comparison of Theorems 5 and 7 with Proposition 5.1 and Corollary 5.1 in [15] indicate that the bias and variance of the spectral estimates using estimates $w_n(b)$ of $b \in \mathcal{L}$ have the same behavior asymptotically as spectral estimates assuming complete knowledge of $b \in \mathcal{L}$.

3. Examples. In this section we give a simulated example. Consider the amplitude modulated (AM) signal model in communication

$$y(t) = \cos(\omega_1 t) Z_1(t) + \cos(\omega_2 t) Z_2(t), \qquad E Z_i(t) = 0, \qquad i = 1, 2, \tag{3.1}$$

where $Z_1(t)$ and $Z_2(t)$ are independent stationary processes with spectral densities $f_1(\lambda)$ and $f_2(\lambda)$ at frequency λ, respectively. Then $\{y(t)\}$ is harmonizable with almost periodic covariance function and spectral support lines on $\lambda = \mu$, $\lambda = \mu \pm 2\omega_1$ and $\lambda = \mu \pm 2\omega_2$. We consider (3.1) with $\omega_1 = \frac{\pi}{4\sqrt{3}}$, $\omega_2 = \frac{\pi}{3\sqrt{2}}$, and $Z_1(t) = \varepsilon_t + 0.5\varepsilon_{t-1}$, $Z_2(t) = \eta_t + 0.3\eta_{t-1}$, where ε_t and η_t are independent standard Gaussian sequences. In this case, $y(t) = \int e^{it\lambda}\,dZ_y(\lambda)$ with $dZ_y(\lambda) = \frac{1}{2}[dz_1(\lambda \mp \omega_1) + dz_2(\lambda \mp \omega_2)]$ and $Z_i(t) = \int e^{it\lambda}\,dz_i(\lambda)$ for $i = 1, 2$. Therefore, the spectrum of $\{y(t)\}$ is given by

$$\begin{aligned}E[dZ_y(\lambda)\overline{dZ_y(\mu)}] &= \tfrac{1}{4}\big\{(f_1(\lambda \mp \omega_1) + f_2(\lambda \mp \omega_2))\delta_{\lambda,\mu} \\ &\qquad + f_1(\lambda \mp \omega_1)\delta_{\lambda,\mu\pm 2\omega_1} + f_2(\lambda \mp \omega_2)\delta_{\lambda,\mu\pm 2\omega_2}\big\}.\end{aligned}$$

Given a realization of $\{y(t)\}$, we are interested in estimating ω_1 and ω_2 and the spectral densities on the lines given by $\lambda = \mu \pm 2\omega_1$ and $\lambda = \mu \pm 2\omega_2$.

As an example, a simulated realization of $\{y(t)\}$ for $t = -512$ to 512 is generated. A stretch of $\{Z_1(t)\}$ and $\{y(t)\}$ are given in Figure 1. The periodogram

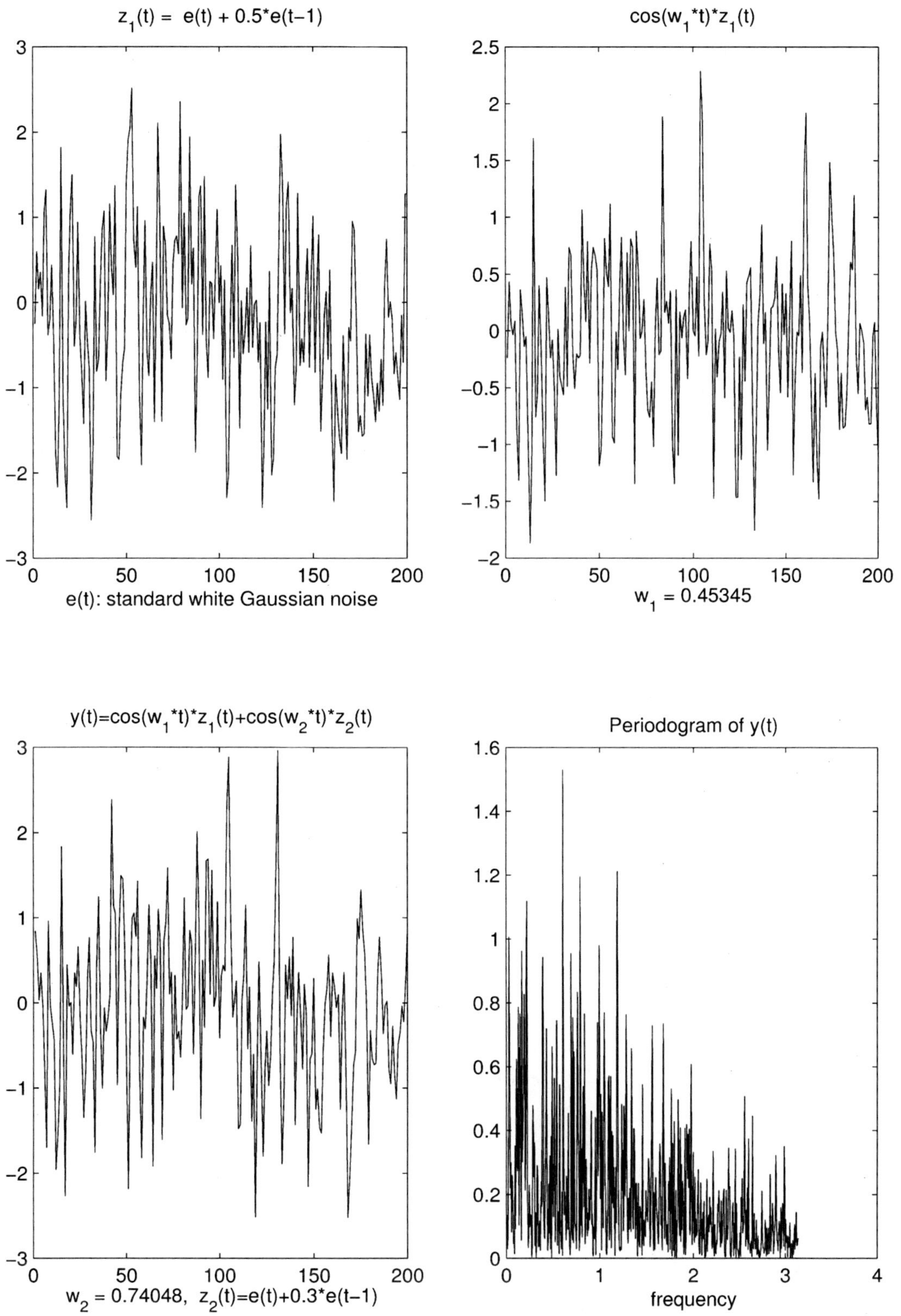

FIG. 1. *Simulated time series with $z_1(t)$, $z_2(t)$ independent stationary time series.*

of $\{y(t)\}$ is given in Figure 1 also. There is no apparent information on ω_1 and ω_2 in the periodogram. The $U_n(b)$ statistic of $\{y(t)\}$ is computed for a range of b's. For $b = b_j = \frac{2\pi j}{1025}$ with $j = 1, 2, \ldots, 512$, $U_n(b)$ and $E(U_n(b))$ are given in Figure 2(a)–(f).

Two spikes are apparent at frequencies near $2\omega_1$ and $2\omega_2$. The largest two b's in $|U_n(b)|$ occurred at $\hat{\omega}_1 = 0.45406$ and $\hat{\omega}_2 = 0.73938$. When the $U_n(b)$ are calculated at a more refined grid with $b = b_j = \frac{2\pi j}{4097}$, $j = 1, 2, \ldots, 2048$, the largest $|U_n(b)|$ occurred at $\hat{\omega}_1 = 0.45329$ and $\hat{\omega}_2 = 0.74015$. These latter values are closer to the true values $\omega_1 = 0.45345$ and $\omega_2 = 0.74048$. These are plotted in Figure 2(g) and (h). The scale in figures (g) and (h) differs from that in figures (e) and (f) since there is leakage at frequencies very close to zero because the computation at (e) and (f) is at a more refined set of frequencies. The zero frequency corresponds to the diagonal line which is the support line of the "power spectrum" of the process. Therefore, the value of $|U_n(0)|$ is relatively large. It is this mass that leaks.

A large value of $|U_n(b)|$ suggests that $\lambda = \mu + b$ is a spectral support line. Figure 3 gives the plots of spectral estimates on these estimated support lines. The dashed line gives the theoretical spectral density if the intercept b is known exactly. In these cases the imaginary part is zero. For the diagonal line $\lambda = \mu$, there is no need to estimate the intercept. The estimated spectral densities on each line when the intercepts are exactly known are given by the solid lines. The dotted lines are spectral estimates when the intercepts are estimated with a crude search, while the dash-dot lines are spectral estimates when the intercepts are estimated by the more refined search.

When the sample size of $\{y(t)\}$ is increased to $t = -2048$ to 2048, we have more resolution as expected.

These graphs show that intercepts of these support lines or frequencies or "periods" of the covariance function can be effectively estimated using $U_n(b)$. These intercepts can be estimated sufficiently accurately to give accurate estimates of the spectral densities on the spectral support lines.

APPENDIX

For the proofs, the following will be useful. Let

$$D_n(x) = \frac{\sin((n+1)/2)x}{\sin(x/2)}, \qquad \widetilde{D}_n(x) = \frac{\sin((n+1)/2)x}{x/2}. \tag{A.1}$$

LEMMA A.1. *If* $|y| \le \pi$,

$$\int_{-\pi}^{\pi} D_n(x+y) D_n(-x)\,dx = \int_{-\pi}^{\pi} \widetilde{D}_n(x+y)\widetilde{D}_n(-x)\,dx + O(\log n), \tag{A.2}$$

with O *uniform in* y.

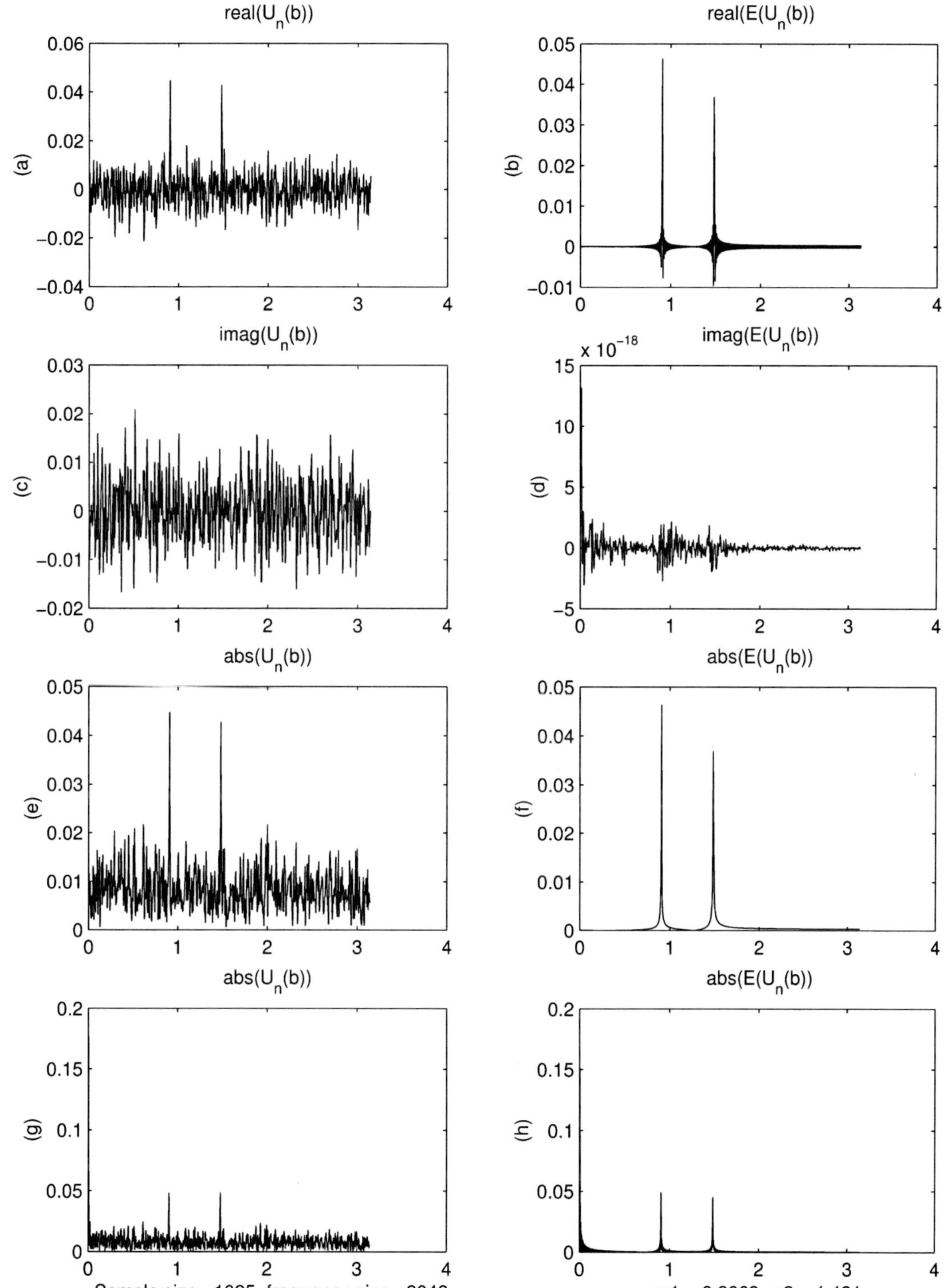

FIG. 2. *$U_n(b)$ statistics of $\{y(t)\}$. Frequencies are computed for increments of $2\pi/1025$ for* (a)–(f) *and for increments of $2\pi/4097$ for* (g)–(h).

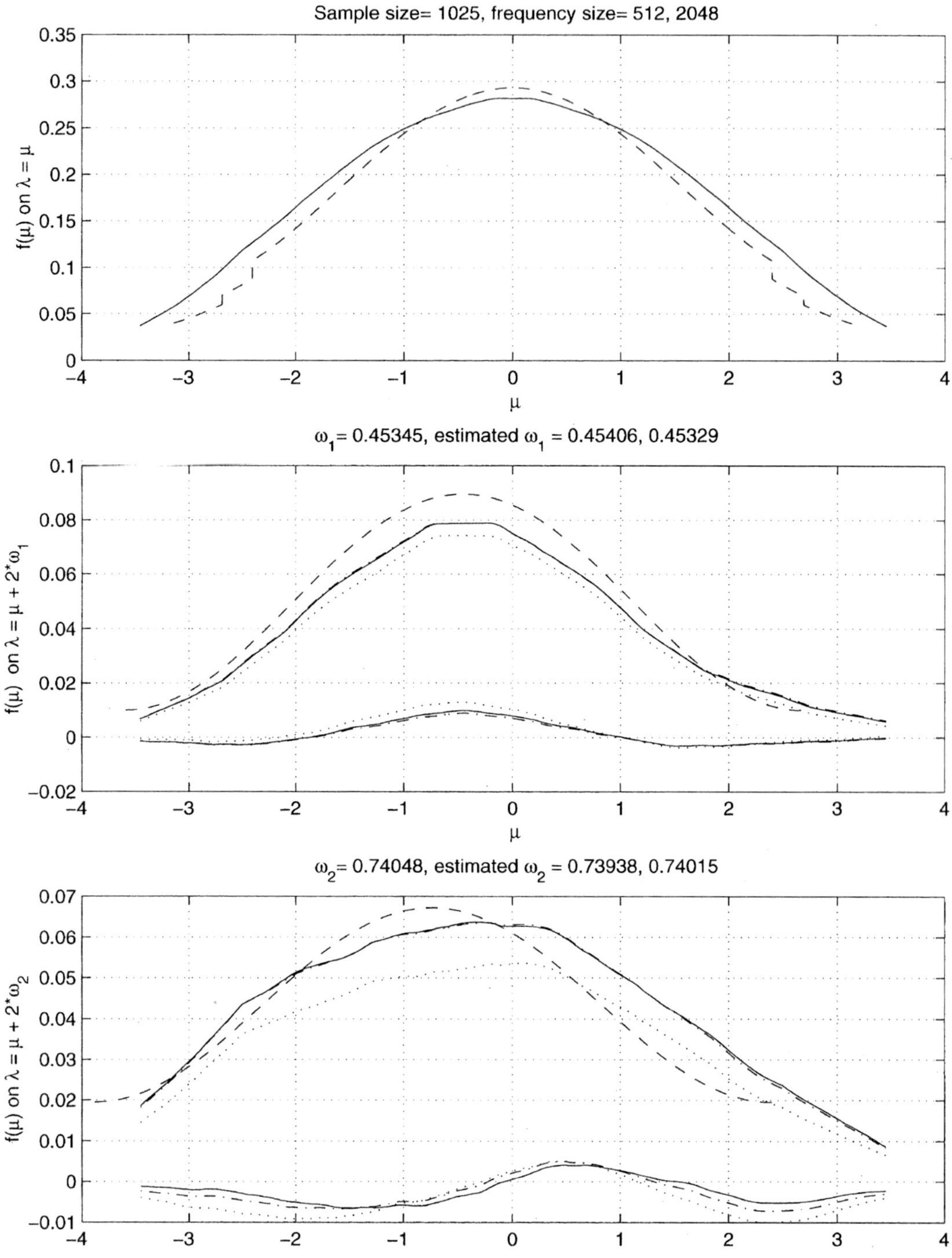

FIG. 3. *Spectral density estimates at different intercepts; the theoretical part is given by the dashed - - - line, estimated part with intercept known is given by the solid line; estimated part with intercept estimated at lower resolution is given by the dotted · · · line; estimated part with intercept estimated at higher resolution is given by dash-dot - · - line. Both real and imaginary parts are given. Theoretical imaginary parts are all zero and they are not plotted.*

The argument given in [15] for Lemma 7.2 holds for this case.

PROOF OF THEOREM 1. Given $X_t, t = -n/2, \dots, n/2$, then

$$\begin{aligned} EI_n(\lambda, \mu) &= \frac{1}{2\pi(n+1)} \sum_{b\in\mathcal{L}} \int_{-\pi}^{\pi} D_n(v + b - \mu - w) D_n(-v + \mu) f_b(v)\, dv \\ &= \sum_{b\in\mathcal{L}} G_1, \end{aligned} \tag{A.3}$$

with $G_1 = \frac{1}{2\pi(n+1)} \int_{-\pi}^{\pi} D_n(v - \mu + y(b, w)) D_n(-v + \mu) f_b(v)\, dv$. The continuous differentiability of f_b implies that

$$f_b(v) = f_b(\mu) + (v - \mu) f_b'(v^*), \tag{A.4}$$

with v^* between μ and v. Then $G_1 = G_2 + G_3$ with [denoting $y(b, w)$ by y]

$$G_2 = \frac{1}{2\pi(n+1)} \int_{-\pi}^{\pi} D_n(v - \mu + y) D_n(-v + \mu)(v - \mu) f_b'(v^*)\, dv,$$

$$G_3 = \frac{1}{2\pi(n+1)} f_b(\mu) \int_{-\pi}^{\pi} D_n(x + y) D_n(-x)\, dx.$$

It is clear that $|G_2| = O(\frac{\log n}{n})$. Also,

$$G_3 = f_b(\mu) \frac{1}{2\pi(n+1)} \int_{-\pi}^{\pi} \widetilde{D}_n(x + y) \widetilde{D}_n(-x)\, dx + O\left(\frac{\log n}{n}\right)$$

using Lemma A.1. To simplify the notation, we define

$$\left\{ \sum_{i=1}^{I} \int_{A_i} \right\} f(x)\, dx = \sum_{i=1}^{I} \int_{A_i} f(x)\, dx.$$

Now

$$\begin{aligned} G_4 &\equiv \frac{1}{2\pi(n+1)} \int_{-\pi}^{\pi} \widetilde{D}_n(x + y) \widetilde{D}_n(-x)\, dx \\ &= \frac{1}{2\pi(n+1)} \\ &\quad \times \left\{ \int_{-\infty}^{\infty} - \int_{|z|>(n+1)\pi} \right\} \frac{\sin(1/2)(z + (n+1)y)}{(z + (n+1)y)/2(n+1)} \frac{\sin(1/2)z}{z/2(n+1)} \frac{dz}{n+1} \\ &= G_5 - G_6, \end{aligned} \tag{A.5}$$

with G_5 and G_6 the two integrals of the previous line. Notice that

$$\begin{aligned} |G_6| \le \frac{1}{2\pi} &\left\{ \int_{(n+1)\pi+1 \ge |z| \ge (n+1)\pi} + \int_{2(n+1)\pi \ge |z| \ge (n+1)\pi+1} + \int_{|z|\ge 2(n+1)\pi} \right\} \\ &\left| \frac{\sin(1/2)(z + (n+1)y)}{(z + (n+1)y)/2} \frac{\sin(1/2)z}{z/2} \right| dz, \end{aligned}$$

with the integrand less than or equal to $1/((n+1)\pi)$ in the first integral, less than or equal to $\frac{1}{(n+1)\pi}(z+(n+1)y)^{-1}$ in the second integral and less than $2/z^2$ in the third integral (because $|y| \le \pi$). These bounds are enough to insure that $G_6 = O(\frac{\log n}{n})$. In evaluating G_5, we note that

$$\frac{\sin(1/2)(z+(n+1)y)}{(z+(n+1)y)/2} = \int_{-\infty}^{\infty} \mathbb{1}_{[-1/2,1/2]}(x) e^{ix(z+(n+1)y)}\, dx. \tag{A.6}$$

The Parseval relation implies that $G_5 = \mathrm{sinc}(y)$. Applying the above bounds for the G_i's proves the theorem. □

PROOF OF THEOREM 2. By a direct calculation, we have four expressions of the same type as that analyzed in Theorem 1. Using (2.8) and (2.10), formula (2.11) is obtained. □

PROOF OF COROLLARY 1. The first equation follows directly from Theorem 2. The order of magnitude statement is obtained by making use of the Schwarz inequality. □

PROOF OF THEOREM 3. First notice that, for distinct indices in the range $-n/2$ to $n/2$,

$$\mathrm{cum}(X_{j_1}^2, X_{j_2}^2, \ldots, X_{j_s}^2) = 2^{s-1} \sum_{\nu} r_{j_1, j_{\alpha_2}} r_{j_{\alpha_2}, j_{\alpha_3}} \cdots r_{j_{\alpha_s}, j_1}, \tag{A.7}$$

where ν is a listing of the $(s-1)!$ irreducible sequences $1 \to \alpha_2 \to \cdots \to \alpha_s \to 1$ with $\alpha_2, \ldots, \alpha_s$ distinct choices of $j_2, \ldots, j_s$. The case in which the indices are not distinct is obtained from (A.7) by making the appropriate identification of identical indices. By (2.5),

$$|r_{j_1, j_2}| \le \sum_{b \in \mathcal{L}} |c_b(j_1 - j_2)|. \tag{A.8}$$

Since the c_b's are uniformly bounded by a summable $c(\cdot) \ge 0$,

$$|r_{j_1, j_{\alpha_2}}, r_{j_{\alpha_2}, j_{\alpha_3}} \cdots r_{j_{\alpha_s}, j_1}| \le q^s c(j_1 - j_{\alpha_2}) c(j_{\alpha_2} - j_{\alpha_3}) \cdots c(j_{\alpha_s} - j_1). \tag{A.9}$$

We shall carry out an argument for $\mathrm{Re}\{V_n(b)\}$ alone since that for $Im\{V_n(w)\}$ is similar. Let σ_n be the variance of $\mathrm{Re}\{V_n(w)\}$. Then

$$\begin{aligned} &\log E \exp[\alpha \,\mathrm{Re}\{V_n(w)\}] - \alpha^2 \sigma_n/2 \\ &\quad \le (n+1) \sum_{s=3}^{\infty} (2q)^s \alpha^s \left(\sum_{j=-\infty}^{\infty} c(j) \right)^s \Big/ s \end{aligned} \tag{A.10}$$

since the sth cumulant of $V_n(b)$ is less than or equal to $(s-1)!(n+1)(2q)^s \times (\sum_j c(j))^s$. If $\alpha > 0$ is sufficiently small,

$$\log E \exp[\alpha \,\mathrm{Re}\{V_n(b)\}] \le (n+1) q^2 \left(\sum_j c(j) \right)^2 (1+\epsilon) \alpha^2 \tag{A.11}$$

since

$$\begin{aligned}0 \le \sigma_n &= \operatorname{var}(\operatorname{Re}\{V_n(w)\})\\ &= \operatorname{var}\left(\sum_{k=-n/2}^{n/2} X_k^2 \cos(kw)\right)\\ &= 2\sum_{j,k=-n/2}^{n/2} r_{j,k}^2 \cos(jw)\cos(kw)\\ &< 2(n+1)q^2\left(\sum_j c(j)\right)^2\end{aligned}$$

if $\{X_k\}$ is not the trivial process that is identically zero. By Lemma 2.1 and Corollary 2.1 of [19] (see also [21], Chapter 10, Theorem 7.28),

$$\sup_w |\operatorname{Re}\{V_n(w)\}| \le \sup_j |\operatorname{Re}\{V_n(u_j)\}|/(1 - 6\pi n R^{-1}), \tag{A.12}$$

with $u_j = 2\pi j/R$, $j = 0, 1, \dots, R-1$ and $R \gg n$. The simple argument for this follows from the fact that $\max_\lambda |p'(\lambda)| \le n \max_\lambda |p(\lambda)|$ for a trigonometric polynomial $p(\lambda) = \sum_{v=-n}^{n} \alpha_v \exp(iv\lambda)$ of order n (Bernstein's inequality, page 11, [21], vol. 2). Then for $\lambda_j = 2\pi j/R$, $j = 0, 1, \dots, R-1$,

$$\max_{|\lambda|\le\pi} |p(\lambda)| \le \max_{|j|\le R} |p(\lambda_j)| + \frac{2\pi}{R}\max_{|\lambda|\le\pi} |p'(\lambda)|. \tag{A.13}$$

Equation (A.13) implies that

$$\left(1 - \frac{2n\pi}{R}\right)\max_{|\lambda|\le\pi} |p(\lambda)| \le \max_{|j|\le R} |p(\lambda_j)|$$

and (A.12) follows. Then

$$\begin{aligned}&E\exp\left\{\alpha \sup_w |\operatorname{Re}\{V_n(w)\}|\right\}\\ &\quad\le E\exp\left\{\frac{\alpha \sup_j |\operatorname{Re}\{V_n(u_j)\}|}{(1-6\pi n R^{-1})}\right\}\\ &\quad\le \sum_j E\exp\left\{\frac{\alpha|\operatorname{Re}\{V_n(u_j)\}|}{(1-6\pi n R^{-1})}\right\}\\ &\quad\le 2\exp\left\{\log R + \frac{(1+\epsilon)(n+1)q^2(\sum c(j))^2\alpha^2}{[(1-6\pi n R^{-1})^2]}\right\}.\end{aligned} \tag{A.14}$$

Our objective is to show by an appropriate choice of a and R that

$$P\left[\sup_w |\operatorname{Re}\{V_n(w)\}| \ge 2a\right] < Cn^{-1-\delta} \tag{A.15}$$

for some constant C. By (A.14) and Chebyshev's inequality, the probability on the left-hand side of (A.15) is less than or equal to

$$\exp\{-2a\alpha\}2\exp\left\{\log R + (1+\epsilon)(n+1)q^2 \times \left(\sum_j c(j)\right)^2 \alpha^2/[(1-6\pi n R^{-1})^2]\right\}. \tag{A.16}$$

Given $\epsilon, \delta > 0$, set

$$a^2 = (1+\epsilon)(2+\delta)q^2(n+1)\log(n+1)\left(\sum_j c(j)\right)^2, \quad R = n\log n \tag{A.17}$$

and

$$\alpha = a(1-6\pi n R^{-1})^2 \Big/ \left[(1+\epsilon)(n+1)q^2\left(\sum_j c(j)\right)^2\right]. \tag{A.18}$$

On inserting these expressions for a, α and R, (A.16) becomes

$$2\exp[-(1+\delta)\log(n+1)] \le Cn^{-1-\delta}. \tag{A.19}$$

The bound $Cn^{-1-\delta}$ has been obtained for sufficiently large n with an appropriate choice for the constant C. Thus,

$$P\left[\sup_w |\mathrm{Re}\{V_n(w)\}|/(n\log n)^{1/2} \ge 2\{(1+\epsilon)(2+\delta)\}^{1/2} q\left(\sum_j c(j)\right)\right] \le Cn^{-1-\delta}. \tag{A.20}$$

By applying the Borel–Cantelli lemma to (A.20), one can see that, for $\eta > 0$,

$$\limsup_{n\to\infty} \sup_w |\mathrm{Re}\{V_n(w)\}|/(n\log n)^{1/2} \le 2^{3/2} q(1+\eta)\left(\sum_j c(j)\right). \tag{A.21}$$

Since the same result holds for $\sup_w |Im\, V_n(w)|$, we obtain

$$\limsup_{n\to\infty} \sup_w |V_n(w)|/(n\log n)^{1/2} \le 2^{5/2}(1+\eta)q\left(\sum_j c(j)\right). \tag{A.22}$$

□

PROOF OF COROLLARY 2. Notice that the probability of (A.15) can be rewritten as $P[\sup_w |\mathrm{Re}\{U_n(w) - EU_n(w)\}| \ge \frac{2a}{n+1}]$. Let $c > 0$ and $a = cn/4$, $R =$

$n \log n$ and $\alpha = n^{-1/2}(\log n)^{1/2}$. Using (A.16), we see that then there is a $C > 0$ such that

$$P\left[\sup |\mathrm{Re}\{U_n(b) - EU_n(b)\}| > \frac{c}{2}\right] \le \exp(-Cn^{1/2}(\log n)^{1/2}).$$

The same type of inequality holds for $Im\{U_n(b) - EU_n(b)\}$. Thus, the corollary holds. □

PROOF OF COROLLARY 3. The arguments are similar but now set $a = cn^{1/2}(\log n)^{1/2}$, $R = n \log n$ and $\alpha = n^{-1/2}(\log n)^{1/2}$ with $c > 0$ fixed but large. □

PROOF OF COROLLARY 4. Trivially,

$$\begin{aligned}\sup_w |U_n(w) - EU_n(w)|^2 &\ge \frac{1}{2\pi}\int_{-\pi}^{\pi} |U_n(w) - EU_n(w)|^2\,dw \\ &= \sum_{j=-n/2}^{n/2} (X_j^2 - EX_j^2)^2 \frac{1}{n+1}.\end{aligned}$$

The covariance of $\frac{1}{n+1}(U_n(w) - EU_n(w))$ and $\frac{1}{n+1}(U_n(\eta) - EU_n(\eta))$ is

$$= \frac{1}{(n+1)^2} \sum_{j,k=-n/2}^{n/2} \mathrm{cov}(X_j^2, X_k^2) e^{-ijw+ik\eta},$$

with $\mathrm{cov}(X_j^2, X_k^2) = 2r_{j,k}^2$ since $\{X_t\}$ is a Gaussian process. But then

$$\begin{aligned}(1) &:= \frac{1}{(n+1)^2} \sum_{j,k=-n/2}^{n/2} \left\{\sum_{b\in\mathcal{L}} e^{ibj} c_b(j-k)\right\}^2 e^{-ijw+ik\eta} \\ &= \frac{2}{(n+1)^2} \\ &\quad \times \sum_{b,b'\in\mathcal{L}} \int_{-\pi}^{\pi} \frac{\sin((n+1)/2)(b+b'-w+\alpha)}{\sin(1/2)(b+b'-w+\alpha)} \frac{\sin((n+1)/2)(\eta-\alpha)}{\sin(1/2)(\eta-\alpha)} \\ &\qquad \times f_b * f_{b'}(\alpha)\,d\alpha,\end{aligned}$$

where "$*$" denotes convolution. If $\eta = w - b - b'$, we essentially get a Fejér kernel centered at this point and so

$$(1) = \frac{4\pi}{n+1} \sum_{b,b'\in\mathcal{L}} \delta(\eta - w + b + b') f_b * f_{b'}(\eta) + O\left(\frac{1}{n^2}\right)$$

if there are a finite number of b in $\mathcal{L}$. Notice that, for $\eta = w$, we get

$$(1) = \frac{2\pi}{n+1} \sum_{b\in\mathcal{L}} f_b * f_{-b}(\eta) + O\left(\frac{1}{n^2}\right).$$

We first show joint asymptotic normality of $U_n(w) - EU_n(w)$ for any finite number of w's. Also, there are an infinite number of w_i in a neighborhood of zero such that $w_i - w_j + b + b' \neq 0$ for any $b, b' \in \mathcal{L}$. For joint asymptotic normality, we have to deal with

$$\sum_{j=1}^{m}[U_n(w_j) - EU_n(w_j)]\alpha_j. \tag{A.23}$$

The argument for estimates of the cumulants of $V_n(b)$ leading to Theorem 3 still holds for the sth cumulant of (A.23) with the additional factor $(\sum|\alpha_j|)^s$. The asymptotic variance is

$$\frac{4\pi}{n+1}\sum_{j,j'}\alpha_j\alpha_{j'}\sum_{b,b'\in\mathcal{L}}\delta(w_j - w_{j'} + b + b')f_b * f_{b'}(w_j) + O\left(\frac{1}{n^2}\right). \tag{A.24}$$

We now consider conditions for positivity of the coefficient of $\frac{4\pi}{n+1}$.

We trivially get asymptotic normality for $\sqrt{n}\sum_{j=1}^{m}\alpha_j[U_n(w_j) - EU_n(w_j)]$ but with a possibly singular distribution because the sth cumulants with $s > 2$ all tend to zero. The asymptotic variance of $U_n(w_j) - EU_n(w_j)$ is $\frac{4\pi}{n+1}\sum_{b\in\mathcal{L}} f_b * f_{-b}(w_j) + O(\frac{1}{n^2})$. Now $f_0 * f_0(0) = \int_{-\pi}^{\pi} f_0(w)^2\,dw > 0$ if f_0 is not identically zero. Further, for $b \in \mathcal{L}$,

$$f_b * f_{-b}(0) = \int_{-\pi}^{\pi} f_b(w)f_{-b}(-w)\,dw = \int_{-\pi}^{\pi} |f_b(w)|^2\,dw \geq 0.$$

Thus, $\sum_{b\in\mathcal{L}} f_b * f_{-b}(0) > 0$ and this implies $\sum_{b\in\mathcal{L}} f_b * f_{-b}(\eta) > \zeta > 0$ for η in some neighborhood of zero. But there are a countable number of $w_i \neq 0$ in this neighborhood such that $w_i - w_j + b - b' \neq 0$ for any $b, b' \in \mathcal{L}$. Therefore, the $\sqrt{n}[U_n(w_j) - EU_n(w_j)]$ are asymptotically jointly normal and independent with variance greater than $\delta > 0$. Therefore,

$$1 = o\Big[\sup_{w_i}|\sqrt{n}[U_n(w_i) - EU_n(w_i)]|\Big] = o\Big[\sup_{w}|\sqrt{n}[U_n(w) - EU_n(w)]|\Big]$$

in probability. We have the corollary. □

PROOF OF COROLLARY 5. The proof is similar to that of Theorem 4 and is omitted. □

PROOF OF THEOREM 5. If $\lambda = \mu + w$, Theorem 1 gives

$$EI_n(\lambda, \mu) = \sum_{\substack{b\in\mathcal{L}\\ |y(b,w)|\leq\pi}} f_b(\mu)\frac{\sin(((n+1)/2)y(b,w))}{((n+1)/2)y(b,w)} + O\left(\frac{\log n}{n}\right).$$

Since

$$\hat{f}_{w_n}(\eta) = \int_{-\pi}^{\pi} I_n(\mu + w_n, \mu)K_n(\mu - \eta)\,du,$$

it follows that

$$E\hat{f}_{w_n(b)}(\eta) = \sum_{b'\in\mathcal{L}} E\left[\frac{\sin((n+1)/2)y(b', w_n(b))}{((n+1)/2)y(b', w_n(b))}\right] \times \int_{-\pi}^{\pi} f_{b'}(\mu)K_n(\mu-\eta)\,d\mu + O\left(\frac{\log n}{n}\right),$$

$$E\left[\frac{\sin((n+1)/2)y(b, w_n(b))}{((n+1)/2)y(b, w_n(b))}\right] = (1) + (2),$$

where

$$(1) := E\left[\frac{\sin((n+1)/2)y(b, w_n(b))}{((n+1)/2)y(b, w_n(b))};\ |n(b - w_n(b))| \le kn^{-1/4}(\log n)^{1/4}\right],$$

$$(2) := E\left[\frac{\sin((n+1)/2)y(b, w_n(b))}{((n+1)/2)y(b, w_n(b))};\ |n(b - w_n(b))| > kn^{-1/4}(\log n)^{1/4}\right],$$

with $E[f(X); A] := \int_A f(x)\,dP(x)$ where $P(\cdot)$ is the distribution of X. Now

$$(1) = E\big[1 - [n(b - w_n(b))]^2/3! + o([n(b - w_n(b))]^2);\ |n(b - w_n(b))| \le kn^{-1/4}(\log n)^{1/4}\big].$$

By Corollary 3, $1 \ge (1) \ge (1 - k^2 n^{-1/2}(\log n)^{1/2})(1 - \frac{1}{n})$, while $|(2)| \le \frac{1}{n}$ if k is large enough. Therefore,

$$\left|E\left[\frac{\sin((n+1)/2)y(b, w_n(b))}{((n+1)/2)y(b, w_n(b))}\right] - 1\right| \le k^2 n^{-1/2}(\log n)^{1/2}.$$

Now if $b' \ne b$ consider

$$E\left[\frac{\sin((n+1)/2)y(b', w_n(b))}{((n+1)/2)y(b', w_n(b))}\right] = (1') + (2'),$$

where

$$(1') := E\left[\frac{\sin((n+1)/2)y(b', w_n(b))}{((n+1)/2)y(b', w_n(b))};\ |n(b - w_n(b))| > c\right],$$

$$(2') := E\left[\frac{\sin((n+1)/2)y(b', w_n(b))}{((n+1)/2)y(b', w_n(b))};\ |n(b - w_n(b))| \le c\right].$$

By Corollary 2 $|(1')| \le \exp(-Cn^{1/2}(\log n)^{1/2}) + O(n^{-2})$, $|(2')| \le C/n$. Notice that $|(\frac{n+1}{2})(b - w_n(b))|^2 \le k \sup|U_n(w) - EU_n(w)|$ except for a set of probability $O(n^{-2})$, which implies

$$P\big[|(n+1)(b - w_n(b))| \le cn^{-1/2}(\log n)^{1/2}\big] \ge 1 - \exp[-c' \log n]$$

for some c' and so $P[|(n+1)(b - w_n(b))| > cn^{-1/2}(\log n)^{1/2}] \le \exp[-c' \log n]$. These estimates imply that, if f_b is twice continuously differentiable, $E\hat{f}_{w_n(b)}(\eta) = f_b(\eta) + \frac{b_n^2}{2} f_b''(\eta) \int x^2 K(x)\,dx + o(b_n^2) + O(n^{-1/2}(\log n)^{1/2})$. □

PROOF OF THEOREM 6. First notice that

$$\begin{aligned}
&\operatorname{cov}(\hat f_{w_n(b)}(\eta), \hat f_{w_n(b')}(\eta')) \\
&\quad = E\big[E\big[\hat f_{w_n(b)}(\eta)\hat f_{w_n(b')}(\eta')|w_n(b)=w_n, w_n(b')=w_n'\big]\big] \\
&\qquad - E\hat f_{w_n(b)}(\eta)E\hat f_{w_n(b')}(\eta') \\
&\quad = E\big[\operatorname{cov}(\hat f_{w_n(b)}(\eta), \hat f_{w_n(b')}(\eta')|w_n(b)=w_n, w_n(b')=w_n')\big] \qquad \text{(A.25)}\\
&\qquad + E\big[E(\hat f_{w_n(b)}(\eta)|w_n(b)=w_n, w_n(b')=w_n') \\
&\qquad\qquad \times E(\hat f_{w_n(b')}(\eta')|w_n(b)=w_n, w_n(b')=w_n')\big] \\
&\qquad - E\hat f_{w_n(b)}(\eta)E\hat f_{w_n(b')}(\eta').
\end{aligned}$$

The estimate for the first term on the right-hand side of (A.25) is obtained by making use of Corollary 1. The remaining terms on the right-hand side of (A.25) can be seen to be

$$\begin{aligned}
&\sum_{b'',b'''\in\mathcal{L}} \operatorname{cov}\left[\frac{\sin((n+1)/2)y(b'', w_n(b))}{((n+1)/2)y(b'', w_n(b))}, \frac{\sin((n+1)/2)y(b''', w_n(b'))}{((n+1)/2)y(b''', w_n(b'))}\right] \\
&\qquad \times \int_{-\pi}^{\pi} f_b(\mu)K_n(\mu-\eta)\,d\mu \int_{-\pi}^{\pi} f_{b'}(\mu)K_n(\mu-\eta')\,d\mu + O\left(\frac{\log n}{n}\right).
\end{aligned}$$

Of course,

$$\begin{aligned}
&|\operatorname{cov}(\operatorname{sinc} y(b'', w_n(b)), \operatorname{sinc} y(b''', w_n(b')))| \\
&\qquad \le \{\operatorname{var}(\operatorname{sinc} y(b'', w_n(b)))\operatorname{var}(\operatorname{sinc} y(b''', w_n(b')))\}^{1/2}.
\end{aligned}$$

Notice that $\operatorname{var}(\operatorname{sinc} y(b', w_n(b)))$ is

$$\begin{aligned}
&E[(\operatorname{sinc} y(b'', w_n(b)))^2] - (E[\operatorname{sinc} y(b'', w_n(b))])^2 \\
&\quad = E[(1+\{(\operatorname{sinc} y(b'', w_n(b))) - 1\})^2] - [E(1+\{\operatorname{sinc} y(b'', w_n(b)) - 1\})]^2 \\
&\quad = E[\{\operatorname{sinc} y(b'', w_n(b)) - 1\}^2] - (E[\{\operatorname{sinc} y(b'', w_n(b)) - 1\}])^2.
\end{aligned}$$

One has to consider the case in which $b = b''$ and that in which $b \neq b''$. Also, one should note that the expression above is less than or equal to $E[\{\operatorname{sinc} y(b'', w_n(b)) - 1\}^2]$. If $b = b''$,

$$\begin{aligned}
&E[\{\operatorname{sinc} y(b, w_n(b)) - 1\}^2; |n(b - w_n(b))| \le kn^{-1/4}(\log n)^{1/4}] \\
&\quad = E\left[\left(\left(\frac{((n+1)/2)(b-w_n(b))}{3!}\right)^2 + o\left(\left(\frac{((n+1)/2)(b-w_n(b))}{3!}\right)^2\right)\right)^2;\right. \\
&\qquad\qquad \left. |n(b-w_n(b))| < kn^{-1/4}(\log n)^{1/4}\right] \\
&\quad \le C\log n/n,
\end{aligned}$$

while, by Corollaries 3 and 4,

$$E\big[\{\operatorname{sinc} y(b, w_n(b)) - 1\}^2; |n(b - w_n(b))| > kn^{-1/4}(\log n)^{1/4}\big] \le c/n$$

if k is large enough. If $b \neq b''$,

$$E\big[\{\operatorname{sinc} y(b'', w_n(b)) - 1\}^2; |n(b - w_n(b))| \le c\big] \le c/n$$

because $w_n(b)$ is close to b and $b'' - w_n(b)$ is large, while

$$\begin{aligned} &E\big[\{\operatorname{sinc} y(b'', w_n(b)) - 1\}^2; |n(b - w_n(b))| > c\big] \\ &\quad \le e^{-Cn^{1/2}(\log n)^{1/2}} + O(n^{-2}) \end{aligned}$$

follows by Corollary 2 as $n \to \infty$. The proof of Theorem 6 is complete. □

PROOF OF THEOREM 7.

$$\begin{aligned} \operatorname{var}(\hat{f}_{w_n(b)}(\eta)) &= E(\operatorname{var}(\hat{f}_{w_n(b)}(\eta)|w_n(b) = w_n)) \\ &\quad + E\big[E(\hat{f}_{w_n(b)}(\eta)|w_n(b) = w_n)^2\big] \\ &\quad - \{E\hat{f}_{w_n(b)}(\eta)\}^2. \end{aligned} \tag{A.26}$$

We first consider the first term on the right-hand side:

$$\begin{aligned} &\operatorname{var}(\hat{f}_{w_n(b)}(\eta)|w_n(b) = w_n) \\ &= \int_{-\pi}^{\pi}\int_{-\pi}^{\pi}\Bigg[\bigg\{\sum_{b\in\mathcal{L}} f_b(\mu' + w_n)\operatorname{sinc}(y(1)) + O\Big(\frac{\log n}{n}\Big)\bigg\} \\ &\qquad \times \bigg\{\sum_{b'\in\mathcal{L}} f_{b'}(-\mu')\operatorname{sinc}(y(2)) + O\Big(\frac{\log n}{n}\Big)\bigg\} \\ &\qquad + \bigg\{\sum_{b\in\mathcal{L}} f_b(-\mu')\operatorname{sinc}(y(3)) + O\Big(\frac{\log n}{n}\Big)\bigg\} \\ &\qquad \times \bigg\{\sum_{b'\in\mathcal{L}} f_{b'}(\mu' + w_n)\operatorname{sinc}(y(4)) + O\Big(\frac{\log n}{n}\Big)\bigg\}\Bigg] \\ &\qquad \times K_n(\mu - \eta)K_n(\mu' - \eta)\,d\mu\,d\mu'. \end{aligned}$$

Making use of the symmetry of K about zero on taking the expectation, we obtain

$$\begin{aligned} &\operatorname{var}(\hat{f}_{w_n(b)}(\eta)|w_n(b) = w_n) \\ &= \frac{2\pi}{nb_n}\Bigg[f_0(\eta + b)f_0(-\eta) \\ &\qquad + \sum_{b'\in\mathcal{L}} \gamma(-2\eta + b' - b)|f_{b'}(-\eta)|^2\Bigg]\int_{-\infty}^{\infty} K^2(x)\,dx + O\Big(\frac{\log n}{n}\Big). \end{aligned}$$

The term

$$E\big[E(\hat{f}_{w_n(b)}(\eta)|w_n(b)=w_n)\big]^2 - \{E\hat{f}_{w_n(b)}(\eta)\}^2$$

can be shown to be $O(\frac{\log n}{n})$ just as in the discussion of the corresponding term in the proof of Theorem 6. □

REFERENCES

[1] ALEKSEEV, V. G. (1988). Estimating the spectral densities of a Gaussian periodically correlated stochastic process. *Problems Inform. Transmission* **24** 109–115. MR0955986

[2] BRILLINGER, D. (1975). *Time Series: Data Analysis and Theory*. Holt, Rinehart and Winston, New York. MR0443257

[3] DANDAWATE, A. V. and GIANNAKIS, G. B. (1994). Nonparametric polyspectral estimators for kth order (almost) cyclostationary processes. *IEEE Trans. Inform. Theory* **40** 67–84.

[4] DEHAY, D. and LESKOW, J. (1996). Functional limit theory for the spectral covariance estimator. *J. Appl. Probab.* **33** 1077–1092. MR1416228

[5] GARDNER, W. A. (1991). Exploitation of spectral redundancy in cyclostationary signals. *IEEE Signal Processing Magazine* **8** (2) 14–36.

[6] GARDNER, W. A., ed. (1994). *Cyclostationarity in Communications and Signal Processing*. IEEE Press, New York.

[7] GERR, N. and ALLEN, J. (1994). The generalized spectrum and spectral coherence of a harmonizable time series. *Digital Signal Processing* **4** 222–238.

[8] GERR, N. and ALLEN, J. (1994). Time-delay estimation for harmonizable signals. *Digital Signal Processing* **4** 49–62.

[9] GLADYSHEV, E. G. (1963). Periodically and almost-periodically correlated random processes with a continuous time parameter. *Theory Probab. Appl.* **8** 173–177.

[10] HURD, H. (1989). Nonparametric time series analysis for periodically correlated processes. *IEEE Trans. Inform. Theory* **35** 350–359. MR0999650

[11] HURD, H. and GERR, N. (1991). Graphical methods for determining the presence of periodic correlation. *J. Time Ser. Anal.* **12** 337–350. MR1131006

[12] HURD, H. and LESKOW, J. (1992). Strongly consistent and asymptotically normal estimation of the covariance for almost periodically correlated processes. *Statist. Decisions* **10** 201–225. MR1183203

[13] HURD, H., MAKAGON, A. and MIAMEE, A. G. (2002). On AR(1) models with periodic and almost periodic coefficients. *Stochastic Process. Appl.* **100** 167–185. MR1919612

[14] LESKOW, J. and WERON, A. (1992). Ergodic behavior and estimation for periodically correlated processes. *Statist. Probab. Lett.* **15** 299–304. MR1192541

[15] LII, K.-S. and ROSENBLATT, M. (2002). Spectral analysis for harmonizable processes. *Ann. Statist.* **30** 258–297. MR1892664

[16] LOÈVE, M. (1963). *Probability Theory*, 3rd ed. Van Nostrand, Princeton, NJ. MR0203748

[17] LUND, R., HURD, H., BLOOMFIELD, P. and SMITH, R. (1995). Climatological time series with periodic correlation. *J. Climate* **8** 2787–2809.

[18] TIAN, C. J. (1988). A limiting property of sample autocovariances of periodically correlated processes with application to period determination. *J. Time Ser. Anal.* **9** 411–417. MR0995617

[19] WOODROOFE, M. B. and VAN NESS, J. W. (1967). The maximum deviation of sample spectral densities. *Ann. Math. Statist.* **38** 1558–1569. MR0216717

[20] YAGLOM, A. M. (1987). *Correlation Theory of Stationary and Related Random Functions* **1**, **2**. Springer, Berlin. MR0893393, MR0915557

[21] ZYGMUND, A. (1959). *Trigonometric Series* **1**, **2**, 2nd ed. Cambridge Univ. Press. MR0107776

DEPARTMENT OF STATISTICS
UNIVERSITY OF CALIFORNIA, RIVERSIDE
RIVERSIDE, CALIFORNIA 92521
USA
E-MAIL: ksl@stat.ucr.edu

DEPARTMENT OF MATHEMATICS
UNIVERSITY OF CALIFORNIA, SAN DIEGO
LA JOLLA, CALIFORNIA 92093
USA
E-MAIL: mrosenblatt@ucsd.edu

The Annals of Statistics
2008, Vol. 36, No. 3, 1508
DOI: 10.1214/07-AOS502
Main article DOI: 10.1214/009053606000000218

CORRECTION

ESTIMATION FOR ALMOST PERIODIC PROCESSES

Ann. Statist. **34** (2006) 1115–1139

BY KEH-SHIN LII AND MURRAY ROSENBLATT

University of California, Riverside and University of California, San Diego

On page 1122 of [1], line 7 from the top, instead of "into blocks of length $1/n$," it should read "into blocks of length $c(n)/n$ [with $\lim_{n\to\infty} c(n) = \infty$ and $c(n) = o(n)$ as $n \to \infty$, e.g., $c(n) = \log n$ or $c(n) = n^{1/2}$]...."

On the same page, line 10, replace $1/n$ with $c(n)/n$.

REFERENCES

[1] LII, K.-S. and ROSENBLATT, M. (2006). Estimation for almost periodic processes. *Ann. Statist.* **34** 1115–1139. MR2278353

DEPARTMENT OF STATISTICS
UNIVERSITY OF CALIFORNIA, RIVERSIDE
RIVERSIDE, CALIFORNIA 92521
USA
E-MAIL: ksl@stat.ucr.edu

DEPARTMENT OF MATHEMATICS
UNIVERSITY OF CALIFORNIA, SAN DIEGO
LA JOLLA, CALIFORNIA 92093
USA
E-MAIL: mrosenblatt@ucsd.edu

Received May 2007; revised May 2007.

Statistics and Probability Letters 78 (2008) 1339–1348

Contents lists available at ScienceDirect

Statistics and Probability Letters

journal homepage: www.elsevier.com/locate/stapro

Prolate spheroidal spectral estimates

K.S. Lii [a], M. Rosenblatt [b,*]

[a] Department of Statistics, University of California at Riverside, Riverside, CA 92521, United States

[b] Department of Mathematics, 0112, University of California at San Diego, La Jolla, CA 92093-0112, United States

ARTICLE INFO

Article history:
Received 28 June 2007
Received in revised form 1 March 2008
Accepted 1 May 2008
Available online 25 May 2008

In memory of David Slepian who gave us so much insight into the prolate spheroidal functions.

MSC:
60
62
62M10
62M15

ABSTRACT

An estimate of the spectral density of a stationary time series can be obtained by taking the finite Fourier transform of an observed sequence $x_0, x_1, \ldots, x_{N-1}$ of sample size N with taper a discrete prolate spheroidal sequence and computing its square modulus. It is typical to take the average K of several such estimates corresponding to different prolate spheroidal sequences with the same bandwidth $W(N)$ as the final computed estimate. For the mean square error of such an estimate to converge to zero as $N \to \infty$, it is shown that it is necessary to have $W(N) \downarrow 0$ with $NW(N) \to \infty$ as $N \to \infty$ and significantly have $K(N) \leq 2NW(N)$ but $K = K(N) \to \infty$ as $N \to \infty$.

0. Introduction

Spectral estimates for stationary time series based on discrete prolate spheroidal sequences (dpss) or discrete prolate spheroidal wave functions (dpwf) have been proposed and utilized in the literature (see Thomson (1990) and Perceival and Walden (1993)). Here we shall derive some of their asymptotic properties and compare them with some of the earlier estimates suggested. Before obtaining some of these results, some of the well known properties of the prolate spheroidal sequences and functions will be given for convenience (see Slepian (1978)). Let $\frac{1}{2} > W > 0$ and N be a positive integer. The integral equation

$$\int_{-W}^{W} \frac{\sin N\pi(f-f')}{\sin \pi(f-f')} \psi(f')\mathrm{d}f' = \lambda\psi(f), \quad -\infty < f < \infty \tag{1}$$

has N positive eigenvalues $\lambda_j(N, W)$, $j = 0, \ldots, N-1$,

$$1 \geq \lambda_0(N, W) > \lambda_1(N, W) > \cdots > \lambda_{N-1}(N, w) > 0.$$

The N real eigenfunctions corresponding to these eigenvalues are $U_0(N, W; f), \ldots, U_{N-1}(N, W; f)$ and they are the dpwf. They are normalized so that

$$\int_{-\frac{1}{2}}^{\frac{1}{2}} |U_k(N, W; f)|^2 \mathrm{d}f = 1$$

* Corresponding author.
E-mail address: mrosenblatt@ucsd.edu (M. Rosenblatt).

0167-7152/$ - see front matter © 2008 Elsevier B.V. All rights reserved.
doi:10.1016/j.spl.2008.05.022

with

$$U_k(N, W; 0) \geq 0, \qquad \frac{dU_k}{df}(N, W; 0) \geq 0,$$

$k = 0, 1, \ldots, N-1$. The dpwf have the double orthogonality property

$$\int_{-W}^{W} U_i(N, W; f) U_j(N, W; f) df = \lambda_i \int_{-\frac{1}{2}}^{\frac{1}{2}} U_i(N, W; f) U_j(N, W; f) df = \lambda_i \delta_{i,j}, \quad i, j = 0, 1, \ldots, N-1.$$

For $k = 0, 1, \ldots, N-1$ the dpss $\{v_n^{(k)}(N, W)\}$ is the real solution of the set of equations

$$\sum_{m=0}^{N-1} \frac{\sin 2\pi W(n-m)}{\pi(n-m)} v_m^{(k)}(N, W) = \lambda_k(N, W) v_n^{(k)}(N, W)$$
$$n = 0, \pm 1, \pm 2, \ldots$$

and normalized so that

$$\sum_{j=0}^{N-1} v_j^{(k)}(N, W)^2 = 1,$$
$$\sum_{j=0}^{N-1} v_j^{(k)}(N, W) \geq 0, \qquad \sum_{j=0}^{N-1} (n - 1 - 2j) v_j^{(k)} \geq 0.$$

The λ_k's and the corresponding dpss $v^{(k)}$ (with subscript limited to the index set $(0, N-1)$) are the eigenvalues and corresponding eigenvectors of the matrix $\rho(N, M)$ with elements

$$\rho(N, W)_{mn} = \frac{\sin 2\pi W(m-n)}{\pi(m-n)}, \quad m, n = 0, 1, \ldots, N-1.$$

The dpss also have a double orthogonality property

$$\sum_{n=0}^{N-1} v_n^{(j)}(N, W) v_n^{(k)}(N, W) = \lambda_j \sum_{n=-\infty}^{\infty} v_n^{(j)}(N, W) v_n^{(k)}(N, W) = \delta_{j,k}.$$

Also,

$$U_k(N, W; f) = e_k \sum_{n=0}^{N-1} v_n^{(k)}(N, W) e^{-i\pi(N-1-2n)f},$$
$$k = 0, 1, \ldots, N-1,$$

with

$$e_k = \begin{cases} 1, & k \text{ even} \\ i, & k \text{ odd}. \end{cases}$$

Notice that U_0 is a Fourier transform of a time limited sequence indexed from 0 to $N-1$ that has maximal L^2 mass in the interval $(-W, W)$. *This is the basic property leading to U_0 as the eigenfunction of (1) with maximal eigenvalue λ_0 and it is what motivates the choice of prolate spheroidal functions in constructing spectral estimates based on them.* Suppose $x_0, \ldots, x_{N-1}$ are observations on a stationary process which we will assume to be Gaussian. For convenience we shall also introduce

$$V_k(f) = \sum_{n=0}^{N-1} v_n^{(k)} e^{-i2\pi nf} = e^{-i\pi(N-1)f} U_k(N, W; -f).$$

The Cramér representation of the stationary $\{x_t\}$ process is

$$x_t = \int_{-\frac{1}{2}}^{\frac{1}{2}} e^{2\pi it\lambda} dz(\lambda)$$

with z the spectral process with orthogonal increments

$$E|dz(\lambda)|^2 = dG(\lambda)$$

and G the spectral distribution function of the process. A finite Fourier transform of $x_0, \ldots, x_{N-1}$ with taper weights the dpss is given by

$$\begin{aligned} y_k(f) &= \sum_{t=0}^{N-1} e^{-2\pi i f t} v_t^{(k)}(N, W) x_t \\ &= \int_{-\frac{1}{2}}^{\frac{1}{2}} V_k(f - f') dz(f'). \end{aligned}$$

An estimate proposed for the spectral density $g(\lambda) = G'(\lambda)$ (assuming G to be continuously differentiable) is

$$\frac{1}{K} \sum_{k=0}^{K-1} |y_k(f)|^2. \tag{2}$$

Spectral estimates based on the prolate spheroidal functions have good bias properties and are thought to be effective in guarding against contamination due to leakage when the spectral density fluctuates over a broad range (see Thomson (1990) and Perceival and Walden (1993)). Estimates for the bias and variance will be given in Section 1. Section 3 will consider the eigenvalue distribution of the integral equation (Eq. (1)) as $NW(N) \to \infty$ and show that for consistency in mean square one must have $NW(N) \to \infty$, $W(N) \downarrow 0$ and $K = K(N) \to \infty$ as $N \to \infty$. Estimates of this sort have been used in the analysis of oceanographic, atmospheric, and astronomical data (Thomson, 1990). Throughout our discussion the eigenvalues and eigenfunctions depend on both N and $W(N)$ even though we have not indicated this explicitly at times so as to avoid cumbersome notation.

1. Asymptotic bias and variance

A relevant question which we will discuss later on is the choice K as $N \to \infty$. The first objective will be to get a bound for the bias of an estimate as $N \to \infty$. It is clear that one will have $W = W(N) \downarrow 0$ as $N \to \infty$. First

$$\begin{aligned} E|y_k(f)|^2 &= \int_{-\frac{1}{2}}^{\frac{1}{2}} |V_k(f - f')|^2 g(f') df' \\ &= \int_{-\frac{1}{2}}^{\frac{1}{2}} |U_k(f - f')|^2 g(f') df' \\ &= \int_{|f-f'|<W} |U_k(f - f')|^2 g(f') df' + \int_{|f-f'|\geq W} \{\ \} df'. \end{aligned}$$

Now

$$\begin{aligned} \int_{|f-f'|\geq W} |U_k(f - f')|^2 g(f') df' &\leq \sup g(f) \int_{|f-f'|\geq W} |U_k(f - f')|^2 df' \\ &\leq \sup g(f)(1 - \lambda_k). \end{aligned}$$

Assuming g to be uniformly continuous and $W(N) \downarrow 0$ as $N \to \infty$

$$\begin{aligned} \int_{|f-f'|<W} |U_k(f - f')|^2 g(f') df' &= g(f)\lambda_k + \int_{|f-f'|<W} |U_k(f - f')|^2 (g(f') - g(f)) df' \\ &= g(f)\lambda_k + o(1)\lambda_k. \end{aligned} \tag{3}$$

If g is twice uniformly continuously differentiable, $E|y_k(f)|^2$ can be written

$$\begin{aligned} &\int_{-\frac{1}{2}}^{\frac{1}{2}} |U_k(f - f')|^2 \left\{ g(f) + \frac{(f - f')^2}{2} g''(f) + o(f - f')^2 \right\} df' \\ &= g(f) + \frac{g''(f)}{2} \int_{-\frac{1}{2}}^{\frac{1}{2}} u^2 |U_k(u)|^2 du + o(W^2) + o(1 - \lambda_k). \end{aligned}$$

Theorem 1. *If g is uniformly continuous the bias of the estimate* (2) *is bounded by*

$$(\sup g(f)) \left\{ 1 - \frac{1}{K} \sum_{k=0}^{K-1} \lambda_k \right\} + o(1).$$

If g is twice uniformly continuously differentiable the bias is bounded by

$$\left(\frac{1}{2}|g''(f)| + o(1)\right)\left(\left\{1 - \frac{1}{K}\sum_{k=0}^{K-1}\lambda_k\right\} + W^2\right).$$

See also Thomson (1982). Now consider

$$\begin{aligned}
\operatorname{cov}(|y_k(f)|^2, |y_j(f')|^2) &= \operatorname{cov}\left[\sum_{t_1,t_2=0}^{N-1} e^{-i2\pi f t_1} e^{i2\pi f t_2} v_{t_1}^{(k)} v_{t_2}^{(k)} X_{t_1} X_{t_2}, \sum_{t_3,t_4=0}^{N-1} e^{-i2\pi f' t_3} e^{i2\pi f' t_4} v_{t_3}^{(j)} v_{t_4}^{(j)} X_{t_3} X_{t_4}\right] \\
&= \int_{-\frac{1}{2}}^{\frac{1}{2}} \sum_{t_1,t_3=0}^{N-1} e^{i2\pi t_1(\alpha - f)} e^{-i2\pi t_3(\alpha + f')} v_{t_1}^{(k)} v_{t_3}^{(j)} g(\alpha)\mathrm{d}\alpha \int_{-\frac{1}{2}}^{\frac{1}{2}} \sum_{t_2,t_4=0}^{N-1} e^{i2\pi t_2(\beta + f)} e^{i2\pi t_4(-\beta + f')} v_{t_2}^{(k)} v_{t_4}^{(j)} g(\beta)\mathrm{d}\beta \\
&\quad + \int_{-\frac{1}{2}}^{\frac{1}{2}} \sum_{t_1,t_4=0}^{N-1} e^{2\pi i t_1(\alpha - f)} e^{2\pi i t_4(-\alpha + f')} v_{t_1}^{(k)} v_{t_4}^{(j)} g(\alpha)\mathrm{d}\alpha \int_{-\frac{1}{2}}^{\frac{1}{2}} \sum_{t_2,t_3=0}^{N-1} e^{2\pi i t_2(\beta + f)} e^{2\pi i t_3(-\beta - f')} v_{t_2}^{(k)} v_{t_3}^{(j)} g(\beta)\mathrm{d}\beta \\
&= \int_{-\frac{1}{2}}^{\frac{1}{2}} V_k(f - \alpha) V_j(\alpha + f') g(\alpha)\mathrm{d}\alpha \int_{-\frac{1}{2}}^{\frac{1}{2}} V_k(-\beta - f) V_j(\beta - f') g(\beta)\mathrm{d}\beta \\
&\quad + \int_{-\frac{1}{2}}^{\frac{1}{2}} V_k(f - \alpha) V_j(\alpha - f') g(\alpha)\mathrm{d}\alpha \int_{-\frac{1}{2}}^{\frac{1}{2}} V_k(-\beta - f) V_j(\beta + f') g(\beta)\mathrm{d}\beta. \qquad (4)
\end{aligned}$$

At this point we remark that the $V_j(\alpha)$'s have the same double orthogonality properties the $U_j(\alpha)$'s have

$$\int_{-W}^{W} V_i(f)\overline{V_j(f)}\mathrm{d}f = \lambda_i \int_{-\frac{1}{2}}^{\frac{1}{2}} V_i(f)\overline{V_j}(f)\mathrm{d}f = \lambda_i \delta_{i,j}$$

and

$$\overline{V_j(f)} = V_j(-f).$$

The Schwarz inequality and (4) imply that

$$\begin{aligned}
&\left|\sum_{j,k=0}^{K-1} \operatorname{cov}(|y_j(f)|^2, |y_k(f)|^2)\right| \\
&\le \left[\iint_{-\frac{1}{2}}^{\frac{1}{2}} \left|\sum_{k=0}^{K-1} V_k(f - \alpha) V_k(-\beta - f)\right|^2 g(\alpha)g(\beta)\mathrm{d}\alpha\mathrm{d}\beta \iint_{-\frac{1}{2}}^{\frac{1}{2}} \left|\sum_{j=0}^{K-1} V_j(\alpha + f') V_j(\beta - f')\right|^2 g(\alpha)g(\beta)\mathrm{d}\alpha\mathrm{d}\beta\right]^{\frac{1}{2}} \\
&\quad + \left[\iint_{-\frac{1}{2}}^{\frac{1}{2}} \left|\sum_{k=0}^{K-1} V_k(f - \alpha) V_k(-\beta - f)\right|^2 g(\alpha)g(\beta)\mathrm{d}\alpha\mathrm{d}\beta\right. \\
&\quad \left.\times \iint_{-\frac{1}{2}}^{\frac{1}{2}} \left|\sum_{j=0}^{K-1} V_j(\alpha - f') V_j(\beta + f')\right|^2 g(\alpha)g(\beta)\mathrm{d}\alpha\mathrm{d}\beta\right]^{\frac{1}{2}}. \qquad (5)
\end{aligned}$$

Now

$$\iint_{-\frac{1}{2}}^{\frac{1}{2}} \left|\sum_{k=0}^{K-1} V_k(f - \alpha) V_k(-\beta - f)\right|^2 g(\alpha)g(\beta)\mathrm{d}\alpha\mathrm{d}\beta \le [\sup g(\alpha)]^2 \iint_{-\frac{1}{2}}^{\frac{1}{2}} \left|\sum_{k=0}^{K-1} V_k(f - \alpha) V_k(-\beta - f)\right|^2 \mathrm{d}\alpha\mathrm{d}\beta$$

and by orthogonality of the V_k's we see that the last bound is less than or equal to

$$[\sup g(\alpha)]^2 K.$$

Using the same sort of bound for the other three integrals on the right side of (5) we see that

$$\left|\operatorname{cov}\left(\frac{1}{K}\sum_{j=0}^{K-1} |y_j(f)|^2, \frac{1}{K}\sum_{k=0}^{K-1} |y_k(f')|^2\right)\right| \le 2[\sup g(\alpha)]^2/K.$$

Theorem 2. *A bound for*

$$\operatorname{cov}\left(\frac{1}{K}\sum_{k=0}^{K-1}|y_k(f)|^2, \frac{1}{K}\sum_{k=0}^{K-1}|y_k(f')|^2\right)$$

is

$$2[\sup g(\alpha)]^2/K.$$

Assume that $N \to \infty$ and $W(N) \downarrow 0$. Let us now make a more detailed analysis of $\operatorname{cov}(|y_k(f)|^2, |y_j(f)|^2)$ with j and k fixed. First consider $j \neq k$ with $f \neq 0, \frac{1}{2}$. Now

$$\int_{-\frac{1}{2}}^{\frac{1}{2}} V_k(f-\alpha)V_j(\alpha+f)g(\alpha)\mathrm{d}\alpha = \int_{|f-\alpha|\le W} + \int_{|f-\alpha|>W} = (i) + (ii) \tag{6}$$

Then

$$|(i)| \le [\sup g(\alpha)]\left\{\int_{|f-\alpha|\le W} |V_k(f-\alpha)|^2\mathrm{d}\alpha \int_{|f-\alpha|\le W} |V_j(\alpha+f)|^2\mathrm{d}\alpha\right\}^{\frac{1}{2}}$$

For sufficiently large N if $|f-\alpha| \le W$ then $|f+\alpha| \ge W$ and so

$$\int_{|f-\alpha|\le W} |V_j(\alpha+f)|^2\mathrm{d}\alpha \le (1-\lambda_j).$$

Thus

$$|(i)| \le [\sup g(\alpha)](1-\lambda_j)^{\frac{1}{2}}.$$

The same argument implies that

$$|(ii)| \le [\sup g(\alpha)](1-\lambda_k)^{\frac{1}{2}}$$

and so

$$\left|\int_{-\frac{1}{2}}^{\frac{1}{2}} V_k(f-\alpha)V_j(f+\alpha)g(\alpha)\mathrm{d}\alpha\right| \le [\sup g(\alpha)]\{(1-\lambda_j)^{\frac{1}{2}} + (1-\lambda_k)^{\frac{1}{2}}\}.$$

The same bound is obtained for

$$\int_{-\frac{1}{2}}^{\frac{1}{2}} V_k(-\beta-f)V_j(\beta-f)g(\beta)\mathrm{d}\beta.$$

The same type of argument gives us the same bounds for

$$\int_{-\frac{1}{2}}^{\frac{1}{2}} V_k(f-\alpha)V_j(\alpha-f)g(\alpha)\mathrm{d}\alpha, \quad \int_{-\frac{1}{2}}^{\frac{1}{2}} V_k(-\beta-f)V_j(\beta+f)g(\alpha)\mathrm{d}\alpha.$$

Hence

$$|\operatorname{cov}(|y_k(f)|^2, |y_j(f)|^2)| \le 2[\sup g(\alpha)]^2\{(1-\lambda_j)^{\frac{1}{2}} + (1-\lambda_k)^{\frac{1}{2}}\}^2 \tag{7}$$

if $j \neq k$ and $W(N) \downarrow 0$ as $N \to \infty$. Now look at the case $j = k, f \neq 0, \frac{1}{2}$. The bounds obtained for (6) and (7) when $j \neq k$ still hold if $j = k$. Set

$$\int_{-\frac{1}{2}}^{\frac{1}{2}} |V_k(f-\alpha)|^2 g(\alpha)\mathrm{d}\alpha = \int_{|f-\alpha|\le W} + \int_{|f-\alpha|>W} = (iii) + (iv).$$

Now

$$\begin{aligned}(iii) &= \int_{|f-\alpha|\le W} |V_k(f-\alpha)|^2\mathrm{d}\alpha g(f) + \int_{|f-\alpha|\le W} |V_k(f-\alpha)|^2\{g(\alpha)-g(f)\}\mathrm{d}\alpha \\ &= \lambda_k g(f) + o(1).\end{aligned} \tag{8}$$

Thus

$$\operatorname{var}(|y_k(f)|^2) = \lambda_k^2 g(f)^2 + o(1) + O(1-\lambda_k).$$

A minor adaptation of our argument shows that one gets the same bound as (7) for $j \neq k$ if $f = 0, \frac{1}{2}$. The term $\lambda_k g(f)$ is replaced by $2\lambda_k g(f)$ in (8) if $f = 0$ or $f = \frac{1}{2}$.

Theorem 3. *If g is uniformly continuous, $W(N) \downarrow 0$ as $N \to \infty$ and K is fixed and finite*

$$var\left(\frac{1}{K}\sum_{k=0}^{K-1}|y_k(f)|^2\right) = \{1+\eta(f)\}\frac{1}{K^2}\sum_{j=0}^{K-1}\lambda_j^2 g(f)^2 + O\left(\frac{1}{K}\sum_{k=0}^{K-1}(1-\lambda_k)\right) + o(1).$$

Here $\eta(f) = 0$ if $f \neq 0, \frac{1}{2}$ mod 1 and one otherwise. If g is twice continuously differentiable $o(1)$ is replaced by $o(W^2)$.

2. Additional remarks

It is clear that the eigenvalues $\lambda_i(N, W)$, $i = 0, 1, \ldots$, are a nonincreasing sequence with $1 \geq \lambda_i(N, W) \geq 0$ (see Slepian (1978)). By Mercer's theorem (Widom, 1969) as applied to the integral equation (1)

$$N = \sum_{j=0}^{N-1}\lambda_j\left|\frac{U_j(N, W; f)}{\sqrt{\lambda_j}}\right|^2, \quad |f| \leq W$$

and integrating over $(-W, W)$ the equation

$$2NW = \sum_{j=0}^{N-1}\lambda_j(N, W) \tag{9}$$

is obtained. If $W = W(N)$ and $NW(N) \to 0$ as $N \to \infty$, it is clear that all the eigenvalues $\lambda_j(N, W) \to 0$ as $N \to \infty$. Clearly this is not of any interest to us since in this case the bias of a corresponding spectral estimate will not tend to zero as $N \to \infty$.

If $NW(N) \to c > 0$ with c a finite value, asymptotically there can by (9) be at most a finite number of λ_j that tend to 1 as $N \to \infty$. But already it is clear by Theorem 3 that in such a case the variance of the spectral estimate need not tend to zero as $N \to \infty$.

If $W(N) \to 0$ as $N \to \infty$ it seems plausible to approximate the operator A

$$Ah(f) = \int_{-W}^{W}\frac{\sin N\pi(f-f')}{\sin\pi(f-f')}h(f')df'$$

by A_1

$$A_1h(f) = \int_{-W}^{W}\frac{\sin N\pi(f-f')}{\pi(f-f')}h(f')df'.$$

Now

$$\frac{\sin N\pi(f-f')}{\sin\pi(f-f')} - \frac{\sin N\pi(f-f')}{\pi(f-f')} = -\sin N\pi(f-f')\left\{\frac{\sin\pi(f-f') - \pi(f-f')}{\sin\pi(f-f')\cdot\pi(f-f')}\right\}. \tag{10}$$

The absolute value of (10) is less than or equal to $\pi|f-f'|$ when $|f|, |f'| \leq W$. Let $A_2 = A - A_1$. Then

$$\begin{aligned}\|A_2h\|^2 &= \int_{-W}^{W}|A_2h(f)|^2 df \\ &\leq \int_{-W}^{W}\left|\int_{-W}^{W}|f-f'||h(f')|df'\right|^2 df\pi^2 \\ &\leq \int_{-W}^{W}\int_{-W}^{W}|f-f'|^2df'\int_{-W}^{W}|h(f')|^2df'df\pi^2 \\ &= \|h\|^2\int_{-W}^{W}\int_{-W}^{W}|f-f'|^2dfdf'\pi^2 \\ &= \frac{8}{3}\pi^2W^4\|h\|^2.\end{aligned}$$

Thus

$$\|A_2\| \leq 3^{\frac{1}{2}}\pi W^2. \tag{11}$$

The operators A, A_1, A_2 are symmetric completely continuous. Let $\mu_p^+, \mu_{1,p}^+, \mu_{2,p}^+$ be the pth nonnegative eigenvalues of A, A_1, A_2 and $\mu_p^-, \mu_{1,p}^-, \mu_{2,p}^-$ be the pth nonpositive eigenvalues of the corresponding operators with the subscript 1 corresponding to the largest eigenvalue in absolute value. The comparison lemma for eigenvalues tells us that

$$\mu_{p+q-1}^+ \leq \mu_{1,p}^+ + \mu_{2,q}^+$$

and

$$\mu_p^+ \geq \mu_{1,p+q-1}^+ + \mu_{2,q}^- \quad \text{(Riesz and Sz. Nagy, 1990).}$$

The bound (11) for $\|A_2\|$ tells us that all the eigenvalues of A_2 are bounded in absolute value by $3^{\frac{1}{2}}\pi W^2$. But then

$$|\mu_p^+ - \mu_{1,p}^+| \leq 3^{\frac{1}{2}}\pi W^2. \tag{12}$$

The integral equation for A_1 paralleling (1) is

$$\int_{-W}^{W} \frac{\sin N\pi(f - f')}{\pi(f - f')}\psi(f')\mathrm{d}f' = \lambda\psi(f)$$

which can be rewritten in the form

$$\int_{-NW}^{NW} \frac{\sin\pi(x - x')}{\pi(x - x')}\psi\left(\frac{x'}{N}\right)\mathrm{d}x' = \lambda\psi\left(\frac{x}{N}\right).$$

Notice that

$$\frac{\sin\pi x}{\pi x} = \int_{-\pi}^{\pi} \mathrm{e}^{\mathrm{i}xy}\frac{\mathrm{d}y}{2\pi},$$

so that $\frac{\sin\pi x}{(\pi x)}$ is the Fourier transform of the function $f(x) = 1$ on $(-\pi, \pi)$ and equal to zero elsewhere. A theorem of Kac et al. (1953) states that if $NW(N) \to \infty$ as $N \to \infty$ then

$$\lim_{N\to\infty}(2NW(N))^{-1}\{\text{number of eigenvalues } \lambda \text{ in } [a, b]\} = (2\pi)^{-1}m\{a < f(x) < b\}$$

where m is the Lebesgue measure, it is understood that $[a, b]$ does not contain the origin and

$$m\{f(x) = a\} = m\{f(x) = b\} = 0.$$

In our case it indicates that

$$\{\text{number of eigenvalues approaching } 1\}/(2NW(N)) \to 1$$

as $N \to \infty$ while the number of eigenvalues λ between ε and $1 - \varepsilon$, $\varepsilon > 0$, is

$$o(NW(N))$$

as $N \to \infty$. By (11) it follows that exactly the same thing holds for the eigenvalues of (1) if $W(N) \downarrow 0$, and $NW(N) \to \infty$ as $N \to \infty$. The remaining eigenvalues of (1) must be zero or tend to zero as $N \to \infty$.

Theorem 4. *If K is fixed and finite, one must have $\lambda_j(N, W(N)) \to 1, j = 0, 1, \ldots, K - 1$ as $W(N) \downarrow 0, N \to \infty$ for asymptotic unbiasedness. For asymptotic consistency of the estimate one must have $K(N) \to \infty$ as $N \to \infty$.*

The fact that $\lambda_j(N, W(N)) \to 1, j = 0, 1, \ldots, K - 1$ for asymptotic unbiasedness follows from Theorem 1 if K is fixed and finite. That $K = K(N) \to \infty$ as $N \to \infty$, $W(N) \downarrow 0$ if one is to have mean square consistency as $N \to \infty$ follows from Theorem 3.

Theorem 5. *If the spectral density g is continuous, $N \to \infty$, $W(N) \to 0$, $NW(N) \to \infty$ and $K = K(N) \to \infty$ as $N \to \infty$ with $K(N) \leq 2NW(N)$, the spectral density estimate (2) is consistent in mean square.*

This follows from Theorems 1 and 2 and our remarks on the eigenvalues λ_j. A more detailed estimate of the rate at which the mean square error of a spectral estimate tends to zero under the assumptions of Theorem 5 would require an improved estimate of the rate at which the largest eigenvalues tend to one as $N \to \infty$.

3. Numerical computations

In this section some computations are given to illustrate the interplay between the bias in Theorem 1 and the variance in Theorem 3 with various choices of sample size N, the value of $NW(N)$ (hence the bandwidth $W = W(N)$ is determined) and the number of averaging terms K.

It is clear from the discussion that leads to Theorem 5 that the number of eigenvalues close to 1 is no bigger than $2NW(N)$. Therefore the choice of K should be no bigger than $2NW(N)$ in view of Theorem 1 where the bias is determined by $B = \{\frac{1}{K}\sum_{k=0}^{K-1}\lambda_k - 1\}$. The computation of the eigenvalues uses the tridiagonal formulation given in Slepian (1978) and discussed in Perceival and Walden (1993). The eigenvalues in the tables, are listed in descending order of magnitude and truncated after a sharp drop to zero.

It is seen from Fig. 1 that there is a sharp drop from 1 around $2NW(N) - 2$ in general. The graph gives a qualitative indication of the drop-off of the eigenvalues while the table gives the quantitative indication of the closeness of the eigenvalues to 1 for small $k < 2NW$ and how fast they drop to 0 when $k > 2NW$.

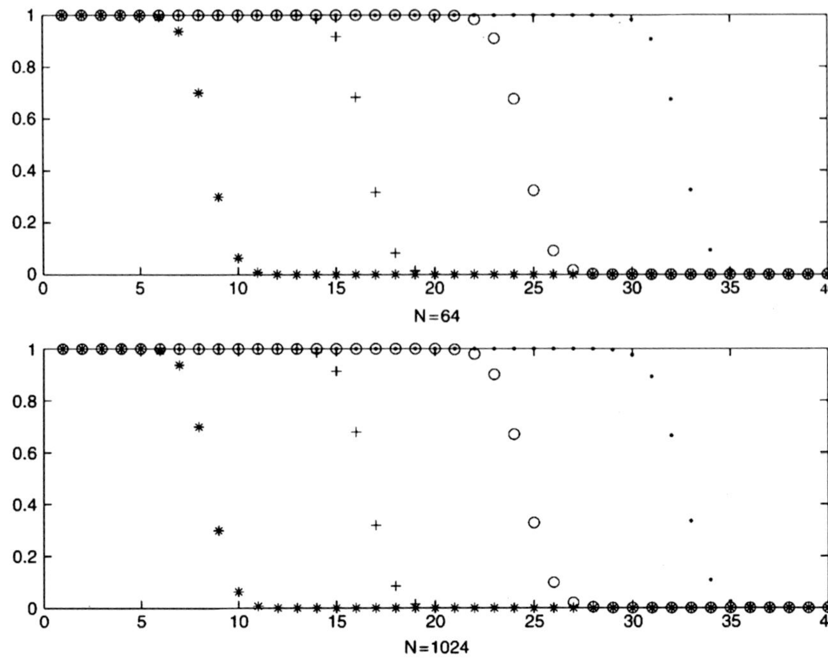

Fig. 1. The first 40 eigenvalues for sample size $N = 64, 1024$ with $NW = 4, 8, 12$ and 16 denoted by *, +, ∘, and . respectively.

```
N=64, NW=4,  W=6.250000000e-02

 9.9999999975e-01,  9.9999997540e-01,  9.9999889519e-01,  9.9996965768e-01
 9.9943654971e-01,  9.9271011593e-01,  9.3746860493e-01,  6.9968507737e-01
 2.9861744341e-01,  6.3390419631e-02,  7.9479055829e-03,  7.2025964440e-04

N=64, NW=8,  W=1.250000000e-01

 1.0000000000e+00,  1.0000000000e+00,  1.0000000000e+00,  1.0000000000e+00
 1.0000000000e+00,  1.0000000000e+00,  9.9999999987e-01,  9.9999999692e-01
 9.9999993815e-01,  9.9999896152e-01,  9.9998535044e-01,  9.9982636034e-01
 9.9828440883e-01,  9.8624672063e-01,  9.1730909795e-01,  6.8266171600e-01
 3.1665152913e-01,  8.2939596727e-02,  1.4057040920e-02,  1.8225016903e-03

N=64, NW=12, W=1.875000000e-01

 1.0000000000e+00,  1.0000000000e+00,  1.0000000000e+00,  1.0000000000e+00
 1.0000000000e+00,  1.0000000000e+00,  1.0000000000e+00,  1.0000000000e+00
 1.0000000000e+00,  1.0000000000e+00,  1.0000000000e+00,  1.0000000000e+00
 1.0000000000e+00,  9.9999999994e-01,  9.9999999891e-01,  9.9999998291e-01
 9.9999976242e-01,  9.9999707620e-01,  9.9996826352e-01,  9.9969814667e-01
 9.9751722771e-01,  9.8288323075e-01,  9.0866103290e-01,  6.7611951975e-01
 3.2360004391e-01,  9.1424001767e-02,  1.7244676102e-02,  2.5343708201e-03

N=64, NW=16, W=2.500000000e-01

 1.0000000000e+00,  1.0000000000e+00,  1.0000000000e+00,  1.0000000000e+00
 1.0000000000e+00,  1.0000000000e+00,  1.0000000000e+00,  1.0000000000e+00
 1.0000000000e+00,  1.0000000000e+00,  1.0000000000e+00,  1.0000000000e+00
 1.0000000000e+00,  1.0000000000e+00,  1.0000000000e+00,  1.0000000000e+00
 1.0000000000e+00,  1.0000000000e+00,  1.0000000000e+00,  1.0000000000e+00
 9.9999999999e-01,  9.9999999987e-01,  9.9999999795e-01,  9.9999997172e-01
```

```
9.9999964723e-01,  9.9999604220e-01,  9.9996026600e-01,  9.9964595666e-01
9.9724077254e-01,  9.8179293460e-01,  9.0610173271e-01,  6.7435370098e-01
3.2564629902e-01,  9.3898267294e-02,  1.8207065404e-02,  2.7592274617e-03
```

N=1024, NW=4, W=3.906250000e-03

```
9.9999999971e-01,  9.9999997233e-01,  9.9999879026e-01,  9.9996756261e-01
9.9941018039e-01,  9.9250530520e-01,  9.3665540821e-01,  6.9883890180e-01
2.9937208026e-01,  6.4238728471e-02,  8.1935666616e-03,  7.5957966448e-04
```

N=1024, NW=8, W=7.812500000e-03

```
1.0000000000e+00,  1.0000000000e+00,  1.0000000000e+00,  1.0000000000e+00
1.0000000000e+00,  9.9999999999e-01,  9.9999999977e-01,  9.9999999490e-01
9.9999990611e-01,  9.9999854539e-01,  9.9998094028e-01,  9.9978872168e-01
9.9803386163e-01,  9.8504547533e-01,  9.1396995463e-01,  6.7987185192e-01
3.1925909354e-01,  8.6324442002e-02,  1.5344220325e-02,  2.1110918855e-03
```

N=1024, NW=12, W=1.171875000e-02

```
1.0000000000e+00,  1.0000000000e+00,  1.0000000000e+00,  1.0000000000e+00
1.0000000000e+00,  1.0000000000e+00,  1.0000000000e+00,  1.0000000000e+00
1.0000000000e+00,  1.0000000000e+00,  1.0000000000e+00,  1.0000000000e+00
9.9999999999e-01,  9.9999999976e-01,  9.9999999653e-01,  9.9999995433e-01
9.9999946178e-01,  9.9999432324e-01,  9.9994656664e-01,  9.9955385850e-01
9.9673740187e-01,  9.7973599628e-01,  9.0102538107e-01,  6.7026844154e-01
3.2916605426e-01,  9.9117137352e-02,  2.0526329047e-02,  3.3808951714e-03
```

N=1024, NW=16, W=1.562500000e-02

```
1.0000000000e+00,  1.0000000000e+00,  1.0000000000e+00,  1.0000000000e+00
1.0000000000e+00,  1.0000000000e+00,  1.0000000000e+00,  1.0000000000e+00
1.0000000000e+00,  1.0000000000e+00,  1.0000000000e+00,  1.0000000000e+00
1.0000000000e+00,  1.0000000000e+00,  1.0000000000e+00,  1.0000000000e+00
1.0000000000e+00,  1.0000000000e+00,  1.0000000000e+00,  9.9999999999e-01
9.9999999989e-01,  9.9999999869e-01,  9.9999998534e-01,  9.9999984930e-01
9.9999857681e-01,  9.9998768999e-01,  9.9990290319e-01,  9.9930664981e-01
9.9557922625e-01,  9.7561620772e-01,  8.9205966296e-01,  6.6402751542e-01
3.3555662036e-01,  1.0802106809e-01,  2.4578335752e-02,  4.5196926372e-03
```

The main term of the variance of the estimate is determined by $V = \frac{1}{K^2}\sum_{j=0}^{K-1}\lambda_j^2$ in Theorem 3.

The following is a table for the bias term B and the variance term V given various choices of N, $NW(N)$, W and the number of averaging terms K. We note that the terms B and V are determined by the eigenvalues and the choice of K and not directly by N. It seems that the B and V may be smaller if we choose larger $NW(N)$ (and K) for the same N. If for different N, $NW(N)$ and K are fixed independent of N, the corresponding B and V are practically the same. The error terms $O(W^2)$ and $o(W^2)$ can at times be comparable to B or V. In applications selection of $W(N)$ and K given sample size N needs to be determined by requirements on control of bias and variance for the problem at hand. This can at times be more complex than the case of classical spectral estimates where asymptotics are determined just by bandwidth and sample size.

N	NW(N)	W	K	B	V
128	4	3.125000000e-02	4	-8.2803085101e-06	2.4999585991e-01
			6	-1.3433910321e-03	1.6622041838e-01
			8	-4.6519768335e-02	1.1484843214e-01
	8	6.250000000e-02	8	-5.9083427040e-10	1.2499999985e-01
			14	-1.1987125972e-03	7.1258441662e-02
			16	-2.6341132401e-02	5.9635224735e-02
	12	9.375000000e-02	12	-4.5630166312e-14	8.3333333333e-02
			22	-1.0463542105e-03	4.5360227285e-02
			24	-1.8689989158e-02	4.0313438974e-02
	16	1.250000000e-01	16	8.8817841970e-16	6.2500000000e-02
			30	-9.1956324501e-04	3.3272629929e-02

			32	-1.4563734465e-02	3.0459759287e-02
512	4	7.812500000e-03	4	-8.4121454063e-06	2.4999579399e-01
			6	-1.3525720581e-03	1.6621737828e-01
			8	-4.6575216198e-02	1.1483681904e-01
	8	1.562500000e-02	8	-6.6334460147e-10	1.2499999983e-01
			14	-1.2239206081e-03	7.1254884374e-02
			16	-2.6451435486e-02	5.9623583927e-02
	12	2.343750000e-02	12	-6.8056671410e-14	8.3333333333e-02
			22	-1.0901916332e-03	4.5356304756e-02
			24	-1.8855910880e-02	4.0301729872e-02
	16	3.125000000e-02	16	2.2204460493e-16	6.2500000000e-02
			30	-9.8376169927e-04	3.3268428113e-02
			32	-1.4786643963e-02	3.0447941589e-02
2048	4	1.953125000e-03	4	-8.4204328419e-06	2.4999578985e-01
			6	-1.3531464737e-03	1.6621718808e-01
			8	-4.6578677068e-02	1.1483609417e-01
	8	3.906250000e-03	8	-6.6811089994e-10	1.2499999983e-01
			14	-1.2254979875e-03	7.1254661803e-02
			16	-2.6458292920e-02	5.9622860174e-02
	12	5.859375000e-03	12	-6.9277916737e-14	8.3333333333e-02
			22	-1.0929317005e-03	4.5356059644e-02
			24	-1.8866157235e-02	4.0301006683e-02
	16	7.812500000e-03	16	-1.3322676296e-15	6.2500000000e-02
			30	-9.8776416716e-04	3.3268166282e-02
			32	-1.4800278406e-02	3.0447218588e-02

Notice that often the variance term is much larger than the bias squared. This suggests increasing the number of tapers used so as to decrease mean square error.

4. Conclusion and comments

We have given the asymptotic properties for bias, variance and mean squared error of spectral density function estimations based on multitaper estimators from prolate spheroidal sequences. It is shown that in order to have good local bias properties the 'bandwidth' W should be small. For a given sample size N a 'small' W would limit the size 2NW which effectively limits the number of tapers that can be used to average in order to reduce the variance. The computation of the prolate spheroidal sequences is a nuisance to compute for large N. Riedel and Sidorenko (1995) have considered a class of orthonormal tapers which minimize the local bias and a class of sinusoidal tapers which give a very close approximation to the minimum local bias tapers. The advantage of the sinusoidal tapers is their simple analytic form and computational efficiency with good local bias properties. Walden et al. (1995) compared effective bandwidth of Slepian and sinusoidal tapers in terms of local bias and variance. A comparison to the usual lag window spectral estimators is also made. The strength of the Slepian prolate spheroidal tapers lies in their small side lobes and hence superior leakage property. For a particular set of data the choice of what kind of multitaper should be used will depend on computational cost, concerns of bias and variance, dynamic range of the spectral density function and its smoothness as well as possible leakage among others.

References

Kac, M., Murdock, W., Szegö, G., 1953. On the eigenvalues of certain hermitian forms. J. Raf. Mech. Anal. 2, 767–800.
Perceival, D., Walden, A., 1993. Spectral Analysis for Physical Applications. Cambridge University Press.
Riedel, K.S., Sidorenko, A., 1995. Minimum bias multiple taper spectral estimation. IEEE Trans. Signal Process. 43 (1), 188–195.
Slepian, D., 1978. Prolate spheroidal wave functions, Fourier analysis, and uncertainty—V: The discrete case. Bell System Tech. J. 57 (5), 1371–1430.
Riesz, F., Sz. Nagy, B., 1990. Functional Analysis. Dover.
Thomson, D.J., 1982. Spectrum estimation and harmonic analysis. Proc. IEEE 1055–1096.
Thomson, D.J., 1990. Time series analysis of Holocene climate data. Philos. Trans. R. Soc. Lond. A330, 601–616.
Walden, A.T., McCoy, E.J., Percival, D.B., 1995. The effective bandwidth of a multitaper spectral estimator. Biometrika 82 (1), 201–214.
Widom, H., 1969. Lectures on Integral Equations. Van Nostrand.

CPSIA information can be obtained at www.ICGtesting.com
228710LV00008B/8/P

9 781441 983381